MICROBIAL MATS
Physiological Ecology of Benthic Microbial Communities

MICROBIAL MATS
Physiological Ecology of Benthic Microbial Communities

Edited by

YEHUDA COHEN
H. Steinitz Marine Biology Laboratory
The Hebrew University of Jerusalem
Eilat, Israel

and

EUGENE ROSENBERG
Department of Microbiology
George S. Wise Faculty of Life Sciences
Tel Aviv University
Ramat Aviv, Israel

AMERICAN SOCIETY FOR MICROBIOLOGY
Washington, D.C.

Library of Congress Cataloging-in-Publication Data

Microbial mats: physiological ecology of benthic microbial communities

"Comprises the scientific presentations at a Bat-Sheva de Rothschild seminar held at the H. Steinitz Marine Biology Laboratory in Eilat, Israel, in September 1987"—Pref.

Includes index.

1. Microbial mats—Congresses. 2. Benthos—Microbiology—Congresses. I. Cohen, Yehuda. II. Rosenberg, Eugene. III. American Society for Microbiology. IV. Batsheva de Rothschild Foundation for the Advancement of Science in Israel.
QR100.8.M37M53 1989 576'.15'0916 88-34241

ISBN 1-55581-002-0

CONTENTS

CONTRIBUTORS

Zeev Aizenshtat • Energy Research Center, Organic Chemistry, and Casali Institute, The Hebrew University of Jerusalem, Jerusalem 91904, Israel

M. Avron • Department of Biochemistry, Weizmann Institute of Science, Rehovot 76100, Israel

Y. Bar-Or • Division of Microbial and Molecular Ecology, Institute of Life Sciences, The Hebrew University of Jerusalem, Jerusalem 91904, Israel

Brad M. Bebout • Institute of Marine Sciences, University of North Carolina at Chapel Hill, Morehead City, North Carolina 28557

Susanne Bekker • Geomicrobiology Division, University of Oldenburg, P.O. Box 2503, D-2900 Oldenburg, Federal Republic of Germany

Shimshon Belkin • Laboratory for Environmental Applied Microbiology, The Jacob Blaustein Desert Research Institute, Ben-Gurion University of the Negev, Sede Boqer Campus 84993, Israel

Y. Bin Zeng • Organic Geochemistry Unit, School of Chemistry, University of Bristol, Cantock's Close, Bristol BS8 1TS, England

Simon Brassell • Department of Geology, Stanford University, Stanford, California 94305-2115

Richard W. Castenholz • Department of Biology, University of Oregon, Eugene, Oregon 97403

Pierre Caumette • Laboratoire de Microbiologie, Faculté des Sciences de Saint-Jérôme, F-13397 Marseille Cedex 13, France

Michael Cheatham • Life Science Division, National Aeronautics and Space Administration, Ames Research Center, Moffett Field, California 94035

Terri Cheatham • Life Science Division, National Aeronautics and Space Administration, Ames Research Center, Moffett Field, California 94035

Peter Bondo Christensen • Institute of Ecology and Genetics, University of Aarhus, Ny Munkegade, DK-8000 Aarhus C, Denmark

Yehuda Cohen • H. Steinitz Marine Biology Laboratory, The Hebrew University of Jerusalem, P.O. Box 469, Eilat 88103, Israel

Sonja E. Cronin • Life Science Division, National Aeronautics and Space Administration, Ames Research Center, N239-4, Moffett Field, California 94035

Elisa D'Antoni D'Amelio • Life Science Division, National Aeronautics and Space Administration, Ames Research Center, N239-4, Moffett Field, California 94035

J. W. de Leeuw • Department of Chemistry and Chemical Engineering, Organic Geochemistry Unit, Delft University of Technology, De Vries van Heystplantsoen 2, 2628 RZ Delft, The Netherlands

Edward F. DeLong • Department of Biology, Institute for Molecular and Cellular Biology, Indiana University, Bloomington, Indiana 47405

Loes de Reus • Laboratory of Microbiology, University of Amsterdam, Nieuwe Achtergracht 127, 1018 WS Amsterdam, The Netherlands

David J. Des Marais • Life Science Division, National Aeronautics and Space Administration, Ames Research Center, N239-4, Moffett Field, California 94035

Ben de Winder • Laboratory of Microbiology, University of Amsterdam, Nieuwe Achtergracht 127, 1018 WS Amsterdam, The Netherlands

Rutger de Wit • Department of Microbiology, University of Groningen, Kerklaan 30, 9751 NN Haren, The Netherlands

Gary Dobson • Organic Geochemistry Unit, School of Chemistry, University of Bristol, Cantock's Close, Bristol BS8 1TS, England

Inka Dor • Oceanography Program and Human Environmental Sciences Division, School of Applied Science and Technology, The Hebrew University of Jerusalem, Jerusalem 91904, Israel

Geoffrey Eglinton • Organic Geochemistry Unit, School of Chemistry, University of Bristol, Cantock's Close, Bristol BS8 1TS, England

Jonathan Erez • Department of Geology, The Hebrew University of Jerusalem, Givat-Ram, Jerusalem 91904, Israel

H. L. Fredrickson • Limnological Institute, "Vijverhof" Laboratory, Rijksstraatweg 6, 3631 AC Nieuwersluis, The Netherlands

E. A. Galinski • Institut für Microbiologie der Rheinischen Friedrich-Wilhelms-Universität, Meckenheimer Allee 168, 5300 Bonn 1, Federal Republic of Germany

William C. Ghiorse • Department of Microbiology, Cornell University, Ithaca, New York 14853

Steven J. Giovannoni • Department of Biology, Institute for Molecular and Cellular Biology, Indiana University, Bloomington, Indiana 47405

Ricardo Guerrero • Department of Genetics and Microbiology, Autonomous University of Barcelona, 08193 Bellaterra (Barcelona), Spain

Heike Heyer • Geomicrobiology Division, University of Oldenburg, P.O. Box 2503, D-2900 Oldenburg, Federal Republic of Germany

Holger W. Jannasch • Biology Department, Woods Hole Oceanographic Institution, Woods Hole, Massachusetts 02543

Barbara Javor • H. Steinitz Marine Biological Laboratory, The Hebrew University of Jerusalem, P.O. Box 469, Eilat 88103, Israel

Bo Barker Jørgensen • Institute of Ecology and Genetics, University of Aarhus, Ny Munkegade, DK-8000 Aarhus C, Denmark

Nachum Kaplan • Department of Microbiology, George S. Wise Faculty of Life Sciences, Tel Aviv University, Ramat Aviv 69978, Israel

M. Kessel • Department of Membrane and Ultrastructure Research, The Hebrew University of Jerusalem-Hadassah Medical School, Jerusalem 91904, Israel

Gary M. King • Darling Marine Center, University of Maine, Walpole, Maine 04573

Andrew H. Knoll • Botanical Museum, Harvard University, Cambridge, Massachusetts 02138

Wolfgang E. Krumbein • Geomicrobiology Division, University of Oldenburg, P.O. Box 2503, D-2900 Oldenburg, Federal Republic of Germany

J. Gijs Kuenen • Department of Microbiology and Enzymology, Delft University of Technology, Julianalaan 67, 2628 BC Delft, The Netherlands

G. F. LaVos • Toegepast-Natuurwetenschappelijk Onderzoek–Centraal Instituut Voedingsmiddelen Onderzoek Institutes, Post Box 360, 3700 AJ Zeist, The Netherlands

Boaz Lazar • H. Steinitz Marine Biological Laboratory, The Hebrew University of Jerusalem, P.O. Box 469, Eilat 88103, Israel

Christopher S. F. Low • Institute for Applied Microbiology, University of Tennessee, 10515 Research Drive, Suite 300, Knoxville, Tennessee 37932-2567

Ronald M. Lynch • Department of Biochemistry, Faculty of Science, The Australian National University, G.P.O. Box 4, Canberra, Australian Capital Territory 2601, Australia

K. C. Marshall • School of Microbiology, University of New South Wales, P.O. Box 1, Kensington, New South Wales 2033, Australia

Jordi Mas • Department of Genetics and Microbiology, Autonomous University of Barcelona, 08193 Bellaterra (Barcelona), Spain

Elaine Munoz • Life Science Division, National Aeronautics and Space Administration, Ames Research Center, N239-4, Moffett Field, California 94035

Luuc R. Mur • Laboratory of Microbiology, University of Amsterdam, Nieuwe Achtergracht 127, 1018 WS Amsterdam, The Netherlands

Hoa Nguyen • Life Science Division, National Aeronautics and Space Administration, Ames Research Center, Moffett Field, California 94035

Lars Peter Nielsen • Institute of Ecology and Genetics, University of Aarhus, Ny Munkegade, DK-8000 Aarhus C, Denmark

Gary J. Olsen • Department of Biology, Institute for Molecular and Cellular Biology, Indiana University, Bloomington, Indiana 47405

John M. Olson • Institute of Biochemistry, Odense University, DK-5230 Odense M, Denmark

Ronald S. Oremland • U.S. Geological Survey, Menlo Park, California 94025

Aharon Oren • Division of Microbial and Molecular Ecology, The Institute of Life Sciences, The Hebrew University of Jerusalem, Jerusalem 91904, Israel

Norman R. Pace • Department of Biology, Institute for Molecular and Cellular Biology, Bloomington, Indiana 47405

Etana Padan • Division of Molecular and Microbial Ecology, Institute of Life Sciences, The Hebrew University of Jerusalem, Jerusalem 91904, Israel

Hans W. Paerl • Institute of Marine Sciences, University of North Carolina at Chapel Hill, Morehead City, North Carolina 28557

Anna C. Palmisano • Life Science Division, National Aeronautics and Space Administration, Ames Research Center, N239-4, Moffett Field, California 94035

Noga Paz • Oceanography Program and Human Environmental Sciences Division, School of Applied Science and Technology, The Hebrew University of Jerusalem, Jerusalem 91904, Israel

Beverly K. Pierson • Biology Department, University of Puget Sound, Tacoma, Washington 98416

Jan Pluis • Laboratory of Physical Geography and Soil Science, University of Amsterdam, Dapperstraat 115, 1093 BS Amsterdam, The Netherlands

Anton F. Post • Laboratory of Microbiology, University of Amsterdam, Nieuwe Achtergracht 127, 1018 WS Amsterdam, The Netherlands

Leslie E. Prufert • Institute of Marine Sciences, University of North Carolina at Chapel Hill, Morehead City, North Carolina 28557

Dave Rensman • Laboratory of Microbiology, University of Amsterdam, Nieuwe Achtergracht 127, 1018 WS Amsterdam, The Netherlands

Niels Peter Revsbech • Institute of Ecology and Genetics, University of Aarhus, Ny Munkegade, DK-8000 Aarhus C, Denmark

W. I. C. Rijpstra • Department of Chemistry and Chemical Engineering, Organic Geochemistry Unit, Delft University of Technology, De Vries van Heystplantsoen 2, 2628 RZ Delft, The Netherlands

Eugene Rosenberg • Department of Microbiology, George S. Wise Faculty of Life Sciences, Tel Aviv University, Ramat Aviv 69978, Israel

Mel Rosenberg • Department of Human Microbiology, Sackler Faculty of Medicine, Tel Aviv University, Ramat Aviv 69978, Israel

Nechemia Sar • Department of Microbiology, George S. Wise Faculty of Life Sciences, Tel Aviv University, Ramat Aviv 69978, Israel

Jentaie Shiea • Department of Chemistry, Montana State University, Bozeman, Montana 59717

M. Shilo • Division of Microbial and Molecular Biology, Institute of Life Sciences, The Hebrew University of Jerusalem, Jerusalem 99104, Israel

Yuval Shoham • Department of Microbiology, George S. Wise Faculty of Life Sciences, Tel Aviv University, Ramat Aviv 69978, Israel

Graham W. Skyring • Division of Water Resources Research, Commonwealth Scientific and Industrial Research Organisation, G.P.O. Box 1666, Canberra, Australian Capital Territory 2601, Australia

Geoffrey D. Smith • Department of Biochemistry, Faculty of Science, The Australian National University, G.P.O. Box 4, Canberra, Australian Capital Territory, 2601, Australia

Lucas J. Stal • Laboratorium voor Microbiologie, Universiteit van Amsterdam, Nieuwe Achtergracht 127, 1018 WS Amsterdam, The Netherlands

Jean-Pierre Sweers • Laboratory of Microbiology, University of Amsterdam, Nieuwe Achtergracht 127, 1018 WS Amsterdam, The Netherlands

A. C. Tas • Toegepast-Natuurwetenschappelijk Onderzoek–Centraal Instituut Voedingsmiddelen Onderzoek Institutes, Post Box 360, 3700 AJ Zeist, The Netherlands

H. G. Trüper • Institut für Mikrobiologie der Rheinischen Friedrich-Wilhelms-Universität, Meckenheimer Allee 168, 5300 Bonn 1, Federal Republic of Germany

Seán Turner • Department of Biology, Institute for Molecular and Cellular Biology, Indiana University, Bloomington, Indiana 47405

Jan van der Greef • Toegepast-Natuurwetenschappelijk Onderzoek–Centraal Instituut Voedingsmiddelen Onderzoek Institutes, Post Box 360, 3700 AJ Zeist, The Netherlands

Andien van der Heuvel • Laboratory of Microbiology, University of Amsterdam, Nieuwe Achtergracht 127, 1018 WS Amsterdam, The Netherlands

Hans van Gemerden • Department of Microbiology, University of Groningen, Kerklaan 30, 9751 NN Haren, The Netherlands

Arnold Veen • Laboratory of Microbiology, University of Amsterdam, Nieuwe Achtergracht 127, 1018 WS Amsterdam, The Netherlands

Marlies Villbrandt • Geomicrobiology Division, University of Oldenburg, P.O. Box 2503, D-2900 Oldenburg, Federal Republic of Germany

David M. Ward • Department of Microbiology, Montana State University, Bozeman, Montana 59717

Roland Weller • Department of Microbiology, Montana State University, Bozeman, Montana 59717

David C. White • Institute for Applied Microbiology, University of Tennessee, 10515 Research Drive, Suite 300, Knoxville, Tennessee 37932-2567

Julian Wimpenny • Department of Microbiology, University College, Newport Road, Cardiff Cf2 1TA, United Kingdom

Tamar Zohary • The Yigal Allon Kinneret Limnological Laboratory, P.O. Box 345, Tiberias 14102, Israel

Gabriel Zwart • Laboratory of Microbiology, University of Amsterdam, Nieuwe Achtergracht 127, 1018 WS Amsterdam, The Netherlands

PREFACE

Microbial mats are probably the oldest form of life on earth, as witnessed by the oldest known microfossils being found in lithified microbial mats: stromatolites which have been dated to over 3.5 billion years old. Stromatolites are the most dominant sedimentary structures in rocks of the Precambrian era, together with vast deposits of the Banded Iron Formation which are often associated with lithified microbial mats at the boundaries of sedimentary basins.

Although the abundance of stromatolites declined during the Cambrian era and later, microbial mats are still presently found in a variety of different biotopes, namely, hypersaline lagoons, alkaline lakes and streams, and hydrothermal vents, both terrestrial and in the deep sea.

Microbial mats are stratified microbial communities that develop in the environmental microgradients established at the interfaces of water and solid substrates. They form a laminated multilayer of biofilms and largely alter the environmental microgradients in this interface as a result of their own communal metabolism. The development of these microbial communities causes steep environmental microgradients and the establishment of a well-defined diffusion boundary layer immediately proximal to the multilayered biofilm.

A steep redox gradient at this interface is expressed in the depletion of oxygen at the mat surface and the buildup of sulfide and often methane with depth in sulfate-containing water bodies. Along with the establishment of the oxygen/sulfide interface, steep gradients of other compounds such as nutrients, low-molecular-weight fatty acids, and available metals are developed. Diffusion of small molecules along these steep microgradients is very fast, allowing rapid internal cycling of carbon, nitrogen, sulfur, and phosphate within the mat to support the high rate of metabolism of microbial mats.

Microbial mats developing in shallow waters are composed primarily of photosynthetic microorganisms of which oxygenic cyanobacteria are dominant. The communal metabolism of the cyanobacterial mats depends on available light for photosynthesis. Steep spectral light gradients are found in these mats. Temporal day-night fluctuations expose the mat communities to vast diurnal fluctuations in both their chemical and optical microenvironments. To cope with these fluctuations, the microorganisms have either to migrate diurnally along with the changing microgradients or to develop physiological flexibility which will allow them to function under both oxidizing and reducing conditions.

Our present knowledge of the structure, function, and dynamics of microbial mats is thanks to the development of suitable microtechniques for the study of mat communities at the microscale, on the one hand, and to physiological experimental research on pure cultures of mat-building microorganisms on the other.

The application of microelectrodes for measurements of oxygen, sulfide, and pH, together with development of micro-optic fibers for the study of spectral light distribution in microbial mats and careful electron microscopy, has brought to light the dynamics of microenvironmental conditions within mat communities. The development of microassays for the study of specific microbial metabolism, including oxygenic photosynthesis, aerobic respiration, and sulfate reduction, at microscale allows an understanding of specific microbial activities and their interrelations within the mat communities.

Stable isotope determination of carbon and sulfur allows the understanding of the integration of microbial activity over time in these systems. Since stable isotope analysis can be applied in both living microbial mats and ancient stromatolites, it can serve as a useful tool in interpreting the ancient sedimentary record from our present knowledge of microbial processes in living mat communities. This comparison is further understood through the use of micropaleontological techniques in relating microfossils to living mats and the use of biomarkers in preserved organic matter. A careful comparison of these data with our present knowledge of the phylogeny of isolated pure cultures of microorganisms, acquired using 16S RNA sequences, may allow us to understand the evolution of microbial mats and stromatolites. Sulfur metabolism emerges from the study as being a primordial metabolic pathway which is of prime importance in present-day microbial mats and probably in Archean stromatolites. Hence the study of sulfur metabolism in pure cultures of mat-forming microorganisms is of prime importance to the understanding of microbial processes in mat communities.

New types of phototrophic bacteria have recently been described in microbial mat communities, some of which are not yet available in pure cultures; examples include marine *Chloroflexus*-like and filamentous *Ectothiorhodospira*-like microorganisms, both of which are abundant in marine microbial mat communities.

Several mat-forming cyanobacteria have been shown to carry out facultative anoxygenic photosynthesis using sulfide, hydrogen, and possibly reduced iron as alternative electron donors.

During the past few years, new nonphotosynthetic sulfide-oxidizing bacteria have been discovered with diverse habitats and physiologies, many of which exhibit metabolic versatility and flexibility with respect to both electron donors and electron receptors.

Beggiatoa mats have been found to be abundant in many marine coastal environments as well as deep-sea hydrothermal beds. Autotrophic CO_2 fixation using sulfide as an electron donor was postulated by Wingradsky over 100 years ago, but was satisfactorily demonstrated only recently in marine *Beggiatoa* isolates. Yet most of the mat-forming *Beggiatoa* spp. from deep-sea hydrothermal vents are not availble in pure cultures.

Our knowledge of the versatility of sulfate-reducing bacteria has also greatly expanded during recent years, and their importance in sulfur metabolism in microbial mats is being emphasized. New species have recently been described, some of which are strict anaerobes while others tolerate oxygen toxicity.

While most of the mat-forming microorganisms belong to the Eubacteria group, several Archaebacteria were recently described, many of which are methane-oxidizing microorganisms. Methane production in marine and hypersaline microbial mats has been found to be an important part of the carbon cycling in these systems as well as in terrestrial hydrothermal mats.

The formation of highly compacted benthic microbial communities requires special characteristics of the cell wall of the microorganisms, namely, the presence of a hydrophotic group at the surface of the cell wall. The hydrophoticity of a mat-forming cyanobacterial cell wall is vital for this group to serve as a skeleton for the adjacent benthic microbial mat community and to allow specific physical interactions with heterotrophic microorganisms. The development of an outer polysaccharide capsule in several marine microorganisms provides attachment to interfaces. The regulation of external polymer production by microorganisms will determine their ability to colonize solid/water interfaces and to form biofilms.

The metabolic diversity and flexibility of a microbial mat community and the metabolic interactions of the various mat-forming microorganisms are the major theme of this book. The book comprises scientific presentations made at a Bat-Sheva de Rothschild seminar held at the H. Steinitz Marine Biology Laboratory in Eilat, Israel, in September, 1987. This meeting followed the first microbial mat meeting held in Woods Hole in 1982, which centered on the interdisciplinary approach to the study of microbial mats and stromatolites.

This book is dedicated to Professor Moshe Shilo of the Hebrew University of Jerusalem, who is a pioneer in many aspects of microbial ecology and who initiated and pursued the development of microbial ecology of microbial mats.

Yehuda Cohen

I. Environments of Depositions

Chapter 1

Hot Spring Microbial Mats: Anoxygenic and Oxygenic Mats of Possible Evolutionary Significance

David M. Ward, Roland Weller, Jentaie Shiea, Richard W. Castenholz, and Yehuda Cohen

INTRODUCTION

Microbiologists interested in interpreting stromatolite microbial communities have mainly studied modern microbial mats which occur in lagoonal settings resembling the coastal environments in which most stromatolites formed (31). The modern lagoonal environment, however, provides quite different biological and chemical features than were present on Earth during the Precambrian era. For example, fossil evidence suggests the absence of animals, which graze upon modern lagoonal mats when salinity is sufficiently low (15, 16, 18). It may also be the case that the eucaryotic cell is of relatively recent origin (27), so that during much of the Precambrian era, mats would have lacked such organisms as diatoms which are common in modern lagoonal mats. Also, unlike modern mats, those of the early Precambrian period may have existed under anoxic conditions (30).

Microbial mats found in modern hot springs provide examples of the range of mat-forming communities which might have existed at different times during the Precambrian era as the biological and chemical environments of Earth changed (28). The elevated temperature of hot spring waters, sometimes in combination with elevated hydrogen sulfide concentration or acidic conditions, eliminates almost all eucaryotic organisms. The variation of these same features restricts microbial diversity so that mats in different springs are formed by distinctly different types of photosynthetic microorganisms (see Fig. 1 and Table 1). Some of these mats were described in a previous review (11). Here we will illustrate hot spring mat types in the context of their possible evolutionary significance. This is especially timely as there is increased information on anoxygenic microbial mats, as well as cyanobacterial mats which might rep-

David M. Ward and Roland Weller • Department of Microbiology, Montana State University, Bozeman, Montana 59717. *Jentaie Shiea* • Department of Chemistry, Montana State University, Bozeman, Montana 59717. *Richard W. Castenholz* • Department of Biology, University of Oregon, Eugene, Oregon 97403. *Yehuda Cohen* • H. Steinitz Marine Biology Laboratory, The Hebrew University of Jerusalem, P.O. Box 469, Eilat 88103, Israel.

Figure 1. (A) Vertical section of the Octopus Spring cyanobacterial mat (55°C, pH 8). (B) Oblique vertical section of the Nymph Creek eucaryotic algal mat (45°C, pH 2.8). (C) Vertical section of the New Pit Spring anoxygenic *Chloroflexus* mat (ca. 55°C, pH 6.5, ca. 30 μM sulfide). (D) Top view of the Roland's Well anoxygenic purple mat (ca. 52°C, pH 6.3, ca. 41 μM sulfide). Note areas where laminated mat has curled away from the rock substrate. Thickness, ca. 3 mm. (E) Vertical section of the New Zealand Travel Lodge Stream *Chlorobium* mat (45°C, pH 5.9, 250 μM sulfide). Diameter of the core tube is ca. 3.5 cm. (F) Top view of Ystihver Spring (Iceland, north of Myvatn) showing orange *Chloroflexus* mat and underlying green *Chlorogloeopsis* mat (58 to 60°C, pH > 8.5, ca. 200 μM sulfide).

Table 1.
Examples of Hot Spring Microbial Mats of Possible Evolutionary Significance

Mat	Temp (°C)	pH	H_2S concn (μM)	Predominant phototrophs				
				Eucaryotic	Cyanobacteria	Photosynthetic bacteria: Green nonsulfur group	Green sulfur group	Purple group
Oxygenic								
Cyanobacterial								
Octopus Spring, YNP[a,b]	40–72	8	ND[c]		*Synechococcus lividus*	*Chloroflexus aurantiacus*		
Eucaryotic algal								
Nymph Creek, YNP	47	3	+[d]	*Cyanidium caldarium*				
Anoxygenic								
New Pit Spring, YNP[e]	52–58	6.3	34			*Chloroflexus* sp.		
Roland's Well, YNP	52–55	6.3	41			*Chloroflexus* sp.		*Chromatium tepidum*
Travel Lodge Stream, New Zealand[f]	45	5.9	250				*Chlorobium* sp.	
Anoxygenic/oxygenic mats with cyanobacteria								
Photosynthetic bacteria above cyanobacteria								
Iceland Spring A[g]	55–61	7.5	40		*Chlorogloeopsis* sp.	*Chloroflexus* sp.		
Cyanobacterial								
Stinky Springs, Utah[h]	42–46	6.1	1,200		*Oscillatoria* spp.	*Chloroflexus* sp.		

[a] And many other hot spring cyanobacterial mats which contain different cyanobacteria, usually with *Chloroflexus* sp. or *Chloroflexus*-like photosynthetic bacteria (6, 11).
[b] YNP, Yellowstone National Park, Wyoming.
[c] ND, Not determined.
[d] Detected in source but not quantified in mat; mat contains some S^0 (see Ward et al., this volume).
[e] And several other *Chloroflexus* mats in Yellowstone National Park, Iceland, and New Zealand between about 50 and 66°C (8, 10, 17).
[f] And other similar New Zealand springs between 45 and 55°C (11a).
[g] From B. B. Jørgensen and D. C. Nelson (19a).
[h] From Cohen et al. (12).

resent a transition from anoxygenic to oxygenic mat types, and as recent molecular phylogenetic work points to the possibility that the earliest microorganisms may have been thermophilic (38). We hope this review will provide the reader interested in microbial evolution (i) an illustration of the variety of microbial mats of modern Earth environments, (ii) an illustration of their physical dimension and structure, (iii) a basic understanding of the major mat-building phototrophs and their relationship to other community members within mats, and (iv) a source of references to more detailed studies of each mat type.

Methods

Red autofluorescence characteristic of chlorophyll *a* was observed using a Leitz

Ortholux II microscope with vertical illumination fluorescence (Leitz B cube excitation/emission filter combination). Phase-contrast photomicrographs were made using a Nikon Optiphot microscope and HFX automatic photomicrography system. In vivo absorption spectra were determined on mat samples homogenized in 60% sucrose, using a Cary model 14 spectrophotometer. Oxygen, sulfide, and pH microelectrode data were obtained as previously described (22).

OXYGENIC MATS BUILT BY CYANOBACTERIA

The most common microbial mats of hot springs are those produced in neutral to alkaline springs by cyanobacteria. As will be detailed below, the communities in such mats include many other procaryotic microorganisms. Above 42 to 50°C, metazoans (35–37) and metaphytes (5) are absent. (Mats are never entirely free of plant and animal contributions, however, an important point for geochemical investigations; see reference 13a and D. M. Ward, J. Shiea, Y. B. Zeng, G. Dobson, S. Brassell, and G. Eglinton, this volume.) Above 55 to 58°C (lower in most springs), eucaryotic microorganisms are absent (5).

A variety of different thermophilic cyanobacteria may produce mats (see reference 6), with distribution controlled by differences in temperature and presumably effluent water chemistry. Discussions of temperature distributions have appeared before (e.g., references 7 and 11). Here we will describe a single intensively studied cyanobacterial mat community, Octopus Spring, Yellowstone National Park (see reference 5 for location), as an example of the hot spring cyanobacterial mat type. The vertical mat structure, community composition, and function will be summarized for a relatively moderate temperature region, 50 to 55°C, where mat development is maximal (14). The reader is referred to reviews for more detailed coverage of the microbiology of this mat (see references 5 and 34). Important distinctions of other hot spring cyanobacterial mats will be pointed out when necessary.

The Octopus Spring mat is well laminated and develops to a thickness of several centimeters (Fig. 1A). The "usual vertical sequence" (as described by Castenholz [11]) of a green cyanobacterial layer above an orange layer of photosynthetic bacteria is observed; however, both microscopic (Fig. 2A) and pigment analyses (see reference 2) suggest that the top layer is composed of both types of phototrophs. The mat is formed by a single cyanobacterium, *Synechococcus lividus*, and a filamentous green nonsulfur bacterium, *Chloroflexus aurantiacus* (Fig. 2A), the latter remaining abundant in the 1 to 2 mm beneath the top green layer where light penetration is sufficient for growth (2, 14). The in vivo absorption spectrum reveals the major photopigments of both phototrophs which absorb at complementary wavelengths (Fig. 2B). (In other hot spring cyanobacterial mats, purple sulfur bacteria are common. This may be a reflection of an additional source of sulfide in such mats resulting from intensive biogenic sulfate reduction [see below].)

The mat grows by upward accretion at rates of 18 to 45 μm/day. Intensive oxygenic photosynthesis by *S. lividus* results in superoxic and alkaline conditions (presumably caused by CO_2 consumption) in the upper mat during daylight (Fig. 2C). This influences *S. lividus* photosynthesis, as such conditions enhance photorespiratory glycolate production and excretion by *S. lividus* (1a). Oxygen is depleted within seconds when the mat is darkened and is replenished rapidly by photosynthesis after return to the light. This certainly means that in addition to diurnal changes, oxygen and other aspects of mat chemistry fluctuate dramatically with changes in cloud cover (24). Superoxic conditions also exclude sulfide from the photic zone at midday light intensities, so that *C. aurantiacus* cannot grow photoautotrophi-

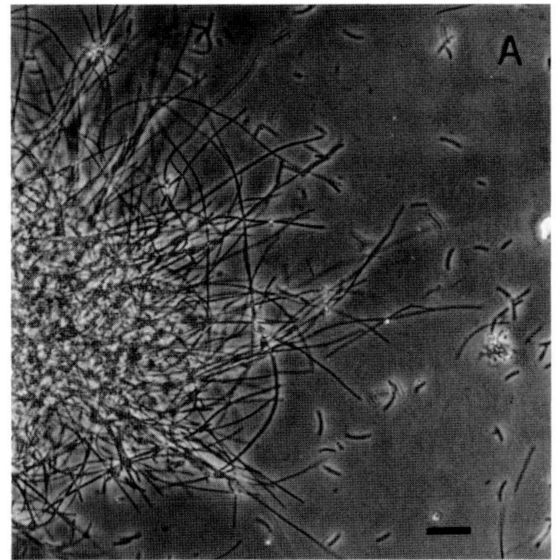

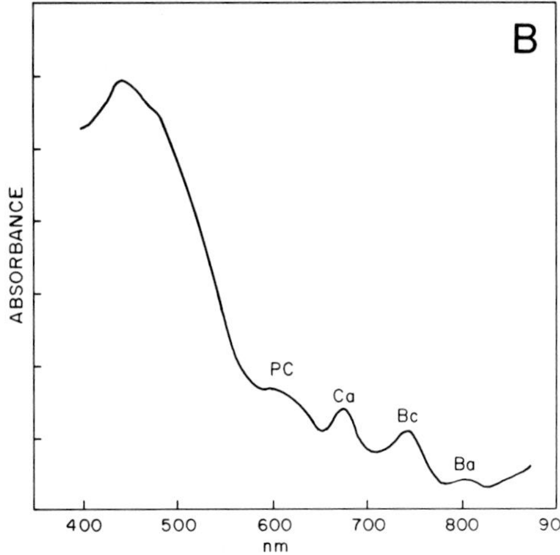

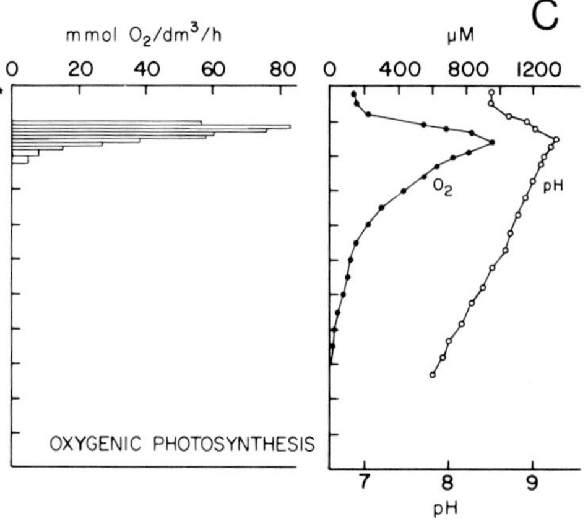

Figure 2. Octopus Spring cyanobacterial mat (50 to 55°C, pH 8) (see Fig. 1A). (A) Phase-contrast photomicrograph of top green layer. Note sausage-shaped cells of the cyanobacterium *S. lividus* and filamentous bacteria, some of which are *Chloroflexus* sp. Bar, 25 µm. (B) In vivo absorption spectrum of mat. Ca, Chlorophyll *a*; PC, phycocyanin; Bc, bacteriochlorophyll *c*; Ba, bacteriochlorophyll *a*. (C) Vertical microprofiles of oxygenic photosynthesis (bars), oxygen, and pH measured in full sunlight (from reference 24).

cally. However, *C. aurantiacus* seems well adapted to the changing environment of the mat since it can grow either photoautotrophically (at low light intensity when sulfide could exist in the photic zone), photoheterotrophically, or by aerobic chemoheterotrophy. Doemel and Brock (14) suggested that movement of *C. aurantiacus*, which can only grow by aerobic means in darkness, toward the oxic mat surface at night may help explain upward mat accretion.

Decomposition, carried out mainly within the top 5 mm of the mat (1, 14) by heterotrophic bacteria, balances accretion so that the mat is nearly in a steady state and there is only slight net accumulation (14). Fermentation recycles reduced carbon (e.g., fermentation products), which helps supply the photoheterotrophic needs of *C. aurantiacus*, but may also cause export of reduced carbon from the mat at night (1, 26, 34). The terminal stages of decomposition lead principally to methane formation in low-sulfate springs such as Octopus Spring. (In high-sulfate springs, the major terminal product is hydrogen sulfide [see references 32 and 33], which may serve as an electron donor for chemolithotrophic bacteria or anoxygenic photosynthetic bacteria, depending on whether there is sufficient oxygen to exclude sulfide from the photic zone [see reference 25].) Since decomposition occurs mainly near the mat surface, most of the organic matter at depth in the mat is presumed to be relatively recalcitrant. In this regard it is interesting that both morphologically (14) and antigenically (29) recognizable remains of *C. aurantiacus* survive at depth. However, the absence of morphologic remains of *S. lividus* beneath the top green layer (14)

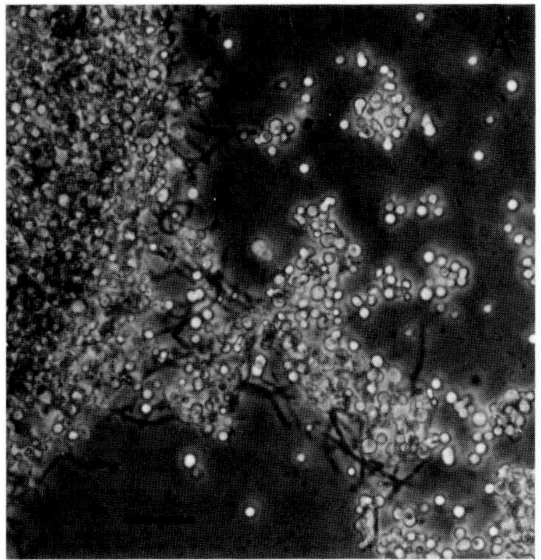

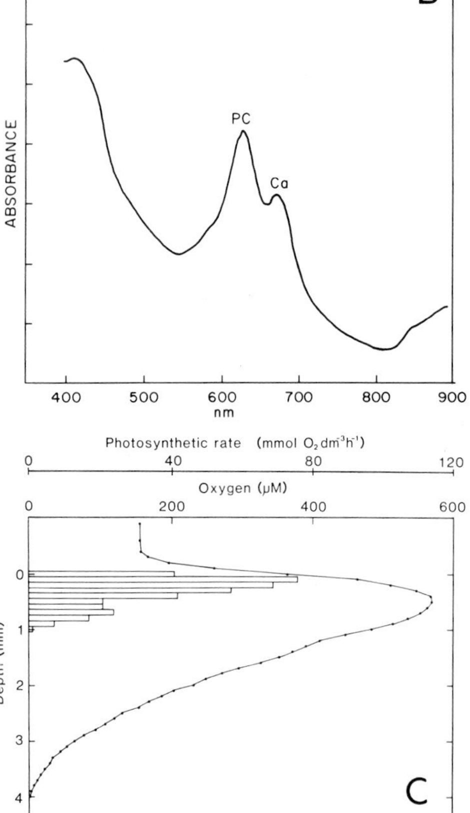

Figure 3. Nymph Creek eucaryotic algal mat (45°C, pH 2.8) (see Fig. 1B). (A) Phase-contrast photomicrograph of top layer. Note refractile spherical cells of the eucaryotic alga *C. caldarium*. Bar, 25 μm. (B) In vivo absorption spectrum of the mat. See legend to Fig. 2B for definition of symbols. (C) Vertical microprofiles of oxygen and oxygenic photosythesis (bars), measured in full sunlight; pH was constant with depth at 2.8 (from reference 23).

provides a vivid example of how microbial decomposition may bias the structures which are ultimately preserved.

OXYGENIC MATS BUILT BY EUCARYOTIC ALGAE

Photosynthetic procaryotes are unable to tolerate pH below about 4.5 (3, 11a). In hot acid environments (pH 1 to 4) such as Nymph Creek, Yellowstone National Park (a small stream flowing into Nymph Lake, located ca. 2 miles northwest of Norris Geyser Basin), mats up to a few centimeters thick (Fig. 1B) develop due to the growth of the oxygenic, eucaryotic alga *Cyanidium caldarium* and possibly related organisms (13) (Fig. 3A). The eucaryotic nature and physiology of this organism have been reviewed elsewhere (5). Such mats exhibit a vertical zonation of color, but are not well laminated, possibly a reflection of the absence of filamentous bacteria as major community members (see also below). Photosynthetic bacteria are also absent, as shown by the lack of bacteriochlorophylls in the in vivo absorption spectrum (Fig. 3B), which is dominated by the major pigments of *C. caldarium*, chlorophyll *a* and phycocyanin. Oxygenic photosynthesis by *C. caldarium* results in superoxic conditions similar to those of cyanobacterial mats (Fig. 3C). Little is known of decomposition in this mat, although a fungus, *Dactylaria gallopava*, and some acidophilic thermophilic bacteria are likely to inhabit the mat (5). While both ancient and modern mats built exclusively by eucaryotic algae are likely rare, such modern environments provide an opportunity to observe the chemical imprint of eucaryotic microorganisms within mats (see Ward et al., this volume).

ANOXYGENIC MATS

Thermophilic cyanobacteria are unable to tolerate combined high temperature and elevated hydrogen sulfide concentration (9, 10). Thus, at temperatures above about 50 to 55°C in springs with primary sulfide, mats are formed exclusively by anoxygenic photosynthetic bacteria. As such springs are rare, such mats are also rare, though they have been observed in North America, Iceland, and New Zealand (8, 11a). In the past few years the discovery and study of anoxygenic mats formed by several different photosynthetic bacteria have extended our knowledge of these types of communities.

Chloroflexus Mats

The best studied anoxygenic mats are those built by an unusual *Chloroflexus* strain which is incapable of growing by aerobic respiration (17). Perhaps the best illustration that a mat built solely by photosynthetic bacteria can be thick and well laminated is shown in cores collected from New Pit Spring (Fig. 1C) (unofficial name for a new spring found about 2 m below the rim of a travertine depression located approximately 100 m west of White Elephant Back Spring on the Upper Terraces Loop of Mammoth Hot Springs in Yellowstone National Park [see map of Castenholz {10}]). Most anoxygenic bacterial mats in the Mammoth Terraces springs are only a few millimeters thick, presumably because travertine deposition within sources limits the longevity of most springs. In New Pit Spring the *Chloroflexus* mat sometimes accumulates to over a centimeter in thickness. *Chloroflexus* mats have been found in neutral to alkaline springs ranging from approximately 50 to 66°C with sulfide concentrations of 30 to 1,000 μM. The principal microbial components are filamentous bacteria (Fig. 4A) which do not autofluoresce red (i.e, lack chlorophyll *a*) and which contain chlorosomes (17). The in vivo absorption spectrum shows the absence of chlorophyll *a* or phycocyanin; the dominant absorption is due to bacteriochlorophyll *c*, the major photopigment of *Chloroflexus* sp. (Fig. 4B). Oxygen microelectrode results show that the mat is devoid of oxygen and oxygenic photosynthesis (17).

Several lines of evidence support the conclusion that mats of this type are produced photoautotrophically by *Chloroflexus* sp. (17). Microelectrode experiments (Fig. 4C) show that sulfide consumption is light dependent. The mats contain a large amount of elemental sulfur (see Ward et al., this volume), some of which adheres to filaments (17). CO_2 fixation is light dependent and is carried out by *Chloroflexus*-like filaments. Photoassimilation of CO_2 is stimulated by sulfide addition, both in the mat and in pure cultures of *Chloroflexus* sp. isolated from the mat, and is not inhibited by restricting irradiation to only infrared light or by addition of the photosystem II inhibitor 3,4-dichlorophenyldimethylurea (8, 17). In the Mammoth Springs area the thin green surface layer of these mats is underlaid by orange laminae which, as in cyanobacterial mats, may contain *Chloroflexus* sp. which probably grow mainly photoheterotrophically on organic compounds excreted by the photoautotrophic *Chloroflexus* sp. of the top layer or supplied through fermentation.

The term "*Chloroflexus* mat" must certainly be an oversimplification, as it is likely that many heterotrophic bacteria carry out nutrient recycling within such mats. This is implied by the fact that *Chloroflexus* mats do not exhibit much net accumulation (i.e., decomposition balances photosynthesis). The rapid accumulation of sulfide when the mat is darkened (Fig. 4C), to a level exceeding the primary sulfide in the spring water, suggests that sulfide is produced within these mats. This could be a result of sulfate reduction (the spring contains millimolar levels of sulfate, and sulfate-reducing bacteria are numerous within the mat [M. M. Bateson and D. M. Ward, unpublished data]) or of reduc-

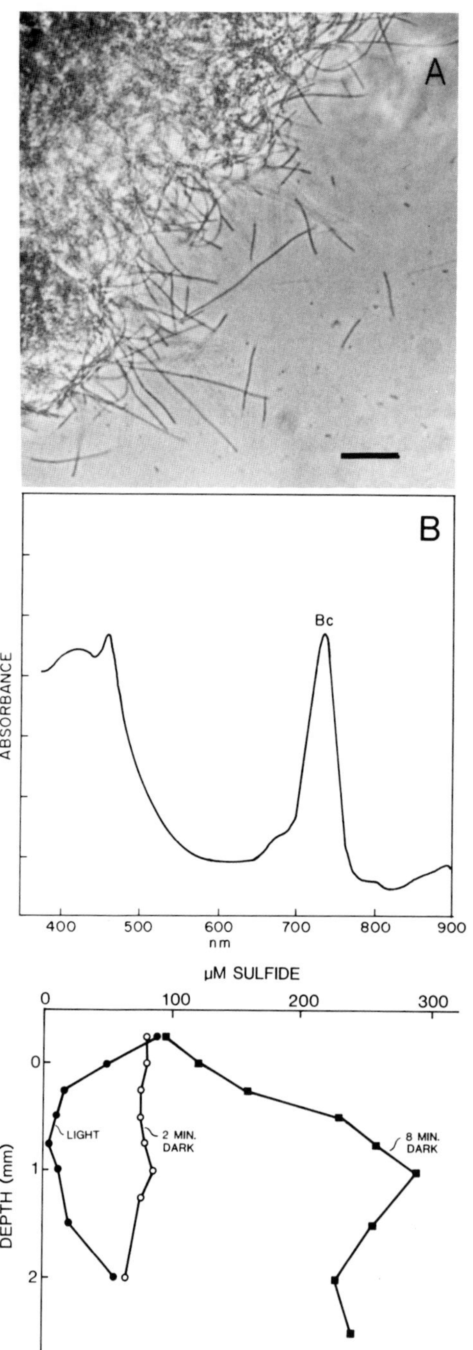

Figure 4. New Pit Spring *Chloroflexus* mat (52 to 58°C, pH 6.3, 34 μM sulfide) (see Fig. 1C). (A) Phase-contrast photomicrograph of top green layer. Bar, 25 μm. (B) In vivo absorption spectrum of mat. See legend to Fig. 2B for definition of symbols. (C) Vertical microprofiles of sulfide in the Painted Pool *Chloroflexus* mat, measured in full sunlight and after shift to darkness (from reference 17). (Results for New Pit Spring mat were similar, but could not be quantified as sulfide concentration was near the detection limit for the microelectrode [Revsbech and Ward, unpublished results].)

tion of S^0 (perhaps even by *Chloroflexus* sp. itself, as speculated by Giovannoni et al. [17]). In any case, the tight coupling between sulfide production and consumption within the mat suggests that biogenic sulfide may be a major source of recycled reduced sulfur used for reducing power in the photoautotrophic mat. In addition to serving as a source of reducing power, the primary sulfide in water above the mat must serve the important additional purpose of shielding the mat from atmospheric oxygen. In an anoxic atmosphere (e.g., during the early Precambrian period) a small net influx of reduced sulfur, and of course light energy, might suffice to drive the synthesis of such a mat. Thus, anoxygenic mats might not have been restricted to high-sulfide environments as they are today, and might have been far more widespread.

Mats Containing *Chromatium* sp.

Many neutral to slightly acid sulfide springs in the Mammoth Terraces Group, Yellowstone National Park, contain mats with *Chromatium* sp. (10). An example is the mat from Roland's Well (unofficial name for a spring found about 40 cm beneath the rim of a small pothole, located approximately 100 m northwest of New Pit Spring) (Fig. 1D). The mat is laminated and has been observed to accumulate to at least 5 mm in thickness. The predominant microorganism within the top purple layer is a short, rod-shaped bacterium which contains granules resembling elemental sulfur. The thin green layer beneath the purple layer is composed of filamentous bacteria (Fig. 5A). The mat is

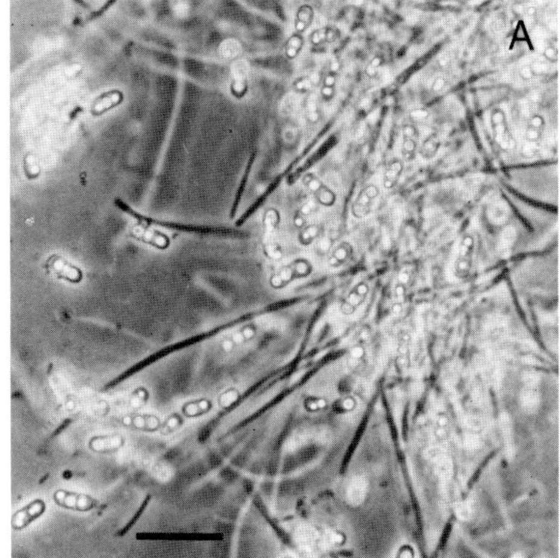

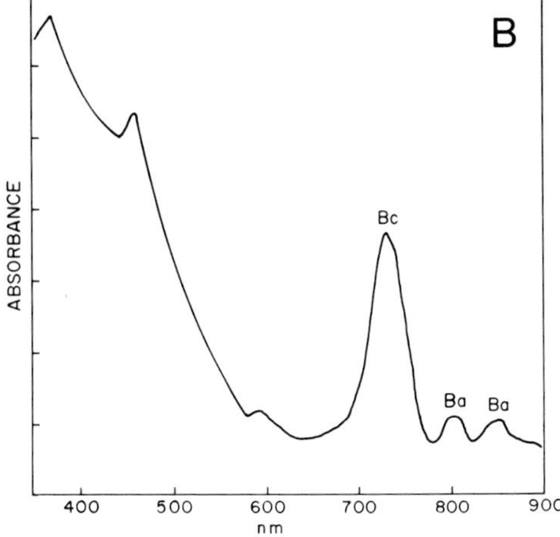

Figure 5. Roland's Well purple mat (52 to 55°C, pH 6.3, 41 μM sulfide) (see Fig. 1D). (A) Phase-contrast photomicrograph of the combined top purple and underlying green layers. Bar, 10 μm. (B) In vivo absorption spectrum of mat. See Fig. 2B for definition of symbols.

devoid of microorganisms that autofluoresce red. The presence of both purple and green photosynthetic bacteria is also suggested by the in vivo absorption spectrum (Fig. 5B): bacteriochlorophylls *a* and *c* are abundant, while chlorophyll *a* and phycobilins are absent. A purple sulfur bacterium, *Chromatium tepidum*, has been isolated from this type of mat (20, 21). These mats are usually found above 50°C (inhibitory to cyanobacteria in these sulfide springs) and below 58°C (the upper temperature limit for *C. tepidum*).

On the basis of the evidence at hand, it is likely that the *Chromatium* sp. grows above a layer of *Chloroflexus* sp. in such mats. It is not clear why the *Chromatium* sp. is restricted to only certain springs and is not present in *Chloroflexus* mats with temperatures below the upper temperature limit of *C. tepidum* (e.g., New Pit Spring). The two phototrophs should not be in competition for light, as they have complementing absorption spectra in the infrared region. Sulfide microelectrode profiles in the Roland's Well mat (N. P. Revsbech and D. M. Ward, unpublished data) resemble those shown in Fig. 4C and suggest that, as in *Chloroflexus* mats, sulfide consumption is light dependent and that the mat community includes sulfate- or sulfur-reducing bacteria, or both, which provide an internal supply of sulfide. Thus, the *Chromatium* sp. and the *Chloroflexus* sp. might not be in competition for the same sulfide pool, as the former could utilize primary sulfide in the spring water (or some other reductant not yet identified) while the latter could be supplied with biogenic sulfide from beneath.

Chlorobium Mats

Recently, Castenholz has described mats formed by *Chlorobium* sp. in New Zealand hot springs (45 to 55°C) which are moderately acid (pH 4.5 to 6) and contain sulfide (200 to 500 μM) (11a) (see Fig. 1E). These mats are up to about 3 mm thick, unlaminated, but slimy in consistency, occurring above a clay sediment in the effluent channel. (As with the *C. caldarium* mats described above, the lack of laminations probably reflects the lack of filamentous microorganisms in this community.) The in vivo absorption spectrum reveals the predominance of bacteriochlorophyll *c* as well as the absence of chlorophyll *a* (Fig. 6); in situ spectral measurements show that this absorption occurs within the upper 0.7 mm of the mat (11a). A moderately thermophilic *Chlorobium* sp. has been isolated from such

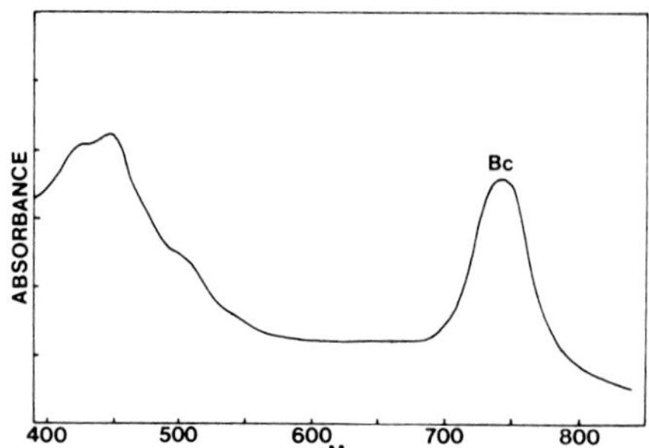

Figure 6. In vivo absorption of the New Zealand Travel Lodge Stream *Chlorobium* mat (45°C, pH 5.9, 250 μM sulfide). See Fig. 1E and also the legend to Fig. 2B for definition of symbols.

mats (M. T. Madigan and R. W. Castenholz, unpublished data).

As with the *Chloroflexus* mats (Fig. 4C), sulfide consumption as measured with microelectrodes was light dependent and sulfide is produced within the mat (11a; R. W. Castenholz, J. Bauld, and B. B. Jørgensen, unpublished data). CO_2 assimilation in the light was stimulated by addition of sulfide and sometimes by sulfite as well. The collective evidence indicates that the mat is built by the photoautotrophic growth of a thermophilic *Chlorobium* sp.

ANOXYGENIC/OXYGENIC MATS WITH CYANOBACTERIA

Though combined high temperature and high sulfide excludes cyanobacteria, it has recently been shown that cyanobacteria can inhabit the subsurface layers of mats which form in high-temperature, high-sulfide springs. For example, some Icelandic springs contain *Chloroflexus* mats which overlie a deeper layer of the cyanobacterium *Chlorogloeopsis* sp. (formerly referred to as "HTF" *Mastigocladus*) (Fig. 1F). Microelectrode studies of a mat of this type revealed the occurrence of oxygenic photosynthesis and resultant accumulation of oxygen in the *Chlorogloeopsis* layer (Fig. 7). Presumably the depletion of sulfide by *Chloroflexus* sp. in the top layer provides a niche for *Chlorogloeopsis* sp., which is sensitive to sulfide (see reference 9).

In high-sulfide springs where temperatures are not extreme, cyanobacteria may carry out anoxygenic photosynthesis themselves (see A. Oren, this volume). One such mat occurs in a slightly acid (pH 6.1), warm (46°C), high-sulfide (1,200 μM) spring called Stinky Springs (located near the south end of Little Mountain, 7 miles west-northwest of Corinne, Utah), where a mat (ca. 1 to 2 mm thick) is produced by *Oscillatoria* spp. (Fig. 8A). Microelectrode experiments (Fig. 8B) show that sulfide is depleted near the mat surface (presumably supporting anoxygenic photosynthesis by *Oscillatoria* spp.; one isolate has been shown to have this capacity in pure culture [12]). Sulfide consumption totally eliminates sulfide within the upper 1,400 μm of the mat, where oxygenic photosynthesis by *Oscillatoria* spp. then occurs.

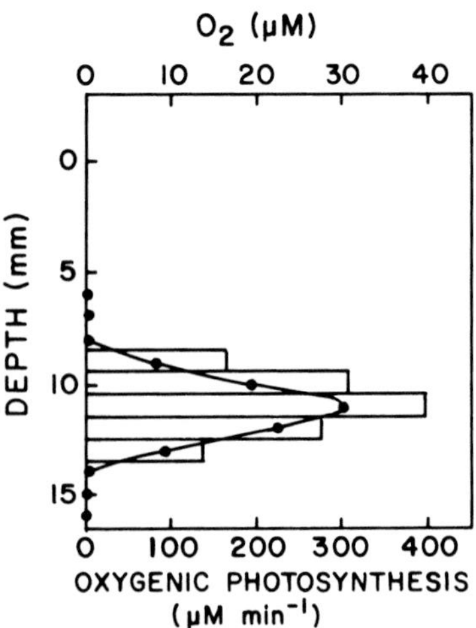

Figure 7. Vertical microprofiles of oxygenic photosynthesis (bars) and oxygen in Icelandic Spring A mat (similar to that shown in Fig. 1F), in which *Chloroflexus* sp. overlies *Chlorogloeopsis* sp. (ca. 60°C, 40 μM sulfide in water above the mat) (from Jørgensen and Nelson [19a]).

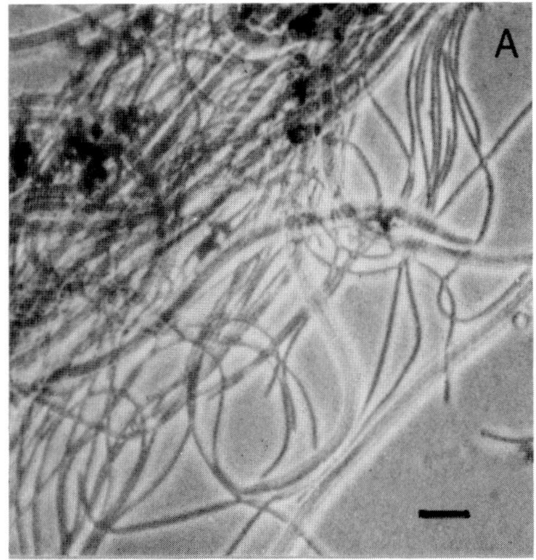

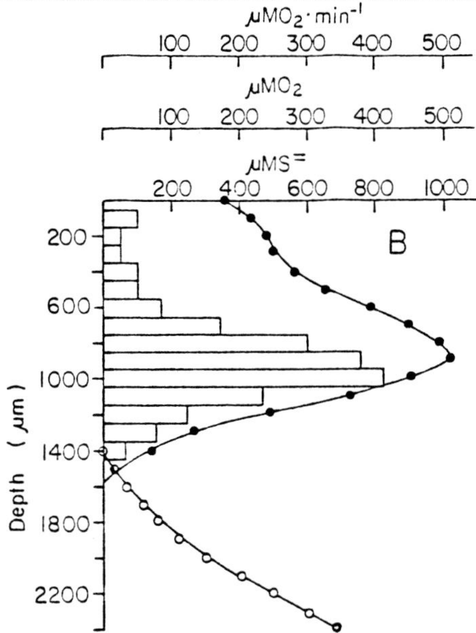

Figure 8. Stinky Springs cyanobacterial mat (46°C, pH 6.1, 1,200 μM sulfide). (A) Phase-contrast photomicrograph of the mat. Bar, 10 μm. (B) Microprofiles of sulfide (○), O_2 (●), and oxygenic photosynthesis (bars) in mat measured in full sunlight. Water above mat contained 1,200 μM sulfide.

POSSIBLE EVOLUTIONARY SIGNIFICANCE

Because of the variety of chemical environments provided by hot springs, thermophilic microbial mats exhibit a great range of types of mat-building microbial communities. A knowledge of these mat types should aid in the interpretation of ancient stromatolitic microbial communities. While the existence of modern mats does not prove the course of evolution of stromatolitic microbial communities, it at least provides evidence of the possible types of mats which may have been important at different times. For example, it is clear that well-laminated mats may be built solely by anoxygenic photosynthetic bacteria. This substantiates the suggestion that stromatolitic communities can be formed in anoxic conditions and in the complete absence of cyanobacteria or their oxygenic ancestors (4). Mats of this type may have preceded the development of cyanobacterial mats in the anoxic early Precambrian Earth (28). Our knowledge of these mats suggests that they may be restricted to specialized environments on the modern Earth mainly for ecologic reasons (i.e., the present atmosphere is oxic); in an anoxic atmosphere (i.e., the Archean or early Proterozoic eras) they might have been widespread.

In some modern mats anoxygenic photosynthesis provides a niche in deeper mat layers for cyanobacteria to carry out oxygenic photosynthesis. Such mats may represent "transitional" environments in which conditions advantageous for the evolution of oxygenic photosynthesis might have existed. It has also been pointed out that these mats do not necessarily oxygenate their external environment, as the oxygen produced is consumed within the mat (19). It is thus possible that stromatolites which formed before the oxygenation of the Earth's atmosphere might have contained oxygenic phototrophs. The removal of oxygen might have been through either abiologic or biologic means (e.g., aerobic respiration).

Modern hot spring microbial mats also provide distinctly different microbial communities which can be exploited for developing and testing hypotheses related to the interpretation of stromatolite microbiotas.

Examples of topics which might be addressed include (i) the role of filamentous bacteria in lamination of mats, (ii) how decomposition biases preservable structural remains, and (iii) whether differences in chemistry reflect differences in key microbial inhabitants (see Ward et al., this volume).

Acknowledgments. We thank the U.S. National Park Service for permission to collect samples in Yellowstone National Park, and Paul McFarlane (Forest Research Institute, Rotorua) for help in processing New Zealand samples.

We also thank the National Science Foundation (grants BSR-8506602 to D.M.W. and BSR-8408179 to R.W.C.) for financial support. Part of this study was supported by a National Research Council associateship (to Y.C.).

LITERATURE CITED

1. **Anderson, K. L., T. A. Tayne, and D. M. Ward.** 1987. Formation and fate of fermentation products in hot spring cyanobacterial mats. *Appl. Environ. Microbiol.* **53**:2343–2352.

1a. **Bateson, M. M., and D. M. Ward.** 1988 Photoexcretion and fate of glycolate in a hot spring cyanobacterial mat. *Appl. Environ. Microbiol.* **54**:1738–1743.

2. **Bauld, J., and T. D. Brock.** 1973. Ecological studies of *Chloroflexis*, a gliding photosynthetic bacterium. *Arch. Mikrobiol.* **92**:267–284.

3. **Brock, T. D.** 1973. Lower pH limit for the existence of blue-green algae: evolutionary and ecological implications. *Science* **179**:480–483.

4. **Brock, T. D.** 1973. Evolutionary and ecological aspects of the cyanophytes, p. 487–500. *In* N. G. Carr and B. A. Whitton (ed.), *The Biology of Blue-Green Algae.* University of California Press, Berkeley.

5. **Brock, T. D.** 1978. *Thermophilic Microorganisms and Life at High Temperatures.* Springer-Verlag, New York.

6. **Castenholz, R. W.** 1969. Thermophilic blue-green algae and the thermal environment. *Bacteriol. Rev.* **33**:476–504.

7. **Castenholz, R. W.** 1973. Ecology of blue-green algae in hot springs, p. 379–414. *In* N. G. Carr and B. A. Whitton (ed.), *The Biology of Blue-Green Algae.* University of California Press, Berkeley.

8. **Castenholz, R. W.** 1973. The possible photosynthetic use of sulfide by the filamentous phototrophic bacteria of hot springs. *Limnol. Oceanogr.* **18**:863–876.

9. **Castenholz, R. W.** 1976. The effect of sulfide on the blue green algae of hot springs. I. New Zealand and Iceland. *J. Phycol.* **12**:54–68.

10. **Castenholz, R. W.** 1977. The effect of sulfide on the blue-green algae of hot springs. II. Yellowstone National Park. *Microb. Ecol.* **3**:79–105.

11. **Castenholz, R. W.** 1984. Composition of hot spring microbial mats: a summary, p. 101–119. *In* Y. Cohen, R. W. Castenholz, and H. O. Halvorson (ed.), *Microbial Mats: Stromatolites.* Alan R. Liss, Inc., New York.

11a. **Castenholz, R. W.** 1988. The green sulfur and nonsulfur bacteria of hot springs, p. 243–255. *In* J. M. Olson, J. G. Ormerod, J. Amesz, E. Stackebrandt, and H. G. Trüper (ed.), *Green Photosynthetic Bacteria.* Plenum Publishing Corp., New York.

12. **Cohen, Y., B. B. Jørgensen, N. P. Revsbech, and R. Poplawski.** 1986. Adaptation to hydrogen sulfide of oxygenic and anoxygenic photosynthesis among cyanobacteria. *Appl. Environ. Microbiol.* **51**: 398–407.

13. **De Luca, P., and A. Moretti.** 1983. Floridosides in *Cyanidium caldarium, Cyanidioschyzon merolae,* and *Galdieria sulphuraria* (Rhodophyta, Cyanidiophyceae). *J. Phycol.* **19**:368–369.

13a. **Dobson, G., D. M. Ward, N. Robinson, and G. Eglinton.** 1988. Biogeochemistry of hot spring environments: extractable lipids of a cyanobacterial mat. *Chem. Geol.* **68**:155–179.

14. **Doemel, W. N., and T. D. Brock.** 1977. Structure, growth, and decomposition of laminated algal-bacterial mats in alkaline hot springs. *Appl. Environ. Microbiol.* **34**:433–452.

15. **Gerdes, G., and W. E. Krumbein.** 1984. Animal communities in recent potential stromatolites of hypersaline origin, p. 59–83. *In* Y. Cohen, R. W. Castenholz, and H. O. Halvorson (ed.), *Microbial Mats: Stromatolites.* Alan R. Liss, Inc., New York.

16. **Gerdes, G., J. Spira, and C. Dimentman.** 1985. The fauna of the Gavish Sabkha and the Solar Lake—a comparative study, p. 322–345. *In* G. M. Friedman and W. E. Krumbein (ed.), *Hypersaline Ecosystems: the Gavish Sabkha.* Springer-Verlag, Heidelberg.

17. **Giovannoni, S. J., N. P. Revsbech, D. M. Ward, and R. W. Castenholz.** 1987. Obligately phototrophic *Chloroflexus*: primary production in anaerobic hot spring microbial mats. *Arch. Microbiol.* 147:80–87.

18. **Javor, B. J., and R. W. Castenholz.** 1984. Invertebrate grazers of microbial mats, Laguna Guerrero Negro, Mexico, p. 85–94. *In* Y. Cohen, R. W.

Castenholz, and H. O. Halvorson (ed.), *Microbial mats: Stromatolites.* Alan R. Liss, Inc., New York.

19. **Jørgensen, B. B., Y. Cohen, and N. P. Revsbech.** 1986. Transition from anoxygenic to oxygenic photosynthesis in a *Microcoleus chthonoplastes* cyanobacterial mat. *Appl. Environ. Microbiol.* **51**:408–417.

19a. **Jørgensen, B. B., and D. C. Nelson.** 1988. Bacterial zonation, photosynthesis, and spectral light distribution in hot spring microbial mats of Iceland. *Microb. Ecol.* **16**:133–147.

20. **Madigan, M. T.** 1984. A novel photosynthetic purple bacterium isolated from a Yellowstone hot spring. *Science* **225**:313–315.

21. **Madigan, M. T.** 1986. *Chromatium tepidum* sp. nov., a thermophilic photosynthetic bacterium of the family *Chromatiaceae. Int. J. Syst. Bacteriol.* **36**:222–227.

22. **Revsbech, N. P., and B. B. Jørgensen.** 1986. Microelectrodes: their use in microbial ecology. *Adv. Microb. Ecol.* **9**:293–352.

23. **Revsbech, N. P., and D. M. Ward.** 1983. Oxygen microelectrode that is insensitive to medium chemical composition: use in an acid microbial mat dominated by *Cyanidium caldarium. Appl. Environ. Microbiol.* **45**:755–759.

24. **Revsbech, N. P., and D. M. Ward.** 1984. Microelectrode studies of interstitial water chemistry and photosynthetic activity in a hot spring microbial mat. *Appl. Environ. Microbiol.* **48**:270–275.

25. **Revsbech, N. P., and D. M. Ward.** 1984. Microprofiles of dissolved substances and photosynthesis in microbial mats measured with microelectrodes, p. 171–188. *In* Y. Cohen, R. W. Castenholz, and H. O. Halvorson (ed.), *Microbial Mats: Stromatolites.* Alan R. Liss, Inc., New York.

26. **Sandbeck, K. A., and D. M. Ward.** 1981. Fate of immediate methane precursors in low-sulfate, hot spring algal-bacterial mats. *Appl. Environ. Microbiol.* **41**:775–782.

27. **Schopf, J. W.** 1978. The evolution of the earliest cells. *Sci. Am.* **239(3)**:110–134.

28. **Schopf, J. W., J. M. Hayes, and M. R. Walter.** 1983. Evolution of earth's earliest ecosystems: recent progress and unsolved problems, p. 361–384. *In* J. W. Schopf (ed.), *Earth's Earliest Biosphere.* Princeton University Press, Princeton, N.J.

29. **Tayne, T. A., J. E. Cutler, and D. M. Ward.** 1987. Use of *Chloroflexus*-specific antiserum to evaluate filamentous bacteria of a hot spring microbial mat. *Appl. Environ. Microbiol.* **53**:1965–1968.

30. **Walker, J. C. G., C. Klein, M. Schidlowski, J. W. Schopf, D. J. Stevenson, and M. R. Walter.** 1983. Environmental evolution of the Archean-early Proterozoic Earth, p. 260–290. *In* J. W. Schopf (ed.), *Earth's Earliest Biosphere.* Princeton University Press, Princeton, N.J.

31. **Walter, M. R.** 1977. Interpreting stromatolites. *Am. Sci.* **65**:562–571.

32. **Ward, D. M., E. Beck, N. P. Revsbech, K. A. Sandbeck, and M. R. Winfrey.** 1984. Decomposition of hot spring microbial mats, p. 191–214. *In* Y. Cohen, R. W. Castenholz, and H. O. Halvorson (ed.), *Microbial Mats: Stromatolites.* Alan R. Liss, Inc., New York.

33. **Ward, D. M., and G. J. Olson.** 1980. Terminal processes in the anaerobic degradation of an algal-bacterial mat in a high sulfate hot spring. *Appl. Environ. Microbiol.* **40**:67–74.

34. **Ward, D. M., T. A. Tayne, K. L. Anderson, and M. M. Bateson.** 1987. Community structure and interactions among community members in hot spring cyanobacterial mats. *Symp. Soc. Gen. Microbiol.* **41**:179–210.

35. **Wickstrom, C. E., and R. W. Castenholz.** 1973. Thermophilic ostracod: aquatic metazoan with the highest known temperature tolerance. *Science* **181**:1063–1064.

36. **Wickstrom, C. E., and R. W. Castenholz.** 1985. Dynamics of cyanobacterial and ostracod interactions in an Oregon hot spring. *Ecology* **66**:1024–1041.

37. **Wiegert, R. G., and R. Mitchell.** 1973. Ecology of Yellowstone thermal effluent systems: intersects of blue-green algae, grazing flies (*Paraceonia, Ephydridae*) and water mites (*Partununiella, Hydrachnellae*). *Hydrobiologia* **41**:251–271.

38. **Woese, C. R.** 1987. Bacterial evolution. *Microbiol. Rev.* **51**:221–271.

Chapter 2

Microbial Mats at Deep-Sea Hydrothermal Vents: New Observations

Shimshon Belkin and Holger W. Jannasch

INTRODUCTION

Until approximately a decade ago, deep-sea microbiology was associated with research on bacterial life at low temperature and high pressure under extremely oligotrophic conditions. Therefore, a term such as "deep-sea microbial mats" would have seemed contradictory: life in the deep sea was believed to be so scarce that it could not include prolific, mat-forming bacterial communities.

This perspective was changed when, by using the research submersible *Alvin*, the existence of dense invertebrate communities was discovered at certain specific sites in the deep ocean, namely, at the immediate vicinities of hydrothermal vents along the fracture zones of submarine spreading centers (3, 13).

The emission of hydrothermal fluid at these vents is the result of seawater circulation through the Earth's crust at tectonic spreading centers. Several kilometers below the sea floor, the water interacts with the basaltic rock at high pressures and temperatures. When it reaches the sea floor surface again, the hot fluid is enriched in metals, hydrogen sulfide, and molecular hydrogen. Depending upon the depth at which the rising fluid mixes with cold, oxidized seawater, it may emerge at either warm or hot vents. In the former case, the mixing occurs below the surface and the exit temperature is up to 20 to 30°C. In the absence of subsurface mixing, the hydrothermal fluid is discharged directly into the bottom seawater at temperatures above 350°C. In these hot vents, calcium and other metal precipitates form characteristic "chimneys" around the vent openings. The typical smokelike appearance of the fluid released from the chimney is caused by particulate metal sulfides and has given rise to the term "black smokers."

The plumes at warm as well as hot vents represent a mix of the anaerobic, highly reduced hydrothermal fluid and the oxygenated bottom seawater. The availability of both oxygen and the geothermally generated reducing power provides an ideal habitat for chemosynthetic bacteria. Indeed, it is assumed that these bacteria provide the organic carbon upon which nutrition of the whole vent community is based (10). Di-

Shimshon Belkin • Laboratory for Environmental Applied Microbiology, The Jacob Blaustein Desert Research Institute, Ben-Gurion University of the Negev, Sede Boqer Campus 84993, Israel. *Holger W. Jannasch* • Biology Department, Woods Hole Oceanographic Institution, Woods Hole, Massachusetts 02543.

verse bacterial populations have been found in three types of ecological niches: in the water itself, in symbiotic associations with vent invertebrates, and on surfaces in the vent surroundings.

Bacteria that densely inhabit the surfaces of lava, bivalve shells, and worm tubes have been described in the past (14). Although they formed a thin (5- to 10-μm) layer in comparison with the *Beggiatoa* mats described below, they appeared to cover all surfaces that were intermittently exposed to the sulfide-containing hydrothermal fluid. Both unicellular and filamentous organisms were observed. Some of the latter, upon isolation, were shown to belong to the genus *Hyphomonas* (18). Other filaments resembled *Beggiatoa*, *Leucothrix*, or *Thiothrix* species, whereas another type exhibited a striking resemblance to species of the cyanobacterium *Calothrix*.

Although these mats have been described earlier and with emphasis on their chemosynthetic function (9), the present discussion concentrates on two new observations: the extremely thermophilic constituents of deep-sea microbial mats and the unusually heavy *Beggiatoa* mats that were discovered at the Guaymas Basin vent site.

SIGNIFICANCE OF SURFACE ATTACHMENT TO VENT MICROORGANISMS

The hydrothermal vent biotope is basically a flowing-water environment. In such surroundings, attachment to a surface can be considered generally advantageous (7), but it would seem to be of particular importance to chemosynthetic bacteria that require both sulfide and oxygen. Unless such bacteria can adhere to a suitable substratum, they are bound to be flushed out of the relatively narrow zone where O_2 and H_2S coexist. The residence time in such zones will depend upon the flow rate of the venting water and upon the degree of turbulance generated. Since flow rates may be very high (2 m s^{-1} in black smokers; 11), a suitable combination of temperature and concentrations of O_2 and sulfide may be encountered only briefly and intermittently.

EXTREMELY THERMOPHILIC BACTERIA

The above limitation applies even more strongly to thermophilic bacteria. Some of the strains isolated from the vents have temperature optima for growth in the 80 to 100°C range (H. W. Jannasch, *in* D. L. Wise, ed., *Biotechnology Applied to Fossil Fuels*, in press). Two such strains, both archaebacteria, temporarily designated S and SY, may belong to the genus *Desulfurococcus* (K. O. Stetter, personal communication); they have temperature optima at 85 and 90°C, respectively. Both have been isolated from the 13°N site in the east Pacific (depth, 2,500 m). A large number of similar isolates have been obtained from the Guaymas Basin vent site (see below). Some extremely thermophilic, anaerobic isolates exhibit a tolerance for free oxygen at temperatures below 65 to 70°C (17). This must be considered an important metabolic characteristic for the survival of extremely thermophilic anaerobes in cold, oxygenated seawater and for their transfer from one vent to another.

One of the well-described organisms, isolated from a black smoker wall at the 21°N site, is a methanogen, *Methanococcus jannaschii*, which grows optimally with a generation time of 28 min at 86°C (15). Another isolate, *Staphylothermus marinus* (6), was obtained from the 11°N vent site and grew at an optimal temperature of 92°C and a maximum of 98°C. Several extreme thermophiles, as yet uncharacterized, were obtained from chimney material collected at the mid-Atlantic vent site at a depth of 3,700 m.

SULFUR REDUCTION AND THERMOPHILIC PHYSIOLOGY

All extreme thermophiles described to date are archaebacteria with optimal growth temperatures above 80°C. The phylogenetic branch to which most of them belong was known in the past as thermoacidophilic. In recent years, this group has justifiably been renamed the sulfur-metabolizing branch of the archaebacteria (17). All members of this diverse group, autotrophic as well as heterotrophic species, are capable of either oxidizing or (in most cases) reducing elemental sulfur.

High-temperature sulfur metabolism, however, is not a property unique to the sulfur-metabolizing archaebacteria. It is also shared by thermophilic methanogens and, more importantly, by newly described eucaryotic extreme thermophiles: *Thermotoga maritima* (8) and *Thermotoga neapolitana* (2, 10a) are very efficient sulfur reducers. Thus, the metabolism of elemental sulfur seems closely related to life at high temperatures. This point is strengthened by the observation that elemental sulfur appears much more amenable to reduction at elevated temperatures (1).

Although the use of elemental sulfur by extreme thermophiles is often facultative (17), in some species it is obligatory. Since elemental sulfur is insoluble in aquatic media, its occurrence would be restricted either to suspended particles or to surface deposits. The need for attachment of thermophiles to a solid substratum may thus be enhanced by the requirement for elemental sulfur.

Since the accumulation of molecular hydrogen is known to inhibit fermentative metabolism, the requirement for elemental sulfur may be interpreted as a mechanism for the removal of hydrogen through the reduction of sulfur (8). This may not apply to the free-flowing vent environment. However, since bacteria inhabiting the chimney itself are exposed to slowly permeating hydrothermal fluid, this suggested role for sulfur reduction may be essential for bacterial survival.

MICROBIAL MATS OF THE GUAYMAS BASIN

The warm and hot vent sites previously discussed, such as the one at the 21°N site (Fig. 1), are all situated along offshore spreading centers and are surrounded by bare basalt. A completely different situation arises along the fracture line as it extends into the Gulf of California (Fig. 1). Unlike the open ocean, this area is characterized by high sedimentation rates (over 1 m/1,000 years) of both terrigenous and pelagic origins (5, 16). As a result, the rift floor at the Guaymas Basin region is covered with sedi-

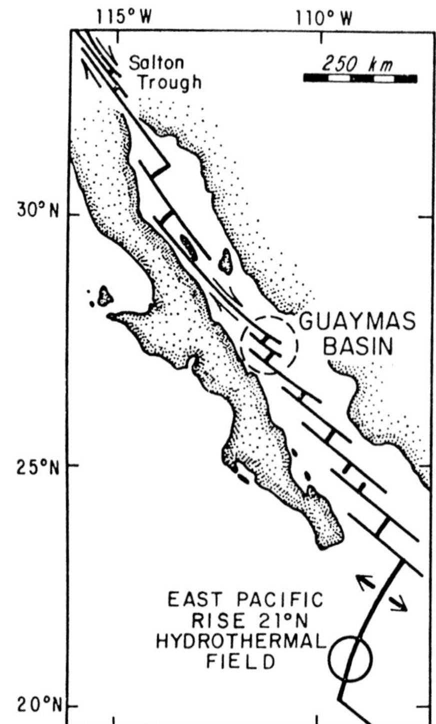

Figure 1. Locations of the Guaymas Basin and 21°N East Pacific Rise vent sites. Reproduced from *Earth and Planetary Science Letters* (16) with permission of the publisher.

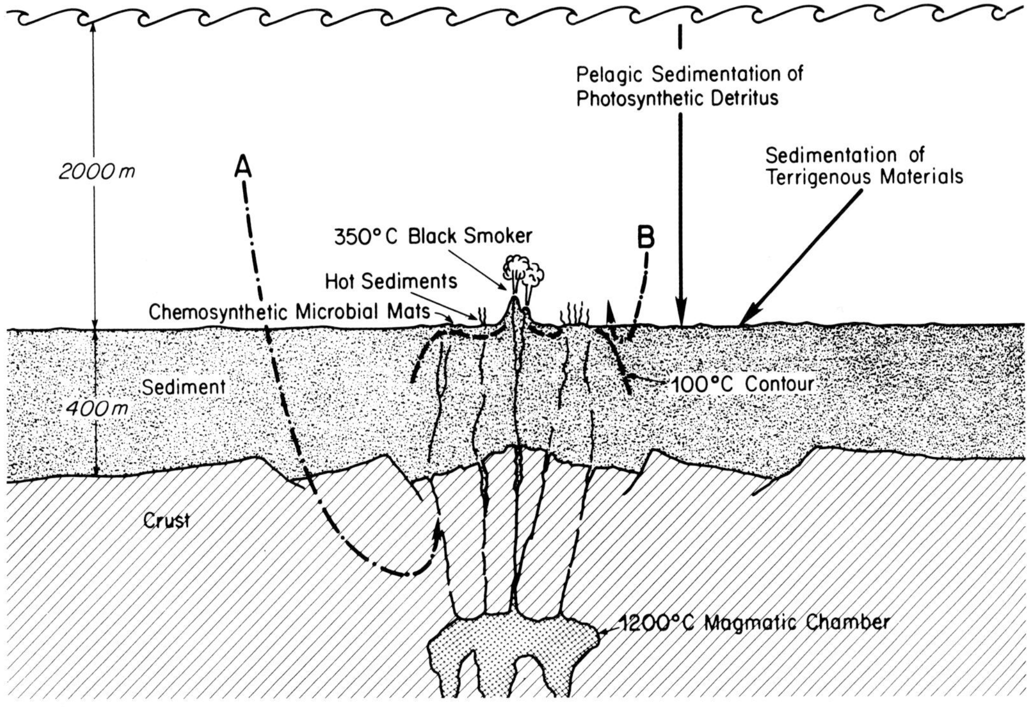

Figure 2. Scheme of the sediment-covered Guaymas Basin vent site. A and B, Seawater circulation systems.

ments of high organic content up to 400 m thick (Fig. 2). The prolonged passage through this layer alters the chemical composition of the venting hydrothermal fluid (Table 1). It is relatively depleted of some of the elements, such as iron and manganese, and enriched in others, such as magnesium and calcium. Ammonia, which is characteristic of the Guaymas Basin emissions, is probably a product of decomposing organic matter.

As in the oceanic vents, the emitted fluids emerge on the basin floor at either warm or hot vents; the latter, with exit temperatures of 155 to 355°C, form the characteristic black smokers. As in other Pacific sites, the most conspicuous invertebrate is the tube worm *Riftia pachyptila*. The most striking phenomena of the basin biota, however, are the thick, white, microbial mats, composed almost entirely of *Beggiatoa*-like filaments. Some of these filaments are

Table 1.
Major Constituents of Hydrothermal Vent Emissions[a]

Constituent (unit) or property	Amt present in samples from[b]:		
	Guaymas Basin	21°N	Ambient seawater
Fe (μM)	56	1,664	0.001
Mn (μM)	139	960	0.001
Ag (nM)	230	38	0.02
Pb (nM)	265	308	0.01
Na (mM)	489	432	464
K (mM)	48.5	28	9.79
Mg (mM)	29	0	52.7
Ca (μM)	29	15.6	10.2
Cl (mM)	601	489	541
SO_4 (mM)	0	0	27.9
Si (mM)	12.9	17.6	0.16
NH_3 (mM)	15.6	0	0
H_2S (mM)	5.82	7.3	0
H_2 (mM)	8–50	8–50	<0.001
CH_4 (mM)	(1–1.6)	1–1.6	<0.001
pH	5.9	3.4	8.0
Alkalinity (meq)	10.6	−0.4	0.16

[a] Depths for Guaymas Basin and 21°N were 2,003 and 2,610 m, respectively.
[b] References 4 and 11.

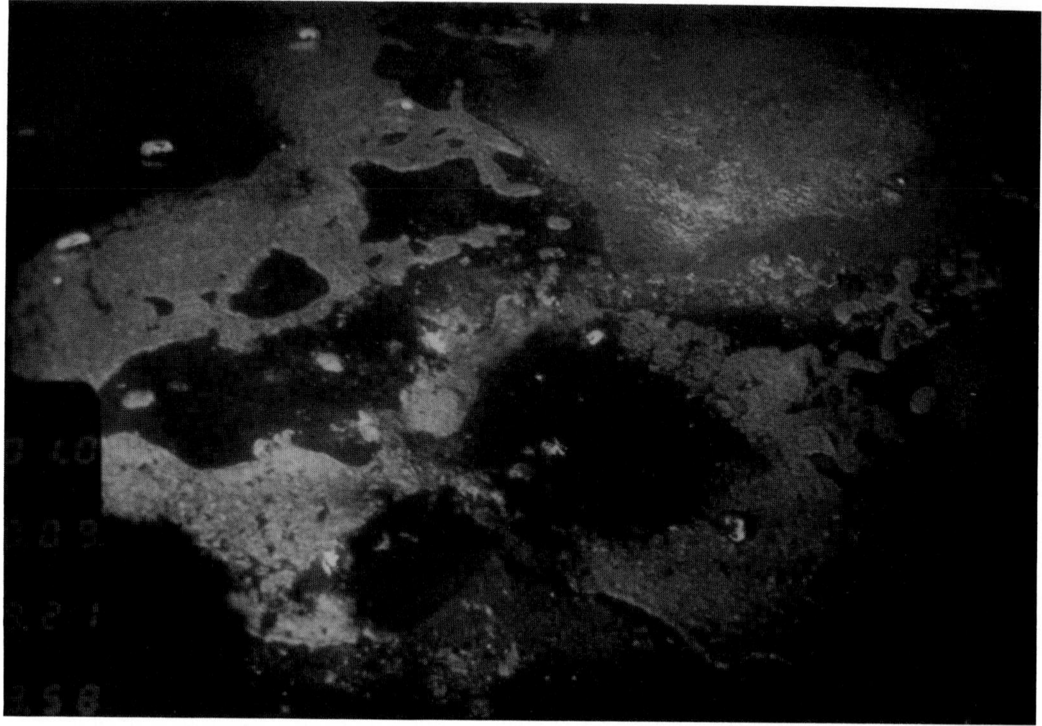

Figure 3. Microbial mats, consisting mainly (about 99%) of *Beggiatoa*-like filaments, at the Guaymas Basin hydrothermal vent site (depth, 2,003 m). The dark, mat-free spot in the lower center is about 1.5 m across (photograph by H. W. Jannasch).

over 100 μm wide (9). They form mats several meters across and 1 to 5 cm thick on the sediment surface (Fig. 3). Mats covering *Riftia* clumps and chimney walls were up to 30 cm thick (D.C. Nelson, C.O. Wirsen, and H.W. Jannasch, manuscript in preparation).

Mat material was collected from the submersible with a device dubbed a slurp gun. In the laboratory on the ship, the filaments settled out to form a close, pink suspension. The cells showed active gliding motility for a few hours, possibly in search of a suitable interface, but attempts to maintain live cultures for prolonged periods have failed so far. $^{14}CO_2$ uptake and ribulose-bisphosphate carboxylase activities were low but measurable, with the latter being optimal at 45°C (Nelson et al., in preparation). The Guaymas Basin, unlike the open ocean environment, is characterized by the presence of organic detritus, and it has yet to be determined whether bacterial chemosynthesis is essential for the community food chain. The presence of *Riftia*, existing in symbiotic association with chemosynthetic bacteria, indicates that chemoautotrophy does indeed play a part.

Like other geothermal areas, the Guaymas Basin region has also proven to be a good source of extreme thermophiles. The need for surface attachment, as discussed above, is more easily met in this sediment-covered environment, and the local thermophiles appear to play an active role in mineralization and transformation processes in the sediment, through which the warm to hot water continuously percolates. Temperature profiles taken from several closely spaced locations in the vicinity of a small black smoker were extremely heterogeneous (Jannasch, in press). The temperatures measured at a depth of 60 cm below the sediment surface varied between 40 and 150°C. This may be explained by the existence of an advective downward flow of pore water in that region (F. L. Sayles, personal communication). It is from these hot sediments that a large number of extreme thermophiles, which are in the process of being characterized, have been isolated.

ACKNOWLEDGMENT. The work presented in this chapter was supported by National Science Foundation grant OCE87-581. This is Woods Hole Oceanographic Institution publication no. 6611.

LITERATURE CITED

1. **Belkin, S., C. O. Wirsen, and H. W. Jannasch.** 1985. Biological and abiological sulfur reduction at high temperatures. *Appl. Environ. Microbiol.* **49:** 1057–1061.
2. **Belkin, S., C. O. Wirsen, and H. W. Jannasch.** 1986. A new sulfur-reducing, extremely thermophilic eubacterium from a submarine thermal vent. *Appl. Environ. Microbiol.* **51:**1180–1185.
3. **Corliss, J. B., J. Dymond, L. I. Gordon, J. M. Edmond, R. P. von Herzen, R. D. Ballard, K. Green, D. Williams, A. Bainbridge, K. Crane, and T. H. van Andel.** 1979. Submarine thermal springs on the Galapagos Rift. *Science* **203:**1073–1083.
4. **Edmond, J. M., and K. L. Von Damm.** 1985. Chemistry of ridge crest hot springs. *Bull. Biol. Soc. Wash.* **6:**43–47.
5. **Einsele, G., J. M. Gieskes, J. Curray, D. M. Moore, E. Aguayo, M.-P. Aubry, D. Fornari, J. Guerrero, M. Kastner, K. Kelts, M. Lyle, Y. Matoba, A. Molina-Cruz, J. Niemitz, J. Rueda, A. Saunders, H. Schrader, B. Simoneit, and V. Vacquier.** 1980. Intrusion of basaltic sills into highly porous sediments, and resulting hydrothermal activity. *Nature* (London) **283:**441–445.
6. **Fiala, G., D. O. Stetter, H. W. Jannasch, T. A. Longworthy, and J. Madon.** 1986. *Staphylothermus marinus* sp. nov. represents a novel genus of extremely thermophilic submarine heterotrophic archaebacteria growing up to 98°C. *Syst. Appl. Microbiol.* **8:**106–113.
7. **Fletcher, M., and K. C. Marshall.** 1982. Are solid surfaces of ecological significance to aquatic bacteria? *Adv. Microb. Ecol.* **6:**199–236.
8. **Huber, R., T. A. Longworthy, H. Konig, M. Thomm, C. R. Woese, U. B. Sleyter, and K. O. Stetter.** 1986. *Thermotoga maritima* sp. nov. represents a new genus of unique extremely thermophilic eubacteria growing up to 90°C. *Arch. Microbiol.* **144:**324–333.
9. **Jannasch, H. W.** 1984. Chemosynthetic microbial mats of deep sea hydrothermal vents, p. 121–131. *In* Y. Cohen, R. W. Castenholz, and H. O. Halvorson (ed.), *Microbial Mats: Stromatolites.* Alan R. Liss, Inc., New York.
10. **Jannasch, H. W.** 1985. The chemosynthetic support of life and the microbial diversity at deep sea hydrothermal vents. *Proc. R. Soc. London Ser. B* **225:**277–297.

10a. **Jannasch, H.W., R. Huber, S. Belkin, and K.O. Stetter.** 1988. *Thermotoga neapolitana* sp. nov. of the extremely thermophilic, eubacterial genus *Thermotoga. Arch. Microbiol.* **150:**103–104.

11. **Jannasch, H. W., and M. J. Mottl.** 1985. Geomicrobiology of deep-sea hydrothermal vents. *Science* **229:**717–725.
12. **Jannasch, H. W., and C. D. Taylor.** 1984. Deep-sea microbiology. *Annu. Rev. Microbiol.* **38:**487–514.
13. **Jannasch, H. W., and C. O. Wirsen.** 1979. Chemosynthetic primary production at East Pacific sea floor spreading centers. *BioScience* **29:**592–598.
14. **Jannasch, H. W., and C. O. Wirsen.** 1981. Morphological survey of microbial mats near deep-sea thermal vents. *Appl. Environ. Microbiol.* **41:**528–538.
15. **Jones, W. J., J. A. Leigh, F. Mayer, C. R. Woese, and R. S. Wolfe.** 1983. *Methanococcus jannaschii* sp. nov., an extremely thermophilic methanogen from a submarine hydrothermal vent. *Arch. Microbiol.* **136:**254–261.
16. **Lonsdale, P. F., J. L. Bischoff, V. M. Burns, M. Kastner, and R. E. Sweeney.** 1980. A high-temperature hydrothermal deposit on the seabed at a Gulf of California spreading center. *Earth Planet. Sci. Lett.* **49:**8–20.
17. **Stetter, K. O., and W. Zillig.** 1985. *Thermoplasma* and the thermophilic sulfur-dependent archaebacteria, p. 85–170. *In* C. R. Woese and R. S. Wolfe (ed.), *The Bacteria*, vol. 8. Academic Press, Inc., New York.
18. **Weiner, R. M., R. A. Devine, D. M. Powell, L. Degasan, and R. L. Moore.** 1985. *Hyphomonas oceanitis* sp. nov., *Hyphomonas hirschiana* sp. nov., and *Hyphomonas jannaschiana* sp. nov. *Int. J. Syst. Bacteriol.* **35:**237–243.

Chapter 3

Photosynthesis in Cyanobacterial Mats and Its Relation to the Sulfur Cycle: a Model for Microbial Sulfur Interactions

Yehuda Cohen

INTRODUCTION

Cyanobacterial mats are organosedimentary structures composed primarily of benthic cyanobacteria together with diverse communities of microorganisms which trap, bind, and precipitate sediment particles. Often, cyanobacterial mats produce laminated sediments, biogenic stromatolites, owing to seasonal changes in the environment of deposition.

Cyanobacterial mats may serve as an ideal model for the study of basic mechanisms of the microbial sulfur cycle and the evolution of the sulfur cycle for several reasons. (i) Cyanobacterial mats develop under environmental conditions that exclude or, at least, limit eucaryotic organisms and grazing metazoans; this allows microbial interactions to be studied with limited bioturbation. (ii) In most microbial mats the organic matter produced is autochthonous, allowing the study of in situ primary production and the coupled processes of mineralization of primary organic carbon from a known source. In most other systems, part or all of the primary organic carbon is transported from other, often unknown, sources and is partially decomposed during transport under unknown conditions. This aspect is particularly important in the study of the decomposition and maturation of microbial mats as a model for oil shale formations (1). (iii) Cyanobacterial mats are the oldest known biogenic sedimentary structures and are found as stromatolites whose origin dates back 3.5 million years. As such, stromatolites are the only fossils which can be found throughout the entire geological era, even though they were most abundant during the Precambrian era (61).

Cyanobacterial mats are confined presently to a restricted range of habitats, including hypersaline coastal marine environments, hot springs, and alkaline lakes. Major marine cyanobacterial mats studied include Solar Lake, Sinai; the Gulf of Aqaba (14, 16, 33, 37, 41, 42); Shark Bay, Western Australia (4, 5, 47, 57); Spencer Gulf, South Australia (4, 5); Laguna Figueroa, Baja California, Mexico (66); Guerrero Negro, Baja California, Mexico (31, 32); and many other mats in Bermuda; the Persian Gulf; Bonair,

Yehuda Cohen • H. Steinitz Marine Biology Laboratory, The Hebrew University of Jerusalem, P.O. Box 469, Eilat 88103, Israel.

Caribbean Ocean; the Bahamas; and the USSR. Cyanobacterial mats in hot springs have been studied primarily in Yellowstone National Park, Wyo. (8, 12), Hunter's Spring, Oreg. (9), and other hot springs in the United States, New Zealand (10), Iceland (10, 62), the Far East (Kamchatka) (26), and Europe. Cyanobacterial mats in alkaline lakes in Ethiopia, the USSR, and various locations in the United States have been described.

All cyanobacterial mats develop in close proximity to microenvironments rich in sulfide (14). (The term sulfide is used to designate the total dissolved sulfide including H_2S, HS^-, and S^{2-}.) The population of species in these mats is pH dependent. Sulfide-rich environments are presently found in geothermal springs supplying dissolved sulfide of geological origin and in organic-rich biotopes where sulfate reduction is a dominant respiratory mechanism. When these sulfurita are exposed to light, four different groups of phototrophic procaryotes may develop: purple phototropic bacteria, green phototropic bacteria, members of the family *Chloroflexaceae*, and cyanobacteria. In trying to understand the different strategies of life under sulfide-rich conditions among the photosynthetic microorganisms, we have to consider the difference between the cyanobacteria and all the other phototrophic bacteria. Cyanobacteria carry out oxygenic photosynthesis with two photosystems in series, in a manner identical to that of eucaryotic phototrophs. Water is the electron donor in oxygenic photosynthesis, and oxygen is the ultimate oxidation product. Cyanobacteria are similar to eucaryotic algae and plants in that they have chlorophyll *a*. In addition, the main light-harvesting pigments are phycobiliproteins, which are also found in eucaryotic red algae together with β-carotene and zeaxanthin as the most common carotenoids (64).

The phototrophic green bacteria and the purple bacteria, in contrast, carry out anoxygenic photosynthesis by using only one photosystem and then require electron donors of redox potential lower than that of water. The suitable donors are reduced-sulfur compounds, simple organic compounds, and molecular hydrogen. The oxidants are organic compounds and CO_2. The photosynthetic pigments are various bacteriochlorophylls, together with a great variety of accessory carotenoids. Although many phototrophic bacteria can grow aerobically in a chemoheterotrophic mode of growth, their long-term photosynthesis is generally restricted to anaerobic, reduced environments where a suitable electron donor can be found. The physiological and ecological consequences of the differences between the oxygenic photosynthesis of the cyanobacteria and the anoxygenic photosynthesis of the other phototrophs are profound. However, cyanobacteria often share the same ecological niche with the other phototrophs.

Cyanobacteria are sometimes found in sulfide-rich environments where exposure to sulfide can be continuous, such as in hot springs with a constant abiogenic sulfide supply, or in sulfurita, where sulfide is produced biogenically and where fluctuations of anaerobic and aerobic conditions are common. These fluctuations may occur seasonally (44), diurnally (22), or even more frequently (33).

Life in sulfide-rich environments exhibits several advantages and some major drawbacks. Sulfide is an effective reducing agent and provides the required low redox potential for growth under anaerobic conditions. It serves as an efficient quencher to potential sulfide toxicity by combining with oxygen chemically or biochemically. In so doing, it reduces or even eliminates photoinhibition, photooxidation, and photorespiration, which are effects of the presence of oxygen. Furthermore, sulfide serves as an electron donor to anoxygenic photosynthesis. It may also serve as an assimilatory sulfur source, especially in organisms unable to assimilate sulfate. However, sulfide is highly toxic for most microorganisms. It reacts with various cytochromes, hemoproteins, and other compounds, and it inhibits the electron transport chain, blocking respiration as well as oxygenic and anoxygenic photosynthesis. Sulfide toxicity is also apparent in organisms which are well adapted to life under sulfide-rich conditions, but the degree of sensitivity to sulfide varies markedly among the different groups of microorganisms. The green sulfur bacterium *Chlorobium* sp. is the most sulfide tolerant of the phototrophic sulfur bacteria. This organism may grow in the presence of as much as 8 mM hydrogen sulfide (54). *Chloroflexus* spp. are found at H_2S concentrations of up to 1.5 mM (10). Purple sulfur bacteria (*Chromatium*, *Thiocystis*, and *Thiocapsa* spp.) may tolerate 0.8 to 4 mM hydrogen sulfide, whereas purple nonsulfur bacteria are less tolerant (0.4 to 2 mM hydrogen sulfide).

Oxygenic photosynthesis of the cyanobacteria in the immediate vicinity of sulfide results in the establishment of a sharp redoxcline of less than 1 mm. Diffusion over these small distances is very rapid, and the redoxcline migrates through the photosynthetically active layer with diurnal changes of light intensity. The cyanobacteria are thus exposed to varying sulfide concentrations for various durations depending on the microenvironmental conditions. The use of microelectrodes for oxygen, sulfide, and pH in various cyanobacterial mats has demonstrated the fast turnover rates of H_2S and oxygen which result in the drastic shifts of the redoxcline in the course of a day (33, 53).

The diurnal oscillations in oxygen and sulfide levels expose the sulfate-reducing bacteria to periodic oxic conditions. Many sulfate-reducing bacteria have been shown to be strict anaerobes (45), yet a new technique for the measurement of in situ sulfate reduction revealed high activities under well-oxygenated microenvironments in the Solar Lake cyanobacterial mats (15). The association of sulfate reduction with primary production activity has also been described by Skyring (63) for cyanobacteria in various Australian mats.

The diurnal migration of the redoxcline within the euphotic zone of the cyanobacterial mats causes periodic release of Fe^{2+} ions from the pool of FeS; these ions serve as a major sink for the sulfides produced by the sulfate-reducing bacteria. Several mat-forming cyanobacterial isolates were capable of efficient CO_2 photoassimilation by using Fe^{2+} ions, which serve as yet another alternative electron donor for photosynthesis in these strains.

The ecological significance of ferrous ion-dependent photosynthesis is currently being studied. It has an important role in the understanding of the evolution of life in the Precambrian era. Banded iron formations were thought to be a result of the oxidation of ferrous ion by the primordial Precambrian oxygen (13) and were taken as a geological proof for the accumulation of oxygen during the Precambrian era (40). Ferrous ion-dependent photosynthesis may be another mechanism for deposition of the banded-iron formation which does not necessarily involve oxygen.

The photosynthetic flexibility of cyanobacteria (65), their tight association with sulfate reduction, and the use of Fe^{2+} by certain cyanobacteria further indicate the antiquity of this group of microorganisms and allow a better understanding of their ecological role in the carbon, sulfur, and iron cycles.

ANOXYGENIC PHOTOSYNTHESIS IN *O. LIMNETICA* AND OTHER CYANOBACTERIA

A flocculant mat, composed mostly of *Oscillatoria limnetica*, as well as the green sulfur bacterium *Prosthecochloris* sp. and the purple sulfur bacteria *Lamprocystis* sp. and *Chromatium* sp., was found at the bottom of the hypolimnion of Solar Lake, Sinai (20). A sulfide concentration of up to 5 mM accumulates in this layer and persists for about 10 months before holomixis occurs. Sulfide is oxidized during holomixis, and the sulfur-dependent phototrophic community disappears, while *O. limnetica* thrives under the oxygen-rich conditions which persist for 2 months before stratification sets in again (17). High primary-production activities of up to 8 g cm^{-2} day^{-1} were measured at Solar Lake under stratification (20), and about 60% of it may be attributed to the activity of *O. limnetica* in the presence of sulfide.

This organism, although capable of oxygenic photosynthesis whenever sulfide is absent, switches off photosystem II (PS II) when sulfide is present and conducts anoxygenic photosynthesis (17, 18). Very low concentrations (0.1 to 0.2 mmol) of sulfide immediately inhibit the oxygenic system (51). However, after 2 h of exposure to light in the presence of a high sulfide concentration (3 mM), photoassimilation reappears and is then insensitive to the PS II inhibitor 3-(3,4-dichlorophenyl)-1,1-dimethylurea (DCMU) (18). The possible participation of PS II in anoxygenic CO_2 photoassimilation has been further excluded (18) by the results of experiments in which PS II was simply not activated (rather than being inhibited) by the use of far-red light. Oxygenic photosynthesis requiring the operation of both photosystems decreased drastically in far-red light (red drop), whereas anoxygenic photosynthesis with sulfide was fully operative under such conditions. Furthermore, if both photosystems could contribute to the reaction, the enhancement in quantum yield of photoassimilation would be predicted with respect to that obtained with only PS I in operation. However, the enhancement phenomenon was observed only for oxygenic photosynthesis (50). The new type of photosynthesis is therefore anoxygenic, independent of PS II, and driven by PS I with sulfide as the electron donor.

Preincubation for 2 h in the presence of sulfide and light is required for anoxygenic photosynthesis, indicating that induction may be involved. Indeed, the protein synthesis inhibitor chloramphenicol inhibits the initiation of anoxygenic photosynthesis (49). To survive sulfide inhibition of PS II, an adaptation which allows the use of sulfide with PS I is necessary. It is suggested that low redox potential elicits a series of events culminating in a modification, possibly a reduction, of an electron carrier(s) that becomes sulfide resistant and is thereby functional in the utilization of sulfide electrons. When sulfide elicits these events, protein synthesis is involved. Addition of sodium dithionate eliminates the lag period observed when *O. limnetica* is transferred to sulfide, possibly by providing the low redox potential needed for the reduction and modification of electron carriers.

Sulfide is oxidized to elemental sulfur by *O. limnetica* according to the following stoichiometric relationship (21):

$$2H_2S + CO_2 \rightarrow H(CH_2O) + 2S + H_2O$$

Elemental sulfur was observed as refractile granules either free in the medium or adhering to the cyanobacterial filament. Otherwise, the appearance of the cells under the electron microscope was the same under both anoxygenic and oxygenic conditions.

If sulfide is removed from sulfide-adapted cells, *O. limnetica* immediately returns to oxygenic photosynthesis (49). This capacity must therefore be constitutively present in the cells, even when PS II is not in operation. The photosynthetic system of *O. limnetica* thus operates facultatively both oxygenically and anoxygenically. When CO_2 is eliminated from the reaction system in *O. limnetica*, sulfide donates electrons for hydrogen evolution (6). This reaction occurs only in cells originally adapted to sulfide in the presence of CO_2, is dependent on light and sulfide, is insensitive to DCMU (like the CO_2 photoassimilation reaction), requires an induction period of 2 h, and is driven by PS

I. However, whereas CO_2 photoassimilation is fully induced after 2 h of incubation in the presence of sulfide, the capacity of hydrogen evolution, although initiated, is very low, and an additional 46 h of incubation in the presence of sulfide is needed for a full induction of hydrogen evolution activity. Sulfide-dependent hydrogen evolution must require a step(s) in addition to that required for sulfide-dependent CO_2 photoassimilation. However, addition of sodium dithionate eliminates the needed induction period, as is the case for anoxygenic photosynthesis.

The concentrations in the cell of the light-harvesting pigments (chlorophyll *a*, phycobilins, and carotenoids) are identical under conditions of both oxygenic and anoxygenic photosynthesis (50). Quantum yield spectra of the two photosynthetic modes show that although only a narrow light band is utilized in the oxygenic mode, the entire absorbed spectrum is used at high quantum efficiency in the anoxygenic mode. The drop in the quantum efficiency of oxygenic photosynthesis at both the blue and red ends of the visible spectrum is marked. This limited range of utilization of the visible-light spectrum in oxygenic photosynthesis of *O. limnetica* is similar to that of other cyanobacteria (46). It is markedly different, however, from that of eucaryotic algae and plants, which also contain chlorophyll *b* in their light-harvesting systems. In these organisms, almost the entire absorbed spectrum is utilized in oxygenic photosynthesis, with the exception of the red drop (27).

A comparison of rates of anoxygenic photosynthesis as a function of sulfide concentration was carried out for 11 strains of cyanobacteria including the mat-forming cyanobacteria *Lyngbia* strain 7104, *Aphanothece halophytica*, and *O. limnetica* (23). In these organisms the pattern of dependence on sulfide concentrations was similar, generating an optimum curve rather than a saturation curve. The drop in the photosynthetic rates at higher sulfide concentrations is caused by sulfide toxicity effects on PS I. The maximal rates of oxygenic and anoxygenic photosynthesis are similar in both *O. limnetica* (1 to 2 μmol of C mg of protein^{-1} h^{-1}) and *A. halophytica* (0.5 to 1 μmol of C mg of protein^{-1} h^{-1}). Although the dependencies on sulfide are similar, both the affinities for sulfide and the tolerances are different. Each strain exhibits a different range of sulfide concentrations at which anoxygenic photosynthesis can be performed. The sulfide concentrations at pH 6.8 are 0.1 to 0.3 mM for *Lyngbia* strain 7104, 0.1 to 1.5 mM for *A. halophytica*, and 0.7 to 9.5 mM for *O. limnetica*. Furthermore, for each of these the concentration range is markedly affected by the pH of the medium (30), which governs the dissociation of hydrogen sulfide. Differences in pH-dependent sulfide concentration ranges are known to constitute a determining factor in the ecology of photosynthetic sulfur bacteria (3, 54). The pattern of sulfide oxidation in anoxygenic photosynthesis of cyanobacteria seems thermodynamically inefficient; photosynthetic bacteria oxidize sulfide to sulfate (54, 55), gaining eight electrons for each H_2S molecule, whereas the cyanobacteria *O. limnetica* and *A. halophytica* remove only two electrons per molecule of sulfur. Elemental sulfur is the only end product of anoxygenic photosynthesis in these cyanobacteria (18). Elemental sulfur may also be a by-product in the oxidation of sulfide by several sulfur phototrophic bacteria (H. G. Trüper, *Abstr. Symp. Prokaryotic Photosynth. Organisms*, p. 160–166, 1973). Finally, other sulfur-containing electron donors which are utilized by photosynthetic bacteria (28, 54, 55) do not seem to serve the photosynthetic system of sulfide-utilizing cyanobacteria. *Anacystis nidulans*, however, can oxidize thiosulfate to sulfate, yet thiosulfate cannot serve as the sole or even as the major electron donor (68).

As in many photosynthetic bacteria and heterocystous cyanobacteria, molecular hydrogen has been shown to serve as an efficient electron donor for CO_2 photoassimilation in an anoxygenic reaction driven by PS I in both *O. limnetica* and *A. halophytica*. Hydrogen evolution dependent on sulfide oxidation in both these strains (6) and heterocystous species (7, 67, 70) has been demonstrated.

In cyanobacteria, sulfide appears to be more toxic to oxygenic photosynthesis than to anoxygenic photosynthesis; in *O. limnetica* the former was inhibited by 0.1 mmol of sulfide (51), whereas the latter was only partially inhibited by 4 mmol (at pH 6.8) (23). These differences have also been observed in other facultatively anoxygenic strains. This difference in toxicity implies that at the higher sulfide concentrations, oxygenic photosynthesis will make only a minor contribution, if any, to photoassimilation of CO_2, whereas at the lower concentrations, oxygenic photosynthesis may occur simultaneously with anoxygenic photosynthesis. An example of such a case has been demonstrated for *O. amphigranulata* (H. C. Utkilen and R. W. Castenholz, *Abstr. 3rd Int. Symp. Photosynth. Prokaryotes*, p. 12, 1979) and occasionally also for the chemocline of Solar Lake (33). Low sulfide concentrations (0.2 mM) are toxic to eucaryotic phototrophs and cyanobacterial strains that do not carry out anoxygenic photosynthesis (10, 30, 39).

The greater sulfide tolerance of anoxygenic photosynthesis operating with PS I confers a selective advantage in sulfide-rich habitats. This advantage is present even if these cyanobacteria cannot grow continuously under strictly anaerobic conditions, for they can at least maintain themselves temporarily in the presence of sulfide. However, some cyanobacteria are capable of permanent growth under these conditions. *O. limnetica* thrives at the hypolimnion of Solar Lake under a 3-m column of sulfide-rich water and grew efficiently in the laboratory with 3.5 mM sulfide at pH 6.8 for several months (50). Facultative anoxygenic photosynthesis is clearly an advantage for life under prolonged periods of exposure to sulfide.

SULFIDE-OXYGEN FLUCTUATION IN CYANOBACTERIAL MATS

The in situ use of microelectrodes for oxygen, sulfide, and pH in various microbial mats demonstrated the establishment of a sharp redoxcline of less than 1 mm as a result of oxygenic photosynthesis of cyanobacteria in the immediate proximity of the sulfide-rich microzone. The first measurements of microgradients of O_2, H_2S, and E_h in microbial mats were carried out in the Flat Shallow cyanobacterial mats at Solar Lake, Sinai (33). This mat, dominated by the cosmopolitan mat-forming cyanobacterium *Microcoleus chthonoplastes*, is situated under 30 cm of oxic water column. Extreme diurnal O_2 and H_2S fluctuations were found, with an O_2 peak of 0.5 mM at 1 to 2 mm below the mat surface during the day and an H_2S peak of 2.5 mM at 2 to 3 mm below the mat at night. The O_2-H_2S interface migrates diurnally from the mat surface at night to 3 mm below the surface at noon. The photic zone extends down to 2.5 mm below the surface, and the H_2S concentration peaked just below the photic zone. H_2S and O_2 were found to coexist at 2.5 mm below the surface over a depth interval of 0.2 to 1 mm, with a turnover rate of less than 1 min.

A later, more detailed study of the microstructures of the various microbial mats in Solar Lake and their O_2, H_2S, and pH microprofiles was carried out (38). The dominant cyanobacterium in the Flat and Blister mats was *M. chthonoplastes*, in association with filamentous flexibacteria, tentatively identified to be *Chloroflexus*-like organisms. Very similar associations were observed in cyanobacterial mats at Sabhat Gavish, Southern Sinai (24), in microbial mats in Laguna Figueroa, Baja California, Mexico (66), and in various microbial mats in Shark Bay, Western Australia, and Spencer

Gulf, South Australia (4). The flat mat from most shallow parts of Solar Lake had a photic zone of merely 0.8 mm, with a maximal photosynthetic activity of 50 μmol of O_2 cm^{-3} h^{-1} at a depth of 0.3 to 0.4 m. In deeper waters the mats were less compacted and the photic zones extended down to over 10 mm with increasing depth of overlying water. In all the different microbial mats examined, a ΔpH of up to 2 pH units developed at the maximal photosynthetically active zone, causing rapid deposition of $CaCO_3$ in this microzone (58).

The microgradients of oxygen, sulfide, and pH in mats from the various hot springs in Yellowstone National Park, Wyo., were measured by Revsbech and Ward (59). These systems exhibit a maximum oxygen concentration of up to 600 μM at concentrations below 2 mM sulfide. The mat is composed of the cyanobacteria *O. terebriformis* and *Synechococcus* sp. At Octopus Springs, Yellowstone National Park, a very thin layer (0.1 mm) of *Synechococcus* sp. is responsible for maximal oxygenic photosynthesis of 150 mmol of O_2 dm^{-3} h^{-1} and a corresponding O_2 production of 47 mmol m^{-2} h^{-1}. At Stinky Springs, Utah, where the sulfide concentration at the source is 1,100 μM, only traces of oxygen can be detected by the activity of *Oscillatoria* sp. which occur together with *Chloroflexus*-like bacteria (19).

In the hypolimnic flocculous mat at Solar Lake, where *O. limnetica* is the dominant cyanobacterium when sulfide is present at 3 mM (42, 43), no O_2 could be detected at the photosynthetically active layer, although this organism can readily produce oxygen at higher light intensities, even in the presence of 3mM sulfide (19).

All cyanobacterial mats examined so far exhibit sharp microgradients of sulfide, oxygen, and pH, which fluctuate diurnally and thus expose both the cyanobacteria and the sulfate-reducing bacteria to conditions which alternate between highly oxygenated water and elevated sulfide concentrations.

OXYGENIC AND ANOXYGENIC PHOTOSYNTHESIS IN MAT-FORMING CYANOBACTERIAL ISOLATES

Axenic cultures of cyanobacteria isolated from various biotopes exposed to varying sulfide concentrations were examined for their capacity to carry out oxygenic and/or anoxygenic photosynthesis under a range of sulfide concentrations. Four different strategies of photosynthetic life under varying degrees of exposures to sulfide can be detected, as follows.

(i) Type 1 photosynthesis is exemplified by irreversible cessation of CO_2 photoassimilation upon brief exposure to sulfide. *A. nidulans* isolated from planktonic blooms without apparent exposure to sulfide is highly sensitive to sulfide toxicity. Exposure to 100 μM sulfide causes 50% inhibition of CO_2 photoassimilation, and at 200 μM sulfide, photoassimilation is blocked completely. When sulfide was removed after 2 h of incubation in the light, no regenerated photoassimilation could be detected. A similar sensitivity to sulfide was found by Schwabe (62) for *Mastigocladus* sp., the thermophilic cosmopolitan cyanobacterium found under very low sulfide concentrations (up to 2 μM) in various hot springs in Iceland, New Zealand, and the United States (8, 10, 11, 62).

(ii) Type 2 photosynthesis involves enhancement of oxygenic photosynthesis upon exposure to sulfide and incapability of utilization of H_2S as an electron donor for anoxygenic photosynthesis. This type of photosynthesis is represented by an *Oscillatoria* sp. isolated from Wilbor Springs, Calif. This organism, which grows under 1 mM sulfide at neutral pH, shows 450% enhancement of oxygenic photosynthesis upon exposure to 800 μM sulfide at neutral pH. With increasing sulfide concentrations, CO_2 photoassimilation was gradually inhibited, yet no anoxygenic, DCMU-insensitive photosynthesis could be detected. Similar activities were reported by Weller et al. (71) for a *Phormidium* sp. isolated from a hot spring at Yellowstone National Park.

(iii) Type 3 photosynthesis involves enhancement of oxygenic photosynthesis at low sulfide concentrations, inhibition of PS II at higher sulfide concentrations, and a concomitant induction of anoxygenic photosynthesis operating in concert with the partially inhibited oxygenic photosynthesis at higher sulfide concentrations. *M. chthonoplastes*, which uses this type of photosynthesis, is a cosmopolitan mat-forming cyanobacterium in hypersaline coastal lagoons. Isolates from Solar Lake, Sabhat Gavish (Sinai), Laguna Figueroa and Guerrero Negro salt pans, Spencer Gulf, and Shark Bay all showed virtually the same photosynthetic activity. The ultrastructures of the cyanobacterial mats dominated by *M. chthonoplastes* from Solar Lake (19, 35), Sabhat Gavish (24), Laguna Figueroa (66), Shark Bay (4, 5), and the Persian Gulf (J. F. Stolz, personal communication) are all extremely similar. The same type of photosynthetic activity was described for *O. amphigranulata* isolated from alkaline hot springs in New Zealand at concentrations of 2.2 mM sulfide (Utkilen and Castenholz, *Abstr. 3rd Int. Symp. Photosynth. Prokaryotes*).

(iv) Type 4 photosynthesis concerns complete reversible inhibition of PS II at low sulfide concentrations and induction of efficient anoxygenic photosynthesis at higher sulfide levels. This type of photosynthesis was initially described by Cohen et al. (18, 21) for *O. limnetica* isolated from Solar Lake and later by Garlick et al. (23) for other cyanobacteria. Oren et al. (51) demonstrated that unlike type 2 and 3 photosynthesis, in type 4 photosynthesis PS II is completely blocked at 100 μM sulfide. When the cyanobacterium is exposed to high sulfide levels, an induction period of 2 h is needed for anoxygenic, PS I-dependent photosynthesis to be fully induced. Recently, this type of activity was also found in a hot-spring isolate, *Oscillatoria* sp., growing in Stinky Springs, Utah, at 1.1 mM sulfide and pH 6.3.

Clearly, cyanobacteria exhibit a high degree of variability in their photosynthesis as a function of varying exposures to sulfide. The four types of photosynthesis described represent increasing degrees of adaptation to photosynthetic life in the presence of sulfide. Prolonged exposures to increasing sulfide concentrations result in the dominance of cyanobacterial strains which can better utilize sulfide, and hence one finds a gradual shift from type 1 to type 4 photosynthesis under conditions of high sulfide concentrations.

Generally, type 1 photosynthesis occurs in cyanobacteria which are either not exposed at all to sulfide or exposed only to vanishingly low concentrations of it. Type 2 photosynthesis is found among cyanobacteria exposed to 1 mM sulfide or less at neutral or high pH. No anoxygenic photosynthesis can be demonstrated in these organisms, and yet their oxygenic activity is enhanced at the low redox potential. This affords protection from photoinhibition and increased efficiency of CO_2 photoassimilation. Types 3 and 4 develop at relatively high sulfide concentrations and neutral or low pH. Sulfide toxicity is pH dependent and is greater at lower pH. Like all weak acids, H_2S penetrates passively through the cell membrane obeying diffusion laws, whereas the ionized forms, namely HS^- and S^{2-}, need active transport mechanisms. Thus, *O. amphigranulata*, which grows at 2.2 mM sulfide and high pH, is a type 3 organism, whereas the *Oscillatoria* sp. from Stinky Springs, which grows under 1.1 mM sulfide but at the lower pH of 6.3, is a type 4 cyanobacterium.

PS II systems of the various types of cyanobacteria show different sensitivities to sulfide inhibition. Exposure to 50 μM sulfide at pH 7.5 induces maximal variable fluorescence of PS II in both *A. nidulans* (type 1) and *O. limnetica* (type 4). Addition of 10^{-4} M DCMU does not further enhance PS II fluorescence, and yet, in the *Oscillatoria* sp. from Wilbor Springs, Calif. (type 2), variable PS II fluoresence is only partially affected, even at 1 mM sulfide (pH 7.5). Further addition of 10^{-4} M DCMU induces the maximal variable fluoresence of PS II (19). Similar results were obtained in cultures of *Synechococcus lividus* from Yellowstone National Park, *Oscillatoria* sp. from Stinky Springs, and the various *M. chthonoplastes* isolates from Solar Lake, Sabhat Gavish, Baja California, and Spencer Gulf.

The ability of the various cyanobacterial mat communities to produce and accumulate oxygen under increasing sulfide concentrations and the efficiency of recovery of oxygenic photosynthesis upon gradual removal of sulfide were measured by introducing microelectrodes for O_2, S^{2-}, and pH into small blocks of various cyanobacterial mats suspended in sulfide-containing media. Given a high enough light intensity, all cyanobacteria of types 2, 3, and 4 are capable of producing and accumulating oxygen in the presence of sulfide. However, the level of sulfide at which oxygen production is detected changes significantly from one type to another. *O. limnetica* (type 4) produces O_2 only at low sulfide concentrations (50 μM); *M. chthonoplastes* (type 3) performs efficient oxygenic photosynthesis at higher sulfide levels (250 μM) at neutral pH; and, finally, *Oscillatoria* sp. from Wilbor Springs (type 2) produces O_2 at much higher sulfide concentrations (millimolar level) (19).

Many, if not all, cyanobacteria of types 3 and 4 occur in close association with *Chloroflexus* or *Chloroflexus*-like organisms. This tight association may enable the community to cope more efficiently with rapid and wide oscillations between high sulfide and high oxygen concentrations. The nature of this intimate coexistence is not understood and is currently being studied.

The various types of photosynthesis are found both in hot sulfur springs and in the various sulfurita. Sulfur springs may serve as a good model for the study of the distribution of different types of cyanobacteria along a sulfide gradient formed downstream from the main source. The degree of exposure to sulfide in sulfurita, on the other hand, depends on the coupling of primary production and sulfate reduction, the only source of H_2S in these biotopes. The importance of sulfate reduction as the major process in the mineralization of the produced organic matter was demonstrated by Jørgensen et al. (37) and Skyring (63).

COUPLING OF PRIMARY PRODUCTION AND SULFATE REDUCTION IN CYANOBACTERIAL MATS

Since the coupling of primary production and sulfate reduction in cyanobacterial mats must occur in close proximity to the photic zone, which is highly oxygenated during most of the day, the usual technique for the measurement of sulfate reduction cannot accurately be applied to these systems. Sulfide undergoes extremely fast turnover owing to its efficient oxidation photosynthetically, chemolithotrophically, heterotrophically, or even chemically in the presence of oxygen. Hence, H_2S produced by the sulfate-reducing bacteria cannot be quantitatively measured by using the technique of injection of $Na_2{}^{35}SO_4$ into sediment cores in proximity to oxygen.

A new method had to be developed to assess the degree of coupling of sulfate reduction to primary production in these systems (15). Silver wires (diameter, 0.15 mm; length, 1 cm) were coated with $Na_2{}^{35}SO_4$ of high specific activity and were introduced into the mat by means of a micromanipulator alongside the microelectrodes for oxygen, sulfide, and pH. The mat was then incubated either in the light or in the dark for 10 min to 2 h after a 2-h preincubation under the same conditions. A series of silver wires were pulled out of the mat at different time intervals and subjected to autoradiography to find the activity of the $Ag_2{}^{35}S$ trapped on the wires after extensive washing to remove all remaining $^{35}SO_4{}^{2-}$. The vertical microprofiles of the sulfate reduction activities were then compared with the microprofiles of oxygen, sulfide, and pH, as well as the profiles of oxygenic primary production. The silver wires were later cut into 1-mm segments, and each segment was counted in a scintillation counter.

This new method presents several major advantages, but also some problems, compared with the conventional method for the determination of sulfate reduction. The new method allows the determination of the $SO_4{}^{2-}$ reduction activity in close proximity to oxic microzones, since the sulfide produced binds to the silver wire at a high affinity and overrides most, if not all, of the concomitant sulfide oxidation process. Thus, even though there may not be an overall sulfide accumulation, since sulfide is immediately oxidized, the sulfide level can still be measured by using this technique; it is impossible, by definition, to determine this activity by using the conventional method. The other advantage is the understanding of the microenvironmental conditions of sulfate reduction and the better insight into the coupling of sulfate reduction to primary production. The disadvantage of this technique is the difficulty in quantification of the sulfate reduction activities, since diffusion of the radioactively labeled $SO_4{}^{2-}$ causes a decrease of specific activity with time of incubation. The label signal decreases with time of incubation owing to isotopic exchange between ^{35}S and ^{32}S: a function of changes in the relative pool sizes of the two isotopes during incubation. Therefore, a time course of the sulfate reduction activity is necessary to quantify the activity of sulfate reduction.

Results from the application of this technique to the Solar Lake cyanobacterial mat demonstrated an extremely tight coupling of sulfate reduction and primary production. Not only can sulfate reduction be detected under the highly oxygenated conditions at the photic microzone of the cyanobacterial mat, but this activity is also enhanced in the light, where oxygen concentrations may accumulate up to 450% saturation of oxygen. Furthermore, the very same light spectra that are most efficient in oxygenic photosynthesis are responsible for the induction of sulfate reduction in the light. Specifically, light wavelengths of 590 to 660 nm, which are absorbed by the major light-harvesting pigment of cyanobacteria, phycocyanin, induce both primary production and sulfate reduction. The sulfate reduction activity under these conditions is fueled directly by photosynthates excreted by the cyanobacteria when performing oxygenic photosynthesis under high oxygen concentrations and high light intensities. The nature of the excretions responsible for the coupling of primary production and sulfate reduction in cyanobacterial mats is currently being studied.

Many sulfate-reducing bacteria have been shown to be strict anaerobes and to be highly sensitive to oxygen toxicity (56), yet in the Solar Lake cyanobacterial mats and probably many other mat systems, these organisms operate well under periodic exposures to high oxygen concentrations. The mechanisms allowing the activity of sulfate reduction under these conditions are not fully understood. Preliminary results may indicate the involvement of hydrogen in the coupling of primary production and $SO_4{}^{2-}$ reduction. Several mat-forming cyanobacteria have been shown to produce hydrogen under CO_2 limitation (S. Belkin, Ph.D. thesis, Hebrew University, Jerusalem, Israel, 1983). Temporary CO_2 limitations may well occur at the photic microzone of the cyanobacterial mats owing to the high specific rate of CO_2 photoassimilation, which creates a temporal high pH of >9.5, resulting in the precipitation of $CaCO_3$ in this calcium-rich system.

Fe^{2+}-DEPENDENT PHOTOSYNTHESIS IN BENTHIC CYANOBACTERIA

The diurnal migration of the redoxcline through the photosynthetic layer ensures the periodic release of Fe^{2+} from the pool of FeS. High concentrations of dissolved Fe^{2+} were observed in the interstitial water of the upper 2 mm of the Solar Lake cyanobacterial mat in the morning hours. Ferrous ion is a good potential electron donor for photosynthesis in cyanobacteria and may operate well thermodynamically at the spectrum of redox potential values of −50 to +50 mV, which is typical for the photic microzone during the daily transition from fully reducing conditions at night to the high oxygen concentration at noon and the opposite transition in the late afternoon. Banded microlayers of iron oxides are not very common in recent cyanobacterial mats, and yet they have been found in several mats in Spencer Gulf and Shark Bay, as well as several coastal lagoons such as the Sippewissett Marsh, Cape Cod, Mass. Iron-dependent photosynthesis in mat-forming cyanobacteria has long been speculated to be responsible, at least in part, for the extensive deposition of the banded-iron formation during the Precambrian era (13). Hartman (29) speculated that this type of photosynthesis was an important step in the evolution of oxygenic photosynthesis.

Several cyanobacteria were examined for their capacity to use Fe^{2+} as an electron donor in photosynthesis. *Oscillatoria* spp. from Wilbor Springs and from Stinky Springs, as well as all *M. chthonoplastes* isolates, have shown Fe^{2+}-dependent CO_2 photoassimilation (19). Fe^{2+} donates electrons primarily to PS II, and this activity is therefore DCMU sensitive. Fe^{2+} initially blocks PS II, as indicated by a temporary sharp decrease in the PS II variable fluorescence in the presence of 10^{-5} M DCMU. However, after a short incubation in the presence of ferrous ion, F_{max} reappears along with an efficient rate of CO_2 photoassimilation and iron oxidation. Iron is oxidized through the intermediate ferritin. The end product of the oxidation, iron oxide, is excreted outside the cell, similar to the excretion of elemental sulfur in sulfide-dependent anoxygenic photosynthesis.

Preliminary investigations have been conducted on the diurnal fluctuation of dissolved ferrous ion, together with the measurement of $\delta^{13}C$ of ΣCO_2 of interstitial water in the upper 10 mm of a sediment core from the San Francisco Marsh. Results showed a disappearance of Fe^{2+} from the photic zone during the day, associated with a second peak of heavy $\delta^{13}C$ at the deepest part of the photic zone at 4 to 6 mm; similar values were observed at the surface, indicating two microzones of autotrophic activities. The upper zone is clearly a result of oxygenic photosynthesis, whereas in the lower layer a possible Fe^{2+}-dependent CO_2 photoassimilation may be inferred.

The importance of Fe^{2+}-dependent photosynthesis as an intermediate between sulfide-dependent activity and oxygenic photosynthesis is not yet understood.

REGULATION OF SULFUR CYCLE IN RECENT CYANOBACTERIAL MATS

Over 99% of CO_2 photoassimilation and SO_4^{2-} reduction in cyanobacterial mats, as well as practically all sulfide oxidation processes, takes place in the upper 5 to 10 mm of the sediment. Any attempt to estimate the rates of processes in this system must be done on a microscale, and analyses of bulk samples must be treated cautiously. In these microdimensions, very sharp gradients of O_2, sulfide, pH, E_h, Fe^{2+}, Fe^{3+}, and other parameters are established. These sharp gradients within a few millimeters are the result of very high specific activities of photosynthesis, sulfate reduction, sulfide oxidation, and heterotrophic microbial activity. The chemical microgradients show a drastic fluctuation on a daily basis and expose the microbial communities to 1,000 μM O_2, a pH of 9 to 10, an E_h of +200 mV, and over 10 mM Fe^{2+} at noon on the one hand and to 5 mM sulfide, a pH of 6.5 to 7.0, and an E_h of −180 mV at night on the other hand. Microorganisms have to adapt to these drastic microenvironmental changes by developing elaborate regulatory mechanisms.

The regulation of photosynthetic activity among mat-forming cyanobacteria has been studied in detail since 1974. Anoxygenic, sulfide-dependent photosynthesis and oxygenic photosynthesis in the presence of sulfide both play a major role in sulfide oxidation processes in these environments. Using our present knowledge, we can predict which type of cyanobacteria will dominate in a given biotope by measuring the diurnal microgradients of oxygen and sulfide.

Since iron serves as a major trap of sulfide, the regulation of the conversion of Fe^{2+} to Fe^{3+} in sediments is closely related to the sulfur cycle. Unfortunately, not enough information on Fe^{2+} photosynthesis in cyanobacteria is presently available to allow a proper assessment of its ecological importance.

When iron is limiting in the environment, sulfide may end up in a pool of polysulfide (16, 33). Polysulfide may attack residual organic matter and cause enrichment of organically bonded sulfur along the sedimentary column. In the Solar Lake sedimentary column, organically bonded sulfur increases from 2% at the surface to 8% at a depth of 60 cm (1, 16). Since sulfide is oxidized to elemental sulfur by the anoxygenic photosynthesis in cyanobacteria, this fraction will contribute to the polysulfide pool.

In the absence of light, sulfur reduction processes prevail via sulfate reduction or via sulfur reduction in cyanobacteria, some phototrophic sulfur bacteria, and *Desulfuromonas* spp. Oxygen and Fe^{2+} efficiently trap most of the sulfide produced. Some reduced sulfur diffuses to the overlying water, but in minute quantities. The efficient trapping and oxidation of sulfide allow the development of diatoms at the mat surface. Diatoms are very sensitive to sulfide toxicity, and their CO_2 photoassimilation process is irreversibly inhibited after exposure to a few micromoles of sulfide. The development of a dense diatom community at the top of the mat requires an efficient removal of sulfide below the diatom layer. By using a microtechnique for the measurement of sulfate reduction, it was shown that this process is very tightly coupled to primary production. The regulatory mechanism of this coupling is not yet fully understood.

PRECAMBRIAN CYANOBACTERIA AND STROMATOLITES

Cyanobacterial mats are an extremely ancient phenomenon; they have been documented in the oldest fossils known, dating back 3.5 million years. For the remaining period of the Archean and through the Proterozoic era, up to 0.57 million years ago, stromatolitic communities of cyanobacterial mats are the most abundant fossils. For the Paleozoic era, the fossil record of stromatolites is limited and is restricted to intertidal and supratidal hypersaline marine environments, thermal springs, and alkaline lakes (2), where they were protected from newly evolved grazing activity.

Another major sedimentary record of the Proterozoic era is the widely spread deposition of finely laminated ferro-ferric oxides known as the banded-iron formation. The understanding of microbial activities in recent cyanobacterial mats may throw light on our understanding of the evolutionary processes during the Precambrian. The recent mat-forming cyanobacteria represent a protocyanobacterial group which differs markedly from modern planktonic cyanobacteria.

LITERATURE CITED

1. **Aizenshtat, Z., G. Lipiner, and Y. Cohen.** 1984. Biogeochemistry of carbon and sulfur cycle in the microbial mats of the Solar Lake (Sinai), p. 281–312. *In* Y. Cohen, R. W. Castenholz, and H. O. Halvorson (ed.), *Microbial Mats: Stromatolites*. Alan R. Liss, Inc., New York.
2. **Awramik, S. M.** 1984. Ancient stromatolites and microbial mats, p. 1–22. *In* Y. Cohen, R. W. Castenholz, and H. O. Halvorson (ed.), *Microbial Mats: Stromatolites*. Alan R. Liss, Inc., New York.
3. **Baas Becking, L. G. M., and E. J. F. Wood.** 1955. Biological processes in the estuarine environment. I–II. Ecology of the sulfur cycle. *Proc. K. Ned. Akad. Wet. Ser. B* **58**:160–181.
4. **Bauld, J.** 1984. Microbial mats in marginal marine environments: Shark Bay, Western Australia, and Spencer Gulf, South Australia, p. 39–58. *In* Y. Cohen, R. W. Castenholz, and H. O. Halvorson (ed.), *Microbial Mats: Stromatolites*. Alan R. Liss, Inc., New York.
5. **Bauld, J., L. A. Chambers, and G. W. Skyring.** 1979. Primary productivity, sulfate reduction and sulfur isotope fractionation in algal mats and sediments of Hamelin Pool, Shark Bay, W.A. *Aust. J. Mar. Freshwater Res.* **30**:753–764.
6. **Belkin, S., and E. Padan.** 1979. Hydrogen metabolism in the facultative anoxygenic cyanobacteria (blue green algae) *Oscillatoria limnetica* and *Aphanotheca halophitica*. *Arch. Microbiol.* **116**:109–111.
7. **Bothe, H., Y. Tennigkeit, and G. Eisbrinner.** 1977. The utilization of molecular hydrogen by the blue green alga *Anabaena cylindrica*. *Arch. Microbiol.* **114**:43–49.
8. **Brock, T. D.** 1978. *Thermophilic Microorganisms and Life at High Temperatures*. Springer-Verlag, New York.
9. **Castenholz, R. W.** 1973. The possible photosynthetic use of sulfide by the filamentous phototrophic bacteria of hot springs. *Limnol. Oceanogr.* **18**:863–876.
10. **Castenholz, R. W.** 1976. The effect of sulfide on the blue green algae of hot springs. I. New Zealand and Iceland. *J. Phycol.* **12**:57–68.
11. **Castenholz, R. W.** 1977. The effect of sulfide on blue green algae of hot springs. II. Yellowstone National Park. *Microb. Ecol.* **3**:79–105.
12. **Castenholz, R. W.** 1984. Composition of hot spring microbial mats: a summary, p. 101–120. *In* Y. Cohen, R. W. Castenholz, and H. O. Halvorson (ed.), *Microbial Mats: Stromatolites*. Alan R. Liss, Inc., New York.
13. **Cloud, P.** 1983. The biosphere. *Sci. Am.* **249(3)**: 176–189.
14. **Cohen, Y.** 1984. The Solar Lake cyanobacterial mats: strategies of photosynthetic life under sulfide, p. 133–148. *In* Y. Cohen, R. W. Castenholz, and H. O. Halvorson (ed.), *Microbial Mats: Stromatolites*. Alan R. Liss, Inc., New York.
15. **Cohen, Y.** 1984. Oxygenic photosynthesis, anoxygenic photosynthesis, and sulfate reduction in cyanobacterial mats, p. 435–441. *In* M. J. Klug and C. A. Reddy (ed.), *Current Perspectives in Microbial Ecology*. American Society for Microbiology, Washington, D.C.
16. **Cohen, Y., Z. Aizenshtat, A. Stoler, and B. B. Jørgensen.** 1980. Microbial geochemistry of Solar Lake Sinai, p. 167–177. *In* P. A. Trudinger and M. R. Walter (ed.), *Biogeochemistry of Ancient and Modern Environments*. Australian Academy of Science, Canberra.
17. **Cohen, Y., M. Goldberg, W. E. Krumbein, and M. Shilo.** 1977. Solar lake (Sinai). 1. Physical and chemical limnology. *Limnol. Oceanogr.* **22**:597–607.
18. **Cohen, Y., B. B. Jørgensen, E. Padan, and M. Shilo.** 1975. Sulphide-dependent anoxygenic photosynthesis in the cyanobacterium *Oscillatoria limnetica*. *Nature* (London) **257**:489–491.
19. **Cohen, Y., B. B. Jørgensen, N. P. Revsbech, and R. Poplawski.** 1986. Adaptation to hydrogen sulfide of oxygenic and anoxygenic photosynthesis among cyanobacteria. *Appl. Environ. Microbiol.* **51**:398–407.
20. **Cohen, Y., W. E. Krumbein, and M. Shilo.** 1977. Solar Lake (Sinai). 2. Distribution of photosynthetic microorganisms and primary production. *Limnol. Oceanogr.* **22**:609–620.
21. **Cohen, Y., E. Padan, and M. Shilo.** 1975. Facultative anoxygenic photosynthesis in the cyanobacterium *Oscillatoria limnetica*. *J. Bacteriol.* **123**:855–861.
22. **Ganf, G. G., and A. B. Viner.** 1973. Ecological stability in a shallow equatorial lake (Lake George, Uganda). *Proc. R. Soc. London Ser. B* **184**:321–346.
23. **Garlick, S., A. Oren, and E. Padan.** 1977. Occurrence of facultative anoxygenic photosynthesis among filamentous and unicellular cyanobacteria. *J. Bacteriol.* **129**:623–629.
24. **Gerdes, G., and W. E. Krumbein.** 1984. Animal communities in recent potential stromatolites of hypersaline origin, p. 59–84. *In* Y. Cohen, R. W. Castenholz, and H. O. Halvorson (ed.), *Microbial Mats: Stromatolites*. Alan R. Liss, Inc., New York.
25. **Giani, D., L. Giani, Y. Cohen, and W. E. Krumbein.** 1984. Methanogenesis in the hypersaline Solar Lake (Sinai). *FEMS Microbiol. Lett.* **25**:219–224.
26. **Gorlenko, V. M., and E. A. Bonch-Osmolovskaya.** 1986. Production and anaerobic destruction of organic matter in cyanobacterial mats of a hydrogen sulphide hot spring in the caldera of Vocano Uzon (Kamchatka).
27. **Govindjee, R., E. Rabinovitch, and R. Govind-**

jee. 1968. Maximum quantum yield and action spectrum of photosynthesis and fluorescence in *Chlorella. Biochim. Biophys. Acta* **162**:539–544.

28. **Gromet-Elhanan, Z.** 1977. Electron transport and photophosphorylation in photosynthetic bacteria, p. 637–662. *In* A. Trebt and M. Avron (ed.), *Encyclopedia of Plant Physiology*. Springer-Verlag KG, Berlin.
29. **Hartman, H.** 1983. The evolution of photosynthesis and microbial mats; a speculation on the banded iron formations, p. 441–454. *In* Y. Cohen, R. W. Castenholz, and H. O. Halvorsen (ed.), *Microbial Mats: Stromatolites*. Alan R. Liss, Inc., New York.
30. **Howlsley, R., and H. W. Pearson.** 1979. pH dependent sulfide toxicity to oxygenic photosynthesis in cyanobacteria. *FEMS Microbiol. Lett.* **6**: 287–292.
31. **Javor, B. J., and R. W. Castenholz.** 1984. Invertebrate grazers of microbial mats, Lagoona Guerrero Negro, Mexico, p. 85–94. *In* Y. Cohen, R. W. Castenholz, and H. O. Halvorson (ed.) *Microbial Mats: Stromatolites*. Alan R. Liss, Inc., New York.
32. **Javor, B. J., and R. W. Castenholz.** 1984. Productivity studies of microbial mats, Lagoona Guerrero Negro, Mexico, p. 149–170. *In* Y. Cohen, R. W. Castenholz, and H. O. Halverson (ed.), *Microbial Mats: Stromatolites*. Alan R. Liss, Inc., New York.
33. **Jørgensen, B. B., and Y. Cohen.** 1977. Solar Lake (Sinai). 5. The sulfur cycle of benthic cyanobacterial mats. *Limnol. Oceanogr.* **22**:657–666.
34. **Jørgensen, B. B., Y. Cohen, and D. J. Des Marais.** 1987. Photosynthetic action spectra and adaptation to spectral light distribution in a benthic cyanobacterial mat. *Appl. Environ. Microbiol.* **53**: 879–886.
35. **Jørgensen, B. B., Y. Cohen, and N. P. Revsbech.** 1986. Transition from anoxygenic to oxygenic photosynthesis in a *Microcoleus chthonoplastes* cyanobacterial mat. *Appl. Environ. Microbiol.* **51**:408–417.
36. **Jørgensen, B. B., Y. Cohen, and N. P. Revsbech.** 1988. Photosynthetic potential and light-dependent oxygen consumption in a cyanobacterial mat. *Appl. Environ. Microbiol.* **54**:176–182.
37. **Jørgensen, B. B., N. P. Revsbech, T. H. Blackburn, and Y. Cohen.** 1979. Diurnal cycle of oxygen and sulfide microgradients and microbial photosynthesis in a cyanobacterial mat sediment. *Appl. Environ. Microbiol.* **38**:46–58.
38. **Jørgensen, B. B., N. P. Revsbech, and Y. Cohen.** 1983. Photosynthesis and structure of benthic microbial mats: microelectrode and SEM studies of four cyanobacterial communities. *Limnol. Oceanogr.* **28**:1075–1093.
39. **Knobloch, K.** 1969. Sulfide oxidation via photosynthesis in green algae, p. 1032–1034. *In* H. Muntzer (ed.), *Progress in Photosynthesis Research*, vol. II. International Union of Biological Sciences, Tübingen, Federal Republic of Germany.
40. **Knoll, A. H.** 1979. Archacean photoautotrophy: some alternatives and limits. *Origins Life* **9**:313–327.
41. **Krumbein, W. E., and Y. Cohen.** 1974. Klastische und evaporitische Sedimentation in einem mesothermen monomiktischen ufernahen See (Golf von Aqaba, Sinai). *Geol. Rundsch.* **63**:1035–1065.
42. **Krumbein, W. E., and Y. Cohen.** 1977. Primary production, mat formation and litification: contribution of oxygenic and facultative anoxygenic cyanobacteria, p. 37–56. *In* E. Flugel (ed.), *Fossil Algae*. Springer-Verlag, New York.
43. **Krumbein, W. E., Y. Cohen, and M. Shilo.** 1977. Solar Lake (Sinai). 4. Stromatolitic cyanobacterial mats. *Limnol. Oceanogr.* **22**:635–656.
44. **Kusnetzov, S. I.** 1970. *The Microflora of Lakes and its Geochemical Activity*. University of Texas Press, Austin.
45. **LeGall, J., and J. R. Postgate.** 1973. The physiology of sulfate reducing bacteria, p. 81–133. *In* H. A. Rose and D. W. Tempest (ed.), *Methods in Microbiology*, vol. 3A. Academic Press, Inc. (London), Ltd., London.
46. **Lemasson, C., N. Tandeau de Marsac, and G. Cohen-Basier.** 1973. Role of allophycocyanin as a light harvesting pigment in cyanobacteria. *Proc. Natl. Acad. Sci. USA* **70**:3130–3133.
47. **Logan, B. W., G. R. Davies, J. F. Read, and D. E. Cebulski.** 1970. Carbonate sedimentation and environments, Shark Bay, Western Australia. *Tulsa Am. Assoc. Petrol. Geol. Mem.* **13**:223.
48. **Margulis, L., S. Ashendorf, S. Banerjee, S. Francis, S. Giovannoni, J. Stolz, E. S. Barghoorn, and O. Chase.** 1980. The microbial community in the layered sediment at Laguna Figueroa, Baja California, Mexico: does it have Precambrian analogues? *Precambrian Res.* **11**:93–123.
49. **Oren, A., and E. Padan.** 1978. Induction of anaerobic photoautotrophic growth in the cyanobacterium *Oscillatoria limnetica*. *J. Bacteriol.* **133**: 558–563.
50. **Oren, A., E. Padan, and M. Avron.** 1977. Quantum yields for oxygenic and anoxygenic photosynthesis in the cyanobacterium *Oscillatoria limnetica*. *Proc. Natl. Acad. Sci. USA* **74**:2152–2156.
51. **Oren, A., E. Padan, and S. Malkin.** 1979. Sulfide inhibition of photosystem 2 in cyanobacteria (blue green algae) and tobacco chloroplasts. *Biochim. Biophys. Acta* **546**:270–279.
52. **Oren, A., and M. Shilo.** 1979. Anaerobic heterotrophic dark metabolism in the cyanobacterium *Oscillatoria limnetica*: sulfur respiration and lactate fermentation. *Arch. Microbiol.* **122**:77–84.
53. **Padan, E., and Y. Cohen.** 1982. Anoxygenic

photosynthesis, p. 215–235. *In* N. C. Carr and B. A. Whitton (ed.), *The Biology of Cyanobacteria*. Blackwell Scientific Publications, New York.

54. **Pfennig, N.** 1975. Phototrophic bacteria and their role in the sulfur cycle. *Plant Soil* 43:1–16.

55. **Pfennig, N.** 1977. Phototrophic green and purple bacteria: a comparative systematic survey. *Annu. Rev. Microbiol.* 31:275–290.

56. **Pfennig, N., F. Widdel, and H. G. Truper.** 1981. The dissimilatory sulfate-reducing bacteria, p. 926–940. *In* M. P. Starr, H. Stolp, H. G. Trüper, A. Balows, and H. G. Schlegel (ed.), *The Prokaryotes*. Springer-Verlag KG, Berlin.

57. **Playford, P. E., and A. E. Cockbain.** 1969. Algal stromatolites: deep water forms in the Devonian of Western Australia. *Science* 165:1008–1010.

58. **Revsbech, N. P., B. B. Jørgensen, T. H. Blackburn, and Y. Cohen.** 1983. Microelectrode studies of the photosynthetics and O_2, H_2S, and pH profiles of a microbial mat. *Limnol. Oceanogr.* 28:1062–1074.

59. **Revsbech, N. P., and D. M. Ward.** 1984. Microprofiles of dissolved substances and photosynthesis in microbial mats measured with microelectrodes, p. 171–188. *In* Y. Cohen, R. W. Castenholz, and H. O. Halvorson (ed.), *Microbial Mats: Stromatolites*. Alan R. Liss, Inc., New York.

60. **Schmidt, T. M., B. Arieli, Y. Cohen, E. Padan, and W. R. Stroll.** 1987. Sulfur metabolism in *Beggiatoa alba*. *J. Bacteriol.* 169:5466–5472.

61. **Schopf, W. J. (ed.).** 1983. *Earth's Earliest Atmosphere: Its Origin and Evolution*. Princeton University Press, Princeton, N.J.

62. **Schwabe, G. H.** 1960. Uber den thermobioten Kosmopoliten *Mastigocladus laminosus* Cohn. Blaugen und Lebensraum. V. Schweeig. *J. Hidrol.* 22:757–792.

63. **Skyring, F. W.** 1984. Sulfate reduction in marine sediments associated with cyanobacterial mats, Australia, p. 265–276. *In* Y. Cohen, R. W. Castenholz, and H. O. Halvorson (ed.), *Microbial Mats: Stromatolites*. Alan R. Liss, Inc., New York.

64. **Stanier, R. Y.** 1974. The origin of photosynthesis in eukaryotes. *Symp. Soc. Gen. Microbiol.* 24:219–240.

65. **Stanier, R. Y.** 1977. The position of the cyanobacteria in the world of phototrophs. *Carlsberg Res. Commun.* 42:77–98.

66. **Stolz, J. F.** 1984. Fine structure of the stratified microbial community at Laguna Figueroa, Baja California, Mexico. II. Transmission electron microscopy as a diagnostic tool in studying microbial communities *in situ*, p. 23–38. *In* Y. Cohen, R. W. Castenholz, and H. O. Halvorson (ed.), *Microbial Mats: Stromatolites*. Alan R. Liss, Inc., New York.

67. **Tel-Or, E., C. W. Luijk, and L. Packer.** 1977. An inductible hydrogenase in cyanobacteria enhances NH_2 fixation. *FEBS Lett.* 78:49–52.

68. **Utkilen, H. C.** 1976. Thiosulphate as electron donor in the blue-green alga *Anacystis nidulans*. *J. Gen. Microbiol.* 95:177–180.

69. **Walter, M. R. (ed.).** 1976. *Developments in Sedimentology*. 20. *Stromatolites*. Elsevier/North Holland Publishing Co., Amsterdam.

70. **Weisman, J. C., and J. R. Benemann.** 1977. Hydrogenase production by nitrogen-starved culture of *Anabena cylindrica*. *Appl. Environ. Microbiol.* 35:123–131.

71. **Weller, D., W. Doemel, and T. D. Brock.** 1975. Requirement of low oxidation-reduction potential for photosynthesis in a blue green alga (*Phormidium* sp.). *Arch. Microbiol.* 104:7–13.

Chapter 4

Multilayered Microbial Communities in Aquatic Ecosystems: Growth and Loss Factors

Ricardo Guerrero and Jordi Mas

INTRODUCTION

In natural environments such as freshwater stratified lakes, phototrophic procaryotes often form multilayered communities (3, 10). These kinds of communities require a physical structure which prevents the different layers from mixing. The nature of this physical structure determines the stability and the persistence of the layering, as well as the scale of the lamination. Phototrophic organisms place themselves at different positions in these structures according to the vertical distribution of environmental factors. Major factors such as light and oxygen, whose levels decrease from the surface to the benthic layers, determine microorganism distribution and abundance. Hydrogen sulfide, whose level, in contrast, increases as a function of depth, is also important for the anoxygenic phototrophic procaryotes (40).

Ecosystems which depend on light as the primary energy source are usually arranged in space in horizontal layers: a consequence of light extinction with depth. Layered communities are widespread among both microbial and plant communities. Tropical forests, microbial mats, and planktonic communities in stratified lakes are considered analogous forms that have evolved to maximize light utilization (Fig. 1). Tropical forests are difficult to study because of the complex trophic relationships involved. The far smaller benthic microbial mats and multilayered planktonic microbial (MPM) communities are far easier systems to analyze. Microbial mats and MPM communities display a far lower degree of complexity and a rather high horizontal uniformity and are composed almost completely of unicellular organisms, most of which are procaryotic.

Both MPM communities and microbial mats require some physical structure to establish and support lamination. Physical support in MPM communities results from thermal or chemical stratification along a vertical axis. Microbial mats, in contrast, are built upon solid substrates by biological accretion. In both cases the active alignment of microorganisms along a light gradient occurs, providing continuous gradients of other environmental factors and making possible the

Ricardo Guerrero and Jordi Mas • Department of Genetics and Microbiology, Autonomous University of Barcelona, 08193 Bellaterra (Barcelona), Spain.

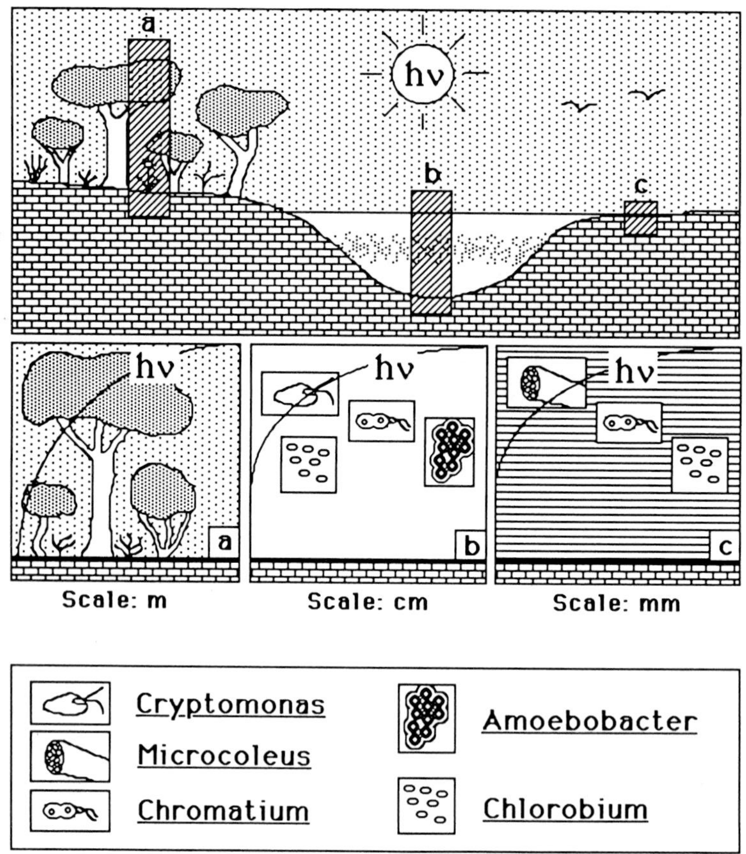

Figure 1. Vertical structuring in several ecosystems in which light is the primary energy source. In the upper panel the following are schematized: (a) a forest ecosystem, in which the photosynthesizing structure extends through several meters; (b) an MPM community, in which the layering of different populations spans in the range of several centimeters; (c) a microbial mat, in which layering of different populations occurs in only a few millimeters. In the lower panel, a key to the main microorganisms of multilayered planktonic communities and microbial mats is presented.

simultaneous existence of different populations situated along them.

The analysis of structure, composition, and ecophysiology of microbial mats has been hampered by the technical problem of sampling complex communities compressed into a few millimeters. The techniques used so far are based on a direct approach: in situ measurements of environmental variables with microprobes (42); use of electron microscopy to examine ultrathin sections of whole transects (45); stable carbon isotope fractionation ($\delta^{13}C$) (2); and analysis of pigments (44), elemental sulfur, acid-volatile sulfide, or nitrogenase activity (L. J. Stal, Ph.D. thesis, 1985) from sliced samples. Although the resolution obtained by use of microprobes is very good (ca. 10 μm), samples obtained after slicing cores are not thin enough (≥1 mm) to give the information necessary to interpret the fine structure of the layering.

On the other hand, MPM communities are far more accessible to detailed studies. This is due to their vertical extension, which expands through centimeters or meters instead of the millimeters used by microbial mats. Detailed results can be obtained with MPM communities by use of relatively simple techniques, thus constituting a good general model for studying other stratified systems or even for understanding more complex microbial ecosystems in which organisms either are not arranged in space in any characteristic pattern or do not display clear fluctuations with time.

Several studies analyzing the vertical structure of the phototrophic community in stratified lakes have been published (3, 10, 16, 23). Variations with time of primary production and, in some cases, of the abundance of the different members of these communities have also been described (5, 6, 22, 33, 36). However, an evaluation of the main factors determining the observed variations in space and time is still lacking.

Although some information about factors related to changes in pigment concentration (39) and biomass (37) in MPM communities has already been assembled, factors determining population loss have seldom been analyzed. As a consequence, our understanding of the mechanisms that regulate the population levels in these communities is limited. The objective of this report is to review our present knowledge about the main factors determining growth and loss in these ecosystems.

MPM communities, owing to the procaryotic nature of most of their constituent species, can almost certainly be considered, along with microbial mats, to have constituted one of the earliest ecosystems on our planet. Therefore, the study of extant MPM communities is of maximal interest for our understanding both of microbial ecology processes and of the mechanisms which have regulated the evolution of ecological relationships throughout the history of the Earth.

MPM COMMUNITIES

MPM communities occur under certain environmental conditions in lakes, lagoons, and narrow estuaries. First, a vertical density gradient, providing both a boundary between aerobic and anaerobic compartments (epilimnion and hypolimnion) and mechanical resistance to wind-mediated mixing, is required. Second, light must be available at the interface to allow phototrophic growth. Finally, hydrogen sulfide must diffuse from the anaerobic to the aerobic compartment. The oxygen, light, and hydrogen sulfide gradients provide a set of different habitats along the density gradient for organisms with different tolerances and affinities for these three environmental factors.

In lakes in temperate climates, in which the density gradient is due to thermal stratification, the layering is transitory and disappears during the cold season. When the density gradient is due to chemical stratification, community stratification may persist year-round, permitting the existence of a permanently layered structure.

The predominant physicochemical conditions in these environments are summarized in Fig. 2, taking as an example Lake Cisó, Spain, where several detailed studies have been already performed (15, 50; C. Pedrós-Alió, J. M. Gasol, and R. Guerrero, *in* F. Megusar, ed., *IV Int. Congr. Microb. Ecol.*, in press). Oxygen, which is usually present in the upper compartment, decreases with depth, reaching zero at the discontinuity layer. Hydrogen sulfide, which is very abundant in the anaerobic compartment, is produced in the sediment through sulfate reduction (and also, in certain amounts, probably enters the system with seepage water). The hydrogen sulfide concentration decreases as it ascends in the water column, until it completely disappears at the interface. Light, which may be the key environmental factor determining the structure of the community, decreases exponentially with depth. If enough light reaches the

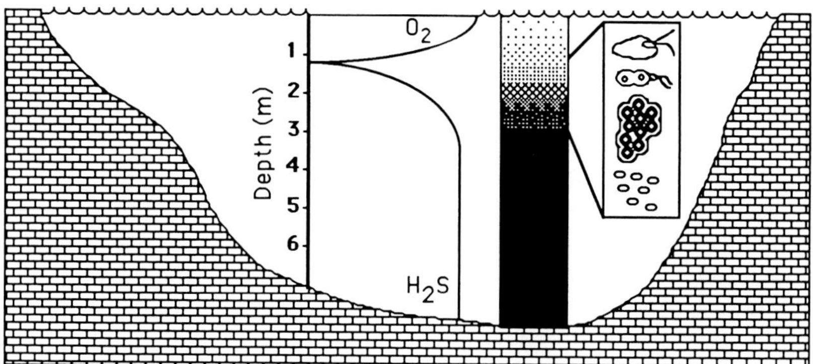

Figure 2. Predominant physicochemical conditions in MPM ecosystems. The figure has been drawn according to the general characteristics of Lake Cisó during summer stratification. Oxygen diffuses into the lake through the surface. Important amounts of hydrogen sulfide are produced in the sediment at the bottom of the lake by the activity of sulfate-reducing bacteria. As a consequence of both processes, gradients of oxygen and hydrogen sulfide are established. At the intersection of and underneath these gradients, a complex community of phototrophs which require, or tolerate, hydrogen sulfide develops. As a consequence of light absorption by this community, a rather steep light gradient is also formed. For a key to the microorganisms, see Fig. 1.

anoxic layers, a complex community formed by aerobic and anaerobic phototrophic microorganisms develops. The composition of this community is variable, but in most cases the relative organization of the different organisms with depth fits the general pattern described here. The description which follows is based mainly on the microbial community of Lake Cisó, Spain.

The upper part of the community consists of aerobic phototrophs, such as diatoms, adapted to the turbulent regimen of the mixed layer. Below, and in the upper part of the gradient, where oxygen is still present and hydrogen sulfide is either very scarce or absent, a population of specialized oxygenic phototrophs develops. This population, which in Lake Cisó is dominated by *Cryptomonas* spp. (38), is formed by several other organisms in other lakes (e.g., cyanobacteria [21, 28], *Chlamydomonas* spp. [24], *Trachelomonas* spp. [43], and *Dunaliella* spp. [32]). In general, these motile phototrophs at the surface layers of these communities tolerate some sulfide. They apparently situate themselves at this position to maximize light and nutrient (phosphorus and nitrogen) conditions.

The next layer is usually inhabited by purple sulfur bacteria. These bacteria have been reported to form highly concentrated populations in different lakes around the world, at depths between 0.5 and 19 m (average depth, 7 m), as reviewed in reference 16, which gives data for 35 lakes. In Lake Cisó, the most abundant of these organisms are *Chromatium* spp. (single, flagellated cells which swim to actively situate themselves at a level where the conditions are optimal for their growth) and *Amoebobacter* spp.(smaller cells which form aggregates surrounded by slime and contain gas vesicles that permit controlled location by floating and sinking in the water column). *Chromatium* spp. are adapted to grow better during the stratification period (summer), whereas *Amoebobacter* spp. are better adapted to conditions of turbulence, during fall and winter (J. Mas, unpublished data).

The *Amoebobacter* strain found in Lake Cisó, initially referred to as M-3 (15), contains okenone as its main carotenoid. Although it was first assigned to the genus *Lamprocystis* (7, 16; Pedrós-Alió et al., in press), its similarity to strains recently isolated from Schleinsee, near Konstanz, Federal Republic of Germany, by Pfennig and co-workers indicates that it should be considered a species of the genus *Amoebobacter* (B. Eichler, University of Konstanz, personal communication).

Purple sulfur bacteria are capable of utilizing light at wavelengths not absorbed by organisms in the upper layers. The conditions prevailing in this zone, low sulfide concentration and occasional exposure to certain amounts of oxygen, correspond to their physiological characteristics. Their sensitivity to high concentrations of sulfide has been reported (40), as well as their tolerance of microaerophilic conditions. Some purple sulfur bacteria use oxygen as an electron acceptor in a chemolithotrophic mode of growth (20; chapter 28 of this volume).

The lowest part of the planktonic community is composed of green sulfur bacteria. These organisms, which tolerate high concentrations of hydrogen sulfide (40), are well suited for the light conditions prevailing at the depth where they are found, i.e., between 2 and 25 m (average, 9.3 m), as reviewed in reference 30, which gives data for 30 lakes. Light penetrating at these depths displays a highly altered spectrum and very low intensity owing to shading by cyanobacteria or algae and by purple sulfur bacteria. Green sulfur bacteria adapt to such conditions by maintaining a high specific content of pigments that absorb mostly in the part of the spectrum not used by the organisms above them (1, 30).

The analogy between MPM communities and microbial mats is strengthened when the anaerobic sediment (Eh, ca. −300 mV) at the bottom of the lake is taken into account. In this sediment, which is the main site of hydrogen sulfide production, a well-developed population of sulfate-reducing bacteria is usually found. Sulfate, which is found in Lake Cisó at concentrations between 10 and 15 mM, comes from the dissolution of gypsum ($CaSO_4 \cdot H_2O$) and enters the lake through seepage. Sulfate reduction in the bottom of the lake, which requires a certain amount of sulfate in the water, is fueled by organic matter (in Lake Cisó it is contributed largely by the sedimenting purple and green sulfur bacteria) from the layers above. This black sediment is analogous to the black layers often found at the bottom of microbial mats. In both cases, the color is due to the accumulation of iron sulfide.

TEMPORAL VARIATIONS IN MPM COMMUNITIES

The multilayered community structure occurs only during periods when lakes are stratified. Stratification, however, is often disturbed, thus removing organisms from their optimal place and promoting the growth of other organisms better suited to the new set of environmental conditions. Even within the stratification period, environmental conditions are, to a certain extent, subject to change. As a consequence, some kind of temporal variations of the populations in the lake can be expected. Integrated cell biovolume measurements of the main components of the planktonic community in Lake Cisó, taken throughout the year, show that variations of this sort do indeed occur (Fig. 3).

Large populations of *Cryptomonas* spp., adapted to relatively high oxygen and low hydrogen sulfide concentrations, are found during summer months when the lake stratifies (Fig. 3A). During the winter, however, a strong decrease in the biomass of *Cryptomonas* spp. is observed, which relates to reduced light availability and continuous exposure to high concentrations of hydrogen sulfide.

Variations in the population of *Amoebobacter* spp. with time (Fig. 3B) are roughly complementary to those displayed by *Cryp-*

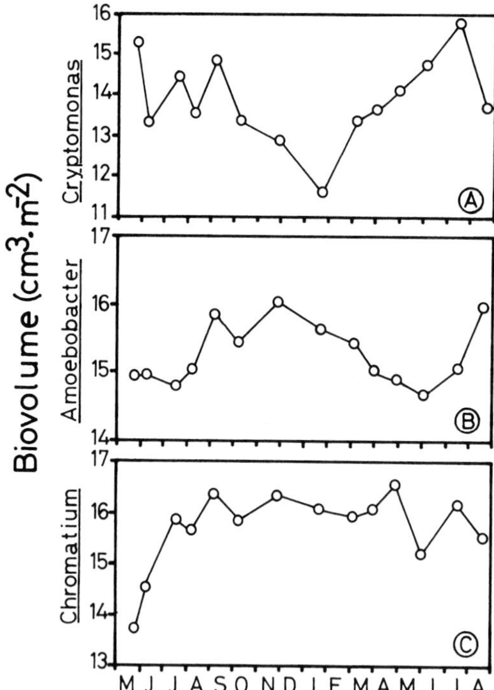

Figure 3. Variation with time of the integrated biovolume for the three most abundant organisms composing the planktonic community of Lake Cisó: the protist *Cryptomonas* (a mastigote) (A) and the purple sulfur bacteria *Amoebobacter* (usually as aggregates) (B), and *Chromatium* (highly motile single cells) (C).

tomonas spp., although they occur within a narrower range. These variations apparently indicate that *Amoebobacter* spp. adapt better than *Cryptomonas* spp. to the turbulent homogeneous conditions prevailing during winter mixing. The physiological basis of this adaptation has not been established.

The variation with time of the integrated biovolume of *Chromatium* spp. is shown in Fig. 3C. The *Chromatium* population increases at the beginning of the first summer, stabilizes around 16 $cm^3 \cdot m^{-2}$, and remains constant thereafter. Variations in the biovolume of *Chromatium* spp., unlike those of *Amoebobacter* and *Cryptomonas* spp., do not appear to have a seasonal basis. In previous studies, however, seasonal variations in *Chromatium* concentration have been described (1, 14, 37, 39) and linked to variations in the concentration of *Chlorobium* spp., which, at that time, was one of the most abundant organisms in the lake.

Specific population changes cannot be interpreted only in relation to environmental factors affecting growth. Variations in population abundance are also a consequence of changes in the intensity of several loss processes. In the MPM community of Lake Cisó these processes are washout, sedimentation, decomposition (death and autolysis), and predation (Fig. 4). A study of the environmental factors affecting growth and loss, as well as an assessment of their impact, are prerequisites for the understanding of the mechanisms regulating microbial populations in nature.

FACTORS AFFECTING GROWTH

Growth of the organisms forming MPM communities depends to a large extent upon light. Usually, a close relationship exists between photosynthetic activity and the depth at which a given layer is found. This relationship is a consequence of light extinction with depth, owing to the combined effect of absorption by water and by planktonic populations (29, 34). Phototrophic microorganisms may modify their light-harvesting capacity through changes in their specific content of pigments (46). Studies performed in natural

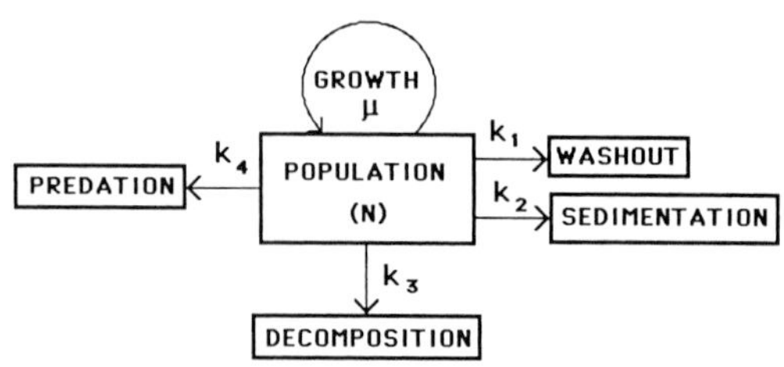

Figure 4. Outline of the main processes affecting population levels of planktonic microorganisms. $dN/dt = (\mu - k_1 - k_2 - k_3 - k_4) \cdot N$, where μ is the specific growth rate and k_1, k_2, k_3, and k_4 are the specific rates of loss due to washout, sedimentation, decomposition, and predation, respectively.

environments have shown that the specific content of pigments changes throughout the year, following a pattern well correlated to the light intensity reaching the layer (39).

Not only the amount but also the spectral composition of light affects the growth of phototrophic organisms. In lakes containing high concentrations of humic substances, which specifically absorb red and blue light, green sulfur bacteria are selected over purple species (35). A similar chromatic effect has been described for MPM communities, in which purple sulfur bacteria absorb wavelengths necessary for the growth of *Chlorobium phaeobacteroides*, thus selecting for *Chlorobium limicola*, which uses a different part of the spectrum (30) (Fig. 5). Phototrophic organisms can compensate somewhat for chromatic effects by means of chromatic adaptation mediated by changes in the composition of their complementary pigments, mainly carotenoids (14).

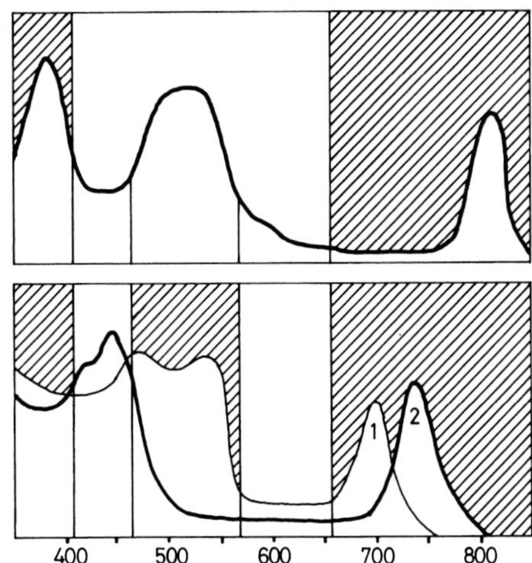

Figure 5. Selective effect of shading by *Chromatium* spp. (upper panel) on the species composition of the layers underneath. The green *Chlorobium* strain (*C. limicola*, absorption spectrum 2) is selected favorably over the brown *Chlorobium* strain (*C. phaeobacteroides*, absorption spectrum 1) owing to the ability of the green bacteria to use the narrow band of the light spectrum not absorbed by *Chromatium* spp. ▨, The part of the spectrum absorbed by either water or *Chromatium* photopigments.

The concentration of hydrogen sulfide plays a fundamental role in MPM communities. In laboratory experiments, very low concentrations of this compound lead to competition between organisms using it (47–49). In nature, however, hydrogen sulfide appears important only at a qualitative level; its availability is what really matters. As long as hydrogen sulfide stays between the limiting and the inhibitory concentrations, phototrophic sulfur bacteria can be found, their abundances and productivities being independent of the actual hydrogen sulfide concentration (37). The main effect of hydrogen sulfide, then, is the exclusion of sensitive species such as most green algae and some groups of cyanobacteria (4). Phototrophic sulfur bacteria in Lake Cisó are not sulfide limited, but are strongly light limited (15). Light limitation is mainly a consequence of the absorption and scattering of light by the upper layers of the phototrophic bacterial population itself, thus suggesting an important regulatory loop mediated by self-shading.

LOSS FACTORS

Light intensity and sulfide concentration play an important role in controlling the growth rate of the populations in the MPM community. However, in most cases this growth rate differs widely from the rate at which population levels actually change. As has been pointed out above, these differences can be accounted for by a certain number of loss processes such as washout, sedimentation, decomposition, and predation (Fig. 4).

Washout

In small bodies of water with a high input of water, population levels of planktonic organisms can be strongly affected by the dilution rate of the system. The volumes of most lakes are so large that dilution rates are negligible when compared with the actual growth rates of the organisms. However, in small lakes such as Lake Cisó (2.2×10^3 m^3), dilution rates can be high enough to have an effect on the planktonic populations. In this lake, where water enters through seepage, the output occurs through a small outlet, where flow determinations can be easily made. Measurements made after periods of heavy rain gave outflow values of 72 $m^3 \cdot day^{-1}$. From this outflow, a dilution rate of 0.033 day^{-1} was calculated. If we express this dilution rate as the time in which the level of a nongrowing population is reduced to one-half, a value of ca. 21 days results, which is short enough to threaten the permanence of the planktonic community. In fact, during this period of high dilution rate, a 10-fold decrease in population levels was observed, thus suggesting an important role for washout as a factor determining variations in the planktonic community.

Sedimentation

Sedimentation of microorganisms has usually been considered negligible. In most cases this assumption is not supported by any direct measurement of the actual sinking speeds, although some studies performed under well-controlled laboratory conditions show that sinking of some bacteria indeed proceeds very slowly (18).

In nature, however, the situation is far more complex. Variations in water turbulence or in the physiological state of the cells can cause unexpected sedimentary behavior. The specific weight of the cells, one of the variables affecting sinking speed, can be greatly altered by accumulation of intracellular inclusions of storage materials such as "elemental sulfur" (probably polysulfide), glycogen, or poly-β-hydroxybutyrate (13, 26, 27). (For clarity, the widely used expression "cell density" should not be used to express variables such as cell concentration or optical density of a culture or a population, but should be used as a synonym of specific weight of the cells [i.e., cell weight divided by cell volume] [13].)

Changes in cell density, however, do not greatly affect sedimentation velocity, unless they are coupled to the presence of gas vesicles. Intracellular inclusions act as ballast, determining either a buoyant or a sedimentary tendency in cells depending on whether the overall density is higher or lower than the density of the surrounding fluid (21). The speed at which cells move (upward or downward) within the fluid is determined largely by changes in size. Variations in the sizes of individual cells, however, have little impact on their speed; the largest changes are observed when clumping occurs.

Determination of the sinking speeds of the main components of the planktonic community in Lake Cisó has revealed several facts. First, sedimentation occurs only during the stratification period, at depths where turbulence is likely to be small (hypolimnion). During the mixing period sedimentation was virtually zero (lower than the lowest theoretical speed predicted by Stokes's Law [Table 1]). These extremely low values were attributed to increased levels of turbulent viscosity under mixing conditions. Second, even during periods of low turbulence, sedimentation was detectable only during short intervals, which, in addition, were different for the various species studied. This last observation strongly suggests that some organism-dependent factor such as its physiological state is involved in the regulation of sinking.

Sinking speeds characteristic of the mixing and stratification periods are shown in Fig. 6 for each of the major organism types in Lake Cisó. Although the values are significantly higher during stratification (1 order of magnitude for *Cryptomonas* spp.), the impact of sedimentation on population dynamics is important only at certain times of the year. In the same figure, below the bars, the time necessary for the organisms to reach the bottom of the lake from the surface (a

Table 1.
Maximum and Minimum Values of Cell Parameters for the Three Most Abundant Organisms in the Planktonic Community of Lake Cisó

Organism	Cell vol (μm³)		Cell density (pg · μm⁻³)		Sinking speed (cm · day⁻¹)	
	Maximum	Minimum	Maximum	Minimum	Maximum[a]	Minimum[b]
Amoebobacter sp.	97.34[c]	31.42[c]	1.070	1.035	11.52	1.61
Chromatium sp.	58.35	31.20	1.160	1.130	18.42	5.95
Cryptomonas sp.	800.00	300.00	1.100	1.100	66.44	20.71

[a] Maximum values of sinking speed were calculated by using the equivalent radius corresponding to the maximum cell volume, the maximum cell density, and the density and viscosity of water at 20°C.
[b] Minimum values of sinking speed were calculated by using the equivalent radius corresponding to the minimum cell volume, the minimum cell density, and the density and viscosity of water at 4°C.
[c] For *Amoebobacter*, since sinking takes place in the form of cell aggregates, the average volume of the aggregates has been taken into account.

distance of 7.5 m) at each of the sinking speeds is shown. As can be seen in this rough approach, bacteria would take at least 100 or 185 days to be eliminated from the planktonic part of the system during the stratification period. During the mixing period this time would be even longer: 332 or 554 days, depending on bacterial type. Therefore, sedimentation is likely to have a low impact in the removal of populations from the water column.

Decomposition

A third mechanism which may be important as a loss factor is decomposition. Several biological processes (e.g., death and autolysis) can contribute to this factor. When energy-generating processes in the cell fall below a certain level, maintenance of the structural and physiological integrity is no longer possible and death occurs. There is no synthesis of new cell material, and degradation of cell components becomes irreversible. Since the ultimate consequence of the degradation of cell components is the disappearance of the cell, the effects of degradation in nature might be difficult to differentiate from those of grazing by other organisms.

Decomposition in Lake Cisó has been analyzed by placing samples in closed containers at different depths and measuring cell concentrations at different times. Parallel samples were processed simultaneously in the laboratory by following a similar procedure in which the samples were incubated at several temperatures. The results of these measurements are summarized in Table 2, in which decomposition rates (k_d) are shown for each organism at three different temperatures. The k_d for *Amoebobacter* spp. was virtually zero in all cases; the k_d for *Chroma-*

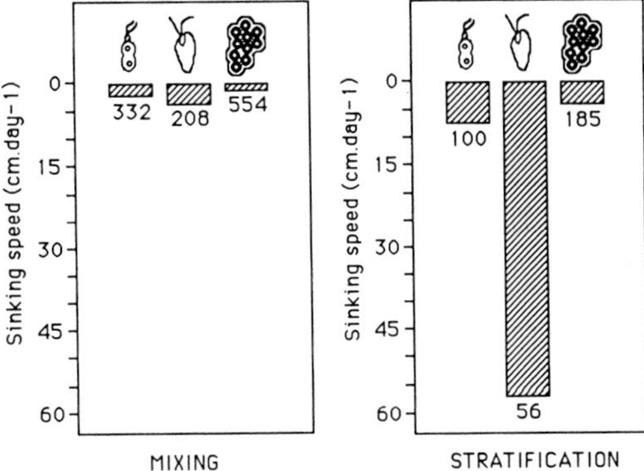

Figure 6. Values of sinking speed representative of mixing and stratification periods for the three most abundant organisms composing the planktonic community of Lake Cisó. For a key to the microorganisms, see Fig. 1. Numbers below each of the bars indicate the time in days that each organism would take to sink from the surface to the bottom of the lake (7.5 m) at each of the specified sinking speeds.

Table 2.
Specific Decomposition Rates for the Three Most Abundant Organisms in the Planktonic Community of Lake Cisó at Different Temperatures

Organism	Decomposition constant[a] (10^3 day^{-1}) at:		
	5°C	10°C	15°C
Amoebobacter	0	0	0
Chromatium	−5	−7	−18
Cryptomonas	−86	−122	−158

[a] *Amoebobacter* organisms form aggregates surrounded by slime, thus hampering the action of decomposing agents. *Chromatium* and *Cryptomonas* organisms have individual cells with large differences in size (Table 1). They do not display aggregation or have a slime envelope. Their decomposition rates are very different, probably owing to the eucaryotic nature of *Cryptomonas* strains.

tium spp. ranged between 5×10^{-3} and 1.8×10^{-2} day^{-1}; and the k_d for *Cryptomonas* spp. was almost 10 times higher, between 8.6×10^{-2} and 1.58×10^{-1} day^{-1}. These differences indicate either the effect of selective predation on *Cryptomonas* spp. (no predators of *Cryptomonas* spp. have yet been demonstrated) or a very high death rate of this organism linked to a very low stability of the dead material. On the other hand, the extremely low *Amoebobacter* decomposition rates are most probably related to the fact that cells of this organism form aggregates surrounded by a thick slime layer, thus rendering access to scavengers or decomposers difficult.

Predation

Although the anoxic nature of many of these systems precludes the presence of important groups of predatory organisms, a few groups of rotifers (11, 28) and protists (7) have been detected which may be grazing on bacterial populations. The predatory impact of these groups, however, has not yet been established. Procaryotic organisms capable of predatory behavior under the conditions prevailing in the lake have been discovered. These organisms, as well as their interaction with their prey, have been characterized (8, 12, 17), and, so far, two main genera have been identified. One of them, tentatively named *Vampirococcus*, attaches to the surfaces of *Chromatium* strains (Fig. 7A), causing alterations visible under transmission electron microscopy as large areas of the cytoplasm where the photosynthetic apparatus has disappeared; thus, the *Vampirococcus* sp. apparently digests its prey while remaining outside of it. This organism, which seems to be strictly anaerobic, has not been yet grown in axenic culture. The second type, *Daptobacter*, which has been grown in axenic culture, has been more extensively studied (12). It consists of a small, actively motile, rod-shaped bacterium, which kills its prey by penetrating its cytoplasm and dividing within it (Fig. 7B). Although predatory bacteria have not yet been found in microbial mats or scums, Stolz, in 1977, observed a bacterium that was very similar to a *Daptobacter* sp. in both morphology and position inside a *Thiocapsa* cell (Fig. 7C) in samples from the surface scums in Laguna Figueroa, Baja California, Mexico (J. Stolz, personal communication).

Both types of predatory bacteria, *Daptobacter* and *Vampirococcus* spp., occur in Lake Cisó; they are present as small buds attached to their prey cells. The importance of predators in population control has not yet been established. Some observations, however, point to an opportunistic scavenging of physiologically altered cells rather than active predation on healthy cells. In Fig. 7D the percentage of prey individuals attacked at different depths is shown. The highest values are found at depths where *Chromatium* spp. probably have serious physiological problems as a consequence of light limitation. Therefore, *Vampirococcus* and *Daptobacter* spp. may be highly specialized scavengers, which are important in degrading decaying populations, but not in promoting the initiation of their decay.

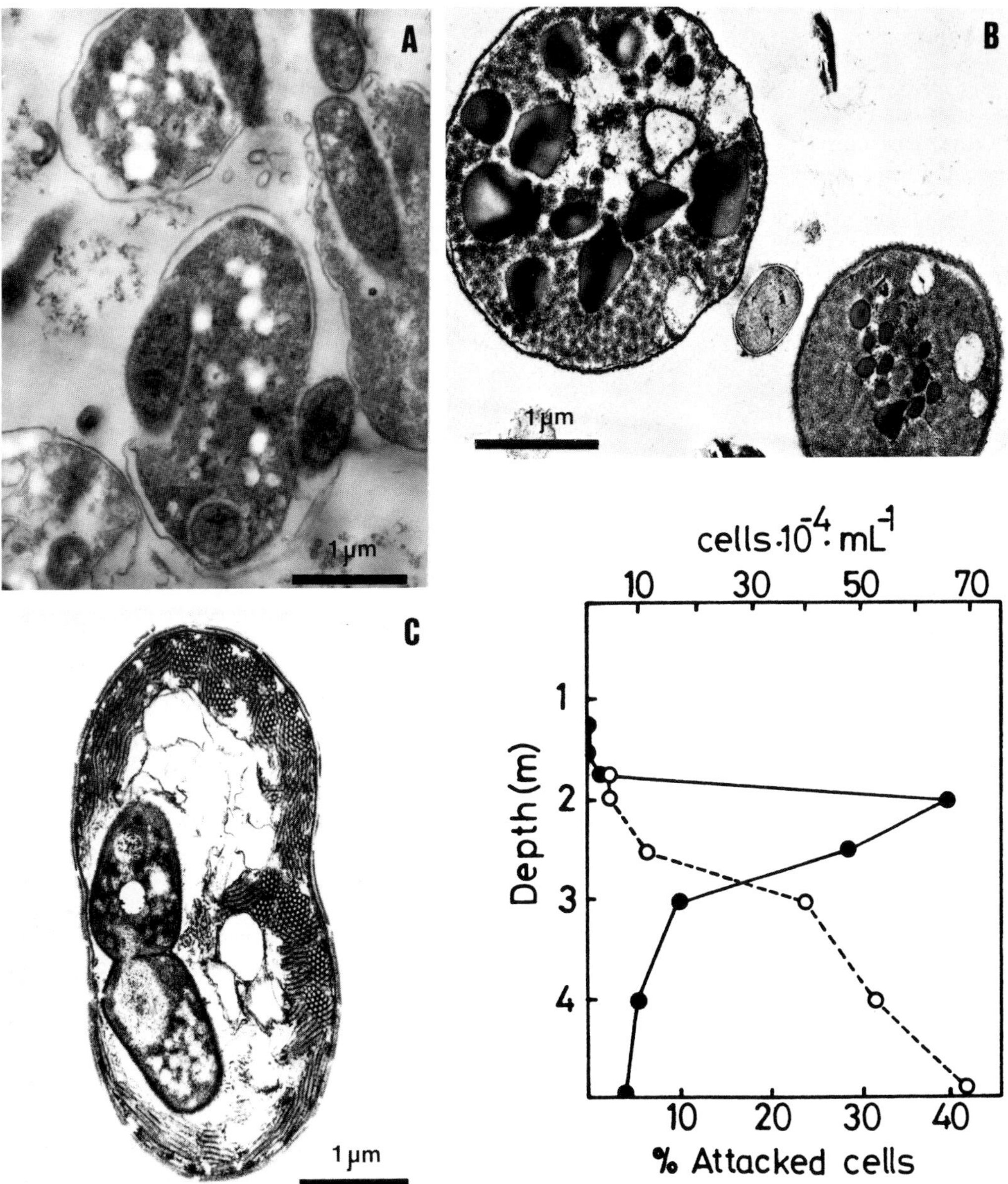

Figure 7. Ultrastructure of several predatory bacteria found feeding on members of the family *Chromatiaceae*. (A) Ultrathin section showing a *Chromatium* cell being attacked by several *Daptobacter* cells in a sample from the planktonic community of Lake Estanya, Huesca, Spain. (B) Transmission electron micrographs of *Vampirococcus* cells attached to *Chromatium* cells in a sample from the planktonic community of Lake Ciső. (C) Ultrathin section showing a *Daptobacter*-like cell inside a purple sulfur bacterium (*Thiocapsa* sp.) in a sample of surface scums from Laguna Figueroa. (D) Vertical profile of *Chromatium* strains (●) in Lake Ciső, showing that the percentage of cells infected by *Vampirococcus* strains (○) increases with depth. (Micrographs in panels A and B are by Isabel Esteve, Autonomous University of Barcelona, Barcelona, Spain; the micrograph in panel C is by John Stolz, University of Massachusetts, Amherst.)

DISCUSSION

Although loss processes could easily bring about the disappearance of some of the populations involved, what actually happens is that a certain equilibrium is reached in which losses by the processes described above are balanced by higher specific growth rates. These increased growth rates can be explained if we assume that the main limiting factor in MPM communities is light and that reduction of population levels results in the availability of more light to the remaining organisms, which, as a consequence, maintain higher specific growth rates and a healthier condition.

Loss processes in microbial mats (i.e., predation, decomposition, and burial), although not yet assessed, are likely to play a significant role. When comparing data available on primary production of microbial mats with data from a number of different systems, it can be seen (Table 3) that production measured in mats is around 4,000 to 5,000 mg of $C \cdot m^{-2} \cdot day^{-1}$, which falls within the range observed in other highly productive ecosystems, such as tropical rain forests, mangrove swamps, hypertrophic lakes, or upwellings in coastal areas. The high production observed indicates that all these systems, each in their own way, maximize energy (light) utilization. Since, in all the cases, light is almost completely absorbed by organisms inside the system, primary production can be expected to be very high and of the same order of magnitude.

One important difference between complex systems such as tropical forests and more simple ones such as microbial mats is the way in which primary production is used. In a study in El Verde, Puerto Rico, net primary production was 6,000 mg of $C \cdot m^{-2} \cdot day^{-1}$ (31). In the same study, soil respiration (corresponding to decomposition of decaying material) accounted for 5,480 mg of $C \cdot m^{-2} \cdot day^{-1}$. Apparently, most of the net primary production was used to replace dead material, whereas only 520 mg of $C \cdot m^{-2} \cdot day^{-1}$ remained for growth purposes or exportation. If we take into account that the total biomass in this system was 11,322 g of $C \cdot m^{-2} \cdot day^{-1}$, a turnover rate (production/biomass) of 4.6×10^{-5} day^{-1} can be calculated, corresponding to a renovation time of ca. 60 years.

Table 3.
Daily Values of Primary Production for Several Ecosystems[a]

Ecosystem	Primary production (mg of $C \cdot m^{-2} \cdot day^{-1}$)	Reference
Tropical rain forest	6,000	25[b]
Mangrove swamp	5,600	25[c]
Lake		
Oligotrophic-distrophic	40–80	25
Mesotrophic-eutrophic	300–3,000	25
Hypertrophic	2,000–5,000	25
Sea		
West Mediterranean	60–500	25
Coastal upwellings (Peru, South Africa)	1,000–4,000	25
Microbial mat[d]		
Shallow flat mat	5,069	19
Deep flat mat	3,830	19
Blister mat	3,873	19
Gelatinous mat	345	19

[a] The ecosystems differ in complexity, degree of organization, and thickness of their photic layers.
[b] Taken from reference 31.
[c] Taken from reference 9.
[d] The original data were expressed as millimoles of oxygen per square meter per hour. Conversion to milligrams of carbon per square meter per day has been performed by assuming a 12-h photoperiod and a molar ratio of 2 between carbon fixed and O_2 released. Respiration during the dark period has not been taken into account. The data are from microbial mats in Solar Lake.

The figures on primary production in microbial mats (Table 3) probably constitute an overestimate, since respiration during the dark period was not taken into account. However, given the very low values of biomass per surface area that are characteristic of these communities, then even if net primary production was 1 order of magnitude lower, the production/biomass ratio would still be much higher than the value observed in the tropical forest quoted above. This high production/biomass ratio, equivalent to a very intense carbon turnover, requires the existence of important loss mechanisms which, to our knowledge, have not yet been identified and whose importance is so far unknown. The study of these mechanisms should be a major topic to be addressed in the future.

It has been suggested that primary production by microbial mats and other communities of procaryotes has been maintained at high levels since the beginning of the Archaean eon (41). Detailed studies of carbon content in shales, formerly organic-rich muds, have been made. These studies of the 3.4-billion-year-old Swaziland System rocks (South Africa) reveal no change in rates of total sedimentation of carbon since that time when the Earth evidently was dominated by phototrophic microbial communities (41).

We conclude that performing detailed analyses of carbon budgets in microbial mats of different characteristics (paying special attention to factors responsible for loss in the productive layers) would give us a deeper insight into the ultimate fate of primary production in these benthic systems. In the meantime, the analysis of growth and loss factors in planktonic systems (such as the MPM communities studied in this article) yields valuable information not only on the mechanisms regulating the laminated microbial communities, but also on the general principles underlying microbial ecology.

Acknowledgments. We thank Carlos Pedrós-Alió, James G. Mitchell, and Lynn Margulis for their stimulating discussions and suggestions during the preparation of this article, and John Stolz for kindly giving us permission to use the electron micrograph in Fig. 7C. The help of David Carlon, Mercè Piqueras, and Peter Knowles with the English language is also acknowledged.

This research was supported by grants from the Comisión Asesora de Investigación Científica y Técnica (Ministerio de Educación y Ciencia), Fondo de Investigaciones Sanitarias de la Seguridad Social (Ministerio de Sanidad y Consumo), and Comissió Interdepartamental de Recerca i Innovació Tecnològica (Generalitat de Catalunya).

LITERATURE CITED

1. **Abellà, C., E. Montesinos, and R. Guerrero.** 1980. Field studies on the competition between purple and green sulfur bacteria for available light. *Dev. Hydrobiol.* 3:173–181.
2. **Aizenshtat, Z., G. Lipiner, and Y. Cohen.** 1984. Biogeochemistry of carbon and sulfur cycle in the microbial mats of the Solar Lake (Sinai), p. 281–312. *In* Y. Cohen, R. W. Castenholz, and H. O. Halvorson (ed.), *Microbial Mats: Stromatolites*. Alan R. Liss, Inc., New York.
3. **Caldwell, D. E., and J. M. Tiedje.** 1975. The structure of anaerobic bacterial communities in the hypolimnia of several Michigan lakes. *Can. J. Microbiol.* 21:377–385.
4. **Cohen, Y.** 1984. The Solar Lake cyanobacterial mats: strategies of photosynthetic life under sulfide, p. 133–148. *In* Y. Cohen, R. W. Castenholz, and H. O. Halvorson (ed.), *Microbial Mats: Stromatolites*. Alan R. Liss, Inc., New York.
5. **Cohen, Y., W. E. Krumbein, and M. Shilo.** 1977. Solar Lake (Sinai). 2. Distribution of photosynthetic microorganisms and primary production. *Limnol. Oceanogr.* 22:609–620.
6. **Culver, D. A., and G. J. Brunskill.** 1969. Fayetteville Green Lake, New York. V. Studies of primary production and zooplankton in a meromictic marl lake. *Limnol. Oceanogr.* 14:862–873.
7. **Dyer, B. D., N. Gaju, C. Pedrós-Alió, I. Esteve, and R. Guerrero.** 1986. Ciliates from a fresh water sulfuretum. *BioSystems* 19:127–135.
8. **Esteve, I., R. Guerrero, E. Montesinos, and C. Abellà.** 1983. Electron microscopy study of the interaction of epibiontic bacteria with *Chromatium minus* in natural habitats. *Microb. Ecol.* 9:57–64.
9. **Golley, F. B., H. T. Odum, and R. F. Wilson.** 1962. The structure and metabolism of a Puerto Rican red mangrove forest in May. *Ecology* 43:9–19.
10. **Gorlenko, V. M., G. A. Dubinina, and S. I. Kuznetsov.** 1983. The ecology of aquatic microorganisms. *Binnengewässer* 28.
11. **Guerrero, R., C. Abellà, and M. R. Miracle.** 1978. Spatial and temporal distribution in a meromictic karstic lake basin: relationships with physicochemical parameters and zooplankton. *Verh. Int. Verein. Limnol.* 20:2264–2271.
12. **Guerrero, R., I. Esteve, C. Pedrós-Alió, and N. Gaju.** 1987. Predatory bacteria in prokaryotic com-

munities: the earliest trophic relationships. *Ann. N.Y. Acad. Sci.* **503**:238–250.

13. **Guerrero, R., J. Mas, and C. Pedrós-Alió.** 1984. Buoyant density changes due to intracellular content of sulfur in *Chromatium warmingii* and *Chromatium vinosum*. *Arch. Microbiol.* **137**:350–356.
14. **Guerrero, R., E. Montesinos, I. Esteve, and C. Abellà.** 1980. Physiological adaptations and growth of purple and green sulfur bacteria in a meromictic as compared to a holomictic lake. *Hydrobiologia* **3**:161–171.
15. **Guerrero, R., E. Montesinos, C. Pedrós-Alió, I. Esteve, J. Mas, H. van Gemerden, P. A. G. Hofman, and J. F. Bakker.** 1985. Phototrophic sulfur bacteria in two Spanish lakes: vertical distribution and limiting factors. *Limnol. Oceanogr.* **30**:919–931.
16. **Guerrero, R., C. Pedrós-Alió, I. Esteve, and J. Mas.** 1987. Communities of phototrophic sulfur bacteria in lakes of the Spanish Mediterranean region. *Acta Acad. Abo.* 47:125–151.
17. **Guerrero, R., C. Pedrós-Alió, I. Esteve, J. Mas, D. Chase, and L. Margulis.** 1986. Predatory prokaryotes: predation and primary consumption evolved in bacteria. *Proc. Natl. Acad. Sci. USA* **83**: 2138–2142.
18. **Jassby, A. D.** 1975. The ecological significance of sinking to planktonic bacteria. *Can. J. Microbiol.* **21**:270–274.
19. **Jørgensen, B. B., N. P. Revsbech, and Y. Cohen.** 1983. Photosynthesis and structure of benthic microbial mats: microelectrode and SEM studies of four cyanobacterial communities. *Limnol. Oceanogr.* **28**:1075–1093.
20. **Kämpf, C., and N. Pfennig.** 1980. Capacity of Chromatiaceae for chemotrophic growth. Specific respiration rates of *Thiocystis violacea* and *Chromatium vinosum*. *Arch. Mikrobiol.* **127**:125–135.
21. **Konopka, A.** 1984. Effect of light-nutrient interactions on buoyancy regulation by planktonic cyanobacteria, p. 41–48. *In* M. J. Klug and C. A. Reddy (ed.), *Current Perspectives in Microbial Ecology*. American Society for Microbiology, Washington, D.C.
22. **Lawrence, J. R., R. C. Haynes, and U. T. Hammer.** 1978. Contribution of photosynthetic green sulphur bacteria to total primary production in a meromictic saline lake. *Verh. Int. Verein. Limnol.* **20**:201–207.
23. **Lindholm, T., and K. Weppling.** 1987. Blooms of phototrophic bacteria and phytoplankton in a small brackish lake on Åland, SW Finland. *Acta Acad. Abo.* 47:45–53.
24. **Lindholm, T., K. Weppling, and H. S. Jensen.** 1985. Stratification and primary production in a small brackish lake studied by close-interval siphon sampling. *Verh. Int. Verein. Limnol.* **22**:2190–2194.
25. **Margalef, R.** 1980. *Ecología*. Ediciones Omega, Barcelona, Spain.
26. **Mas, J., C. Pedrós-Alió, and R. Guerrero.** 1985. Mathematical model for determining the effects of intracytoplasmic inclusions on volume and density of microorganisms. *J. Bacteriol.* **164**:749–756.
27. **Mas, J., and H. van Gemerden.** 1987. Influence of sulfur accumulation and composition of sulfur globule on cell volume and buoyant density of *Chromatium vinosum*. *Arch. Microbiol.* **146**:362–369.
28. **Miracle, M. R., and E. Vicente.** 1983. Vertical distribution and rotifer concentration in the chemocline of meromictic lakes. *Hydrobiologia* **104**:259–267.
29. **Montesinos, E., and I. Esteve.** 1984. Effect of algal shading on the net growth and production of phototrophic sulfur bacteria in lakes of the Banyoles karstic area. *Verh. Int. Verein. Limnol.* **22**: 1102–1105.
30. **Montesinos, E., I. Esteve, C. Abellà, and R. Guerrero.** 1983. Ecology and physiology of the competition for light between *Chlorobium limicola* and *Chlorobium phaeobacteroides* in natural habitats. *Appl. Environ. Microbiol.* 46:1007–1016.
31. **Odum, H. T. (ed.).** 1970. *A Tropical Rain Forest. A study of Irradiation and Ecology at El Verde, Puerto Rico.* Division of Technical Information, Atomic Energy Commission, Oak Ridge, Tenn.
32. **Oren, A., and M. Shilo.** 1982. Population dynamics of *Dunaliella parva* in the Dead Sea. *Limnol. Oceanogr.* 27:201–211.
33. **Parker, R. D., and U. T. Hammer.** 1983. A study of Chromatiaceae in a saline meromictic lake in Saskatchewan, Canada. *Int. Rev. Gesamten Hydrobiol.* **68**:839–851.
34. **Parkin, T. B., and T. D. Brock.** 1980. Photosynthetic bacterial production in lakes: the effects of light intensity. *Limnol. Oceanogr.* **25**:711–718.
35. **Parkin, T. B., and T. D. Brock.** 1980. The effects of light quality on the growth of photosynthetic bacteria in lakes. *Arch. Microbiol.* **125**:19–27.
36. **Parkin, T. B., and T. D. Brock.** 1981. Photosynthetic bacterial production and carbon mineralization in a meromictic lake. *Arch. Hydrobiol.* **91**:366–382.
37. **Pedrós-Alió, C., C. Abellà, and R. Guerrero.** 1984. Influence of solar radiation, water flux and competition on biomass of photosynthetic bacteria in Lake Cisó, Spain. *Verh. Int. Verein. Limnol.* **22**:1097–1101.
38. **Pedrós-Alió, C., J. M. Gasol, and R. Guerrero.** 1987. On the ecology of a *Cryptomonas phaseolus* population forming a metalimnetic bloom in Lake Cisó, Spain: annual distribution and loss factors. *Limnol. Oceanogr.* **32**:285–298.
39. **Pedrós-Alió, C., E. Montesinos, and R. Guer-**

rero. 1983. Factors determining annual changes in bacterial photosynthetic pigments in holomictic Lake Cisó, Spain. *Appl. Environ. Microbiol.* **46**:999–1006.

40. **Pfennig, N.** 1978. General physiology and ecology of photosynthetic bacteria, p. 3–18. *In* R. K. Clayton and W. R. Sistrom (ed.), *The Photosynthetic Bacteria.* Plenum Publishing Corp., New York.
41. **Reimer, T. O., E. S. Barghoorn, and L. Margulis.** Primary productivity in an early Archaean microbial ecosystem. *Precambrian Res.* **9**:93–104.
42. **Revsbech, N. P., and D. M. Ward.** 1984. Microprofiles of dissolved substances and photosynthesis in microbial mats measured with microelectrodes, p. 171–181. *In* Y. Cohen, R. W. Castenholz, and H. O. Halvorson (ed.), *Microbial Mats: Stromatolites.* Alan R. Liss, Inc., New York.
43. **Reynolds, C. S.** 1984. *The Ecology of Freshwater Phytoplankton.* Cambridge University Press, Cambridge.
44. **Stal, L. J., H. van Gemerden, and W. E. Krumbein.** 1984. The simultaneous assay of chlorophyll and bacteriochlorophyll in natural microbial communities. *J. Microbiol. Methods* **2**:295–306.
45. **Stolz, J. F.** 1984. Fine structure of the stratified microbial community at Laguna Figueroa, Baja California, Mexico. II. Transmission electron microscopy as a diagnostic tool in studying microbial communities in situ, p. 23–38. *In* Y. Cohen, R. W. Castenholz, and H. O. Halvorson (ed.), *Microbial Mats: Stromatolites.* Alan R. Liss, Inc., New York.
46. **Takahashi, M., K. Shiokawa, and S. Ichimura.** 1972. Photosynthetic characteristics of a purple sulfur bacteria grown under different light intensities. *Can. J. Microbiol.* **18**:1825–1828.
47. **van Gemerden, H.** 1974. Coexistence of organisms competing for the same substrate: an example among the purple sulfur bacteria. *Microb. Ecol.* **1**:104–119.
48. **van Gemerden, H.** 1981. Coexistence of *Chlorobium* and *Chromatium* in a sulfide-limited continuous culture. *Arch. Microbiol.* **129**:32–34.
49. **van Gemerden, H.** 1983. Physiological ecology of purple and green bacteria. *Ann. Inst. Pasteur Microbiol.* **134B**:73–92.
50. **van Gemerden, H., E. Montesinos, J. Mas, and R. Guerrero.** 1985. Diel cycle of phototrophic sulfur bacteria in Lake Cisó (Spain). *Limnol. Oceanogr.* **30**:932–943.

Chapter 5

Cyanobacterial Hyperscums of Hypertrophic Water Bodies

Tamar Zohary

INTRODUCTION

A hyperscum is a crusted buoyant cyanobacterial mat, often decimeters thick, in which the organisms are so densely packed that free water is not evident (38). The phenomenon is an undesirable consequence of excessive eutrophication of fresh waters and is caused by the accumulation of surface cyanobacterial blooms in wind-protected sites on lee shores, where, depending on weather conditions, they may persist for weeks to months (38).

Hyperscums differ from scums or water blooms in the magnitude of their spatial and temporal dimensions and cell concentrations. Whereas relevant dimensions for scums would usually be millimeters (for thickness), hours (for duration), and 10^2 to 10^3 μg of chlorophyll liter^{-1} (indicator of algal biomass), those for hyperscums would be decimeters, weeks or months, and 10^5 μg liter^{-1}, respectively. These differences justify the new term hyperscum. It was felt that the use of the alternative term neustonic mats, previously used to describe similar phenomena (27, 32), was unjustified. Neuston, by definition, refers to organisms adapted to life at the air/water interface (36). In hyperscums, however, the bulk of the population is below the surface, and cells at the interface die of photooxidation and dehydration (see below).

Surface scums of the dominant gas-vacuolated, bloom-forming genera *Anabaena*, *Aphanizomenon*, and *Microcystis* (34) occur throughout the world in a wide range of freshwater bodies (Table 1). Similar phenomena in the marine environment are surface blooms of the cyanobacterium *Oscillatoria erithraea (Trichodesmium)*, which appear in windrows that can be several miles long (7, 9).

In comparison with the large volume of scientific literature dealing with cyanobacterial scums, little is known about hyperscums. A major reason is the infrequent occurrence of the latter (Table 1). Published accounts of hyperscum occurrence include the neustonic mats from the Dnieper reservoirs, USSR (32), and hyperscums from Hartbeespoort Dam, South Africa (26, 38). Surface scums of decimeters in thickness were recorded in Lakes Suwa and Kasumigaura, Japan (27, 37), and Lake Brielle, The Netherlands (16), but the data provided were insufficient to indicate whether their classification as hyper-

Tamar Zohary • The Yigal Allon Kinneret Limnological Laboratory, P.O. Box 345, Tiberias 14102, Israel.

Table 1.
Records of Occurrence of Hyperscums in Eutrophic Water Bodies Where Cyanobacterial Blooms Occur Regularly

Water body	Trophic status	Dominant bloom genera	Mention of hyperscums	Description of scums or hyperscums	Reference
Fishponds, Israel	Eutrophic	*Microcystis, Oscillatoria, Spirulina*	No	Surface blooms in early mornings disperse in afternoons	33
King Talal Reservoir, Jordan	Hypertrophic	*Microcystis*	Yes	Hyperscum, ca. 40 cm thick, persisted for 4 months in summer 1986	F. Hashwa, personal communication
Lake Brielle, The Netherlands	Hypertrophic	*Microcystis*	?	"Drijflagen" (floating layers)	16
Lake Akersvatn, Norway	Eutrophic	*Microcystis*	No	No hyperscums, but 5 to 10 g (dry wt) of algae liter^{-1} in lee shore aggregates	5
Kremenchug Reservoir, USSR	Hypertrophic	*Microcystis, Anabaena*	Yes	"Dry crusts," "neustonic mats," "neustonic phase"	32
Lake Mendota, Wis.	Eutrophic	*Aphanizomenon*	No	"Massive blooms"	6
Potomac River, Washington, D.C.	Eutrophic	*Microcystis, Oscillatoria*	No	"Green paste"	15
Lake George, Uganda	Eutrophic	*Microcystis*	No	Surface accumulations in early mornings disperse in afternoons	10
Hartbeespoort Dam, South Africa	Hypertrophic	*Microcystis*	Yes	Crusted buoyant cyanobacterial mats of decimeter thickness	26, 38
Lake Suwa, Japan	Eutrophic	*Microcystis*	No	No hyperscums, but 95 mg of chlorophyll liter^{-1} recorded	37
Lake Kasumigaura, Japan	Hypertrophic	*Microcystis, Anabaena*	?	"Heavy neustonic mats of decimeters thickness"	27
Ciénaga Grande Lagoon, Colombia	Eutrophic	*Microcystis*	No	"Mass accumulation"	12
Queensland Coast, Pacific Ocean		*Trichodesmium*	No	Bloom occupying up to 52,000 km^2	9

scums was appropriate. However, unpublished observations indicate that hyperscums do occur elsewhere, e.g., in the King Talal Reservoir, Jordan (F. Hashwa, personal communication; Table 1).

In this paper I summarize the current state of knowledge about the hyperscum phenomenon. The data set originates from work carried out on Hartbeespoort Dam samples during 1983 to 1987, and mostly from intensive field studies during the 1984 (8 May to 24 August) and 1986 (15 May to 30 August) hyperscum "seasons."

STUDY SITE

Hartbeespoort Dam is a warm, monomictic, man-made lake, located some 50 km north of Johannesburg, South Africa (25°43′ S, 27°51′ E). At full capacity the lake covers 20 km^2 and has a maximum depth of 32.5 m, a mean depth of 9.6 m, and a capacity of 195 × 10^6 m^3. During summer stratification, surface temperatures exceed 25°C and an anaerobic hypolimnion develops, which may extend upward to within 8 m of the surface (24). Following autumn overturn (March or April), the water temperature declines to a minimum of about 12°C in winter (25). Solar radiation is high throughout the year (about 13 MJ m^{-2} day^{-1} in winter and 30 MJ m^{-2} day^{-1} in summer) owing to the subtropical latitude, the high elevation (1,162 m at sea level), and the climate, with winter days being mostly cloudless. In winter the midday photosynthetically active radiation is usually >1,000 microeinsteins m^{-2} s^{-1}, comparable with midsummer photosynthetically active radiation levels in many temperate lakes in the northern hemisphere. Wind speeds over the lake are low, especially in winter. The annual mean wind speed in 1982 to 1983 was 1.6 m s^{-1} (25), and the wind speed exceeded 3.7 m s^{-1} (the speed required to initiate turbulent vertical mixing [11]) only 13% of the time in 1984 and 12% of the time in 1985 (T. Zohary, Ph.D. thesis, University of Natal, Pietermaritzburg, South Africa, 1987).

The lake is hypertrophic owing to high nutrient loads originating from domestic and industrial effluents from northern Johannesburg. Total phosphorus concentrations of about 0.5 mg $liter^{-1}$ and soluble inorganic nitrogen concentrations of 1 to 2 mg $liter^{-1}$ are maintained throughout the year (4). With nutrients being in excess of algal growth requirements, the lake contains dense phytoplankton populations (euphotic-zone chlorophyll *a* concentration of up to 400 mg m^{-3} [25]). The phytoplankton is virtually unialgal for up to 10 months of the year, when the cyanobacterium *Microcystis aeruginosa* Kütz. emend Elenkin makes up >90% of the total phytoplankton volume (25). Phytoplankton primary production is among the highest measured in freshwater lakes, with hourly euphotic-zone values exceeding 3 g of C m^{-2} h^{-1} and annual production values exceeding 2 kg of C m^{-2} (22, 23).

HYPERSCUM OCCURRENCE IN HARTBEESPOORT DAM

The main accumulation site of large hyperscums in Hartbeespoort Dam occurs against the dam wall. The wall is situated in a narrow gorge in which winds are channeled in two opposing directions. The prevailing southeast winds blow toward the wall, transporting drifting blooms from the main basin toward the accumulation site. The wall shelters an area of 1 to 2 ha, depending on wind speed and water level, from low-speed winds in the opposite direction. Drifting blooms that are transported into the sheltered area become trapped for weeks to months, until the wind regime changes.

Figure 1 illustrates the seasonal changes of monthly mean total chlorophyll *a* content, the proportion of the total chlorophyll *a* that was periodically contained in hyperscums, and the monthly mean wind speeds. It is evident that hyperscums formed in late autumn and winter (May to July), when wind activity was lowest, and dispersed in August to September, when the wind activity increased. In 1983 a hyperscum that covered more than 1 ha contained up to 2 tonnes of chlorophyll *a* and up to 70% of the total chlorophyll *a* content of the lake for more than 4 months. In 1984 a hyperscum was present that contained up to 1.25 tonnes and 30% of the total chlorophyll for nearly 4 months (Fig. 1). The hyperscum chlorophyll *a* content was related to the total chlorophyll *a* content of the lake preceding hyperscum formation: the larger the standing stock, the larger the hyperscum. In 1985, when the total chlorophyll *a* content of the lake in May was considerably smaller than in previous years, a hyperscum did not form (Fig. 1).

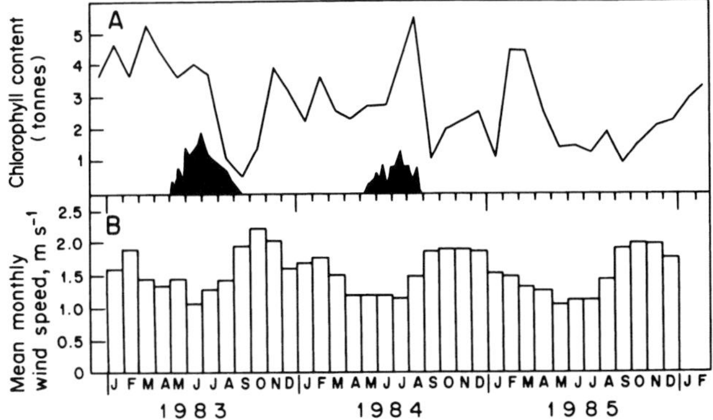

Figure 1. (A) Seasonal variations in estimates of total chlorophyll *a* content in Hartbeespoort Dam, derived by volume-weighted extrapolation of weekly chlorophyll depth profiles at three pelagic and four littoral stations plus the chlorophyll that was periodically contained in hyperscums. Solid areas show the proportion contained in hyperscums. (B) Seasonal changes in mean monthly wind speeds recorded on the southeastern shore of the lake.

Large hyperscums (>1 ha) formed again in the winter of 1986 and late autumn of 1987 (data not shown).

HYPERSCUM FORMATION

Favorable Conditions

Reynolds and Walsby (21) pointed out that cyanobacterial surface scums occur only when three preconditions coexist: a preexisting cyanobacterial population, positive buoyancy in a significant proportion of the organisms, and turbulence that is too weak to negate the tendency of buoyant organisms to float. On the basis of the data from Hartbeespoort Dam, Zohary and Breen (T. Zohary and C. M. Breen, *Hydrobiologia*, in press) concluded that these preconditions are essential but not sufficient for hyperscum formation. The additional requirements are as follows. (i) The preexisting standing stock of positively buoyant cyanobacteria must be large (>100 mg of chlorophyll m^{-2} and 10^2 to 10^3 kg of chlorophyll per lake). This requirement limits hyperscum occurrence to hypertrophic water bodies. (ii) Calms (no winds) or low-speed winds (<3.7 m s^{-1}) must occur continuously over long periods (weeks) and be uninterrupted by storms. (iii) Shore morphometry must be such that there are wind-protected accumulation sites in the lee of the prevailing winds. (iv) High solar radiation levels must be present (see below).

The co-occurrence of all the above-mentioned preconditions is apparently rare, making hyperscums an unusual sight. However, with increasing hypertrophy throughout the Third World (23), the frequency and distribution of hyperscum occurrence are likely to increase.

Algal Accumulation

Figure 2 illustrates the accumulation of chlorophyll *a* (extracted, analyzed and corrected for phaeopigments by the method of Nusch [17]) in the upper 1 m of water near the dam wall of Hartbeespoort Dam over 28 days. On 1 May 1984 the surface chlorophyll *a* concentration was 185 mg m^{-3}, similar to concentrations that are often recorded in the main basin (22, 25). By 8 May *M. aeruginosa* colonies at the top 20 cm were so densely packed that the chlorophyll *a* concentration

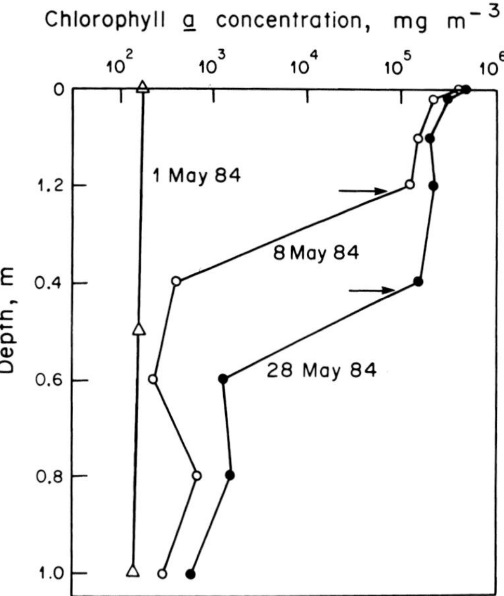

Figure 2. Time series of depth profiles of chlorophyll *a* concentration (note the log scale) at a hyperscum accumulation site next to the dam wall of Hartbeespoort Dam.

between 10 and 20 cm exceeded 100,000 mg m^{-3} and no free water was visible between colonies. Owing to the high colony concentration, samples had to be cored like sediments. A 1-m Perspex corer with holes drilled at 2-cm intervals and taped externally was used. Samples from desired depths were obtained with a syringe fitted with a 16-gauge needle. The highly buoyant colonies tended to float and became compacted as they got closer to the surface, so that the chlorophyll *a* concentration at the surface was even higher (>400,000 mg m^{-3}). By 28 May chlorophyll concentrations exceeding 100,000 mg m^{-3} extended to a depth of 42 cm. A steep vertical gradient (2 to 3 orders of magnitude) in chlorophyll *a* concentration was recorded at the bottom of the hyperscum (the depth where free water between colonies was first evident).

Crust Formation

The crust covering the hyperscum probably forms as a consequence of photooxidative death and subsequent dehydration of the densely packed cells at the surface. Both processes are enhanced by high insolation levels. Photooxidative conditions are characterized by high solar radiation levels, oxygen supersaturation, and CO_2 depletion and are lethal to cyanobacteria (2, 8). The coexistence under high light intensities of oxygen supersaturation and CO_2 depletion at the surface of a new hyperscum, prior to crust formation, was demonstrated using oxygen and pH microelectrodes (20). Figure 3 shows representative microgradients of oxygen concentrations and photosynthetic activity over the uppermost 2 mm in a core sample from a green, noncrusted region in a newly formed hyperscum (ca. 2 to 3 days old). The core was exposed to light of 625 microeinsteins m^{-2} s^{-1} for 1 h before measurements were taken. As a result of photosynthetic oxygen production (maximum rate recorded, 24.7 mmol of O_2 liter^{-1} h^{-1}), the oxygen concentration reached a peak of 409 μM, or 163% saturation, at a depth of 0.2

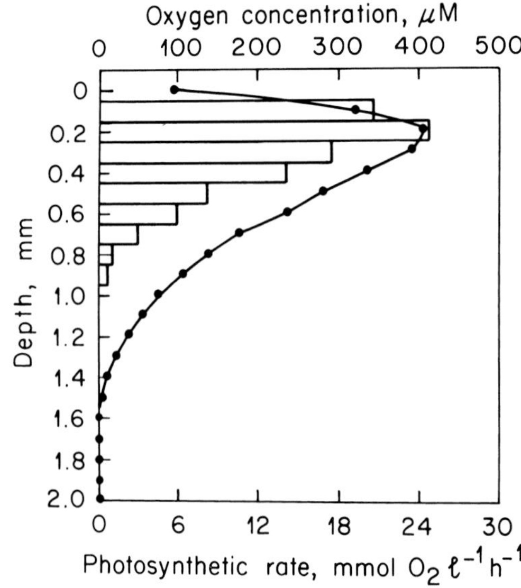

Figure 3. Microprofiles of oxygen (●) and photosynthetic rates (bars) over the upper 2 mm of a newly formed hyperscum (ca. 2 to 3 days) after 1 h of exposure to light of 625 microeinsteins m^{-2} s^{-1}.

mm. Owing to the attenuation of light by the densely packed colonies, photosynthetic activity and hence oxygen concentrations diminished rapidly with depth (Fig. 3). At a depth of 1 mm, net photosynthesis could not be detected, and oxygen disappeared at 1.7 mm. Deeper in the core the hyperscum was aphotic and anaerobic.

High rates of photosynthesis could be sustained as long as CO_2 supplies lasted. With time in the light, photosynthesis became CO_2 limited, as indicated indirectly by an increase in the pH over the top 0.2 mm in the same core from 8.1 in the dark to above 10.5 after 1 h in the light (1,000 microeinsteins m^{-2} s^{-1}).

The high insolation levels in Hartbeespoort Dam ensured the development of photooxidizing conditions within hours at the surface of newly accumulated hyperscums. In addition, exposure of the densely packed colonies at the surface to high solar radiation levels caused their dehydration. Unlike photooxidation, dehydration was a slow process, so that a fully developed dry crust was established within 4 to 6 weeks.

HYPERSCUM STRUCTURE

The hyperscum community was a cyanobacterial-bacterial association in which *M. aeruginosa* was the dominant organism, making up >95% of the cyanobacterial-plus-bacterial volume. Bacteria, however, exceeded *Microcystis* organisms in cell numbers (Table 2) and exhibited a wide range of sizes and morphological types.

Three structural zones could be distinguished in the mature hyperscum (i.e., a hyperscum with a well-developed crust) (Fig. 4). These were a crust ca. 2 mm thick; a compact layer, ca. 5 to 10 mm thick, of partially dehydrated colonies below the crust; and the less compact bulk of the hyperscum, extending to a depth of several decimeters. The changes from one layer to the underlying layer were gradual, but for convenience the three layers are treated separately. Below the hyperscum lay about 20 m of oxygenated, freely circulating lake water.

Scanning electron microscopy of the crust material revealed tightly packed, dehydrated *Microcystis* cells that were heavily

Table 2.
Biomass-Related Parameters in the Three Hyperscum Zones of Fig. 4[a]

Hyperscum layer and depth	Chlorophyll *a* concn (g liter^{-1})	Phaeopigments (% of chlorophyll + phaeopigments)	Dry wt/(algae + interstitial water wt) (%)	10^9 No. of *Microcystis* cells ml^{-1}	*Microcystis* cell diam (μm)[b]	10^9 Bacterial cells ml^{-1}
Crust, 0–2 mm	3.07 ± 33%, 1.14–4.33 (9)	35.5 ± 47%, 6.88–63.96 (9)	85.9 ± 24%, 81.2–88.5 (4)	ND[c]	ND	102 ± 73%, 13.3–238 (9)
Compact, 2–10 mm	1.00 ± 31%, 0.53–1.63 (10)	16.9 ± 67%, 5.68–42.81 (10)	22.8 ± 11%, 15.4–37.5 (5)	4.86 ± 51%, 0.87–8.50 (9)	4.3 ± 8.2, 3.3–4.9 (9)	33.1 ± 53%, 12.5–65.2 (9)
Bulk, 10 cm	0.33 ± 27%, 0.22–0.52 (10)	13.6 ± 37%, 3.14–19.96 (8)	4.3 ± 0.5%, 3.6–5.1 (7)	3.10 ± 54%, 0.51–4.98 (9)	4.8 ± 5.5%, 3.3–5.7 (9)	3.65 ± 67%, 0.8–7.37 (9)

[a] Data are shown as mean ± coefficient of variation, range, and number of cases (in parentheses) for determinations on weekly or biweekly samples collected between 3 June and 2 September 1986.
[b] Means and coefficients of variation are of the medians of nine cases. Medians are of at least 20 cells.
[c] ND, No data.

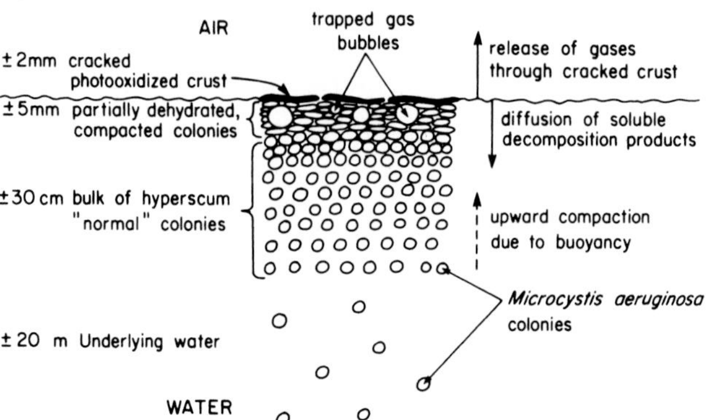

Figure 4. Schematic depth profile through a hyperscum of *M. aeruginosa.*

colonized by bacteria (Fig. 5d). It was impossible to count *Microcystis* cells in the crust because of their partially decomposed state, but chlorophyll *a* concentrations gave an indication of the ultimate degree of cell compaction. The maximum recorded concentration was 4.3 g of chlorophyll liter^{-1} (Table 2), or 0.43% of the crust weight. The average dry weight of the crust was 85.9% of its wet weight; thus, the maximum chlorophyll concentration was 0.37% of the dry weight. This value approached the range of 0.78 to 1.52% reported by Kappers (F. I. Kappers, Ph.D. thesis, University of Amsterdam, Amsterdam, The Netherlands, 1984) to be the axenic *Microcystis* chlorophyll *a* content expressed as a percentage of the dry weight, demonstrating how close the crust was to a dry pellet of axenic *M. aeruginosa.*

With time, the chlorophyll in the crust degraded to phaeopigments. For example, on 3 June 1984, phaeopigments made up 6.9% of the total chlorophyll-derived pigments. By 27 August, the proportion had increased to 64%. The chlorophyll *a* concentration declined over the same period from 4.3 to 1.1 g liter^{-1}.

Just below the crust, colony compaction was still extremely high (average of 1.0 g of chlorophyll liter^{-1} and 4.9×10^9 *Microcystis* cells ml^{-1} [Table 2]). Partial dehydration of the cells was evident by scanning electron microscopy (Fig. 5c) and in a reduced cell diameter compared with those of cells in the bulk of the hyperscum (Table 2) or in the main basin (Fig. 5a and b). The dry weight made up 23% of the total (cells plus interstitial water) weight (Table 2); this proportion approaches the reported dry/wet weight ratios for *Oscillatoria agardhii* (3). Conspicuously large gas bubbles, often 10 mm in diameter and containing end products of anaerobic decomposition (see below), were trapped within this layer (Fig. 4).

Deeper, at a depth of 10 cm within the bulk layer (Fig. 4), colony compaction was 1 order of magnitude lower than in the crust and one-third of that in the compact layer (Table 2). Still, *Microcystis* cell numbers (up to 5×10^9 ml^{-1}) and chlorophyll *a* concentrations (maximum, 516 mg liter^{-1}) in this layer were higher than any previously published values. Scanning electron microscopy revealed spherical, nondistorted cells, many in various stages of cell division, embedded in a mucilagenous sheath and similar in appearance and diameter to cells in control lake populations (Fig. 5a and b).

PROCESSES WITHIN THE HYPERSCUM

The mature crust acted as a physical barrier between the underlying colonies and

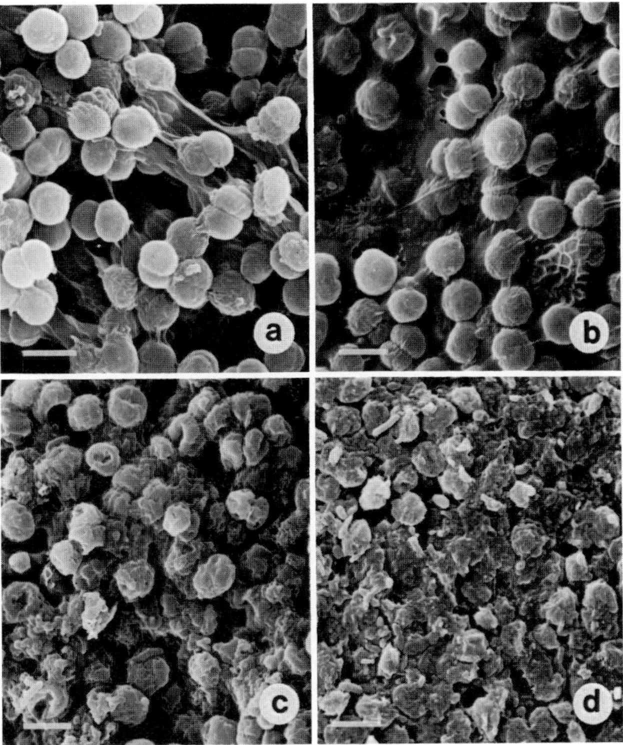

Figure 5. Scanning electron micrographs of planktonic and hyperscum populations of *M. aeruginosa* in Hartbeespoort Dam on 20 August 1986. (a) Planktonic, main basin (depth, 3 m); (b) hyperscum (depth, 10 cm); (c) hyperscum, compact layer beneath crust (depth, ca. 5 mm); (d) hyperscum, crust. Bars, 5 μm.

the atmosphere, drastically reducing free gas exchange and absorbing (owing to its high pigment content) all the incident light. As the crust shrank with time owing to continued dehydration, it cracked, exposing underlying colonies in the cracks to photooxidizing conditions and to a secondary process of crust formation. Just below the crust, photosynthetic oxygen production stopped owing to the absence of light, but respiratory oxygen consumption continued. With the high concentration of respiring organisms, oxygen consumption exceeded the available supplies, causing anoxia.

Under the aphotic, anaerobic conditions in the hyperscum, growth of obligate phototrophs, such as *M. aeruginosa* (31), would be disadvantaged, and a prevalence of anaerobic decomposition processes could be expected. Evidence that decomposition indeed prevailed was plentiful. Analysis by gas chromatography of the chemical composition of the trapped gas bubbles (Fig. 4) revealed that they contained no oxygen, 28% methane, 19% CO_2, 53% N_2, and traces of H_2, which are clearly end products of bacterial anaerobic degradation of organic matter (35). A range of volatile fatty acids, the intermediate products in such processes (35), were found (by using high-pressure liquid chromatography) in the interstitial water. Acetic acid was found at the highest concentrations, which increased with time from 0.2 mM in a newly formed hyperscum to 1.0 mM in a mature one. In addition, pH values at a depth of 10 cm declined with time (possibly owing to the accumulation of organic acids) from 6.6 in a new hyperscum to 5.9 in a mature one.

As the hyperscum aged with time, the interstitial water (collected by dialysis for 24 h [38]) became enriched with nutrients in their reduced form (e.g., ammonia and sulfide), while the oxidized forms were kept at relatively low concentrations. Figure 6 illustrates the changes with time in concentrations of nitrogen from ammonia, soluble organic nitrogen, and soluble reactive phosphorus in the interstitial water of the 1984 hyperscum at a depth of 10 cm. The concentration of nitrogen from ammonia increased from 0.45 mg liter^{-1} on 28 May to 119 mg liter^{-1} on 22 August, while nitrate plus nitrite concentrations remained below 0.3 mg liter^{-1}. The soluble organic nitrogen concentration increased over the same period, from 0.27 to 12.7 mg liter^{-1}, and that of soluble reactive phosphorus also increased, from 2.8 to 83.3 mg liter^{-1}. These concentrations declined to ambient concentrations in the lake in late August, when the hyperscum dispersed.

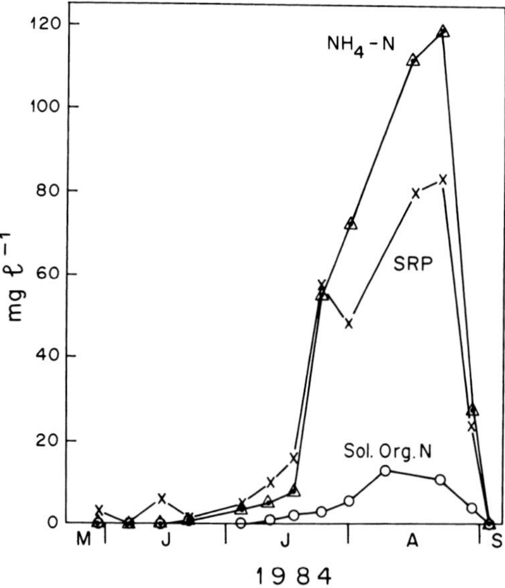

Figure 6. Changes with time in ammonia (NH_4-N), soluble reactive phosphorus (SRP), and soluble organic nitrogen (Sol. Org. N) concentrations in hyperscum interstitial water at a depth of 10 cm.

Despite the evidence for the prevalence of decomposition processes, *M. aeruginosa* in the bulk of the hyperscum (Fig. 4) did not seem to be decomposing for at least 3 months of the hyperscum season. Unlike cells in the crust and compact layers, cells at a depth of 10 cm maintained their structure and integrity (Fig. 5). The proportion of phaeopigments at 10 cm was maintained below 20% (Table 2). Zohary (38) demonstrated that after hyperscum *M. aeruginosa* was reacclimatized to euphotic, aerobic conditions in the main basin (24 h of incubation in dialysis bags at a depth of 0.5 m), it exhibited photosynthetic rates per unit of biomass that were not statistically different from those of control populations in the main basin. Some deterioration was, however, noticed over the fourth (last) month. Similar results were obtained when the experiments were repeated at weekly intervals over the hyperscum season during the following year (Zohary, thesis).

A possible explanation of this apparent contradiction is that *Microcystis* organisms decomposed in the compact layer beneath the crust, where conditions were more stressful (see below), whereas they could survive longer in the bulk of the hyperscum. Products of decomposition near the surface diffused downward along concentration gradients, causing the increases with time in ammonia and soluble reactive phosphorus concentrations at a depth of 10 cm (Fig. 6). As colonies near the surface decomposed and disintegrated, they were replaced by other colonies from below, while new colonies were always added to the hyperscum at the bottom.

The stressful conditions in the compact layer were a consequence of its proximity to the surface. Although protected from photooxidation by the crust, this layer was still subject to evaporative loss of water and to extreme diel fluctuations in temperature, ranging at times from as low as 1°C before sunrise to >20°C at midday. Deeper in the hyperscum, diel temperature fluctuations were moderated by the underlying lake water. In addition, if the compact layer was indeed the source of the nutrient increases at 10 cm, it is likely that accumulated ammonia, sulfide, and other decomposition products reached potentially toxic concentrations (1, 14, 18) much earlier than at deeper layers.

That *M. aeruginosa* survived for several months under dark, anaerobic conditions deeper in the hyperscum is not surprising. Cyanobacteria are known to withstand long periods of dark, anaerobic conditions while maintaining their photosynthetic capacity (28–30). *M. aeruginosa*, in particular, is known to overwinter on dark, hypolimnetic sediments of temperate lakes. Some filamentous nonplanktonic cyanobacteria are capable of heterotrophic growth in the dark (13, 19), but *M. aeruginosa* is not a known heterotroph (31). The question of how *M. aeruginosa* maintains itself for several months under dark, anaerobic conditions both in hyperscums and in hypolimnetic sediments remains unanswered. Does it survive solely by endogenous respiration of stored carbohydrate, or is it capable of generating energy via alternative metabolic pathways?

ECOLOGICAL ROLE OF HYPERSCUMS

The ecological significance of hyperscums is not yet clearly understood. Are they merely a consequence of a coincidental combination of environmental conditions, or do they play an important role in the annual cycle and population dynamics of *M. aeruginosa* in hypertrophic lakes? Topachevskiy et al. (32) claimed that the role of the dry crusts is similar to that of overwintering benthic stocks of *M. aeruginosa* and of resting spores or cysts of other organisms, i.e., providing the inoculum for growth when conditions become favorable. They based their conclusion on observations demonstrating that some of the cells retain their viability and enzymatic activity for long periods and return to their active state when rehydrated.

In Hartbeespoort Dam an inoculum per se is not important, because planktonic populations of *M. aeruginosa* organisms remain in the water column throughout the year. However, in years when hyperscums formed in winter (1982, 1983, 1984, and 1986), *M. aeruginosa* remained the dominant phytoplankton species (>80% by volume) all winter, whereas in years when hyperscums did not form (1985) or were flushed downstream during the initial formation period (1987), the planktonic populations of this species declined dramatically in late autumn. It therefore seems that the to-and-fro transport of *M. aeruginosa* between the lake and the hyperscum during winter, when water temperatures severely limit the growth rate of this cyanobacterium, contributed to its prolonged dominance among the phytoplankton populations.

ACKNOWLEDGMENTS. This is contribution no. 60 to the Hartbeespoort Dam Ecosystem Programme. The work was supported by the Foundation for Research Development, CSIR.

I thank A. M. Pais Madeira for considerable contributions to the field and laboratory work, Y. Cohen for carrying out the microelectrode studies, R. I. Mackie and K. Westlake for gas sample collection and chromatography and for volatile fatty acid analysis, L. M. Sephton for bacterial counts, the analytical division of the Division of Water Technology, CSIR, for N and P determinations, and C. M. Breen, A. Oren, and R. D. Robarts for critical assessment of the draft manuscript. The Department of Water Affairs provided the meteorological data.

LITERATURE CITED

1. **Abeliovich, A., and Y. Azov.** 1976. Toxicity of ammonia to algae in sewage oxidation ponds. *Appl. Environ. Microbiol.* 31:801–806.

2. **Abeliovich, A., and M. Shilo.** 1972. Photooxidative death in blue-green algae. *J. Bacteriol.* **111:** 682–689.
3. **Ahlgren, G.** 1983. Comparison of methods for estimation of phytoplankton carbon. *Arch. Hydrobiol.* **98:**489–508.
4. **Ashton, P. J.** 1985. Nitrogen transformations and the nitrogen budget of a hypertrophic impoundment (Hartbeespoort Dam, South Africa). *J. Limnol. Soc. South. Afr.* 11:32–42.
5. **Berg, K., W. W. Carmichael, O. M. Skulberg, C. Benestad, and B. Underdal.** 1987. Investigation of a toxic water-bloom of *Microcystis aeruginosa* (Cyanophyceae) in Lake Akersvatn, Norway. *Hydrobiologia* 144:97–103.
6. **Brock, T. D.** 1985. *A Eutrophic Lake: Lake Mendota, Wisconsin,* p. 101–105. Springer-Verlag, New York.
7. **Devassy, V. P., P. M. A. Bhattathiri, and S. Z. Quazim.** 1978. *Trichodesmium* phenomenon. *Indian J. Mar. Sci.* 7:168–186.
8. **Eloff, J. N., Y. Steinitz, and M. Shilo.** 1976. Photooxidation of cyanobacteria in natural conditions. *Appl. Environ. Microbiol.* **31:**119–126.
9. **Ferguson Wood, E. J.** 1965. *Marine Microbial Ecology.* Chapman & Hall, Ltd., London.
10. **Ganf, G. G.** 1974. Diurnal mixing and the vertical distribution of phytoplankton in a shallow equatorial lake (Lake George, Uganda). *J. Ecol.* **62:**611–629.
11. **George, D. G., and R. W. Edwards.** 1976. The effect of wind on the distribution of chlorophyll *a* and crustacean plankton in a shallow eutrophic reservoir. *J. Appl. Ecol.* **13:**667–690.
12. **Hoppe, H.-G., K. Gocke, D. Zamorano, and R. Zimmermann.** 1983. Degradation of macromolecular organic compounds in a tropical lagoon (Ciénaga Grande, Colombia) and its ecological significance. *Int. Rev. Gesamten Hydrobiol.* **68:**811–824.
13. **Khoja, T., and B. A. Whitton.** 1971. Heterotrophic growth of blue-green algae. *Arch. Microbiol.* **79:**280–282.
14. **Knobloch, K.** 1969. Sulfide oxidation via photosynthesis in green algae, p. 1032–1034. *In* H. Muntzer (ed.), *Progress in Photosynthesis Research.* International Union of Biological Sciences, Tübingen, Federal Republic of Germany.
15. **Krogmann, D. W., R. Butalla, and J. Sprinkle.** 1986. Blooms of cyanobacteria on the Potomac River. *Plant Physiol.* **80:**667–671.
16. **Meijer, M. L., and H. van der Honing.** 1986. Drijflagen van blauwalgen in het Brielse Meer. *H_2O* **19(5):**90–94.
17. **Nusch, E. A.** 1980. Comparison of different methods for chlorophyll and phaeopigment determination. *Arch. Hydrobiol.* 14:14–36.
18. **Oren, A., E. Padan, and S. Malkin.** 1979. Sulfide inhibition of photosystem 2 in cyanobacteria (blue-green algae) and tobacco chloroplasts. *Biochim. Biophys. Acta* **546:**270–279.
19. **Raboy, B., E. Padan, and M. Shilo.** 1976. Heterotrophic capacities of *Plectonema boryanum. Arch. Microbiol.* **110:**77–85.
20. **Revsbech, N. P., and B. B. Jørgensen.** 1986. Microelectrodes: their use in microbial ecology. *Adv. Microb. Ecol.* **9:**293–351.
21. **Reynolds, C. S., and A. E. Walsby.** 1975. Water blooms. *Biol. Rev. Camb. Philos. Soc.* **50:**437–481.
22. **Robarts, R. D.** 1984. Factors controlling primary production in a hypertrophic lake (Hartbeespoort Dam, South Africa). *J. Plankton Res.* **6:**91–105.
23. **Robarts, R. D.** 1985. Hypertrophy, a consequence of development. *Int. J. Environ. Stud.* **25:**167–175.
24. **Robarts, R. D., P. J. Ashton, J. A. Thornton, H. J. Taussig, and L. M. Sephton.** 1982. Overturn in a hypertrophic, warm, monomictic impoundment (Hartbeespoort Dam, South Africa). *Hydrobiologia* 97:209–224.
25. **Robarts, R. D., and T. Zohary.** 1984. *Microcystis aeruginosa* and underwater light attenuation in a hypertrophic lake (Hartbeespoort Dam, South Africa). *J. Ecol.* **72:**1001–1017.
26. **Robarts, R. D., and T. Zohary.** 1986. The influence of a cyanobacterial hyperscum on the heterotrophic activity of planktonic bacteria in a hypertrophic lake. *Appl. Environ. Microbiol.* **51:**609–613.
27. **Seki, H., M. Takahashi, Y. Hara, and S. Ichimura.** 1980. Dynamics of dissolved oxygen during algal bloom in Lake Kasumigaura, Japan. *Water Res.* **14:**179–183.
28. **Sentzova, O. Y., K. A. Nikitina, and M. V. Gusev.** 1975. Oxygen exchange of the obligate-phototrophic blue-green alga *Anabaena variabilis* in the darkness. *Mikrobiologiya* 44:283–288. (In Russian.)
29. **Sirenko, L. A.** 1972. *Physiological Basis of Multiplication of Blue-Green Algae in Reservoirs,* p. 88–162. Naukova Dumka, Kiev. (Translation no. RTS 8132 from the British Library lending division.)
30. **Stanier, R. Y., and G. Cohen-Bazire.** 1977. Phototrophic prokaryotes: the cyanobacteria. *Annu. Rev. Microbiol.* 31:225–274.
31. **Stanier, R. Y., R. Kunisawa, M. Mandel, and G. Cohen-Bazire.** 1971. Purification and properties of unicellular blue-green algae (order *Chroococcales*). *Bacteriol. Rev.* **35:**171–205.
32. **Topachevskiy, A. V., L. P. Braginskiy, and L. A. Sirenko.** 1969. Massive development of blue-green algae as a product of the ecosystem of a reservoir. *Hydrobiol. J.* **5(6):**1–10.
33. **van Rijn, J., and M. Shilo.** 1985. Carbohydrate fluctuations, gas vacuolation, and vertical migration

of scum-forming cyanobacteria in fishponds. *Limnol. Oceanogr.* **30**:1219–1228.

34. **Walsby, A. E.** 1975. Gas vesicles. *Annu. Rev. Plant Physiol.* **36**:427–439.
35. **Ward, D. M., E. Beck, N. P. Revsbech, K. A. Sandbeck, and M. R. Winfrey.** 1984. Decomposition of hot spring microbial mats, p. 191–214. *In* Y. Cohen, R. W. Castenholz, and H. O. Halvorson (ed.), *Microbial Mats: Stromatolites.* Alan R. Liss, Inc., New York.
36. **Wetzel, R. G.** 1975. *Limnology*, p. 300. The W. B. Saunders Co., Philadelphia.
37. **Yamagishi, H., and K. Aoyama.** 1972. Ecological studies on dissolved oxygen and bloom of *Microcystis* in Lake Suwa. I. Horizontal distribution of dissolved oxygen in relation to drifting of *Microcystis* by wind. *Bull. Jpn. Soc. Sci. Fish.* **38**:9–16.
38. **Zohary, T.** 1985. Hyperscums of the cyanobacterium *Microcystis aeruginosa* in a hypertrophic lake (Hartbeespoort Dam, South Africa). *J. Plankton Res.* 7:399–409.

Chapter 6

Photosynthetic and Heterotrophic Benthic Bacterial Communities of a Hypersaline Sulfur Spring on the Shore of the Dead Sea (Hamei Mazor)

Aharon Oren

INTRODUCTION

As a result of the recent drop in the water level of the Dead Sea (15), a series of hypersaline sulfur springs has become exposed on the western shore of the lake, about 4 km south of Ein Gedi (Hamei Mazor). Figure 1 shows an overview of the site. The water has a strong smell of sulfide. On its way to the Dead Sea, the water of the springs forms a series of shallow pools and streams, which are covered with green and red microbial mats.

I have been studying several aspects of the microbiology of the springs during the past 5 years, and in this chapter I present a summary of the results of these studies.

Aharon Oren • Division of Microbial and Molecular Ecology, The Institute of Life Sciences, The Hebrew University of Jerusalem, Jerusalem 91904, Israel.

EXPERIMENTAL SURVEY OF SPRING WATER AND MATS

Spring Water

Physical and Chemical Characterization

The specific gravity of the water was estimated in the field by using a hydrometer and measured later in the laboratory by weighing a known volume of water. The pH of the water was measured in the laboratory with a calibrated glass electrode within 2 h of sampling; completely filled sample bottles were used. The content of total dissolved salts was determined by drying a known volume of water at 150°C under vacuum to constant weight. Na and K were assayed by flame photometry; Ca and Mg were determined colorimetrically (14, 28); Cl plus Br was determined by titration with $AgNO_3$; and sulfide was found by addition of an excess of iodine, acidification, and titration of residual iodine with thiosulfate, in a pro-

Figure 1. View of the Hamei Mazor sulfur springs, with the Dead Sea in the background.

cedure enabling the dissolved oxygen concentration to be determined simultaneously (12). Gas analyses were made by using a Packard model 427 gas chromatograph equipped with a thermal conductivity detector and a glass column (0.6 by 180 cm) containing Molecular Sieve 5A (80/100 mesh). Argon served as the carrier gas.

Microbial Mats

CO_2 Photoassimilation

A sample of the cyanobacterial mat was divided into equal portions, which were put into 100-ml glass bottles. After 60 μCi of $NaH^{14}CO_3$ had been added to all the bottles and 10^{-5} M 3-(3,4-dichlorophenyl)-1,1-dimethylurea (DCMU) had been added to some of the bottles, they were incubated in the light or in the dark (wrapped in aluminium foil) in the spring for 1.5 h, whereafter the reaction was stopped by the addition of 1 ml of 100% trichloroacetic acid. The bottles were transferred to the laboratory, their contents were filtered on glass fiber filters (Whatman GF/C), and 2 drops of 10% acetic acid were added to each filter. After the filters had dried overnight at room temperature, 8 ml of Instagel scintillation cocktail (Packard) was added to each filter, and the radioactivity retained on the filters was assayed in a scintillation counter.

Microscopy

Microbial mat samples were examined microscopically with a Zeiss standard microscope equipped with phase-contrast optics.

Enrichment Cultures and Growth Experiments

Enrichment cultures for photosynthetic purple bacteria and anaerobic halophilic chemoorganotrophic bacteria were set up, and growth experiments were performed as specified.

CHARACTERISTICS OF SPRING WATER AND MATS

Spring Water

Physical and Chemical Properties

The properties of the spring water are summarized in Table 1. The water is characterized by a high salinity (total dissolved salts, ca. 170 g/liter), a high sulfide content (2.5 mM at the source), a low pH (around 5.2), and a temperature of 39°C at the source. The sulfide is of geothermal origin, and no significant bacterial sulfate reduction in situ was demonstrated. Gas bubbles rise from the pools of the spring, consisting of about 4% methane and 96% nitrogen. Analysis of inorganic nitrogen compounds showed the presence of about 2 mg of ammonia nitrogen per liter; no nitrite or nitrate was detected (S. Diab, personal communication). The physical and chemical properties of the water were found to be constant over the 5 years of sampling and from one season to the next. Table 1 also shows data on the composition of Dead Sea water and water from Hamei Yesha, a 30-m-deep well with saline sulfide-containing water which is approximately 1 km south of our sampling site and whose published salt composition (17) is very similar to that indicated by our data for Hamei Mazor. A comparison of the composition of the spring water with that of the Dead Sea brines shows that the spring water is less saline than the Dead Sea and also differs significantly in salt composition. Notably, the ratio of divalent to monovalent cations is high, although lower than in the Dead Sea; the biological implications of this property are discussed elsewhere (20). Both Dead Sea water and the water of the spring are acidic. Dead Sea brines are known to show a rapid increase in pH upon dilution with distilled water (1) (Table 1), and this effect was attributed to a shift in the bicarbonate-carbonate equilibrium: in strong brines the dissociation of bicarbonate is depressed (1). No such significant increase in pH was observed when spring water was diluted.

The sulfide content of the water (initial concentration, 2.5 mM) decreased rapidly with distance from the source; e.g., I measured (6 June 1987) 2.51 mM at a distance of 2 m (temperature, 38°C), 1.54 mM at 7 m (36.5°C), and 0.59 mM at 20 m (33.5°C). This decrease in sulfide concentration is due to volatilization, to biological oxidation by photosynthetic procaryotes, and possibly also to chemolithotrophic oxidation (see below) and abiotic oxidation.

Microbial Mats

Cyanobacterial Flora

The pools formed by the spring and their outlets to the Dead Sea are covered with benthic microbial mats (Fig. 2), varying in thickness from a few millimeters to 1 cm. The nature of the biota of this mat varies with distance from the source: in the pools near the springs, where the sulfide concentration is the highest (around 2.5 mM), the mat consists almost entirely of long, thick (4-μm) green filaments of *Oscillatoria* sp. (Fig. 2A and 3). The only other forms of life observed in this area were a few long, slender, rod-shaped bacteria and short, motile bacteria. The cyanobacterium displays a highly hydrophobic behavior: when a suspension of filaments is shaken with hexadecane (10), all the cells move to the organic phase.

The *Oscillatoria* strain from the spring grew relatively well in the laboratory on plates of Chull medium (27a), prepared in water from the spring. The halophilic nature of this organism was demonstrated in an experiment in which plates containing different NaCl concentrations were prepared: optimal growth was observed in NaCl concentrations of between 60 and 80 g/liter, with some growth at concentrations of up to 180 g/liter; no growth occurred at NaCl

Table 1.

Physical and Chemical Properties of the Water of the Hamei Mazor Sulfur Spring Compared with Dead Sea Water

Location	Temp (°C)	pH			Specific gravity at 25°C	Total dissolved salts (g/liter)	Concn (mol/liter) of:									Divalent/ monovalent cations
		Undiluted	Diluted 1:1	Diluted 1:9			Na^+	K^+	Mg^{2+}	Ca^{2+}	Cl^-	Br^-	SO_4^{2-}	HCO_3^-	Sulfide (mmol/liter)	
Hamei Mazor	38–40	5.2 ± 0.3	5.52	6.10	1.113 ± 0.003	169.2	1.25	0.08	0.59	0.303	2.98	ND[a]	ND	ND	2.5	0.67
Hamei Yesha[b]	39	ND	ND	ND	ND	ND	1.20	0.09	0.75	0.30	3.32	ND	0.02	1.8	1.17	0.81
Dead Sea[c]	18–37[d]	5.9	6.65	7.96	1.23	339.6	1.74	0.196	1.81	0.429	6.34	0.066	0.004	ND	0	1.17

[a] ND, Not determined.
[b] Data on the chemical composition of Hamei Yesha water (sampled from a 30-m-deep well in 1963) were derived from reference 17.
[c] Data on the composition of Dead Sea water (March 1977, average values) were derived from reference 3.
[d] Reference 22.

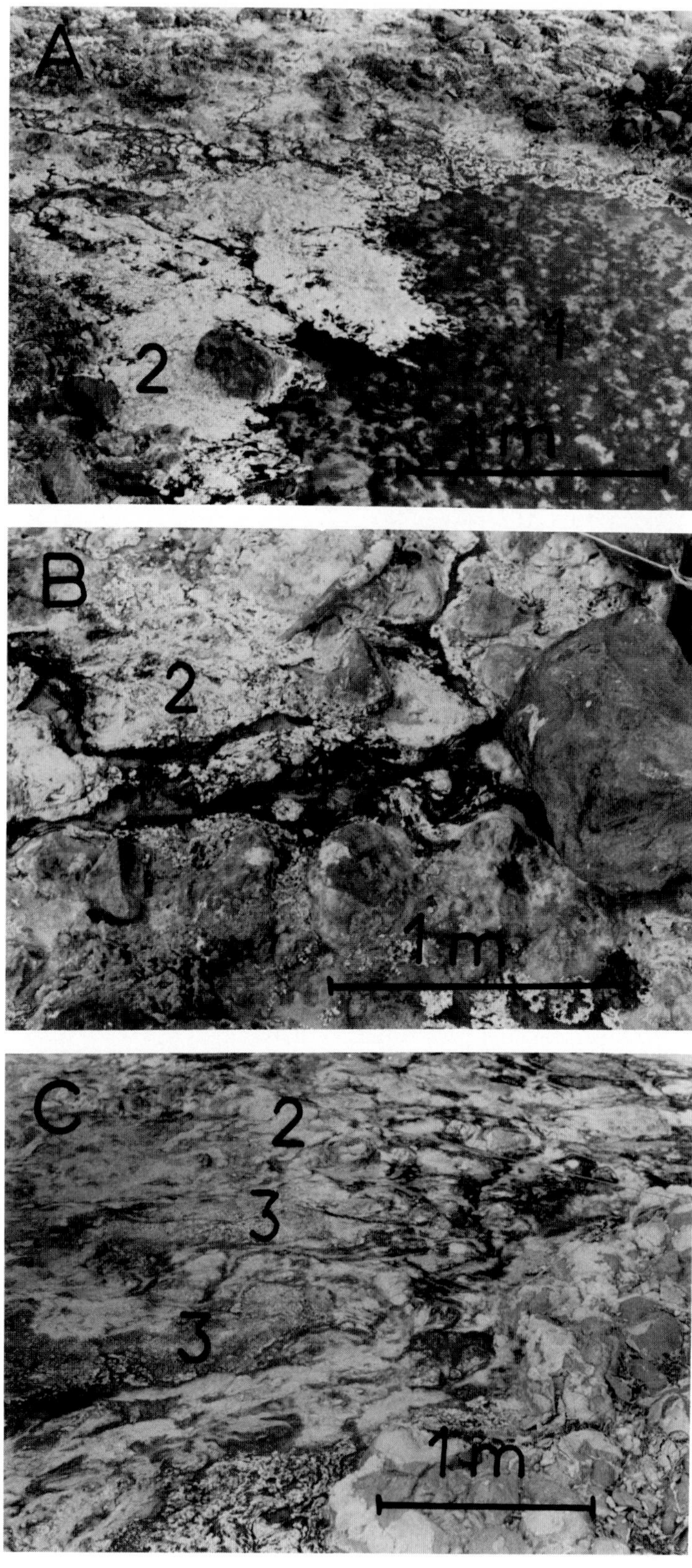
A
2
1
1m
B
2
1m
C
2
3
3
1m

concentrations of 40 g/liter and lower. As with many other halophilic cyanobacteria (26), the cells accumulate glycine-betaine as an intracellular osmotic solute (R. H. Reed, personal communication).

Oscillatoria-type filaments were also abundant farther downstream, where the water, on its way to the Dead Sea, forms a series of shallow mud flats (Fig. 2B and C), but other photosynthetic procaryotes were also found there (see below). Parts of the mats downstream, where sulfide is scarce or absent, are brownish and slimy. In these mats, unicellular cyanobacteria (*Aphanothece halophytica*) dominate, together with a rich bacterial flora (Fig. 4B).

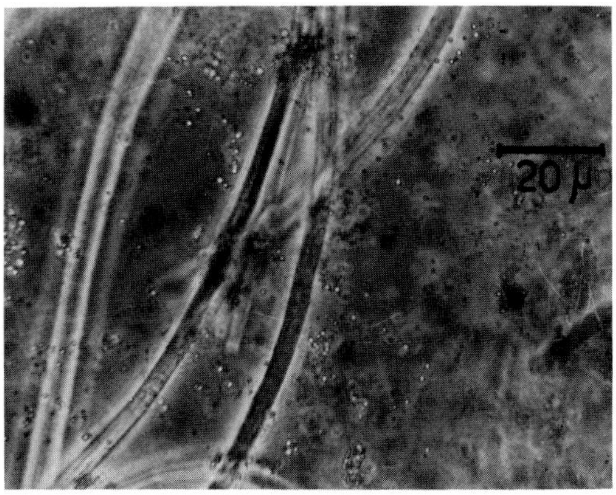

Figure 3. Filaments of *Oscillatoria* sp., covered with sulfur granules, from the microbial mat close to the spring (Fig. 2A).

Other Microbial Communities

Although the area near the spring, where sulfide concentrations are highest, is covered with a microbial mat consisting almost entirely of filamentous cyanobacteria, farther downstream a more varied microbial flora is found (Fig. 4). In this area, characterized by lower temperatures and lower sulfide concentrations, pink patches are seen, which consist of purple photosynthetic bacteria. Microscopically motile forms (*Chromatium* type) are seen, as well as clumps of bright-red nonmotile cells (Fig. 4A), sometimes referred to as "*Thiopolycoccus ruber*" (24); this type of organism was never isolated in pure culture and probably should be classified as a *Thiocapsa* sp. The purple areas contain bacteriochlorophyll *a*, as is evident from the absorption peak at 760 to 770 nm in a 1:1 (vol/vol) methanol-acetone extract of material from the mat. The distribution of the purple sulfur bacteria is probably determined more by the sulfide concentration in situ than by the temperature: red patches in the mat were found at temperatures as high as 35°C, but never at sulfide concentrations above 1.5 mM. Attempts to isolate purple sulfur bacteria of the *Chromatium* type from the site have not succeeded. Only recently, the first isolation of an obligately halophilic *Chromatium* sp., growing at NaCl concentrations of up to 8.5% in Hamelin Pool, Shark Bay, Australia, was reported (2).

In the area where sulfide concentrations are reduced, a variety of amoeboid and ciliate protozoa predating on the cyanobacteria and the purple and colorless bacteria are found (Fig. 5). Amoebae were often found loaded with purple sulfur bacteria (Fig. 5a to c). Similar protozoa were isolated from a hypersaline lagoon in Western Australia (25). Although no protozoa were observed in the Dead Sea during our studies on the biology of the lake from 1980 onwards (20, 22), the occurrence of a dimastigamoeba in the Dead Sea was reported as early as 1943

Figure 2. Benthic microbial mats of the Hamei Mazor sulfur springs near the source of the water (39°C, 2.5 mM sulfide) (A), at a drainage channel 2 m from the source (B), and at a distance of 7 m from the source (sulfide concentration, ca. 1.5 mM) showing green and red patches (C). 1, Green cyanobacteria; 2, white sulfur deposit; 3, pink patches of purple sulfur bacteria.

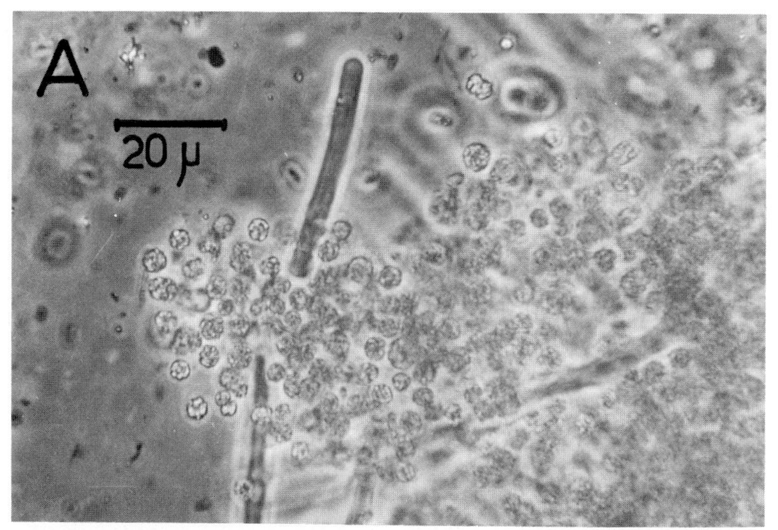
A
20 μ

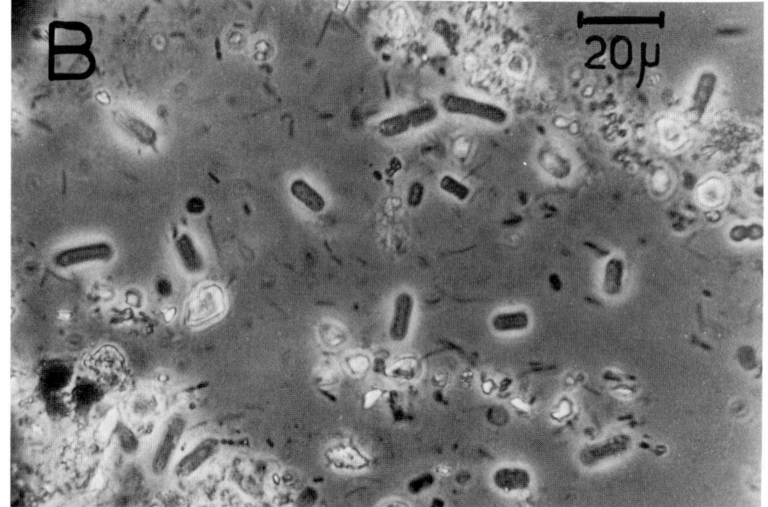
B
20 μ

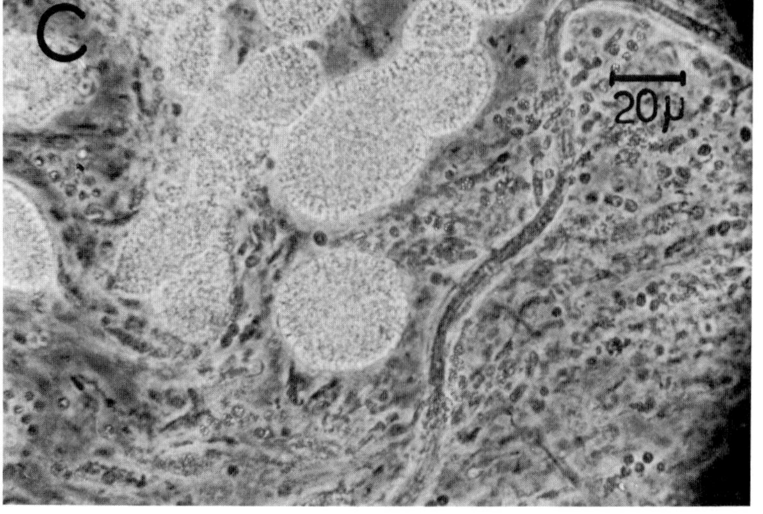
C
20 μ

(8); other types of protozoa were reported from the Dead Sea in 1944 (9, 30). It is possible that the protozoa previously observed in the Dead Sea are unable to live in the Dead Sea nowadays, but are derived from less saline springs on the shore of the lake.

Occasionally, unicellular green algae (*Dunaliella* sp.) were observed microscopically, and on one occasion an unknown, colony-forming, colorless organism was seen (Fig. 4C). Metazoa predating on the mats were generally absent, but on one occasion in early spring, ephydrid flies were seen in great masses, predating on the mat and depositing their eggs (C. Dimentman, personal communication).

Anoxygenic Photosynthesis of Oscillatoria sp. In Situ

The fact that the cyanobacterial mat is found in water with a high sulfide content and devoid of oxygen suggests the possibility that the *Oscillatoria* sp. performs an anoxygenic type of photosynthesis with sulfide as the electron donor (5–7, 11). This suggestion was strengthened by the observation of sulfur granules around the cyanobacterial filaments (Fig. 3).

To test this hypothesis, CO_2 photoassimilation experiments were performed in the presence and absence of DCMU, which inhibits electron flow between photosystem II and photosystem I and thus enables oxygenic and anoxygenic photosynthesis to be distinguished (7, 11). Approximately equal pieces of *Oscillatoria* mat were incubated in 100-ml bottles with water from the spring, with 60 μCi of $NaH^{14}CO_3$, and with or without 10^{-5} M DCMU, for 1.5 h in the light or in the dark, whereafter the amount of radioactive label incorporated by the cells was determined. No significant inhibition of light-dependent CO_2 assimilation by DCMU was observed (Fig. 6), suggesting an anoxygenic type of photosynthesis with sulfide as an electron donor. This conclusion was confirmed in a laboratory experiment, performed in collaboration with Y. Cohen, involving the use of an oxygen microelectrode to measure oxygenic photosynthesis in the *Oscillatoria* mat (5, 27): no light-dependent oxygen evolution could be demonstrated. Since the *Oscillatoria* sp. of Hamei Mazor is able to grow aerobically on plates, we have here another example of a cyanobacterium displaying facultatively anoxygenic photosynthesis.

Other Halophilic Bacteria

For some time I have been studying obligately anaerobic bacteria from the Dead Sea and other hypersaline environments, and I have characterized a novel group of halophilic eubacteria that are not related to any of the previously recognized subgroups within the eubacterial kingdom (for a review, see reference 18). Enrichment cultures for anaerobic halophilic chemoorganotrophs, with material from the microbial mat of Hamei Mazor as the inoculum, yielded facultatively anaerobic halophiles (*Vibrio costicola* and motile rods) at NaCl concentrations below 100 g/liter; however, at NaCl concentrations of between 140 and 200 g/liter, long, slender, obligately anaerobic, fermentative rods, resembling *Halobacteroides halobius* (isolated previously from sediments of the Dead Sea [23]), developed.

Since methane was identified in the gas bubbles that rise from the pools, attempts were made to test for the presence of halophilic methanogenic bacteria. A methanogenic coccus was isolated which grew optimally at 120 g of NaCl per liter; utilized

Figure 4. Microbial types observed in the Hamei Mazor microbial mats. (A) Nonmotile purple sulfur bacteria around a cyanobacterial filament; (B) unicellular cyanobacteria (*A. halophytica*) and bacteria from a brown, slimy part of the mat; (C) colorless colonies of an unknown microorganism.

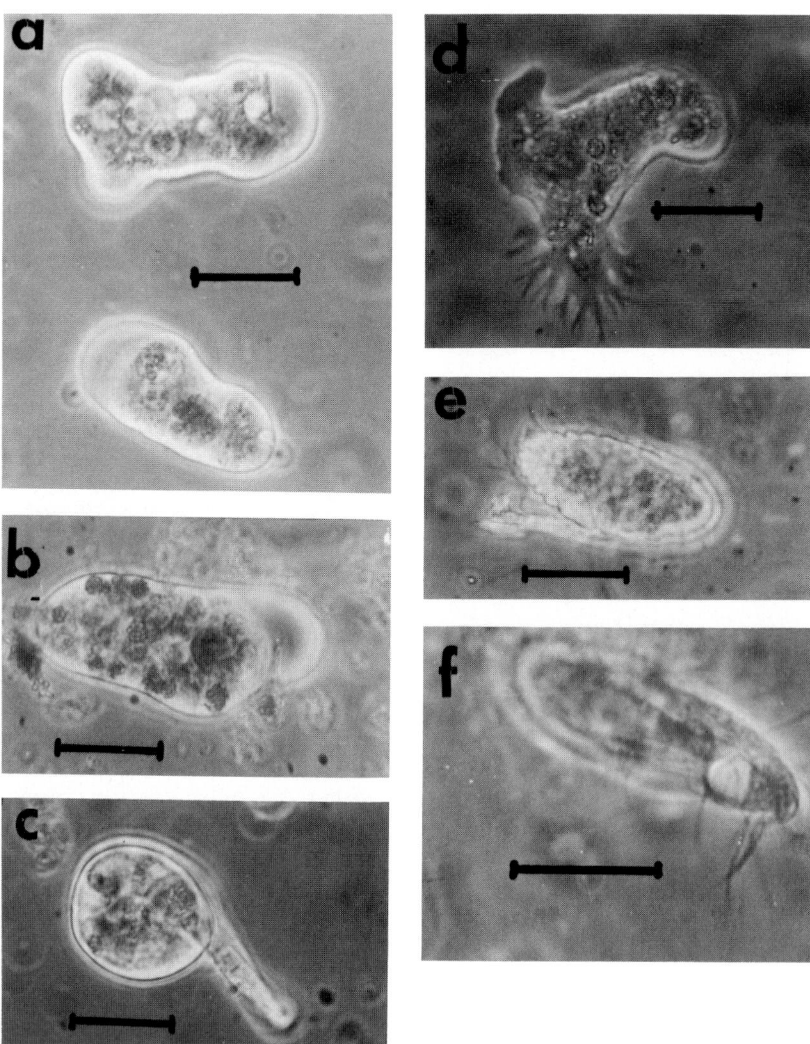

Figure 5. Protozoa observed in the Hamei Mazor microbial mats. (a to d) Amoeboid protozoa, in part loaded with purple sulfur bacteria; (e and f) different types of ciliates. Bars, 20 μm.

trimethylamine and methanol as carbon and energy sources; and was unable to utilize H_2 plus CO_2, dimethylamine, formate, or acetate (U. Marchaim, personal communication).

The possibility that part of the decrease in sulfide concentration was due to biological oxidation by aerobic chemolithotrophic bacteria was investigated by enriching for halophilic chemolithotrophs, with thiosulfate as the energy source in enrichment cultures. On several occasions small, motile, rod-shaped bacteria were observed which produced elemental sulfur from thiosulfate (possibly a *Thiobacillus* sp.); all attempts to maintain these cultures have failed.

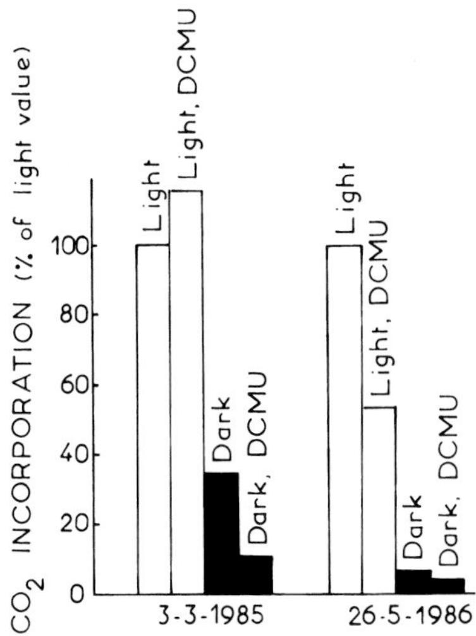

Figure 6. Anoxygenic photosynthesis of *Oscillatoria* sp. in the Hamei Mazor microbial mats under in situ conditions.

Isolation of Photosynthetic Purple Bacteria

As stated above, attempts to isolate the halophilic purple sulfur bacteria that are abundant in the spring have not yet succeeded, but enrichment cultures designed for their isolation yielded a novel type of photosynthetic purple bacterium: an irregularly rod-shaped motile *Ectothiorhodospira*-like organism (Fig. 7A). The organism developed in enrichment cultures consisting of a medium of neutral pH containing spring water enriched with bicarbonate, sulfide (1 mM), and acetate. The bacterium is motile owing to a bundle of four or five polar flagella. The organism (temporarily designated strain EG-1) was obtained in pure culture and is routinely grown at 35°C in a medium containing 100 g of NaCl per liter, 0.1 g of yeast extract per liter (not required for growth, but highly stimulatory), salts including phosphate and ammonium salts, trace elements, acetate, carbonate, and 0.08 g of sodium dithionite per liter (pH 6.8). The range of carbon sources, in addition to acetate, that can be used is limited: succinate, malate, fumarate, and pyruvate supported good growth, and fair growth was obtained on propionate or high concentrations of yeast extract. A source of reduced sulfur is required for growth (sulfide, cysteine, or dithionite), and all attempts to grow the organism aerobically or in the absence of reduced-sulfur compounds have failed. The strain grows optimally at NaCl concentrations of between 40 and 100 g/liter, at temperatures between 35 and 40°C, and at neutral pH (6 to 8); growth below pH 5.5 is very poor (Fig. 8).

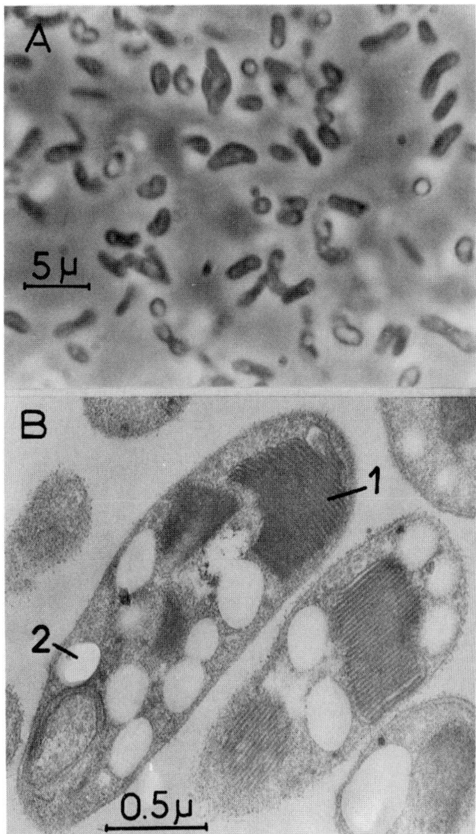

Figure 7. Novel type of halophilic purple bacterium (strain EG-1) isolated from the Hamei Mazor microbial mat. (A) Phase-contrast micrograph. (B) Thin section observed in the electron microscope (courtesy of M. Kessel): 1, stacks of photosynthetic membranes; 2, granules of poly-β-hydroxybutyrate.

The photosynthetic system of strain EG-1 is located on stacks of intracellular

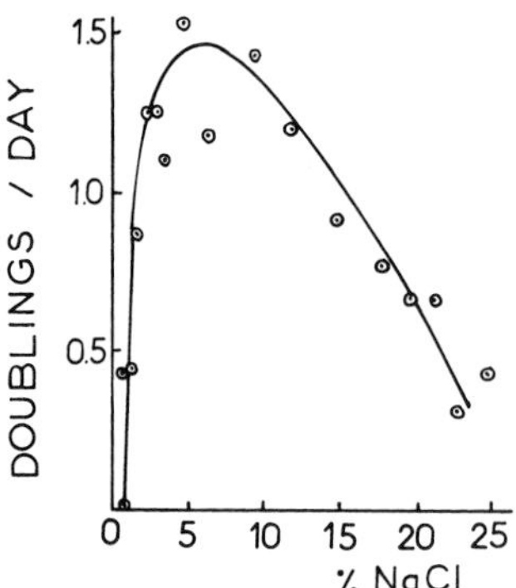

Figure 8. Effect of NaCl concentration on the growth of the purple bacterium EG-1. Cells were grown at 35°C in media of different NaCl concentrations. After different incubation periods, the turbidity of the cultures was measured at 600 nm, and the growth rates were calculated from the slopes of the exponential part of the growth curves.

membranes (Fig. 7B), resembling the photosynthetic system of the members of the genus *Ectothiorhodospira* (purple sulfur bacteria) (29). The in vivo absorption spectrum shows peaks at 380, 490 to 495, 515 and 550, 591, 796, and 860 nm, with a shoulder at 880 nm, suggesting the presence of bacteriochlorophyll *a* and carotenoids of the spirilloxanthin series.

Photoautotrophic growth of strain EG-1 was recently demonstrated, and a formal description of the strain as a new *Ectothiorhodospira* species has been submitted (A. Oren, M. Kessel, and E. Stackebrandt, submitted for publication).

DISCUSSION

The special properties of the biota of the Hamei Mazor sulfur springs are determined mainly by three factors: a high sulfide concentration, low pH values, and a high salinity.

Development of cyanobacteria in the presence of sulfide is a common phenomenon, and the mechanisms enabling certain cyanobacteria to live and grow under anaerobic conditions in the presence of sulfide are well established (5–7, 11, 21): specific adaptations include (i) the potency of using sulfide as the electron donor in an anoxygenic type of photosynthesis driven by photosystem I only and (ii) the absence of polyunsaturated fatty acids from the cellular lipids (an essential feature for organisms growing in the absence of molecular oxygen, since oxygen is required in the biosynthesis of polyunsaturated fatty acids). A special feature of cyanobacterial development in the spring is the combination of high sulfide concentrations (2.5 mM) and a low pH (around 5.2). Sulfide is toxic, even to microorganisms that utilize it as an electron donor, and this toxicity depends primarily on the concentration of undissociated H_2S, which freely passes through biological membranes. Different cyanobacteria show various degrees of adaptation to high sulfide concentrations (5, 11), and *Oscillatoria limnetica* from Solar Lake, Sinai, Egypt, shows the highest sulfide tolerance of all: optimal CO_2 photoassimilation occurs at 3 mM sulfide and neutral pH, i.e., about 1.5 mM undissociated H_2S (5, 6). In Stinky Springs, Utah, an *Oscillatoria* strain thrives at a similar concentration of undissociated sulfide (total sulfide concentration, 1.1 mM [pH 6.3]) (5). At the low pH (5.2) of the Hamei Mazor spring, essentially all sulfide appears as undissociated H_2S, which suggests that its halophilic *Oscillatoria* sp. is one of the most sulfide-tolerant phototrophs known.

Although it is widely assumed that the cyanobacteria are highly adaptable and are very tolerant of environmental extremes, they are conspicuously absent from acidic environments and show a preference for alkaline conditions (4). Even in mildly acidic waters (pH values of 5 to 6), cyanobacteria are uncommon. With a very few exceptions (e.g., the occurrence of a variety of filamentous cyanobacteria in a microbial mat developing in a Swedish lake at a pH of 4.3 to 4.7 [16]), the lowest pH at which cyanobacteria thrive seems to be 4.8 to 5 (4), and most of the common species are already severely inhibited below pH 6 (13). Thus, it appears that in Hamei Mazor the cyanobacterial mat develops not only at an extremely high sulfide concentration, but also at an uncommonly low pH value for cyanobacteria.

The salt concentration of the spring water is only about half of that of the Dead Sea itself, and all organisms living in the spring appear to be well adapted to salt concentrations of up to 170 g/liter, the concentration found in the spring. Since the salt concentrations (and, more specifically, the divalent cation concentrations) in the Dead Sea are supraoptimal for the development of *Dunaliella* spp., the main or only primary producer in the lake (19, 22), a *Dunaliella* bloom can develop in the Dead Sea only when the salinity of the upper water layers is temporarily reduced as a result of massive winter floods. Such a bloom was observed in the summer of 1980, but not during any of the subsequent 7 years (20), and since no *Dunaliella* cells could be demonstrated in the Dead Sea water column in the periods preceding and following the bloom, the question arose as to the source of the *Dunaliella* cells serving as the inoculum when conditions become favorable (22). The finding of *Dunaliella* cells in the Hamei Mazor springs (and possibly at additional sites around the Dead Sea), at the lowered salinities that support their development, may explain the rapid development of *Dunaliella* blooms in the Dead Sea as soon as the physical and chemical conditions become suitable.

Summarizing, the varied biota of the Hamei Mazor sulfur springs develop in an interesting combination of extremes of salt concentration, low pH values, and high sulfide concentrations, and the study of their properties broadens our understanding of microbial life in extreme environments.

ACKNOWLEDGMENTS. I thank Y. Cohen, S. Diab, C. Dimentman, M. Kessel, U. Marchaim, E. Raz, and R. H. Reed for their contributions to this work and for stimulating discussions.

This work was supported in part by a grant from Houston Lighting & Power Co., Houston, Tex., under a university participation program administered by Dynatech R/D Co.

LITERATURE CITED

1. **Amit, O., and Y. K. Bentor.** 1971. pH-dilution curves of saline waters. *Chem. Geol.* 7:307–313.
2. **Bauld, J., J. L. Favinger, M. T. Madigan, and H. Gest.** 1987. Obligately halophilic *Chromatium vinosum* from Hamelin Pool, Shark Bay, Australia. *Curr. Microbiol.* 14:335–339.
3. **Beyth, M.** 1980. Recent evolution and present stage of Dead Sea brines, p. 155–165. *In* A. Nissenbaum (ed.), *Hypersaline Brines and Evaporitic Environments*. Elsevier Scientific Publishing Co., Amsterdam.
4. **Brock, T. D.** 1973. Lower pH limit for the existence of blue-green algae: evolutionary and ecological implications. *Science* 179:480–483.
5. **Cohen, Y.** 1984. Oxygenic photosynthesis, anoxygenic photosynthesis, and sulfate reduction in cyanobacterial mats, p. 435–441. *In* M. J. Klug and C. A. Reddy (ed.), *Current Perspectives in Microbial Ecology*. American Society for Microbiology, Washington, D.C.
6. **Cohen, Y., B. B. Jørgensen, E. Padan, and M. Shilo.** 1975. Sulphide-dependent anoxygenic photosynthesis in the cyanobacterium *Oscillatoria limnetica*. *Nature* (London) 257:489–492.
7. **Cohen, Y., E. Padan, and M. Shilo.** 1975. Facultative anoxygenic photosynthesis in the cyanobacterium *Oscillatoria limnetica*. *J. Bacteriol.* 123:855–861.
8. **Elazari-Volcani, B.** 1943. A dimastigamoeba in the bed of the Dead Sea. *Nature* (London) 152:301–302.
9. **Elazari-Volcani, B.** 1944. A ciliate from the Dead Sea. *Nature* (London) 154:335.
10. **Fattom, A., and M. Shilo.** 1984. Hydrophobicity as the adhesion mechanism of benthic cyanobacteria. *Appl. Environ. Microbiol.* 47:135–143.
11. **Garlick, S., A. Oren, and E. Padan.** 1977. Occurrence of facultative anoxygenic photosynthesis among filamentous and unicellular cyanobacteria. *J. Bacteriol.* 129:623–629.
12. **Ingvorsen, K., and B. B. Jørgensen.** 1979. Combined measurement of oxygen and sulfide in water samples. *Limnol. Oceanogr.* 24:390–393.
13. **Kallas, T., and R. W. Castenholz.** 1983. Internal

pH and ATP-ADP pools in the cyanobacterium *Synechococcus* sp. during exposure to growth-inhibiting low pH. *J. Bacteriol.* **149**:229–236.

14. **Kerr, J. R. W.** 1960. The spectrophotometric determination of microgram amounts of calcium. *Analyst* **85**:867–870.
15. **Klein, C.** 1981. The influence of rainfall over the catchment area on the fluctuations of the level of the Dead Sea since the 12th century. *Isr. Meteor. Res. Pap.* **3**:29–57.
16. **Lazarek, S.** 1980. Cyanophytan mat communities in acidified lakes. *Naturwissenschaften* **67**:97–98.
17. **Mazor, E., E. Rosenthal, and J. Ekstein.** 1969. Geochemical tracing of mineral water sources in the south western Dead Sea basin. *Isr. J. Hydrol.* **7**:246–275.
18. **Oren, A.** 1986. The ecology and taxonomy of anaerobic halophilic eubacteria. *FEMS Microbiol. Rev.* **39**:23–29.
19. **Oren, A.** 1986. Dynamics of *Dunaliella* in the Dead Sea, p. 351–359. *In* Z. Dubinsky and Y. Steinberger (ed.), *Environmental Quality and Ecosystem Stability*, vol. IIIA. Bar Ilan University Press, Ramat Gan, Israel.
20. **Oren, A.** 1988. The microbial ecology of the Dead Sea, p. 193-229. *In* K.C. Marshall (ed.), *Advances in Microbial Ecology,* vol.10. Plenum Publishing Corp., New York.
21. **Oren, A., A. Fattom, E. Padan, and A. Tietz.** 1985. Unsaturated fatty acid composition and biosynthesis in *Oscillatoria limnetica* and other cyanobacteria. *Arch. Microbiol.* **141**:138–141.
22. **Oren, A., and M. Shilo.** 1982. Population dynamics of *Dunaliella parva* in the Dead Sea. *Limnol. Oceanogr.* **27**:201–211.
23. **Oren, A., W. G. Weisburg, M. Kessel, and C. R. Woese.** 1984. *Halobacteroides halobius* gen. nov., sp. nov., a moderately halophilic anaerobic bacterium from the bottom sediments of the Dead Sea. *Syst. Appl. Microbiol.* **5**:58–69.
24. **Pfennig, N., and H. G. Trüper.** 1974. The photosynthetic bacteria, p. 24–60. *In* R. E. Buchanan and N. E. Gibbons (ed.), *Bergey's Manual of Determinative Bacteriology*, 8th ed. The Williams & Wilkins Co., Baltimore.
25. **Post, F. J., L. J. Borowitzka, M. A. Borowitzka, B. Mackay, and T. Moulton.** 1983. The protozoa of a Western Australian hypersaline lagoon. *Hydrobiologia* **105**:95–113.
26. **Reed, R. H., L. J. Borowitzka, M. A. Mackay, J. A. Chudek, R. Foster, S. R. C. Warr, D. J. Moore, and W. D. P. Stewart.** 1986. Organic solute accumulation in osmotically stressed cyanobacteria. *FEMS Microbiol. Rev.* **39**:51–56.
27. **Revsbech, N. P., B. B. Jørgensen, T. H. Blackburn, and Y. Cohen.** 1983. Microelectrode studies of the photosynthesis and O_2, H_2S, and pH profiles of a microbial mat. *Limnol. Oceanogr.* **28**:1062–1074.

27a. **Stanier, R. Y., R. Kunisawa, M. Mandel, and G. Cohen-Bazire.** 1971. Purification and properties of unicellular cyanobacteria (order *Chroococcales*). *Bacteriol. Rev.* **35**:171–205.

28. **Taras, M.** 1948. Photometric determination of magnesium in water with brilliant yellow. *Anal. Chem.* **20**:1156–1158.
29. **Trüper, H. G, and N. Pfennig.** 1981. Characterization and identification of the anoxygenic phototrophic bacteria, p. 229–312. *In* M. P. Starr, H. Stolp, H. G. Trüper, A. Balows, and H. G. Schlegel (ed.), *The Prokaryotes. A Handbook of Habitats, Isolation, and Identification of Bacteria*, vol. 1. Springer-Verlag KG, Berlin.
30. **Volcani, B. E.** 1944. The microorganisms of the Dead Sea, p. 71–85. *In Papers Collected Commemorate the 70th Anniversary of Dr. Chaim Weizmann.* Collective volume. Daniel Sieff Research Institute, Rehovoth.

Chapter 7

Characterization of a Cyanobacterial, Algal Crust in the Coastal Dunes of The Netherlands

Ben de Winder, Jan Pluis, Loes de Reus, and Luuc R. Mur

INTRODUCTION

Algal crusts are layers of microorganisms which are subjected to periods of dryness. They are a geographically widespread phenomenon (2, 3, 8, 9, 13, 17). During drying, they become rigid. These rigid crusts play an important role in erosion processes, especially in arid zones. When a crust desiccates, its surface becomes hydrophobic, which will cause a surface runoff during rain. This in turn leads to an increase in water erosion. On the other hand, wind erosion is reduced. It was shown in a simple laboratory experiment that the amount of sand blown away from a well-developed dry crust was 100 times smaller than that blown from an uncolonized surface (4, 20).

In most cases, cyanobacteria are the organisms responsible for the initial formation of aggregates with the sand. The cyanobacteria are surrounded by mucilage, which acts as a cementing agent (4, 5, 15). While the crust is being formed, the algae are capable of increasing the total amount of organic matter. An increase in organic matter and a decrease in wind erosion favor the subsequent colonization by mosses and annuals.

In the crust system described here, the algae are present in so-called blowouts (7). Blowouts are erosional depressions that often develop on south-exposed slopes of sand dunes. Figure 1 shows a schematic view of a longitudinal section of a blowout (20). The cyanobacteria and algae colonize on the lee side. In this chapter the results of microscopic and experimental investigations of sand-alga aggregates from blowouts in the coastal dunes in The Netherlands are presented. A comparison with algal crusts in other areas is made. This study forms a part of a larger project, which focuses on the role of algae in the stabilization of blowouts.

METHODS

Algal crusts were sampled in a dune area called Meijendel, north of The Hague.

Ben de Winder, Loes de Reus, and Luuc R. Mur • Laboratory of Microbiology, University of Amsterdam, Nieuwe Achtergracht 127, 1018 WS Amsterdam, The Netherlands. *Jan Pluis* • Laboratory of Physical Geography and Soil Science, University of Amsterdam, Dapperstraat 15, 1093 BS Amsterdam, The Netherlands.

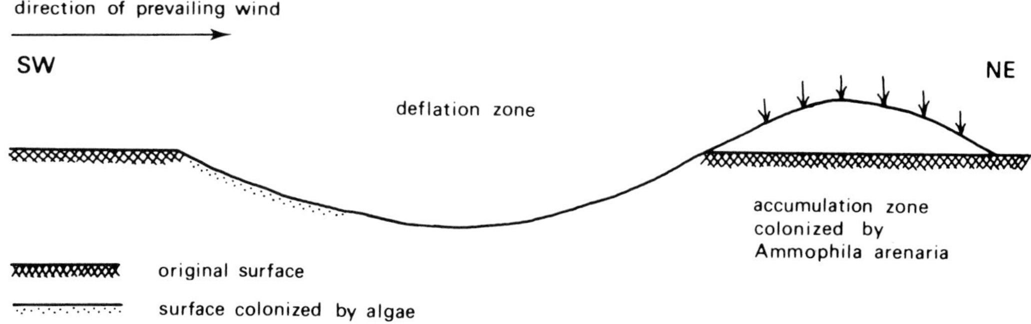

Figure 1. Schematic view of a longitudinal section of a blowout (adapted from reference 20).

This area is situated directly behind the strip of foredunes along the coast and is part of the Younger Dunes, which were formed between the 12th and 19th centuries (6).

Isolation and Growth

Organisms were isolated by suspending and subsequently diluting the samples in BG-11 (14). After 5 days of growth, culture material was plated on BG-11 solidified with 1% Difco Bacto-Agar. Cultures were grown photoautotrophically at 20°C under continuous light (provided by cold white fluorescent lamps at a light intensity of 40 microeinsteins $\cdot$ $m^{-2} \cdot s^{-1}$). Stock cultures were maintained under photoautotrophic growth conditions at a light intensity of 10 microeinsteins $\cdot$ $m^{-2} \cdot s^{-1}$. Unicyanobacterial cultures were obtained by picking up single filaments or small colonies of unicellular species with sterile pipettes. A dissecting microscope was used for this work.

SEM

Scanning electron microscope (SEM) studies were performed with an ISI-DS130 dual-stage SEM (International Scientific Instruments, Inc., Tokyo). Two preparation methods were used. In the first, samples were air dried, sputtered with gold, and stored in a dark, cool place prior to examination. In the second, water-saturated samples were fixed in 4% glutaraldehyde in 0.1 M cacodylate buffer (pH 6.8). The samples were dehydrated in an ethanol series, critical point dried, and sputtered with gold.

Photosynthesis

Photosynthesis was measured with a Clark-type oxygen microelectrode (12). Experiments were done with water-saturated samples kept in a core with a diameter of 10 cm and at 20°C. Samples were not older than 4 days. The net production of oxygen on the surface was measured at light intensities of 400 microeinsteins $\cdot$ $m^{-2} \cdot s^{-1}$ provided by a cold-halogen lamp.

FIELD OBSERVATIONS

The crusts on the lee side of the blowouts were investigated. The organisms were found in the uppermost 5 mm of the sandy surface throughout the year. The sand consisted of rounded silica grains with a size of 150 to 400 μm. The amount of lime, present mainly as shell fragments with a maximum diameter of 500 μm, was 2 to 4% (wt/wt). The porosity of the substrate was 62%.

Obviously, no growth will occur during dry periods. The period from January to May 1987 comprised the driest months. Biomass, expressed as the amount of chlorophyll, was at its lowest level (10 to 30 mg · m^{-2}) during this period. From August to November 1986, i.e., the months with a maximum amount of precipitation, the chlorophyll content was 5 times as high. During this period, large areas of freshly colonized surfaces were also found. In these areas the chlorophyll content was 1 to 20 mg · m^{-2}, with maxima of up to 100 mg · m^{-2}.

COMPOSITION OF CRUST

Light-microscopic studies revealed that the crusts are built by the following two groups of microorganisms. The cyanobacteria are the initial colonizers; they are present in the uppermost 4 mm of the sand. At the initial stage of colonization, the sand is still moved by the wind and water and no real crust is formed. *Synechococcus*, *Gloeocapsa*, *Oscillatoria* strains 01 through 05, LPP-A strain 13, LPP-B strain 08, and LPP-B (*Microcoleus*) strains 14 to 17 were observed and isolated. *Oscillatoria* strain 05 was the most frequently found cyanobacterium. The names of these cyanobacteria are those given by Rippka et al. (14). LPP is a group consisting of *Lyngbia*, *Plectonema*, and *Phormidium* spp. When no strain number is mentioned, the species has not yet been isolated. The isolated filamentous cyanobacteria from the crust were all motile. Motility is an asset for organisms

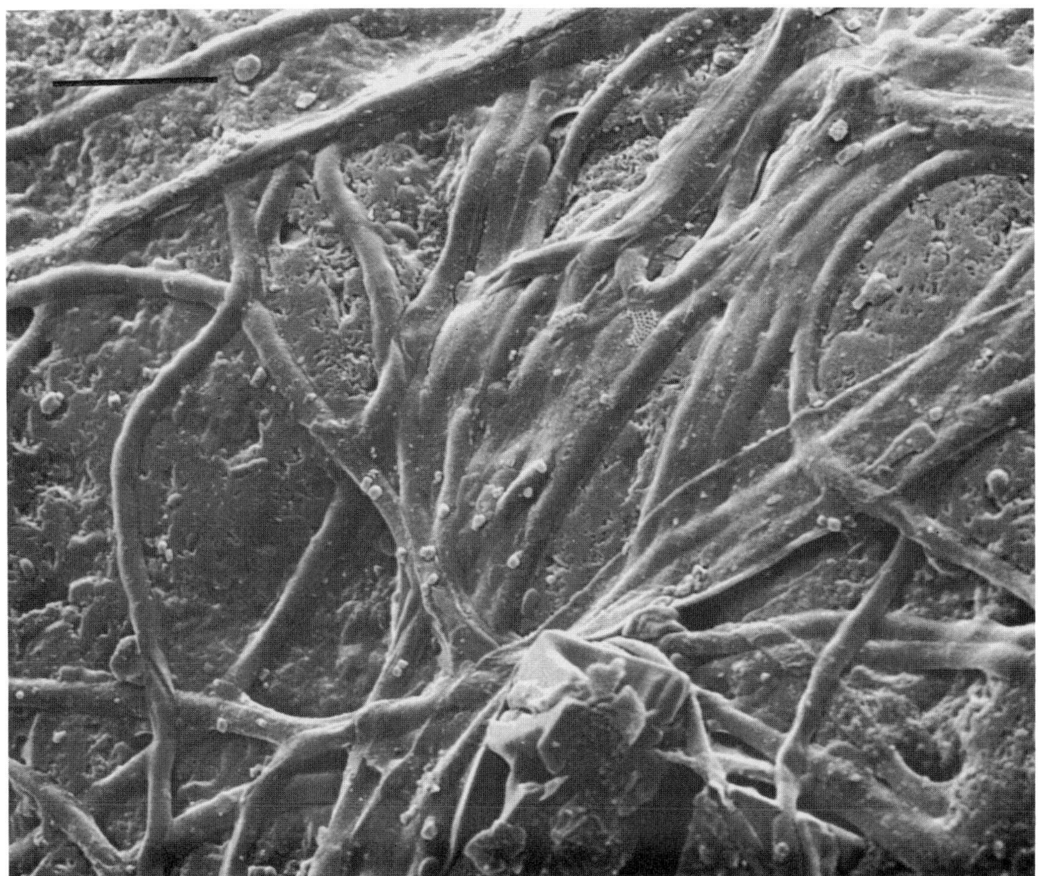

Figure 2. SEM of a dry-crust sample, showing epipsamnic cyanobacteria of the LPP group on the surface of the sand grain. Bar, 10 μm.

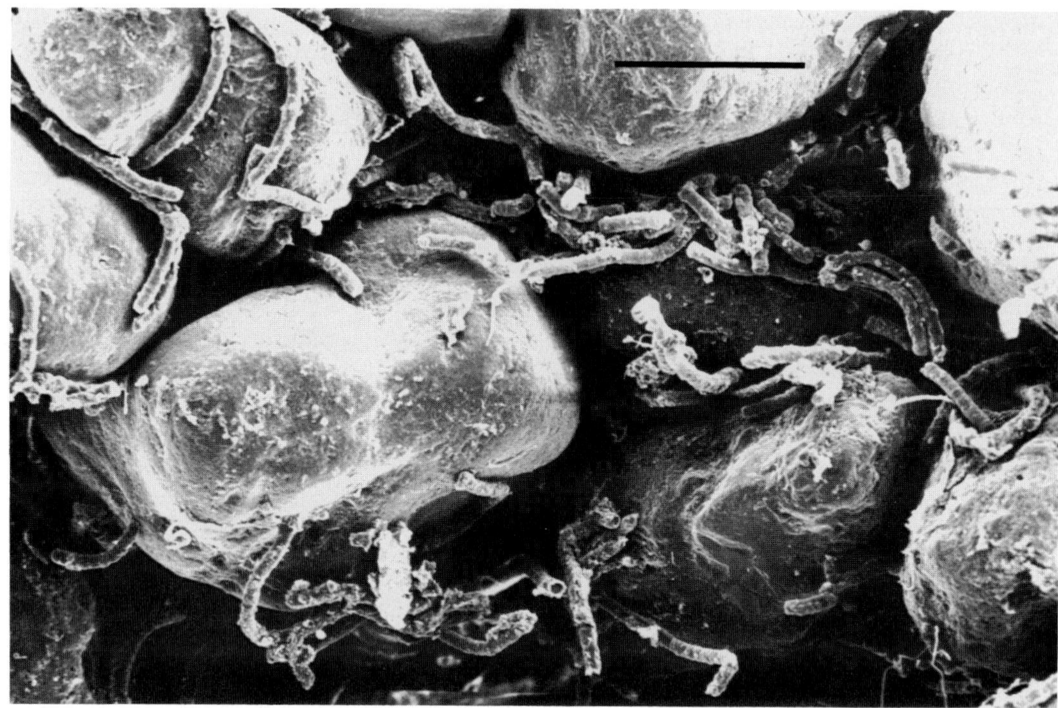

Figure 3. SEM of a transfer section of the top layer of a dry crust. The crust is dominated by *K. flaccidum*. Bar, 100 μm.

which are the first colonizers under conditions where they may be buried in the sand. Later, when the sand becomes aggregated, less-motile organisms become dominant. In this study we found that the green alga *Klebsormidium flaccidum* (Kützing) became dominant. When *K. flaccidum* became dominant, a rigid crust was formed.

STRUCTURE OF CRUST

The microscopic structure of the crust was studied with an SEM. Figure 2 shows the surface of a sand grain occupied by thin filaments of LPP species. The presence of a slime layer is evident from Fig. 2 as well as from additional phase-contrast microscopy. This slime layer probably enabled the organisms to adhere tightly to the surface. Figure 3 shows an electron micrograph taken of a dry sample of a *Klebsormidium*-dominated crust. The filaments form a continuous network and thus entangle the sand grains. This *Klebsormidium* species also possesses a slime layer, which contributes to the formation of aggregates.

The distribution pattern of the contributing organisms as derived from SEM and light-microscopic observations is shown in a schematic model of the crust in Fig. 4. The uppermost zone (A) was a network of filamentous algae, predominantly *Klebsormidium* and sometimes a few *Microcoleus* bundles. This layer is approximately 1 mm thick. Underneath layer A is a layer (B) with shorter filaments that are sometimes clumped with slime remnants. Below this is a layer (C) with short filaments that are clumped and surrounded by slimy organic matter. In the lowest zone (D), recognizable algal filaments are rare. In the interstitial spaces of the crust, particularly, there are clumps of algal residues. The transition from one layer to the

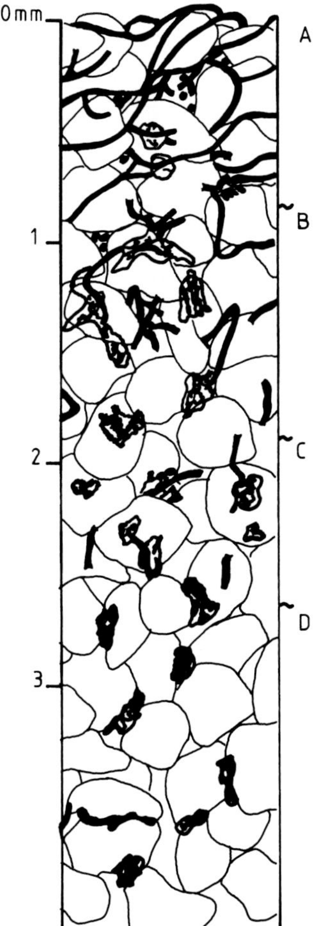

Figure 4. Model of an algal crust dominated by *K. flaccidum*. The model was drawn from SEM and light-microscopic observations.

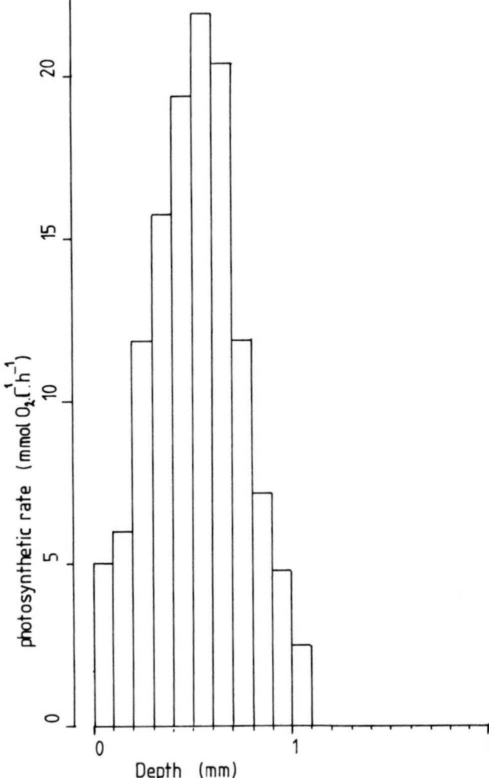

Figure 5. Photosynthetic rate in a well-developed *K. flaccidum*-dominated crust, measured under water-saturated conditions. Porosity was estimated to be 62%. The temperature was 20°C, and the light intensity at the surface was 400 microeinsteins · $m^{-2} \cdot s^{-1}$.

next is gradual. From the model, the crust system represents one compact structure, tightly bound together by the organisms and their residues.

PHOTOSYNTHETIC CAPACITY

For technical reasons it was not possible to measure oxygen evolution under non-water-saturated conditions. Under water-saturated conditions, the maximum net oxygen production was always recorded at depths of between 0.3 and 0.5 mm. Below this depth the profile shows a sharp decrease in oxygen production. No net production was detected below 1.2 mm. The maximum oxygen production in the crust was 130 mmol of $O_2 \cdot m^{-2} \cdot h^{-1}$. The net production is calculated as millimoles of oxygen per square meter by taking into account a porosity of 62%.

If the production data (Fig. 5) are compared with the model constructed from the SEM studies of the same crust samples (Fig. 4), it can be seen that the production area is located in the same layer as the one in which the algae are filamentous and spread over the surface of the sand grains. Apparently, structure and function coincide in the algal crust.

DISCUSSION

The pioneer colonizers of the sand surface in the blowouts, the cyanobacteria, have to cope with conditions of moving sand. In sand under these conditions, light intensities decrease exponentially with depth. At least two properties of the filamentous cyanobacteria allow the organisms to survive burial in the sand: (i) phototactic movements and (ii) a higher net growth yield than green algae (because of their low maintenance energy requirements), thus allowing the cyanobacteria to be active in extremely low light intensities (21).

When the colonization by the cyanobacteria is not disturbed by environmental factors such as heavy winds, the sand becomes more tightly bound; the green alga *K. flaccidum* then becomes the main stabilizer of the sand. It is not clear why this organism becomes so dominant. It is thought that the ample supply of nitrogen in the atmosphere promotes its growth. The whole coastal area has a high nitrogen input from wet and dry deposition. Levels of ammonium and nitrate together are estimated to be 3.8 g $\cdot$ m^{-2} $\cdot$ $year^{-1}$ (report from Dune Water Works, The Hague, The Netherlands). This high nitrogen input might also be the reason for the absence of heterocystous cyanobacteria and for the fact that no nitrogenase activity could be detected (unpublished data).

Desert crusts built by *Microcoleus* sp. become thicker, because under favorable conditions, i.e., at water saturation, the filaments are pulsed out of their common sheath and afterwards moved via phototaxis, leaving behind a lot of sheath material and slime. This material is responsible for the increase in thickness of *Microcoleus* crusts in the desert (2). It is hypothesized that although *Klebsormidium* sp. is not capable of moving, the slime excreted by this organism contributes to an increase in the thickness of the dune crust. Sand particles, which are transported by wind, may be glued to the surface by the slime, whereafter the organisms have to grow up again and the process will repeat. This would mean that the algal remnants in the deepest layers shown in Fig. 4 are inactive organisms buried by the moving sand.

Water-saturated conditions seldom occur in situ, since water quickly percolates through the sand. Only production data collected under water-saturated conditions are reliable, since the sediment/gas interphase causes more severe interference with photosynthetic measurements than the gas/water interphase does, owing to a higher degree of oxygen diffusion in gas (see also reference 11). However, it has been shown (1, 10, 18) that physiological activities, in particular photosynthesis, are possible at lowered matrix water potentials. This means that algal crust development still proceeds under non-water-saturated conditions. Continued colonization with algae and cyanobactria improves water availability (4), which favors longer periods of activity, and this, in turn, triggers colonization by moss protonema and higher plants. Colonization of the sand results in an increase in surface resistance and therefore results in less wind erosion. By decreasing wind erosion, the crusts may contribute to stabilization of barren soils.

Further studies, with emphasis on the mechanism of crust formation and the activities of the organisms at a lowered matrix water potential, are in progress.

Acknowledgments. This work was supported by grant 77-112 from the Dutch Organization for Pure Research.

We thank Bert van Groen for skillful help in building the oxygen microelectrodes.

LITERATURE CITED

1. **Brock, T. D.** 1975. Effect of water potential on a *Microcoleus* (Cyanophyceae) from a desert crust. *J. Phycol.* 11:316–320.
2. **Campbell, S. E.** 1979. Soil stabilization by a prokaryotic desert crust: implications for precambrian land biota. *Origins Life* 9:335–348.

3. **Dulieu, D., A. Gaston, and J. Darley.** 1977. La dégradation des pâturages de la region N'Djamena (Republique du Tchad) en relation avec la présence de cyanophycées psamnophiles. *Rev. Elev. Med. Vet. Pays Trop.* **30:**181–190.
4. **Forster, S. M., and T. H. Nicolson.** 1981. Microbial aggregation in a maritime dune succession. *Soil Biol. Biochem.* **13:**205–208.
5. **Graebner, P.** 1910. Pflanzen leben auf den Dünen, p. 183–296. *In* F. Solger, P. Graebner, J. Thienemann, P. Speiser, and F. W. O. Schulze (ed.), *Dünenbuch.* Enke Verlag, Stuttgart, Federal Republic of Germany.
6. **Jelgersma, J. S., J. de Jong, W. H. Zagwijn, and J. F. van Regteren Altena.** 1970. The coastal dunes of the western Netherlands; geology, vegetation history and archeology. *Meded. Rijks Geol. Dienst* **21:**93–167.
7. **Jungerius, P. D., A. T. J. Verheggen, and J. Wiggers.** 1981. The development of blow-outs in "De Blink" a coastal dune area near Noordwijkerhout, the Netherlands. *Earth Surf. Proc. Landf.* **6:**375–396.
8. **Klubek, B., and J. Skujins.** 1981. Heterotrophic N_2-fixation in arid soil crusts. *Soil Biol. Biochem.* **12:**229–236.
9. **MacEntee, F. J., and H. C. Bold.** 1978. Some microalgae from sand. *Tex. J. Sci.* **130:**167–173.
10. **Potts, M., and E. I. Friedman.** 1981. Effect of water stress on cryptoendolithic cyanobacteria from hot desert rocks. *Arch. Microbiol.* **130:**267–271.
11. **Revsbech, N. P., and B. B. Jørgensen.** 1983. Photosynthesis of benthic microflora measured with high spatial resolution by the oxygen microprofile method: capabilities and limitations of the method. *Limnol. Oceanogr.* **28:**749–756.
12. **Revsbech, N. P., and D. M. Ward.** 1983. Oxygen microelectrode that is insensitive to medium chemical composition: use in an acid microbial mat dominated by *Cyanidium caldarium. Appl. Environ. Microbiol.* **45:**755–759.
13. **Reynaud, P. A., and P. A. Roger.** 1981. Variation saisonnières de la flore algale et de l'activité fixatrice d'azote dans un sol engorgé de bas de dune. *Rev. Ecol. Biol. Sol* **18:**9–27.
14. **Rippka, R., J. Deruelles, J. B. Waterbury, M. Herdman, and R. Y. Stanier.** 1979. Generic assignments, strain histories and properties of pure cultures and properties of cyanobacteria. *J. Gen. Microbiol.* **111:**1–61.
15. **Robins, R. J., D. O. Hall, D. J. Shi, R. J. Turner, and M. J. C. Rhodes.** 1986. Mucilage acts to adhere cyanobacteria and cultured plant cells to biological and inert surfaces. *FEMS Microbiol. Lett.* **34:**155–160.
16. **Roger, P. A., and P. A. Reynaud.** 1982. Free-living blue-green algae in tropical soils, p. 147–168. *In* Y. R. Dommergues and H. G. Diem (ed.), *Microbiology of Tropical Soils and Plant Productivity.* Martinus Nijhoff/Dr. W. Junk, The Hague, The Netherlands.
17. **Shields, L. M., C. Mitchell, and F. Drouet.** 1957. Alga- and lichen-stabilized surface crusts as soil nitrogen sources. *Am. J. Bot.* 44:489–497.
18. **Smith, D. W., and T. D. Brock.** 1973. The water relations of the alga Cyanidium caldarium in soil. *J. Gen. Microbiol.* **79:**219–231.
19. **Stewart, W. D. P., G. P. Fitzgerald, and R. H. Burris.** 1968. Acetylene reduction by nitrogen-fixing blue-green algae. *Arch. Microbiol.* **62:**336–348.
20. **van den Ancker, J. A. M., P. D. Jungerius, and L. R. Mur.** 1985. The role of algae in the stabilization of coastal dune blow-outs. *Earth Surf. Proc. Landf.* **10:**189–192.
21. **Van Liere, L., and L. R. Mur.** 1979. Growth kinetics of Oscillatoria agardhii in continuous culture, limited in its growth by the light energy supply. *J. Gen. Microbiol.* **115:**153–160.

Chapter 8

Total Alkalinity in Marine-Derived Brines and Pore Waters Associated with Microbial Mats

Boaz Lazar, Barbara Javor, and Jonathan Erez

TOTAL ALKALINITY AS A TRACER OF MICROBIAL ACTIVITY

Total alkalinity (A_T) is defined as the acid-neutralizing capacity (20) of a solution. A_T is defined as the number of equivalents of a strong acid needed to reach the titration equivalence point (20). The equivalence point of H_2CO_3 is commonly used for natural waters.

According to this definition, an aqueous solution containing only pure weak acids and salts which have no common anions with the acids has zero alkalinity. To explain this, consider an aqueous solution which consists of NaCl, K_2SO_4, and CO_2. The charge balance in the solution requires the following (where brackets denote concentrations):

$$[Na^+] + [K^+] + [H^+] = [Cl^-] + 2[SO_4^{2-}] + [HCO_3^-] + 2[CO_3^{2-}] + [OH^-] \quad (1)$$

Boaz Lazar and Barbara Javor • H. Steinitz Marine Biological Laboratory, The Hebrew University of Jerusalem, P.O. Box 469, Eilat 88103, Israel. ***Jonathan Erez*** • Department of Geology, The Hebrew University of Jerusalem, Givat-Ram, Jerusalem 91904, Israel.

By rearranging equation 1 to have all the species of weak acids on the right side of the equation, we arrive at the following expression for A_T:

$$[Na^+] + [K^+] - [Cl^-] - 2[SO_4^{2-}] = [HCO_3^-] + 2[CO_3^{2-}] + [OH^-] - [H^+] = A_T \quad (2)$$

The right part of equation 2 is the total alkalinity (acid-neutralizing capacity) as defined above. In this case $A_T = 0$, as proved by inspecting the left part of equation 2.

Conservation of A_T is its most important characteristic. This characteristic is shown by inspecting the left part of equation 2 for the case $A_T \neq 0$ (true for most natural waters). This part stays constant if normalized to the water concentration of the solution (moles of H_2O per kilogram of solution). Thus, changes in A_T serve as tracers for nonconservative processes in the solution (see examples below).

A_T has a typical value of ca. 2.5 $meq \cdot kg^{-1}$ in normal oxic marine waters. The major components of A_T in normal marine waters are the anions, i.e., bicarbonate, carbonate, and borate. Minor contributions from ions such as fluoride, phosphate, silicate, and other trace ions exist. There are special aquatic environments in which inorganic and biochemical processes may completely change the A_T. An important example of such environments is that of sediment pore waters in which numerous microbially catalyzed redox reactions exist. In normal marine sediments, the alkalinity of pore waters is generally higher than that of the overlying seawater (3, 4, 14). This is because seawater contains a high concentration of sulfate, which is susceptible to microbial reduction (see the explanation for reaction 5 below).

The most common biologically mediated reactions which involve changes in the weak acids and bases of the solution are presented below.

(i) The first reaction is oxygenic CO_2 photoassimilation. This process can be expressed schematically as follows:

$$CO_2 + H_2O \rightarrow CH_2O + O_2 \qquad (3)$$

For this reaction $*A_T = 0$, where $*A_T = [A_T\text{ (products)} - A_T\text{(reactants)}]/\text{(moles of total } CO_2 \text{ produced)}$. The reverse process of reaction 3 describes aerobic respiration, which also results in $*A_T = 0$.

(ii) The second reaction is nitrification, which may be written schematically as follows:

$$2CO_2 + NH_4^+ + H_2O \rightarrow 2CH_2O + NO_3^- + 2H^+$$

It has $*A_T = (-1/-2) = 1/2$.

(iii) The third reaction is denitrification, which is expressed schematically as follows:

$$5CH_2O + 4NO_3^- + 4H^+ \rightarrow 7H_2O + 2N_2 + 5CO_2$$

It has $*A_T = 4/5$. This means that for every 5 mol of total carbon produced by denitrification, the total alkalinity increases by 4 eq.

(iv) The fourth reaction is anoxygenic CO_2 photoassimilation with H_2S as an electron donor and sulfate reduction. If CO_2 photoassimilation produces only elemental sulfur, it is expressed by the following reaction:

$$CO_2 + 2H_2S \rightarrow CH_2O + H_2O + 2S^0 \qquad (4)$$

It has $*A_T = 0$. If the reaction proceeds all the way to sulfate, it takes the following form:

$$2CO_2 + H_2S + 2OH^- \rightarrow 2CH_2O + SO_4^{2-} \qquad (5)$$

It has $*A_T = (-2/-2) = 1$. The reverse reaction is sulfate reduction, which also has $*A_T = 2/2 = 1$.

(v) The fifth reaction is methane production and oxidation. Reduction of carbohydrate to methane is expressed as follows:

$$2CH_2O \rightarrow CH_4 + CO_2$$

It has $*A_T = 0$. Oxidation by the reverse reaction results in the same $*A_T$. Methane oxidation with sulfate (6) is expressed as follows:

$$CH_4 + SO_4^{2-} \rightarrow CO_2 + 2H_2O + S^{2-}$$

It results in $*A_T = 2$.

(vi) The sixth reaction is a fermentation, expressed as follows:

$$3CH_2O + H_2O \rightarrow 2CH_3OH + CO_2$$

It results in $*A_T = 0$. Other fermentation reactions involving the conversion of carbohydrate into weak organic acids, such as acetic acid and lactic acid, also result in $*A_T = 0$.

(vii) The seventh reaction is that of precipitation and dissolution. Precipitation of bases such as calcite, aragonite, magnesite, pyrite, and siderite reduces the A_T. For example, the precipitation of aragonite, which is expressed as follows:

$$Ca^{2+} + CO_3^{2-} \rightarrow CaCO_3 \quad (6)$$

results in $^*A_T = (-2/-1) = 2$. When the dissolution involves oxidation, the A_T may drop markedly, such as for pyrite oxidation (20):

$$4FeS_2(s) + 15O_2 + 14H_2O \rightarrow 4Fe(OH)_3(s) + 8SO_4^{2-} + 16H^+$$

which results in a drop in A_T of 4 eq · mol of pyrite oxidized^{-1}.

(viii) The eighth reaction is oxidation with oxygen. Oxidation of dissolved sulfide with oxygen may proceed nonbiologically or biologically (18), according to the following reaction:

$$2H_2S + O_2 \rightarrow 2H_2O + 2S^0 \quad (7)$$

It results in no change in A_T. If the reaction proceeds to sulfate, as follows:

$$S^{2-} + 2O_2 \rightarrow SO_4^{2-} \quad (8)$$

then A_T decreases by 2 eq · mol of sulfide oxidized^{-1}.

The few basic examples above demonstrate the usefulness of A_T as a tracer for biologically mediated processes in natural waters.

Organic acids are not thought to play a major role in controlling the pH of marine pore waters (3). Acetic acid rarely (15) and amino acids never (9) reach millimolar concentrations in coastal marine sediment pore waters.

It was demonstrated (4) that the total alkalinity in fjord sediments increased from 2.2 meq · liter^{-1} in surface waters to 76.5 meq · liter^{-1} in sediment at a depth of 62 to 68 cm. Although sulfate reduction could account for the increase in A_T until sulfate was depleted at a depth of about 27 cm, ammonia accumulation (up to 11.4 mM) could account for the continued alkalinity increase with depth.

In hypersaline marine-derived brines, A_T is influenced by the effects of $CaCO_3$ precipitation (equation 6) and solubility (11). High concentrations of sulfate, particulate organic matter, and dissolved organic matter in hypersaline brines associated with microbial mats should provide a large reservoir of substrates to drive diverse reactions (see above) capable of changing drastically the A_T of the pore waters. The alkalinity changes may trigger other processes in the interstitial water. For example, an elevation of pore water alkalinity could cause local precipitation of various carbonate minerals.

SAMPLING AND ANALYTICAL CONSIDERATION

The purpose of this study was to demonstrate that A_T is a useful additional tracer for redox reactions which involve the consumption and/or production of a weak acid or base species. To do this, we evaluated the relationship between A_T, salinity, and sulfide concentration in marine-derived brines collected from the sediments and the evaporation pans of the Israel Salt Co., Eilat.

Most of the sampling for this work was done in pan 203. The A_T of brines and pore waters (where noted) was determined by titration with a two-electrode system and a Radiometer model pHM 84 pH meter. For samples with sulfide, an additional KCl bridge with a ceramic frit was inserted on the reference electrode to prevent any sulfide precipitation at the electrode. Surface brines were filtered through a Whatman GF/C filter before titration. C_T (total dissolved carbon) was measured manometrically on a vacuum line during extraction of the CO_2 for $\delta^{13}C$. The $\delta^{13}C$ was determined with a VG-602 mass spectrometer. Br^- was measured with an ion chromatograph (Dionex 2010i).

Sediments were collected from the salt pans in plastic corers (diameter, 8.5 cm). They were then extruded and sliced into 2-mm sections under air and rapidly loaded to screw-cap centrifuge tubes; the tubes were filled until no headspace remained. We think that sulfide oxidation during that short period would not drastically affect our results, because the half time for sulfide removal from O_2-saturated seawater is on the order of a few hours (20). Pore waters were collected by centrifuging the sediments at 10,000 × *g* for 15 or 20 min. For sulfide measurements, samples of the pore waters (5 to 200 μl) were immediately precipitated with zinc acetate (200 μl of 2% [wt/wt] zinc acetate, 200 μl of 0.5 N NaOH) and analyzed by the methylene blue method of Gilboa-Garber (7). The total alkalinity in 5-ml samples of the sulfide-containing pore waters was determined within several hours of pore water extraction. During that time the samples were kept in closed vials filled with N_2. Salinity was measured with an Atago-28 refractometer.

To measure the possible alkalinity change due to sulfide oxidation by oxygen (reaction 8), microbial mats and underlying sediments were collected in 24-mm plastic corers, sliced, and centrifuged in air. In this case, pore water sulfide was allowed to oxidize in air. All supernatants from the same depth were combined to a total volume of >2 ml. An alkalinity titration was conducted on 2-ml samples after further centrifugation to remove suspended elemental sulfur.

A Winogradsky column experiment was performed to drive the redox reactions to maximum. Four black-walled columns (diameter, 10 cm; height, 40 cm), made of polyvinyl chloride, were preconditioned in seawater for 2 weeks before use to wash out any possible contaminant. The columns were filled to a depth of about 30 cm with a mixture containing the top 5 cm of microbial mats (1 part) and underlying sediments from pan 202 (two columns) or pan 203 (two columns) (3 parts). The initial salinities for pan 202 and 203 sediments were ca. 130 and 150‰, respectively. In one column of each salinity, ca. 50 g of coral aragonite powder (particle size, <60 μm) was added to ensure that the pore waters would not reach supersaturation with respect to $CaCO_3$ during the experiment. This was done to provide a control on the alkalinity changes.

Sediments were covered with brines from the salt pans, and the columns were covered with clear glass plates. The columns were incubated outdoors for 6 months (March to August 1987) in a running-seawater bath (21 to 26°C). If the brine overlying the sediments evaporated, it was replaced by fresh brine of approximately the same salinity. Pore waters were collected and processed in completely filled screw-cap centrifuge tubes as described above.

A single light and dark bell jar experiment was conducted on 2 March 1987 in pan 203 (144‰ salinity), where the sediments are covered by a microbial mat dominated by *Aphanothece halophytica*. The two bell jars were made from Plexiglas tubing (diameter, 20 cm). Two special sampling ports equipped with a luer-type adapter were placed at the top of each bell jar. The bell jars were carefully inserted into the sediment, trapping ca. 2 liters of brine. After 4.3 h, brines were collected through the sampling ports. The brines were analyzed for oxygen by using a modified Winkler method (19). The analyses of total CO_2 and $\delta^{13}C$ are described above.

$\delta^{13}C$, ALKALINITY, AND SULFIDE IN BRINES AND PORE WATERS

As surface brines increase in salinity or degree of evaporation (DE; defined as the molal ratio of Br^- in the brine to Br^- in mean ocean water), the total alkalinity increases in a predictable manner (Fig. 1). Below a DE of ca. 6, A_T behaves nonconservatively owing to $CaCO_3$ precipitation (11). The $\delta^{13}C_T$ (the $\delta^{13}C$ of the total CO_2) in the evaporation pans decreases with increasing salinity (Fig. 2) until the point of massive gypsum precipitation (about 200‰ salinity). This minimum in $\delta^{13}C_T$ coincides with the

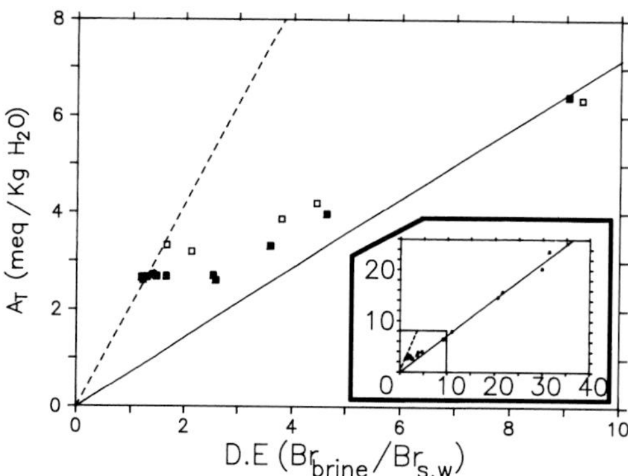

Figure 1. Total alkalinity as a function of DE during the evaporation of seawater in the solar salt pans of Israel Salt Co. Symbols: ■, August, 1986; □, December, 1986. The inset shows the general behavior (up to DE of ca. 35), and the blown-up portion is marked with a line. -----, Conservative slopes for total alkalinity; ——, conservative behavior of A_T after depletion due to $CaCO_3$ precipitation. Note that for DE < 6 the alkalinity in December is always higher than in August. This reflects the retrograde solubility of $CaCO_3$ as a function of temperature (the average temperature in August and December was 30 and 20°C, respectively).

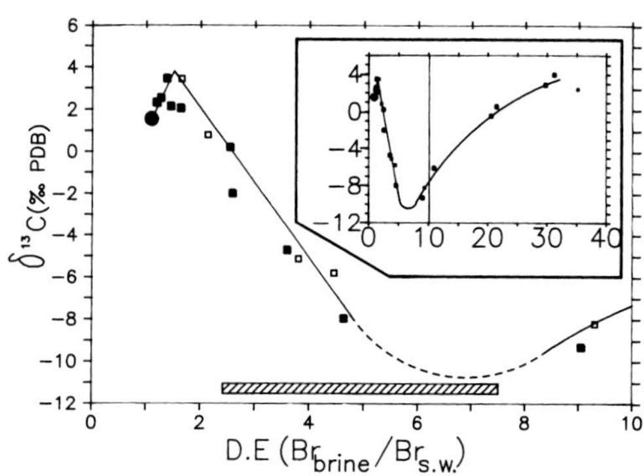

Figure 2. $\delta^{13}C$ of total CO_2 as a function of DE during the evaporation of seawater. The inset shows the general behavior (up to DE of ca. 35), and the blown-up portion is marked with a line. The symbols are as for Fig. 1; ●, open-sea water. The decrease of $\delta^{13}C$ and its minimum (-----) coincide with algal mats covering the bottom of the pans (▨). We suggest that the bulk of this depletion is caused by kinetic fractionation during CO_2 invasion from the atmosphere.

maximum salinities permitting microbial mat development.

The results of the bell jar experiment show that the gross CO_2 demand for primary production was 0.47 μmol · cm^{-2} · h^{-1} and that the level of CO_2 added to the dark bell jar was 0.27 μmol · cm^{-2} · h^{-1}. This translates into a net CO_2 transport from the sediments of 0 to 0.8 μmol · cm^{-2} · day^{-1} (depending on the average light/dark time ratio in Eilat). It means that the sediment either is balanced or reflects slight net degradation of organic matter. Gross photosynthetic O_2 evolution was 0.18 μmol · cm^{-2} · h^{-1}. The differences between the O_2 and CO_2 data suggest that anoxygenic photosynthesis accounts for some of the CO_2 removal.

The A_T and total soluble sulfide (S_T) of different layers in cores taken from *Aphanothece* mats and associated sediments are presented in Fig. 3. The data show that there is a large peak in A_T just below the 4-mm-deep photosynthetic zone. According to reaction 8, it is possible to estimate the maximum contribution of sulfide to A_T throughout the section. The potential role of sulfide in A_T is relatively small, and the addition of all the equivalent amount of sulfide to the alkalinity of the oxidized samples will not change the shape of that curve drastically (Fig. 3).

Analyses of A_T and S_T of pore waters from cores cut on a millimeter scale (Fig. 3)

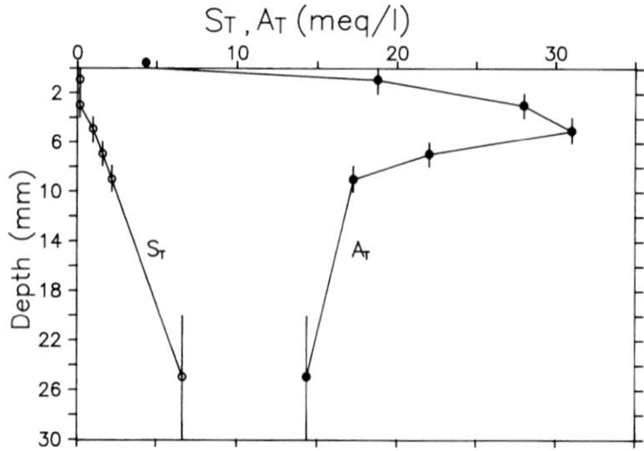

Figure 3. Total alkalinity and total sulfide (both expressed in milliequivalents per liter) profiles from an *Aphanothece* mat in the solar salt pans of the Israel Salt Co. Symbols: ●, A_T of oxidized pore waters; ○, S_T (equivalent units) in freshly collected pore waters with no oxidation.

Table 1.
Salinity, Sulfide, and Total Alkalinity in Pan 203 and Winogradsky Column Sediments

Sample	Sediment source	Depth (cm)	Salinity (‰)	Sulfide concn (mM)	A_T (meg · liter^{-1})
1	Pan 203	Surface	176	0	4.26
2	Pan 203	0–1	169	1.78	5.96
3	Pan 203	1–2	158	2.21	9.26
4	Pan 203	2–3	154	3.27	11.5
5	Pan 203	3–4	153	2.09	12.0
6	Pan 203	4–5	153	4.20	12.4
7	Pan 203	5–6	152	3.47	12.6
8	Pan 203	6–7	152	2.03	12.1
Winogrodsky columns					
I-1	Pan 202[a]	0–1	144	8.67	12.2
I-2	Pan 202[a]	4–5	137	28.4	30.3
I-3	Pan 202[a]	9–10	133	76.9	73.0
I-4	Pan 202[a]	19–20	129	126	84.8
II-1	Pan 202[b]	0–1	142	42.9	63.0
II-2	Pan 202[b]	4–5	156	48.5	65.0
II-3	Pan 202[b]	9–10	150	81.6	91.0
II-4	Pan 202[b]	19–20	164	89.8	95.8
III-1	Pan 203[a]	0–1	152	35.1	52.0
III-2	Pan 203[a]	4–5	173	67.2	81.0
III-3	Pan 203[a]	9–10	173	59.6	94.0
III-4	Pan 203[a]	19–20	164	84.9	109
IV-1	Pan 203[b]	0–1	194	45.7	48.4
IV-2	Pan 203[b]	4–5	192	52.0	58.4
IV-3	Pan 203[b]	9–10	185	66.1	83.4
IV-4	Pan 203[b]	19–20	168	65.3	135

[a] Controls; no addition.
[b] Aragonite added.

and from sediments cut on a centimeter scale (Table 1) show relatively small variations in A_T. The analyses from the Winogradsky column experiments (Table 1) show much larger variations and hence give a clue about the potential behavior of A_T versus S_T. The larger variations in the results of the Winogradsky column experiments are most probably due to the rich source of organic matter (introduced by mixing fresh microbial mats with reducing sediments) coupled with the high sulfate content of the brines and the presence of solid sulfate (gypsum). The data show that A_T is independent of DE in the presence of sulfide (Fig. 4 and Table 1). The normalized A_T values deviate systematically

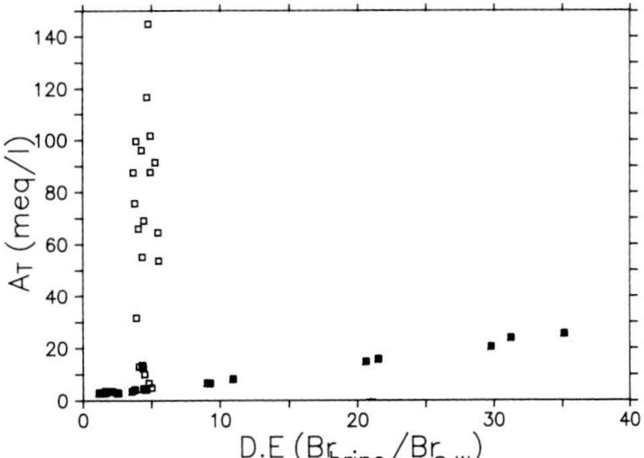

Figure 4. Total alkalinity versus DE of brines from the solar salt pans (■) and pore waters (□) from freshly collected cores and Winogradsky column sediments. Note that there is no correlation between DE and A_T in the interstitial waters.

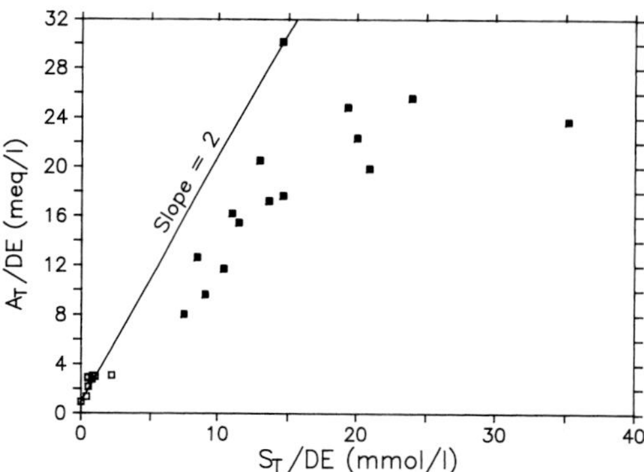

Figure 5. Total alkalinity as a function of total sulfide normalized to the DE of freshly collected cores (□) and Winogradsky column sediments (■). The line indicates the predicted behavior of total alkalinity versus total sulfide during sulfate reduction ($A_T/S_T = 2$). Note that the *y* intercept is the A_T of the overlying brines.

from the predicted slope of 2 meq of A_T generated per mmol of S_T added (Fig. 5) (the reverse of reaction 5). Only when both S_T and A_T values were low (in the fresh sediment core) did the measured values conform to the predicted slope. Once the S_T/DE ratio reached about 20 mmol · liter^{-1}, there was no further increase in A_T with increases in S_T. Only one set of values from the Winogradsky column sediment experiments fell on the predicted curve.

X-ray diffraction analyses of the composition of pan 203 sediments showed the presence of calcite, high-magnesium calcite, and gypsum, along with quartz, small amounts of dolomite, and unidentified minor minerals. The calcite-to-quartz ratio showed a significant increase with depth in the columns for samples III and IV.

ALKALINITY DEPLETION VERSUS $\delta^{13}C$ DEPLETION IN THE BRINES

$CaCO_3$ precipitation depletes the $\delta^{13}C_T$ of the solution, making it ca. 1.8‰ lighter than the solid precipitated $CaCO_3$ (13). On this basis, we have used the alkalinity depletion curve (Fig. 1) to estimate the $\delta^{13}C$ depletion due to $CaCO_3$ precipitation (11) in the evaporation pans. The calculated depletion is only ca. 4‰, which is only a small portion of the observed depletion (Fig. 2). Thermodynamic calculations based on the carbonate system data (8, 11) show that up to a DE of ca. 3.5, the brines are depleted up to 20% in C_T with respect to equilibrium with atmospheric CO_2. This depletion indicates that the brine is in a state of dynamic balance, rather than in a state of thermodynamic equilibrium. This calculation, as well as the results from the bell jar experiment (which show potentially high CO_2 demand by the mats), suggests that the average rate of CO_2 photoassimilation (at least up to a DE of ca. 3.5) is higher than the rate of CO_2 gain by organic-matter degradation. Thus, it is unlikely that the major source of isotopically light C_T in the brines overlying the microbial mats is isotopically light CO_2 from sulfate reduction in the sediments (see the reverse of reaction 5). This is because sulfate reduction would require a C_T excess rather than deficit. The C_T deficit implies that CO_2 should continuously invade the brines from the atmosphere. Herczeg and Fairbanks (10) found that the CO_2 demand by a bloom of cyanobacteria in a soft-water lake resulted in a C_T which is ca. 13‰ isotopically lighter than atmospheric CO_2. Measurements of the isotopic composition of CO_2 absorbed by a hydroxide solution showed an isotopic depletion of 15‰ relative to the isotopic composition of CO_2 gas (2, 5). Such kinetic fractionation could explain the depletion of the carbon isotopic composition observed in the brines. A small part of the light isotopic signal is due to $CaCO_3$ precipitation (13), sulfate reduction in the sediments, and methane oxidation in the oxic zone (16). Gas bubbles from nearby sediments in the same pan contained 40% methane and 60% N_2 (A. Oren, personal communication).

A_T IN PORE WATERS: EVIDENCE FOR REDOX REACTIONS

Figure 1 suggests that the brines are depleted in A_T up to 60% relative to seawater. This is due to aragonite precipitation (11). Pore waters from *Aphanothece* mats cut into 2-mm slices show completely different A_T behavior. The profile has an alkalinity maximum of ca. 30 meq · liter^{-1} at a depth of about 5 mm (Fig. 3), which is ca. sixfold the A_T in the overlying brine. If the maximum possible sulfide contribution to alkalinity (Fig. 3, ○) were added to the alkalinity measured (Fig. 3, ●), the shape of the profile would remain basically the same. The very high A_T values in and just below the photosynthetic zone in the presence of relatively low sulfide concentrations suggest that components other than sulfide (such as ammonia [4]) may be important in generating alkalinity in the organic-rich matrix of the microbial mats near the sediment/brine interface. Alternatively, any process that removes sulfide without affecting the alkalinity might explain the data. Such processes include upward diffusion of H_2S, anoxygenic photosynthesis (reaction 4), and nonbiological or biological oxidation with oxygen (reaction 7). Upward diffusion of H_2S is likely to occur in sulfate reduction environments because the pK_1 of H_2S is ca. 1 unit higher than the pK_1 of H_2CO_3 (20). This implies that even at relatively high pH (where the major carbon species is bicarbonate), the major sulfide species is still H_2S.

Sediments sliced on a centimeter scale lose the fine A_T structure of the upper part (Table 1; Fig. 5, □), but their composition follows a general pattern of sulfate reduction (the reverse of reaction 5); i.e., these samples follow a line with an A_T/S_T ratio of 2 (Fig. 5). In the Winogradsky column experiment (in which the interstitial reactions reached the maximum potential extent), A_T/S_T deviates from the predicted value of 2 and reaches zero with increasing sulfide concentrations (Fig. 5). Thus, once S_T/DE values reached and exceeded ca. 20 mmol · liter^{-1} (corresponding to sulfide concentrations of >80 mM), no further increase in A_T was detected. Reduction of such large quantities of sulfate (comparable to the total sulfate content of original seawater) from a gypsum-saturated solution requires the dissolution of solid gypsum. The dissolution maintains the state of saturation of the interstitial solution with respect to solid sulfate. No estimates of rates of reduction or dissolved-sulfate pool sizes were made, because much of the reduced sulfate was probably bound as FeS, FeS_2, and polysulfides, and the exact amount of gypsum in the sediments is unknown. The opposite behavior was found in fjord sediments (where solid gypsum does not exist), in which A_T increased from the predicted A_T versus S_T values (based on sulfate concentrations) even after sulfate was completely used (4). In that case, the A_T increase was caused by ammonia generation (4).

MICROBIAL MATS AS A POTENTIAL DOLOMITE-PRECIPITATING ENVIRONMENT

In the interstitial solutions of the Winogradsky columns, solid gypsum serves as a buffer for alkalinity increase. Dissolution of gypsum releases Ca^{2+}, which (in a $CaCO_3$-saturated solution) causes calcium carbonate precipitation:

$$2CH_2O + CaSO_4 \cdot 2H_2O \rightarrow CaCO_3 + H_2S + CO_2 + 3H_2O \quad (9)$$

Such reaction results in $*A_T = 0/1 = 0$. This means that $CaCO_3$ precipitation could account for the increase in S_T without any A_T change (Fig. 5).

Such brines, containing very high magnesium concentrations, should be a potential environment for the deposition of dolomite. A reaction similar to reaction 9 will precipitate dolomite:

$$4CH_2O + 2CaSO_4 \cdot 2H_2O + Mg^{2+} \rightarrow CaMg(CO_3)_2 + 2H_2S + 2CO_2 + 4H_2O + Ca^{2+}$$

This reaction would lower the A_T/S_T ratio in the same manner as in reaction 9.

The presence of protodolomite in sediments of similar microbial mats in environments such as Solar Lake was demonstrated (1). In the geological record, laminae of bituminous dolomite are fairly common features in gypsum or anhydrite evaporite formations (17). Such carbonate precipitation is believed to have been triggered by a period of brine freshening accompanied by the influx of organic matter. Similarly, Magaritz (12) found bituminous dolomitic laminae in gypsum beds of the Mohilla Formation with light $\delta^{13}C$ values in the carbonates (down to $-14‰$). The precipitation of light carbonate was attributed to the influx of fresh continental water floating on the dense body of brine. The data from the present investigation suggest that brine freshening is not a prerequisite for bituminous dolomite precipitation in brines saturated with respect to gypsum. The in situ generation of organic matter followed by its isolation and degradation by sulfate reducers within the sediment would produce such laminae of bituminous carbonates. The negative carbon isotopic composition is explained by the light initial composition of the overlying brine (Fig. 2). This is especially applicable to shallow hypersaline environments, where mixing of less salty brine with brine from the gypsum facies should result in $CaCO_3$ precipitation (Fig. 1).

Even when our Winogradsky columns reached a maximum A_T (Fig. 5), significant additional dolomite did not precipitate in 6 months of incubation. X-ray diffractions suggest that calcite probably did precipitate. However, the high Mg-calcite ($Ca_{\sim 0.88}$, $Mg_{\sim 0.12}$ CO_3) level found in pan 203 mat sediments and the protodolomite detected in the hypersaline sediments of Solar Lake (1) strongly support the mechanism suggested above. Further experimentation with hypersaline microbial mat sediments under a variety of conditions would demonstrate whether primary dolomitization can occur in situ in the mats.

Acknowledgments. We thank E. Gilboa and A. Ravizky, Israel Salt Co., for their assistance and Y. Cohen and D. Des Marais for critical reviews and useful discussions.

LITERATURE CITED

1. **Aharon, P., Y. Kolodny, and E. Sass.** 1977. Recent hot brine dolomitization in the Solar Lake, Gulf of Elat: isotopic, chemical and mineralogical study. *J. Geol.* **85**:27–48.
2. **Baertschi, P.** 1952. Die Fraktionierung der Hohlenstoff-isotope bei der Absorption von Kohlendioxid. *Helv. Chim. Acta* **35**:1030–1036.
3. **Ben-Yaakov, S.** 1973. pH buffering of porewater of recent anoxic marine sediments. *Limnol. Oceanogr.* **18**:86–94.
4. **Berner, B. J., and I. R. Kaplan.** 1968. Carbonate alkalinity in pore waters of anoxic marine sediments. *Limnol. Oceanogr.* **13**:544–549.
5. **Craig, H.** 1953. The geochemistry of the stable carbon isotopes. *Geochim. Cosmochim. Acta* **3**:53–92.
6. **Fenchel, T., and T. H. Blackburn.** 1979. *Bacteria and Mineral Cycling.* Academic Press, Inc., New York.
7. **Gilboa-Garber, N.** 1971. Direct spectrophotometer determination of inorganic sulfide in biological materials and other complex mixtures. *Anal. Biochem.* **43**:129–133.
8. **Harvie, C. E., N. Moller, and J. H. Weare.** 1984. The prediction of mineral solubilities in natural waters: the Na-K-Mg-Ca-H-Cl-SO_4-OH-HCO_3-CO_3-CO_2-H_2O system to high ionic strengths at 25°C. *Geochim. Cosmochim. Acta* **48**:723–751.
9. **Henrichs, S. M., J. W. Farrington, and C. Lee.** 1984. Peru upwelling region sediments near 15°S. 2. Dissolved free and total hydrolyzable amino acids. *Limnol. Oceanogr.* **29**:20–34.
10. **Herczeg, A. L., and R. G. Fairbanks.** 1987. Anomalous carbon isotope fractionation between atmospheric CO_2 and dissolved inorganic carbon induced by intense photosynthesis. *Geochim. Cosmochim. Acta* **51**:895–899.
11. **Lazar, B., A. Starinsky, A. Katz, E. Sass, and S. Ben-Yaakov.** 1983. The carbonate system in hy-

persaline solutions: alkalinity and $CaCO_3$ solubility of evaporated seawater. *Limnol. Oceanogr.* **28**:978–986.

12. **Magaritz, M.** 1984. Carbon isotope composition of an Upper Triasssic evaporite section in Israel: evidence for meteoric water influx. *Am. Assoc. Pet. Geol. Bull.* **68**:37.
13. **Mook, W. G., J. C. Bommerson, and W. H. Staverman.** 1974. Carbon isotope fractionation between dissolved bicarbonate and gaseous carbon dioxide. *Earth Planet. Sci. Lett.* **22**:169–176.
14. **Presley, B. J., and I. R. Kaplan.** 1968. Changes in dissolved sulfate, calcium and carbonate from interstitial water of near-shore sediments. *Geochim. Cosmochim. Acta* **32**:1037–1048.
15. **Sansone, F. J., and C. S. Martens.** 1982. Volatile fatty acid cycling in organic-rich marine sediments. *Geochim. Cosmochim. Acta* **46**:1575–1589.
16. **Schidlowski, M.** 1985. Early life and mineral resources. *Nat. Resour.* **21**:127–134.
17. **Sonnenfeld, P.** 1984. *Brines and Evaporites.* Academic Press, Inc., New York.
18. **Stanier, R. Y., E. A. Adelberg, and J. L. Ingraham.** 1976. *The Microbial World*, 4th ed. Prentice-Hall, Inc., Englewood Cliffs, N.J.
19. **Strickland, J. D. H., and T. R. Parsons.** 1972. *A Practical Handbook of Seawater Analysis*, 2nd ed. Bulletin 167. Fisheries Research Board of Canada, Ottawa.
20. **Stumm, W., and J. J. Morgan.** 1981. *Aquatic Chemistry*, 2nd ed. John Wiley & Sons, Inc., New York.

II. Structure and Function of Benthic Microbial Communities

Chapter 9

Comparative Functional Ultrastructure of Two Hypersaline Submerged Cyanobacterial Mats: Guerrero Negro, Baja California Sur, Mexico, and Solar Lake, Sinai, Egypt

Elisa D'Antoni D'Amelio, Yehuda Cohen, and David J. Des Marais

INTRODUCTION

Hypersaline cyanobacterial mats are presently found scattered throughout the world, mostly in remote areas where arid climatic conditions prevail along the coastlines. Two major types of cyanobacterial mats develop in these areas: (i) supralittoral mats, which are regularly exposed, often by daily tidal fluctuations, and are desiccated, and (ii) submerged mats, which, in contrast, may be exposed only seasonally (2, 21). The first type of mat is dominated by either the filamentous cyanobacterium *Lyngbia* sp. or the colonial unicellular cyanobacterium *Entophysalis* sp., whereas the submerged mat is invariably dominated by the cosmopolitan mat-forming cyanobacterium *Microcoleus chthonoplastes* (2). This type of hypersaline cyanobacterial mat was first found in 1770 on the Danish coast of the Wadden Sea (7) and has since been found on every continent.

The depositional environment of submerged cyanobacterial mats allows the accretion of well-laminated microbial communities which bear the record of seasonal changes in the environment. Anaerobic conditions immediately below the surface of these mats allow the preservation of several components of the living microbial community which may serve as meaningful modern analogs to Precambrian stromatolites (1). Most of the functional morphological information about the living microbial communities disappears during the processes of degradation, calcification, and lithification, which transform the living laminated cyano-

Elisa D'Antoni D'Amelio and David J. Des Marais • Life Science Division, National Aeronautics and Space Administration, Ames Research Center, N239-12, Moffett Field, California 94035. **Yehuda Cohen** • H. Steinitz Marine Biology Laboratory, The Hebrew University of Jerusalem, P.O. Box 469, Eilat 88103, Israel.

bacterial mat into aged lithified stromatolites. The study of functional morphology of recent cyanobacterial mats and the environment of deposition of these mats may provide a key to understanding Precambrian microbial life.

The best-studied submerged microbial mats are those of the Persian Gulf (21); Shark Bay, Western Australia (2); Solar Lake, Sinai (2, 18); and Baja California, Mexico (14). In this paper we have chosen to compare the functional ultrastructure of two submerged *M. chthonoplastes*-dominated which are geographically separated by more than 10,000 km and are found in the salt pans in Guerrero Negro (GN), Baja California Sur, Mexico, and in Solar Lake (SL). This comparison demonstrates the universal nature of these microbial communities and emphasizes the differences between the two mats which can be detected only through detailed, comparative, ultrastructural study.

The two sites are at similar geographical longitudes, and the climatic conditions in the two areas are dry and arid. The sites are fed by evaporated seawater of 50 to 70‰ salinity in GN and 41‰ in SL. One major difference between the two study sites is that the salinity of the overlying water in GN is kept at 60 to 95‰, whereas that in SL ranges widely between 45 and 180‰; this results in seasonal precipitation of gypsum in SL (18), which is not found in GN. Another major difference between the two sites is the depth of the overlying water column, which fluctuates from a few centimeters to 70 cm in SL (17), whereas in the GN mat it ranges between 60 and 120 cm.

METHODS FOR ELECTRON MICROSCOPY AND CHEMICAL MICROPROFILE DETERMINATIONS

The mat samples used in this study were collected from Pond 5 in GN in April 1985 and from the Shallow Flat Mat of SL in June 1986. All the samples were collected in the early afternoon (2 p.m.). They were cut in blocks of approximately 0.5 cm in width and 1.0 cm in length and prefixed in situ with 2.5% glutaraldehyde made up in prefiltered pond water. After being postfixed in 0.1 M OsO_4 and dehydrated in a graded alcohol series, the samples were dissected into 0.5-mm pieces from the top down to a depth of 2 mm. Entire samples (2 mm) were cut as well. Each sample was then oriented to show vertical sections, labeled, and embedded in a low-viscosity resin, Quetol 651 (19). Semithin (thickness, 0.25 to 0.5 μm) sections, stained with toluidine blue, were used for the light microscopy study. Ultrathin (thickness, 60 to 90 nm) sections, stained with barium salts (13) or uranyl acetate-lead citrate (24), were used for the transmission electron microscopy (TEM) study. The location and distribution of the main organisms were found by light microscopy. The corresponding ultrathin section was subsequently examined under TEM.

Chemical microprofiles of dissolved oxygen, sulfide, and pH, as well as microprofiles of oxygenic photosynthesis, were determined for the two cyanobacterial mats on samples collected from the same area as those used for electron microscopy, and the chemical analyses were carried out on the same days that sampling for microscopy was done. The sensing tips of the microelectrodes used in this study were 4 μm for O_2, 120 μm for sulfide, and 60 μm for the pH microelectrode. Measurements were carried out near the sampling site by using a light source providing "white" light of 1,000 microeinsteins m^{-2} s^{-1} at the sediment surface, and analyses were carried out as previously described (16, 17, 23).

SPECIES COMPOSITION AND MICROBIAL ZONATION

Generally, both the species composition and microbial zonation in both mats are very similar (Fig. 1). They are both dominated by the filamentous forms of *M. chthonoplastes*, and both communities are embedded in a polysaccharide matrix excreted by this organism and other members of the benthic community. In general, the SL mat is more dense than the GN microbial community and zonation in the SL mat is more defined, resulting in a more distinct lamination of the SL mat.

Several morphological species found in the GN mat are absent in the SL sample. Most notable is the occurrence of two types of the major mat-building cyanobacterium in the GN sample, namely, the "big" and the "small" *M. chthonoplastes* (defined below), with the small form deeper in the mat. The big *M. chthonoplastes* is absent in the SL mat. Of the chemosynthetic bacteria, both *Beggiatoa alba* and *Thioploca nigra* were found only in the GN mat. The cyanobacteria *Oscillatoria limnetica*, *Oscillatoria salina*, and *Spirulina* sp. were all previously described for the SL mat (18), but were not found in the Shallow Flat Mat analyzed here. The unicellular cyanobacterium *Synechococcus* sp. is found in both mats; it is restricted to the upper 0.5 mm in the SL mat, whereas in the GN sample it can be found to at least 1.5 mm.

The similarities between the mats of GN and those of SL are expressed in virtually identical chemical microprofiles of oxygen, sulfide, pH, and oxygenic photosynthesis (Fig. 2). Figure 2 presents average values measured in triplicate sediment cores from the two sites of study.

In both mats the photic zone was restricted to the upper 800 μm of the mat, and oxygenic photosynthesis could be detected down to 600 μm when the mat was exposed to white light of 1,000 microeinsteins m^{-2} s^{-1}. The chemocline, as defined by the microzone where oxygen and sulfide overlap, was found under experimental illumination conditions at 600 to 1,000 μm. Exponentially increasing levels of sulfide, generated by the high rate of microbial sulfate reduction in the mats, were measured below the chemocline (15).

THE GN MAT

The GN mat surface was flat to slightly undulating. A relatively small community of diatoms (*Nitzschia* and *Navicula* spp.) appeared at the top (Fig. 3). Filamentous cyanobacteria, namely, *Oscillatoria* spp. and *Spirulina labyrynthiformis* (9, 25), as well as colonies of unicellular cyanobacteria, most of them *Synechococcus* sp., were observed in the upper 300 μm of the photic zone (Fig. 3 to 5). *Synechococcus* sp. was a conspicuous component of the mat. The cells of this cyanobacterium measured about 3 μm and had groups of lamellae, usually near the periphery, a few carboxysomes, and several lacunalike inner spaces. These spaces were partially filled with deposits of medium to high electron density (Fig. 5). *Synechococcus* sp. was represented by groups of several heavily ensheathed cells that appeared with no defined pattern of distribution throughout and below the entire photic zone (see Fig. 23 and 24). Clusters of small, rectangular, unicellular organisms which measured 0.75 by 1.3 μm were observed between the top and a depth of 300 μm. These organisms, located in the vicinity of mucopolysaccharide sheaths, each had a seemingly rigid cell wall, a dense cytoplasm with a wavy rim, and one to three lighter cytoplasmic spots that showed stacks of up to 25 faintly stained lamellae (Fig. 6). The cosmopolitan filamentous cyanobacterium *M. chthonoplastes*, the dominant component of the mat, grew mainly in bundles of variable numbers of trichomes enveloped in a common mucopolysaccharide sheath. *M. chthonoplastes* started to form layers at a depth of 50 μm and was found throughout the entire photic zone and below it to about 1,200 μm. The region of its highest frequency was located in a microlayer that extended between 300 and 700 μm from the surface.

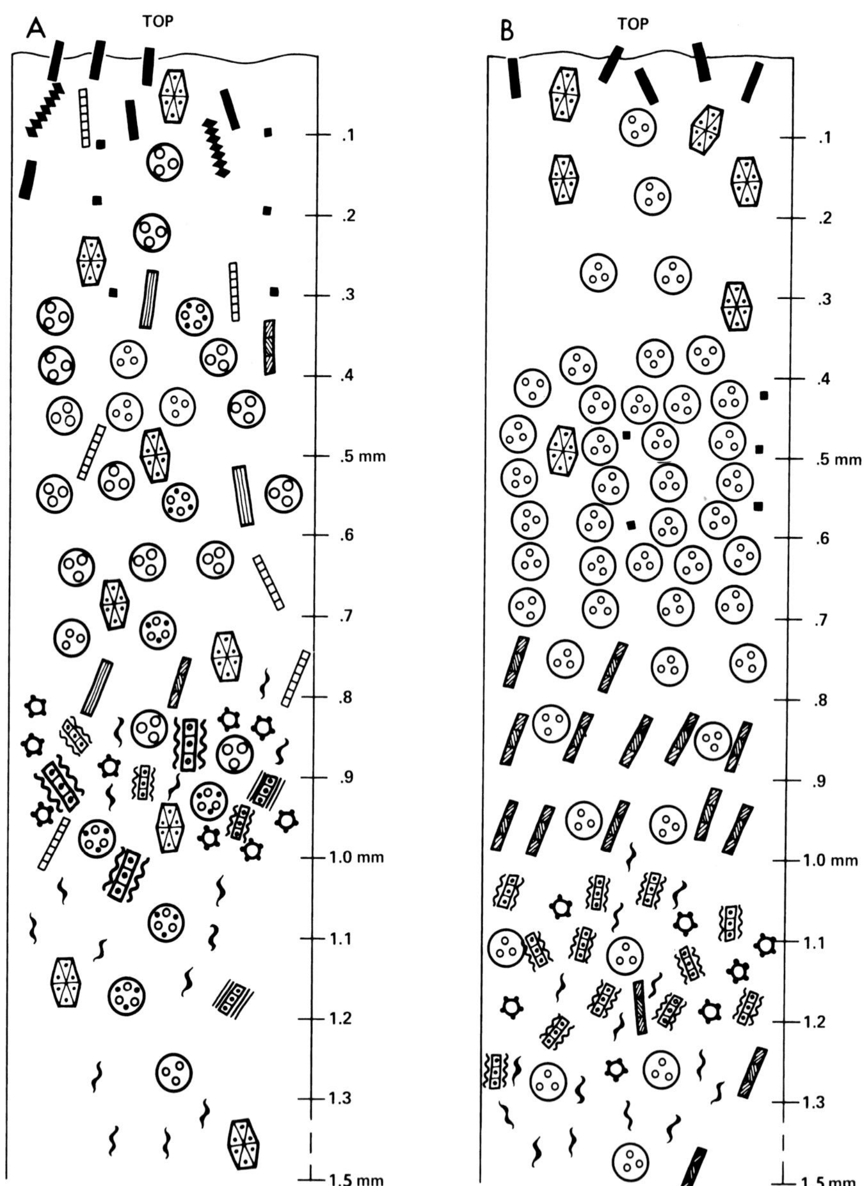

Figure 1. Schematic summary of the vertical distribution of organisms through the top 1.5 mm of the two cyanobacterial mats studied. (A) GN; (B) SL. On facing page are schematic assignments of the different morphologically identifiable mat-forming microorganisms.

Fine Structure of *M. chthonoplastes*

Cross sections of *M. chthonoplastes* trichomes under TEM at high magnification showed a gram-negative cell wall with a distinct peptidoglycan layer (5). Organisms exhibited the typical radial arrangement of thylakoids and a nucleoplasmic region containing carboxysomes and large numbers of ribosomes (Fig. 7). Longitudinal sections studied under TEM at high magnification

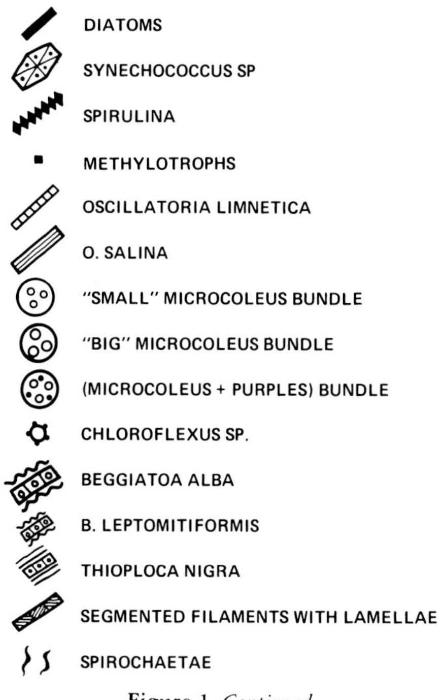

Figure 1. *Continued.*

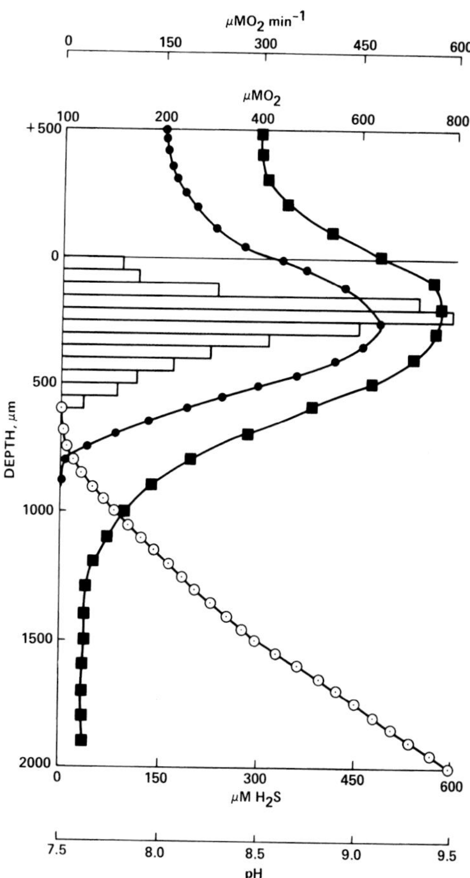

Figure 2. Chemical microprofiles of dissolved oxygen (●), sulfide (⊙), pH (■), and oxygenic photosynthesis (bars) in the upper 2 mm of both cyanobacterial mats from GN and SL. Measurements were carried out with specific microelectrodes for oxygen, sulfide, and pH as described in the text.

showed the detailed location and features of the phycobilisomes. They appeared in the interthylakoidal spaces and extended over the external surface of the thylakoids (10). The phycobilisomes were semicircular at this cutting angle and measured 50 by 25 nm, with a center-to-center distance of 75 nm (Fig. 8).

Two types of *M. chthonoplastes* with different trichome diameters were observed in the mat. One type, the small *M. chthonoplastes*, had trichomes of 2.7 to 3.5 μm in diameter (Fig. 9). The other type, the big *M. chthonoplastes*, had trichomes with a diameter of 3.5 to 6.0 μm (Fig. 10). Each type was observed in separate bundles. The number of trichomes per bundle ranged from 1 to 90. Both types of *M. chthonoplastes* bundles harbored variable numbers of an unidentified filamentous bacterium (Fig. 10). This association accounted for up to 15% of the total *M. chthonoplastes* bundles in the community and attained its highest frequency at a depth of 800 to 1,000 μm. The unidentified filamentous bacterium found inside the *M. chthonoplastes* bundles had a gram-negative cell wall (8), was 0.8 to 0.9 μm in width, and exhibited stacks of lamellae oriented at sharp angles with respect to the cell wall. The structure and arrangement of the lamellae resemble those of the purple bacterium *Ectothiorhodospira* sp. (22). The nucleoplasmic region exhibited PHB-like granules. This unidentified filamentous organism was suggested to be a new type of gliding, filamentous, purple phototroph (6). Filamentous organisms with features similar to those

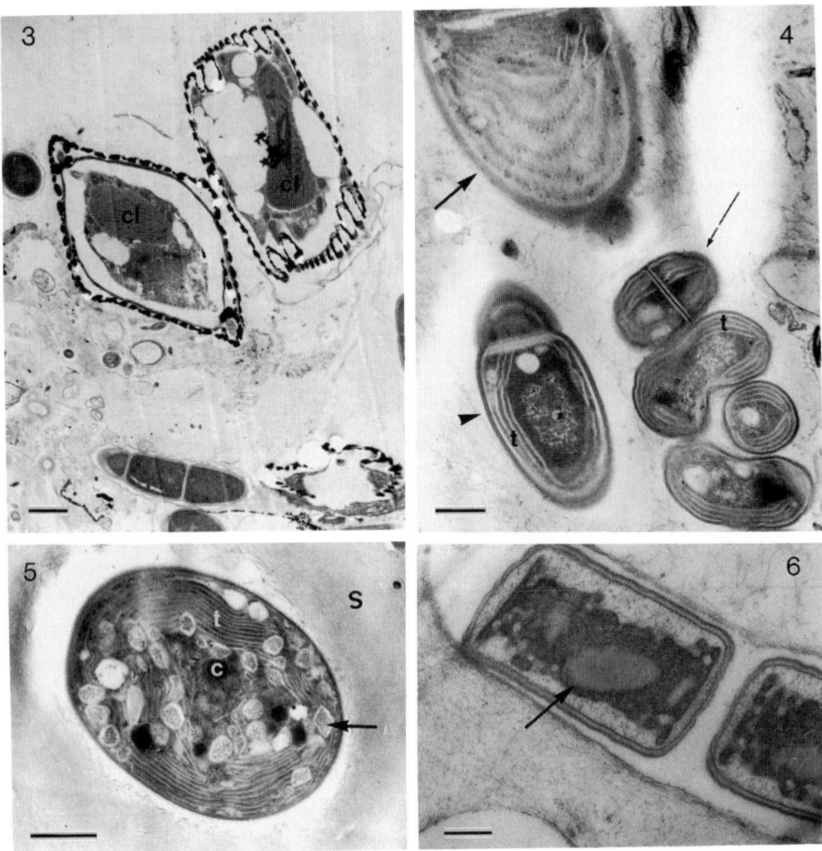

Figure 3. Top of the mat consisting of diatoms (note the silicous frustules and chloroplasts) and filaments of *O. limnetica*. cl, Chloroplast. Bar, 1 μm.

Figure 4. View of the mat at 200 μm from the top, showing the assembly of filamentous cyanobacteria found at this depth. *Oscillatoria* sp. (arrow), *O. limnetica* (arrowhead), and *Spirulina labyrynthiformis* (small arrow) are shown. t, Thylakoids. Bar, 0.5 μm.

Figure 5. The unicellular cyanobacterium *Synechococcus* sp., a conspicuous member of the microbial community. Note thylakoids (t), carboxysomes (c), lacunalike spaces containing a grainy deposit (arrow), and mucopolysaccharide sheath (s). Depth, 300 μm. Bar, 0.2 μm.

Figure 6. Unidentified microorganism frequently found in the upper regions of the photic zone, showing stacks of lamellae (arrow) resembling those of methylotrophic bacteria. Depth, 300 μm. Bar, 0.2 μm.

of the ones inside *M. chthonoplastes* bundles were also found free in the mat. They were slightly wider, measuring about 1.3 μm in diameter, and were surrounded by a thin sheath. Details of their fine structure are depicted in Fig. 11 and 12.

Other Mat-Building Cyanobacteria

Other filamentous cyanobacteria were found in the photic zone of the mat, within the same region that was dominated by *M. chthonoplastes*. *O. limnetica* appeared throughout the photic zone, seemingly at random (Fig. 3, 4, and 9). The filaments of *O. limnetica* were 1.2 to 1.4 μm in diameter. Each segment in the filament was about 2.5 μm in length in longitudinal section, and stacks of photosynthetic lamellae, parallel to the cell wall, made up the nucleoplasmic region. Arrays of gas vesicles located near the septa

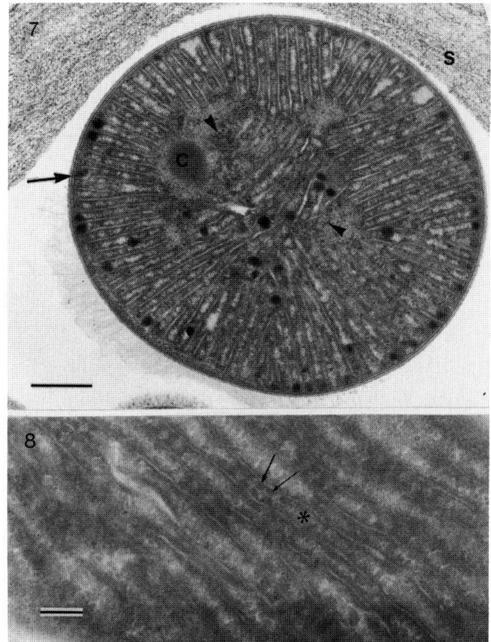

Figure 7. Cross section of *M. chthonoplastes*, the dominant mat-building cyanobacterium, showing the typical radial arrangement of the photosynthetic lamellae. A carboxysome (c) and abundant ribosomes (arrowheads) appear in the nucleoplasm. Arrow, Peptidoglycan layer of the cell wall. S, Mucopolysaccharide sheath. Depth, 500 μm. Bar, 0.5 μm.

Figure 8. Longitudinal section of *M. chthonoplastes* at 600 μm from the top. Rows of phycobilisomes (arrows) are attached to the external surface of the thylakoids. *, Interthylakoidal space. Bar, 0.2 μm.

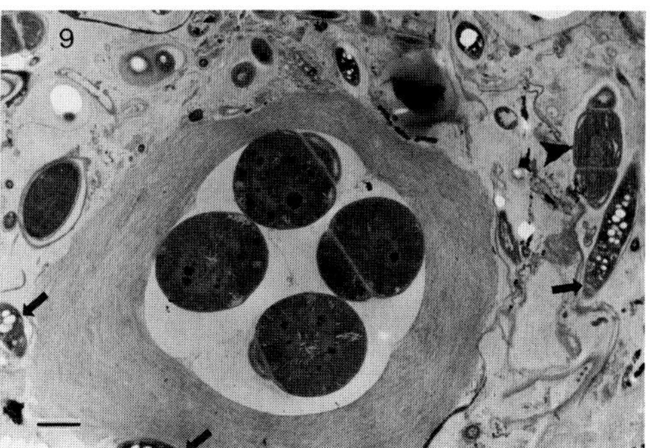

Figure 9. View of the mat at a depth of 200 μm, showing a cross-sectioned bundle of small *M. chthonoplastes*, *O. limnetica* (arrowhead), and filaments of an unidentified organism with round inclusions (arrows). Bar, 1.5 μm.

between two cells were observed in this species (Fig. 13). Filaments of *O. salina* and *Oscillatoria* sp., both remarkably less abundant than the previously described cyanobacteria, were observed within a region that extended between 300 and 500 μm from the surface (Fig. 4; see also Fig. 23).

Other Components of the Photic Zone

Clusters of presumably heterotrophic, unicellular microorganisms, 0.7 to 0.9 μm in size and containing PHB granules, were found in close proximity to *M. chthonoplastes* sheaths at a depth of 300 to 600 μm (Fig. 14). Also regularly found in the photic zone of the cyanobacterial community was a filamentous, segmented organism of about 1 μm in width and several micrometers in length. This unidentified organism was surrounded by a thin capsule and had a well-structured cell wall with a noticeable peptidoglycan layer. Round deposits of variable size (30 to 90 nm) and medium to high electron density, located near the cell periphery, were contained in the light paths. These paths traversed the cytoplasm and appeared to be connected (Fig. 9, 15, and 16). Detailed TEM analyses at high magnification revealed the existence in the mat of

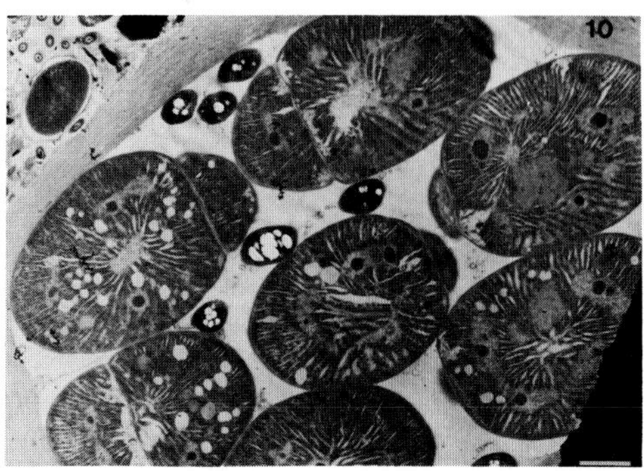

Figure 10. Cross section of a big *M. chthonoplastes* bundle harboring several unidentified filamentous purple bacteria. Depth, 500 μm. Bar, 1 μm.

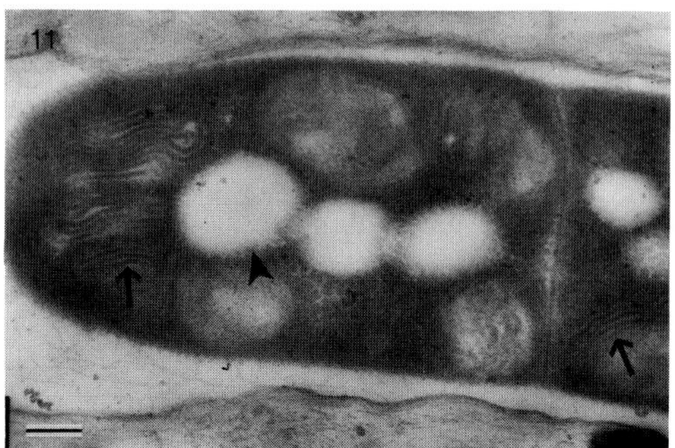

Figure 11. Filamentous bacterium found free in the mat. This bacterium has features similar to those of the filamentous purple bacterium observed inside *M. chthonoplastes* bundles. Arrowhead, PHB-like deposits. Arrows, Stacks of lamellae. Bar, 0.2 μm.

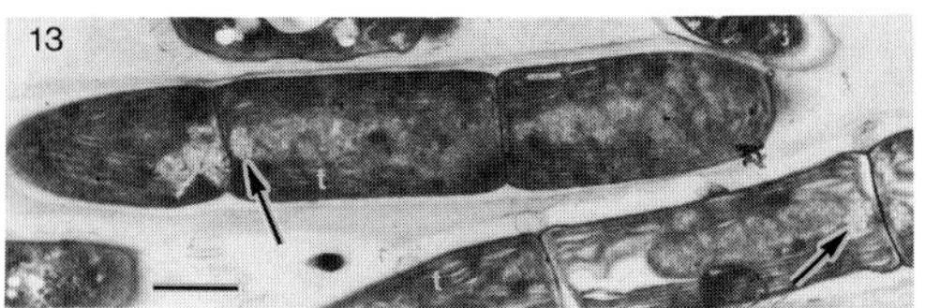

Figure 13. *O. limnetica*, a conspicuous member of the cyanobacterial community, shown in longitudinal section. Arrows denote gas vesicles. t, Thylakoids. Depth, 500 to 600 μm. Bar, 1.0 μm.

two rather similar types of unidentified flexibacteria. They were basically segmented filaments which had a dense nucleoplasm encircled by lamellae. The fine structure of these flexibacteria is shown in Fig. 17 through 19.

Chemocline of the Microbial Mat

The microbial community showed a well-defined microlayer 200 to 300 μm thick. This microlayer was located at 800 to 1,000 μm from the mat surface, at a depth that corresponds to the chemocline of the microecosystem, where oxygen and sulfide coexist at midday. Although there was some overlapping of components of the photic zone (see Fig. 23), the chemocline of the community was dominated by the extensive development of members of the family *Beggiatoaceae* and large numbers of the filamentous green phototroph *Chloroflexus* sp.

Figure 20 depicts several filaments of *Chloroflexus* sp. The organisms were 0.55 to 0.65 μm in width and had a continuous subcortical layer of chlorosomes (20). These characteristic features of green bacteria measured 40 to 55 nm by 130 to 145 nm and were orientated with their longest axis par-

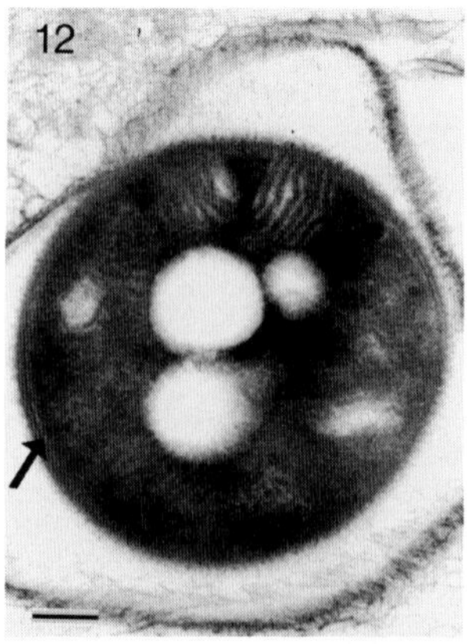

Figure 12. The same organism as in Fig. 11, in cross section. The arrow indicates the peptidoglycan layer of the cell wall. Bar, 0.25 μm.

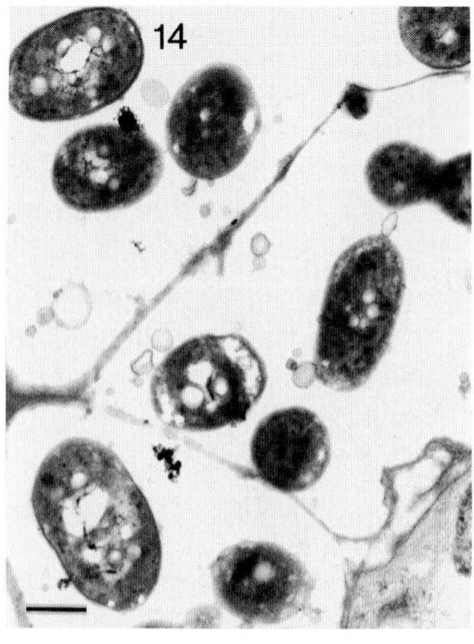

Figure 14. Group of presumably heterotrophic organisms observed near *M. chthonoplastes* sheaths. PHB-like deposits are visible. Bar, 0.5 μm.

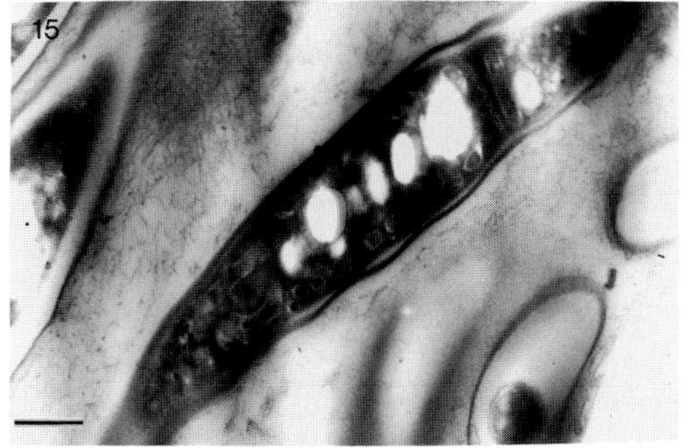

Figure 15. Longitudinal section of an unidentified filamentous organism at a depth of 600 μm which is often found within the photic zone. Note the round deposits contained in light cytoplasmic paths. Bar, 0.2 μm.

allel to the cell wall. The other major components of the chemocline community were nonphotosynthetic sulfur bacteria that belonged to the family *Beggiatoaceae.* Three species of the family *Beggiatoaceae* were identified in the chemocline: *Beggiatoa alba*, with free filaments of 3 to 5 μm in diameter (Fig. 21); *B. leptomitiformis*, also with free filaments that were 1.0 to 1.3 μm wide (Fig. 21); and *Thioploca nigra* (Fig. 22), with ensheathed bundles of straight trichomes that measured about 3 μm in diameter (28). All three species had a multilayered cell wall with the basic arrangement of a gram-negative cell wall plus two slightly undulating external layers of low electron density (12). The cytoplasm contained sulfur deposits

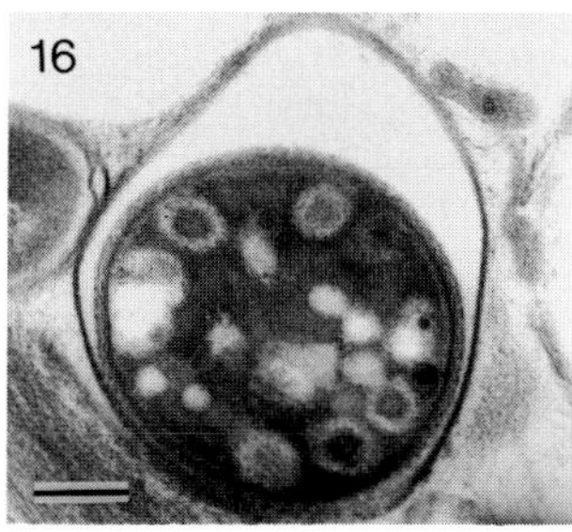

Figure 16. Cross section of the organism in Fig. 15 with round deposits, which are located near the cell periphery. Bar, 0.2 μm.

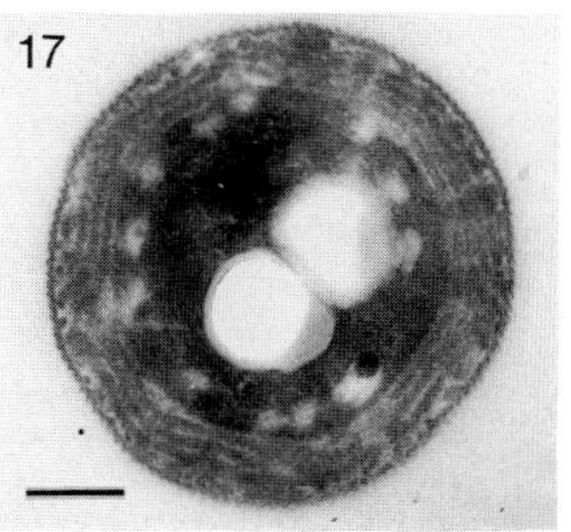

Figure 17. Cross section of a presumably photosynthetic filamentous component of the photic zone of the mat. The nucleoplasm is encircled by lamellar structures. Bar, 0.25 μm.

which appear as empty spaces bordered by a single unit membrane, as well as PHB deposits of medium electron density (26, 27).

Mat Community below 1 mm from the Surface

Below 1 mm from the surface, the cyanobacterial population was drastically reduced to a few *M. chthonoplastes* bundles, some of them showing an association with the unidentified purple bacterium, and some ensheathed colonies of *Synechococcus* sp. (Fig. 23 and 24). Beggiatoans and large numbers of spirochetes and other heterotrophic organisms also appeared between 1,000 and 1,500 μm of depth (Fig. 25).

THE SL MAT

The surface of the SL cyanobacterial community was nearly flat. Naviculate dia-

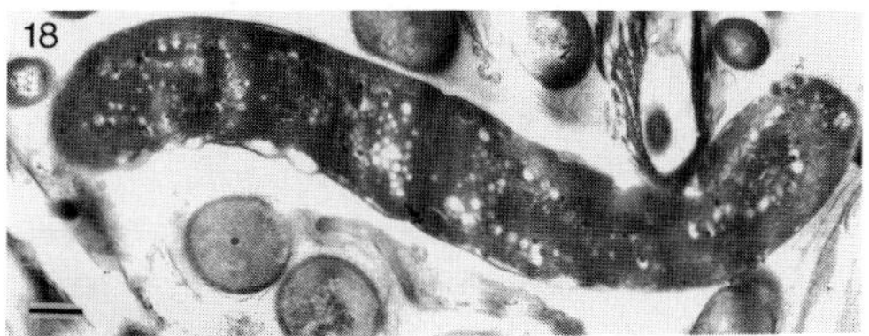

Figure 18. Longitudinal section of the organism depicted in Fig. 17. Bar, 0.5 μm.

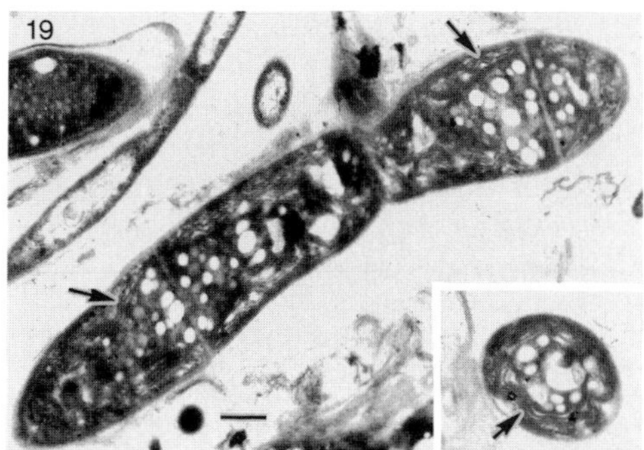

Figure 19. Filamentous segmented organism containing subperipheral lamellae that run almost parallel to the cell wall. Inset, The same organism in cross section. Arrows indicate lamellae. Depth, 800 μm. Bar, 0.5 μm.

toms appeared at the very top. The cyanobacteria *Johannesbaptistia* sp. and *Synechococcus* sp., both ensheathed, and very few bundles of *M. chthonoplastes*, also heavily ensheathed, were observed in the uppermost layer, which extended down 200 to 300 μm (Fig. 26 and 27). *M. chthonoplastes*, the dominant component of the mat, started layering, although a bit loosely, between a depth of 300 and 400 μm. It appeared in bundles of variable numbers of trichomes surrounded by a thick mucopolysaccharide envelope (Fig. 28). In cross section, the trichomes had a diameter of 2.8 to 3.34 μm (Fig. 28 and 29). The thylakoids, exhibiting the typical radial arrangement for this species, filled almost all the cell lumen. The main region of *M. chthonoplastes* in the Shallow Flat Mat was located between 450 and 700 μm from the surface. Clusters of organisms with features resembling those of methylotrophic bacteria were observed at a depth of 400 to 600 μm (Fig. 30). Round to oval ensheathed microorganisms, apparently heterotrophic, were often found near *M. chthonoplastes* sheaths (Fig. 31).

Filamentous organisms with an extensive intracytoplasmic lamellar system started to appear in a microlayer located at 700 to 750 μm below the surface (Fig. 32). These organisms measured 0.9 to 1.4 μm in diam-

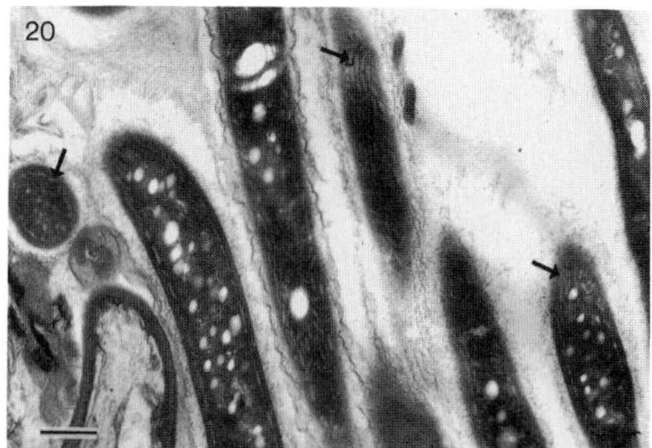

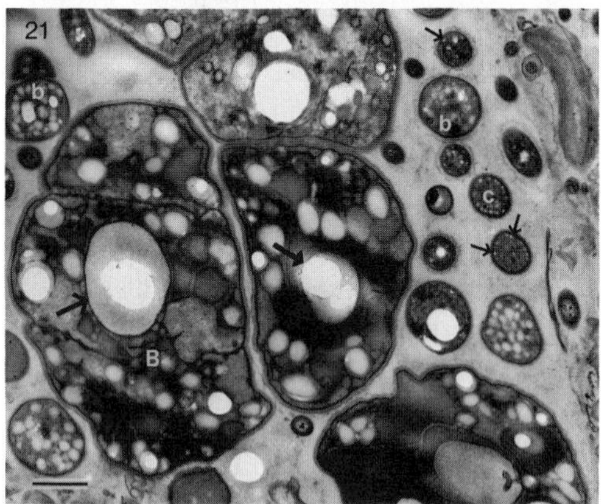

Figure 20. One cross section and several longitudinal sections of *Chloroflexus*-like organisms. These filamentous gliding phototrophs are found at a microzone that corresponds to the chemocline of the community. Arrows denote chlorosomes. Depth, 800 μm. Bar, 0.5 μm.

Figure 21. The nonphotosynthetic, chemotrophic sulfur bacteria *B. alba* (B) and *B. leptomitiformis* (b), together with several *Chloroflexus* spp. (c) in cross section, at a depth of 800 to 900 μm. Arrows show sulfur granules. Small arrows show chlorosomes. Bar, 1 μm.

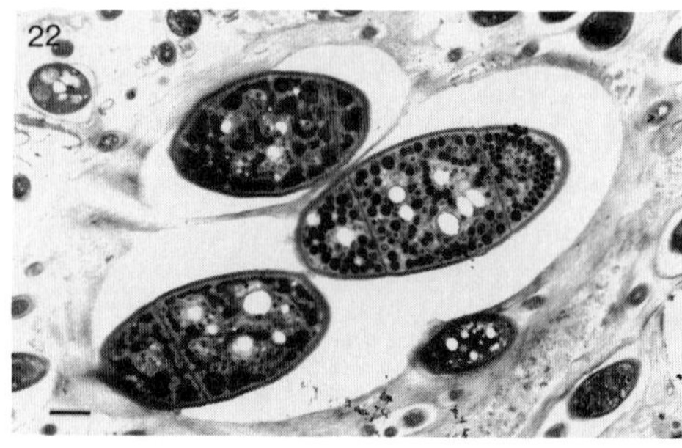

Figure 22. The ensheathed filamentous sulfur bacterium *Thioploca nigra*, found at a depth of 900 μm. Bar, 1 μm.

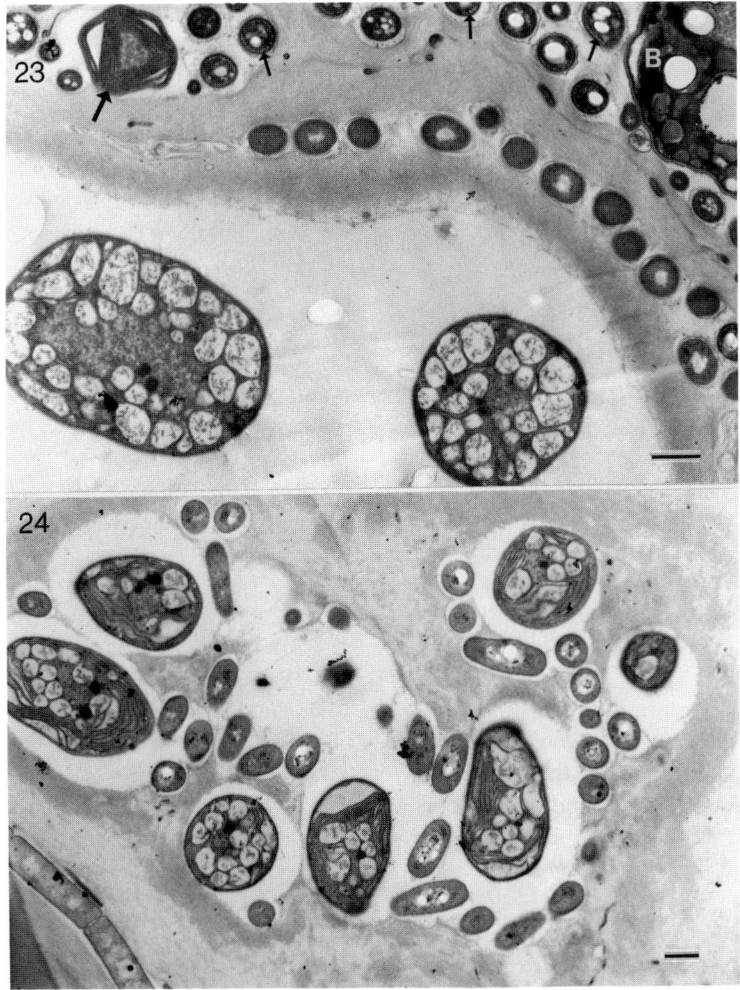

Figure 23. Microbial community at a depth of 800 μm, showing part of a colony of *Synechococcus* sp. with a few spirochetes and other heterotrophic organisms lodged in its mucopolysaccharide envelope. A beggiatoan (B), several *Chloroflexus*-like organisms (both characteristic components at this depth), and an *Oscillatoria* sp. in cross section (arrow) are also shown. Small arrows denote chlorosomes. Bar, 1.0 μm.

Figure 24. Ensheathed colony of *Synechococcus* sp. found at a depth of 1,200 μm, where the environment is anaerobic. Note the distended thylakoids in some of the cyanobacterial cells. Several unidentified rods are seen associated with the *Synechococcus* colony. Bar, 1.0 μm.

eter and were surrounded by a thin, fibrous sheath. Some of them appeared immersed in the *M. chthonoplastes* polysaccharide envelopes. Differences in the arrangement of the inner lamellar system were observed in these filaments. In some, the lamellae appeared in stacks, oriented at sharp angles to the cell wall (Fig. 32 and 33A); in others, the lamellae were grouped in a single central stack that ran almost parallel to the cell wall (Fig. 32 and 33B). Organisms with these types of lamellar arrangements were the most abun-

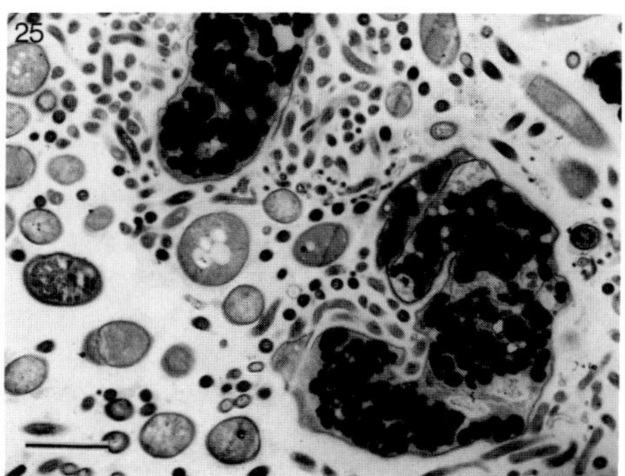

Figure 25. *B. alba* containing large numbers of highly electron-dense deposits, presumably polyphosphate, and scarce sulfur or PHB granules at 1,200 μm from the surface. Numerous spirochetes appear at this microzone, which is well within the sulfide-rich layer of the mat. Bar, 2.0 μm.

dant. A few filaments had the lamellae concentrically arranged, surrounding the cytoplasm (Fig. 32). Also observed in this microlayer were filaments with a dense nucleoplasmic region containing very abundant round deposits which exhibited various degrees of electron density (Fig. 33C). These organisms appeared in small numbers, interspersed among the ones with inner membranes.

As the depth increased, the *M. chthonoplastes* population diminished, although a microlayer located between 800 and 1,000 μm from the top was clearly dominated by these filamentous organisms.

The mat became more complex with increasing depth. The nonphotosynthetic sulfur bacterium *B. leptomitiformis* started layering at a microzone located at 1,000 to 1,100 μm from the surface. The filaments of this bacterium were 1.0 to 1.1 μm wide. The cytoplasm contained sulfur granules and large numbers of PHB deposits (Fig. 34). At a depth of 1,100 to 1,200 μm, *B. leptomitiformis* was a major component of the community. Filaments of the photosynthetic green bacterium *Chloroflexus* sp., exhibiting their distinctive chlorosomes, appeared at this microzone. In cross section, the *Chloroflexus* sp. measured 0.60 to 0.65 μm and had a continuous subperipheral layer of chlorosomes (Fig. 35). *B. leptomitiformis*, *Chloroflexus* sp., and the filaments with lamellae overlapped at a depth between 1,200 and 1,500 μm. *M. chthonoplastes* was found in this microlayer as well, although it was scarce. Spirillae and spirochetes were frequently found within the mat. They were particularly abundant between 1,000 and 1,500 μm. Figure 35 depicts a trichome of *M. chthonoplastes*, heavily ensheathed. Several groups of *Beggiatoa* spp., *Chloroflexus* spp., and numerous spirillae and spirochetes are lodged toward the periphery of the *M. chthonoplastes* sheath.

DISCUSSION

The results of the microscopy studies revealed a finely laminated cyanobacterial community in the samples from both localities. Both communities were clearly dominated by *M. chthonoplastes*.

The vertical distribution of organisms followed a basically similar pattern in both mats. Communities were heterogeneous in the photic zone. Eucaryotic phototrophs (diatoms) and colonies of unicellular cyanobacteria, most of them *Synechococcus* spp., appeared in the upper microlayers, whereas ensheathed bundles of *M. chthonoplastes* extended through almost the entire profile of the photic zone (Fig. 1). Segmented filaments with an intracytoplasmic lamellar system devoid of phycobilisomes were also components of the photic zones of both mats (Fig. 10, 11, 12, 17, 18, 19, 32, 33A, and 33B). Other comparable components were clusters of apparently heterotrophic organisms observed in the near vicinity of *M. chthonoplastes* polysaccharide envelopes (Fig. 14 and 31) and groups of microorganisms that had stacks of lamellae similar to those of methylotrophic bacteria (Fig. 6 and 30). Methane bacteria were enriched from the SL cyanobacterial mats (11), and their autofluo-

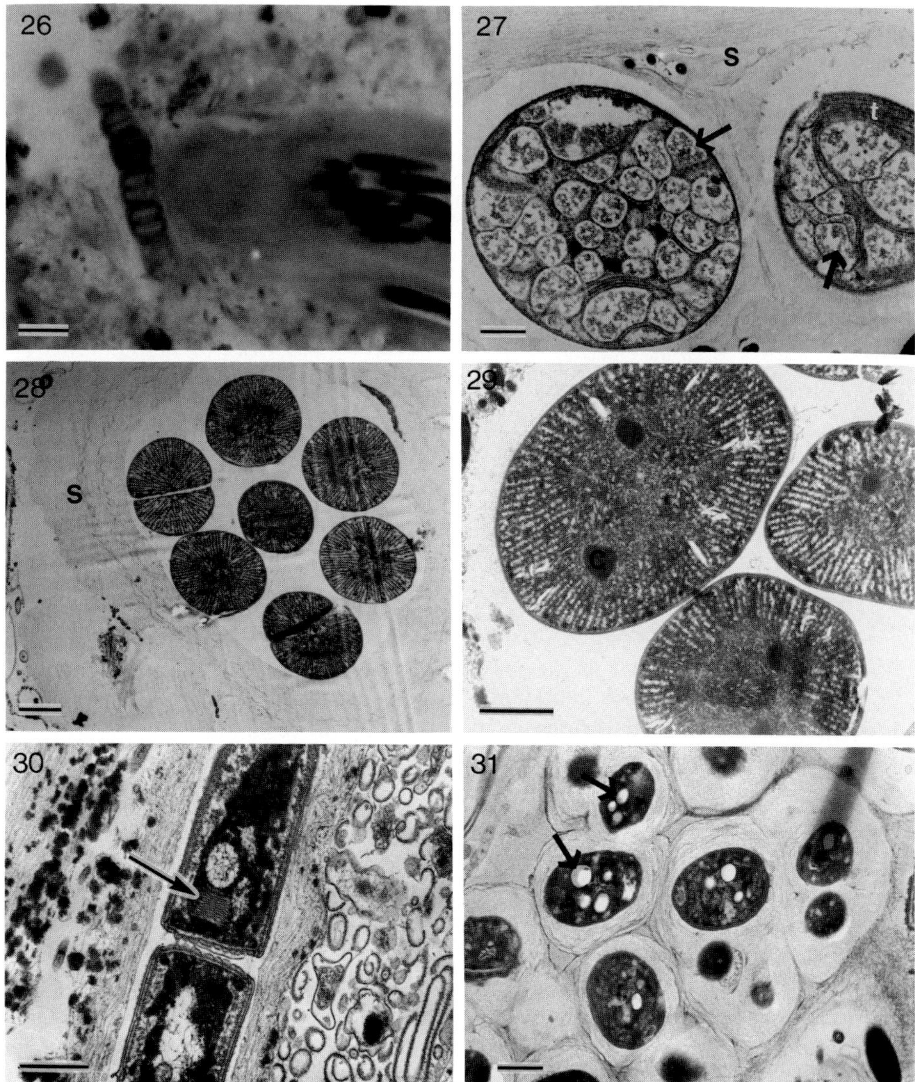

Figure 26. The ensheathed cyanobacteria *Johannesbaptistia* sp. and *M. chthonoplastes* at a depth of 50 μm. Bar, 10 μm.

Figure 27. The unicellular cyanobacterium *Synechococcus* sp. at a depth of 200 μm. S, Mucopolysaccharide sheath; t, thylakoids. Arrows indicate lacunae with grainy deposits. Bar, 0.5 μm.

Figure 28. Ensheathed bundle of *M. chthonoplastes*, the dominant component of the mat, at 200 μm below the surface. S, Mucopolysaccharide envelope. Bar, 1.5 μm.

Figure 29. *M. chthonoplastes* at a higher magnification. Note the radial photosynthetic lamellae and dense nucleoplasm containing carboxysomes (c). Bar, 1 μm.

Figure 30. Microorganism with lamellae (arrow), resembling methylotrophic bacteria, at 500 to 600 μm from the mat surface. Bar, 0.5 μm.

Figure 31. Group of presumably heterotrophic organisms, found inside an empty sheath of *M. chthonoplastes*. Arrows indicate PHB-like deposits. Bar, 0.5 μm.

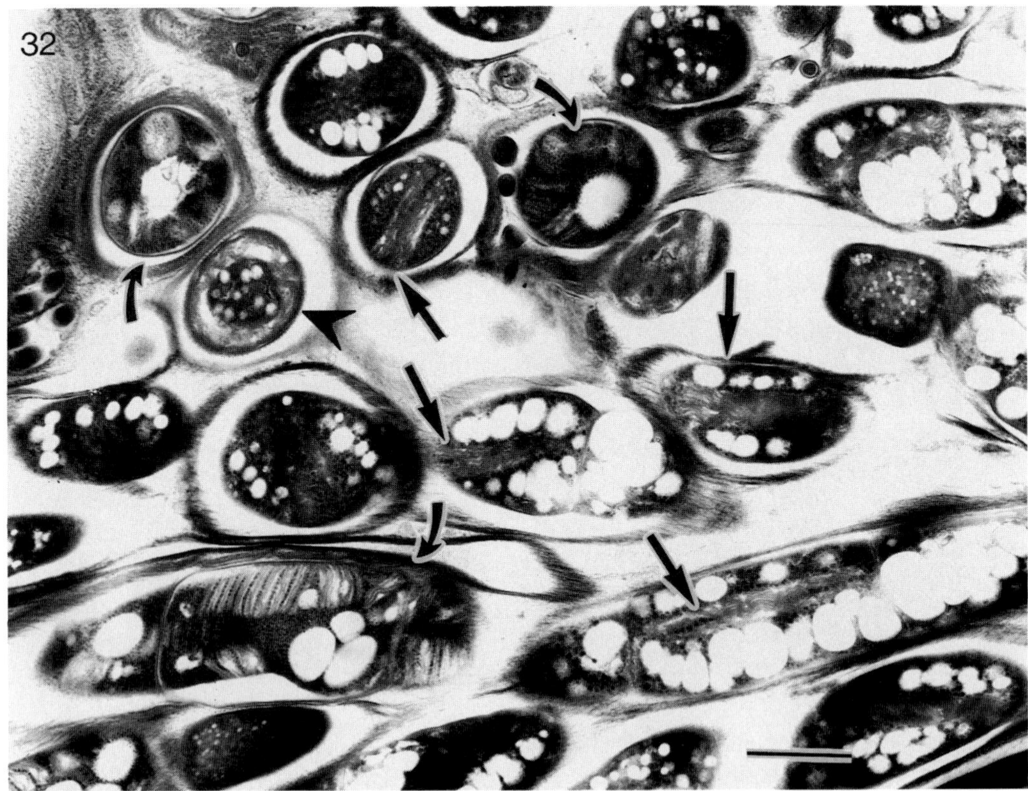

Figure 32. Organisms found in the Shallow Flat Mat at 900 μm from the surface. These organisms are filamentous and show extensive intracytoplasmic membrane systems. Three types of lamellar arrangement can be distinguished: grouped in stacks oriented at sharp angles to the cell wall (curved arrows), concentrically arranged surrounding the cytoplasm (arrowhead), and sandwiched by the cytoplasm (arrows). Bar, 1.0 μm.

rescence was detected in SL sediment samples. Both apparently phototrophic and heterotrophic communities showed a microlayer located at the depth of their correspondent chemoclines, in which the distinctive organisms were filaments of the green phototroph *Chloroflexus* sp. and members of the *Beggiatoaceae* family of nonphotosynthetic sulfur bacteria (Fig. 21 and 35).

This study also revealed some differences in the texture of both mats, as well as in the arrangement and degree of association between some components of the photic zone. As mentioned above, colonies of *Synechococcus* sp., often heavily ensheathed, were components of the upper regions of the photic zone in GN and in SL. This unicellular cyanobacterium was almost totally restricted to these layers, about 300 μm from the surface, in the SL mat, whereas in GN it was observed through the entire section, well below 1 mm from the surface.

In the GN mat, *M. chthonoplastes* started layering at 50 μm from the surface, but remained loosely knit even in regions of its highest frequency. This loose texture allowed other cyanobacteria, particularly *Synechococcus* sp. and, to a lesser degree, *Oscillatoria* spp., to occur in considerable numbers within the layers dominated by *M. chthonoplastes*. The main cyanobacterium itself exhibited three types of bundles: small and big *M. chthonoplastes* and *M. chthonoplastes* plus purple organisms. Most of the purple segmented filamentous organisms exhibiting an intracytoplasmic lamellar system were ob-

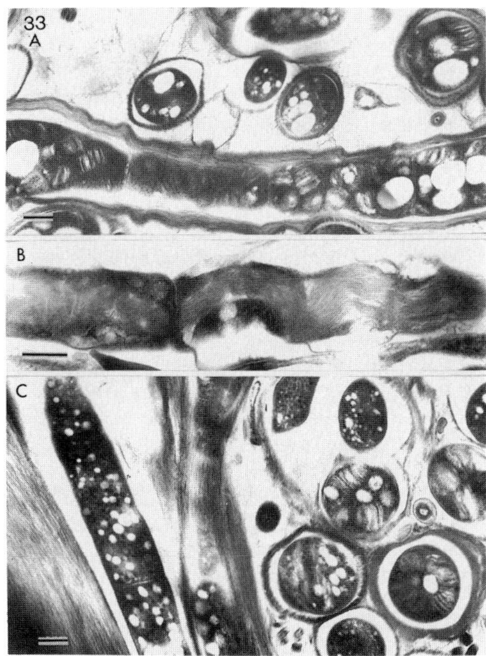

Figure 33. (A) Longitudinal section of the segmented filament with extensive stacks of lamellae oriented at sharp angles to the cell wall. Note the PHB-like deposits and abundant ribosomes. The filament is enveloped in a fibrous sheath. A similar microorganism is sectioned at the upper right-hand corner. Depth, 900 μm. Bar, 0.5 μm. (B) View in longitudinal section of the filament with large numbers of lamellae grouped in a central single stack. Depth, 1,000 μm. Bar, 0.5 μm. (C) Filamentous, segmented organism with a very dense cytoplasm filled with unidentified round deposits of different electron density. Depth, 800 μm. Bar, 0.5 μm.

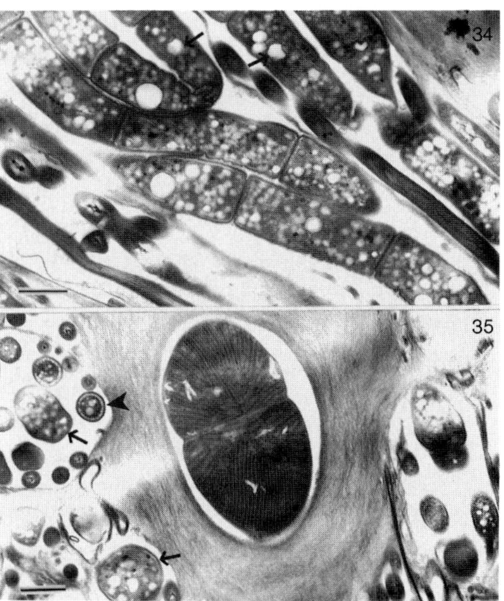

Figure 34. The nonphotosynthetic sulfur bacterium *B. leptomitiformis* at 1,100 to 1,200 μm from the mat surface. Arrows indicate sulfur granules. Bar, 1.0 μm.
Figure 35. Semioblique sectioned trichome of heavily ensheathed *M. chthonoplastes*. *Beggiatoa* (arrows) and *Chloroflexus* (arrowhead) filaments are lodged toward the periphery of the *M. chthonoplastes* sheath. Depth, 1,200 μm. Bar, 1.0 μm.

served inside *M. chthonoplastes* bundles, in a very close association with the cyanobacterium, which is well adapted to sulfide (4), whereas those observed free in the mat appeared in small numbers. *Synechococcus* colonies were observed harboring unidentified heterotrophic rods (Fig. 24). Within the GN chemocline community, the *Beggiatoaceae* family of nonphotosynthetic sulfur bacteria was represented by two genera and three species (Fig. 21 and 23).

In the SL mat, the upper 300 μm of the photic zone contained most of the unicellular cyanobacteria (*Johannesbaptistia* sp., *Synechococcus* sp.) and very few *M. chthonoplastes* bundles. *M. chthonoplastes* started layering at a deeper microlayer than it did in the GN mat (about 300 to 350 μm from the surface) and produced a cohesive population. In fact, *M. chthonoplastes* bundles with trichomes of only one type were arranged very close to each other, particularly at the layers of their highest incidence, where *M. chthonoplastes* was almost the sole cyanobacterial component (see Fig. 1). The distinctive feature of the SL mat was the presence in large numbers of filamentous, segmented organisms exhibiting an elaborate system of intracytoplasmic lamellae. One of these organisms, with lamellae arranged in stacks oriented at sharp angles to the cell wall (Fig. 32 and 33A), resembled the organism found inside *M. chthonoplastes* bundles in the GN mat. There were also similarities between the filament depicted in Fig. 19 and the organism indicated by an arrowhead in Fig. 32.

The filamentous organisms with inner lamellae had a very dense, granular nucleoplasm and exhibited large PHB-like deposits. No phycobilisomes were observed in these organisms, which were always found free in the SL mat and clearly outnumbered *M. chthonoplastes* below 800 to 900 μm from the mat surface. The SL chemocline community was defined by the extensive development of members of the families *Chloroflexaceae* and *Beggiatoaceae*, with the latter represented by one species, *B. leptomitiformis* (Fig. 34).

In summary, the GN mat exhibited a rather loose texture with a diversified cyanobacterial community, whether referring to the species found in the community or to the morphological types of the dominant mat-building organism. The major components of the mat, e.g., *M. chthonoplastes* and *Synechococcus* sp., showed an intimate degree of association with other members of the community (Fig. 10 and 24). In contrast, the SL mat had a compact texture that was exhibited not only by the dominant *M. chthonoplastes*, but also by the filamentous, segmented organism with inner lamellae and by the constituents of the chemocline community. It was, therefore, a mat with a high degree of cohesiveness.

Finally, overlapping of distributions of some typical members of the photic zone was observed in the chemoclines of both mats (Fig. 23 and 35), and the morphologically preserved microbial communities in the permanently reduced layers of the mats are quite similar.

Despite the large geographical distance between the two mats, they are very similar. The distinct differences between the two mats is primarily an outcome of the differences in the environment of deposition of the two sites.

LITERATURE CITED

1. **Awramik, S. M.** 1984. Ancient stromatolites and microbial mats, p. 1–22. *In* Y. Cohen, R. W. Castenholz, and H. O. Halvorson (ed.), *Microbial Mats: Stromatolites.* Alan R. Liss, Inc., New York.
2. **Bauld, J.** 1984. Microbial mats in marginal marine environments: Shark Bay, Western Australia, and Spencer Gulf, South Australia, p. 39–58. *In* Y. Cohen, R. W. Castenholz, and H. O. Halvorson (ed.), *Microbial Mats: Stromatolites.* Alan R. Liss, Inc., New York.
3. **Cohen, Y.** 1984. The Solar Lake cyanobacterial mats: strategies of photosynthetic life under sulfide, p. 133–148. *In* Y. Cohen, R. W. Castenholz, and H. O. Halvorson (ed.), *Microbial Mats: Stromatolites.* Alan R. Liss, Inc., New York.
4. **Cohen, Y., B. B. Jorgensen, N. P. Revsbech, and R. Poplawski.** 1986. Adaptation to hydrogen sulfide of oxygenic and anoxygenic photosynthesis among cyanobacteria. *Appl. Environ. Microbiol.* **51**:398–407.
5. **Costerton, J. W., J. Ingram, and K. J. Cheng.** 1974. Structure and function of the cell envelope of gram-negative bacteria. *Bacteriol. Rev.* **38**:87–104.
6. **D'Amelio, E. D., Y. Cohen, and D. J. Des Marais.** 1987. Association of a new type of gliding filamentous phototrophic bacterium inside bundles of *Microcoleus chthonoplastes* in hypersaline cyanobacterial mats. *Arch. Microbiol.* **147**:528–534.
7. **Debe, G. C.** 1770. *Flora Danica*, vol. 3, Table 1385. Nicolaus Moller, Royal Bookbinder, Copenhagen.
8. **Drews, G., J. Weckesser, and H. Mayer.** 1978. Cell envelopes, p. 61–77. *In* R. K. Clayton and W. R. Sistrom (ed.), *The Photosynthetic Bacteria.* Plenum Publishing Corp., New York.
9. **Erlich, A., and I. Dor.** 1985. Photosynthetic microorganisms of the Gavish Sabkha, p. 296–331. *In* G. M. Friedman and W. E. Krumbein (ed.), *Hypersaline Ecosystems. The Gavish Sabkha.* Springer-Verlag KG, Berlin.
10. **Gantt, E.** 1980. Structure and function of phycobilisomes: light-harvesting pigment complexes in red and blue-green algae. *Int. Rev. Cytol.* **66**:45–80.
11. **Giani, D., L. Giani, Y. Cohen, and W. E. Krumbein.** 1984. Methanogenesis in the hypersaline Solar Lake (Sinai). *FEMS Microbiol. Lett.* **25**:219–224.
12. **Glauert, A. M., and M. J. Thornley.** 1969. The topography of the bacterial cell wall. *Annu. Rev. Microbiol.* **23**:159–196.
13. **Grove, S. N., C. E. Bracker, and D. J. Morre.** 1968. Cytomembrane differentiation in the endoploasmic reticulum–Golgi apparatus–vesicle complex. *Science* **161**:171–173.
14. **Javor, B. J., and R. W. Castenholz.** 1984. Productivity studies of microbial mats, Lagoona Guerrero Negro, Mexico, p. 149–170. *In* Y. Cohen, R. W. Castenholz, and H. O. Halvorson (ed.), *Microbial Mats: Stromatolites.* Alan R. Liss, Inc., New York.

15. **Jørgensen, B. B., and Y. Cohen.** 1977. Solar Lake (Sinai). 5. The sulfur cycle of benthic cyanobacterial mats. *Limnol. Oceanogr.* **22:**657–666.

16. **Jørgensen, B. B., N. P. Revsbech, and Y. Cohen.** 1979. Diurnal cycle of oxygen and sulfide microgradients and microbial photosynthesis in a cyanobacterial mat sediment. *Appl. Environ. Microbiol.* **38:**46–58.

17. **Jørgensen, B. B., N. P. Revsbech, and Y. Cohen.** 1983. Photosynthesis and structure of benthic microbial mats: microelectrode and SEM studies of four cyanobacterial communities. *Limnol. Oceanogr.* **28:**1075–1093.

18. **Krumbein, W. E., Y. Cohen, and M. Shilo.** 1979. Solar Lake (Sinai). 4. Stromatolitic cyanobacterial mats. *Limnol. Oceanogr.* **22:**635–656.

19. **Kushida, H.** 1974. A new method for embedding with a low viscosity resin, Quetol 651. *J. Electron Microsc.* **213:**197–204.

20. **Pierson, B. K., and R. W. Castenholz.** 1974. A phototrophic gliding filamentous bacterium of hot springs, *Chloroflexus aurantiacus*, gen. and sp. nov. *Arch. Microbiol.* **100:**5–24.

21. **Purser, B. H. (ed.).** 1973. *The Persian Gulf.* Springer-Verlag KG, Berlin.

22. **Remsen, C. C., S. W. Watson, J. B. Waterbury, and H. G. Truper.** 1968. Fine structure of *Ectothiorhodospira mobilis. J. Cell Bacteriol.* **19:**2374–2392.

23. **Revsbech, N. P., B. B. Jørgensen, T. H. Blackburn, and Y. Cohen.** 1983. Microelectrode studies of the photosynthesis and O_2, H_2S, and pH profiles of a microbial mat. *Limnol. Oceanogr.* **28:**1062–1074.

24. **Reynolds, E. S.** 1963. The use of lead citrate as an electron opaque stain in electron microscopy. *J. Cell Biol.* **17:**208–213.

25. **Rippka, R., J. B. Waterbury, and R. Y. Stanier.** 1981. Provisional generic assignments for cyanobacteria in pure culture, p. 247–256. *In* M. P. Starr, H. Stolp, H. G. Trüper, A. Balows, and H. G. Shlegel (ed.), *The Prokaryotes.* Springer-Verlag KG, Berlin.

26. **Schmidt, T. M., B. Arieli, Y. Cohen, E. Padan, and W. R. Stroll.** 1987. Sulfur metabolism in *Beggiatoa alba. J. Bacteriol.* **169:**5466–5472.

27. **Shively, J. M.** 1974. Inclusion bodies of prokaryotes. *Annu. Rev. Microbiol.* **28:**167–187.

28. **Wiessner, W.** 1981. The family Beggiatoaceae, p. 380–389. *In* M. P. Starr, H. Stolp, H. G. Trüper, A. Balows, and H. G. Shlegel (ed.), *The Prokaryotes.* Springer-Verlag KG, Berlin.

Chapter 10

Temporal and Spatial Distribution of Mat Microalgae in the Experimental Solar Ponds, Dead Sea Area, Israel

Inka Dor and Noga Paz

INTRODUCTION

Experimental solar ponds constructed and operated in the Dead Sea area by Solmat Systems Ltd. Israel for applied purposes offer researchers a unique opportunity to investigate young, mat-forming communities of microalgae under controlled field conditions. These hypersaline, meromictic water bodies have an artificially supported stable halocline, with diluted Dead Sea water at the surface and saturated brine at the bottom. An inverted thermal stratification develops naturally and results in a steep increase of the temperature with depth.

The pond slopes are richly populated by benthic communities of diatoms and mat-forming cyanobacteria (10, 13–15), which change composition with the seasons and vertical gradients of salinity and temperature. *Euglena* sp. appears as a stable component of this benthic community (10). *Dunaliella parva* Lerche is the only eucaryotic phytoplankter present in the upper water mass and occasionally forms blooms (12). Some gas-vacuolated strains of *Aphanothece halophytica* Fremy also live suspended in the water column (16).

Apart from a few protozoan species, grazers are absent, since they have been eliminated by the high salinity and unusual ionic composition of Dead Sea water (25). In such all-producer communities thriving under the extreme environmental conditions, the species diversity is much reduced and the distribution of various forms depends primarily on the ability to cope with the abiotic stresses (26). However, for species having overlapping salinity and temperature ranges, the interspecific competition for space, light, and nutrients becomes an important factor, promoting selection and establishment of the microalgal community (13). Under multiple physicochemical and competitive stresses and with the lack of significant grazing, distribution patterns of various benthic forms delineate their optimal growth ranges.

This chapter describes the physicochemical characteristics of the experimental solar pond ecosystem and provides environmental ranges for several predominant dia-

Inka Dor and Noga Paz • Oceanography Program and Human Environmental Sciences Division, School of Applied Science and Technology, The Hebrew University of Jerusalem, Jerusalem 91904, Israel.

toms and cyanobacteria and one euglenide alga. Competitive relationships between various species and some strategies for adaptation to this hypersaline and thermal habitat are also discussed.

SAMPLING PROCEDURES

All seven solar ponds investigated had the same general structure: a depth of about 2 m with a convective upper layer of diluted Dead Sea water (thickness, 40 to 60 cm), a nonconvective gradient layer, and Dead Sea brine at the bottom. Fresh water was added continuously to compensate for evaporation and to maintain a constant salinity (29). Pond sizes ranged from 1,000 to 210,000 m^2. Various ponds were constructed from 1978 to 1982; the investigations took place from 1981 to 1984.

Samples of benthic communities were collected from the pond slopes during different seasons, at 20- to 30-cm intervals from close to the shore line down to about 60 to 80 cm, where the temperature usually exceeded 48°C and the salinity was higher than 210‰. Below this level microalgae were absent. The sampled littoral zone was 3 to 5 m wide. Samples either were collected directly from the bottom sediment for species identification and relative abundance estimation (a total of 116 samples) or were collected using a plastic frame (8 by 25) for the estimation of cell concentration per area unit (a total of 73 samples). Material was preserved in 5% formaldehyde. Small subsamples were homogenized and counted with a hemacytometer. In identifying the cyanobacterial species, Geitler's taxonomy was followed. Data on the relative abundances (see Fig. 4 to 8) are adapted from Dor (10).

Parallel with the collection of the microalgae from the pond slopes, temperature and density profiles were taken. Salinity was extrapolated from the density by the method of Dor and Hornoff (15). Several times during the investigation period P_i was determined in the water column (28). Light penetration was measured with a Lambda Li-Cor model 185-A Quantum Meter, always between 10:00 and 11:00 a.m.

DISTRIBUTION OF MICROALGAE IN SOLAR PONDS

Pond water salinities ranged from 30 to 140‰ in the upper, convective water layer and up to 320‰ in the deep brine. Temperatures ranged from ambient values in the top layer and up to 60 to 98°C in the heat-storing water mass at the bottom. Figure 1 shows typical profiles of temperature and density in the two representative solar ponds. Seasonal changes in light intensity, measured in the air and at a depth of 60 cm, are shown in Fig. 2. Concentrations of P_i in the upper water layer were very low throughout the year (Fig. 3). An increase was regularly recorded in the gradient layer in each pond, particularly at the beginning of winter. This seasonal enrichment resulted from the windborne desert dust, which contains considerable amounts of phosphate in the north Dead Sea area (10).

Figures 4 to 10 show the relative abundances and seasonal concentrations of the most common microalgae as a function of salinity and temperature. *A. halophytica* Fremy (Fig. 4) appeared at a wide range of temperatures and salinities, being common to abundant at 16 to 48°C and 40 to 180‰, respectively. Its maximal seasonal concentrations, exceeding 10^{11} cells m^{-2}, were found in summer. *Phormidium hypolimneticum* Campbell (Fig. 5) showed optimal growth within a slightly more extended range of salinities (up to 200‰) but a more limited temperature range (up to 40°C), reaching maximal concentrations (10^{10} filaments m^{-2}) in the fall and winter. The remaining species, *Chroococcidiopsis* sp. (Fig. 6), *Phormidium hypersalinum* Campbell (Fig. 7), and *Spirulina subsalsa* Oersted (Fig. 8), were less abundant,

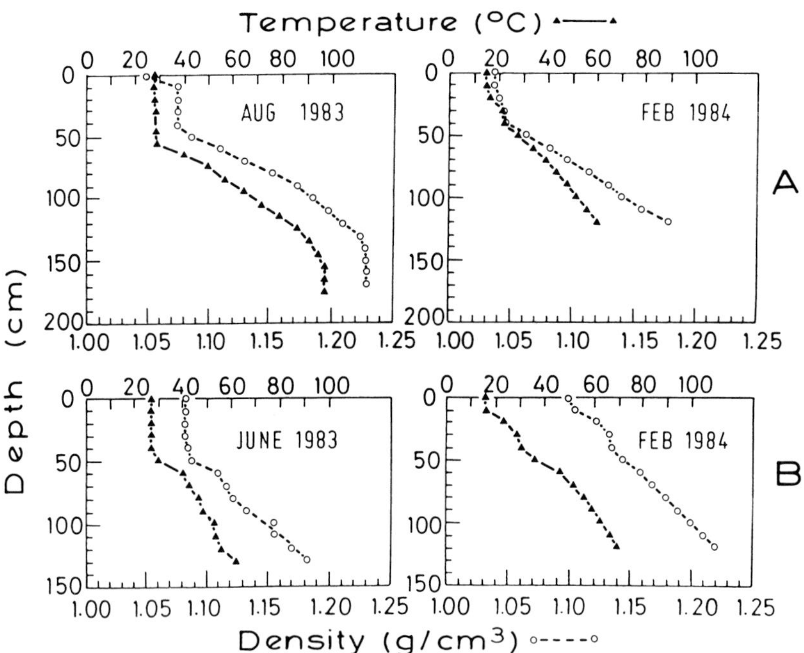

Figure 1. Profiles of temperature and density in two representative solar ponds, a 40-dunam pond (A) and a 210-dunam pond (B), Beth Arava, north Dead Sea (dunam = 1,000 m^2). Densities of 1.05, 1.10, 1.15, and 1.20 g cm^{-3} correspond approximately to salinities of 72, 127, 211, and 250‰, respectively (15).

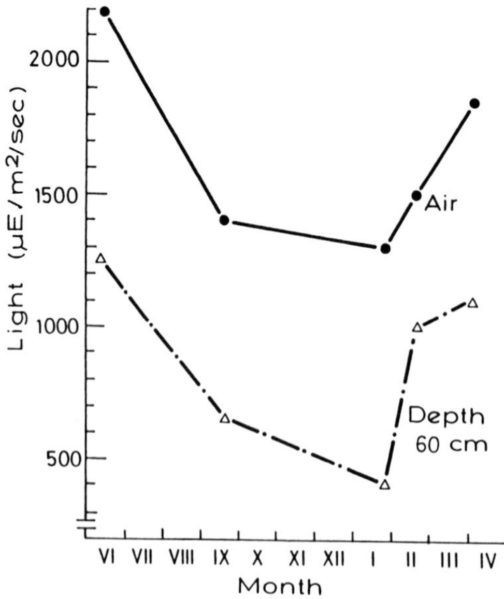

Figure 2. Seasonal changes in light intensity (in microeinsteins [μE] per square meter per second) measured in the air and at a depth of 60 cm in the 40-dunam pond between 10:00 and 11:00 a.m. Similar light intensities were also recorded in the other ponds investigated.

had more restricted temperature and salinity ranges, and showed maximal growth at different seasons. Diatoms (Fig. 9) were abundant within temperature and salinity ranges of 15 to 35°C and 40 to 100‰, respectively, and their levels peaked in the winter at 10^{11} cells m^{-2}. *Euglena* sp. (Fig. 10) exhibited an optimal growth at salinities of 70 to 100‰ and temperatures not exceeding 35°C. Maximal concentrations of 10^{10} cells m^{-2} were recorded in the fall. Figure 11 summarizes distribution patterns of the dominant microalgae on the solar pond littoral.

MICROALGAL GROWTH IN SOLAR PONDS

The experimental solar ponds of the type described here constitute an extreme but stable environment. As explained above, the salinities of the upper, convective water

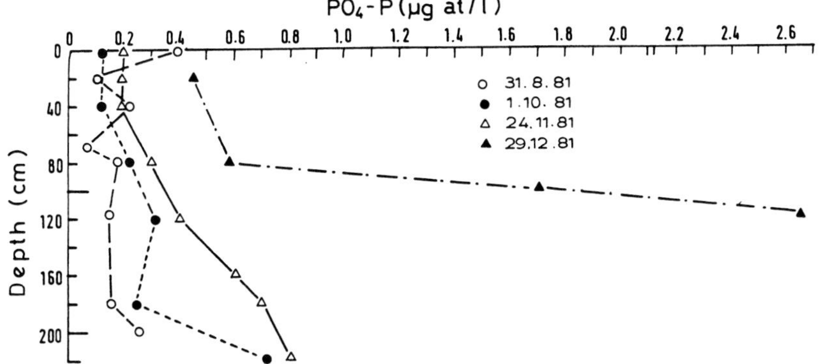

Figure 3. Concentrations of P_i in the 40-dunam pond. Similar values and seasonal changes were also found in the other ponds investigated.

layer are monitored and kept at the required level by the addition of fresh water. As a result of an established halocline, an inverted thermocline develops and the gradient layer, where the temperature and salinity increase gradually with depth, remains stable for prolonged time intervals. Under these strictly controlled conditions and in the absence of grazers, the preferences of various species for salinity, temperature, and light are maximally expressed and determine the pattern of their spatial and temporal distribution.

Available P_i in the ponds was scarce and, according to our laboratory bioassays (10), appeared to be the limiting nutrient in this ecosystem. As a result of this limitation, very little phytoplankton thrived in the ponds and the water column remained quite transparent. The light intensity of 400 microeinsteins m^{-2} s^{-1} measured at a depth of 60 cm, near the lower limits of the microalgal belt, is higher by at least a factor of 10

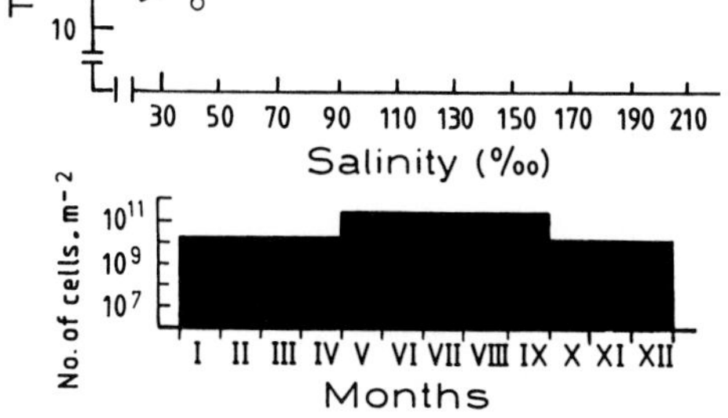

Figure 4. *A. halophytica* Fremy. The upper graph shows relative abundance (●, common to abundant; ○, rare) as a function of temperature and salinity; the number of points gives some approximation of the species frequency in the littoral samples. The lower graph shows seasonal changes in cell concentrations per unit area of the littoral within the optimal environmental range.

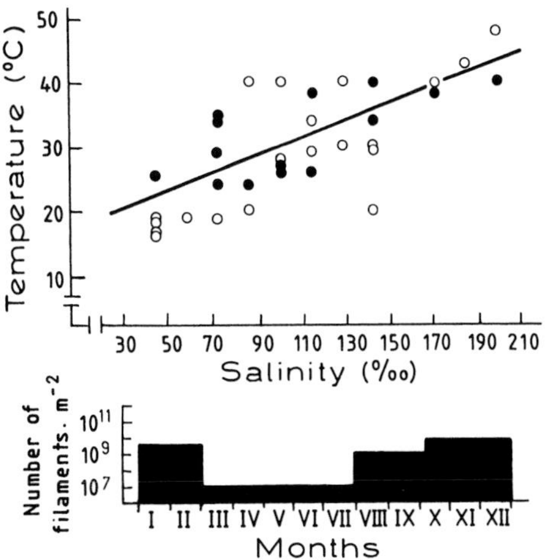

Figure 5. *P. hypolimneticum* Campbell. For a description of the graphs, see the legend to Fig. 4.

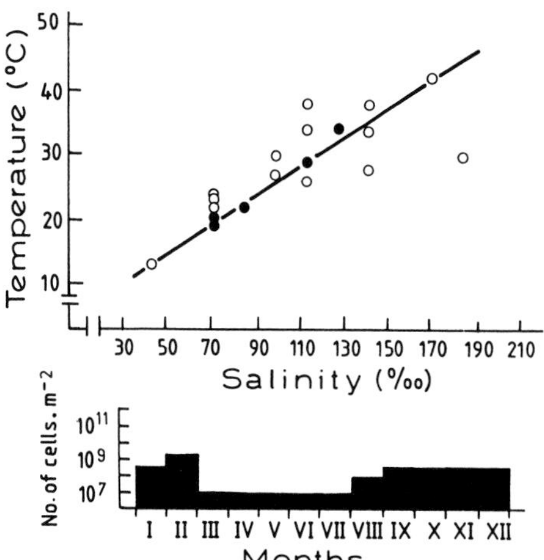

Figure 6. *Chroococcidiopsis* sp. For a description of the graphs, see the legend to Fig. 4.

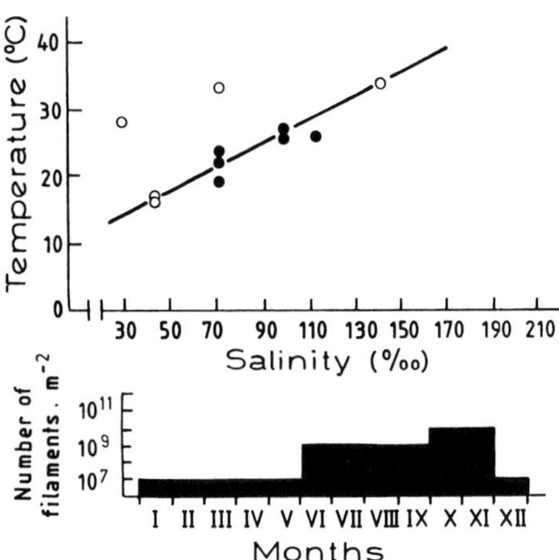

Figure 8. *S. subsalsa* Oersted. For a description of the graphs, see the legend to Fig. 4.

than the optimal light intensity used for laboratory growth of solar pond cyanobacteria (15, 16). Since the arid Dead Sea area is characterized by strong radiation and almost year-round cloudless weather, light is presumably an inhibitory rather than a limiting factor of the microalgal growth on the shallow pond littoral.

At the time of the present study, the experimental solar ponds were 1 to 6 years old and were evidently in the early stages of establishing their littoral communities. Multilayered mats with the underlying sulfuretum, expected to be present in such habitats (6, 7), were not yet formed. The diatoms, cyanobacteria, and a euglenide alga inhabiting the pond littoral showed definite vertical zonation and seasonal fluctuations.

The uppermost zone, down to a depth of 30 to 40 cm, with prevailing high light intensities and comparatively low temperature and low salinity, was evidently well

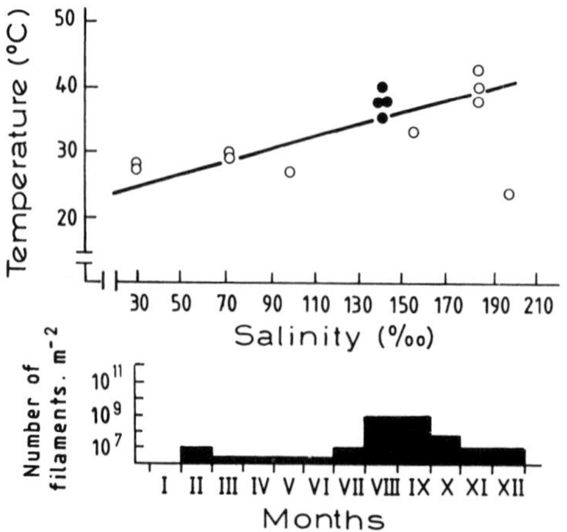

Figure 7. *P. hypersalinum* Campbell. For a description of the graphs, see the legend to Fig. 4.

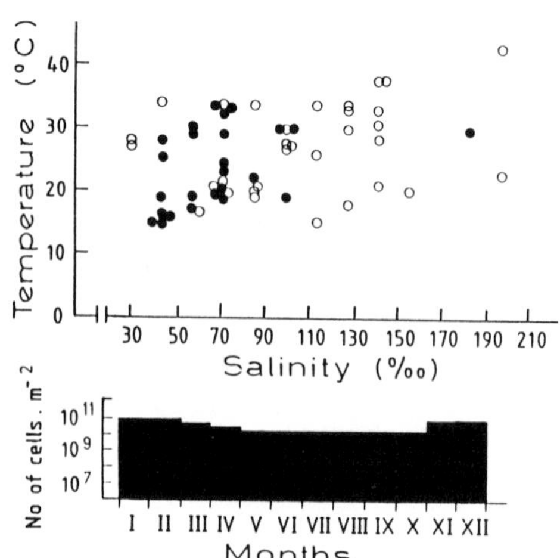

Figure 9. Diatoms: *Nitzschia* sp. aff. *N. rostellata* Husted and several species of *Navicula* (combined). For a description of the graph, see the legend to Fig. 4.

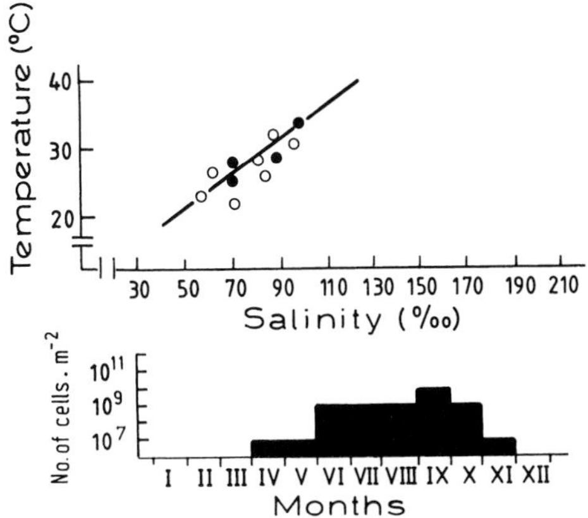

Figure 10. *Euglena* sp. For a description of the graphs, see the legend to Fig. 4.

suited to the environmental requirements of diatoms and *A. halophytica* among the cyanobacteria (Fig. 11). The mixed community appeared there as single cells, small flocks, or gelatinous masses on the bottom sediment. Diatoms, represented mostly by *Nitzschia* and *Navicula* spp., constituted about 40% of the total cell count in this shallow belt in the summer and reached 80 to 100% in the winter to early spring, outcompeting cyanobacteria in this season. *A. halophytica*, a predominant form, always flourished in the summer, when the temperatures increased, evidently providing a competitive advantage

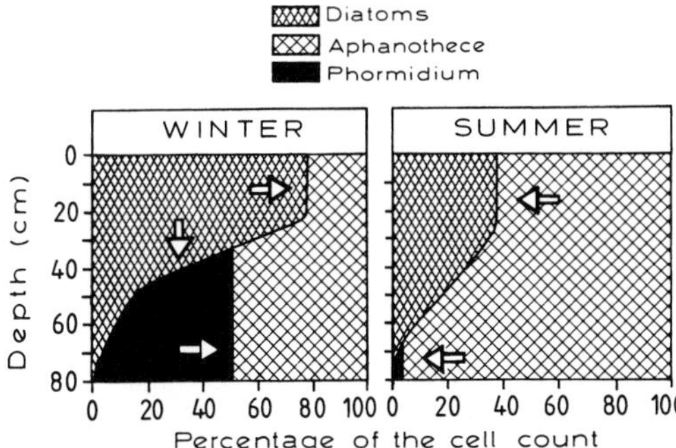

Figure 11. Schematic presentation of the temporal and spatial distribution of the dominant microalgae (*A. halophytica*, *P. hypolimneticum*, and diatoms) on the solar pond littoral. Arrows indicate directions of the assumed competitive pressures. For details, see the text.

for this cyanobacterium while restricting the growth of diatoms. The summer bloom of *A. halophytica* also suggests that it tolerates high light intensity; this was experimentally proved, at least for one littoral strain isolated (16). The deeper zone, where temperatures exceeded 35°C and salinities were higher than 72‰, was populated predominantly by cyanobacteria, which formed soft, gelatinous films of various thicknesses that persisted throughout the year. Also at this lower belt of the littoral, *A. halophytica* constituted a prominent component of the community. The unicellular cyanobacteria tentatively classified by Hof and Fremy into the taxon *A. halophytica* (20) have been repeatedly found (sometimes reported as *Synechococcus*, *Coccochloris*, or "coccoid cyanobacteria") in a variety of saline and hypersaline habitats (1, 3, 4, 9, 22). Our recent experimental studies revealed that strains of *A. halophytica*, although being extremely euryhaline and eurythermal, differ in their optimal environmental ranges. A similar conclusion, based on field data, was reached by Golubic (18). These strains increase their cell size and shape with increasing salinity, but differ in the extent of their polymorphism (16). An increase in cell size as a response to increased salinity was also reported for *A. halophytica* isolated from the Great Salt Lake, Utah (30). At higher temperatures (46°C as compared with 42°C), *A. halophytica* grown in the laboratory showed increased amounts of carbohydrate cell wall materials (15), which can explain the appearance of the thick, gelatinous films in the lower, warm zone of the ponds inhabited by these cyanobacteria. Various strains of *A. halophytica* have in common a coupling of salinity and temperature requirements: to sustain optimal growth, any increase in temperature must be compensated for by a concomitant increase in salinity. However, although the strains investigated were similar in their general response, their salinity and temperature optima differed. Such a phenomenon was also recently recorded for

Dunaliella (10) and *Staphylococcus* (21) species. The physiological adaptation of a single *A. halophytica* strain and the existence of numerous strains under overlapping environmental conditions explain the worldwide distribution of this group of halophilic organisms in the variable and unstable hypersaline habitats.

S. subsalsa grew mostly in the summer. This filamentous cyanobacterium, common in saline and hypersaline habitats (17, 27), grew on the top of the gelatinous mass of *A. halophytica* in the middle zone of the littoral, where salinities were 70 to 110‰ and temperatures were not higher than 28°C. After reaching a thickness of a few millimeters, the entangled mass of filaments was detached from the underlying *A. halophytica* mass by entrapped bubbles of photosynthetic oxygen and floated to the surface. Owing to this mechanism of spatial separation, direct competition for light and nutrients between the two species was evidently much reduced.

P. hypolimneticum, although abundant and widely distributed, appeared as single filaments; in these young ponds it did not form the continuous carpet characteristic of its growth in Solar Lake, Sinai (coast of the Red Sea) (5, 7). This species grew abundantly mostly in the deeper belt of the littoral, and its maximum growth occurred in the winter, which indicates a tolerance of high salinity and temperature, as well as a preference for lower light intensities. It should also be noted that the winter peak of *P. hypolimneticum* growth coincided with a considerable increase in the P_i content in the deeper water mass (Fig. 3). This deeper population of *P. hypolimneticum* appeared to be separated spatially from the diatoms blooming in the same season on the shallow littoral and to be separated temporally from *A. halophytica* inhabiting the same level, whose growth peaked in a different season (Fig. 11). Similar spatial divisions of the shallow littoral between the surface diatoms and subsurface *A. halophytica* were recently also reported to occur in Shark Bay and Spencer Gulf, Australia (2). The winter increase in concentration of filamentous cyanobacteria was also observed in Solar Lake (24). It may be assumed that at a later stage of maturation of the solar ponds, when a sulfuretum develops under the aerobic community on the accumulated organic matter, *P. hypolimneticum* will take advantage of its anoxygenic photosynthetic abilities in high concentrations of sulfide (6, 8) and become predominant in the deeper litoral zone.

The lower graphs in Fig. 4 and 5 provide seasonal counts of *A. halophytica* and *P. hypolimneticum*, respectively; together, these organisms constitute the dominant biomass of cyanobacteria on the solar pond littoral. However, it must be remembered that for *A. halophytica* single cells were counted, whereas for *P. hypolimneticum* multicellular filaments of various lengths were counted. Therefore, these counts do not give comparable quantitative information about the contribution of the two components to the total biomass.

P. hypersalinum (Fig. 7) was unique in this group of species by having quite a narrow salinity and temperature range for optimal growth, which was strongly shifted toward an increased salinity (130 to 150‰) and temperature (34 to 40°C). *Chroococcidiopsis* sp. constituted a consistent component of the cyanobacterial community, growing abundantly at 70 to 130‰ salinity and temperatures of 20 to 35°C. Its seasonal maximum in fall and winter indicates a preference for lower light intensity (Fig. 6). Representatives of this little-known genus are mostly aerophytic inhabitants of caves, soils, desert rocks, and stones (I. Friedman, personal communication). Our strain, although growing successfully in fresh water and at low temperatures (unpublished data), exhibited considerable halotolerance and thermotolerance in the solar ponds. However, under conditions of increasing salinity, the relative number of vegetative cells in the population decreased, with the concomitant appearance of encysted unicells and sporangia. A parallel reduction in cell size and a decline in metabolic activity were recorded. This passive resistance strategy evidently helps *Chroococcidiopsis* spp. to endure prolonged periods of excessive osmotic water stress while preserving viability. Diatoms were restricted to the shallow belt of the solar pond littoral; their growth peaked in winter, and the optimal conditions for growth were a salinity of 30 to 72‰ and a temperature not exceeding 30°C. *Euglena* sp., which occurred at greatest abundance within the salinity range of 70 to 100‰, is probably among the most halotolerant representatives of this genus (Fig. 10). Some unidentified *Euglena* species have also been reported in Spain and Germany, growing at salinities exceeding 50 and 100‰, respectively (19). However, in our ponds, near its upper salinity limits, *Euglena* cells round up and are immobilized.

CONCLUSIONS

The microalgal community of the solar ponds in the Dead Sea area has a greatly reduced species diversity as compared with similar communities in marine littoral (27), marginal marine environments (2, 6, 11, 17, 23), or inland saline lakes (19) of equivalent salinity. *Lyngbia*, *Microcoleus*, *Schizothrix*, and *Entophysalis* spp., which are present in the above environments as common components of microbial mats, are absent in the solar ponds of the Dead Sea area. Moreover, species recorded in this study show more restricted salinity ranges than those in the other hypersaline habitats. For example, *A. halophytica* and *S. subsalsa* were common in Gavish Sabkha up to salinities of 250 to 330 and 205‰, respectively. In the solar ponds the upper limits were only 210 and 150‰, respectively.

Similarly, diatoms tolerated higher salinities in Gavish Sabkha than in the solar ponds (13, 17). Although the species diversity is presumably limited by the particular ionic composition of the Dead Sea water (25), the narrowing of the salinity ranges evidently occurs because the salinity gradients in the solar ponds are always coupled with increasing temperature gradients. Thus, the combined effect of both factors determines the species distribution in the solar pond environment.

ACKNOWLEDGMENT. These studies were supported by a grant from Solmat Systems Ltd. Israel.

LITERATURE CITED

1. **Bauld, J.** 1981. Occurrence of benthic microbial mats in saline lakes. *Hydrobiologia* **81**:87–111.
2. **Bauld, J.** 1984. Microbial mats in marginal marine environments: Shark Bay, Western Australia, and Spencer Gulf, South Australia, p. 39–58. *In* Y. Cohen, R. W. Castenholz, and H. O. Halvorson (ed.), *Microbial Mats: Stromatolites.* Alan R. Liss, Inc., New York.
3. **Borowitzka, L. J.** 1981. The microflora: adaptations to life in extremely saline lakes. *Hydrobiologia* **81**:33–46.
4. **Brock, T. D.** 1976. Halophilic blue-green algae. *Arch. Microbiol.* **107**:109–111.
5. **Campbell, S. E., and S. Golubic** 1985. Benthic cyanophytes (cyanobacteria) of Solar Lake (Sinai). *Arch. Hydrobiol.* (Suppl. 71) **38/39**:311–329.
6. **Cohen, Y.** 1984. The Solar Lake cyanobacterial mats: strategies of photosynthetic life under sulfide, p. 133–148. *In* Y. Cohen, R. W. Castenholz, and H. O. Halvorson (ed.), *Microbial Mats: Stromatolites.* Alan R. Liss, Inc., New York.
7. **Cohen, Y., W. E. Krumbein, and M. Shilo.** 1977. Solar Lake (Sinai). 2. Distribution of photosynthetic microorganisms and primary production. *Limnol. Oceanogr.* **22**:609–620.
8. **Cohen, Y., E. Padan, and M. Shilo.** 1975. Facultative anoxygenic photosynthesis in the cyanobacterium *Oscillatoria limnetica. J. Bacteriol.* **123**:855–861.
9. **Dor, I.** 1967. Algues des sources thermales de Tiberiade. *Sea Fish. Res. Stn. Haifa Bull.* **48**:3–29.
10. **Dor, I.** 1983. *Research on Algae of Solar Ponds.* Report no. 1 to Solmat Systems Ltd. The Hebrew University of Jerusalem, Jerusalem, Israel. (In Hebrew.)
11. **Dor, I.** 1984. Epiphytic blue-green algae (cyanobacteria) of the Sinai Mangal: considerations on vertical zonation and morphological adaptations, p.

35–54. *In* F. D. Por and I. Dor (ed.), *Hydrobiology of the Mangal.* Dr. W. Junk Publishers, The Hague, The Netherlands.

12. **Dor, I.** 1985. Long lasting effect of dilution on the cell volume, motility, division rate and vertical distribution of *Dunaliella parva* Lerche. *J. Exp. Mar. Biol. Ecol.* **91**:183–197.
13. **Dor, I., and A. Ehrlich.** 1986. Cyanobacteria-diatom competition for the littoral zone of the experimental solar ponds, Dead Sea area, Israel, p. 361–367. *In* Z. Dubinski and Y. Steinberger (ed.), *Environmental Quality and Ecosystem Stability*, vol. 3A. Bar-Ilan University Press, Ramat Gan, Israel.
14. **Dor, I., and A. Ehrlich.** 1987. The effect of salinity and temperature gradients on the distribution of littoral microalgae in experimental solar ponds, Dead Sea area, Israel. *Mar. Ecol.* **8(3)**:193–205.
15. **Dor, I., and M. Hornoff.** 1985. Salinity—temperature relations and morphotypes of a mixed population of coccoid cyanobacteria from a hot, hypersaline pond in Israel. *Mar. Ecol.* **6(1)**:13–25.
16. **Dor, I., and M. Hornoff.** 1985. Studies on *Aphanothece halophytica* Fremy from a solar pond: comparison of two isolates on the basis of cell polymorphism and growth response to salinity, temperature and light conditions. *Bot. Mar.* **28**:389–398.
17. **Ehrlich, A., and I. Dor.** 1985. Photosynthetic microorganisms of the Gavish Sabkha, p. 296–321. *In* G. M. Friedman and W. E. Krumbein (ed.), *Hypersaline Ecosystems: the Gavish Sabkha.* Springer-Verlag KG, Berlin.
18. **Golubic, S.** 1980. Halophily and halotolerance in cyanophytes. *Origins Life* **10**:169–183.
19. **Hammer, U. T.** 1986. *Saline Lake Ecosystems of the World.* Dr. W. Junk Publishers, Dordrecht, The Netherlands.
20. **Hof, T., and P. Fremy.** 1933. On myxophyceae living in strong brines. *Recl. Trav. Bot. Neerl.* **30**:140–162.
21. **Hurst, A., A. Hughes, and R. Pontefract.** 1980. Mechanism of salts on *Staphylococcus aureus. Can. J. Microbiol.* **26**:511–515.
22. **Imhoff, J. F., F. Hashwa, and H. G. Trüper.** 1978. Isolation of extremely halophilic, phototrophic bacteria from the alkaline Wadi Natrun Egypt. *Arch. Hydrobiol.* **84(3)**:381–388.
23. **Javor, B. J., and R. W. Castenholz.** 1984. Productivity studies of microbial mats, Laguna Guerrero Negro, Mexico, p. 149–170. *In* Y. Cohen, R. W. Castenholz, and H. O. Halvorson (ed.), *Microbial Mats: Stromatolites.* Alan R. Liss, Inc., New York.
24. **Krumbein, W. E., Y. Cohen, and M. Shilo.** 1977. Solar Lake (Sinai). 4. Stromatolitic cyanobacterial mats. *Limnol. Oceanogr.* **22**:635–656.
25. **Nissenbaum, A.** 1975. The microbiology and biogeochemistry of the Dead Sea. *Microb. Ecol.* **2**:139–161.
26. **Por, F. D.** 1980. A classification of hypersaline waters based on trophic criteria. *Mar. Ecol.* **1**:121–131.
27. **Stal, L. J., and W. E. Krumbein.** 1985. Isolation and characterisation of cyanobacteria from a marine microbial mat. *Bot. Mar.* **28**:351–365.
28. **Strickland, J. D. H., and T. R. Parsons.** 1972. *A Practical Handbook of Seawater Analysis*, 2nd ed. Bulletin 167. Fisheries Research Board of Canada, Ottawa.
29. **Tabor, H.** 1981. Solar ponds. *Sol. Energy* **27**:171–194.
30. **Yopp, J. H., D. R. Tindall, D. M. Miller, and W. E. Schmid.** 1978. Isolation, purification and evidence for the halophilic nature of the blue-green alga *Aphanothece halophytica* Fremy (Chroococcales). *Phycologia* **17**:172–178.

Chapter 11

Light Penetration, Absorption, and Action Spectra in Cyanobacterial Mats

Bo Barker Jørgensen

INTRODUCTION

The thick cyanobacterial mats which grow in shallow hypersaline ponds are frequently laminated as a result of seasonal variations in light, temperature, salinity, etc. (1, 13, 22, 25). The alternating colored layers are inhabited by phototrophs with different pigment compositions and thus with different patterns of spectral utilization of the available light. Organisms in the top layer are illuminated directly by daylight with a broad spectral composition. Organisms in the deeper layers within the photic zone receive light which has been filtered through the overlying pigments. The available light therefore has a lower intensity and is spectrally altered relative to the incident light.

These cyanobacterial mats are ideal systems for use in the study of several aspects of spectral adaptation in natural populations of phototrophs. In contrast to planktonic systems, most photosynthetic microorganisms within the mats will continuously and actively move in response to varying light intensities or chemical gradients (4). Since the photic zone is often less than 1 mm deep, the study of photosynthetic activity within individual mat layers has been greatly facilitated by the use of oxygen microelectrodes (29). The advantage of these electrodes is not only their high spatial resolution of about 0.1 mm, but also their speed of measurement. Since each determination of photosynthetic rate takes only a few seconds (followed by a reequilibration period of 10 to 60 s), the microelectrodes are ideal for ecophysiological studies of, e.g., light saturation or action spectra (17, 30). Such studies, however, require information on the spectral composition and intensity of the available light in situ. We therefore developed a fiber-optic microprobe which also allowed spectral light measurements to be made at a resolution of <0.1 mm (19). The goal was to concurrently analyze photosynthetic rates and spectral quantum flux within small clusters of cells. This required a more detailed study of the optical properties of the mats, including the effects of light scattering.

The cyanobacterial mats investigated in this work were located in large, hypersaline ponds at Guerrero Negro, Baja California, Mexico (13, 14). The mats were dominated by the filamentous cyanobacterium *Microcoleus chthonoplastes*, which grows in dense

Bo Barker Jørgensen • Institute of Ecology and Genetics, University of Aarhus, Ny Munkegade, DK-8000 Aarhus C, Denmark.

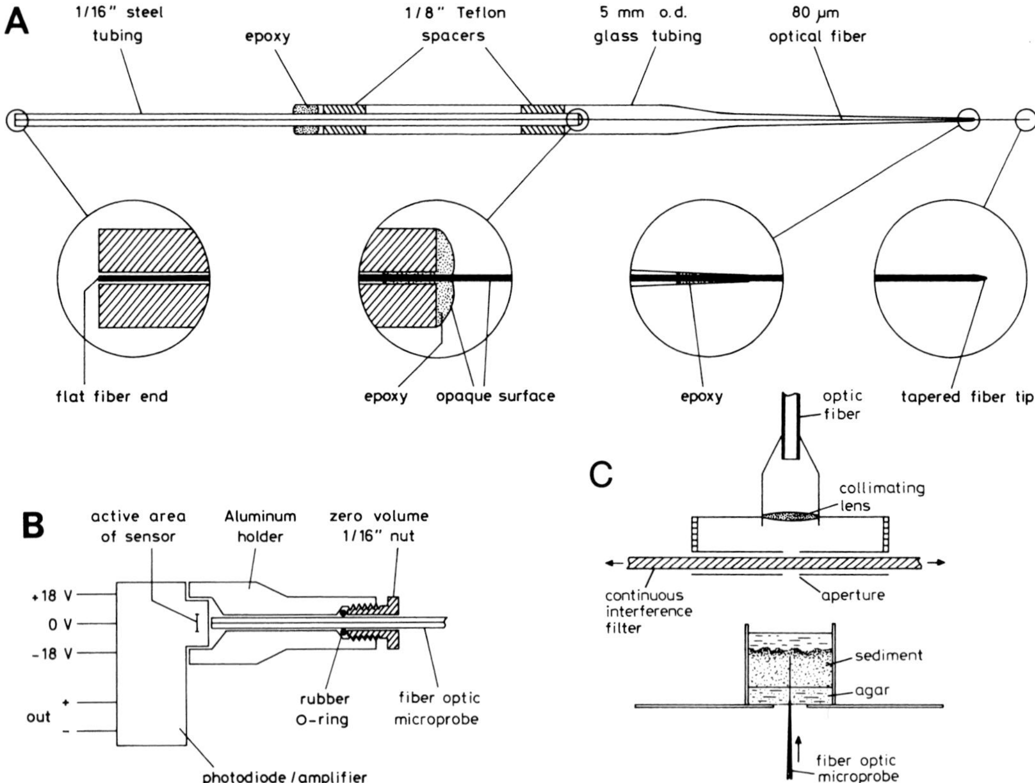

Figure 1. Spectral radiance measurements with a fiber-optic microprobe. (A) The light probe with a single-stranded optical fiber, tapered at the end. (B) Attachment of the probe to a hybrid photodiode/amplifier detector. (C) Experimental setup for measurement of spectral radiance gradients in cyanobacterial mats. Reproduced from *Limnology and Oceanography* (19) with permission of the publisher.

bundles surrounded by common sheaths. *Oscillatoria* and *Phormidium* spp. were also abundant in the mats, as were purple sulfur bacteria and green sulfur bacteria, mostly *Chloroflexus* spp. On the surface was a thin, fluffy film of diatoms, mostly *Nitzschia* sp. The mats grew on the bottom of artificial evaporation ponds, 0.5 to 1 m deep, which contained seawater concentrated to a brine of 60 to 120‰. Water temperatures were 15 to 23°C. The main results of some recent studies of these mats are summarized and discussed below.

FIBER-OPTIC MICROPROBES

The simple fiber-optic microprobe, its coupling to a light detector, and the experimental setup for measurements of light penetration into the cyanobacterial mats are shown in Fig. 1. The probe consisted of a single-stranded optical fiber of 80 µm in diameter built into a shaft of steel and glass (Fig. 1A). The fiber was tapered at the end and had a rounded sensing tip of 20 to 30 µm in diameter. This tip shape allowed a spatial resolution of about 50 µm. During measurements, the rounded tip reduced the mechanical disturbance of the mat due to the advancing fiber. It also modified the numerical aperture of the fiber (Fig. 2). The acceptance half-angle of the fibers was 25 to 30° in air (about 20° in water) and increased slightly with wavelength. The light probe was optically coupled to an ultra-low-noise photodiode detector, which was attached directly

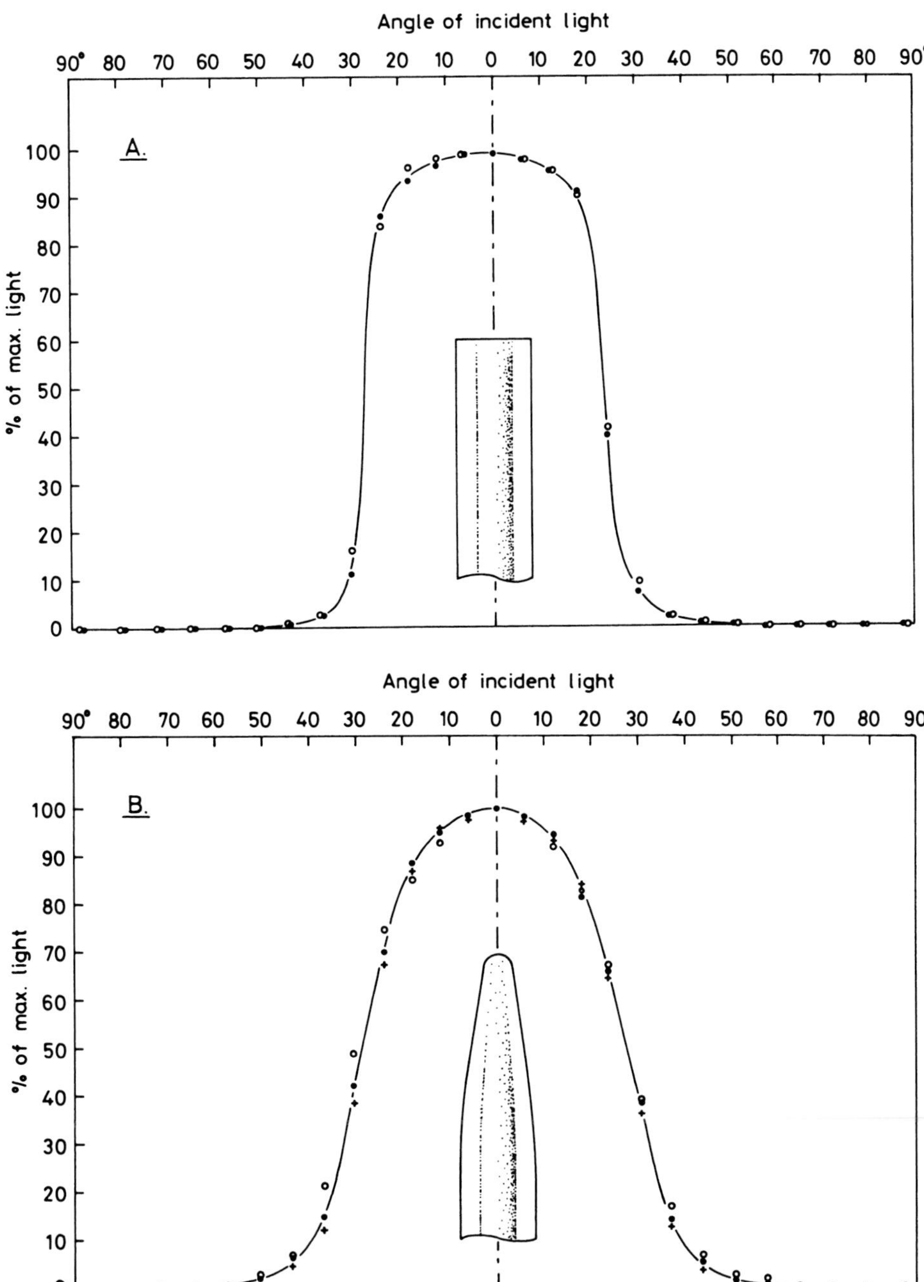

Figure 2. Examples of two fiber tips and their acceptance angles for light in air. (A) Flat fiber end; the fiber was rotated 90° between the first (○) and the second (●) set of measurements of white light to check axial symmetry. (B) Tapered and rounded fiber end; measurements were made in monochromatic light at 400 nm (○), 700 nm (●), and 1,000 nm (+). Reproduced from *Limnology and Oceanography* (20a).

to the end of the fiber (Fig. 1B). Despite the narrow aperture of the fiber tip, 0.0002 mm^2, the probe was very sensitive. The detectability of white light was about 0.001 W m^{-2} or 0.01 microeinstein m^{-2} s^{-1}.

Measurements of downwelling light in the cyanobacterial mats require that the fiber penetrate the mat from below. Small mat segments were cored with plastic tubing which was then sealed at the bottom with an agar plug and positioned over a hole in a plastic plate (Fig. 1C). The mat surface was illuminated vertically by monochromatic light, and the spectrum was scanned at 10-nm intervals through the visible range or at 25-nm intervals through the visible plus near-infrared range. The light microprobe was attached to a micromanipulator and was advanced up through the mat. Light spectra were measured at depth increments of 50 to 100 μm and recalculated for each wavelength as a percentage of the surface value. Measurements were made with the fiber pointing directly toward the light source (0° angle) as well as at other angles.

RADIANCE, IRRADIANCE, AND SCALAR IRRADIANCE

The fiber-optic microprobe has a directional sensitivity (Fig. 2). It will measure the radiant flux, Φ, in the direction of its axis. This direction can be defined relative to the horizontal mat surface by its zenith, θ, and azimuth, ϕ, angles. The optical property, which is measured by the light microprobe, is a radiance, $L(\theta,\phi)$. The radiance is the radiant flux per unit area, A, and per unit solid angle (steradian), ω (Fig. 3), and is expressed as follows:

$$L(\theta, \phi) = d^2\Phi/dA \cdot d\omega$$

The probe detects only a small part of the total light field in the mat. When positioned vertically, it measures the attenuation of the collimated light beam below the mat surface.

Most of the earlier studies of light penetration into mats and sediments involved the use of large irradiance sensors, over which mat slices of various thicknesses were placed (10, 21, 33). These sensors have a cosine collector which measures the total downward irradiance, E_d, by integrating the radiance from all directions over the upper hemisphere (solid angle = 2π) onto a horizontal surface. The equation for E_d is as follows:

$$E_d = \int_{2\pi} L(\theta,\phi) \cdot \cos\theta \cdot d\omega$$

Irradiance measurements have the advantage of including both the collimated light and the forward (downward)-scattered light. The vertical attenuation of downward irradiance is therefore a more relevant parameter than downward radiance for describing how much light reaches the different mat layers.

Much of the available light within the cyanobacterial mats is due to scattering rather than to the direct light beam. The light is scattered both forward and backward. The most relevant measure of available light for an algal cell is therefore not the downward irradiance but, rather, the scalar irradiance, E_0, i.e., the integrated radiance distribution at the point of the cell over all directions, which is given by the following equation:

$$E_0 = \int_{4\pi} L(\theta,\phi) \cdot d\omega$$

As a measure of light, the scalar irradiance is similar to the spherical irradiance, E_s, which is the total radiant flux onto a spherical surface divided by the area of the surface. The two quantities differ by a factor of 4: $E_0 = 4E_s$. They both describe the total flux of quanta from all directions which reaches a cell at a given point within the mat.

The scalar irradiance could, in theory, be measured for a given point or depth by pointing the fiber toward all directions until

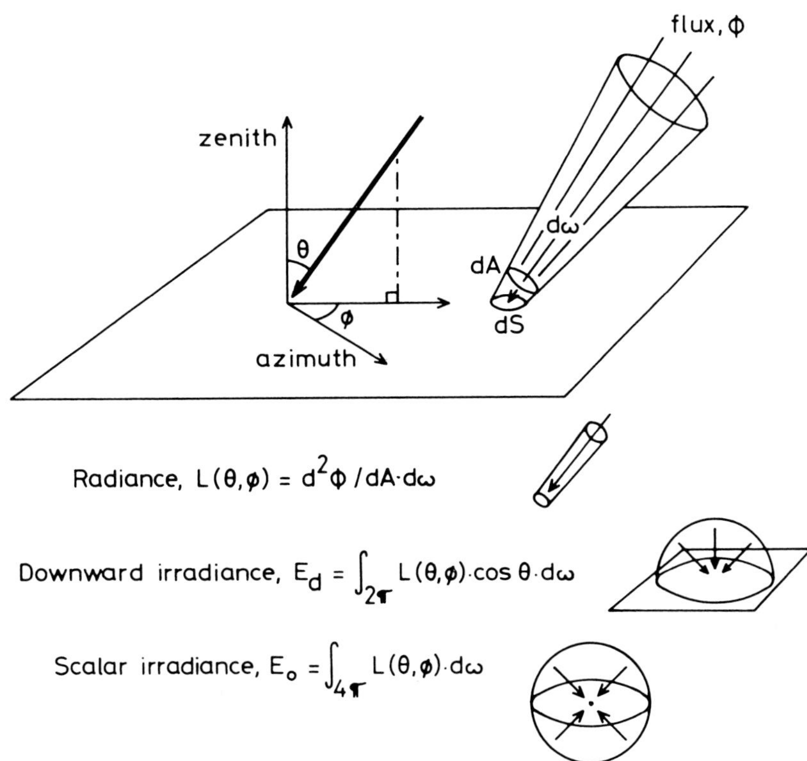

Figure 3. Radiance. The radiance, $L(\theta,\phi)$, is defined as the energy or quantum flux within the solid angle, $d\omega$, through the area, dA, perpendicular to the flux. The direction is defined by the spherical coordinates, θ and ϕ, whereas $d\omega$ and dA represent infinitesimal quantities. The projected area of dA on a horizontal surface is dS, and thus $dA = dS \cdot \cos\theta$. The radiances onto the horizontal surface per unit area are thus given by $L(\theta,\phi) \cdot \cos\theta \cdot d\omega$. When the radiances are integrated over all solid angles of the upper hemisphere, which has a solid angle of 2π, the downward irradiance, E_d, is obtained. When the radiances are integrated over the whole sphere, the scalar irradiance is obtained. Since the scalar irradiance is not defined relative to a plane but relative to a point, it does not have the cosine dependence.

the acceptance cone had covered the whole sphere. This, however, is hardly feasible in practice. Instead, one can use the fact that the light field has axial symmetry around the vertical axis when the mat is illuminated by collimated light which is incident perpendicular to the surface. The radiance is independent of the azimuth angle. The fiber therefore needs only to be pointed in enough directions relative to the vertical (zenith angles) to cover the half-circle, 0 to 180°. This is illustrated in Fig. 4. The directions covered were 0, 30, 60, 120, and 155° of light-fiber angle. For each direction, the measured radiance was then multiplied by the surface area of the spherical zone, which it represents. The areas of the spherical zones are only weighting factors for each direction of measurement. Since the sphere has unit radius, the areal sum of the spherical zones is equal to the surface area of the whole sphere, 4π. This way of summing the radiance distributions from all directions corresponds to theoretically sweeping the acceptance cone over the inner surface of the sphere during one rotation for each fixed

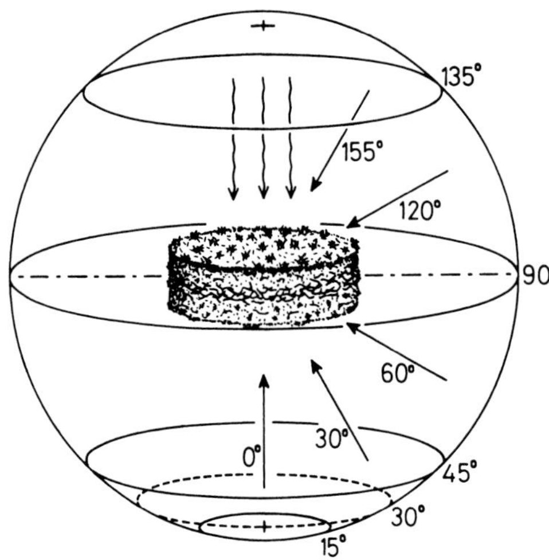

Figure 4. Calculation of scalar irradiance from radiance measurements. The mat segment is illuminated vertically from above. The straight arrows indicate five different orientations of the fiber-optic microprobe. A sphere of unit radius is drawn around the mat. Each spherical zone represents a fiber orientation. The surface area of the spherical zone is then a weighing factor by which the radiance in that direction is multiplied. Reproduced from *Limnology and Oceanography* (20a).

angle. If this is done with the probe fixed at 0, 30, 60°, etc., of light-fiber angle, the whole sphere will be covered.

More detailed discussions of the optical properties of scattering media and aquatic environments, as well as of definitions and units, are presented elsewhere (9, 11, 12, 15, 23, 24, 27).

RADIANCE GRADIENTS

The downwelling radiance in cyanobacterial mats is attenuated by a combination of absorption and scattering. Scattering increases rather monotonously from the longer toward the shorter wavelengths. Absorption in the mats is due mostly to the pigments of the dominant phototrophs, and the radiance spectra at different depths clearly show the absorption peaks from these pigments. Figures 5, 6, and 7 show three different ways of presenting the radiance gradients in cyanobacterial mats from the hypersaline ponds at Guerrero Negro: by tracing the attenuation of individual wavelengths with depth (Fig. 5); by measuring complete spectra at individual depths (Fig. 6); and by combining the first two ways to form three-dimensional plots (Fig. 7).

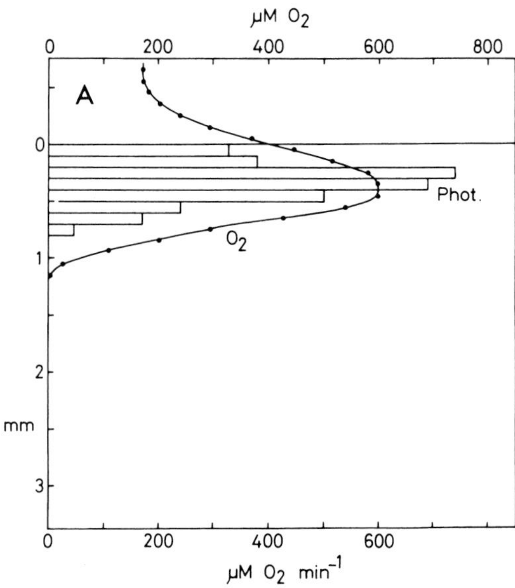

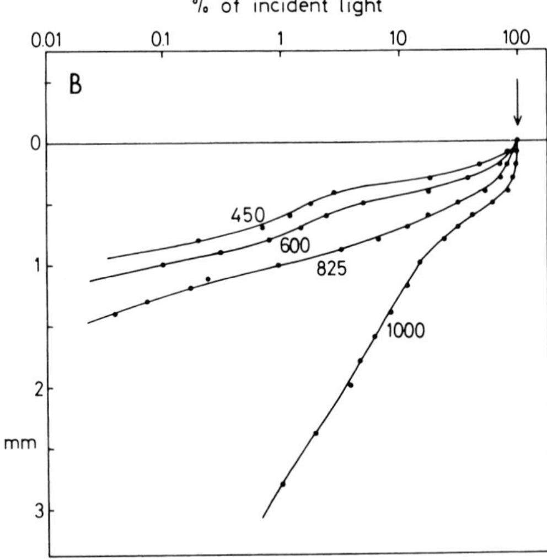

Figure 5. Depth distribution of (A) photosynthesis (Phot.) and oxygen (O_2) and (B) light in a *Microcoleus* mat. The euphotic zone was 0.8 mm deep. Numbers on light gradients indicate wavelengths (in nanometers) at which the downwelling radiance was measured. Reproduced from *Applied and Environmental Microbiology* (17).

Figure 5 shows the depth distribution of photosynthesis relative to the light gradients. The rates of oxygenic photosynthesis were measured with oxygen microelectrodes by using the light-dark shift technique (29). The euphotic zone was only 0.8 mm deep. The lower boundary of detectable photosynthesis coincided with the depth where visible light fell below 1% of the surface intensity. Only infrared light, especially of the longer wavelengths, penetrated deeper into the mat. In contrast to the near-infrared light of 825 nm, light of 1,000 nm is not absorbed by bacteriochlorophylls in these mats.

This spectral effect is shown more clearly in Fig. 6, which presents results for a very gelatinous mat. Owing to copious production of polysaccharide sheath material, the density of phototrophs was low and the euphotic zone reached a depth of over 4 mm. Oxygen produced by the cyanobacteria reached only slightly deeper, to 5.5 mm, by diffusion away from the oxygen maximum. Hydrogen sulfide, which was produced by sulfate-reducing bacteria in the mat layers below, diffused upward almost far enough to make contact with the oxygen. At the interface between the oxic and sulfidic zones

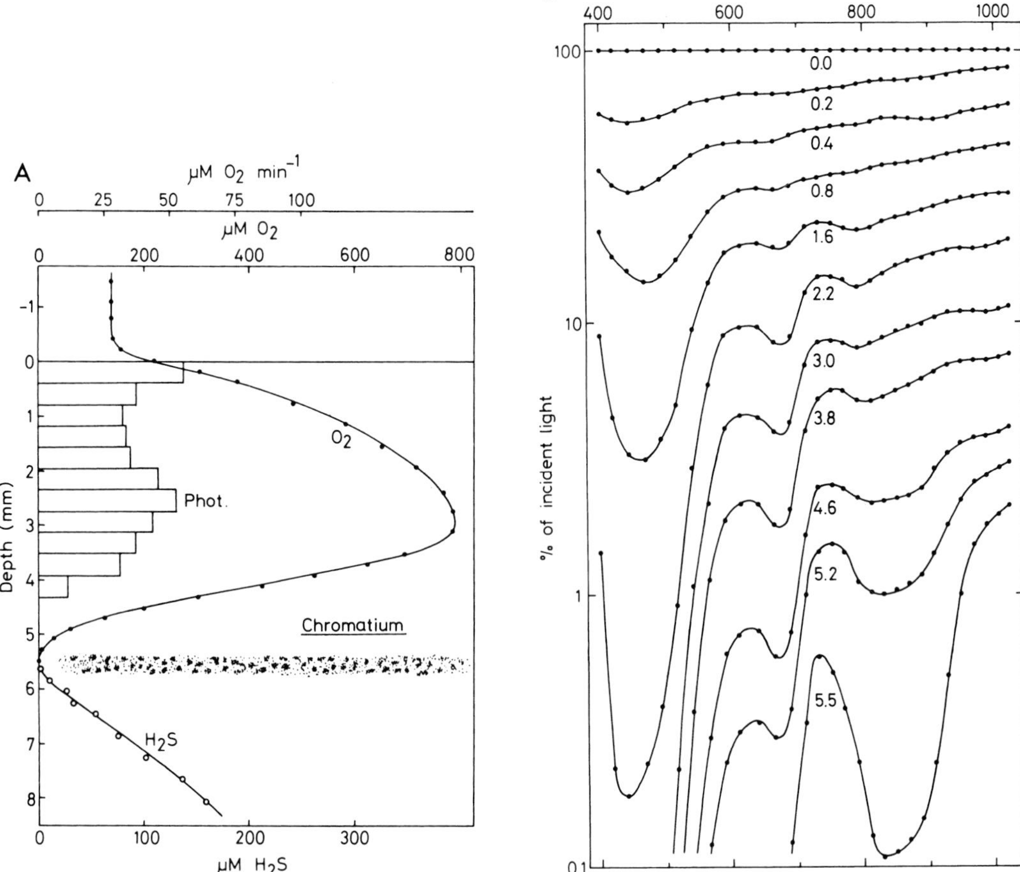

Figure 6. Vertical distribution of (A) photosynthesis (Phot.), oxygen (O_2), and sulfide (H_2S) and (B) light in a relatively translucent mat. A dense, pink band of purple sulfur bacteria was positioned between oxygen and sulfide at a depth of 5.5 mm. Spectra of downwelling radiance were measured at the depths (in millimeters) indicated. Reproduced from *FEMS Microbiology Ecology* (20) with permission of the publisher (A) and from *Limnology and Oceanography* (20a) (B).

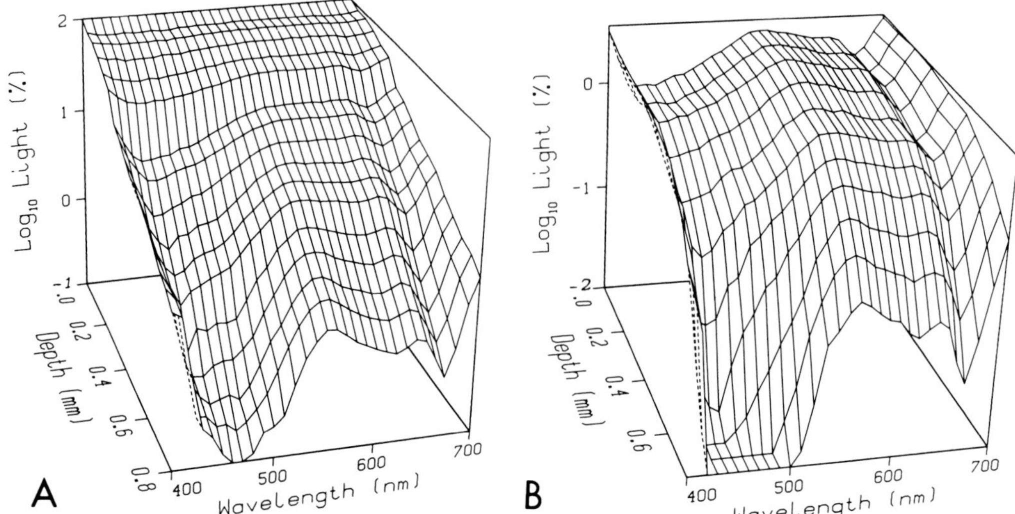

Figure 7. (A) Downward (0°) and (B) upward (145°) spectral radiance in a *Microcoleus* mat (Pond 5). The radiances were calculated as percentages of the collimated light incident on the mat surface for each wavelength. The downward radiance was therefore 100% ($\log_{10}$ = 2) at the surface. The upward radiance was around 1% ($\log_{10}$ = 0) at the surface. Broken lines are used where the back side of the topographic surface is visible. Reproduced from *Limnology and Oceanography* (20a).

there was a sharp, pink band of purple sulfur bacteria. Although visible light had been attenuated below the threshold for oxygenic photosynthesis at a depth of 5.5 mm, there was evidently enough near-infrared light for the phototrophic bacteria to grow. Indeed, the radiance spectra showed that although all visible light had been attenuated below 1% at 4.6 mm, a small percentage of the infrared light remained. At 5.5 mm, in the middle of the pink band, the attenuation of light at a wavelength of 800 to 900 nm was very intensive owing to absorption by bacteriochlorophyll *a* in the purple bacteria. Light in the range of 700 to 800 nm, which can be utilized by the green sulfur bacteria, was not strongly absorbed, and neither was light of >900 nm. In the visible range, the most distinct spectral feature was the absorption of blue light by carotenoids.

When analyzed at higher spectral resolution, the absorption by phycobilins and chlorophyll a (Chl *a*) was detectable. In Fig. 8, a radiance spectrum of the *M. chthonoplastes* mat from Baja California is compared

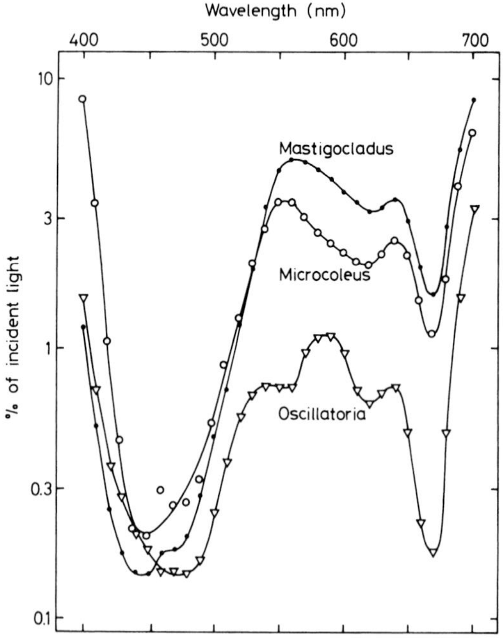

Figure 8. Comparison of downward radiance spectra in the lower euphotic zones of three cyanobacterial mats (*Microcoleus*, *Mastigocladus*, and *Oscillatoria* mats). Reproduced from *Microbial Ecology* (20b); *Oscillatoria* data from R. W. Castenholz, J. Bauld, and B. B. Jørgensen, manuscript in preparation.

with those of two cyanobacterial mats from hot springs: a *Mastigocladus laminosus* mat from Iceland and an *Oscillatoria boryanum* mat from New Zealand. The spectra were measured in the lowest euphotic zone of each mat. Abundant carotenoids in all the mats combined with the blue absorption peak of Chl *a* caused the strong attenuation of blue light at wavelengths between 430 and 500 nm. Also, the red absorption peak of Chl *a* at 675 nm was very distinct, whereas absorption by phycocyanin at 620 nm was less pronounced. The radiance spectra of the *M. laminosus* and *M. chthonoplastes* mats were quite similar, whereas that of the *O. boryanum* mat had an additional absorption band at 550 to 570 nm due to phycoerythrin.

A complete data set of downward spectral radiance is shown in Fig. 7A in a three-dimensional plot. Light intensities were measured at 50-μm depth increments from the surface to 0.8 mm for every 10 nm. The light intensities are again expressed as a percentage of the surface radiance for each wavelength and plotted logarithmically. The cyanobacterial mat was covered by a 0.2-mm-thick diatom film with a relatively low light attenuation and spectral modulation. The dense cyanobacterial layer below the diatoms showed strong absorption due to Chl *a*, phycocyanin, and carotenoids. The blue light intensity had already dropped to 1% of the surface radiance at 0.8 mm, which illustrates the necessity of submillimeter resolution in light measurements of such compact photosynthetic communities.

The backscattered light had a relative spectral composition similar to that of the downward radiance (Fig. 7B). The upward radiance at the mat surface was only 1 to 2% of the downward radiance. The relative intensity and spectral composition of this upward radiance determine the apparent visual brightness and color of the mat. Since relatively more of the scattered light in the range of 550 to 650 nm escapes absorption and is radiated back across the surface, the mats generally have an orange-brown appearance.

SCALAR IRRADIANCE

The scalar irradiance comprises light from all directions and was calculated from spectral radiance gradients measured under five representative light-fiber angles. Both the downward radiance (0° light-fiber angle) and the backscattered light were rather uniformly attenuated with depth (Fig. 7). The forward-scattered light, in contrast, showed a more complex spectral depth distribution, as shown in Fig. 9 for the radiance spectra for a 60° light-fiber angle. In the water above the mat, only a little forward-scattered light was detected, since the vertical, collimated light fell outside the 20° acceptance cone of the optical fiber. As the collimated beam penetrated down into the mat, the light was gradually scattered by multiple refraction and reflection and was also absorbed by pigments. As a result, the forward-scattered light reached a maximum at 0.05 to 0.1 mm below the mat surface. The light penetrated most deeply in spectral regions with the least absorption. Below the maximum, the 60° light decreased exponentially with depth.

The combination of absorption and scattering effects described above gives the mat surface optical properties which may be intuitively unexpected and which are very important for the light flux to the microalgae. In short, there is more light available at the surface than we may have expected, and it has a spectral composition which is different from that of the light source.

This phenomenon is illustrated by two examples in Fig. 10. Scalar irradiance spectra were calculated at 50- to 100-μm depth intervals for two *Microcoleus* mats. One of the mats (Pond 5) had a very dense population of cyanobacteria and very few diatoms, whereas the other (Pond 8) had a relatively lower density of cyanobacteria, with more sheath material and a diatom film on the surface. The scalar irradiances are expressed as a percentage of the downward scalar irradiance, i.e., of the light intensity at which the mats were illuminated. In the Pond 5 mat,

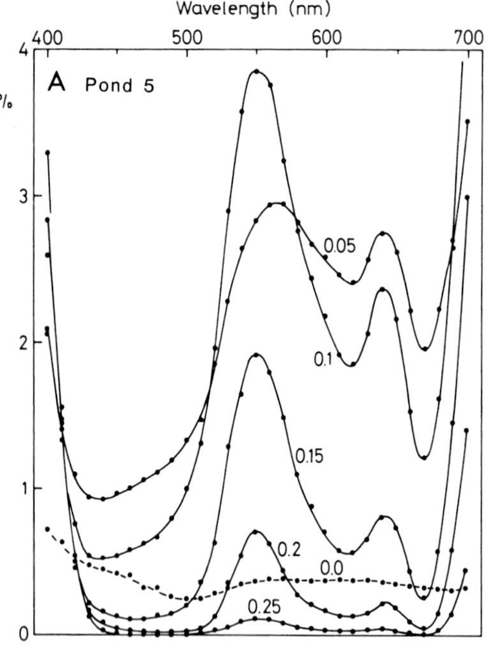

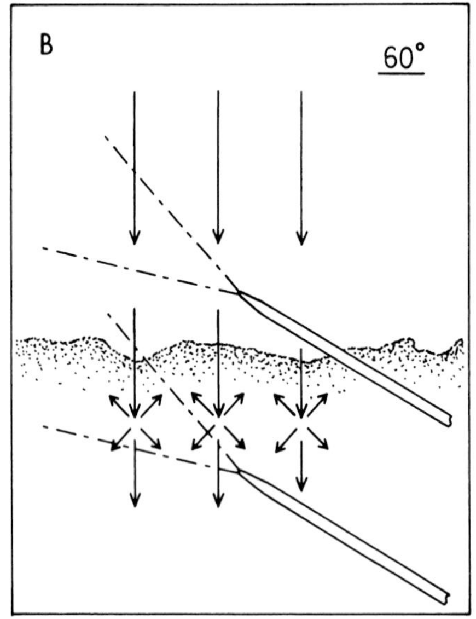

Figure 9. Forward-scattered light in a *Microcoleus* mat. (A) Radiance spectra were measured at a 60° light-fiber angle. Numbers on curves indicate depths (in millimeters) below the mat surface. Notice that the forward-scattered light at this angle reaches its maximum level below the mat surface. (Reproduced from *Limnology and Oceanography* [20a].) (B) Light microprobe at 60° above and below the mat surface. The figure illustrates the build-up of forward-scattered light to a maximum below the surface.

the pigment concentration was very high and absorption was dominant relative to scattering. In the Pond 8 mat, there was relatively more scattering, which added to the light of the collimated beam near the surface. As a result, the scalar irradiance reached up to 190% of the downward scalar irradiance at a depth of 0.1 to 0.2 mm. At 0.2 mm, there was a threefold-higher level of orange-red light (570 to 650 nm) than blue light (430 to 500 nm) relative to the levels in the light source.

These strong effects of light scattering shown for Pond 8 are very important in understanding light saturation and action spectra in compact photosynthetic communities. Similar optical properties must exist not only in other microbial mats but also in any type of phototrophic cell population in illuminated sediments, as well as in epiphytic films, thalli of macroalgae, leaves of higher plants, etc. (31, 34). Yet, to my knowledge, no simultaneous measurements have been made of photosynthesis and scalar irradiance at this resolution.

ACTION SPECTRA

It is known from earlier studies, either in situ or in pure cultures, that planktonic and benthic algae adapt to the ambient light intensity and spectral composition (6, 7, 28, 32, 36, 37). Shade-adapted organisms living in the lower part of the photic zone develop more antenna pigment per reaction center and become light saturated at a lower light intensity. Also, a spectral adaptation to the prevailing blue-green light at greater depths is often observed. Many algae increase the amount of accessory pigments such as carotenoids, and some cyanobacteria produce more phycoerythrin as a chromatic adaptation (3).

It is obvious from the present studies of spectral scalar irradiance gradients that the phototrophic organisms living on the mat surface have access to a wider spectral range

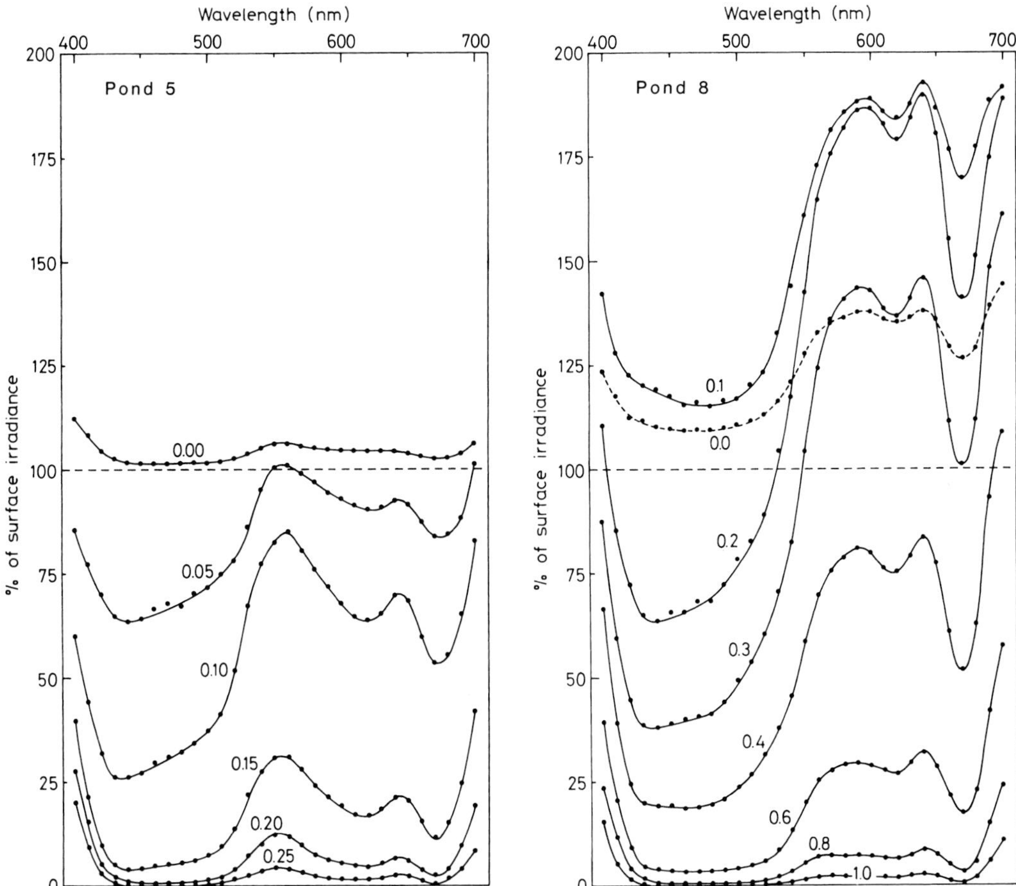

Figure 10. Scalar irradiance spectra at different depths (in millimeters) of two *Microcoleus* mats. The spectra were normalized for each wavelength as a percentage of the downward scalar irradiance at the mat surface. Reproduced from *Limnology and Oceanography* (20a).

of light than those living deeper in the euphotic zone. It is therefore interesting to know whether they also utilize this wider range. In several undisturbed mats, in situ action spectra for oxygen evolution were measured for the two dominant phototrophic communities: the diatom film on the surface and the dense *Microcoleus* band below. The mats were illuminated by monochromatic light which was shifted through the visible spectrum at 10-nm intervals. Photosynthesis rates were measured with an oxygen microelectrode at each wavelength and normalized to a light intensity of 20 microeinsteins m^{-2} s^{-1}. Owing to the high spatial resolution of the technique, it can separate the photosynthetic activities of cell populations spaced only 0.1 mm apart.

The action spectrum of diatoms growing on the mat surface showed the highest activity at the blue and red absorption maxima of Chl *a* (Fig. 11A). A broad shoulder in the green light of 480 to 530 nm was due to photosynthetically active carotenoids. Another shoulder around 620 nm was due to phycocyanin in unicellular cyanobacteria which occurred scattered within the diatom film. Apart from this cyanobacterial interference, Fig. 11A represents a typical diatom action spectrum.

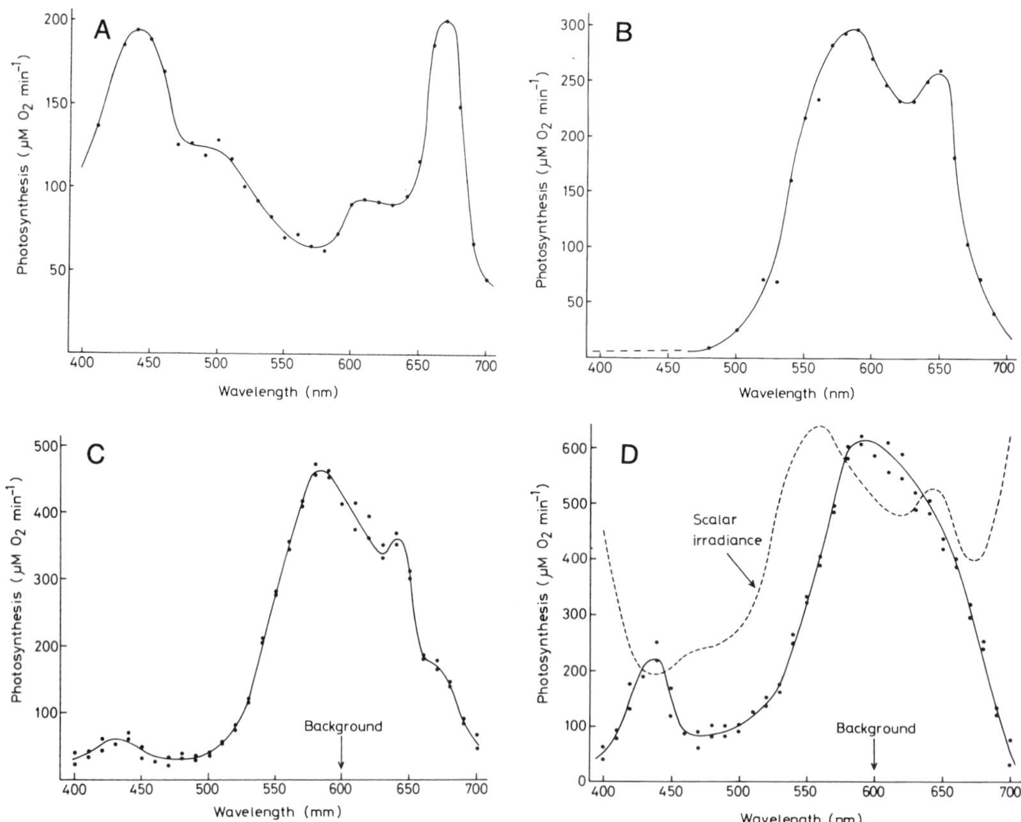

Figure 11. Action spectra for oxygen evolution measured in situ in a cyanobacterial mat. Photosynthesis rates were normalized to 20 microeinsteins m^{-2} s^{-1} for each wavelength. (A) Diatom film on mat surface. (B to D) Dense *Microcoleus* layer at a depth of 0.3 to 0.8 mm, measured without background illumination (B), measured with background illumination at 600 nm to excite PS II (C), or recalculated to account for the altered scalar irradiance spectrum within the mat (D). Reproduced from *Applied and Environmental Microbiology* (17).

The dense *Microcoleus* layer showed an action spectrum almost complementary to that of the diatoms (Fig. 11B). There was insignificant activity due to Chl *a* or carotenoids and a broad activity maximum in the absorption range of phycobilins associated with photosystem (PS) II (8). The photosynthesis rates peaked at 580 and 650 nm. The former wavelength falls between the typical absorption maxima of phycoerythrin or phycoerythrocyanin (570 to 580 nm) and of phycocyanin (610 to 630 nm). The latter wavelength corresponds to the absorption maximum of allophycocyanin.

Oxygenic photosynthesis requires simultaneous excitation of and electron transport through both PS I and PS II. Since the cyanobacteria showed high activity of the phycobilins, which are associated with PS II, the phycobilins can evidently pass energy efficiently to both PS II and PS I. The reverse, however, was not the case. Antenna Chl *a*, which is associated with PS I, showed very little activity. This phenomenon, called blue and red drop, is well known from studies of cyanobacteria exposed to monochromatic light (2, 16, 26). The action spectrum was therefore measured again with a continuous background illumination of 600-nm light to excite PS II (Fig. 11C). There was now a detectable, but still surprisingly low, activity at the blue and red absorption maxima of Chl *a*.

The low PS I activity could be due to low light intensity in the blue and red range. These action spectra were measured before the optical fiber technique had been developed to a level which allowed the calculation of scalar irradiance spectra. The action spectra had therefore been normalized to the spectral intensity of incident light at the mat surface, not to that of the in situ light. To compensate for this, the action spectrum in Fig. 11C was corrected for the relative scalar irradiance spectrum at a depth of 0.1 mm in the same mat (but measured 1 year later). The result is shown in Fig. 11D. Activity at the blue Chl *a* peak was enhanced, but was still relatively low. The shoulders in the red range have been smoothed out, and there was no activity maximum at the red Chl *a* peak. Since the action spectrum in Fig. 11D probably represents an overcorrection, the true action spectrum should be intermediate between those in Fig. 11C and D.

The lack of significant Chl *a* activity was unexpected, because the light attenuation spectra showed abundant Chl *a* in the cyanobacterial layer. Previous studies of action spectra of cyanobacteria in pure cultures have been done on planktonic species. The benthic species tend to clump and are therefore less suited to culture and physiological experimentation. The planktonic species generally show quite significant Chl *a* activity (see, e.g., reference 16). However, action spectra of low-Chl *a Anacystis* mutants have been recorded that are similar to those presented here (35). It is an interesting possibility that the spectral scalar irradiance in cyanobacterial mats, which is totally different from that in the water column, selects against Chl *a*-based photosynthesis. This could be an adaptation to the predominantly orange light below the mat surface by a mechanism different from the well-known chromatic adaptation caused by variations in the phycocyanin-to-phycoerythrin ratio (3). Studies of benthic cyanobacteria in pure culture are required to answer this question.

Only the role of light intensity and spectral composition for the zonation of phototrophs has been discussed here. It should be remembered that other factors, such as the toxicity of sulfide, may play an equally important role (5, 18). The photosynthetic activity of the diatoms, which live exclusively on the mat surface, is fully inhibited when they are exposed to low levels of sulfide. Many of the cyanobacteria, e.g., *M. chthonoplastes*, are well adapted to sulfide, which is used as an electron donor for anoxygenic photosynthesis. Since sulfide accumulates within the mat at night, the diatoms are restricted to the surface and are probably even dependent on the cyanobacteria to remove sulfide diffusing up from below. Thus, the vertical zonations in the mats result from a dynamic balance and interaction between the physicochemical gradients and the constant migrations and growth of phototrophic microorganisms.

CONCLUSIONS

Fiber-optic microprobes are a new tool for the study of optical properties and light distribution in cyanobacterial mats. The sensing tips of the single-stranded fibers are 20 to 30 μm in diameter, and the detectability of white light is 0.01 microeinstein m^{-2} s^{-1}. Spectral radiance gradients can be measured at a spatial resolution of 50 to 100 μm, similar to the resolution of photosynthesis measurements with oxygen microelectrodes. By measuring radiance gradients in several representative directions, it is possible to calculate the scalar irradiance, i.e., the total quantum flux to a point from all directions. This is considered to be the most relevant light parameter for microbial photosynthesis.

Scalar irradiance spectra of *Microcoleus* mats from Baja California, Mexico, showed that the light available for photosynthesis on an illuminated mat surface is both quantitatively and qualitatively different from the light source. There is more light available than expected from the downwelling irradiance, especially in mats with high scattering relative to absorption. The spectral composition of the light is also altered relative to that of the light source. As a result, the phototrophs below the mat surface are exposed to relatively more orange light of 550 to 650 nm, whereas blue light is particularly scarce.

On the basis of the action spectra, the dominant phototrophs seem to be adapted to this spectral situation. Diatoms growing on the mat surface utilize the blue and red light, which is absorbed by Chl *a* and by carotenoids. Cyanobacteria living below utilize the intermediate region of the spectrum between 550 and 650 nm, which passes through the diatom layer and which is absorbed by the phycobilins. Surprisingly, very little activity associated with PS I in the cyanobacteria was detectable, although the attenuation spectra indicated strong absorption by Chl *a*.

The results demonstrate the importance of making simultaneous measurements of photosynthesis and of scalar irradiance. Such combined measurements have not yet been carried out.

Acknowledgments. This research was done while I held a National Research Council Associateship at the National Aeronautics and Space Administration, Ames Research Center, Moffett Field, Calif.

I thank David J. Des Marais and Yehuda Cohen for inspiring an enjoyable collaboration and Barbara J. Javor for helpful field guidance.

LITERATURE CITED

1. **Bauld, J.** 1984. Microbial mats in marginal marine environments: Shark Bay, Western Australia, and Spencer Gulf, South Australia, p. 39–58. *In* Y. Cohen, R. W. Castenholz, and H. O. Halvorson (ed.), *Microbial Mats: Stromatolites.* Alan R. Liss, Inc., New York.
2. **Blinks, L. R.** 1960. Action spectra of chromatic transients and the Emerson effect in marine algae. *Proc. Natl. Acad. Sci. USA* **46**:327–333.
3. **Bogorad, L.** 1975. Phycobiliproteins and complementary chromatic adaptation. *Annu. Rev. Plant Physiol.* **26**:369–401.
4. **Castenholz, R. W.** 1982. Motility and taxes, p. 413–439. *In* N. G. Carr and B. A. Whitton (ed.), *The Biology of Cyanobacteria.* University of California Press, Berkeley.
5. **Cohen, Y.** 1984. The Solar Lake cyanobacterial mats: strategies of photosynthetic life under sulfide, p. 133–148. *In* Y. Cohen, R. W. Castenholz, and H. O. Halvorson (ed.), *Microbial Mats: Stromatolites.* Alan R. Liss, Inc., New York.
6. **Dring, M. J.** 1981. Chromatic adaptation of photosynthesis in benthic marine algae: an examination of its ecological significance using a theoretical model. *Limnol. Oceanogr.* **26**:271–284.
7. **Dring, M. J, and K. Lüning.** 1985. Emerson enhancement effects and quantum yield of photosynthesis for marine macroalgae in simulated underwater light fields. *Mar. Biol.* (Berlin) **87**:109–117.
8. **Glazer, A. N.** 1981. Photosynthetic accessory proteins with bilin prosthetic groups, p. 51–97. *In* M. D. Hatch and N. K. Boardman (ed.), *The Biochemistry of Plants*, vol. 8. *Photosynthesis.* Academic Press, Inc., New York.
9. **Grum, F., and R. J. Becherer.** 1979. *Optical Radiation Measurements*, vol. 1. *Radiometry.* Academic Press, Inc., New York.
10. **Haardt, H., and G. A. E. Nielsen.** 1980. Attenuation measurements of monochromatic light in marine sediments. *Oceanol. Acta* **3**:333–338.
11. **Højerslev, N. K.** 1986. Optical properties of sea water, p. 383–462. *In* J. Sünderman (ed.), *Oceanography*, vol. V/3. Landolt-Börnstein, Springer-Verlag KG, Berlin.
12. **IAPSO.** 1985. The international system of units (SI) in oceanography. *UNESCO Tech. Pap. Mar. Sci.*, vol. 45.
13. **Javor, B. J., and R. W. Castenholz.** 1981. Laminated microbial mats, Laguna Guerrero Negro, Mexico. *Geomicrobiol. J.* **2**:237–273.
14. **Javor, B. J, and R. W. Castenholz.** 1984. Productivity studies of microbial mats, Laguna Guerrero Negro, Mexico, p. 149–170. *In* Y. Cohen, R. W. Castenholz, and H. O. Halvorson (ed.), *Microbial Mats: Stromatolites.* Alan R. Liss, Inc., New York.
15. **Jerlov, N. G.** 1976. Marine optics. *Elsevier Oceanogr. Ser.* **14**:1–232.
16. **Jones, L. W., and J. Myers.** 1964. Enhancement in

the blue-green algae, *Anacystis nidulans*. *Plant Physiol.* 39:938–946.

17. **Jørgensen, B. B., Y. Cohen, and D. J. Des Marais.** 1987. Photosynthetic action spectra and adaptation to spectral light distribution in a benthic cyanobacterial mat. *Appl. Environ. Microbiol.* 53:879–886.
18. **Jørgensen, B. B., Y. Cohen, and N. P. Revsbech.** 1986. Transition from anoxygenic to oxygenic photosynthesis in a *Microcoleus chthonoplastes* cyanobacterial mat. *Appl. Environ. Microbiol.* 51:408–417.
19. **Jørgensen, B. B., and D. J. Des Marais.** 1986. A simple fiber-optic microprobe for high resolution light measurements: application in marine sediment. *Limnol. Oceanogr.* 31:1374–1381.
20. **Jørgensen, B. B., and D. J. Des Marais.** 1986. Competition for sulfide among colorless and purple sulfur bacteria in a cyanobacterial mat. *FEMS Microbiol. Ecol.* 38:179–186.

20a. **Jørgensen, B. B., and D. J. Des Marais.** 1988. Optical properties of benthic photosynthetic communities: fiber-optic studies of cyanobacterial mats. *Limnol. Oceanogr.* 33:99–113.

20b. **Jørgensen, B. B., and D. C. Nelson.** 1988. Bacterial zonation, photosynthesis, and spectral light distribution in hot spring microbial mats of Iceland. *Microb. Ecol.* 16:133–147.

21. **Jørgensen, B. B., N. P. Revsbech, T. H. Blackburn, and Y. Cohen.** 1979. Diurnal cycle of oxygen and sulfide microgradients and microbial photosynthesis in a cyanobacterial mat sediment. *Appl. Environ. Microbiol.* 38:46–58.
22. **Jørgensen, B. B., N. P. Revsbech, and Y. Cohen.** 1983. Photosynthesis and structure of benthic microbial mats: microelectrode and SEM studies of four cyanobacterial communities. *Limnol. Oceanogr.* 28:1075–1093.
23. **Kirk, J. T. O.** 1983. *Light and Photosynthesis in Aquatic Ecosystems.* Cambridge University Press, Cambridge.
24. **Kishino, M., C. R. Booth, and N. Okami.** 1984. Underwater radiant energy absorbed by phytoplankton, detritus, dissolved organic matter, and pure water. *Limnol. Oceanogr.* 29:340–349.
25. **Krumbein, W. E., Y. Cohen, and M. Shilo.** 1977. Solar Lake (Sinai). 4. Stromatolitic cyanobacterial mats. *Limnol. Oceanogr.* 22:635–656.
26. **Larkum, A. W. D., and S. K. Weyrauch.** 1977. Photosynthetic action spectra and light-harvesting in *Griffithsia monilia* (Rhodophyta). *Photochem. Photobiol.* 25:65–72.
27. **Morel, A.** 1978. Available, usable, and stored radiant energy in relation to marine photosynthesis. *Deep Sea Res.* 25:673–688.
28. **Prézelin, B. B.** 1981. Light reactions in photosynthesis. *Can. Bull. Fish. Aquat. Sci.* 210:1–43.
29. **Revsbech, N. P., and B. B. Jørgensen.** 1983. Photosynthesis of benthic microflora measured with high spatial resolution by the oxygen microprofile method: capabilities and limitations of the method. *Limnol. Oceanogr.* 28:749–756.
30. **Revsbech, N. P., and B. B. Jørgensen.** 1985. Microelectrodes: their use in microbial ecology. *Adv. Microb. Ecol.* 9:293–352.
31. **Seyfried, M., and L. Fukshansky.** 1983. Light gradients in plant tissue. *Appl. Opt.* 22:1402–1408.
32. **SooHoo, J. B., D. A. Kiefer, D. J. Collins, and I. S. McDermid.** 1986. *In vivo* fluorescence excitation and absorption spectra of marine phytoplankton. I. Taxonomic characteristics and responses to photoadaptation. *J. Plankton Res.* 8:197–214.
33. **Taylor, W. R., and C. D. Gebelein.** 1966. Plant pigments and light penetration in intertidal sediments. *Helgol. Wiss. Meeresunters.* 13:229–237.
34. **Vogelmann, T. C., and L. O. Björn.** 1984. Measurement of light gradients and spectral regime in plant tissue with a fiber optic probe. *Physiol. Plant.* 60:361–368.
35. **Wang, R. T., C. L. R. Stevens, and J. Myers.** 1977. Action spectra for photoreactions I and II of photosynthesis in the blue-green alga *Anacystis nidulans*. *Photochem. Photobiol.* 25:103–108.
36. **Whitney, D. E., and W. M. Darley.** 1983. Effect of light intensity upon salt marsh benthic microalgal photosynthesis. *Mar. Biol.* (Berlin) 75:249–252.
37. **Wood, A. M.** 1985. Adaptation of the photosynthetic apparatus of marine ultraphytoplankton to natural light fields. *Nature* (London) 316:253–255.

Chapter 12

Distribution and Survival of Lipophilic Pigments in a Laminated Microbial Mat Community near Guerrero Negro, Mexico

Anna C. Palmisano, Sonja E. Cronin, Elisa D. D'Amelio, Elaine Munoz, and David J. Des Marais

INTRODUCTION

The microbial mats occurring in a salina near Guerrero Negro, Baja California Sur, Mexico, have emerged as one of the most promising systems for examining highly laminated phototrophic mat communities (11). Similar to the better-known microbial mats in Solar Lake, Sinai (15, 18), and Shark Bay, Australia (1), the Guerrero Negro mats are based on the filamentous cyanobacterium *Microcoleus chthonoplastes*. In Guerrero Negro, *M. chthonoplastes* is found in association with unicellular cyanobacteria, *Oscillatoria* sp., diatoms, *Chloroflexus* spp., and purple bacteria (6). Nonphototrophic bacterial components of the Guerrero Negro mats include *Beggiatoa* sp. and sulfate-reducing bacteria. Jørgensen and Des Marais (13) examined the vertical zonation of light, O_2, H_2S, and pH and their effects on the distribution of colorless and purple sulfur bacteria in two types of Guerrero Negro mats (Ponds 5 and 6). Using a fiber-optic probe to measure spectral radiance and oxygen microelectrodes, Jørgensen et al. (12) further examined the photosynthetic action spectra within Guerrero Negro mats (Pond 5).

Our research has focused on highly laminated communities because of their putative homology with fossil stromatolites, the earliest record of ancient microbial life. In trying to understand the formation of such laminated microbial mat communities on early Earth, two approaches have prevailed. First, stromatolites have been examined for the presence of microfossils, i.e., the microscopic remains of microbial life with morphologies resembling cyanobacteria (16, 34). Schopf and Packer (29) reported a microfossil of possible cyanobacterial origin dating back to 3.5 billion years ago. A second approach, developed by Eglinton and co-

Anna C. Palmisano, Sonja E. Cronin, Elisa D. D'Amelio, Elaine Munoz, and David J. Des Marais • Life Science Division, National Aeronautics and Space Administration, Ames Research Center, N239-4, Moffett Field, California 94035.

workers (7, 8), is to look for chemical fossils, i.e., organic compounds whose structure, in part, has survived in the fossil record. Analytical techniques continue to develop rapidly, increasing our sensitivity of detection of such compounds.

Summons and Powell (32) reported a diagenetic product of isorenieratene (a carotenoid from members of the family *Chlorobiaceae*) in Silurian and Devonian rocks. Lipophilic pigments (carotenoids and chlorophylls) are important cellular constituents of phototrophic microorganisms in modern microbial mats. Their potential for survival in the fossil record and their critical roles in the photophysiology and ecology of living microbial mats have prompted us to examine the distribution, abundance, and source organisms of lipophilic pigments in Guerrero Negro mats.

INVESTIGATION PROCEDURES

Sample Collection

Our study site was located in a commercial salina, Exportadora de Sal, near Guerrero Negro, Baja California Sur, Mexico, located on the Pacific coast at 28° N, 114° W (Fig. 1). Microbial mats dominated by *M. chthonoplastes* were found in the concentrating ponds at salinities between 65 and 128‰; freshwater input into these mats is minimal. Samples were collected from Ponds 5 and 8 in November 1985 (salinities, 66 and 115‰, respectively) and June 1986 (salinities, 82 and 128‰, respectively). Water temperatures at midday were 20.2 to 20.8°C in November 1985 and 24.5 to 27.0°C in June 1986. We established two shaded quadrats in Pond 5 in November 1985 by using 1 m^2 of fiber glass screening suspended approximately 0.5 m above the mat surface on wooden frames. Cores were taken with 30-ml-cutoff plastic syringes, sliced into 0.5- or 1.0-mm sections, stored frozen in the dark, and returned to Ames Research Center. It should be noted that the mat samples were collected at midday; mat microorganisms, including *M. chthonoplastes*, *Chloroflexus* spp., and *Beggiatoa* sp., are capable of migrating in response to changing environmental gradients at night (14, 15).

Pigment Analysis

Frozen mat sections were lyophilized, homogenized, extracted in 90% acetone overnight at 0°C in the dark, and then reextracted in fresh 90% acetone for 1 to 2 h. Pigment extracts were filtered through a Whatman GF/C filter and then refiltered through a membrane filter (pore size, 0.5 μm; FH, Millipore Corp.); these extracts were stored under N_2 in the dark on ice and chromatographed within several hours of final extraction.

Pigments were separated in a Hewlett-Packard 1084B high-performance liquid chromatography instrument on a C-18 Hypersil column (250 by 4.6 mm) with 5-μm packing (Alltech Associates, Inc.) protected by a Whatman GSK guard column filled with C-18 Co:Pel (30- to 38-μm packing). The UV-visible-wavelength absorbance detector with a built-in scanning mode (250 to 550 nm) was set at 440 nm (chlorophylls and carotenoids), 360 nm (bacteriochlorophyll *a* and bacteriophaeophytin *a*), or 410 nm (phaeophytin *a*).

Samples were chromatographed by using a program described by Paerl et al. (23). The mobile phase of solvent A (methanol-acetonitrile, 90:10, vol/vol) and solvent B (100% acetone) was regulated at a flow rate of 1.5 $ml \cdot min^{-1}$ by using the following program: 0 to 7 min at 100% solvent A, 7 to 11 min for a linear increase to 60% solvent B, 11 to 20 min at 60% solvent B, and 20 to 28 min for a linear decrease to 100% solvent A. This program was modified to improve the separation of more-polar carotenoids by addition of deionized water to solvent A for a final concentration of 9% H_2O, 9% acetonitrile, and 82% methanol.

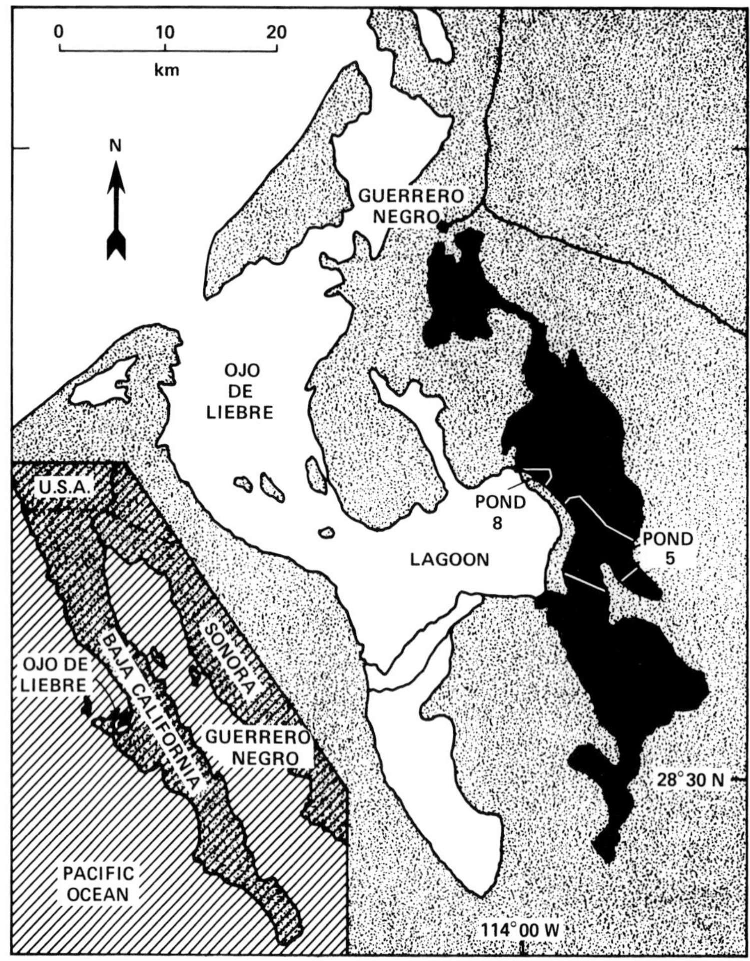

Figure 1. Map of the study site near Guerrero Negro. Microbial mats were collected from two concentrating ponds (Pond 5 and Pond 8) within a salina.

Pigments from microbial mats were identified by characteristic absorption maxima in organic solvents by UV-visible spectroscopy and by cochromatography with standards obtained from Sigma Chemical Co. and Hoffmann-La Roche Inc. or isolated from pure cultures of *M. chthonoplastes*, *Navicula saprophyta* (diatom), and *Chloroflexus aurantiacus*.

Pure-Culture Experiments

To test the effects of salinity on the pigment composition of the dominant cyanobacteria in the Guerrero Negro mats, we grew cultures of a *Microcoleus* sp. (isolated from Spencer Gulf, Australia) and a unicellular cyanobacterium (isolated from Guerrero Negro Pond 8) at 30, 60, 90, and 120‰ in ASN III medium, pH 7.5 (35), at approximately 100 microeinsteins $\cdot$ $m^{-2} \cdot s^{-1}$ and 35°C. Cells were harvested by centrifugation, lyophilized, and analyzed for lipophilic pigments, as described above.

LAMINA MICROSTRUCTURE

Microbial mats from Ponds 5 and 8 were finely laminated to a depth of 5 to 7 cm

Figure 2. Photograph of the highly laminated microbial mat community found in Pond 5 near Guerrero Negro. Individual laminae have a distinct pigmentation.

(Fig. 2). Individual laminae had distinct pigmentation (Fig. 3). These laminae varied between Ponds 5 and 8 and between the November 1985 and June 1986 samples. The surface yellow-orange layer in Pond 5, June 1986, and the surface pink layers in Pond 8 were composed primarily of unicellular cyanobacteria with some pennate diatoms. The dark-green layer corresponded to a dense, cohesive population of the filamentous cyanobacterium *M. chthonoplastes*. D'Amelio et al. (6) have reported a new filamentous purple gliding bacterium within 5 to 15% of the bundles of the *M. chthonoplastes* population. Populations of *Chloroflexus* spp. and *Beggiatoa* sp. were most abundant in the light-green layer just below the dark-green *M. chthonoplastes* layer. Below a depth of 2 mm it was more difficult to identify individual species components by light microscopy; however, microorganisms resembling cyanobacteria and flexibacteria have been observed at depths of 5 to 6 mm. Artificial shading of the Pond 5 community resulted in a sharp decrease in the lamination of the mat (Fig. 3).

PIGMENTS

Chromatography

A group of chromatograms of lipophilic pigments from the *M. chthonoplastes* layer in Pond 8, November 1985, is shown in Fig. 4. All carotenoids and chlorophyll *a* were estimated from chromatograms at 440 nm, the wavelength at which these pigments have a peak in absorbance. Bacteriochlorophyll *a* and bacteriophaeophytin *a*, however, absorb maximally at 360 nm, whereas phaeophytin *a* absorbs maximally at 410 nm in our solvent program.

Distribution with Depth

The distribution of chloropigments (chlorophyll *a*, bacteriochlorophyll *a*, and their corresponding phaeophytins) with depth in Pond 5 (Fig. 5) and Pond 8 (Fig. 6) in November 1985 is shown. The chlorophyll *a* concentration was maximal in the surface section (depth, 0 to 0.5 mm), reaching 2.7 μg/mg (dry weight) in Pond 5 and 0.7

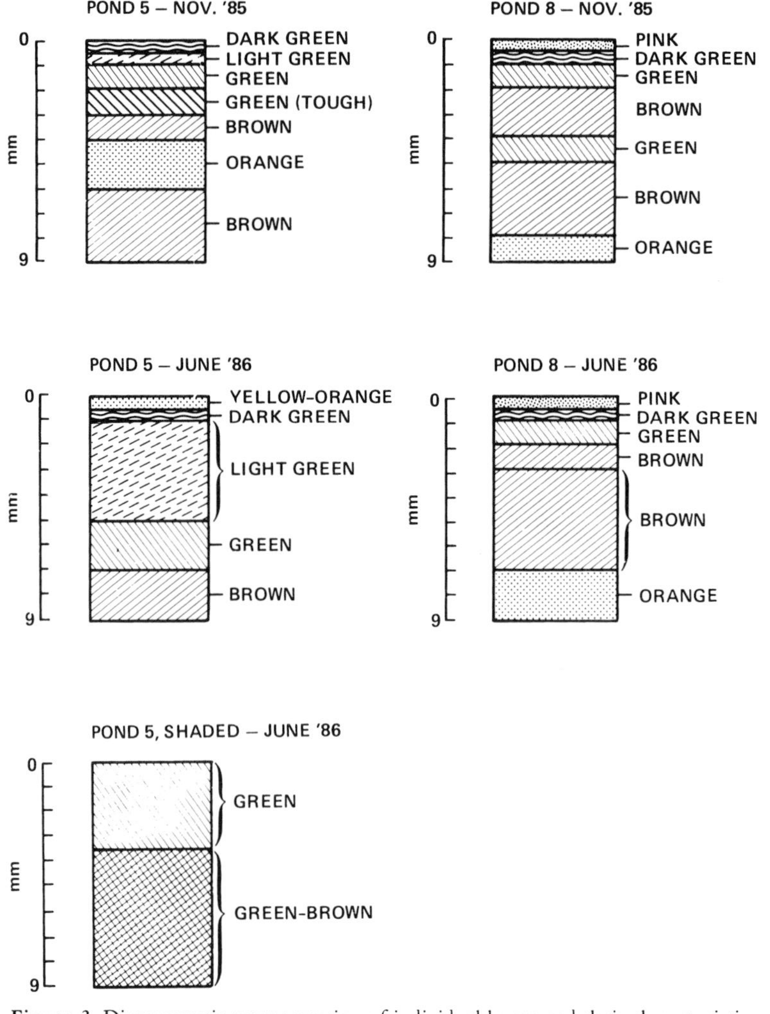

Figure 3. Diagrammatic representation of individual layers and their characteristic pigmentation in Ponds 5 and 8 in November 1985 and June 1986 and in a shaded quadrat of Pond 5 in June 1986.

μg/mg (dry weight) in Pond 8. Bacteriochlorophyll *a* concentrations showed one peak at a depth of 0.5 to 1 mm just below the chlorophyll *a* maximum and a second peak deeper within the mat between 5 and 8 mm. Phaeophorbides, which may be formed as a result of grazing activity (30), were not detected. Phaeophytin *a* concentrations never exceeded 4% of the chlorophyll *a* concentration in either pond. By contrast, bacteriophaeophytin *a* concentrations reached as high as 50% of the bacteriochlorophyll *a* concentration. Chlorophyll *c* was found in small amounts (<2 ng/mg [dry weight]) in the surface layer of the mat. Bacteriochlorophyll *c* was present, but in most cases was below the limits of sensitivity of our method; bacteriochlorophyll *c* chromatographed as four separate peaks and required >42 ng for accurate detection (23a).

The distribution of six carotenoids (myxoxanthophyll, fucoxanthin, zeaxanthin,

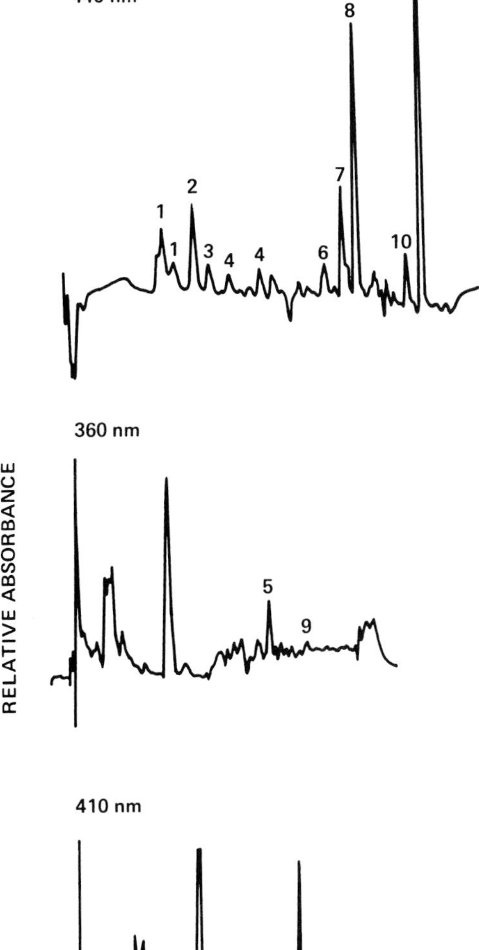

Figure 4. High-performance liquid chromatogram of lipophilic pigments from the *M. chthonoplastes* layer in Pond 8, November 1985, at 440, 360, or 410 nm. Numbered peaks correspond to the following pigments: 1, myxoxanthophyll; 2, zeaxanthin; 3, canthaxanthin; 4, bacteriochlorophyll *c*; 5, bacteriochlorophyll *a*; 6, chlorophyll *a'*; 7, echinenone; 8, chlorophyll *a*; 9, bacteriophaeophytin *a*; 10, γ-carotene; 11, β-carotene; 12, phaeophytin *a*; 13, phaeophytin *a'*.

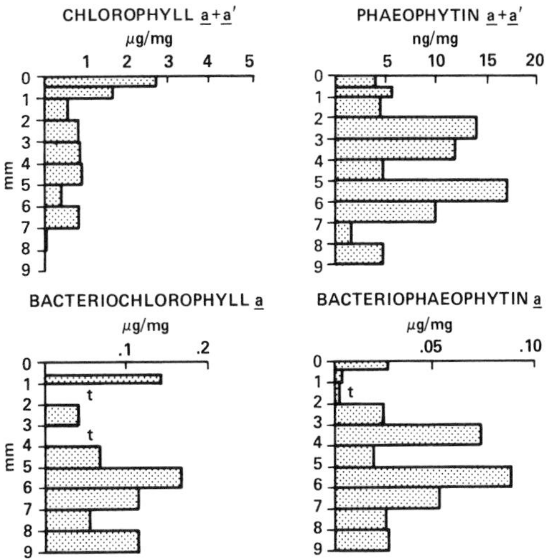

Figure 5. Distribution of four chloropigments with depth in a laminated microbial mat from Pond 5 in November 1985.

echinenone, β-carotene, and γ-carotene) with depth is shown in Fig. 7 (Pond 5) and Fig. 8 (Pond 8). The cyanobacterial pigments myxoxanthophyll, zeaxanthin, and echinenone were high in the topmost 0- to 0.5-mm mat layers in both ponds. In Pond 8, a second and larger peak of these three carotenoids was found at a depth of 7 to 9

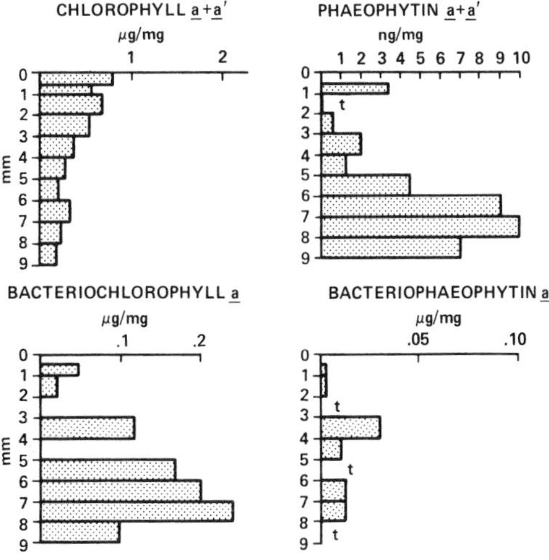

Figure 6. Distribution of four chloropigments with depth in a laminated microbial mat from Pond 8 in November 1985.

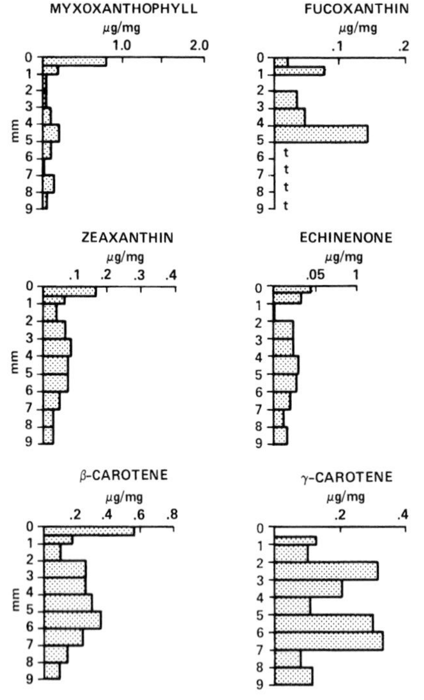

Figure 7. Distribution of six carotenoids with depth in a laminated microbial mat from Pond 5 in November 1985.

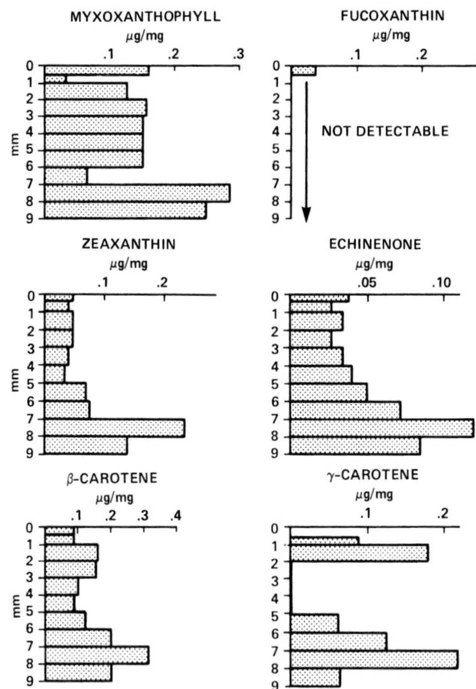

Figure 8. Distribution of six carotenoids with depth in a laminated microbial mat from Pond 8 in November 1985.

mm. Canthaxanthin (not shown) was present as a minor component of the cyanobacterial carotenoids. Fucoxanthin from diatoms occurred in the surface 0 to 0.5 mm and was not detectable below 5 mm in Pond 5 or below 0.5 mm in Pond 8. γ-Carotene had a bimodal distribution in both ponds, with peaks between 1 and 3 mm and between 5 and 8 mm. The γ-carotene concentration was highest at the surface 0 to 0.5 mm in Pond 5 and between 7 and 8 mm in Pond 8.

In November 1985, carotenoid/chlorophyll *a* ratios (by weight) showed a steady increase with depth, ranging from 0.5 to 7.0 in Pond 5 and from 3 to 27 in Pond 8 (Fig. 9). Similarly, in June 1986 in Pond 8 (Fig. 9), carotenoid/chlorophyll *a* ratios increased from 1 to 12. In June 1986 in Pond 5, however, a somewhat anomalous distribution was found, with a peak in carotenoid/chlorophyll *a* at a depth of 6 mm decreasing below to 9 mm.

Total Pigments in Mat Surface

Chlorophyll *a* and carotenoids per unit area integrated over the top 9 mm of mat are shown in Table 1 for Ponds 5 and 8 in November 1985 and June 1986. The individual carotenoids in each pond are also expressed as a percentage of the total carotenoids. The Pond 5 mat accreted about 10 mm between November 1985 and June 1986, while Pond 8 accreted about 4 mm during this same period. The chlorophyll *a* concentration was consistently higher in Pond 5 than in Pond 8 in both November 1985 and June 1986. Among the six carotenoids examined, the biggest difference was seen in a dramatic increase in myxoxanthophyll in both ponds between November

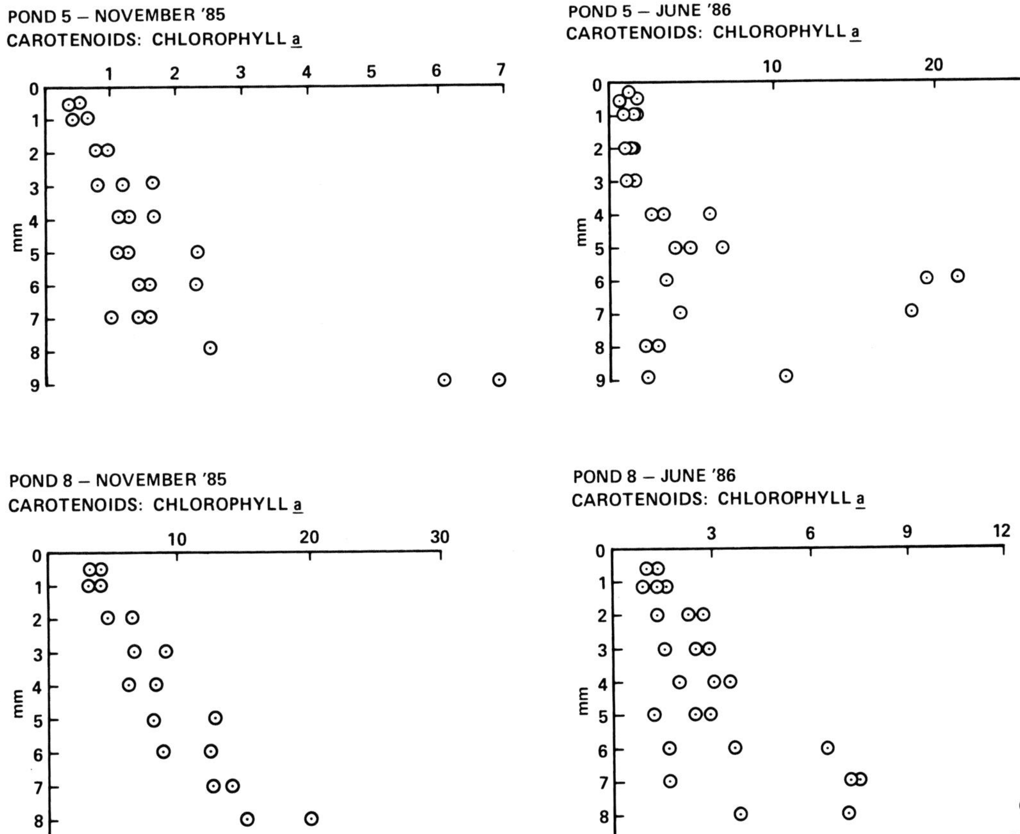

Figure 9. Carotenoid/chlorophyll *a* ratios (by weight) in mats from Pond 5 and Pond 8 in November 1985 and June 1986.

1985 and June 1986. Levels of myxoxanthophyll jumped 6- to 10-fold in both ponds, with myxoxanthophyll accounting for an average of 62% of the total carotenoids in June 1986 compared with an average of 33% in November 1986.

EFFECTS OF SALINITY ON PURE CULTURES

Ratios of four carotenoids to chlorophyll *a* (by weight) in two cyanobacterial isolates from mats at Guerrero Negro were examined over a range of salinities from 30 to 120‰ (Table 2). The unicellular cyanobacterium had a 5- to 10-fold-higher carotenoid/chlorophyll *a* ratio than the *Microcoleus* sp. did at a given salinity. More than 70% of the carotenoids in the unicellular cyanobacterium were contributed by myxoxanthophyll. No distinct trends were evident in carotenoids in this isolate with changes in salinity. Carotenoids from *Microcoleus* sp. were distributed in approximately equal portions among myxoxanthophyll, zeaxanthin, and β-carotene, and echinenone was not detected. Individual and total carotenoid/chlorophyll *a* ratios in *Microcoleus* sp. showed an overall two- to threefold increase when the salinity increased from 30 to 120‰.

Table 1.
Lipophilic Pigments per Unit Area[a]

Location and date	Chlorophyll *a*	Amt ($\mu g \cdot cm^{-2}$) (% of total carotenoids) of: Myxoxanthophyll	Zeaxanthin	Echinenone	β-Carotene	γ-Carotene	Total carotenoids
Nov 1985							
Pond 5							
Core 1	104.4	21.7 (28.8)	7.4 (9.9)	3.4 (4.5)	25.1 (33.4)	17.2 (22.9)	75.0
Core 2	64.9	15.3 (24.0)	5.6 (8.8)	4.0 (6.3)	21.8 (34.2)	17.1 (26.8)	63.8
Pond 8							
Core 1	43.4	34.4 (36.9)	10.6 (11.3)	9.5 (10.1)	27.5 (29.5)	11.2 (12.0)	93.2
Core 2	53.1	33.3 (43.5)	7.8 (10.3)	6.5 (8.5)	20.4 (26.7)	8.5 (11.0)	76.5
June 1986							
Pond 5							
Core 1	154.9	258.8 (65)	27.4 (6.9)	13.7 (3.5)	50.0 (12.7)	44.7 (11.3)	394.6
Core 2	204.7	271.9 (52.8)	45.1 (8.8)	23.9 (4.7)	82.3 (16.0)	89.5 (17.5)	512.8
Core 3	142.2	164.2 (57.6)	23.7 (8.3)	11.5 (4.0)	48.6 (17.0)	37.0 (13.0)	285.0
Pond 8							
Core 1	105.6	118.0 (60.4)	18.2 (9.3)	12.4 (6.3)	30.7 (15.7)	16.2 (8.2)	195.5
Core 2	101.0	194.0 (66.2)	21.0 (7.2)	15.9 (5.4)	47.5 (16.2)	14.7 (5.0)	293.2
Core 3	75.1	181.5 (68.3)	16.2 (6.1)	13.6 (5.1)	38.3 (14.4)	16.0 (6.0)	265.6

[a] Levels are integrated over the top 9 mm of microbial mat.

EFFECTS OF SHADING ON MAT STRUCTURE

Artificial shading of 1 m^2 of mat in Pond 5 for 7 months (November 1985 to June 1986) resulted in a redistribution of lipophilic pigments compared with those in an unshaded control community. In June 1986, the control had a subsurface (0.5 to 1 mm) maximum of chlorophyll *a*. γ-Carotene showed a bimodal distribution as in November 1985, with peaks at 1 to 2 and 5 to 6 mm

Table 2.
Effect of Salinity on Lipophilic Pigment Composition in Cyanobacterial Isolates from Microbial Mat Communities

Isolate and salinity (‰)	Ratio of following carotenoid to chlorophyll *a* (% of total carotenoids): Myxoxanthophyll	Zeaxanthin	Echinenone	β-Carotene	Total carotenoids
Unicellular cyanobacterium					
30	2.85 (77.0)	0.05 (1.4)	0.10 (2.7)	0.42 (11.3)	3.70
60	1.38 (70.0)	0.10 (5.1)	0.08 (4.1)	0.47 (23.8)	1.97
90	3.37 (83.6)	0.04 (1.0)	0.08 (2.0)	0.43 (10.7)	4.03
120	2.06 (82.1)	0.02 (0.08)	0.07 (2.8)	0.38 (15.1)	2.51
Microcoleus sp.					
30	0.13 (37.1)	0.10 (28.6)	0	0.12 (34.2)	0.35
60	0.13 (56.5)	0.07 (30.4)	0	0.09 (39.1)	0.23
90	0.12 (22.6)	0.14 (26.4)	0	0.27 (50.9)	0.53
120	0.41 (40.2)	0.30 (29.4)	0	0.31 (30.4)	1.02

(Fig. 10A). In the shaded quadrat, the subsurface chlorophyll *a* and γ-carotene peaks were coincident at the surface (0 to 0.5 mm); a second, greater, γ-carotene peak occurred at a depth of 6 to 7 mm (Fig. 10B). Overall levels of myxoxanthophyll were slightly lower in the shaded quadrat than in the control community.

DISCUSSION

The distribution of lipophilic pigments in the Guerrero Negro microbial mats reflected both the source organisms and the relative preservation of pigments in this environment. Possible source organisms for individual pigments found in the Guerrero Negro mats are listed in Table 3. These microorganisms have been observed in mats from Ponds 5 and 8 by extensive use of light microscopy and transmission electron microscopy (6).

Fucoxanthin and trace amounts of chlorophyll *c* were probably derived from diatoms which were found together with unicellular cyanobacteria in the surface layer of the mat. Diatoms, a minor component of the mat community, may be limited to the mat surface by their sensitivity to sulfide (24). Fucoxanthin and chlorophyll *c* also occur in chrysophytes, haptophytes, and phaeophytes (3); however, these organisms are not typically found in mat communities. Myxoxanthophyll, a glycosidic carotenoid found exclusively in cyanobacteria (19), was the most abundant carotenoid in the mats. Zeaxanthin is found in organisms other than cyanobacteria, including green algae and some higher plants (10). Echinenone and canthaxanthin are ketocarotenoids whose natural distribution is limited to cyanobacteria and lower marine animals (10). Of the phototrophs present in the microbial mats at Guerrero Negro, however, cyanobacteria are the most likely sources of these four carotenoids. Chlorophyll *a*, its stereoisomer chlorophyll *a*′, phaeophytin *a*, and its stereoisomer phaeophytin *a*′ were contributed primarily by cyanobacteria, which dominated the community, with a much smaller amount coming from diatoms. Assuming a typical fucoxanthin/chlorophyll *a* ratio (by weight) of 1 in diatoms (4), diatoms accounted for only 0.7 and 4% of the chlorophyll *a* in the surface mat layer in Pond 5 and Pond 8, respectively.

In most photosynthetic bacteria, bacteriochlorophyll *a* is part of the photosynthetic reaction center (25); both bacteriochlorophyll *a* and bacteriophaeophytin *a* in the Guerrero Negro mats were probably derived from *Chloroflexus* spp. and purple bacteria. Bacteriochlorophyll *c* is the primary light-harvesting pigment in *Chloroflexus* spp. (26), and observations by D'Amelio et al. (E. D'Amelio, Y. Cohen, and D. Des Marais, manuscript in preparation) suggest that there are as many as four morphological types of *Chloroflexus* spp. in the Guerrero Negro mats. γ-Carotene is one of the dominant carotenoids found in *Chloroflexus* spp. (26). The occurrence of γ-carotene in nature is limited to *Chloroflexus* spp. and fungi (10); fungi have not been observed in the Guerrero Negro microbial mats. Trace amounts of γ-carotene have been reported in the cyanobacterium *Chloroglea* sp. and in two species of the dinoflagellate *Peridinium* (10), which are also absent from these mats. Therefore, *Chloroflexus* spp. are the most likely source of γ-carotene in the Guerrero Negro mats. By contrast, β-carotene is widely distributed in nature and is produced by many of the phototrophic organisms found in the Guerrero Negro mats, including cyanobacteria, diatoms, and *Chloroflexus* spp. Despite the observation of purple bacteria in the Guerrero Negro mats (6), carotenoids such as spirilloxanthin and rhodopin have not yet been detected.

Lipophilic pigments have at least three possible roles in phototrophic organisms in microbial mats: light harvesting, photoprotection, and membrane stabilization. First, chlorophylls and carotenoids can harvest

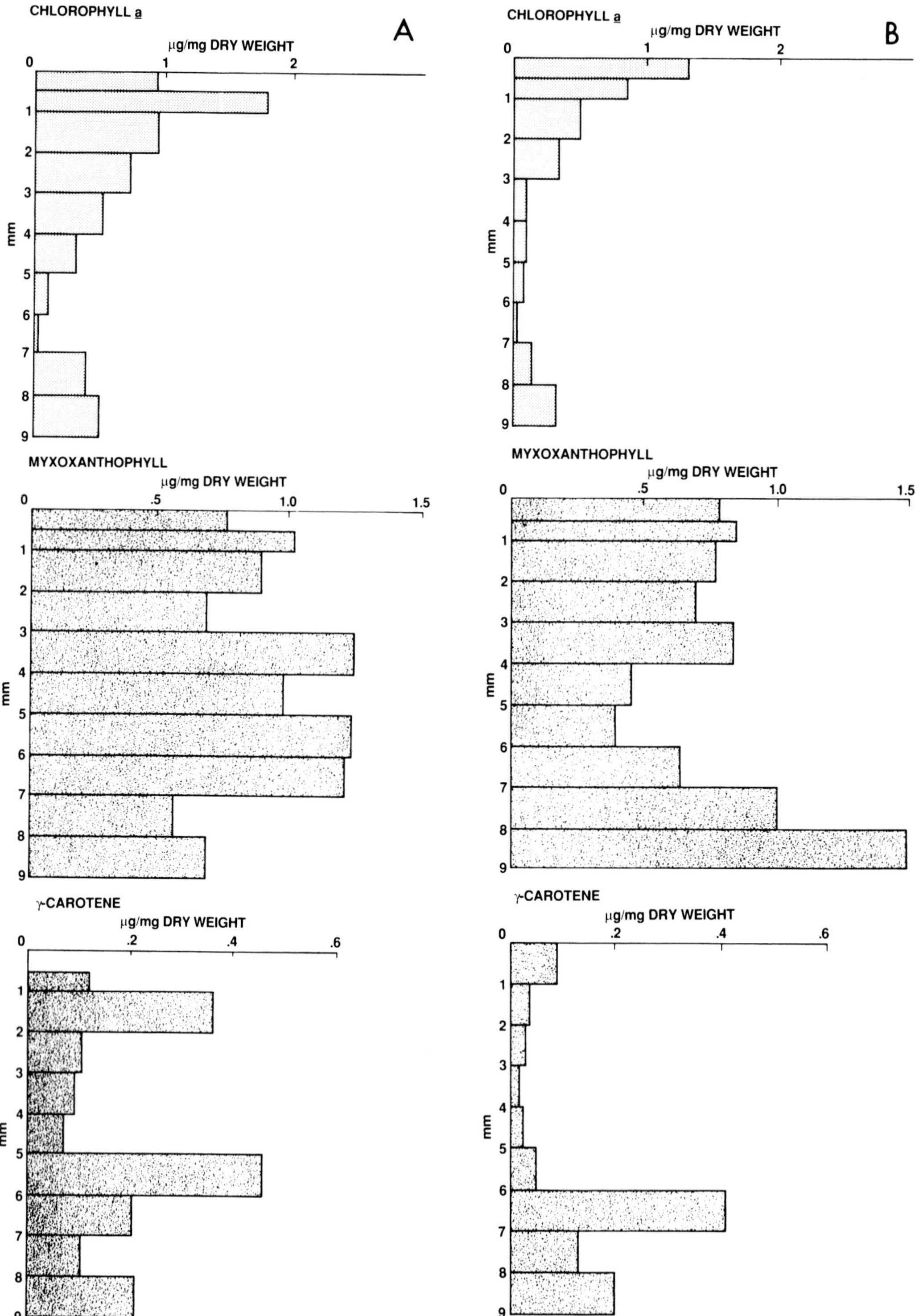

Figure 10. Distribution of chlorophyll *a*, myxoxanthophyll, and γ-carotene in (A) an unshaded control mat in Pond 5 in June 1986 and (B) an artificially shaded quadrat in Pond 5 in June 1986.

Table 3.
Possible Source Organisms for Lipophilic Pigments in Microbial Mats from Guerrero Negro

Pigment	Possible source
Chlorophyll *c*	Diatoms
Fucoxanthin	Diatoms
Myxoxanthophyll	Cyanobacteria
Zeaxanthin	Cyanobacteria
Canthaxanthin	Cyanobacteria
Bacteriochlorophyll *c*	*Chloroflexus* spp.
Bacteriochlorophyll *a*	Photosynthetic bacteria including *Chloroflexus* spp.
Chlorophyll *a'*	Cyanobacteria, diatoms
Echinenone	Cyanobacteria
Chlorophyll *a*	Cyanobacteria, diatoms
Bacteriophaeophytin *a*	Photosynthetic bacteria including *Chloroflexus* spp.
γ-Carotene	*Chloroflexus* spp.
β-Carotene	Cyanobacteria, diatoms, photosynthetic bacteria
Phaeophytin *a*	Cyanobacteria, diatoms
Phaeophytin *a'*	Cyanobacteria, diatoms

light, transferring energy to a reaction center, where it may be used to drive photosynthetic carbon fixation (5, 27). In the Guerrero Negro mats, photosynthetic action spectra of the surface films of unicellular cyanobacteria and diatoms showed maximal activity for wavelengths absorbed in vivo by chlorophyll *a* (430 and 670 nm) and by carotenoids (440 to 500 nm [12]). The photosynthetic activity of the underlying *M. chthonoplastes* layer, however, was dependent on light harvesting by the water-soluble phycobiliproteins between 550 and 660 nm, despite the relatively high abundance of chlorophyll *a* and carotenoids in the *M. chthonoplastes* layer. In *Chloroflexus* spp., γ-carotene is capable of energy transfer to reaction center bacteriochlorophyll *a* with a 40% efficiency (33). However, Jørgensen et al. (12) demonstrated that the blue light (450 nm) absorbed by carotenoids was attenuated 10 times more strongly than red light (600 nm) at a depth of 2.2 mm in the mat. *Chloroflexus* spp., lying below the *M. chthonoplastes* layer, would receive mostly red light, which is absorbed most efficiently by bacteriochlorophylls. Second, carotenoids protect cells against the harmful effects of visible radiation by quenching the triplet state of chlorophyll, inactivating singlet oxygen, and serving as an oxidizable substrate to protect other molecules (17). Using an inhibitor of carotenoid synthesis, diphenylamine, Paerl (22) demonstrated the importance of carotenoids in maintaining cyanobacterial blooms by increasing photoprotection as well as photosynthetic performance. Photoprotection is critical to organisms at the mat surface which may be exposed to surface downwelling irradiance as high as 1,000 microeinsteins $\cdot$ $m^{-2} \cdot s^{-1}$. Third, Ourisson and Rohmer (21) have suggested that hydrophilic carotenoids found in procaryotes may stabilize membranes in a manner similar to hopanoids and sterols.

Environmental factors influence the relative abundance of pigments in mat communities by affecting species composition through differential growth or chemotaxis. Artificial shading of the Pond 5 mat resulted in a restructuring of the mat surface, with *M. chthonoplastes* and *Chloroflexus* spp. displacing unicellular cyanobacteria, leading to a redistribution of chlorophyll *a* and γ-carotene in the top 1 mm. In the unshaded control, *M. chthonoplastes* absorbed wavelengths of light not absorbed by the overlying unicellular cyanobacteria and diatoms. With a reduction in irradiance, *M. chthonoplastes* responded by migrating to the surface, where photon flux was greatest and where the full spectrum of light was available, no longer "filtered" by overlying organisms.

The abundance of lipophilic pigments in microbial mats is determined in part by environmental factors affecting their biosynthesis by phototrophic organisms. Carotenoid synthesis has been shown to be affected by light intensity (20), spectral composition of light (9), growth phase (4), salinity (2, 20), and nitrogen availability (2). In our study, changes in salinity had little effect on the relative abundance of pigments in a unicellular cyanobacterial isolate; however, the *Microcoleus* sp. showed an increase in the carotenoid/chlorophyll *a* ratio with increased salinity.

Carotenoids have been proposed as possible biomarkers for phototrophic organisms (19). For an organic compound to be useful as a biomarker in a microbial mat, it must (i) be specific to a certain taxon or group of microorganisms, (ii) be sufficiently abundant to detect and identify with confidence, and (iii) have a relatively high preservation potential in this environment. In the Guerrero Negro mat, lipophilic pigment biomarkers for cyanobacteria might include myxoxanthophyll, echinenone, zeaxanthin, and canthaxanthin, while γ-carotene might serve as a useful biomarker for *Chloroflexus* spp. Fucoxanthin is a poor biomarker because, although specific to a group of organisms, it is poorly preserved owing to the presence of an epoxide ring (10, 28). β-Carotene is clearly too widely distributed to be useful as a biomarker for a given taxon of microorganisms.

The overall increase in carotenoid/chlorophyll *a* ratios with depth resulted from a preservation of carotenoids with degradation of chlorophyll *a* over a time scale of months. Only small amounts of phaeophytin *a* (<4% of the total chlorophyll *a* concentration) and no phaeophorbide *a* or chlorophyllide *a* were detected. Our results contrast with those of Stal et al. (31), who found a phaeophytin *a* concentration of >60% of the total chlorophyll *a* concentration in microbial mats from the intertidal zone of the North Sea Island of Mellum after a phase separation of lipophilic pigments. In general, the conditions of the deeper layers of the mat (dark, reducing conditions, a moderate pH, and an absence of grazers) are conducive to pigment preservation, especially of carotenoids (7). Edmunds and Eglinton (7) examined carotenoid preservation in a microbial mat from Solar Lake to a depth of 70 cm (about 2,500 years old). Carotenoid concentrations decreased with depth by 1.5 orders of magnitude down to the first 20 mm and then leveled off. The accretion rate at Solar Lake is much slower than at Guerrero Negro; therefore, degradation could be monitored over a much longer time frame. In the Guerrero Negro mats, carotenoids were preserved over a 7-month period during which about 10 mm of mat accreted in Pond 5 and about 4 mm of mat accreted in Pond 8. The cyanobacterial carotenoids, myxoxanthophyll, zeaxanthin, and echinenone, showed generally similar distribution; however, γ-carotene had a bimodal distribution in all 12 cores examined from Ponds 5 and 8. We have yet to determine whether this bimodality represents (i) the presence of a viable, subsurface population of microorganisms containing γ-carotene or (ii) a record of episodic growth of *Chloroflexus* spp.

Carotenoids in microbial mats may be considered potential chemical fossils; for this reason it is critical to understand their distribution, abundance, and source organisms in modern mats. Carotenoids showed relatively high survival potential upon burial, and some, such as myxoxanthophyll and γ-carotene, contained useful chemotaxonomic information. More research is needed, however, on the factors affecting both biosynthesis and degradation of these compounds in microbial mats.

Acknowledgments. We thank Barbara Javor for invaluable guidance in the field and the personnel of Exportadora de Sal for their cooperation. Hoffmann-La Roche Inc. generously supplied carotenoid standards, B. Pierson provided cultures of *C. aurantiacus*, and D. Robinson and M. Lizotte provided cultures of *N. saprophyta* for carotenoid isolations.

This study was supported by a National Research Council fellowship to A.C.P. and by a grant from the National Aeronautics and Space Administration Planetary Biology Program to D.J.D.

LITERATURE CITED

1. **Bauld, J. B.** 1984. Microbial mats in marginal marine environments: Shark Bay, Western Australia, and Spencer Gulf, South Australia, p. 39–58. *In* Y. Cohen, R. W. Castenholz, and H. O. Halvorson (ed.), *Microbial Mats: Stromatolites*. Alan R. Liss, Inc., New York.
2. **Ben-Amotz, A., and M. Avron.** 1983. On the factors which determine massive β-carotene accumulation in the halotolerant *Dunaliella bardawil. Plant Physiol.* 72:593–597.
3. **Bold, H. C., and M. J. Wynne.** 1978. *Introduction to the Algae: Structure and Reproduction*. Prentice-Hall, Inc., Englewood Cliffs, N.J.
4. **Carreto, J. I., and J. A. Catoggio.** 1976. Variations in pigment contents of the diatom *Phaeodactylum tricornutum* during growth. *Mar. Biol.* (Berlin) **36**:105–112.
5. **Cogdell, R. J.** 1978. Carotenoids in photosynthesis. *Philos. Trans. R. Soc. London Ser. B* **284**: 569–579.
6. **D'Amelio, E. D., Y. Cohen, and D. J. Des Marais.** 1987. Association of a new type of gliding, filamentous, purple phototrophic bacterium inside bundles of *Microcoleus chthonoplastes* in hypersaline cyanobacterial mats. *Arch. Microbiol.* **147**:213–220.
7. **Edmunds, K. L. H., and G. Eglinton.** 1984. Microbial lipids and carotenoids and their early diagenesis in the Solar Lake laminated microbial mat sequence, p. 343–389. *In* Y. Cohen, R. W. Castenholz, and H. O. Halvorson (ed.), *Microbial Mats: Stromatolites*. Alan R. Liss, Inc., New York.
8. **Eglinton, G.** 1973. Chemical fossils: a combined organic geochemical and environmental approach. *Pure Appl. Chem.* **34**:611–632.
9. **Fiksdahl, A., P. Foss, and S. Liaaen-Jensen.** 1983. Carotenoids of blue-green algae. 11. Carotenoids of chromatically-adapted cyanobacteria. *Comp. Biochem. Physiol.* **76B**:599–601.
10. **Goodwin, T. W.** 1980. *The Biochemistry of Carotenoids*. Chapman & Hall, London.
11. **Javor, B. J., and R. W. Castenholz.** 1981. Laminated microbial mats, Laguna Guerrero Negro, Mexico. *Geomicrobiol. J.* **2**:237–273.
12. **Jørgensen, B. B., Y. Cohen, and D. J. Des Marais.** 1987. Photosynthetic action spectra and adaptation to spectral light distribution in a benthic cyanobacterial mat. *Appl. Environ. Microbiol.* **53**:879–886.
13. **Jørgensen, B. B., and D. J. Des Marais.** 1986. Competition for sulfide among colorless and purple sulfur bacteria in cyanobacterial mats. *FEMS Microbiol. Ecol.* **38**:179–186.
14. **Jørgensen, B. B., N. P. Revsbech, T. H. Blackburn, and Y. Cohen.** 1979. Diurnal cycle of oxygen and sulfide microgradients and microbial photosynthesis in a cyanobacterial mat sediment. *Appl. Environ. Microbiol.* **38**:46–58.
15. **Jørgensen, B. B., N. P. Revsbech, and Y. Cohen.** 1983. Photosynthesis and structure of benthic microbial mats: microelectrode and SEM studies of four cyanobacterial communities. *Limnol. Oceanogr.* **28**:1075–1093.
16. **Knoll, A. H.** 1985. Exceptional preservation of photosynthetic organisms in silicified carbonates and silicified peats. *Philos. Trans. R. Soc. London Ser. B* **311**:111–122.
17. **Krinsky, N. I.** 1978. Non-photosynthetic functions of carotenoids. *Philos. Trans. R. Soc. London Ser. B* **284**:581–590.
18. **Krumbein, W. E., Y. Cohen, and M. Shilo.** 1977. Solar Lake (Sinai). 4. Stromatolitic cyanobacterial mats. *Limnol. Oceanogr.* **22**:635–656.
19. **Liaaen-Jensen, S.** 1979. Carotenoids—a chemosystematic approach. *Pure Appl. Chem.* **51**:661–675.
20. **Loeblich, L. A.** 1982. Photosynthesis and pigments influenced by light intensity and salinity in the halophile *Dunaliella salina* (Chlorophyta). *J. Mar. Biol. Assoc. U.K.* **62**:293–308.
21. **Ourisson, G., and M. Rohmer.** 1982. Prokaryotic polyterprenes: phylogenetic precursors of sterols. *Curr. Top. Membr. Transp.* **17**:153–182.
22. **Paerl, H. W.** 1984. Cyanobacterial carotenoids: their roles in maintaining optimal photosynthetic production among aquatic bloom forming genera. *Oecologia* **61**:143–149.
23. **Paerl, H. W., J. Tucker, and P. T. Bland.** 1983. Carotenoid enhancement and its role in maintaining blue-green algal (*Microcystis aeruginosa*) surface blooms. *Limnol. Oceangr.* **29**:847–857.

23a. **Palmisano, A. C., S. E. Cronin, and D. J. Des Marais.** 1988. Analysis of lipophilic pigments from a phototrophic microbial mat community by high performance liquid chromatography. *J. Microbiol. Methods* **8**:209–217.

24. **Patrick, R.** 1977. Ecology of freshwater diatoms—diatom communities, p. 284–332. *In* D. Werner (ed.), *The Biology of Diatoms*. University of California Press, Berkeley.
25. **Pfennig, N.** 1978. General physiology and ecology of photosynthetic bacteria, p. 3–18. *In* R. K. Clayton and W. R. Sistrom (ed.), *The Photosynthetic Bacteria*. Plenum Publishing Corp., New York.
26. **Pierson, B. K., and R. W. Castenholz.** 1974. Studies of pigments and growth in *Chloroflexus aurantiacus*, a phototrophic filamentous bacterium. *Arch. Microbiol.* **100**:283–305.
27. **Prezelin, B. B.** 1981. Light reactions in photosynthesis. *Can. Bull. Fish. Aquat. Sci.* **210**:1–43.
28. **Repeta, D. J., and R. B. Gagosian.** 1984. Transformation reactions and recycling of carotenoids

and chlorins in the Peru upwelling region (15°S, 75°W). *Geochim. Cosmochim. Acta* **48**:1265–1277.

29. **Schopf, J. W., and B. M. Packer.** 1987. Early Archean (3.3-billion to 3.5-billion-year-old) microfossils from Warrawoona Group, Australia. *Science* **237**:70–73.
30. **Shuman, F. R., and C. J. Lorenzen.** 1975. Quantitative degradation of chlorophyll by a marine herbivore. *Limnol. Oceanogr.* **20**:580–586.
31. **Stal, L. J., H. van Gemerden, and W. E. Krumbein.** 1984. The simultaneous assay of chlorophyll and bacteriochlorophyll in natural microbial communities. *J. Microbiol. Methods* **2**:295–306.
32. **Summons, R. E., and T. G. Powell.** 1986. Chlorobiaceae in Palaeozoic seas revealed by biological markers, isotopes and geology. *Nature* (London) **319**:763–765.
33. **Vasmel, H., R. J. van Dorssen, G. J. de Vos, and J. Amesz.** 1986. Pigment organization and energy transfer in the green photosynthetic bacterium *Chloroflexus aurantiacus. Photosynth. Res.* **7**:281–294.
34. **Walter, M. R.** 1976. *Stromatolites.* Elsevier/North-Holland Publishing Co., Amsterdam.
35. **Waterbury, J. B., and R. Y. Stanier.** 1981. Isolation and growth of cyanobacteria from marine and hypersaline environments, p. 221–223. *In* M. P. Starr, H. Stolp, H. G. Trüper, A. Balows, and H. G. Schlegel (ed.), *The Prokaryotes*, vol. 1. Springer-Verlag KG, Berlin.

Chapter 13

Microelectrode Analysis of Photosynthetic and Respiratory Processes in Microbial Mats

Niels Peter Revsbech, Peter Bondo Christensen, and Lars Peter Nielsen

INTRODUCTION

It is difficult to study the photosynthetic generation of organic matter in the topmost few millimeters of a microbial mat and the subsequent degradation processes within and below the photic zone without the use of microsensors. Microsensors are, however, available only for light (9) and for a relatively small spectrum of chemical species (16), and our understanding of the microbially mediated reactions in microbial mats may thus be improved by the development of microsensors for other ions and molecules.

The quantitatively important autotrophic photosynthetic reactions use either water or reduced sulfur compounds as electron donors. Microsensors are available for oxygen, which is the product of the oxidation of water, and for hydrogen sulfide, which is probably the quantitatively most important sulfur compound used for anoxygenic photosynthesis. In principle, we are therefore able to measure gradients and dynamics of oxygen and hydrogen sulfide in sediments and to deduce rates of oxygenic and anoxygenic photosynthesis from these data. Photosynthesis is, however, tightly coupled with respiratory reactions, and a thorough understanding of the photosynthetic reactions in microbial mats therefore requires knowledge about the simultaneously occurring respiratory processes.

The energetically most favorable respiratory processes use oxygen as the electron acceptor. The oxygen consumption of a microbial mat can be deduced from oxygen profiles as measured by oxygen microsensors. Oxygen is, however, present only within the uppermost fraction of a millimeter in most dark-incubated microbial mats, and the respiratory processes in deeper layers must therefore be based on other electron acceptors. Next to oxygen, nitrate is the most favorable electron acceptor. The most common pathway of nitrate respiration is denitrification, in which nitrate is reduced to gaseous nitrogen through a reduction se-

Niels Peter Revsbech, Peter Bondo Christensen, and Lars Peter Nielsen • Institute of Ecology and Genetics, University of Aarhus, Ny Munkegade, DK-8000 Aarhus C, Denmark.

quence. Nitrous oxide is an intermediate in this reduction. Acetylene is known to inhibit the reduction of nitrous oxide, and denitrification can therefore be estimated as the rate of nitrous oxide accumulation in an acetylene-inhibited system. By use of a nitrous oxide microelectrode, it is now possible to observe the microdistribution of denitrification. Ferric and manganese hydroxides and oxides are reduced when nitrate is absent, but at present we are not able to analyze these processes by using microsensors, although it may be possible to construct microsensors for Mn^{2+} and Fe^{2+}. Even with microelectrodes for these reduced metal ions, it would be difficult to perform quantitative calculations based on the measured microprofiles, since only a minor fraction of the total pool of these ions is found dissolved in the pore water. Below the zone where nitrate and the transition metals are reduced, sulfate is the principal electron acceptor. Reduction of sulfate results in the production of hydrogen sulfide, and it is therefore possible to determine the sulfate-reducing activity by using sulfide microelectrodes. Only carbon dioxide can be used as the electron acceptor at depths where no more sulfate is present. However, although carbon dioxide microsensors do exist, it would be extremely difficult to obtain much information about methanogenesis by using them.

In this chapter we discuss the capacities and limitations of microsensor analysis in investigating photosynthesis and respiration in microbial mats.

PHOTOSYNTHESIS

Oxygenic Photosynthesis

Microsensor analysis of oxygenic photosynthesis in microbial mats has been described in great detail (15, 16, 19). The oxygen microsensors (21) used for these measurements are only 2 to 10 μm wide at the tip and have 90% response times of only 0.2 to 0.5 s to changes in oxygen concentration. The analysis is based on the measurement of the decrease in oxygen concentration after darkening at some specific depth within the mat. The initial rate of decrease in oxygen concentration after darkening corresponds to the rate of photosynthesis in that specific layer. That photosynthesis can be determined this way may not seem very obvious. After the mat has been illuminated for some time, the oxygen concentration at each depth reaches a steady state because the photosynthetic oxygen production equals the amount removed by respiration and diffusion. When light is extinguished, the production of oxygen stops, but the processes removing oxygen continue at the same rate, so that the oxygen concentration therefore starts to decrease at a rate equaling the rate at which oxygen was previously supplied to the layer, i.e., the photosynthetic rate. One limitation of the method is that at least 1 s of incubation in the dark is necessary to observe the rate of decrease in oxygen concentration, and the oxygen molecules diffuse around within that 1 s, resulting in an integration of the photosynthetic activity within ca. 0.1 mm from the sensor tip (15). A mathematical formulation of the spatial resolution of the method has also been published (20). The spatial resolution is, however, fine enough for most applications. Figure 1 shows a profile of oxygen and photosynthesis within a cyanobacterial mat from a shallow locality in Limfjorden, Denmark. A high activity was found within a ca. 0.2-mm-wide zone at the surface, but photosynthesis could be detected down to the oxic/anoxic interface at a depth of 0.9 mm depth. Photosynthetic activity could be measured in the water 0.05 mm above the surface of the mat, where there were no organisms and hence no real activity, and it should also have been possible to measure a low activity 0.1 mm above the mat because of the ca. 0.1-mm spatial resolution described above. The water 0.1 mm above the mat was part of the diffusive boundary layer (11), but

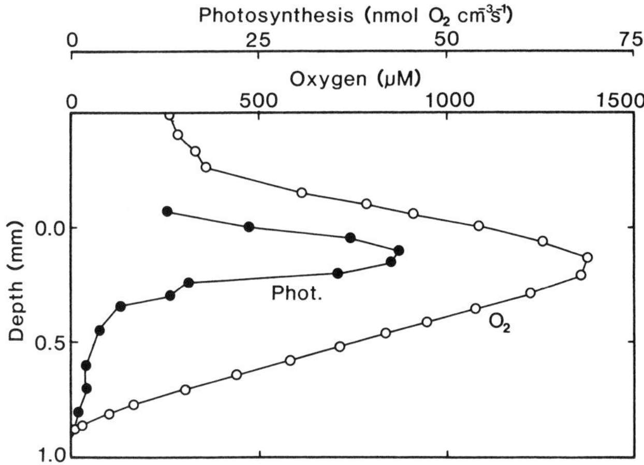

Figure 1. Profiles of oxygen and oxygenic photosynthesis in a microbial mat. The mat was illuminated at 310 microeinsteins m^{-2} s^{-1}.

some laminar flow must have occurred, since the oxygen reading fluctuated so much that a low photosynthetic activity could not be detected. The photosynthetic activity integrated over all layers amounted to 1.20 nmol of O_2 m^{-2} s^{-1}. The high rate of photosynthesis resulted in an oxygen concentration of 1,376 μmol $liter^{-1}$ within the most active layer. The finding of oxygenic photosynthesis down to the oxic/anoxic interface indicates that sufficient light might have been present in the uppermost anoxic layers to allow some anoxygenic photosynthesis. Light utilization by the photosynthetic bacteria found below the oxic/anoxic interface can be studied by using light microsensors (7, 10).

Anoxygenic Photosynthesis

Anoxygenic, sulfide-consuming photosynthesis within microbial mats can be measured by using sulfide microelectrodes. During this procedure, the algal mat should be illuminated until a steady-state sulfide profile is developed. After the light has been extinguished, the sulfide concentration should theoretically start to increase at a rate corresponding to the former photosynthetic rate. There are, however, many problems associated with such an approach. First, the response of the sulfide microelectrode is dependent on the pH, and pH measurements close to the tip of the sulfide microelectrode must therefore be conducted simultaneously (18). Even when the pH is known, the millivolt output from the sulfide-sensing circuit is difficult to transform to an accurate sulfide concentration, since the response is logarithmic, slow, and nonideal at concentrations below ca. 100 μmol $liter^{-1}$. The sulfide concentrations in the surface layers of many microbial mats are often significantly below 100 μmol $liter^{-1}$. In addition to these problems, sulfide and sulfide-utilizing photosynthetic organisms exhibit features which further complicate interpretation. Hydrogen sulfide forms polysulfides with elemental sulfur (2), and changes in the pool of dissolved sulfide may be masked by this reaction. Photosynthetic sulfur bacteria are able to compete with the precipitation reaction between hydrogen sulfide and ferrous iron (H. van Gemerden, Ph.D. thesis, University of Leiden, Leiden, The Netherlands, 1967), and they may consequently also be able to assimilate some of the sulfide bound as FeS. The use of hydrogen, elemental sulfur, and thiosulfate as electron donors (22) also complicates the picture. It has also been shown that some purple sulfur bacteria are able to oxidize hydrogen sulfide to elemental sulfur very rapidly at the beginning of a light period (25), with a further oxidation of this sulfur when the supply of hydrogen sulfide becomes limiting. An assumption of a constant hydrogen sulfide consumption corresponding to a predictable carbon dioxide fixation during illumination of anoxygenic photosynthetic communities may thus be far from reality. The simplest case to analyze is the anoxygenic photosynthesis of cyanobacteria, which can utilize only sulfide as the electron donor. Figure 2 shows data from such an experiment. Sulfide disappeared at a rate of ca. 65 μmol liter^{-1} s^{-1} after the start of illumination, which should correspond to the photosynthetic activity in this layer. Hydrogen sulfide was, however, soon depleted from the mat, and the cyanobacteria shifted to oxygenic photosynthesis. Steady-state anoxygenic photosynthesis was consequently not obtained in this system, and similar conditions may prevail in many communities with the capacity for anoxygenic photosynthesis. The electron donor for oxygenic photosynthesis, water, is always present in adequate concentrations (ca. 55 M) in the aquatic environment. Dissolved sulfide is, on the contrary, rarely present at concentrations above a few millimolar, and it is actually poisonous at higher concentrations, even for the organisms utilizing sulfide in their metabolism. Given enough light, a dense population of anoxygenic phototrophs will decrease the sulfide concentration to near zero, and their growth is therefore regulated by the diffusion flux of sulfide. In the modern world, only a few microbial mat environments are found where the depletion of sulfide after the start of illumination is not followed by oxygenic photosynthesis in the topmost layers (4; chapter 1 of this volume). The anoxygenic phototrophs in these layers must therefore be able to tolerate high oxygen concentrations or must migrate to deeper, anoxic layers. It has recently been demonstrated that purple sulfur bacteria may also continue to perform anoxygenic photosynthesis under oxic conditions if a

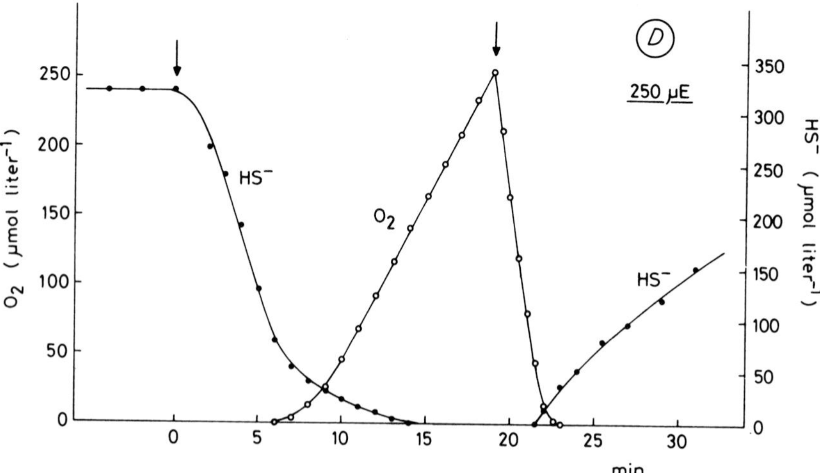

Figure 2. Concentrations of oxygen and hydrogen sulfide at a depth of 0.75 mm in the Solar Lake microbial mat during light-dark cycles. The light was turned on at the time indicated by the first arrow and turned off again at the time indicated by the second arrow. The light intensity at the mat surface was 400 microeinsteins (μE) m^{-2} s^{-1}. Reproduced from *Limnology and Oceanography* (8) with permission of the publisher.

reduced-sulfur compound is available (chapter 26 of this volume). It has also been shown that many purple sulfur bacteria are able to grow chemolithoautotrophically in the dark at the expense of reduced-sulfur compounds and oxygen (12, 13).

RESPIRATION

Oxygen Respiration

The flux of oxygen to and from individual layers of a microbial mat can be calculated from Fick's first and second laws of diffusion (3). According to Fick's first law of diffusion, the diffusion flux, J, can be calculated from the equation $J = \Phi D_s(dC/dx)$, where Φ is the porosity, D_s is the diffusion coefficient of oxygen within that particular layer of the mat, and dC/dx is the slope of the oxygen profile with depth at depth x. The steady-state oxygen profile in Fig. 1 can be used to demonstrate how such calculations can be performed. Immediately above the mat surface, the oxygen profile is linear through the diffusive boundary layer, which covers the mat surface like a blanket (13). The slope of the profile is 32,914 nmol cm^{-4}. The diffusion coefficient in 20‰ salinity seawater at 20°C is 2.0×10^{-5} cm^2 s^{-1}, and the porosity is 1. When these values are inserted into the above equation, a diffusion flux from the mat to the overlying water of 0.66 nmol cm^{-2} s^{-1} is obtained. The flux of oxygen at a depth of 0.50 mm can be calculated from a slope of 21,333 nmol cm^{-2} s^{-1}, a diffusion coefficient of 1.41 cm^2 s^{-1} (20), and a porosity of 1, to give a flux to deeper layers of 0.29 nmol cm^{-2} s^{-1}. The photosynthetic activity in the mat down to a depth of 0.50 mm is 1.13 nmol cm^{-2} s^{-1}, and 0.66 + 0.29 nmol cm^{-2} s^{-1} of this oxygen is thus exported from the photosynthetic layer. The oxygen consumption (respiration) within the 0- to 0.50-mm layer is thus 0.18 nmol cm^{-2} s^{-1}, which is low compared with the photosynthetic activity. A very high rate of oxygen consumption occurs at the oxic/anoxic interface. The flux downward at a depth of 0.86 mm can be calculated to be 0.18 nmol cm^{-2} s^{-1}, and this oxygen was consumed within the next 0.03-mm layer. This corresponds to a specific oxygen consumption of 60 nmol cm^{-3} s^{-1}, whereas the specific consumption in the 0- to 0.50-mm layer was only 3.6 nmol cm^{-3} s^{-1}. The very high rate of oxygen consumption at the oxic/anoxic interface is due to the oxidation of sulfide (6) and other reduced inorganic and organic species diffusing up from deeper layers.

Oxygen consumption rates can also be calculated from non-steady-state oxygen microprofiles by the use of Fick's second law of diffusion:

$$\delta C(x,t)/\delta t = \delta C(x,t)/\delta x^2 - R(x,t) + P(x,t)$$

where $C(x,t)$ is the concentration of oxygen at depth x at time t, $R(x,t)$ is the rate of oxygen consumption, and $P(x,t)$ is the rate of oxygen production. Figure 3 shows plots of oxygen concentration versus time after darkening at various depths in a mat similar to that analyzed in Fig. 1. Darkening resulted in a very fast disappearance of the oxygen peak, and a new steady-state profile was approached after only 80 s. A computer simulation based on an extended version of Fick's second law of diffusion (20) showed that the profile of respiratory rate illustrated in Fig. 4 could explain the disappearance rate of oxygen (Fig. 3) at all depths. The respiration was highest in the photosynthetically most active layer just below the surface of the mat and at the oxic/anoxic interface. The peak of respiration in the photosynthetically most active layer is not surprising, since this layer contains a dense population of metabolically very active microorganisms. The peak at the oxic/anoxic interface was due to oxidation of reduced compounds as described above.

The major problem of using oxygen microelectrode data to calculate respiration

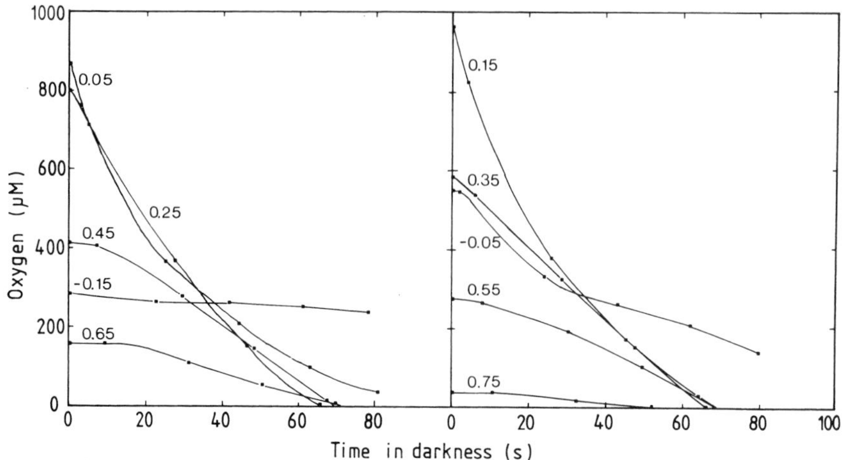

Figure 3. Oxygen concentrations at various depths in a microbial mat as a function of time after the light was extinguished. The light intensity before being extinguished was 250 microeinsteins m^{-2} s^{-1}. Reproduced from *Limnology and Oceanography* (20) with permission of the publisher.

in microbial mats and similar communities is the small-scale heterogeneity of most natural communities, causing invalidation of the applied one-dimensional diffusion models. The

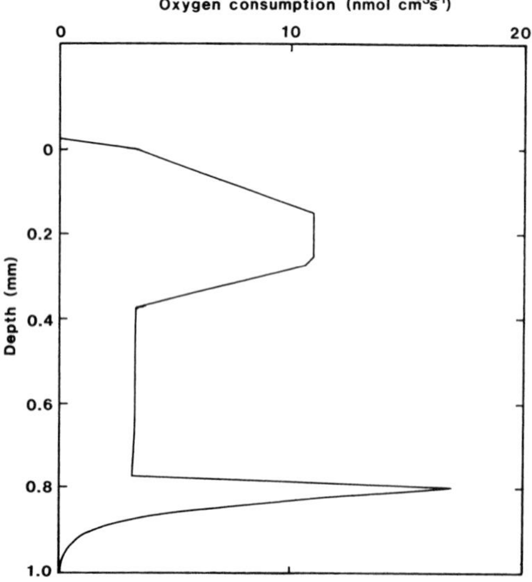

Figure 4. Profile of oxygen consumption calculated by computer simulation of data in Fig. 3. Reproduced from *Limnology and Oceanography* (20) with permission of the publisher.

measurements and calculations should be performed at several locations within the same sediment core and preferably also in several sediment cores to obtain representative values. Local and temporal variations in the turbulence pattern in the water above the substrate may also cause variations in the thickness of the diffusive boundary layer and thus cause fluctuating oxygen readings. Burrowing meiofauna and macrofauna within the substrate cause similar disturbances. It is difficult to perform any calculations with such temporally variable oxygen profiles, and so it is often necessary to look for an especially suitable microbial mat when highly reproducible calculations of oxygen fluxes are essential. Measurement of rates of photosynthesis by using oxygen microelectrodes does not require such highly stable oxygen microprofiles, but extreme instability may necessitate several readings at the same depth and a subsequent calculation of the mean rate. Rates of photosynthesis representative of the locality (ecosystem) should always be obtained as the mean of

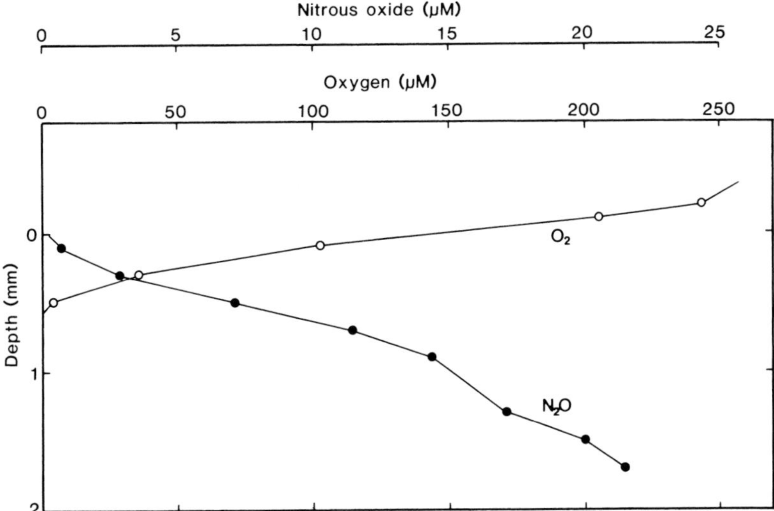

Figure 5. Profiles of oxygen and nitrous oxide in a biofilm incubated with acetylene. The 2-mm-thick biofilm was collected in a small, moderately eutrophic stream.

many photosynthesis profiles measured at random positions.

Nitrate Respiration

Nitrate is probably not a quantitatively important electron acceptor in most naturally occurring microbial mats, but our knowledge of denitrification in these systems is not very extensive. Nitrate is, however, a very important electron acceptor in microbial films from sewage treatment plants and also in biofilms and sediments from freshwater and estuarine environments, where nitrate concentrations higher than 1 mmol liter^{-1} may be found in the overlying water. We have just developed a microsensor for simultaneous measurement of O_2 and N_2O which makes it possible to observe the microdistribution of denitrifying activity in such environments (20a). The microsensor has a tip diameter of ca. 20 μm and can detect both O_2 and N_2O down to concentrations of 1 μmol liter^{-1}. There are no agents that interfere significantly, except for sulfide, which in high concentrations rapidly deactivates the N_2O sensor. The 90% response times are 10 to 20 s for the N_2O sensor and 1 to 2 s for the O_2 sensor. The electrode can be used to study the natural N_2O concentrations in sediments and soils. Low N_2O concentrations are often found in nature, since both nitrification and denitrification can result in the formation of N_2O. Figure 5 shows profiles of O_2 and N_2O in a biofilm growing on a piece of plastic collected in a small creek near Aarhus, Denmark. The water phase above the biofilm was kept at 10% acetylene saturation for 1 h prior to the measurements. This acetylene concentration efficiently inhibits the reduction of N_2O (14), thus causing an N_2O outflux from the biofilm proportional to the denitrifying activity within the biofilm. Oxygen disappeared at 0.6 mm below the surface of the biofilm, and below this depth denitrification caused a production of N_2O which reached a concentration of 21.5 μmol liter^{-1} at a depth of 1.65 mm in the biofilm. The slope of the N_2O profile at the oxic/anoxic interface was 230 μmol cm^{-4}, and, assuming a diffusion coefficient of 1.5×10^{-5} cm^2 s^{-1} within the biofilm (1, 20) and a porosity of 1, a flux of N_2O of 0.0033 nmol cm^{-2} s^{-1} can then be calculated from Fick's first law of diffusion. One molecule of N_2O is formed for every two nitrate ions reduced in the denitrification, so that this flux corresponds to a denitrifying activity (nitrate disappearance rate) of 0.0066 nmol cm^{-2} s^{-1}. This is a very low activity compared with the oxygen consumption, which also can be estimated from Fig. 5. Assuming that the diffusion coefficient at a depth of 0 mm is identical with the one for pure water, 2.1×10^{-5} cm^2 s^{-1}, the downward flux of oxygen is 0.10 nmol cm^{-2} s^{-1}, or about 15 times that of nitrate. More detailed microprofiles obtained for a more homogeneous biofilm would have allowed calculations of microprofiles of both oxygen respiration and denitrification. Such calculations have been performed for other substrates, and significant denitrifying activity always seemed to be associated with anoxic conditions. The denitrifying zone was often very narrow, since the availability of nitrate was a limiting factor in most substrates. The limiting factor in the biofilm analyzed in Fig. 5 was degradable organic matter, since addition of yeast extract caused a 10-fold increase in the denitrifying activity (data not shown).

The use of an N_2O microelectrode results in an excellent spatial resolution in the determination of denitrifying activity, but it does not solve the methodological problems (14a) which may occur in the interpretation of data obtained by the acetylene inhibition technique. One of these problems is the inhibition of nitrification by acetylene. Denitrification is often tightly coupled to nitrification, since nitrification produces the substrate for denitrification. The acetylene inhibition technique is therefore most applicable in microbial communities with a large net import of nitrate from outside and with little nitrification within the substrate. Bacteria may also reduce nitrate to ammonia, which does not result in production of N_2O and is hence not included in the rates of nitrate reduction obtained by the acetylene inhibition technique.

Sulfate Reduction

Respiration in which sulfate is reduced to hydrogen sulfide is a very important process in many microbial mats. It is usually quantified by using radioactively labeled sulfate; this method is very sensitive, since even a small proportion of radiolabeled sulfide in the sample can easily be detected. Such an approach does, however, have its limitations when oxidized or even oxic layers are found close to the zones with sulfate reduction. A significant proportion of radiolabeled sulfide formed during incubation may then be reoxidized during the incubation, and the sulfate-reducing activity will therefore be underestimated.

Sulfide microelectrodes may be used to measure microprofiles of sulfide from which sulfate-reducing activity can be calculated, but such measurements are complicated by the analytical difficulties described above. The measurements are further complicated by the binding of sulfide to transition metals such as iron, but that is mostly a problem in bioturbated sediments in which the iron and chemically bound sulfide are oxidized at intervals by burrowing fauna (17). Figure 6 shows oxygen and sulfide profiles in the microbial mats of Solar Lake, Sinai, in the dark. The sulfide diffused up from deeper layers and was oxidized in a ca. 0.25-mm-thick zone at the oxic/anoxic interface. The flux of sulfide just below the oxic/anoxic interface must equal the integrated rate of sulfate reduction in the layers below, assuming a steady-state situation. The flux was calculated (18) to be 0.43 mmol m^{-2} h^{-1}, or 0.012 nmol cm^{-2} s^{-1}. This is only 15% of the flux which would be expected from measurements of the sulfate reduction in the same mat by using radiolabeled sulfate (5). The reason for this discrepancy cannot be explained by burial of sulfide as FeS and FeS_2, since these minerals accounted for only 1.5‰ of the sulfide production as measured by the radiotracer method (5). Comparisons with the primary productivity of the

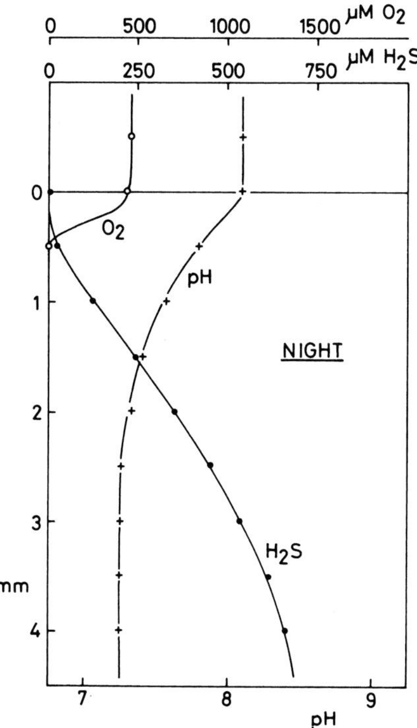

Figure 6. Profiles of oxygen, pH, and dissolved sulfide in the microbial mat from Solar Lake. Reproduced from *Limnology and Oceanography* (18) with permission of the publisher.

mat did, however, indicate that the radiotracer method yielded an overestimate of the rate of sulfate reduction (5). It should also be possible to calculate the depth distribution of sulfate reduction from a profile of dissolved sulfide, but such deep profiles have never been measured in undisturbed microbial mats.

LITERATURE CITED

1. **Broecker, W. S., and T. H. Peng.** 1974. Gas exchange rates between air and sea. *Tellus* **26:**21–35.
2. **Chen, K. Y., and S. K. Gupta.** 1973. Formation of polysulfides in aqueous solution. *Environ. Lett.* **4:**187–200.
3. **Crank, J.** 1983. *The Mathematics of Diffusion.* Oxford University Press, London.
4. **Giovannoni, S. J., N. P. Revsbech, D. M. Ward, and R. W. Castenholz.** 1987. Obligately phototrophic *Chloroflexus*: primary production in anaerobic hot spring microbial mats. *Arch. Microbiol.* 147:80–87.
5. **Jørgensen, B. B., and Y. Cohen.** 1977. Solar Lake (Sinai). 5. The sulfur cycle of the benthic cyanobacterial mats. *Limnol. Oceanogr.* 22:657–666.
6. **Jørgensen, B. B., Y. Cohen, and D. J. Des Marais.** 1987. Photosynthetic action spectra and adaptation to spectral light distribution in a benthic cyanobacterial mat. *Appl. Environ. Microbiol.* **53:**879–886.
7. **Jørgensen, B. B., Y. Cohen, and N. P. Revsbech.** 1986. Transition from anoxygenic to oxygenic photosynthesis in a *Microcoleus chthonoplastes* cyanobacterial mat. *Appl. Environ. Microbiol.* **51:**408–417.
8. **Jørgensen, B. B., and D. J. Des Marais.** 1986. A simple fiber-optic microprobe for high resolution light measurements. *Limnol. Oceanogr.* **31:**1376–1383.
9. **Jørgensen, B. B., and D. J. Des Marais.** 1986. Competition for sulfide among colorless and purple sulfur bacteria in cyanobacterial mats. *FEMS Microb. Ecol.* **38:**179–196.
10. **Jørgensen, B. B., and N. P. Revsbech.** 1983. Colorless sulfur bacteria, *Beggiatoa* spp. and *Thiovolum* spp., in O_2 and H_2S microgradients. *Appl. Environ. Microbiol.* **45:**1261–1270.
11. **Jørgensen, B. B., and N. P. Revsbech.** 1985. Diffusive boundary layers and the uptake of sediment and detritus. *Limnol. Oceanogr.* **30:**111–122.
12. **Kämpf, C., and N. Pfennig.** 1986. Chemoautotrophic growth of *Thiocystis violacea*, *Chromatium gracile*, and *C. vinosum* in the dark at various O_2 concentrations. *J. Basic Microbiol.* **26:**517–531.
13. **Kämpf, C., and N. Pfennig.** 1986. Isolation and characterization of some chemoautotrophic chromatiaceas. *J. Basic Microbiol.* **26:**507–515.
14. **Knowles, R.** 1982. Denitrification. *Microbiol. Rev.* **46:**43–70.

14a. **Koike, I., and J. Sørensen.** 1988. Nitrate reduction and denitrification in marine sediment, p. 251–273. *In* T. H. Blackburn and J. Sørensen (ed.), *Nitrogen Cycling in Coastal Marine Environments: Scope 33.* John Wiley & Sons, Inc., New York.

15. **Revsbech, N. P., and B. B. Jørgensen.** 1983. Photosynthesis of benthic microflora measured with high spatial resolution by the oxygen microprofile method: capabilities and limitations of the method. *Limnol. Oceanogr.* **28:**749–756.
16. **Revsbech, N. P., and B. B. Jørgensen.** 1986. Microelectrodes: their use in microbial ecology. *Adv. Microb. Ecol.* **9:**293–352.
17. **Revsbech, N. P., B. B. Jørgensen, and T. H. Blackburn.** 1980. Oxygen in the seabottom measured with microelectrode. *Science* **207:**1355–1356.

18. **Revsbech, N. P., B. B. Jørgensen, T. H. Blackburn, and Y. Cohen.** 1983. Microelectrode studies of photosynthesis and O_2, H_2S and pH profiles of a microbial mat. *Limnol. Oceanogr.* **28:**1062–1074.
19. **Revsbech, N. P., B. B. Jørgensen, and O. Brix.** 1981. Primary production of microalgae in sediments measured by oxygen microprofile, $H^{14}CO_3^-$ fixation, and oxygen exchange methods. *Limnol. Oceanogr.* **26:**717–730.
20. **Revsbech, N. P., B. Madsen, and B. B. Jørgensen.** 1986. Oxygen production and consumption in sediments determined at high spatial resolution by computer simulation of oxygen microelectrode data. *Limnol. Oceanogr.* **31:**293–304.
20a. **Revsbech, N. P., L. P. Nielsen, P. B. Christensen, and J. Sørensen.** 1988. Combined oxygen and nitrous oxide microsensor for denitrification studies. *Appl. Environ. Microbiol.* **54:**2245–2249.
21. **Revsbech, N. P., and D. M. Ward.** 1983. Oxygen microelectrode that is insensitive to medium chemical composition: use in an acid microbial mat dominated by *Cyanidium caldarium*. *Appl. Environ. Microbiol.* **45:**755–759.
22. **Trüper, H. G., and U. Fischer.** 1982. Anaerobic oxidation in sulfur compounds as electron donors for bacterial photosynthesis. *Philos. Trans. R. Soc. London Ser. B* **298:**529–542.
23. **van Gemerden, H.** 1974. Coexistence of organisms competing for the same substrate: an example among the purple sulfur bacteria. *Microb. Ecol.* **1:**104–119.

Chapter 14

Manganese and Iron as Physiological Electron Donors and Acceptors in Aerobic-Anaerobic Transition Zones

William C. Ghiorse

INTRODUCTION

My purpose is to discuss a historically controversial topic: the role of Fe and Mn as physiological electron donors (energy sources) and electron acceptors (energy sinks) in neutral-pH environments. I will discuss this topic in the context of the biogeochemistry of processes that occur at aerobic-anaerobic (oxic-anoxic) transition zones, primarily in aquatic systems. Detailed discussions of the underlying experimental work and more complete lists of references can be found in two review papers (26, 27a).

BIOGEOCHEMISTRY OF Fe AND Mn

Several excellent sources discuss the thermodynamics and kinetics of Fe and Mn reactions in a biogeochemical context (21, 44, 46, 54). The salient thermodynamic and kinetic considerations can be summarized as follows. (i) Thermodynamic stabilities of Fe and Mn phases ultimately control the distribution of these metals on a global scale. However, because they do not give reaction rate information, thermodynamic equilibrium considerations (e.g., E_h-pH stability field diagrams) can be used only as rough guides for predicting the forms of these metals under given natural conditions. On the other hand, thermodynamic calculations can be used to predict a sequence of Fe and Mn redox process (54) that will operate in geochemical stratified environments and to show that Fe and Mn oxidation and reduction may provide physiologically useful energy to support microbial growth. (ii) Kinetic considerations are very important for understanding both the biogeochemistry and physiological roles of these metals as electron donors and acceptors. The rate of abiotic Fe oxidation can be so high in oxygenated natural waters at neutral pH that it may be difficult for biological processes to compete, whereas the rate of abiotic Mn oxidation is low enough under these conditions for biological processes to be very important. Fe and Mn reduction kinetics in natural systems are not well understood; however, it has been demonstrated that both metals can

William G. Ghiorse • Department of Microbiology, Cornell University, Ithaca, New York 14853.

be reduced rapidly by both biological and abiotic processes (21, 54).

Despite a long history of speculation and the accumulation of a large body of literature, the study of microbial effects on the biogeochemistry of Fe and Mn is still in its developmental stages. For example, the influence of Fe and Mn cycles in controlling the electron balance and chemical dynamics in stratified systems such as those that occur at aerobic/anaerobic interfaces is not widely recognized. Nealson (47) asked several important questions about Fe and Mn biogeochemistry that are yet to be answered completely. (i) What is the microbial contribution to Fe and Mn redox reactions? (ii) How are microbial Fe and Mn cycles coupled to other biogeochemical cycles? (iii) What are the mechanisms of catalysis of Fe and Mn redox reactions?

The first two of these questions have been addressed in numerous investigations of freshwater (7, 10, 14, 25, 28, 35, 38, 39, 43, 45, 57–59), marine (8, 9, 11, 12, 18, 23, 49, 50, 55, 56), and groundwater treatment (6, 13, 24, 31–33; B. D. Rundell and S. J. Randtke, personal communication) systems. Some answers are clear. As predicted by thermodynamic and kinetic considerations, when O_2 is present in neutral-pH environments Mn^{2+} oxidation is greatly influenced by microbial activity; in some instances Mn^{2+} oxidation is almost entirely microbial. On the other hand, Fe^{2+} oxidation at neutral pH is affected by microbial activity only when O_2 levels are very low, and even then the extent of the microbial influence is uncertain. Fe and Mn reduction under both aerobic and anaerobic conditions also can be coupled to oxidation of organic compounds, but the extent of microbial versus abiotic Fe and Mn reduction is still not certain because of difficulties in distinguishing direct enzymatic attack on iron and manganese oxides from indirect abiotic reduction by microbial metabolites (27a, 51).

Fe^{2+} AND Mn^{2+} AS PHYSIOLOGICAL ELECTRON DONORS

As mentioned above, thermodynamic calculations (20) show that oxidation of Fe^{2+} or Mn^{2+} at neutral pH by reaction 1 or 2 is exergonic:

$$2Fe^{2+} + 0.5O_2 + 5H_2O \rightarrow 2Fe(OH)_3 + 4H^+$$
$$\Delta G'^{\circ} = -8.6 \text{ kcal} \cdot \text{mol}^{-1} \quad (-36 \text{ kJ} \cdot \text{mol}^{-1}) \qquad (1)$$

$$Mn^{2+} + 0.5O_2 + H_2O \rightarrow MnO_2 + 2H^+$$
$$\Delta G'^{\circ} = -16.3 \text{ kcal} \cdot \text{mol}^{-1} \quad (-68.2 \text{ kJ} \cdot \text{mol}^{-1}) \qquad (2)$$

In theory, neutral-pH Fe- and Mn-oxidizing chemolithotrophic bacteria should grow in enrichment cultures provided with Fe^{2+} or Mn^{2+} and a suitable carbon source as well as other essential nutrients and the proper redox conditions. The fact that such organisms have not been readily enriched or isolated from samples taken from Fe- or Mn-rich environments shows the difficulties in demonstrating that these metallic species can serve as energy sources.

Problems with such enrichment cultures include rapid abiological Fe^{2+} oxidation and Mn^{2+} toxicity (26, 27, 34, 41, 47). In some instances when culture problems have been understood and overcome, bacteria previously thought to be Mn or Fe chemolithotrophs, such as *Leptothrix discophora*, have been shown not to use the energy available from metal oxidation to support growth (1). *Gallionella ferruginea* is still the only neutral-pH "iron bacterium" for which reasonably good evidence of chemoautotrophy exists (26, 27, 34). In addition, recent work supports the idea that Mn^{2+} oxidation can supply at least part of the energy needs of certain marine bacteria. Ehrlich (19, 20, 22) has demonstrated that ATP synthesis is

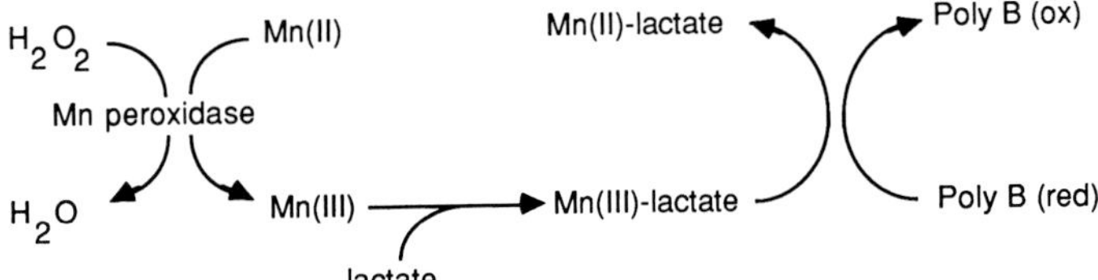

Figure 1. Indirect oxidation of the polymeric dye poly B by an extracellular manganese peroxidase of *P. chrysosporium* showing Mn(III) lactate as an obligatory intermediate. Redrawn from reference 30, Fig. 6.

coupled to Mn^{2+} oxidation by bacterial membrane vesicles from at least three gram-negative marine isolates. Also, Kepkay et al. (40, 41) have shown that Mn^{2+} stimulates the growth and CO_2 fixation of marine *Pseudomonas* strain S-36 in chemostat cultures. The data suggest that this organism may be capable of growing mixotrophically or even chemoautotrophically with Mn^{2+} as an energy source.

The ability to deposit iron oxides is probably universal among microorganisms that produce anionic exopolymers to which colloidal iron oxides can bind (29). However, at neutral pH, biological Fe^{2+} oxidation may compete with abiological oxidation only under low-O_2 conditions. Except for *G. ferruginea* (34), the significance of Fe^{2+} oxidation as an energy source for bacteria in microaerobic environments has been largely neglected.

The ability to oxidize Mn^{2+} and deposit manganese oxide extracellularly is a common property of heterotrophic bacteria and fungi (26). These organisms do not appear to derive direct energy benefits from Mn^{2+} oxidation. Instead, they may derive other benefits from the deposition of manganese oxide outside their cells. For example, the deposition of extracellular oxide may serve to relieve Mn^{2+} or other metal toxicity or to dissipate the effects of toxic oxygen species such as H_2O_2, which is rapidly oxidized by manganese oxide in a catalaselike reaction. A more clear-cut role of heterotrophic Mn^{2+} oxidation has been demonstrated for the lignin-degrading white rot fungus *Phanerochaete chrysosporium*, which produces an extracellular manganese peroxidase that is thought to be important in the lignin-degrading system of the fungus (42). The peroxidase catalyzes the oxidation of phenol red, *o*-dianisidine, and polymeric dyes (30) indirectly by oxidizing Mn(II) to Mn(III), which complexes with lactate (or other α-hydroxy acids). The Mn(III) complex in turn oxidizes the dye (Fig. 1). If a complexing agent such as lactate is omitted, the dye-oxidizing activity is lost and an insoluble manganese oxide is formed (30). It appears that several white rot fungi may use similar manganese peroxidase enzymes in their extracellular lignin-degrading systems, as witnessed by the accumulation of manganese oxides in wood decayed by white rot fungi (4).

Peroxidatic Mn oxidation has also been reported to occur in *Leptothrix pseudoochracea* and several other Mn-oxidizing bacteria (26), but no direct coupling of the bacterial activity to oxidation of phenolic or other organic compounds has been demonstrated. Interestingly, the controversial fungal symbiont *Metallogenium symbioticum* is one of the organisms reported to oxidize Mn in this way (17). It now seems possible that structures resembling *M. symbioticum* are extracellular products of certain fungi, possibly related to their extracellular manganese peroxidase activity. The connection between Mn oxidation and extracellular manganese peroxidase activity of fungi is an area ripe for further investigation.

Mechanisms of extracellular Mn oxidation by heterotrophic bacteria probably are different from those of fungi. Mn-oxidizing bacteria in the *Sphaerotilus-Leptothrix* group have been studied extensively in the past (26), and *L. discophora* SS-1 (28) has been the subject of several recent studies (1–3, 5). In this strain, which has lost its ability to form a sheath (1, 2), Mn^{2+} oxidation occurs extracellularly in association with complex protein-lipid-polysaccharide exopolymers (3, 5). SS-1 excretes at least one Mn-oxidizing protein (molecular weight, 100,000) which can be harvested along with exopolymers from spent growth medium. The Mn-oxidizing activity in the concentrated supernatant obeys Michaelis-Menten kinetics. It is inhibited by a variety of protein-denaturing treatments (e.g., heat, pronase, and detergents) and metabolic inhibitors (e.g., azide, *o*-phenanthroline, $HgCl_2$, and other metals). An active Mn-oxidizing protein band can be separated from all other bands by sodium dodecyl sulfate-polyacrylamide gel electrophoresis, but isolation of the protein in sufficient quantities for chemical analysis has not yet been possible. The product of oxidation reaction 2 appears to be a mixed Mn(III,IV) oxide (L. F. Adams, Ph.D. thesis, Cornell University, Ithaca, N.Y., 1986). On the basis of currently available data, it can be concluded that extracellular Mn-oxidizing activity of SS-1 behaves like an Mn^{2+}-oxidase.

Other *Leptothrix* strains and several other Mn-oxidizing heterotrophic bacteria (e.g., *Pedomicrobium* sp. [29]) also excrete proteinaceous extracellular Mn-oxidizing factors into the surrounding medium (A. Ress and W. Ghiorse, manuscript in preparation). Further studies of the extracellular Mn-oxidizing systems of these bacteria are expected to reveal the similarities of such systems among bacteria. At this point it seems likely that exopolymer-associated Mn-oxidizing systems like that of SS-1 are very common. Given the abundance of Fe- and Mn-depositing bacteria and exopolymers found in particles in natural systems (11, 12, 25, 26, 58), the significance of this type of system in nature may be quite high.

Another type of bacterial Mn-oxidizing system, which appears to be common among aerobic endosporeforming bacteria (26), has been found in a marine *Bacillus* sp. and described in detail (15, 36, 48). The surfaces of endospores of this *Bacillus* sp. bind and oxidize Mn^{2+} to MnO_2 by reaction 2. A spore coat protein is suspected to be involved in the catalytic activity. It may be environmentally significant that vegetative cells of this organism also reduce MnO_2 by an enzyme-linked process similar to that found in other Mn-reducing species (16, 27a), suggesting that this organism, like other bacteria reported earlier (26, 27a), can catalyze both oxidation and reduction of Mn at different stages in its life cycle.

Fe AND Mn AS PHYSIOLOGICAL ELECTRON ACCEPTORS

Biologically produced iron and manganese oxides probably play a larger role in oxidation of naturally occurring organic compounds than previously has been thought. These oxides may be highly reactive in oxidizing a number of organic compounds (37, 51–53). Manganese oxides also may react rapidly to oxidize inorganic species such as H_2S and H_2O_2. Because of possible abiological reactions, the concept of Fe and Mn as physiological electron acceptors must be viewed cautiously. I have reviewed this topic recently (27a); therefore, only major conclusions are summarized here.

The literature on microbial Mn and Fe reduction clearly shows that these processes probably are biologically mediated in anaerobic and, to a lesser extent, aerobic environments. Physiological studies of bacterial Fe and Mn reduction show that respiratory enzyme systems can be involved, but many of these studies are compromised by the often overlooked possibility that abiotic reductions mediated by metabolic end products are involved. Such reductions are likely to occur simultaneously with reduction coupled to electron transport systems in the cell (27a, 51). Recent work in which abiotic reactions were taken into account (8, 9, 38, 39, 43, 50) shows that Fe and Mn oxides can be reduced in sediments by processes that are coupled biologically to anaerobic oxidation of organic matter, but the nature of the coupling mechanisms and the biochemistry involved have not been studied as yet. More work is needed to investigate the physiology and biochemistry of Fe and Mn reduction with respect to their geochemistry in natural systems.

ACKNOWLEDGMENTS. Research in my laboratory has been supported by the National Science Foundation, the U.S. Environmental Protection Agency, and the U.S. Department of Energy, Office of Health and Environmental Research.

The secretarial help of Patti Lisk is gratefully acknowledged.

LITERATURE CITED

1. **Adams, L. F., and W. C. Ghiorse.** 1985. Influence of manganese on growth of a sheathless strain of *Leptothrix discophora*. *Appl. Environ. Microbiol.* **49**: 556–562.
2. **Adams, L. F., and W. C. Ghiorse.** 1986. Physiology and ultrastructure of *Leptothrix discophora* SS-1. *Arch. Microbiol.* **145**:126–135.
3. **Adams, L. F., and W. C. Ghiorse.** 1987. Characterization of extracellular Mn^{2+}-oxidizing activity and isolation of an Mn^{2+}-oxidizing protein from *Leptothrix discophora* SS-1. *J. Bacteriol.* **169**:1279–1285.
4. **Blanchette, R. A.** 1984. Manganese accumulation in wood decayed by white-rot fungi. *Phytopathology* 74:725–730.
5. **Boogerd, F. C., and J. P. M. deVrind.** 1987. Manganese oxidation by *Leptothrix discophora*. J. Bacteriol. **169**:489–494.
6. **Boudou, J. P., P. Kaiser, and J. M. Phillipot.** 1985. Elimination du fer et du manganese: interet des procedes biologique. *Water Supply* **3B**:151–155.
7. **Brannon, J. M., D. Gunnison, R. M. Smart, and R. L. Chen.** 1984. Effects of organic matter on iron and manganese redox systems in sediment. *Geomicrobiol. J.* 3:319–341.
8. **Burdige, D. J., and K. H. Nealson.** 1985. Microbial manganese reduction by enrichment cultures from coastal marine sediments. *Appl. Environ. Microbiol.* **50**:491–497.
9. **Burdige, D. J., and K. H. Nealson.** 1986. Chemical and microbiological studies of sulfide-mediated manganese reduction. *Geomicrobiol. J.* 4:361–387.
10. **Chapnick, S. D., W. S. Moore, and K. H. Nealson.** 1982. Microbially mediated manganese oxidation in a freshwater lake. *Limnol. Oceanogr.* **27**:1004–1014.
11. **Cowan, J. B., and K. W. Bruland.** 1985. Metal deposits associated with bacteria: implications for Fe and Mn biogeochemistry. *Deep Sea Res.* **32**:253–272.
12. **Cowan, J. B., and M. W. Silver.** 1984. The association of iron and manganese with bacteria on marine microparticulate material. *Science* **224**:1340–1342.
13. **Czekalla, C., W. Mevius, and H. Hanert.** 1985. Quantitative removal of iron and manganese by microorganisms in rapid sand filters (*in situ* investigations). *Water Supply* **3B**:111–123.
14. **Davison, W., and C. Woof.** 1984. A study of the cycling of manganese and other elements in a seasonally anoxic lake, Rostherne Mere, U.K. *Water Res.* **18**:727–734.
15. **deVrind, J. P. M., E. W. deVrind-deJong, J.-W. H. deVoogt, P. Westbroek, F. C. Boogerd, and R. A. Rosson.** 1986. Manganese oxidation by spores and spore coats of a marine *Bacillus* species. *Appl. Environ. Microbiol.* **52**:1096–1100.
16. **deVrind, J. P. M., F. C. Boogerd, and E. W. deVrind-deJong.** 1986. Manganese reduction by a marine *Bacillus* species. *J. Bacteriol.* **167**:30–34.
17. **Dubinina, G. A.** 1979. Mechanism of the oxidation of divalent iron and manganese by iron bacteria growing at neutral pH of the medium. *Microbiology* (English translation) 47:471–478.
18. **Edenborn, H. M., Y. Paquin, and G. Chateauneuf.** 1985. Bacterial contribution to manganese oxidation in a deep coastal sediment. *Estuarine Coastal Shelf Sci.* **21**:801–815.
19. **Ehrlich, H. L.** 1976. Manganese as an energy source for bacteria, p. 633–644. *In* J. O. Nriagu (ed.), *Environmental Biogeochemistry*, vol. 2. *Metals Transfer and Ecological Mass Balances*. Ann Arbor Science, Ann Arbor, Mich.
20. **Ehrlich, H. L.** 1978. Inorganic energy sources for chemolithotrophic and mixotrophic bacteria. *Geomicrobiol. J.* 1:65–83.
21. **Ehrlich, H. L.** 1981. *Geomicrobiology*. Marcel Dekker, Inc., New York.

22. **Ehrlich, H. L.** 1983. Manganese-oxidizing bacteria from a hydrothermally active area on the Galapagos rift. *Ecol. Bull. NFR* 35:357–366.
23. **Emerson, S., S. Kalhorn, L. Jacobs, B. M. Tebo, K. H. Nealson, and R. A. Rosson.** 1982. Environmental oxidation rate of manganese(II): bacterial catalysis. *Geochim. Cosmochim. Acta* **46**:1073–1079.
24. **Frischherz, H., F. Zibuschka, H. Jung, and W. Zerobin.** 1985. Biological elimination of iron and manganese. *Water Supply* **3B**:125–136.
25. **Ghiorse, W. C.** 1984. Bacterial transformations of manganese in wetland environments, p. 615–622. *In* M. J. Klug and C. A. Reddy (ed.), *Current Perspectives in Microbial Ecology*. American Society for Microbiology, Washington, D.C.
26. **Ghiorse, W. C.** 1984. Biology of iron- and manganese-depositing bacteria. *Annu. Rev. Microbiol.* **38**:515–550.
27. **Ghiorse, W. C.** 1986. Biology of *Leptothrix*, *Gallionella*, and *Crenothrix*, relationship to plugging, p. 97–108. *In* R. D. Cullimore (ed.), *Proceedings of the International Symposium on Biofouled Aquifers: Prevention and Restoration*. American Water Resources Association, Urbana, Ill.
27a. **Ghiorse, W. C.** 1988. Microbial reduction of manganese and iron, p. 305–331. *In* A. J. B. Zehnder (ed.), *Biology of Anaerobic Microorganisms*. John Wiley & Sons, Inc., New York.
28. **Ghiorse, W. C., and S. D. Chapnick.** 1983. Metal-depositing bacteria and the distribution of manganese and iron in swamp waters. *Ecol. Bull.* NFR **35**:367–376.
29. **Ghiorse, W. C., and P. Hirsch.** 1979. An ultrastructural study of iron and manganese deposition associated with extracellular polymers of *Pedomicrobium*-like budding bacteria. *Arch. Microbiol.* **123**:213–226.
30. **Glenn, J. K., L. Akileswaran, and M. H. Gold.** 1986. Mn(II) oxidation is the principal function of the extracellular Mn-peroxidase from *Phanerochaete chrysosporium*. *Arch. Biochem. Biophys.* **251**:688–696.
31. **Gottfreund, E., I. Gerber, and R. Schweisfurth.** 1985. Verteilung verschiedener physiologischer Gruppen von Bakterien in einem oxidierten und einem reduzierten Grundwasserleiter. *Landwirtsch. Forsch.* **38**:72–79.
32. **Gottfreund, E., J. Gottfreund, I. Gerber, G. Schmitt, and R. Schweisfurth.** 1985. Occurrence and activities of bacteria in the unsaturated and saturated underground in relation to the removal of iron and manganese. *Water Supply* **3A**:109–115.
33. **Gottfreund, E., J. Gottfreund, and R. Schweisfurth.** 1985. Mikrobiologische Grundlage der Subterrestrischen Enteisenung und Entmanganung. *Forum Städte-Hyg.* **36**:178–183.
34. **Hanert, H. H.** 1981. The genus *Gallionella*, p. 509–515. *In* M. P. Starr, H. Stolp, H. Trüper, A. Balows, and H. G. Schlegel (ed.), *The Prokaryotes*, vol. 1. Springer-Verlag, New York.
35. **Hart, B. T., and M. J. Jones.** 1984. Oxidation of manganese(II) in island billabong water. *Environ. Technol. Lett.* **6**:87–92.
36. **Hastings, D., and S. Emerson.** 1986. Oxidation of manganese by spores of a marine bacillus: kinetic and thermodynamic considerations. *Geochim. Cosmochim. Acta* **50**:1819–1824.
37. **Jauregui, M. A., and H. M. Reisenauer.** 1982. Dissolution of oxides of manganese and iron by root exudate components. *Soil. Sci. Soc. Am. J.* **46**:314–317.
38. **Jones, J. G., S. Gardener, and B. M. Simon.** 1983. Bacterial reduction of ferric iron in a stratified eutrophic lake. *J. Gen. Microbiol.* **129**:131–139.
39. **Jones, J. G., S. Gardener, and B. M. Simon.** 1984. Reduction of ferric iron by heterotrophic bacteria in lake sediments. *J. Gen. Microbiol.* **130**:45–51.
40. **Kepkay, P. E., D. J. Burdige, and K. H. Nealson.** 1984. Kinetics of bacterial manganese binding and oxidation in the chemostat. *Geomicrobiol. J.* **3**:245–262.
41. **Kepkay, P. E., and K. H. Nealson.** 1987. Growth of a manganese oxidizing *Pseudomonas* sp. in continuous culture. *Arch. Microbiol.* **148**:63–67.
42. **Kirk, T. K., and R. L. Farrell.** 1987. Enzymatic "combustion": the microbial degradation of lignin. *Annu. Rev. Microbiol.* **41**:465–505.
43. **Lovley, D. R., and E. J. P. Phillips.** 1986. Organic matter mineralization with reduction of ferric iron in organic sediments. *Appl. Environ. Microbiol.* **51**: 683–689.
44. **Lundgren, D. G., and W. Dean.** 1979. Biogeochemistry of iron, p. 211–251. *In* P. A. Trudinger and D. J. Swaine (ed.), *Biogeochemical Cycling of Mineral-Forming Elements*. Elsevier/North-Holland Publishing Co., Amsterdam.
45. **Madsen, E. L., M. D. Morgan, and R. E. Good.** 1986. Simultaneous photoreduction and microbial oxidation of iron in a stream in the New Jersey Pinelands. *Limnol. Oceanogr.* **31**:832–838.
46. **Marshall, K. C.** 1979. Biogeochemistry of manganese minerals, p. 253–292. *In* P. A. Trudinger and D. J. Swaine (ed.), *Biogeochemical Cycling of Mineral-Forming Elements*. Elsevier/North-Holland Publishing Co., Amsterdam.
47. **Nealson, K. H.** 1983. Microbial oxidation and reduction of manganese and iron, p. 459–487. *In* P. Westbroek and E. W. deJong (ed.), *Biomineralization and Biological Metal Accumulation*. D. Reidel, Boston.
48. **Rosson, R. A., and K. H. Nealson.** 1982. Man-

ganese binding and oxidation by spores of a marine bacillus. *J. Bacteriol.* **151**:1027–1034.

49. **Rosson, R. A., B. M. Tebo, and K. H. Nealson.** 1984. Use of poisons in determination of microbial manganese binding rates in seawater. *Appl. Environ. Microbiol.* **47**:740–745.
50. **Sørensen, J.** 1982. Reduction of ferric iron in anaerobic, marine sediment and interaction with reduction of nitrate and sulfate. *Appl. Environ. Microbiol.* **43**:319–324.
51. **Stone, A. T.** 1987. Microbial metabolites and the reductive dissolution of manganese oxides: oxalate and pyruvate. *Geochim. Cosmochim. Acta* **51**:919–925.
52. **Stone, A. T., and J. J. Morgan.** 1984. Reduction and dissolution of manganese(III) and manganese(IV) oxides by organics. 1. Reaction with hydroquinone. *Environ. Sci. Technol.* **18**:450–456.
53. **Stone, A. T., and J. J. Morgan.** 1984. Reduction and dissolution of manganese(III) and manganese(IV) oxides by organics. 2. Survey of the reactivity of organics. *Environ. Sci. Technol.* **18**:617–624.
54. **Stumm, W., and J. J. Morgan.** 1981. *Aquatic Chemistry*, 2nd ed. John Wiley & Sons, Inc., New York.
55. **Tebo, B. M., and S. Emerson.** 1985. Effect of oxygen tension, Mn(II) concentration, and temperature on the microbially catalyzed Mn(II) oxidation rate in a marine fjord. *Appl. Environ. Microbiol.* **50**:1268–1273.
56. **Tebo, B. M., K. H. Nealson, S. Emerson, and L. Jacobs.** 1984. Microbial mediation of Mn(II) and Co(II) precipitation at the O_2/H_2S interfaces in two anoxic fjords. *Limnol. Oceanogr.* **29**:1247–1258.
57. **Tipping, E.** 1984. Temperature dependence of Mn(II) oxidation in lakewaters: a test for biological involvement. *Geochim. Cosmochim. Acta* **48**:1353–1356.
58. **Tipping, E., J. G. Jones, and C. Woof.** 1985. Lacustrine manganese oxides: Mn oxidation states and relationship to "Mn depositing bacteria." *Arch. Hydrobiol.* **105**:161–175.
59. **Tipping, E., D. W. Thompson, and W. Davison.** 1984. Oxidation products of Mn(II) in lake waters. *Chem. Geol.* **44**:359–383.

Chapter 15

Quantitative Relationships between Carbon, Hydrogen, and Sulfur Metabolism in Cyanobacterial Mats

Graham W. Skyring, Ronald M. Lynch, and Geoffrey D. Smith

INTRODUCTION

Cyanobacterial mats are useful in quantifying the interaction between the carbon and sulfur cycles because primary productivity is entirely local and several investigators have shown that sulfate reduction in these communities generally occurs at very high rates (up to 104 mmol m^{-2} day^{-1}) in close proximity to, or even within, the phototrophic zone of the cyanobacterial mat (5, 6, 12, 13, 17, 20, 27, 27a, 28). In cyanobacterial mats in which the photic-aphotic transition occurs within 1 mm or less (5, 27, 28), the complete set of biogeochemical reactions concerning carbon and sulfur metabolism occurs within or very close to the phototrophic zone of the mat. Quantitative relationships between primary productivity and sulfate reduction were first calculated for Solar Lake cyanobacterial mats by Jørgensen and Cohen (13). They found that 14 to 20% of the daily primary productivity (0.7 to 1.0 mol of C m^{-2} day^{-1}) was oxidized by the sulfate-reducing bacteria (SRB). Skyring et al. (28) showed that for the smooth cyanobacterial mat from Spencer Gulf, it was possible that 100% of the organic carbon synthesized by the cyanobacteria in the mat was eventually oxidized by the SRB. They estimated that the molar ratio between photosynthetically fixed carbon and reduced sulfate was 2:1 (±20%). They also showed that sulfate reduction rates correlated positively with primary productivity of the *Microcoleus* mat when the productivity data were transformed by a time series equation which accounted for the delay in the decomposition of the photosynthate. Chambers (4) also showed that there were close spatial and temporal relationships between the production of low-molecular-weight organic compounds (mainly acetate) and the oxidation of these compounds by the SRB in *Microcoleus* mats from Hamelin Pool, Shark Bay, Western Australia. The reliability of these calculations is constrained by the assumption that all energy substrates for the SRB are derived

Graham W. Skyring • Division of Water Resources Research, Commonwealth Scientific and Industrial Research Organisation, G.P.O. Box 1666, Canberra, Australian Capital Territory 2601, Australia. ***Ronald M. Lynch and Geoffrey D. Smith*** • Department of Biochemistry, Faculty of Science, The Australian National University, G.P.O. Box 4, Canberra, Australian Capital Territory 2601, Australia.

from C-linked photosynthetic reactions. However, hydrogen production by cyanobacteria in axenic or pure culture is well known (11, 14, 16), and if significant quantities of H_2 were produced by the cyanobacteria in a manner not linked to photosynthetic CO_2 fixation, then the calculated quantitative relationships between primary productivity and sulfate reduction rates would have to be modified. The present communication describes the results of experiments on the H_2-producing capacity of various types of intact cyanobacterial mats. These experiments were prompted by a discussion of the possible constraints on the calculation of quantitative relationships between carbon and sulfur metabolism in cyanobacterial mats and the possible role for a significant quantitative contribution by H_2 as an energy substrate for the SRB (27, 28, 32).

MEASUREMENT OF ACETYLENE REDUCTION AND H_2 PRODUCTION

Cyanobacterial mats were collected from Hamelin Pool; within 3 to 4 days they were placed in a shallow tank containing aerated Hamelin Pool water and maintained under a constant photon irradiance of ca. 8 microeinsteins m^{-2} s^{-1} (photosynthetically active radiation, 400 to 700 nm). Sections (5 by 1.2 cm; thickness, 2 ± 1 mm) of the mats were incubated in Pierce vials (volume, 8.8 ml) fitted with silicone rubber septa. The vials were filled with a defined medium (1; all components at one-eighth strength except phosphate and $MgSO_4$) supplemented with $NaHCO_3$ (3 mM), NaCl (3.5%, wt/vol), and other components as indicated below. $NiSO_4$ was omitted because the nickel content of the smooth mat was 7 ppm (μg/g) (wt/dry wt), which would be adequate for hydrogenase activity (7). A headspace of air or argon (at atmospheric pressure) was formed by withdrawing liquid (4.4 ml) from the vials, which were then flushed with the appropriate gas for at least 10 min. Acetylene (5 or 10%, vol/vol) and H_2 (4.5%, vol/vol) were injected into the headspace for nitrogenase (31) and H_2 uptake measurements, respectively. The vials were shaken on their sides at 90 oscillations per min. When illumination was required, the photon irradiance was constant within the range of 40 to 70 microeinsteins m^{-2} s^{-1} (photosynthetically active radiation, 400 to 700 nm). Nitrogenase activity was measured by the production of ethylene, which, together with H_2, N_2, and O_2, was measured by gas chromatography (15).

MEASUREMENT OF SULFATE REDUCTION RATES

The samples for sulfate reduction rate measurements were put into Pierce vials and prepared and incubated under the same conditions as for the H_2 production experiments. $Na_2{}^{35}SO_4$ (0.16 μmol) at 0.87 GBq $mmol^{-1}$ was added to all vials, and 0.5 ml of 20% $ZnCl_2$ was added to each vial at the end of the incubation period. The samples were then frozen until required for distillation analyses. Five slices of mat, representing an area of 30 cm^2, were used for each experimental condition.

The $H_2{}^{35}S$ formed in the samples during incubation was recovered by the following procedure. The samples were thawed and transferred to a reaction flask which was continuously gassed with oxygen-free nitrogen. The sample port was closed, and gassing with nitrogen continued for 5 min. Two drops of antifoam (Dow Corning Antifoam C emulsion) and 50 ml of 5 N HCl were added, and the H_2S thus released from acid-volatile sulfide in the mats was carried by the stream of N_2 into two 2 N NaOH traps. When this reaction was completed, the NaOH traps were replaced and 5 g of granulated tin and 30 ml of 10 N HCl containing 20% $SnCL_2$ were added to the reaction flask. The remainder of the sample was then boiled under a reflux condenser for 1 h.

Previous experiments showed that this period was sufficient to recover non-acid-volatile sulfide in similar sediments. The second reaction with Sn was required to recover the ^{35}S which was present in pyrite or other forms of non-acid-volatile sulfide. Replicate samples (0.5 ml) from the NaOH traps were placed in scintillation vials containing 4.0 ml of H_2O and 5.0 ml of Instagel (Packard Instrument Co.) and counted in a Packard Scintillation Counter with the spectralyzer channels set between 4 and 167. Around 90% of the S expected from 25 mg of authentic pyrite (kindly supplied by T. H. Donnelly) was recovered by the tin distillation procedure.

The sulfate reduction rates were calculated by the method of Skyring et al. (28) from the sum of the acid-volatile and tin-reducible sulfur that was formed during incubation.

H_2 PRODUCTION AND CONSUMPTION

Production

The rates of net H_2 production by smooth mat, tufted mat, and pustular mat are given in Fig. 1a, b, and c, respectively. Data are recorded for H_2 production in each mat ecosystem in the light and the dark in the presence of molybdate and also 3-(3,4-dichlorophenyl)-1,1-dimethylurea (DCMU). Sodium molybdate, a specific inhibitor of the SRB (34), was added to determine whether the SRB were consuming H_2. DCMU was added as a specific inhibitor of oxygenic photosynthesis. All of the cyanobacterial mat ecosystems produced H_2 under an atmosphere of argon (Fig. 1a to c); however, in the tufted mat this was evident only in the presence of molybdate (see below). Patterns of H_2 production under an atmosphere of air were similar, but quantitatively the production was one to several orders of magnitude lower. The rates of H_2 consumption could not be shown by headspace H_2 accumulation measurements alone. However, when slices of smooth and pustular mat were incubated under a 4.5% H_2–9.5% argon atmosphere, the H_2 consumption rate in the light exceeded that in the dark: H_2 consumption rates in the tufted mat in the dark and in the light were comparable (Fig. 1d to f). Thus, H_2 produced either by the photosynthetic components of the mat or by the nonphotosynthetic heterotrophic bacterial population is consumed by the photosynthesizing cyanobacterial population.

The effect of molybdate was to enhance H_2 production in the dark but not in the light in smooth and tufted mats (Fig. 1a and b). Because molybdate inhibits sulfate reduction by the SRB, it may be expected that substrates which are oxidized by the SRB would accumulate in the ecosystem. Thus, it appears that H_2 may be a quantitatively significant substrate for the SRB in the dark. However, the H_2 appears to be preferentially consumed by the photosynthetic cyanobacteria or other photosynthetic components of the ecosystem in the light. Molybdate significantly reduced the rates at which H_2 was consumed in the smooth mat (Fig. 1d), and this indicated that the SRB are quantitatively important components of the cyanobacterial mat community. However, there was no effect of molybdate on the H_2-producing capacity of the pustular mat ecosystem (Fig. 1c), and this is consistent with the previous observations of Skyring (27) that sulfate reduction occurs in smooth and tufted mat communities but that pustular mat communities generally do not support an active SRB component. DCMU did not appear to significantly affect H_2 accumulation in any of the mat ecosystems, suggesting that the H_2 is produced by fermentation and not by the photoclastic split of water.

Consumption

A rapid light-dependent H_2 uptake was shown for the three mat types (Fig. 1d to f). Since the sulfate reduction rates were similar

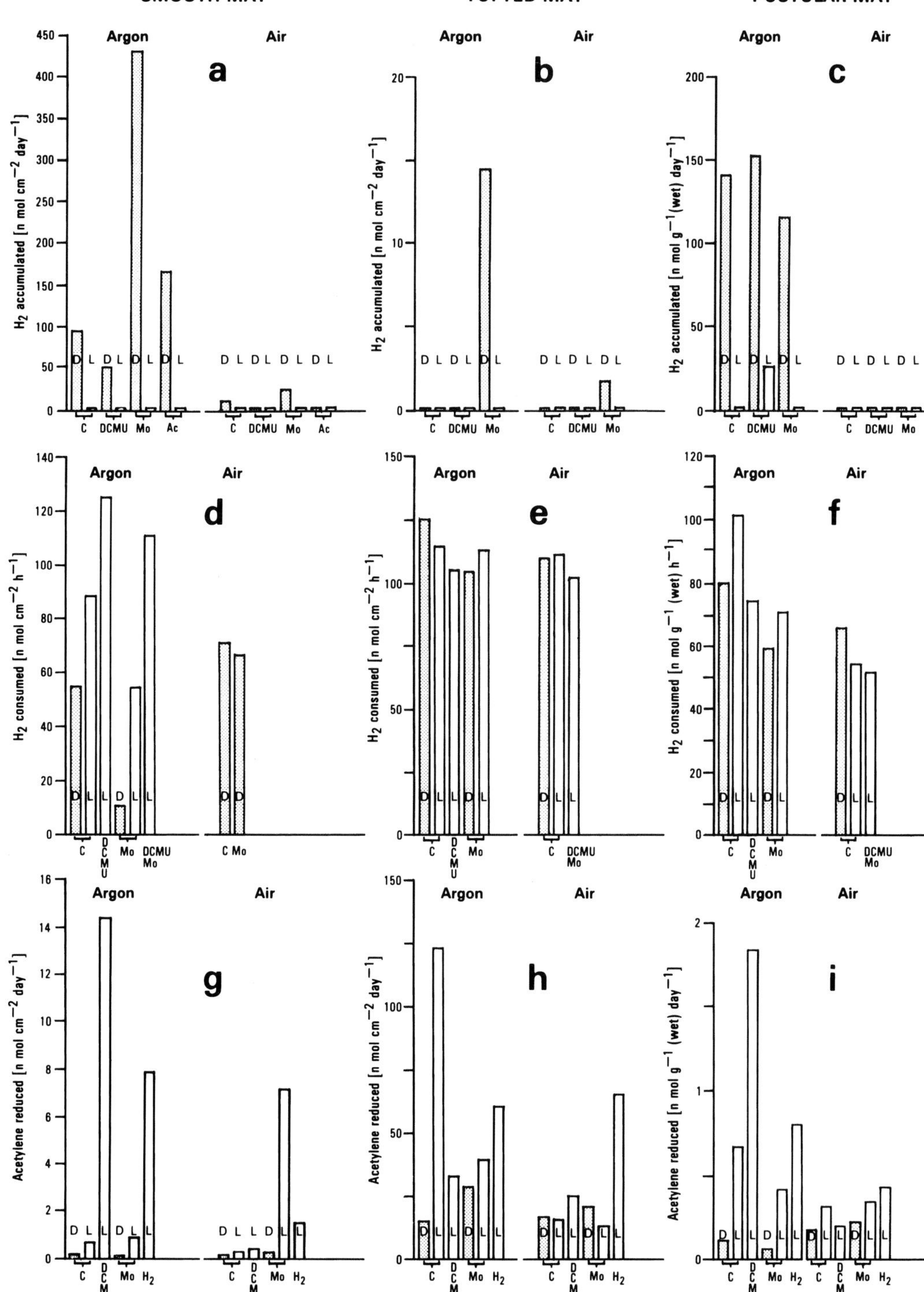

Figure 1. Hydrogen production and consumption and acetylene reduction by smooth, tufted, and pustular cyanobacterial mats from Hamelin Pool. The mats were cut into sections as described in the text and were incubated for 24 h in the presence of 1 mM sulfate in the light (L) or dark (D) with the specific chemicals added at the following concentrations: DCMU, 50 μM; sodium molybdate (Mo), 10 mM; sodium acetate (Ac), 1 mM. The control rates are denoted C. The gas phase initially contained argon or air. For H_2 consumption measurements, sections of the mat were incubated for 24 h under a gas phase of argon or air, supplemented with an initial partial pressure of H_2 of 4.5% (vol/vol).

Table 1.
Sulfate Reduction Rates and Net H_2 Production Rates in Smooth Mats from Spencer Gulf

Conditions	Sulfate reduction rate (nmol cm^{-2} day^{-1}) by:			H_2 production rate (nmol cm^{-2} day^{-1})[a]
	AVS[b]	NAVS[b]	Total[c]	
Dark	540	441	983	3
Light	645	377	1,022	2
Dark (Mo)			5	872
Light (Mo)			5	6

[a] $n = 4$ for all conditions.
[b] Abbreviations: AVS, acid-volatile sulfide; NAVS, non-acid-volatile sulfide.
[c] $n = 9$ for all conditions.

in the dark and the light for the Spencer Gulf smooth mats (Table 1), it may be inferred that fermentative processes, which are the sources of substrates for the SRB, also occur at similar rates in the light and in the dark. Therefore, it appears that there are three distinct types of H_2 uptake processes: light-, oxygen-, and sulfate-dependent uptake. The fact that there were measurable effects of molybdate on H_2 metabolism in smooth and tufted mats indicated that the SRB were quantitatively important consumers of H_2 in these communities.

SOURCES OF H_2

To determine the stoichiometry of carbon and sulfur metabolism in cyanobacterial mat ecosystems, it was also important to determine the metabolic sources or mechanisms by which the H_2 was produced. Possible mechanisms are (i) fermentation of photosynthetically produced organic matter, (ii) H_2 production as an inevitable consequence of nitrogenase activity, and (iii) photoclastic (biophotolysis) split of water into H_2 and O_2 (in which the H_2 production is independent of carbon metabolism). We therefore measured the rates of acetylene reduction (nitrogenase activity) for smooth, tufted, and pustular mats from Hamelin Pool. Microscopic examinations showed that cyanobacteria were the major phototrophs in these mat communities. A more detailed account of this work with smooth-mat communities is given by Skyring et al. (28a). The rates of acetylene reduction in the light and the dark in the presence of DCMU, molybdate, and H_2 are given in Fig. 1g to i. The results showed that nitrogenase activity was greater in the light than in the dark in the three mat communities and that it was greatly enhanced by DCMU in the light in smooth and pustular mats. These results indicated that the nitrogenase of the smooth and pustular communities was sensitive to O_2 and that it occurred predominantly within photosynthetic organisms, presumably the dominant mat-constructing organisms, *Microcoleus* or *Entophysalis* spp. The nitrogenase activity in the tufted mat was not detected in the *Lyngbya* component (isolated from the Hamelin Pool tufted mat), indicating that in this community, nitrogenase activity appeared to be due to the *Microcoleus* component. For each mat type, the rates of acetylene reduction in the dark were similar under argon and air, suggesting that the nitrogenase which was responsible for this activity was not as oxygen sensitive as the light-dependent nitrogenase was. However, it was not possible to attribute the nitrogenase that was active in the dark to any specific group of microorganisms. The findings reported here were reproducible in mats collected at various times of the year, although there was some degree of quantitative variability. The magnitude of the effect of DCMU on acetylene reduction, in particular, was variable, and on some occasions an inhibitory rather than a stimulatory effect was observed. We previously speculated that this may be due to seasonal variation of the intracellular carbohydrate component of the cyanobacteria or of the cyanobacterial composition of the mats (28a). Stal et al. (29) also showed that DCMU stimulated nitrogenase activity in a *Microcoleus-Oscillatoria-Spirulina-Gloeocapsa* mat and attributed this to O_2 sensitivity of the nitrogenase.

Nitrogen fixation by heterocystous cyanobacteria has been long established and is well documented (9, 10, 30). Nonheterocystous cyanobacteria have also been shown to be nitrogen fixers. Potts and Whitton (24), Pearson et al. (22), and Stal et al. (29) have shown that mats composed of nonheterocystous *Microcoleus* spp. had nitrogenase activity which was maximal in the light. Pearson et al. (23) showed that axenic cultures of *Microcoleus chthonoplastes* fixed N_2 when grown under continuous illumination, but that when the cultures were exposed to a light-dark regimen, maximal nitrogenase activity occurred in the dark. In the present investigation it was shown that intact mats with *Microcoleus* sp. as the major photosynthetic component had an active light-dependent nitrogenase system and that pustular mats, composed almost entirely of another nonheterocystous cyanobacterium, *Entophysalis* sp., also contained a light-dependent nitrogenase which appeared to be O_2 sensitive. Nitrogenase activity in the latter cyanobacterial communities has not been previously reported.

Cyanobacteria are known to produce H_2 via either hydrogenase or nitrogenase (16). The stoichiometry of the relationship between N_2 fixation and H_2 evolution by nitrogenase is given by the following equation:

$$N_2 + 8H^+ + 8e^- + 16ATP \rightarrow 2NH_3 + H_2 + 16ADP + 16\ P_i$$

(8, 19, 21, 26). Although it is clearly possible that significant H_2 production is associated with N_2 fixation in these ecosystems (especially the tufted mat), it could not account for the total quantity of H_2 that is produced in the cyanobacterial mats.

Light-dependent nitrogenase activity was lower for all of the mat types in air rather than argon, indicating that the nitrogenase systems were O_2 sensitive. Molybdate significantly enhanced the nitrogenase activity in smooth and tufted mats in the light but not in the dark, and added H_2 also stimulated light-dependent nitrogenase activity in both argon and air. These results indicate that the inhibition of the SRB enhances the amount of H_2 available to support nitrogenase activity by the photosynthetic microorganisms. On the other hand, H_2 inhibition of N_2 fixation but not of acetylene reduction is well known (25), and it may be that the rapid consumption of fermentative H_2 prevents the inhibition of N_2 fixation in cyanobacterial communities. The amount of H_2 released in the dark in smooth and tufted mats in the presence of molybdate was high (14 to 425 nmol cm^{-2} day^{-1}), suggesting that the H_2 formed by fermentative bacteria in the dark may be a substantial proportion of the energy substrates for the SRB in the mat.

STOICHIOMETRY BETWEEN C AND S, AND H METABOLISM

Since the original work which prompted these investigations was done with a *Microcoleus* smooth mat from Spencer Gulf (28), sulfate reduction rates and H_2 production rates in a set of replicate samples of smooth mat from Spencer Gulf were determined simultaneously.

The sulfate reduction rates and H_2 production rates in replicate samples of smooth mat from Spencer Gulf are given in Table 1. There was no significant difference between sulfate reduction rates in the dark and light, and molybdate completely inhibited sulfate reduction under all conditions. The patterns of H_2 accumulation were similar to those observed previously for smooth and tufted mats from Hamelin Pool. The H_2 which accumulated in the presence of molybdate in the dark accounted for about 20% of the oxidizable substrates that would have been required by the SRB had they not been inhibited by molybdate. Since the experiments described above showed that the SRB did not effectively compete with the cyanobacterial component of the mat for H_2 in the light, it may be inferred that in this particular instance, H_2 contributes about 10% daily (assuming 12 h light and 12 h dark) of the oxidizable substrates for the SRB.

The sulfate reduction rates which were determined for the present samples of smooth mat from Spencer Gulf were in the range previously determined for similar smooth mats in both Spencer Gulf and Hamelin Pool (28, 28a). These investigations and those of Jørgensen and Cohen (13), together with the results of Skyring and Cohen (Table 2; unpublished results), clearly indicate that the SRB are important microbial components of cyanobacterial mats in environments where fine-grained sediments permit the establishment of anoxic conditions. The latter situation generally excludes the pustular mat, which appears to prefer to colonize loose, coarse-grained sediments that are not easily waterlogged or made anaerobic. The observations that sulfate reducers are active in smooth and tufted mats are enigmatic because all of the SRB which have been isolated are obligate anaerobes and, although they may tolerate several minutes to several hours of exposure to O_2, they are unable to reduce sulfate even in the presence of small quantities of O_2 (27a). Also, the cyanobacteria in these communities are actively producing O_2 during daylight hours. However, there is clearly a close spatial relationship between photosynthesis, diagenesis, and mineralization of carbon by the SRB.

Westrich and Berner (33) calculated that the kinetics of sulfate reduction, in which the organic source was planktonic carbon, could be described by the sequential decomposition of two organic fractions with decay constants of 8 ± 1 year^{-1} and 0.94 ± 0.25 year^{-1}. In marked contrast to this, the data of Jørgensen and Cohen (13) and Skyring et al. (28) indicated that organic matter from cyanobacterial mats has a decay constant equivalent to that of the rapidly metabolized organic fraction described by Westrich and Berner (33) and also that the SRB have an important role in its diagenesis. One possible reason for this is the close spatial relationship between the phototrophs and SRB in cyanobacterial mat communities. It is known that some of the low-molecular-weight organic material produced during photosynthesis is immediately available to the microbial populations of the mat ecosystem and that this fraction of organic matter may account for 1 to 6% of the primary productivity (2). After these studies, Chambers (4) showed that acetate was a significant component of this dissolved organic fraction and concluded from molybdate inhibition studies that the acetate produced during photosynthesis was immediately oxidized by the SRB of smooth and tufted mat communities from Shark Bay. Other studies provide data which are consistent with a rapid transfer between producers and consumers. Howarth and Marino (12) found a fivefold increase in sulfate reduction in cyanobacterial mats (Great Sippewissett Marsh, Mass.)

Table 2.
Sulfate Reduction Rates in a Smooth Mat from Solar Lake, Sinai

Depth (mm)	Sulfate reduction rate (nmol cm^{-2} day^{-1})[a]
1	11
2	10
3	7
5	7
10	5

[a] $n = 2$ for all depths.

on cold sunny days as opposed to warm cloudy days; they suggested that this indicated a short-term coupling of mat photosynthesis and sulfate reduction. A diurnal effect on sulfate reduction in a subtropical sea grass ecosystem was also observed by Moriarty et al. (18).

CONCLUDING REMARKS

Ward (32) briefly discussed various aspects of the decomposition of microbial mats. He suggested that underestimations of the primary productivity by cyanobacterial mats may have resulted in overestimation of the quantities of organic matter estimated to be oxidized to CO_2 by the SRB (28). However, work by Bauld et al. (2, 3) on the primary productivity of cyanobacterial ecosystems from Spencer Gulf and Hamelin Pool suggests that the original estimates of primary productivity for the Spencer Gulf smooth mat were probably not grossly underestimated. Both Skyring (27) and Ward (32) discussed the possibility that H_2 whose production was not coupled to carbon metabolism in cyanobacterial mats was a significant oxidizable substrate for the SRB residing in the mats. The results of the present investigation suggest that this does not occur, and it appears that the original suggestion of Skyring et al. (27) and Cohen et al. (6), i.e., that the SRB within cyanobacterial mat ecosystems are responsible for the oxidation to CO_2 of most of the organic carbon synthesized by phototrophs, was correct. We have shown that there was potential for H_2 production as a consequence of N_2 fixation in all of the mat types investigated, but that in smooth and tufted mats, where sulfate reduction occurred, the quantity of H_2 produced by this mechanism was small compared with that produced by fermentative processes. Our investigations also indicated that the photoclastic split of water was not a significant H_2 source in these cyanobacterial mats.

Acknowledgments. We thank the Commonwealth Scientific and Industrial Research Organization/Australian National University Collaborative Research Fund, which supported this project. The Baas Becking Geobiological Laboratory, Canberra (at which G.W.S. conducted some of this research), was supported by the Bureau of Mineral Resources, the Commonwealth Scientific and Industrial Research Organisation, and the Australian Mineral Industries Research Association. We also sincerely thank the B. De Rothschild Foundation for the Advancement of Science in Israel, Inc., for support in presenting this paper at the Bat Sheva De Rothschild seminar entitled *Microbial Mats: Ecological Physiology of Benthic Microbial Communities* at Eilat, Israel, 13 to 20 September 1987.

We sincerely thank Andrea Blanks for assistance in collecting and maintaining the mats and in determining the sulfate reduction rates.

LITERATURE CITED

1. **Allen, M. B., and D. I. Arnon.** 1955. Studies on nitrogen-fixing blue-green algae. I. Growth and nitrogen fixation by *Anabaena cylindrica* Lemm. *Plant Physiol.* **30:**366–372.
2. **Bauld, J.** 1984. Microbial mats in marginal marine environments: Shark Bay, Western Australia, and Spencer Gulf, South Australia, p. 39–58. *In* Y. Cohen, R. W. Castenholz, and H. O. Halvorson (ed.), *Microbial Mats: Stromatolites.* Alan R. Liss, Inc., New York.
3. **Bauld, J., L. A. Chambers, and G. W. Skyring.** 1979. Primary productivity, sulfate reduction and sulfur isotope fractionation in algal mats and sediments in Hamelin Pool, Shark Bay, W.A. *Aust. J. Mar. Freshwater Res.* **30:**753–764.
4. **Chambers, L. A.** 1985. Biogeochemical aspects of the carbon metabolism of microbial mat communities, p. 371–376. *In* C. Gabrie, J. L. Toffart, and B. Salvat (ed.), *Proceedings of the 5th International Coral Reef Congress*, vol. 3. Antenne Museum-Ethe, Moorea, French Polynesia.
5. **Cohen, Y.** 1984. The Solar Lake cyanobacterial mats: strategies of photosynthetic life under sulfide,

p. 133–148. *In* Y. Cohen, R. W. Castenholz, and H. O. Halvorson (ed.), *Microbial Mats: Stromatolites.* Alan R. Liss, Inc., New York.

6. **Cohen, Y., Z. Aizenshtat, A. Stoler, and B. B. Jørgensen.** 1980. The microbial geochemistry of Solar Lake, Sinai, p. 167–172. *In* P. A. Trudinger, M. R. Walter, and B. J. Ralph (ed.), *Biogeochemistry of Ancient and Modern Environments.* Australian Academy of Science, Canberra.
7. **Daday, A., and G. D. Smith.** 1983. The effect of nickel on the hydrogen metabolism of the cyanobacterium *Anabaena cylindrica. FEMS Microbiol. Lett.* **20:**327–330.
8. **Emerich, D. W., R. V. Hageman, and R. H. Burris.** 1981. Interactions of dinitrogenase and dinitrogenase reductase. *Adv. Enzymol. Relat. Areas Mol. Biol.* **52:**1–22.
9. **Haselkorn, R.** 1978. Heterocysts. *Annu. Rev. Plant Physiol.* **29:**319–344.
10. **Haselkorn, R.** 1986. Organization of the genes for nitrogen fixation in photosynthetic bacteria and cyanobacteria. *Annu. Rev. Microbiol.* **40:**525–547.
11. **Houchins, J. P.** 1984. The physiology and biochemistry of hydrogen metabolism in cyanobacteria. *Biochim. Biophys. Acta* **768:**227–255.
12. **Howarth, R. W., and R. Marino.** 1984. Sulfate reduction in salt marshes, with some comparisons to sulfate reduction in microbial mats, p. 254–263. *In* Y. Cohen, R. W. Castenholz, and H. O. Halvorson (ed.), *Microbial Mats: Stromatolites.* Alan R. Liss, Inc., New York.
13. **Jørgensen, B. B., and Y. Cohen.** 1977. Solar Lake (Sinai). 5. The sulfur cycle of benthic cyanobacterial mats. *Limnol. Oceanogr.* **22:**657–666.
14. **Kumazawa, S., and A. Mitsui.** 1985. Comparative amperometric study of uptake hydrogenase and hydrogen photoproduction activities between heterocystous cyanobacterium *Anabaena cylindrica* B296 and nonheterocystous cyanobacterium *Oscillatoria* sp. strain Miami BG7. *Appl. Environ. Microbiol.* **50:**287–291.
15. **Lambert, G. R., and G. D. Smith.** 1980. Hydrogen metabolism by filamentous cyanobacteria. *Arch. Biochem. Biophys.* **205:**36–50.
16. **Lambert, G. R., and G. D. Smith.** 1981. The hydrogen metabolism of cyanobacteria (blue-green algae). *Biol. Rev. Camb. Philos. Soc.* **56:**589–660.
17. **Lyons, W. B., M. E. Hines, and H. E. Gaudette.** 1983. Major and minor element pore water geochemistry of modern marine sabkhas: the influence of cyanobacterial mats, p. 411–424. *In* Y. Cohen, R. W. Castenholz, and H. O. Halvorson (ed.), *Microbial Mats: Stromatolites.* Alan R. Liss, Inc., New York.
18. **Moriarty, D. J. W., P. Boon, J. Hansen, W. G. Hunt, I. R. Pointer, P. C. Pollard, G. W. Skyring, and D. C. White.** 1984. Microbial biomass and productivity in seagrass beds. *Geomicrobiol. J.* 4:21–51.
19. **Mortenson, L. E.** 1978. The role of dihydrogen and hydrogenase in nitrogen fixation. *Biochimie* **60:**219–223.
20. **Nedwell, D. B., and J. W. Abram.** 1978. Bacterial sulfate reduction in relation to sulfur geochemistry in two contrasting areas of salt marsh sediment. *Estuarine Coastal Mar. Sci.* **6:**341–351.
21. **Newton, W., J. R. Postgate, and C. Rodriguez-Barrueco.** 1977. *Recent Developments in Nitrogen Fixation.* Academic Press, Inc. (London), Ltd., London.
22. **Pearson, H. H., R. Howsley, C. K. Kjeldsen, and A. E. Walsby.** 1979. Aerobic nitrogenase activity associated with a non-heterocystous filamentous cyanobacterium. *FEMS Microbiol. Lett.* **5:**163–167.
23. **Pearson, H. W., G. Malin, and R. Howsley.** 1981. Physiological studies on in vitro nitrogenase activity by anexic cultures of the blue-green alga *Microcoleus chthonoplastes. Br. Phycol. J.* **16:**139.
24. **Potts, M., and B. A. Whitton.** 1979. Nitrogen fixation by blue-green algal communities in the intertidal zone of the lagoon of Aldabra Atoll. *Oecologia* **27:**275–283.
25. **Robson, R. L., and J. R. Postgate.** 1980. Oxygen and hydrogen in biological nitrogen fixation. *Annu. Rev. Microbiol.* **34:**183–207.
26. **Simpson, F. B., and R. H. Burris.** 1984. A nitrogen pressure of 50 atmospheres does not prevent evolution of hydrogen by nitrogenase. *Science* **224:**1095–1097.
27. **Skyring, G. W.** 1984. Sulfate reduction in marine sediments associated with cyanobacterial mats in Australia, p. 265–275. *In* Y. Cohen, R. W. Castenholz, and H. O. Halvorson (ed.), *Microbial Mats: Stromatolites.* Alan R. Liss, Inc., New York.

27a. **Skyring, G. W.** 1987. Sulfate reduction in coastal ecosystems. *Geomicrobiol. J.* **5:**295–374.

28. **Skyring, G. W., L. A. Chambers, and J. Bauld.** 1983. Sulfate reduction in sediments colonized by cyanobacteria, Spencer Gulf, South Australia. *Aust. J. Mar. Freshwater Res.* **34:**359–374.

28a. **Skyring, G. W., R. M. Lynch, and G. D. Smith.** 1988. Acetylene reduction and hydrogen metabolism by a cyanobacterial/sulfate-reducing bacterial mat ecosystem. *Geomicrobiol. J.* **6:**25–31.

29. **Stal, L. J., S. Grossberger, and W. E. Krumbein.** 1984. Nitrogen fixation associated with the cyanobacterial mat of a marine laminated microbial ecosystem. *Mar. Biol.* (Berlin) **82:**217–224.
30. **Stewart, W. D. P.** 1980. Some aspects of structure and function in N_2-fixing cyanobacteria. *Annu. Rev. Microbiol.* **34:**497–536.
31. **Stewart, W. D. P., G. P. Fitzgerald, and R. H.**

Burris. 1967. In situ studies on N_2-fixation using the acetylene reduction technique. *Proc. Natl. Acad. Sci. USA* **58**:2071–2078.

32. **Ward, D. M.** 1984. Decomposition of microbial mats—discussion, p. 277–280. *In* Y. Cohen, R. W. Castenholz, and H. O. Halvorson (ed.), *Microbial Mats: Stromatolites*. Alan R. Liss, Inc., New York.

33. **Westrich, J. T., and R. A. Berner.** 1984. The role of sedimentary organic matter in bacterial sulfate reduction: the G model tested. *Limnol. Oceanogr.* **29**:236–249.

34. **Wilson, L. G., and R. S. Bandurski.** 1958. Enzymatic reactions involving sulfate, sulfite, selenate and molybdate. *J. Biol. Chem.* **233**:975–981.

Chapter 16

Methanogenesis in Hypersaline Environments

Ronald S. Oremland and Gary M. King

INTRODUCTION

Methanogenesis in certain ecosystems such as the rumen, sludge, and both freshwater and marine sediments has been well studied. A number of reviews have summarized microbiological, physiological, and ecological characteristics of methane production in these habitats (12a, 27, 29). However, environments representing biological extremes in terms of temperature, pH, and/or salinity have not been extensively considered. The impetus for examining such habitats has been, in part, the phylogenetic linkages between methanogenic isolates and other representatives of the archaebacteria. Thus, a number of moderate and extreme halophiles have been isolated and characterized (11, 18, 32, 33). However, this approach cannot address other basic questions including the mechanisms of methane formation and its significance to carbon mineralization in hypersaline environments. Furthermore, because methane can also arise from nonbiological sources associated with the process of petroleum formation (5, 26), the geochemical characteristics of the methane encountered in hypersaline environments (e.g., stable carbon isotopic composition) must be carefully assessed in conjunction with microbial activities to determine whether the methane present is truly of a current bacterial origin or is derived from older biological or nonbiological (thermogenic) sources present in the vicinity. Because many hypersaline environments are located in geological settings where thermal processes also occur (e.g., the Red Sea brines and Mono Lake), the origins of the methane in these environments may be diverse and complex (1, 15a).

Ronald S. Oremland • U.S. Geological Survey, Menlo Park, California 94025. ***Gary M. King*** • Darling Marine Center, University of Maine, Walpole, Maine 04573.

BIOGEOCHEMICAL CHARACTERISTICS OF HYPERSALINE ENVIRONMENTS

Some aspects of the water chemistry, as well as the geochemical properties of biogenic methane found in some hypersaline environments, are given in Table 1. The environments show great variability with respect to sulfate content, pH, salinity, and the abundance of methane within their anoxic waters and sediments. Three moderately hypersaline lakes (Big Soda, Soap, and Mono Lakes) occurring in the Great Basin of the

Table 1.
Characteristics of the Aquatic Chemistry and Geochemical Properties of Methane Found in the Anoxic Waters and Sediments from Various Hypersaline Environments

Site	Anoxic water column				Bottom sediments				References
	Salinity (‰)	pH	SO_4^{2-} (mM)	CH_4 (μM)[a]	CH_4 (μM)[a]	$\delta^{13}CH_4$ (‰)[b]	$C_1/(C_2 + C_3)$[c]	Methanogenic activity	
Big Soda Lake, Nev.	89	9.8	68	57	418	−74	1,736	Yes	13, 14
Mono Lake, Calif.	90	9.8	130	57	3,000	−86	750	Yes	6, 15a
Soap Lake, Wash.	100	9.8	221	1,000	ND[d]	ND	3,415[e]	Yes	21; L. Miller and R. Oremland, unpublished
Orca Basin, Gulf of Mexico	250	ND	38	899	4,500	−105	1,040	No	20, 28
Atlantis II Deep, Red Sea	156[f]	ND	ND	0.003	3	ND	1,040	ND	1
Westend Pond, St. Croix	180	7.2	100	ND	5	ND	ND	Yes	10
Solar Lake, Sinai	180	6.9	120	ND	0.02	ND	132	Yes	2, 4; R. Oremland, unpublished
Great Salt Lake, Utah	333	7.7	28	25	ND	ND	ND	Yes	19, 22, 31
Dead Sea, Israel	323	6.1	5	ND[g]	ND	ND	ND	ND	12

[a] Methane concentrations are the highest concentrations reported in the literature.
[b] $\delta^{13}CH_4$ values are the most negative values reported in the literature.
[c] $C_1/(C_2 + C_3)$ ratios are the highest ratios reported in the literature.
[d] ND, Not determined.
[e] Value for water column.
[f] As chloride.
[g] Methane present but not investigated (A. Nissenbaum, personal communication).

western United States are characteristically highly alkaline (pH 9.8), whereas the extremely hypersaline brines have pH values closer to 7 (Table 1). The sulfate content varies considerably, with the Dead Sea having only 5 mM sulfate and both the Great Salt Lake and the Orca Brine having lower levels than would intuitively be anticipated for concentrated seawater. This is due to removal of sulfate from solution as the brines become more concentrated. Sulfate precipitates to form minerals such as mirabilite ($Na_2SO_4 \cdot 10H_2O$) under these conditions. By contrast, environments with <200 g of dissolved solids liter^{-1} typically have very high sulfate concentrations (68 to 221 mM).

Methane is present in all the hypersaline environments examined to date (Table 1). In all the environments except the Dead Sea, the methane appears to be of biogenic origin. This is based on its stable carbon isotopic composition ($\delta^{13}CH_4$), relative hydrocarbon abundances ($C_1/[C_2 + C_3]$, where C_1 is CH_4, C_2 is C_2H_6, and C_3 is C_3H_8), and/or the presence of methanogenic activity in sediment samples. It is generally considered that methane formed by bacteria is enriched in ^{12}C relative to ^{13}C and thus exhibits $\delta^{13}CH_4$ values more negative than about −55% (24). Clearly, methane from Mono Lake and Big Soda Lake sediments, as well as from the Orca Brine, appears biogenic. In addition, $C_1/(C_2 + C_3)$ ratios above ca. 100 are usually indicative of bacterially formed gases (24), and thus the above three environments, as well as Soap Lake and Atlantis II Deep, fall into this category. Evidence for a biogenic origin of methane in Westend Pond consists solely of detection of bacterial methanogenic activity in sediments (10). Curiously, preliminary investigations of this type did not yield discernible activity for the Orca Brine (see below). Finally, although traces of methane are present in the Dead Sea, it is not clear whether it is from a biogenic or thermogenic source; more work must be done in that environment.

SOURCES OF METHANOGENIC SUBSTRATES

Methanogenic bacteria, as well as sulfate respirers, occupy a terminal niche in the degradation of organic matter in anaerobic ecosystems. Complex organic compounds are degraded to simpler molecules by the various nonmethanogenic denizens of the microbial food web (Fig. 1). Recognized substrates include hydrogen plus carbon dioxide, acetate, formate, methanol, methylated amines, and dimethyl sulfide. There is considerable diversity among methanogens with respect to substrate utilization. Most species (but not all) use hydrogen plus carbon dioxide, and a few can metabolize formate as well. A restricted few can grow by using the energy released from acetate cleavage, whereas some have a methylotrophic affinity in that they can grow on methylated amines, dimethyl sulfide, or methanol (for reviews, see references 12a, 27, and 29). Obligately methylotrophic methanogens, such as *Methylococcoides methylutens* (25), are unusual in that they can grow only on methylated C_1 compounds and do not use hydrogen. These organisms occur in estuarine (6, 25) as well as in hypersaline (11, 14, 18, 32, 33) environments.

Trimethylamine and dimethyl sulfide are derived from the degradation of compounds such as glycine betaine and dimethylsulfoniopropionate (6, 8, 9). Aquatic plants and phytoplankton use these compounds as compatible solutes as part of a molecular strategy for survival in osmotically stressed environments. Thus, these compounds are quite common in hypersaline environments. Methanol is associated with the degradation of plant pectins (23) and, perhaps, lignins and does not reflect osmotic stress. In environments rich in plant biomass

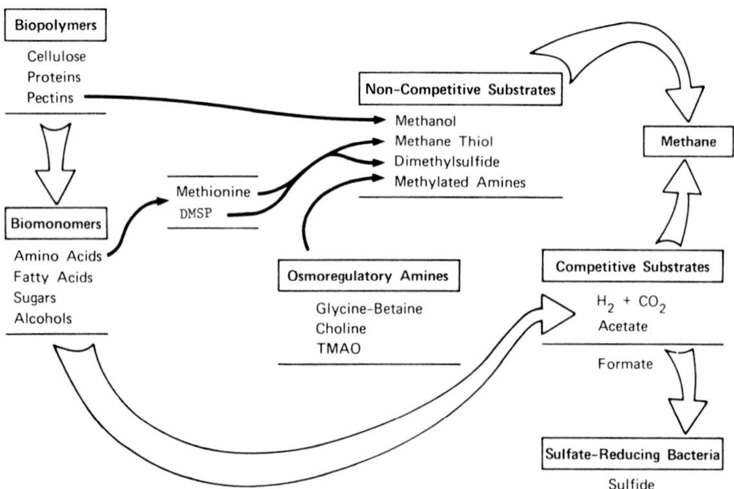

Figure 1. Anaerobic food web for formation of competitive and noncompetitive methanogenic substrates in an aquatic ecosystem. Adapted from Oremland (12a).

Table 2.
Levels of Methane during Incubation of Sediment Slurries Collected from the Orca Brine Sediments[a]

Conditions	No. of expts (n)	Methane (μmol/125 ml of slurry)[b] after:		
		2 days	4 days	43 days
H_2 (4°C)	3	7.4 (1.0)	7.3 (1.0)	7.5 (1.5)
H_2 + CCl_4 (4°C)	3	9.2 (0.2)	9.4 (0.8)	9.3 (0.7)
H_2 + autoclaved control (4°C)	2	0.59 (0.11)	0.61 (0.07)	0.65 (0.05)
N_2 (4°C)	2	24.6 (0.3)	24.3 (0)	24.2 (0.2)
H_2 (20°C)	2	11.1 (0.8)	11.2 (1.4)	11.5 (1.8)

[a] Samples were collected from a water depth of ca. 2,000 m on 22 March 1978. Sediments (12 ml) were slurried with 113 ml of anoxic brine and incubated in sealed conical flasks (total volume, 260 ml) with shaking. Incubations were started aboard ship. Variations in methane contents of headspaces were caused by variability in headspace gas flushing time intervals.
[b] Parentheses indicate 1 standard deviation (n = 3) or spread of values from average (n = 2).

(such as salt marshes), methanol, trimethylamine, and dimethyl sulfide, as well as their molecular precursors, are present at relatively high concentrations (15). Similarly, high levels of both glycine betaine and trimethylamine have been detected in the microbial mats of hypersaline Westend Pond (10). However, there is a general paucity of information about the abundance of these compounds in hypersaline environments.

SUBSTRATE AND INHIBITOR ADDITIONS TO SEDIMENT SLURRIES

One experimental procedure for detecting methanogenic activity is to add high concentrations of substrates or inhibitors to sediment slurries and to look for stimulation or inhibition of methane formation relative to that of an unamended control. This procedure has been used extensively in the study of both freshwater and marine sediments to gain insight into carbon and electron flow in those communities (15, 17, 28, 30). In a preliminary experiment with sediment slurries from the Orca Brine, no evidence for methanogenic activity was noted (Table 2). For all three experimental conditions tested, the levels of methane formed by slurries under H_2 or N_2 at 4°C or H_2 at 20°C did not increase over a 43-day incubation period. These were the same results as observed in the chloroform-inhibited or autoclaved controls. Furthermore, no discernible consumption of H_2 occurred, which contrasts with results for marine sediment slurries (15, 17). In another experiment, production of $^{14}CH_4$ by sediments was not observed after 19 days of incubation with ^{14}C-labeled formate, acetate, or bicarbonate (data not shown). These negative results raised the question of how the biogenic geochemical signals with regard to methane (Table 1) were detected in the Orca Brine without the presence of apparent bacterial activity in the sediments. Possible explanations include the following. (i) The activity is present at deeper horizons in the sediments (and at lower salinities) than the near-surface samples assayed. (ii) The activity occurred earlier in time, when the migrating brine first entered the Orca depression, and is not occurring now because of inhibition of methanogenesis by the high salinity. (iii) Methanogenic activity takes place by mechanisms other than CO_2 reduction, cleavage of acetate, or metabolism of formate.

To answer the above questions, we chose a site which would not entail the logistical difficulties of the Orca Basin; Big

Soda Lake, Nev. (Table 1), fit these requirements. Sediment slurries from this environment produced small quantities of methane, and production was enhanced by methanol, methionine, or trimethylamine, but not by H_2, formate, or acetate (14; Fig. 2). Further examination by using estuarine sediments revealed that sulfate was the key factor in restricting methanogenesis from acetate or hydrogen, but had no effect on the conversion of methionine, methanol, or trimethylamine to methane (16). The last three compounds have been referred to as noncompetitive methanogenic substrates (8, 16) because sulfate reducers do not have as strong an affinity for them as they do for acetate or hydrogen; this allows methanogenesis and sulfate reduction to operate concurrently (15). Recently, dimethyl sulfide was found to behave like a noncompetitive substrate, and its addition to Mono Lake sediments enhanced methanogenesis (6). Thus, methionine is not attacked directly by methanogens, but by other anaerobes, which remove the methiol group to yield dimethyl sulfide (7). Ethylated analogs of dimethyl sulfide (e.g., diethyl sulfide and ethane thiol) were found to enhance the formation of ethane by methanogenic bacteria in Mono Lake sediments (17a).

These results indicate that methanogenesis does indeed occur in hypersaline environments and that perhaps the failure to detect activity in the Orca Basin was due to option (iii) above, i.e., the wrong choice of substrates. Subsequently, addition of methylated amines resulted in stimulation of methanogenesis in slurries from hypersaline Solar Lake (4) and Westend Pond (10). Moderate and extremely halophilic methanogens which grow only on methylated C_1 compounds (obligate methylotrophs) have also been isolated from hypersaline environments (11, 14, 18, 32, 33), indicating that there is an exclusive niche for methanogens which live solely on methylated, noncompetitive C_1 compounds.

It should be borne in mind, however, that experiments with sediment slurries suffer from considerable limitations with respect to certain conclusions about in situ methanogenesis. First, enhancement of

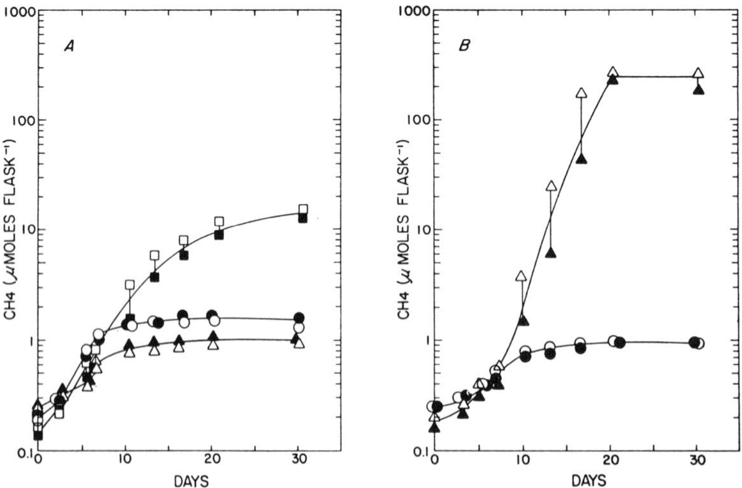

Figure 2. Formation of methane during incubation of sediment slurries from Big Soda Lake. Duplicates are indicated by open and solid symbols. (A) Symbols: △,▲, no amendments; ○,●, 10 mM formate; □,■, 10 mM methionine. (B) Symbols: ○,●, acetate; △,▲, methanol. Slurry volume, 75 ml per flask. Reproduced from reference 14.

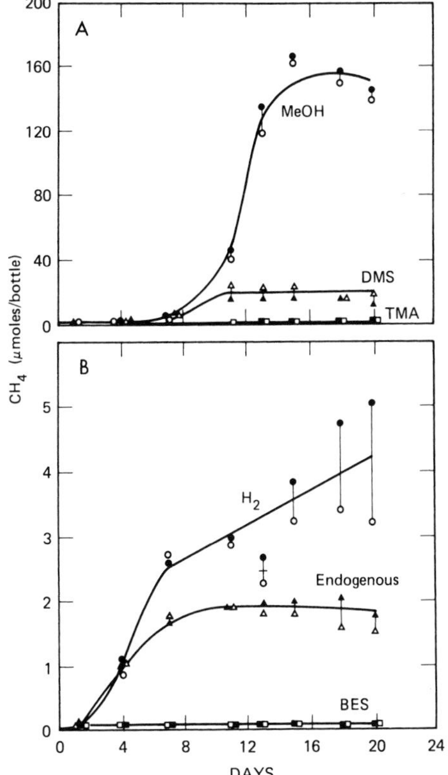

Figure 3. Formation of methane during incubation of Mono Lake sediment slurries (30 ml) in the presence of substrates or inhibitors. Duplicates are indicated by open and solid symbols. (A) Symbols: ○,●, 10 mM methanol; △,▲, 10 mM dimethyl sulfide; □,■, 10 mM trimethylamine. (B) Symbols: △,▲, no additions; ○,●, hydrogen gas phase; □,■, 20 mM 2-bromoethanesulfonic acid. Note the scale differences between panels A and B.

methanogenesis by a given compound does not mean that the compound is important in that ecosystem. Analyses with radioisotopes and determinations of substrate pool sizes are necessary for estimates of in situ methane formation from various precursors (see below). In addition, sediment slurries can sometimes yield ambiguous results. For example, when amended with 10 mM methanol or dimethyl sulfide, Mono Lake slurries showed enhanced methane formation (Fig. 3), had only slightly enhanced activity with H_2, and were inhibited by 2-bromoethanesulfonic acid (a specific inhibitor of methanogens) (12b). However, 10 mM trimethylamine also entirely inhibited methanogenesis in these slurries (Fig. 3). It is not clear why this occurred, since trimethylamine was previously reported to be a substrate for methanogenesis in estuarine (9) and hypersaline (10) sediments, but it was not due to pH changes, because these waters are strongly buffered at pH 9.8. On the basis of these results, one could conclude (incorrectly) that trimethylamine is not a substrate in this environment and may act to impede methanogenesis. However, work with radioisotopes at concentrations closer (15 µM) to the in situ pool sizes than can be achieved with slurries demonstrated that some trimethylamine is indeed converted to methane (see Table 4). Therefore, it is important to view all work with substrate addition to slurries in hypersaline systems with a "grain of salt."

STUDIES WITH RADIOISOTOPES

The noncompetitive-substrate theory for methanogenesis in sulfate-containing environments was refined by investigations with radioisotopes. For methanol and dimethyl sulfide, the $^{14}CH_4/^{14}CO_2$ ratios achieved in sediment incubations were considerably less than the 3:1 ratio predicted for pure cultures (6, 8). Addition of molybdate to inhibit sulfate-respiring bacteria resulted in increased ratios (Fig. 4). These observations indicated that sulfate-respiring bacteria will outcompete methanogens for either methanol or dimethyl sulfide when these compounds are added at micromolar concentrations to sediments (radioassays), but that the reverse is true for the millimolar concentrations used in substrate amendment experiments. However, trimethylamine appeared to be the one compound for which sulfate reducers had little affinity at any concentration (9). Thus, it is a true noncompetitive methanogenic substrate, at least for the estuarine environments tested. These

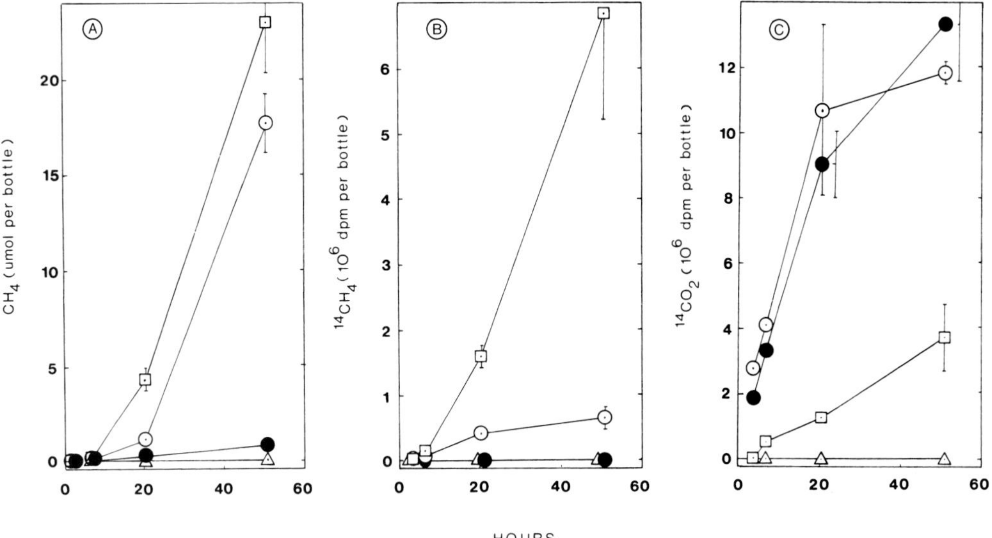

Figure 4. Production of methane (A), $^{14}CH_4$ (B), and $^{14}CO_2$ (C) by San Francisco Bay salt marsh slurries incubated with [^{14}C]dimethyl sulfide (5 μCi/50 ml of slurry). Results represent the mean and one standard deviation for three samples. Symbols: ⊙, no additions; ⊡, 20 mM molybdate; ●, 40 mM 2-bromoethanesulfonic acid; △, autoclaved. Adapted from *Applied and Environmental Microbiology* (7).

results indicated that the mechanisms of methanogenesis in hypersaline systems had to be scrutinized closely by using radioisotopes.

Preliminary studies of this sort were conducted to determine whether hypersaline sediments had the capacity to form methane over short incubation periods. Formation of $^{14}CH_4$ from [^{14}C]methionine or [^{14}C]methanethiol was observed with sediments from the Great Salt Lake (31). Addition of $^{14}CH_3OH$ to Mono Lake sediments resulted in the formation of both $^{14}CH_4$ and $^{14}CO_2$ over a time course, indicating that the sediments had an active methanogenic flora capable of metabolizing methylotrophic substrates (Table 3). The low $^{14}CH_4/^{14}CO_2$ ratio (ca. 1.25) indicated that sulfate-respiring bacteria also oxidized this substrate. The issue which needs further clarification is that of the relative contributions of competitive and noncompetitive substrates to methanogenesis in various hypersaline ecosystems.

RATES OF METHANE FORMATION FROM VARIOUS PRECURSORS IN HYPERSALINE ENVIRONMENTS

Profiles of the rates of sulfate reduction and methanogenesis in Mono Lake sediments are presented in Table 4. Sulfate levels decreased, but did not disappear, with depth, whereas methane increased to saturating levels. Sulfate reduction rates greatly exceeded those of methanogenesis at the top of the sediments; however, rates for the two processes were roughly equivalent at depths beneath 5 cm. Most of the methane was formed via CO_2 reduction. The methyl groups of acetate and dimethyl sulfide accounted for only about 1.4 and 0.2% of the methane evolved at the top of the sediments, respectively, and for even less at greater depths. Trimethylamine, however, accounted for as much as 33% of the activity at the top of the sediments. It should be borne in mind that for acetate, trimethylamine, and dimethyl sulfide, these rates are probably

Table 3.
Formation of ^{14}C-Labeled Gases from $^{14}CH_3OH$ Addition to Intact Sediment Cores from Mono Lake[a]

Product	Amt of radiolabeled product (10^3 dpm) after:		
	0 h	4 h	8 h
$^{14}CH_4$	0	2,225	5,550
$^{14}CO_2$	1,950	2,500	4,450

[a] Cores were taken from the top 11 cm of Mono Lake bottom sediments. The cores were retrieved with an Eckman dredge, and plastic subcores (volume, 25 ml; 11 by 2.5 cm) were taken. The cores were injected with radioisotopes (20 μCi) and incubated in the dark at 6°C. Individual cores were frozen at various times to arrest activity. Radiolabeled products were extracted from the cores by extrusion into a conical flask containing a saturated NaCl solution. Radioactive gases were analyzed by gas proportional counting (3).

overestimates, because addition of isotope greatly exceeded the in situ pool sizes. In addition, these substrates yielded very low $^{14}CH_4/^{14}CO_2$ ratios (Table 4), indicating that other microorganisms (e.g., sulfate reducers) metabolized them as well. Thus, the only reliable estimate is for CO_2 reduction.

In contrast to the study with the pelagic sediments of Mono Lake, the shallow (2 m), warm (40 to 55°C) Westend salt pond sediments exhibited a very different pattern of methanogenesis. These laminated sediments were covered with a cyanobacterial mat composed primarily of *Spirulina* sp., and oxygen penetrated down to a depth of only about 2 to 4 mm, depending upon illumination conditions (10). Methane concentrations, as well as the rate of methane formation by incubated (55°C) sediment slurries, were highest in the top 1 cm of the sediments and then decreased with depth (Fig. 5). Glycine betaine accounted for about 20% of the total nitrogen in these sediments, with the highest values also occurring in the top 1 cm (10). A similar profile was observed for sediment concentrations of trimethylamine and ammonia (Fig. 6). Addition of trimethylamine (or glycine betaine) to slurries greatly enhanced methanogenesis, whereas addition of acetate or hydrogen did not, which strongly suggests that trimethylamine was the primary precursor of methane formed within the mat (10). Of course, radiotracer studies are necessary to confirm these observations.

DISCUSSION

The above results with pelagic sediments from Mono Lake and those from a shallow cyanobacterial mat represent the only detailed studies conducted thus far which can address the mechanism of meth-

Table 4.
Estimated Rates of Methanogenesis and Sulfate Reduction in Pelagic Sediments of Mono Lake[a]

Depth (cm)	SO_4^{2-} (mM)	CH_4 (mM)	SO_4^{2-} reduction rate[b]	Rate of methanogenesis from[b]:			
				HCO_3^-	Acetate	DMS	TMA
1		0.3		1.65	0.034	0.0050	0.825
3	133	0.5	225.0	1.80	0.039	0.0006	0.113
5		0.7		1.73	0.019	0.0005	0.072
7	115	1.0	5.9	6.08	0.005	0.0003	0.074
17.5	95	1.3	4.2	2.03	0.003	0.0001	0.055
32.5	67	2.1	2.1	1.14	0.003	0.0001	0.020
47.5	60	2.5	2.0	1.14	0.003	0.0001	0.002

[a] The water depth was 26 m. Methane and sulfate concentrations were determined in August 1986 (15a). Data for the sulfate reduction rates were determined by using ^{35}S tracers in a core taken during August 1986 (R. L. Smith, unpublished data); data for methanogenic precursors were determined in November 1986. Added pools (concentrations achieved by adding radioisotope to sediments) were as follows: HCO_3^-, 150 μM; acetate, 20 μM; dimethyl sulfide, 2.5 μM; trimethylamine, 15 μM. The estimated in situ pools were as follows: HCO_3^-, 300,000 μM; acetate, 1 μM; dimethyl sulfide, 0.1 μM; trimethylamine, 1 μM. $^{14}CH_4/^{14}CO_2$ ratios were as follows: acetate, <0.06; dimethyl sulfide, <0.1; trimethylamine, <0.6.

[b] Rates are expressed in micromoles per liter of sediment per day. DMS, Dimethyl sulfide; TMA, trimethylamine.

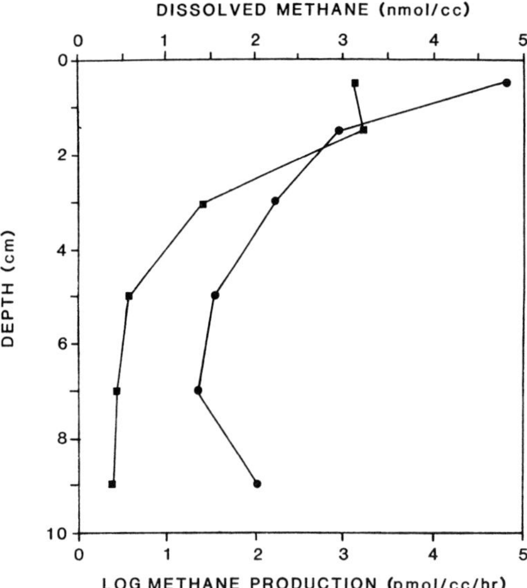

Figure 5. Dissolved methane (■) and methane production rates (●) with depth in the sediments of Westend Pond during April 1986. Reproduced from reference 10.

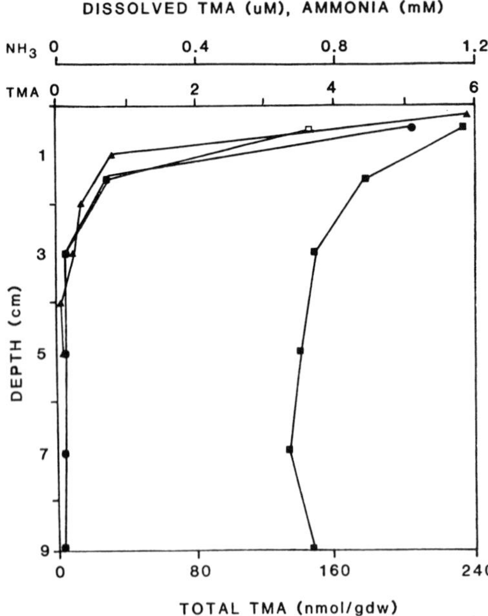

Figure 6. Dissolved trimethylamine (TMA) in March 1985 (▲) and April 1986 (●) and dissolved ammonia in March 1985 (□) and April 1986 (■) with depth in the sediments of Westend Pond. Reproduced from reference 10.

ane formation in hypersaline environments. Westend Pond sediments clearly use trimethylamine as a precursor of methane, and the fact that trimethylamine (as well as its glycine betaine precursor) was abundant in the sediments and displayed similar depth profiles (Fig. 5 and 6) strongly suggests that it is the primary substrate for methanogenesis. These results are consistent with mechanisms of methanogenesis via noncompetitive substrates in high-sulfate environments (Fig. 1). In Mono Lake, however, most of the methane is derived from hydrogen-mediated reduction of carbon dioxide (Table 3). Trimethylamine accounts for <33% of the methane formed at the sediment surface, and its importance declines with depth. Neither dimethyl sulfide nor acetate appears to be a significant precursor. Thus, although sulfate reduction and methanogenesis operate concurrently and at equal rates below a depth of 5 cm, this situation does not appear to occur via metabolism of noncompetitive substrates by methanogens, but rather by competition for hydrogen between methanogens and sulfate reducers. For some reason, the affinity of sulfate reducers for hydrogen is impaired with depth in these sediments.

The striking differences between the methane metabolism displayed by these two hypersaline, sulfate-rich sediments can be explained by a proximity to inputs of fixed carbon by the primary producers of the communities. In the Westend Pond algal mat, the cyanobacteria were located directly above the anaerobic decomposition zone (less than 0.5 cm), and thus a rapid mineralization of dead cells occurred which was manifested by the release of their osmoregulatory solute glycine betaine (Fig. 6). In contrast, Mono Lake sediments were located in deep water (26 m), the lowest 8 m of which was anoxic. Sulfate reduction rates in the anoxic water column were typically about 1 to 3 μmol liter^{-1} day^{-1}, which accounted for half the integrated total sulfate reduction (ca. 25 mmol m^{-2} day^{-1}) taking place in the anoxic waters and sediments of the lake (R. L. Smith, unpublished data). Thus, sinking phytoplankton were apparently effectively mineralized in the anoxic waters and within the surface of the sediments. This explains the low trimethylamine pools (<1 μM) in Mono Lake sediments compared with Westend Pond (>240 μM). The absence of significant levels of noncompetitive substrates in the Mono Lake sediments, as well as the suggested ability of sulfate-respiring bacteria to mineralize these compounds (as manifested by the low $^{14}CH_4/^{14}CO_2$ ratios [Table 4]), indicated that their lack of availability limited their significance as a methane precursor. In contrast, hydrogen appeared to be the most important precursor, which would indicate that both sulfate respirers and methanogens had similar apparent K_s values for this substrate.

CONCLUSIONS

Methanogenic bacteria, as well as methanogenic activity, occur in hypersaline ecosystems. In algal mat communities, methanogenesis takes place at the expense of noncompetitive substrates (e.g., trimethylamine), whereas in pelagic communities, competitive substrates (e.g., hydrogen) are of greater importance. Because of these disparities, it would be prudent for an investigator not to make a priori conclusions concerning the nature of methanogenic pathways in a particular hypersaline system. Experimental tests must be conducted and must go beyond the sediment slurry amendment approach. In situ activity estimates made with radioisotopes (when possible) in conjunction with substrate pool sizes are vital for a comprehensive understanding.

LITERATURE CITED

1. **Burke, R. A., Jr., J. M. Brooks, and W. M. Sackett.** 1981. Light hydrocarbons in Red Sea brines and sediments. *Geochim. Cosmochim. Acta* 45:627–634.
2. **Cohen, Y., W. E. Krumbein, M. Goldberg, and M. Shilo.** 1977. Solar Lake (Sinai). 1. Physical and chemical limnology. *Limnol. Oceanogr.* 22:597–608.
3. **Culbertston, C. W., A. J. B. Zehnder, and R. S. Oremland.** 1981. Anaerobic oxidation of acetylene by estuarine sediments and enrichment cultures. *Appl. Environ. Microbiol.* 41:396–403.
4. **Giani, D., L. Giani, Y. Cohen, and W. E. Krumbein.** 1984. Methanogenesis in the hypersaline Solar Lake (Sinai). *FEMS Microbiol. Lett.* 25:219–224.
5. **Hunt, J. M.** 1979. *Petroleum Geochemistry and Geology*. W. H. Freeman & Co., San Francisco.
6. **Kiene, R. P., R. S. Oremland, A. Catena, L. G. Miller, and D. Capone.** 1986. Metabolism of reduced methylated sulfur compounds by anaerobic sediments and a pure culture of an estuarine methanogen. *Appl. Environ. Microbiol.* 52:1037–1045.
7. **Kiene, R. P., and P. T. Visscher.** 1987. Production and fate of methylated sulfur compounds from methionine and dimethylsulfoniopropionate in anoxic salt marsh sediments. *Appl. Environ. Microbiol.* 53:2426–2434.
8. **King, G. M.** 1984. Utilization of hydrogen, acetate and "non-competitive" substrates by methanogenic bacteria in marine sediments. *Geomicrobiol. J.* 3:275–306.
9. **King, G. M.** 1984. On the metabolism of trimethylamine, choline and glycine betaine by sulfate-reducing and methanogenic bacteria in marine sediments. *Appl. Environ. Microbiol.* 48:719–725.
10. **King, G. M.** 1988. Methanogenesis from methylated amines in a hypersaline algal mat. *Appl. Environ. Microbiol.* 54:130–136.
11. **Mathrani, I. M., and D. R. Boone.** 1985. Isolation and characterization of a moderately halophilic methanogen from a solar saltern. *Appl. Environ. Microbiol.* 50:140–143.
12. **Nissenbaum, A.** 1975. The microbiology and biogeochemistry of the Dead Sea. *Microb. Ecol.* 2:139–161.

12a. **Oremland, R. S.** 1988. Biogeochemistry of methanogenic bacteria, p. 641–705. *In* A. J. B. Zehnder (ed.), *Biology of Anaerobic Microorganisms*. John Wiley & Sons, Inc., New York.

12b. **Oremland, R. S., and D. G. Capone.** 1988. Use of "specific" inhibitors in biogeochemistry and microbial ecology. *Adv. Microb. Ecol.* 10:285–383.

13. **Oremland, R. S., and D. J. Des Marais.** 1983. Distribution, abundance and carbon isotopic composition of gaseous hydrocarbons in Big Soda Lake, Nevada: an alkaline, meromictic lake. *Geochim. Cosmochim. Acta* 47:2107–2114.
14. **Oremland, R. S., L. Marsh, and D. J. Des Marais.**

1982. Methanogenesis in Big Soda Lake, Nevada: an alkaline, moderately hypersaline desert lake. *Appl. Environ. Microbiol.* **43**:462–468.

15. **Oremland, R. S., L. M. Marsh, and S. Polcin.** 1982. Methane production and simultaneous sulfate reduction in anoxic, salt marsh sediments. *Nature* (London) **296**:143–145.

15a. **Oremland, R. S., L. G. Miller, and M. J. Whiticar.** 1987. Sources and flux of natural gases from Mono Lake, California. *Geochim. Cosmochim. Acta* **51**:2915–2929.

16. **Oremland, R. S., and S. Polcin.** 1982. Methanogenesis and sulfate reduction: competitive and noncompetitive substrates in estuarine sediments. *Appl. Environ. Microbiol.* **44**:1270–1276.

17. **Oremland, R. S., and B. F. Taylor.** 1978. Sulfate reduction and methanogenesis in marine sediments. *Geochim. Cosmochim. Acta* **42**:209–214.

17a. **Oremland, R. S., M. J. Whiticar, F. E. Strohmaier, and R. P. Kiene.** 1988. Bacterial ethane formation from reduced, ethylated sulfur compounds in anoxic sediments. *Geochim. Cosmochim. Acta* **52**:1895–1904.

18. **Paterek, J. R., and P. H. Smith.** 1985. Isolation and characterization of a halophilic methanogen from Great Salt Lake. *Appl. Environ. Microbiol.* **50**:877–881.

19. **Post, F. J.** 1981. Microbiology of the Great Salt Lake north arm. Hydrobiologia **81**:59–69.

20. **Sackett, W. M., J. M. Brooks, B. B. Bernard, C. R. S. Schwab, and H. Chung.** 1979. A carbon inventory for Orca Basin brines and sediment. *Earth Planet. Sci. Lett.* **44**:73–81.

21. **Sanchez, A. L., J. W. Murray, W. R. Schell, and L. G. Miller.** 1986. Fallout plutonium in two oxic-anoxic environments. *Limnol. Oceanogr.* **31**: 1110–1121.

22. **Schink, B., F. S. Lupton, and J. G. Zeikus.** 1983. Radioassay for hydrogenase activity in viable cells and documentation of aerobic hydrogen-consuming bacteria living in extreme environments. *Appl. Environ. Microbiol.* **45**:1491–1500.

23. **Schink, B., and J. G. Zeikus.** 1980. Microbial methanol formation: a major endproduct of pectin metabolism. *Curr. Microbiol.* **4**:387–389.

24. **Schoell, M.** 1983. Genetic characterization of natural gases. *Am. Assoc. Pet. Geol. Bull.* **67**:2225–2238.

25. **Sowers, K. R., and J. G. Ferry.** 1983. Isolation and characterization of a methylotrophic marine methanogen, *Methylococcoides methylutens* gen. nov., sp. nov. *Appl. Environ. Microbiol.* **45**:684–690.

26. **Tissot, B. P., and D. H. Welte.** 1978. *Petroleum Formation and Occurrence.* Springer-Verlag, New York.

27. **Ward, D. M., and M. R. Winfrey.** 1985. Interactions between methanogenic and sulfate-reducing bacteria in sediments. *Adv. Aquat. Microbiol.* **3**:141–179.

28. **Wiesenburg, D. A., J. M. Brooks, and B. B. Bernard.** 1985. Biogenic hydrocarbon gases and sulfate reduction in the Orca Basin brine. *Geochim. Cosmochim. Acta* **49**:2069–2080.

29. **Winfrey, M. R.** 1984. Microbial production of methane, p. 153–219. *In* R. M. Atlas (ed.), *Petroleum Microbiology.* Macmillan Publishing Co., New York.

30. **Winfrey, M. R., and J. G. Zeikus.** 1977. Effect of sulfate on carbon and electron flow during microbial methanogenesis in freshwater sediments. *Appl. Environ. Microbiol.* **33**:275–281.

31. **Zeikus, J. G.** 1983. Metabolic communication between biodegradative populations in nature. *Soc. Gen. Microbiol. Symp.* **34**:423–462.

32. **Zhilina, T. N.** 1983. New obligate halophilic methane producing bacterium. *Mikrobiologiya* **40**: 674–680.

33. **Zhilina, T. N., and G. A. Zavarzin.** 1987. *Methanohalobium evastigatus* N-Gen N-Sp—extremely halophilic methane-forming archaebacterium. *Dokl. Akad. Nauk. SSSR Ser. Biol.* **293**:464–468.

Chapter 17

Carbon Isotopic Trends in the Hypersaline Ponds and Microbial Mats at Guerrero Negro, Baja California Sur, Mexico: Implications for Precambrian Stromatolites

David J. Des Marais, Yehuda Cohen, Hoa Nguyen, Michael Cheatham, Terri Cheatham, and Elaine Munoz

INTRODUCTION

Microbial mats growing in hypersaline marine environments are modern homologs of at least some of the environmentally diverse Precambrian stromatolite-forming benthic communities. Because stromatolites constitute the oldest and most widespread evidence of life in the Precambrian era, it is important to examine the microbial processes in mats to understand the impact of these processes on the geologic record.

Studies of organic compounds in mats and stromatolites are directly relevant to paleobiology; however, the ravages of time and temperature frequently have altered the structure of Precambrian organic matter to an extent to which biological marker compounds have been either altered beyond recognition or else destroyed. Fortunately, the stable carbon isotopic composition of organic matter is more resistant to thermal alteration (14) and therefore may be a more useful indicator of biospheric evolution. Accordingly, we have examined the distribution of carbon isotopes in the microbial mats and brines at Guerrero Negro, Baja California Sur, Mexico, to provide base-line information for the interpretation of Precambrian stromatolites.

The study site is located in the seawater evaporation ponds of a solar salt works operated by Exportadora de Sal, S. A., a Mexican corporation which produces 5.5×10^6 metric tons of sodium chloride per year, making it the world's largest single salt works (15). The total evaporation pond area exceeds 250 km^2, of which about 100 km^2 is

David J. Des Marais, Hoa Nguyen, Michael Cheatham, Terri Cheatham, and Elaine Munoz • Life Science Division, National Aeronautics and Space Administration, Ames Research Center, N239-12, Moffett Field, California 94035. *Yehuda Cohen* • H. Steinitz Marine Biology Laboratory, The Hebrew University of Jerusalem, Eilat 88103, Israel.

underlain by microbial mats. The mats were shown to us by Barbara Javor, who had previously described the flora and fauna of these ponds (16, 17). The microbial mats growing within these ponds offer the following advantages for biogeochemical research: (i) the mat communities in the ponds grow in permanently submerged, relatively stable environments whose salinities are regulated by salt company operations; (ii) the company maintains continuous records of water flow, salinity, temperature, evaporation rate, and major inorganic ion concentrations; (iii) the brine comes from unpolluted, evaporated seawater which, because the ponds are above the local hydrologic gradient, is also uncontaminated by groundwater; (iv) local annual rainfall is typically less than 30 mm and thus exerts a negligible effect upon the brines; (v) essentially no inorganic detritus is admixed to the pond sediments, and thus the mats are almost purely organic deposits; (vi) at salinities above 70‰ virtually no vascular plants coexist with the mats, and therefore the chemical contributions of plants are trivial; (vii) even populations of planktonic photosynthetic microorganisms are relatively minor in the salinity range between 80 and 120‰; and (viii) the shallow (less than 1 m), wide (kilometers in scale) geometry of the ponds amplifies the chemical effects of the mats upon the brines. It should be possible to monitor mat activity by measuring chemical changes in the brines as they flow through the ponds.

SAMPLING SITES

Water sampling sites were selected to monitor the chemical changes in the brine as it moved from the lagoon through the first 10 seawater-concentrating ponds. The inlet site is located in the water channel between the lagoon and the pumping station (Fig. 1, arrow between lagoon and Pond 1). The

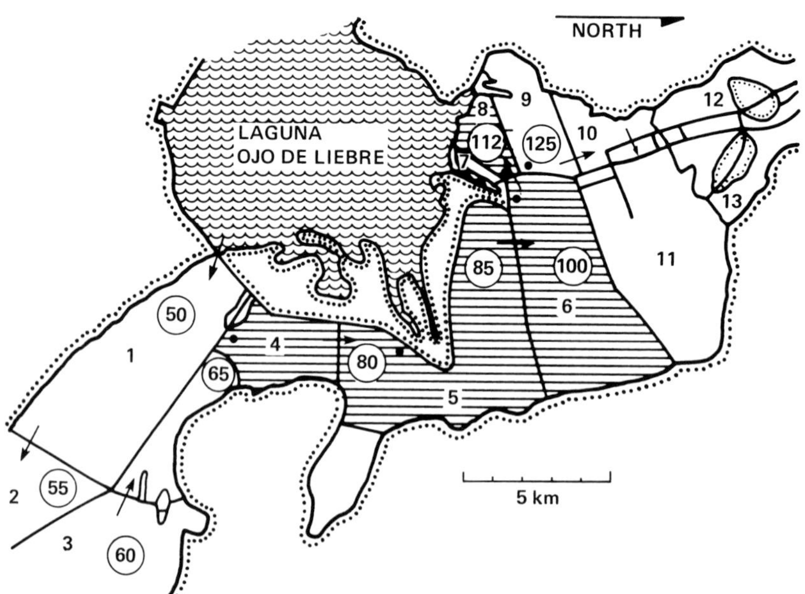

Figure 1. Seawater concentrating ponds of Exportadora de Sal. Water is pumped into Pond 1 and flows to other ponds through water gates (as shown by arrows). Circled numbers give average salinities (per mille) for each pond. Pond areas covered by mats are indicated by horizontal lines. Mat sampling sites are shown by the black dots. Water sampling locations are described in the text.

Pond 1 station samples Pond 1 waters at the gate where they flow into Pond 2. The Pond 2 station is located in the northernmost corner of that pond. The Pond 3 (high salt) station is located along the dike between Pond 3 and Pond 4, 100 m north-northeast of the water gate between Ponds 3 and 4. This station samples a current which flows from the southeastern half of Pond 3. Brine flows from Pond 2 to Pond 3 through two breaches in the dike; one is located within 100 m of the northeastern end of the dike, and the second is at the opposite, southeastern end of the dike. Evaporation increases the salinity of the brine which circulates through the southeastern end of these two ponds before arriving at the gate between Ponds 3 and 4. This brine mixes with the brine which has traveled a shorter route from Pond 1, thus delivering brine to Pond 4 which is somewhat less saline than that sampled at the Pond 3 (high salt) station. The Pond 4 site is located near the gates between Ponds 4 and 5. The Pond 5 site is located at the Pond 5 mat sampling site (Fig. 1, black dot). The Pond 6 site is midway between the gates between Ponds 5 and 6 (Fig. 1, arrow) and Ponds 6 and 7. This site is also used to sample mats. The Pond 8 site is midway along the dike between Ponds 6 and 8, also at a mat sampling site. The Pond 9 and Pond 10 sites are situated at the southeastern corners of those ponds. That the brines in the corners of Ponds 9 and 10 reflect the overall chemistry of these ponds was confirmed by reconnaissance sampling with a salinity refractometer.

The sampling sites for the mats are indicated by black dots in Fig. 1. The sites were chosen to sample permanently submerged mats growing under the widest possible range of salinities. Each site is at least 50 m from the nearest dike and thus avoids both detritus from the dike and the effects of relatively shallower water. The mat samples taken at each site are judged to be representative of the mats in the surrounding pond area.

ANALYSIS OF SAMPLES

Salinity, Temperature and pH

Measurements of salinity, temperature, and pH were made during sample collection. Salinity was measured with a Kahlsico salinity refractometer calibrated by density measurements. Temperature was measured with a mercury thermometer. The pH was recorded with an Orion model 231 field pH meter, whose electrode was calibrated against an Orion Ross electrode having a known response to salinity effects (sodium error, etc.).

DIC Abundance and Stable Isotopic Composition

Approximately 50- to 80-ml brine samples were immediately filtered through Whatman GF/F filters into preevacuated 130-ml Wheaton bottles by penetrating their silicone stopper seals with a 22-gauge syringe needle. The bottles were kept in the dark and chilled to below 4°C within 3 h of collection. In the laboratory, 5 ml of brine was transferred by syringe to another evacuated 130-ml Wheaton bottle, wherein it was acidified with 0.2 ml of "102%" phosphoric acid (also used by us for carbonate isotopic analyses). The phosphoric acid was prepared by evaporating reagent-grade 85% phosphoric acid under vacuum until it contained a small concentration of anhydrous phosphoric acid. After the acidified brine had equilibrated for 20 min, gases in the headspace of the bottle were drawn into a vacuum line through a cold trap (−196°C) fashioned from four 15-cm oblong loops of stainless steel tubing (outer diameter, 1/8 in. [0.32 cm]). Once the CO_2 and water in the trap were free of nonvolatile constituents, the CO_2 was purified from water by passing this mixture through a variable-temperature cold trap (7) held at −125°C and then trapping the CO_2 in a −196°C trap. The CO_2 was measured by using a mercury manometer and flame sealed in a 6-mm Pyrex (Corning Glass Works) tube for isotopic analysis. Carbon isotopic measurements were performed with a Nuclide 6-60RMS mass spectrometer modified for small-sample analysis (13). For the overall procedure, the 95% confidence limits were about 0.2‰. All isotopic results are reported as per mille values, relative to the Peedee belemnite carbonate standard, as follows: $\delta^{13}C = [(R_{sample}/R_{standard}) - 1]1{,}000$, where $R = {}^{13}C/{}^{12}C$ and $R_{standard} = 0.0112372$.

Organic Carbon Abundance and Isotopic Composition

Microbial mat samples were chilled to below 4°C within 3 h of collection and frozen to below −2°C within 1 day of collection at the field site. Upon return to the laboratory 2 days later, samples were stored at −60°C until lyophilized to dryness and then rinsed with distilled water to remove soluble salts. Dried samples about 1 to 10 mg in size were combusted by the bomb method (10, 11). Control measurements with unrinsed lyophilized mat samples yielded $\delta^{13}C$ values identical to those from rinsed samples; however, the salt in the unrinsed samples frequently caused the quartz combustion bombs to rupture during combustion. The CO_2 was separated from N_2 and water by passing the combustion products through a −125°C cold trap (7), trapping the CO_2 in a cold trap at −196°C, and recovering the N_2 elsewhere with a mercury Toepler pump. The CO_2 was quantified, packaged, and measured isotopically by using the same procedures described above for DIC.

DOC Levels

Brine samples (15 ml) were acidified with concentrated phosphoric acid to pH 2 and sparged for 10 min with purified O_2 to remove inorganic carbon. Control experiments with purified N_2 indicate that sparging with O_2 does not affect the dissolved organic carbon (DOC) content of the sample. Samples were analyzed by using a Dohrmann model DC-80 Total Organic Carbon analyzer. In this method, the DOC of a sample is oxidized in a persulfate solution which is sparged with pure O_2 and subjected to UV irradiation. The abundant halides in the sample can slow the rate of DOC oxidation, and therefore an $HgCl_2$-$Hg(NO_3)_2$ reagent described in the Dohrmann DOC analyzer manual is added to complex the halides and overcome this effect. The performance of the DOC analyzer was checked by using standards consisting of glucose, glutamic acid, or acetic acid dissolved in brine (120‰ salinity). A 100% CO_2 recovery with the standards was confirmed by using a Horiba model PIR-2000 infrared analyzer, which monitors the effluent O_2 gas stream. Depending on the salinity and DOC content, between 100 μl (for high-DOC, high-salinity samples) and 1 ml (for relatively dilute samples) of brine was injected into the analyzer. The 95% confidence limits of the measurements were equal to approximately 10% of the DOC values obtained.

Microelectrode Measurements

Depth distributions of oxygen concentration and photosynthesis were measured by oxygen microelectrode measurements as described by Revsbech and Jørgensen (25). The mats were under constant illumination of 1,000 microeinsteins $m^{-2}\ s^{-1}$ at the mat surface, and readings were taken at 100-μm depth intervals with a Clark-type oxygen microelectrode attached to a micromanipulator. Oxygenic photosynthesis was measured by the light-dark shift technique (24).

Microcoleus Cultures

The axenic cultures of *Microcoleus chthonoplastes* we used were purified from a crude culture isolated by John Bauld from mats in Spencer Gulf, South Australia. These cells were grown in ASN III medium (34) at 80‰ salinity. Sodium bicarbonate was added to three sets of flasks to prepare media containing between 65 and 70 mM, 20 and 25 mM, and 5 and 7 mM bicarbonate. The pH was adjusted to 7.5. To prevent exchange with the atmosphere, the flasks were firmly stoppered, yet also fitted with ports to permit samples to be obtained aseptically during incubation. The flasks were incubated in a New Brunswick incubator-shaker at 30°C and illuminated by cool white fluorescent lights at 185 microeinsteins m^{-2} s^{-1}. Samples were taken after 1, 5, and either 8 or 10 days, as required to obtain sufficient DIC for isotopic analysis.

Table 1.
General Composition of Mats Collected during November 1985

Constituent	Amt (% of total wet wt) in:	
	Pond 5	Pond 8
Water	83.2	82.6
Soluble salt[a]	5.8	9.4
Organic carbon	1.9	1.6
Organic matter[b]	4.8	4.0
Other[c]	6.2	4.0

[a] Calculated as anhydrous salts, assuming that Pond 5 and 8 salinities are 66 and 115‰, respectively (see Table 3).
[b] Assumes the stoichiometry for organic matter to be CH_2O.
[c] Includes waters of salt hydration, reduced sulfur and nitrogen, siliceous debris, etc.

GENERAL TRENDS IN BIOLOGICAL ACTIVITY WITH INCREASING SALINITY

Ponds 1 through 3 (Fig. 1) contain green algal blooms which form floating or suspended masses in the water column. The sediments of these ponds are covered with algal debris, below which is a black sulfidic organic-rich sediment. The winds prevailing from the northwest have driven large quantities of organic matter into the southeastern portions of Ponds 1 through 4. The largest accumulations of floating organic matter occur at the southeastern ends of Ponds 2 and 3, occupying perhaps one-third or more of their surface areas. Organic decomposition is very probably the dominant biological process in these areas. The southeastern end of Pond 4 contains unconsolidated, black sulfidic sediment populated every few tens of meters by colonies (diameter, 2 to 3 m) of *Ruppia* seagrass. Toward the northwest end of the pond, this community is replaced by the first permanent microbial mats, dominated by the filamentous cyanobacterium *M. chthonoplastes* and *Oscillatoria* spp. Mat thicknesses vary from a few millimeters to as much as 4 cm. The floors of Ponds 5 through 8 are essentially completely covered with cyanobacterial mats, varying in thickness from 3 to 10 cm. These mats consist almost entirely of organic matter and brine (Table 1); few, if any, inorganic minerals have been observed microscopically. The cyanobacterium *M. chthonoplastes* dominates throughout, but populations of unicellular cyanobacteria (e.g., *Synechococcus* and *Aphanothece* spp.) become more abundant at higher salinities. Gypsum is deposited periodically in Pond 9 and thus prevents thick organic mats from accumulating. Only wave-dispersed mats a few millimeters thick occur. No intact organic mats occur in Pond 10; however, photosynthetic bacteria do inhabit the substantial gypsum deposits which line the pond floor.

CHEMICAL TRENDS IN THE BRINES

It is useful to examine the changes in the dissolved carbon species to monitor the effects of biological activity in the ponds. Data for pH and total DIC are given in Tables 2 and 3 and illustrated in Fig. 2. Consistent trends with salinity were observed during the five field trips. As brine moves from the inlet through Pond 1 to Pond 2, its pH increases by almost a whole unit (e.g., typically from 8.2 to 8.9), whereas

Table 2.
Water and Mat Data for the April and June 1985 Field Trips

Date and site	Salinity (‰)	[Halide] (mmol/kg)	Temp (°C)	pH	[DIC] (mmol/kg)	$\delta^{13}C_{DIC}$	$\delta^{13}C_{mat}$
Apr 1985							
Inlet	39	0.61	24	8.4	2.5	+1.1	
1	51	0.80	22	8.7			
3 (high salt)	66	1.04		8.9	1.8	−4.8	
4	76	1.19	22		1.9	−5.6	
5	91	1.43	21	8.8	2.0	−5.7	−11.3
6	119	1.87	21		2.3	−5.4	−13.4
8	129	2.03	22	8.3			−12.6
9	136	2.14		8.3	2.3	−5.7	−12.5
June 1985							
Inlet	41	0.64	25	8.3	2.4	+0.4	
1	43	0.68	25	8.3			
2	50		25		1.6	−1.7	
3 (high salt)	79		28		1.9	−5.7	
4	71	1.11	28	8.6	1.8	−5.4	−10.6
5	83	1.30	28	8.6	1.7	−6.1	−12.9
6	117	1.84	27	8.6	2.4	−5.4	−13.0
8	118	1.85	27	8.4	2.5	−5.4	

DIC concentrations (denoted by [DIC]) decline substantially (e.g., typically from 3.0 to 2.0 millimolal [mmol/kg]). These trends indicate that CO_2 is withdrawn from the water, presumably owing to inorganic carbon assimilation by the algal blooms in Pond 1. These pH and DIC trends disappear by the time the brine reaches Pond 4, indicating that essentially no net transport of inorganic carbon into and out of the brine occurs in Pond 4. At higher salinities, the pH slowly declines toward 8.0. This decline may indicate that the balance between the rate of biological uptake of CO_2 and its rate of resupply from the atmosphere is shifting in favor of resupply.

A proper evaluation of the [DIC] trends with salinity must compensate for concentration increases associated merely with evaporation of the brines. This can be achieved by normalizing [DIC] values to concentrations of some dissolved species, for example, Cl^-, which is conserved in the evaporating brines in Ponds 1 through 10. Indeed, halide concentrations in the first nine ponds (Table 2) increase with salinity at the rate predicted for dissolved species conserved during evaporation. Data for both [DIC] and [DOC] are normalized to the total halide concentrations of the brines and plotted versus salinity in Fig. 3. Consistent with the trend in Fig. 2, Fig. 3 confirms that [DIC] declines from the inlet to Pond 3. However, this [DIC] decline continues in Ponds 4 through 8 (salinities between 60 and 120‰), presumably owing to DIC assimilation by the cyanobacterial mats. Only above salinities of 130‰ do [DIC] values approach a steady-state balance between supply from the atmosphere and loss to organisms and sediments.

The trend of [DOC] with increasing salinity indicates that a net addition of organic carbon to the brines occurs (Table 3; Fig. 3). The rate of addition is consistently highest in Ponds 2 and 3. Very probably the decomposition of the large accumulations of windblown organic detritus observed in these two ponds is adding DOC to the brine. Note that station 3 (high salt) (Table 3), which samples brine flowing from the south-

Table 3.
Water and Mat Data for the November 1985 and January and June 1986 Field Trips

Date and site	Salinity (‰)	Temp (°C)	pH	[DIC] (mmol/kg)	$\delta^{13}C_{DIC}$	[DOC] (mmol/kg)	$\delta^{13}C_{mat}$
Nov 1985							
Inlet	46		8.1	3.0	−0.2	0.35	
1	50		8.7	2.4	+0.2	0.62	
3 (high salt)	69		8.7	2.1	−4.0	1.3	
4	56	22		2.4	−4.8	1.3	−11.8
5	66	19	8.8	2.2	−6.0	1.6	−12.5
6	97	20	8.6	2.4	−7.2	2.9	−12.9
8	115	19	8.3	2.8	−6.3	4.1	−14.7
9	138	17	8.3	2.9	−6.1	4.9	
10	156	18	8.0	3.9	−4.1	6.6	
Jan 1986							
Inlet	38	18	8.1	3.0	−0.2	0.34	
1	44	18	8.6	2.4	−0.9	0.9	
3 (high salt)	60	21	8.8	2.0	−5.0	1.9	
4	53	20	9.0	1.7		1.3	
5	63	21	9.0	1.9	−9.1	2.3	−13.1
6	85	21	8.6	2.4	−7.2	2.8	
8	102	23	8.2	2.9	−6.0	4.1	−14.5
9	142	20	8.2	3.0	−4.8	5.4	
10	170	20	8.0	3.8	−3.7	6.2	
June 1986							
Inlet	46	24	8.4	2.4	+1.0	0.6	
1	55	24	9.0	1.9	−0.8	0.8	
3 (high salt)	83	25	8.9	1.9	−7.3	2.0	
4	70	30	8.9	1.8	−7.4	1.8	−10.1
5	82	29	9.0	1.7	−9.0	2.5	−12.5
6	114	32		1.9	−9.2	3.3	−13.2
8	128	31	8.6	2.1	−8.4	3.8	−14.8
9	150	30	8.5	2.5	−8.0	4.4	
10	167	31	8.3	2.8	−7.4	5.1	

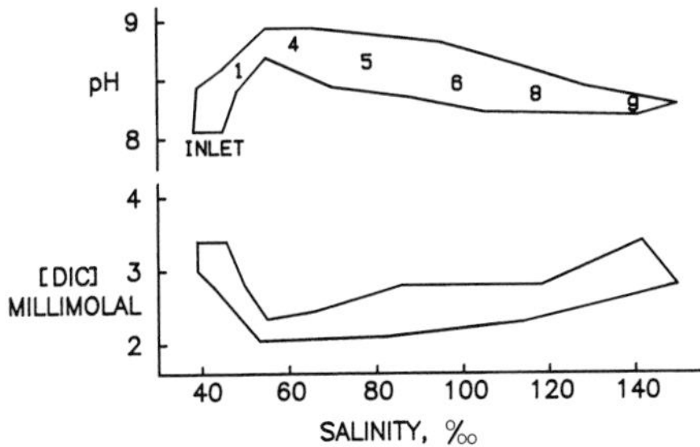

Figure 2. Trends in pH and total [DIC] with salinity. The lines enclose all datum points (Tables 2 and 3) measured during the four field trips. Numbers in the figure for pH give approximate positions (pond number; see Fig. 1) for the data from each of the ponds.

eastern end of Pond 3, had, on two occasions (January and June, 1986), an even higher [DOC] than did Pond 4 brine. A much lower, but detectable, [DOC] increase is observed from Ponds 4 through 8. This increase reflects contributions either from the cyanobacterial mats or from decomposing particulate organic carbon which has been carried into these ponds from further upstream. Net additions of DOC to the brine are trivial at salinities above 130‰ (Fig. 3).

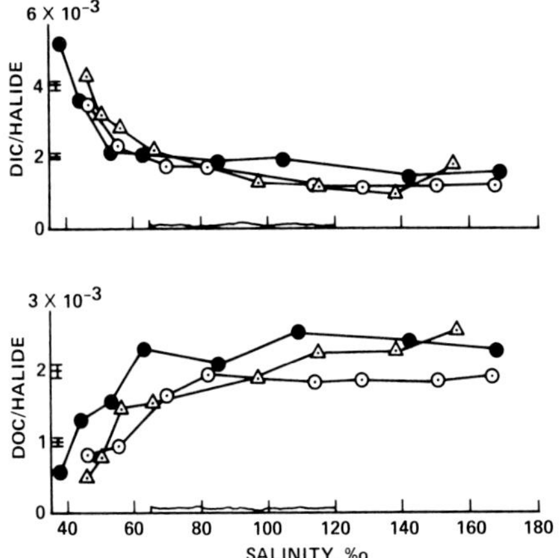

Figure 3. Trends in [DIC]/[halide] and [DOC]/[halide] versus salinity. Datum point symbols designate data from each field trip, as follows: △, November, 1985; ●, January, 1986; ⊙, June, 1986. Error bars just to the right of the vertical axis depict 95% confidence intervals for the data. The uneven line along the horizontal axis indicates the salinity range within which abundant mats occur (65 to 120‰).

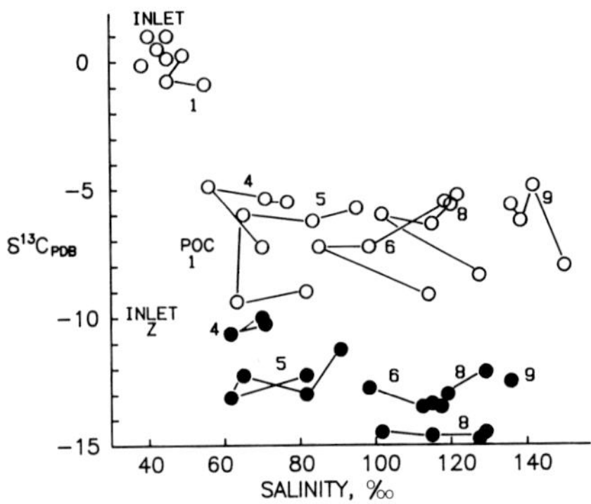

Figure 4. Carbon isotopic composition of DIC (○) and microbial mats (●) as a function of salinity. Numbers adjacent to datum points designate the ponds from which the samples were obtained. The value for particulate organic carbon from Pond 1 is represented by POC. The value for Zostera grass collected from the inlet channel is represented by Z.

CARBON ISOTOPES

Carbon Isotopic Trends with Salinity

Three features in Fig. 4 merit attention, namely the $\delta^{13}C$ trend with salinity observed for the DIC ($\delta^{13}C_{DIC}$), the magnitude of the carbon isotopic difference between the DIC and the microbial mats, and the $\delta^{13}C$ trend with salinity for the mat organic matter ($\delta^{13}C_{mat}$).

Carbon Isotopes in DIC

At the inlet, $\delta^{13}C_{DIC}$ values are within 1‰ of 0 and are typical for seawater (5). As the brine flows through Ponds 2, 3, and 4, however, its $\delta^{13}C_{DIC}$ decreases sharply to values typically between −5 and −8‰ (Tables 2 and 3, compare values for Pond 1 with those for Pond 3 [high salt]). This decline coincides with the presence in these ponds of abundant decaying organic matter, an increase in brine [DOC], and a diminishing rate of decline in [DIC]. All of these observations are consistent with the interpretation that decomposing organic matter in Ponds 2 and 3 is injecting ^{13}C-depleted DIC into the brine. A similar pattern is observed in the open ocean, where organic decomposition below the photic zone causes DIC in the deep ocean to become ^{13}C depleted, relative to surface water DIC (19). Downstream from Pond 4, little further change in $\delta^{13}C_{DIC}$ was observed, indicating either that relatively little additional organically derived DIC is added or that the $\delta^{13}C$ of any added DIC is similar to that of the total DIC pool in these brines.

Carbon Isotopes in Mat Organic Matter

The $\delta^{13}C_{mat}$ values measured (Tables 1 and 2; Fig. 4) are much more positive than the $\delta^{13}C$ values typically found in marine organic matter (e.g., −17 to −31‰ for plankton [6]). The difference in $\delta^{13}C$ between DIC and organic matter ($\Delta\delta^{13}C_{mat-DIC}$) is only −4 to −6‰ for Pond 4 and 5 mats and −6 to −8‰ for Pond 6 and 8 mats. Similar elevated $\delta^{13}C_{mat}$ values have been reported for recent hypersaline mats elsewhere (1–3, 29, 31). It is useful to explore the cause of such high $\delta^{13}C_{mat}$ values.

The carbon isotopic composition of a typical biological community is controlled principally by isotopic discrimination associated with DIC uptake by the primary producers of organic matter in that community. This is because isotopic discrimination by other members further along the organic-matter food chain either is relatively small (4, 6) or has a minor effect upon the total pool of accumulated organic matter. To date, plants have received more attention than other organisms regarding isotope discrimination during photosynthetic carbon assimilation (9, 22). Sharkey and Berry (33) applied the basic principles used in the plant work to cultures of green algae. It is useful to explore their applicability to microbial mats.

The assimilation of inorganic carbon by a mat cyanobacterium can be described by the multistep process shown in Fig. 5, where the arrows inside the mat indicate net carbon flows. Processes associated with the transport of inorganic carbon to the cell cytoplasm do not discriminate strongly against ^{13}C (22), whereas the enzyme ribulose-1,5-bisphosphate carboxylase (RUBISCO) prefers to assimilate ^{12}C at a rate which is between 2 and 3% (20 to 30‰) faster than the rate at which it assimilates ^{13}C. Thus, given an unlimited supply of CO_2, an autotrophic cell using RUBISCO will produce organic matter which is 20 to 30‰ depleted in ^{13}C relative to its CO_2 source. However, if the supply rate of DIC becomes so limited

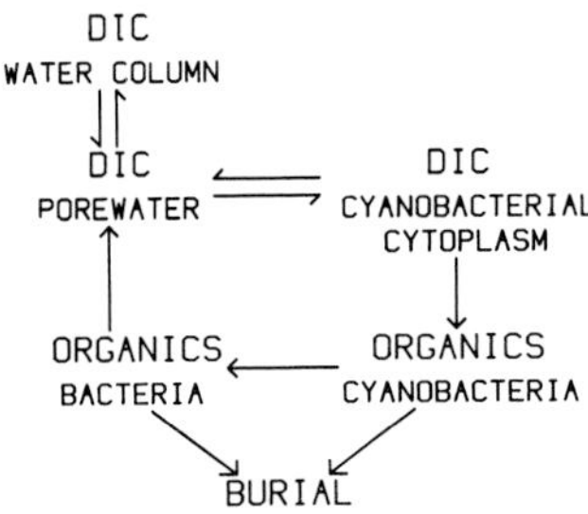

Figure 5. Schematic showing major carbon flows in a microbial mat. Carbon flows involving organic matter are shown as arrows depicting net flows.

that this enzyme must assimilate all of the CO_2 reaching the cytoplasm, then the organic matter synthesized will be isotopically identical to the DIC source of the cell. Accordingly, if carbon is completely recycled within the microbial mat with no net import or export (that is, no exchange with a large external reservoir of DIC [Fig. 5]), then no net isotopic discrimination can occur. However, exchange with an external inorganic carbon reservoir obviously does happen; therefore, it is the dynamics of this exchange process which controls the magnitude of isotopic discrimination. For the purposes of this discussion, Fig. 5 can be simplified to the two-step process shown below.

$$\text{External DIC} \underset{F_{\text{leak}}}{\overset{F_{\text{import}}}{\rightleftarrows}}$$

$$\text{DIC in mat} \xrightarrow{F_{\text{net fix.}}} \text{Mat organics}$$

The first step is the uptake or loss by the mat of DIC from the overlying water column. The second step is the net fixation (conversion) of DIC to organic matter. The following relationship describes semiquantitatively the net carbon isotopic discrimination by this system:

$$D_{\text{overall}} \approx (F_{\text{leak}}/F_{\text{import}})(D_e + D_r) \quad (1)$$

where D_{overall} is the overall discrimination (per mille) by the mat against ^{13}C, D_e is the equilibrium isotope effect associated with the conversion of HCO_3^- to CO_2, D_r is the kinetic isotopic discrimination associated with RUBISCO, F_{leak} is the flux of DIC from the mat back into the water column, and F_{import} is the flux of DIC into the mat. Note also that

$$F_{\text{leak}} = F_{\text{import}} - F_{\text{fix.}} \quad (2)$$

where $F_{\text{fix.}}$ is the net rate of DIC fixation to organic matter. If the rate of import of DIC (by diffusion, etc.) is very high relative to its rate of conversion to organic matter, then the fraction F_{leak}/F_{import} will approach unity ($F_{import} - F_{fix.}$ is large) and the overall discrimination will approach the sum of the equilibrium isotope effect (e.g., −7.4‰ at 30°C [21]) and the maximum possible discrimination by RUBISCO (−29.2% in green plants [27]). As the supply of DIC relative to its rate of conversion to organic matter becomes smaller, F_{leak}/F_{import} decreases ($F_{import} - F_{fix.}$ declines) and the overall discrimination decreases toward zero.

This relationship between isotopic discrimination and DIC availability is illustrated by the experiment summarized in Table 4. Three sets of *M. chthonoplastes* cultures were grown at different bicarbonate concentrations (67, 23, and 6 mM), and the $\delta^{13}C$ changes in DIC and cells were monitored with time. The growth rates for the different sets differed by less than a factor of 2. Note that for the cultures grown at lower bicarbonate concentrations, the $\delta^{13}C$ values of both DIC and cells increased more rapidly. For example, after 10 days of growth, $\delta^{13}C$ values of cells grown at 67 and 23 mM bicarbonate increased to −29.3 and −14.9, respectively (Table 4); cells grown at 6 mM bicarbonate already reached a $\delta^{13}C$ value of −8.4 after only 8 days. This is because, as the remaining DIC became depleted and more ^{13}C enriched, the cells grown at lower bicarbonate concentrations had to assimilate relatively more of this enriched DIC. Despite these differences in ^{13}C enrichment, all of the cultures were discriminating against ^{13}C to the same degree (see fractionation factors in Table 4). However, the last fractionation factor listed in this column is lower. This is probably because, as [DIC] declined even further, the rate of import of DIC from the culture medium into the microenvironment of the cells (F_{import}) declined relative to its rate of fixation into organic matter ($F_{fix.}$) (equations 1 and 2).

The microenvironment of the photic zone within the microbial mat very probably attains very low [DIC] during the daytime, as evidenced by the very high pH values attained (26). Therefore, a large fraction of DIC which enters the mat is assimilated; that is, F_{leak}/F_{import} and $F_{import} - F_{fix.}$ are small. This circumstance promotes the minimal isotopic discrimination which is observed.

Schidlowski et al. (31) also noted extreme ^{13}C enrichments in hypersaline microbial mats from Solar Lake, Sinai, Egypt, and

Table 4.
Carbon Isotope Data for *Microcoleus* Cultures[a]

Growth time (days)	[DIC] (mM)	$\delta^{13}C_{DIC}$	$\delta^{13}C_{mat}$	Fractionation factor[b]
5	66	0		
10	61.3	+1.9	−29.3	1.026
5	21.8	0		
10	13.5	+12.1	−14.9	1.026
1	5.5	0		
5	4.5	+4.9	−22.9	1.025
8	1.1	+26.5	−8.4	1.019

[a] Cultures were grown at 30°C and pH 7.5 to 8 under illumination of 100 microeinsteins m^{-2} s^{-1}.

[b] Fractionation factor expressing the rate of $^{12}CO_2$ fixation relative to the rate of $^{13}CO_2$ fixation. Calculated from DIC abundance and isotope data (see reference 20 for method).

explained this enrichment by invoking low CO_2 availability in the microbial mats. They attributed this low availability to the decline in CO_2 solubility in brines at elevated salinities and temperatures. Their hypothesis predicts, therefore, that as salinity and/or temperature increases, the $\delta^{13}C$ difference between DIC and mat organic matter should decline. Schidlowski et al. (31) proposed that Precambrian stromatolitic organic matter has much lower $\delta^{13}C$ values (30) because many of these ancient communities grew at the lower salinities characteristic of the open ocean.

The salinity-temperature hypothesis of Schidlowski et al. (31) can be tested directly by measuring the $\delta^{13}C$ values of organic matter from microbial mats grown at various salinities (Tables 2 and 3; Fig. 4). As noted above, the difference between DIC and organic matter $\delta^{13}C$ values ($\Delta\delta^{13}C_{mat - DIC}$) is only −4 to −6‰ for Pond 4 and 5 mats (grown at 65 to 85‰ salinities) and −6 to −8‰ for Pond 6 and 8 mats (grown at 100 to 120‰ salinities). Thus, isotopic discrimination actually increases slightly at higher salinities, which is opposite to the trend predicted by Schidlowski et al. (31). Greater discrimination can occur if F_{import} increases, as would happen if brine [CO_2] increased. The [DIC] does not increase appreciably between Pond 4 and Pond 8, which occupy opposite extremes in the salinity range within which mats accumulate (see [DIC] data in Tables 2 and 3). The pH of the brines does decline somewhat, thus favoring greater [CO_2], but this relatively small pH trend may be caused by increases in organic acid concentrations associated with higher [DOC] (Table 3). Furthermore, the [CO_2]-enhancing effect of lower pH is opposed by the salinity-related decline in CO_2 solubility (31).

An increase in carbon isotopic discrimination at higher salinities can also be caused by a decline in the rate of DIC conversion to organic matter in the mats. Such a decline would cause both $F_{import} - F_{fix.}$ and F_{leak}/F_{import} to increase (see equations 1 and 2 above). Indeed, a decline in DIC uptake rates by the mats at higher salinities is evidenced by a marked decrease in the rates of oxygenic photosynthesis (Fig. 6). This decrease could be caused by the increasing proportion of unicellular cyanobacteria, which form a looser, less dense mat texture. Alternatively, higher salinities might slow rates of photosynthesis in both *M. chthonoplastes* and the unicellular cyanobacteria. Regardless of the specific cause, the effect of lowered rates of photosynthetic carbon uptake is to increase the availability of DIC relative to its rate of conversion to organic matter. This increase favors greater carbon isotopic discrimination by the mats.

The evidence presented here indicates that stable isotopic differences between the organic constituents of modern mats and ancient stromatolites cannot be attributed to the restriction of modern mats to hypersaline environments. However, it must be emphasized that other environmental factors (e.g., atmospheric and oceanic [CO_2]) which controlled the availability of DIC to the mats could have caused this isotopic trend. In addition, any parameter which caused rates of photosynthesis in Precambrian mats to have been lower than they are today must be considered.

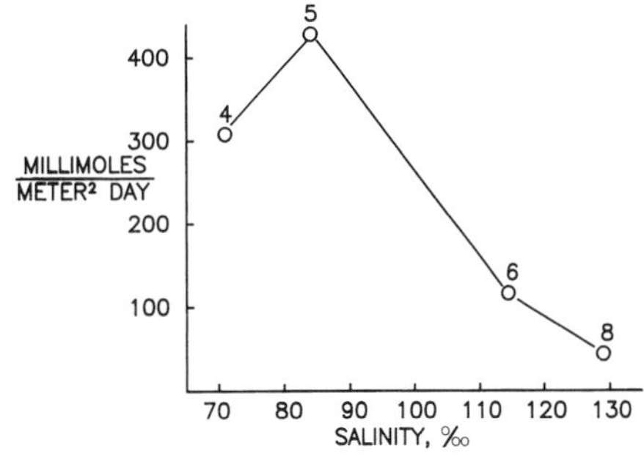

Figure 6. Rates of oxygen production in the mats in different ponds. Pond numbers are indicated next to the datum points. The data were obtained by using oxygen microelectrodes (see text).

Several investigators have proposed that the Precambrian era witnessed elevated atmospheric [CO_2]. Higher concentrations could have been supported in part by higher rates of volcanic CO_2 outgassing (8). Sagan and Mullen (28) recognized that because the early Sun was cooler, the Earth's atmosphere must have exerted a stronger greenhouse effect to prevent the oceans from freezing. Carbon dioxide apparently was the most suitable greenhouse gas (12, 23). Kasting et al. (18) calculated that an atmospheric [CO_2] approximately 600 times greater than the present concentration was required to sustain early Archean ocean temperatures similar to those of today. A factor of 600 would have greatly increased the expression of carbon isotopic discrimination by *M. chthonoplastes* (Table 4), a cyanobacterium whose ancestors possibly produced mats as old as 3.5×10^9 years (32).

ACKNOWLEDGMENTS. We thank Exportadora de Sal, S. A., for their permission to study the salt ponds and for their support of our field efforts. We are indebted to Barbara Javor for introducing us to the site and for guiding us on our early trips.

This research was supported by a grant from the National Aeronautics and Space Administration Planetary Biology Program.

LITERATURE CITED

1. **Aizenshtat, Z., G. Lipiner, and Y. Cohen.** 1984. Biogeochemistry of carbon and sulfur cycle in the microbial mats of the Solar Lake (Sinai), p. 281–312. *In* Y. Cohen, R. W. Castenholz, and H. O. Halvorson (ed.), *Microbial Mats: Stromatolites.* Alan R. Liss, Inc., New York.
2. **Barghoorn, E. S., A. H. Knoll, H. Dembicki, Jr., and W. G. Meinschein.** 1977. Variations in stable carbon isotopes in organic matter from the Gunflint Iron Formation. *Geochim. Cosmochim. Acta* **41:** 425–430.
3. **Behrens, E. W., and S. A. Frischman.** 1971. Stable carbon isotopes in blue-green algal mats. *J. Geol.* **79:**94–100.
4. **Blair, N., A. Leu, E. Munoz, J. Olson, and D. J. Des Marais.** 1985. Carbon isotopic fractionation by *Escherichia coli* K-12. *Appl. Environ. Microbiol.* **50:**996–1001.
5. **Broecker, W. S., and T.-H. Peng.** 1982. *Tracers in the Sea*, p. 306–311. Lamont-Doherty Geological Observatory, Palisades, N.Y.
6. **Deines, P.** 1980. The isotopic composition of reduced organic carbon, p. 329–406. *In* P. Fritz and J. C. Fontes (ed.), *Handbook of Environmental Isotope Geochemistry,* vol. 1. Elsevier Scientific Publishing Co., Amsterdam.
7. **Des Marais, D. J.** 1978. Variable temperature trap for the separation of gas mixtures. *Anal. Chem.* **48:**1651–1652.
8. **Des Marais, D. J.** 1985. Carbon exchange between the mantle and the crust, and its effect upon the atmosphere: today compared to Archean time. *Geophys. Monogr.* **32:**602–611.
9. **Farquhar, G. D., M. H. O'Leary, and J. A. Berry.** 1982. On the relationship between carbon dioxide discrimination and the intracellular carbon dioxide concentration in leaves. *Aust. J. Plant Physiol.* **9:**121–137.
10. **Frazier, J. W.** 1962. Simultaneous determination of carbon, hydrogen, and nitrogen, part II. *Mikrochim. Acta* **1962:**993–999.
11. **Frazier, J. W., and R. Crawford.** 1963. Modifications in the simultaneous determination of carbon, hydrogen and nitrogen. *Mikrochim. Acta* **1963:** 561–566.
12. **Hart, M. H.** 1978. The evolution of the atmosphere of the earth. *Icarus* **33:**23–39.
13. **Hayes, J. M., D. J. Des Marais, D. W. Peterson, D. A. Schoeller, and S. P. Taylor.** 1977. High precision stable isotope ratios from microgram samples. *Adv. Mass Spectrom.* **7:**475–480.
14. **Hayes, J. M., I. R. Kaplan, and K. W. Wedeking.** 1983. Precambrian organic geochemistry, preservation of the record, p. 93–104. *In* J. W. Schopf (ed.), *Earth's Earliest Biosphere, Its Origin and Evolution.* Princeton University Press, Princeton, N.J.
15. **Holser, W. T., B. Javor, and C. Pierre.** 1981. *Geochemistry and Ecology of Salt Pans at Guerrero Negro, Baja California.* Field Guidebook, Geological Society of America Cordilleran Section. Geological Society of America, Boulder, Colo.
16. **Javor, B.** 1983. Planktonic standing crop and nutrients in a saltern ecosystem. *Limnol. Oceanogr.* **28:**153–159.
17. **Javor, B. J.** 1983. Nutrients and ecology of the Western Salt and Exportadora de Sal Saltern Brines, p. 195–205. *In Sixth International Symposium on Salt,* vol. 1. Salt Institute, Alexandria, Va.
18. **Kasting, J. F., J. B. Pollack, and D. Crisp.** 1984. Effects of high CO_2 levels on surface temperature and atmospheric oxidation state of the early earth. *J. Atmos. Chem.* **1:**403–408.
19. **Kroopnick, P., W. G. Deuser, and H. Craig.**

1970. Carbon-13 measurements on dissolved inorganic carbon at the North Pacific (1969) GEOSECS station. *J. Geophys. Res.* 75:7668–7671.

20. **Melander, L., and W. H. Saunders.** 1980. *Reaction Rates of Isotopic Molecules*, p. 95–113. John Wiley & Sons, Inc., New York.
21. **Mook, W. G., J. C. Bommerson, and W. H. Staverman.** 1974. Carbon isotope fractionation between dissolved bicarbonate and gaseous carbon dioxide. *Earth Planet. Sci. Lett.* 22:169–176.
22. **O'Leary, M. H.** 1981. Carbon isotopic fractionation in plants. *Photochemistry* 20:553–567.
23. **Owen, T., R. D. Cess, and V. Ramanathan.** 1979. Enhanced CO_2 greenhouse to compensate for reduced solar luminosity on early Earth. *Nature* (London) 277:640–642.
24. **Revsbech, N. P., and B. B. Jørgensen.** 1983. Photosynthesis of benthic microflora measured with high spatial resolution by the oxygen microprofile method: capabilities and limitations of the method. *Limnol. Oceanogr.* 28:749–756.
25. **Revsbech, N. P., and B. B. Jørgensen.** 1986. Microelectrodes: their use in microbial ecology. *Adv. Microb. Ecol.* 9:293–352.
26. **Revsbech, N. P., B. B. Jørgensen, T. H. Blackburn, and Y. Cohen.** 1983. Microelectrode studies of the photosynthesis and O_2, H_2S, and pH profiles of a microbial mat. *Limnol. Oceanogr.* 28:1062–1074.
27. **Roesske, C. A., and M. H. O'Leary.** 1984. Carbon isotope effect on the enzyme catalyzed carboxylation of ribulose 1,5 bisphosphate. *Biochemistry* 23:6275–6284.
28. **Sagan, C., and G. Mullen.** 1972. Earth and Mars: evolution of atmospheres and surface temperatures. *Science* 177:52–56.
29. **Schidlowski, M.** 1985. Carbon isotope discrepancy between Precambrian stromatolites and their modern analogs: inferences from hypersaline microbial mats of the Sinai Coast. *Origins Life* 15:263–277.
30. **Schidlowski, M., J. M. Hayes, and I. R. Kaplan.** 1983. Isotopic inferences of ancient biochemistries: carbon, sulfur, hydrogen and nitrogen, p. 149–186. *In* J. W. Schopf (ed.), *Earth's Earliest Biosphere, Its Origin and Evolution.* Princeton University Press, Princeton, N.J.
31. **Schidlowski, M., U. Matzigkeit, and W. E. Krumbein.** 1984. Superheavy organic carbon from hypersaline microbial mats. *Naturwissenschaften* 71:303–308.
32. **Schopf, J. W., and B. M. Packer.** 1987. Early Archean (3.3-billion to 3.5-billion-year-old) microfossils from Warrawoona Group, Australia. *Science* 237:70–73.
33. **Sharkey, T. D., and J. A. Berry.** 1984. Carbon isotopic fractionation of algae as influenced by an inducible CO_2 concentrating mechanism, p. 389–401. *In* W. J. Lucas and J. A. Berry (ed.), *Inorganic Carbon Uptake by Aquatic Photosynthetic Organisms.* American Society of Plant Physiologists, Rockville, Md.
34. **Waterbury, J. B., and R. Y. Stanier.** 1981. Isolation and growth of cyanobacteria from marine and hypersaline environments, p. 221–223. *In* M. P. Starr, H. Stolp, H. G. Trüper, A. Balows, and H. G. Schlegel (ed.), *The Prokaryotes,* vol. 1. Springer-Verlag KG, Berlin.

III. Regulation of Adhesion and Hydrophobicity of Cell Surfaces in the Formation of Microbial Mats

Chapter 18

The Unique Characteristics of Benthic Cyanobacteria

M. Shilo

Many cyanobacteria are benthic, endolithic, epiphytic, or symbiotic and are therefore attached to submerged solid surfaces or internal cavities within other organisms. The first question I would like to address is whether there is a common mechanism of nonspecific adherence of all these organisms to nonanimated as well as animated surfaces which is not found for the planktonic cyanobacteria. The second is the special physiological adaptations that are required that have evolved as a consequence of this attached existence (28). Among the unique adaptive processes found in attached cyanoabacteria are (i) mechanisms for detachment to ensure spread and dispersion to new colonization sites; (ii) adaptation to cope with the rapidly changing environment (this requires metabolic flexibility and rapid shifts between alternative pathways); (iii) adaptation to resist high, toxic H_2S concentrations; (iv) adaptation to low light levels and restricted spectrum quality or to high light intensities; and (v) resistance to periodic desiccation.

All cyanobacteria adherent to surfaces, benthic as well as epilithic and symbiotic organisms, share the property of having a hydrophobic cell surface. All planktonic cyanobacteria that have been tested, on the other hand, have a cell surface which is highly hydrophilic (2, 12). This conclusion was established by using many of the conventional methods for measurement of hydrophobicity, including partitioning between hexadecane and water (24), drop-angle measurements between a cell layer and a drop of water or an air bubble, and adhesion to phenyl- or octyl-Sepharose beads. In *Phormidium* strain J-1, hydrophobic sites are scattered around the entire cell surface; this was demonstrated microscopically by the nature of their adherence to an oil/water interface (12).

All attached microorganisms, including the benthic and epilithic cyanobacteria, must be able to detach themselves from their substratum at some stage in their life cycle. This is an absolute requirement for any further spread and colonization of new suitable surfaces. In other aquatic microorganisms such as the caulobacters (22), *Nevskia* sp., or *Rhodo-hyphomicrobium* sp., alternations between mature cells adherent to interfaces and freely mobile flagellated cells in the aquatic environment are well known.

Two different mechanisms for attach-

M. Shilo • Division of Microbial and Molecular Ecology, Institute of Life Sciences, The Hebrew University of Jerusalem, Jerusalem 99104, Israel.

ment and desorption have been found to operate in cyanobacteria; both involve a change of the cell surface properties. Primarily, these mechanisms involve a transition of the hydrophobic cell exterior to hydrophilicity. The first mechanism is alternation between different cell types in the growth cycle. This involves the formation of special cells, such as baeocytes in the pleurocapsalean cyanobacteria, or the production of hormogonia in many attached filamentous cyanobacteria. The notion that this transition involves the change from hydrophobicity to hydrophilicity was demonstrated for several hormogonium-producing cyanobacteria (12).

It was relatively easy to obtain synchronously produced hormogonia by growing the cells in a column of glass beads to which the hydrophobic mature filaments adhere, releasing the hydrophilic hormogonia, which could be washed off the column and collected.

The development and transition of the hydrophilic *Plectonema hormogonia* to mature hydrophobic filaments with a change of the cell surface properties take place within 24 to 48 h and require both photosynthetic energy and de novo synthesis of protein(s). Darkness, 3-(3,4-dichlorophenyl)-1,1-dimethylurea (DCMU), or chloramphenicol completely blocked the transition (12). Baeocyte formation in the pleurocapsalean cyanobacteria may involve an analogous modulation of the cell surface from hydrophobicity to hydrophilicity, but this has not yet been demonstrated experimentally (31).

A second mechanism, found in cyanobacteria which lack a biphasic life cycle which includes hormogonia, is one in which the hydrophobic mature cells produce and excrete an amphiphilic, high-molecular-weight compound which covers the cell surface and temporarily masks its hydrophobicity (2, 14). One such heteropolymer was shown to be a polysaccharide with emulsifying properties and was therefore called emulcyan (14). Encapsulation, slime production, and change of surface hydrophobicity have similarly been shown to occur in the oil-degrading organism *Acinetobacter calcoaceticus* (25, 32).

Experiments with *Phormidium* strain J-1 have shown that the external masking layer accumulates on the cell exterior as a function of the age of the culture (2) and is easily removed by washing, by centrifugation, or by mechanical stirring. Sar and Rosenberg have recently shown (27) that several hydrophilic bacterial isolates from fish skin also became more hydrophobic after washing.

Chemical, enzymatic, and mechanical treatments of the cell surface, together with electron-microscopic studies, have been used to uncover the composition and architecture of the surface layers of benthic cyanobacteria and to understand the nature of the surface modulation processes. On the basis of these studies, a structural scheme has been proposed which represents the present-day knowledge of the cell surface of *Phormidium* strain J-1 (2), the complex relationship between the different components of the cell surface, and the effect of the different treatments.

The nature of the macromolecules responsible for masking of the cell surface and for detachment from the substratum has been deduced from the effects of the above-mentioned treatments, as well as from the analysis of purified emulcyan preparations from *Phormidium* strain J-1. The mechanism of detachment by surface masking may occur more widely than just in some of the cyanobacteria; chemically different amphiphilic macromolecules formed by different organisms, such as emulsan produced by *A. calcoaceticus*, may have an analogous function (25).

Recently in the analysis of *Phormidium* strain J-1 emulcyan it was shown that the crude preparations could be resolved on acrylamide agarose columns (LKB Ultrogel AcA 5:4; molecular weight, 5,000 to 70,000) into a high-molecular-weight emulsifying substance and a low-molecular-weight surface-active component (N. Hershkowitz and Y. Elkana, unpublished data).

The second general property characterizing the attached cyanobacteria and separating them from their free-floating planktonic counterparts is their versatile metabolism and their ability to shift rapidly between alternative metabolic pathways.

Many of the planktonic cyanobacteria are capable of buoyancy regulation due to their interior gas vacuoles and the formation and degradation of granules of reserve material of relatively high specific gravity. This allows them to change their position in the water column as environmental conditions change and thus to always face a nearly constant external milieu (29, 30).

The sessile benthic cyanobacteria, however, remain attached to the water/sediment interface, an environment which shows rapid and drastic fluctuations of light, pH, E_h, H_2S, and oxygen concentrations, as has been clearly demonstrated in situ with the aid of microelectrodes (24). These organisms, therefore, have to change the patterns of their metabolism, rather than change their location in the water column. Furthermore, benthic cyanobacteria often remain viable under aphotic conditions for many years, buried under the layers which sediment above them annually (16). This again requires unique adaptive metabolic properties and maintenance mechanisms.

One of these flexible metabolic patterns is the shift found only in these cyanobacteria, from oxygenic to anoxygenic photosynthesis, from a photosystem I (PS I) plus PS II mechanism to photosynthesis driven by PS I only with H_2S and not water as the electron donor (8, 10, 19–21). Another example of alternative metabolic pathways is the ability of *Oscillatoria limnetica* to shift in the dark from aerobic (CO_2-producing) to anaerobic (S to H_2S) respiration and/or fermentation of polyglucose to lactate (18). An additional example of the metabolic versatility is the ability to use the same photosynthetic electron pathway alternatively for photoassimilation of CO_2, for production of hydrogen under overreduced conditions, or for the fixation of atmospheric nitrogen (5–7). After a shift from anaerobic to aerobic conditions, a rapid induction of superoxide dismutase, up to a level 8- to 10-fold higher than under O_2-depleted conditions, occurs (15). A schematic representation of the metabolic shifts in *O. limnetica* after a shift from light to dark conditions and from aerobic to anaerobic conditions is shown in Fig. 1.

Different modes of adaptation to the presence of H_2S in the environment are found in the cyanobacteria, allowing them to resist its toxic effects. Cohen et al. (9) have shown that cyanobacteria have four types of adaptation to high H_2S concentration. Although all planktonic cyanobacteria photosynthesize only with a coupled PS I plus PS

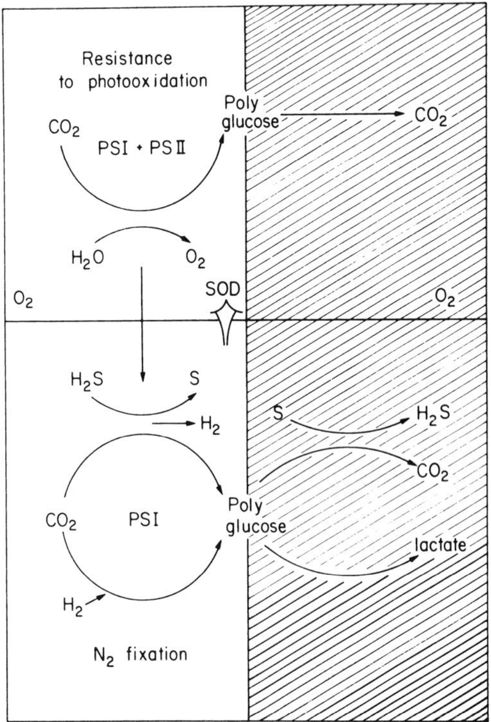

Figure 1. Schematic representation of metabolic shifts in *O. limnetica* after a shift from light to dark and from aerobic to anaerobic conditions. SOD, Superoxide dismutase.

II system and stop photosynthetic activity in the presence of H_2S, various resistance mechanisms exist among benthic cyanobacteria (Table 1): (i) resistance of PS II to H_2S but no anoxygenic photosynthesis; (ii) partial resistance of PS II to H_2S with activation of PS I by H_2S as the electron donor in anoxygenic photosynthesis; and (iii) a mechanism like that in *O. limnetica*, for which in the presence of the H_2S anoxygenic PS I, only photosynthesis is found, with complete inhibition of PS II.

Another unique feature of the benthic cyanobacteria is their ability to adapt to the quality and low intensities of light reaching the benthic interface. This involves phototactic gliding to the sediment surface to prevent the burying of the benthic mat by sedimentation, chromatic adaptation and enhanced synthesis of photosynthetic pigments, and an increase in the number or size of reaction centers. We also showed (13, 28) that some cyanobacteria are able to produce a macromolecule which can cause flocculation of particular matter; this may enable the benthic cyanobacterial mats to secure sufficient light for photosynthesis, by clarifying turbid waters. The macromolecular bioflocculant formed by *Phormidium* strain J-1 can be separated from the emulcyan by precipitation in ethanol; emulcyan remains in the ethanol-soluble fraction (13).

Several different filamentous cyanobacteria share the property of bioflocculant production; however, different strains form chemically different acidic macromolecular bioflocculants. Some produce sulfated polysaccharides, whereas others contain heteropolysaccharides rich in pyruvate or uronic acids. It thus seems probable that by polyphyletic convergent evolution, different molecules have evolved to carry out a similar function (3).

Many epiphytic and epilithic cyanobacteria are periodically, at low tide, exposed to extremely high light intensities and have to protect themselves from lethal photooxidation (1, 11). Some benthic, epilithic, or saprophytic cyanobacteria are periodically exposed to extreme dryness; this may be in a diurnal tide cycle or for longer periods in seasonal cycles. *Phormidium autumnale* flourishes in the drainage basins in the sewage disposal plant of the Tel Aviv area, in which flooding and drying occur in regular alternation; the totally dried cyanobacterial mats retain full viability and return to full photosynthetic activity within minutes after becoming wet (17). The epiphytic *Scytonema* spp. live on the aerial roots of mangrove (*Avicenia*) trees, which are affected by the tidal cycle. Every drying and rewetting episode involves the inducible resynthesis of a new nitrogenase cycle (23).

Cell surface hydrophobicity, which is characteristic of all attached cyanobacteria, including benthic, epilithic, epiphytic, and even many symbiotic cyanobacteria (such as *Anabaena* spp., which live in *Azolla* leaf cavities), is not a permanent constant fea-

Table 1.
Adaptation of Cyanobacteria to H_2S[a]

Organism	H_2S sensitive (oxygenic PS only)	H_2S resistant or dependent (oxygenic PS, no anoxygenic PS)	H_2S resistant (oxygenic and anoxygenic PS)	H_2S dependent (PS I only) in DCMU
Anacystis nidulans	+	−	−	−
Phormidium strain Yellow Spring	−	+	−	−
Microcoleus chthonoplastes	−	−	+	−
Oscillatoria limnetica	−	−	−	+

[a] Reproduced from *Applied and Environmental Microbiology* (9).

ture, but can change and fluctuate as a function of cell age, growth conditions, and growth rate.

There are indications that nutrient limitation in periods of starvation (or nutrient flux) is involved in the triggering of changes in the cell surface properties and thus affects attachment or detachment of the organisms. Experimentally, such situations can be simulated in the cyanobacterium *Phormidium* strain J-1 by aging of the cells, by mechanical stirring, by washing, or by centrifugation. In *Phormidium* strain J-1, inhibition of de novo protein synthesis, starvation, and the action of enzymes and detergents have been shown to affect cell surface properties (Table 2) (2, 3, 9).

Belas et al. (4) have made the first attempt to understand the genetic regulation of the transformation of cell surface on attachment. Their elegant work has shown that the signal that triggered the change in the cell surface properties of *Vibrio parahaemolyticus* was physical rather than chemical. Contact with solid surfaces or viscous liquids induced the formation of the lateral hydrophobic flagella (or the emission of light for cells in which the lux genes were introduced into the middle of the lateral flagellum gene).

Table 2.
Modulation of Cell Surface Hydrophobicity in Cyanobacteria

Condition and strain	Hydrophobicity (%)
Genetic alterations	
Phormidium strain J-1, wild type	80–100
Phormidium strain J-1, spontaneous mutant	65
Phormidium strain J-1, phage resistant	55
Spirulina platensis, wild type	0
Spirulina platensis, AcA^r (clumping)	20
Spirulina platensis, 5FT^r (clumping)	40
Aphanothece halophytica, wild type	+
Aphanothea halophytica phototaxis-negative mutant	–
Cell cycle stages	
Phormidium strain J-1, young	80–100
Phormidium strain J-1, old	40–50
Plectonema boryanum, vegetative	95
Plectonema boryanum, hormogonia	10
Metabolic inhibitors	
Phormidium strain J-1, CAP treated	10
Phormidium strain J-1, DCMU treated	30
Phormidium strain J-1, dark incubation	20
Mechanical or chemical modifications	
Phormidium strain J-1, Omnimixer treated	40
Phormidium strain J-1, sonicated	15
Phormidium strain J-1, DSD-protease	10

In conclusion, not only does adherence of cyanobacteria involve cell surface properties connected with the attachment per se, but also, as a consequence of adhesion, the organisms have to face unique environmental conditions. Benthic cyanobacteria must thus have far-reaching physiological adaptations to allow their detachment for spread and dispersal, to be capable of shifts between alternative metabolic pathways to survive rapidly changing environmental conditions, to resist highly toxic H_2S concentrations, and to adapt to low light intensities.

It thus is appropriate to speak of a benthic way of life, which includes all the different and unique properties shared by these organisms.

Acknowledgments. The results and ideas summarized in this paper are the product of a group effort in the Division of Microbial and Molecular Ecology. The collaborating scientists include Etana Padan, Yehudit Elkana, Devorah Friedberg, Yehuda Cohen, and Aaron Oren, as well as several present and former Ph.D. students: Ali Fattom, Yeshayahu Bar-Or, Nizan Hershkowitz, Jaap Van Rijn, and Shimshon Belkin. I am grateful to all of them for their individual contributions and for their enthusiastic and friendly collaboration.

LITERATURE CITED

1. **Abeliovich, A., and M. Shilo.** 1972. Photooxidative death in blue-green algae. *J. Bacteriol.* **111:** 682–689.
2. **Bar-Or, Y., M. Kessel, and M. Shilo.** 1985. Modulation of cell surface hydrophobicity in the benthic cyanobacterium *Phormidium* J-1. *Arch. Microbiol.* 142:21–27.
3. **Bar-Or, Y., and M. Shilo.** 1987. Characterization of macromolecular flocculants produced by *Phormidium* J-1 and by *Anabaenopsis circularis* PCC 6720. *Appl. Environ. Microbiol.* **53:**2226–2230.
4. **Belas, R., M. Simon, and M. Silverman.** 1986. Regulation of lateral flagella gene transcription in *Vibrio parahaemolyticus*. *J. Bacteriol.* **167:**210–218.
5. **Belkin, S., and E. Padan.** 1978. Hydrogen metabolism in the facultative anoxygenic cyanobacteria (blue-green algae) *Oscillatoria limnetica* and *Aphanothece halophytica*. *Arch. Microbiol.* **116:**109–111.
6. **Belkin, S., and E. Padan.** 1978. Sulfide-dependent hydrogen evolution in the cyanobacterium *Oscillatoria limnetica*. *FEBS Lett.* **94:**291–294.
7. **Belkin, S., and E. Padan.** 1983. Na-dithionite promotes photosynthetic sulfide utilization by the cyanobacterium *Oscillatoria limnetica*. *Plant Physiol.* 72:825–828.
8. **Cohen, Y., B. B. Jørgensen, E. Padan, and M. Shilo.** 1975. Sulphide-dependent anoxygenic photosynthesis in the cyanobacterium *Oscillatoria limnetica*. *Nature* (London) **257:**489–492.
9. **Cohen, Y., B. B. Jørgensen, N. P. Revsbech, and R. Poplawski.** 1986. Adaptation to hydrogen sulfide oxygenic and anoxygenic photosynthesis among cyanobacteria. *Appl. Environ. Microbiol.* **51:** 398–407.
10. **Cohen, Y., E. Padan, and M. Shilo.** 1975. Facultative anoxygenic photosynthesis in the cyanobacterium *Oscillatoria limnetica*. *J. Bacteriol.* **123:**855–861.
11. **Eloff, J. N., Y. Steinitz, and M. Shilo.** 1976. Photooxidation of cyanobacteria in natural conditions. *Appl. Environ. Microbiol.* **31:**119–126.
12. **Fattom, A., and M. Shilo.** 1984. Hydrophobicity as an adhesion mechanism of benthic cyanobacteria. *Appl. Environ. Microbiol.* **47:**135–143.
13. **Fattom, A., and M. Shilo.** 1984. *Phormidium* J-1 bioflocculant: production and activity. *Arch. Microbiol.* **139:**421–426.
14. **Fattom, A., and M. Shilo.** 1985. Production of emulcyan by *Phormidium* J-1: its activity and function. *FEMS Microbiol. Ecol.* **31:**3–9.
15. **Friedberg, D., M. Fine, and A. Oren.** 1979. Effect of oxygen on the cyanobacterium *Oscillatoria limnetica*. *Arch. Microbiol.* **123:**311–313.
16. **Jørgensen, B. B., Y. Cohen, and N. P. Revsbech.** 1988. Photosynthetic potential and light-dependent O_2 consumption in a benthic cyanobacterial mat. *Appl. Environ. Microbiol.* **54:**176–182.
17. **Katznelson, R.** 1986. Cyanobacterial mats in groundwater recharge basins of the Dan Region wastewater reclamation project, p. 933–939. *In* Z. Dubinsky and Y. Steinberger (ed.), *Environmental Quality and Ecosystem Stability*, vol. IIIB. Bar-Ilan University Press, Ramat Gan, Israel.
18. **Oren, A., and M. Shilo.** 1979. Anaerobic heterotrophic dark metabolism in the cyanobacterium *Oscillatoria limnetica*: sulfur respiration and lactate fermentation. *Arch. Microbiol.* **122:**77–84.
19. **Padan, E.** 1979. Facultative anoxygenic photosynthesis in cyanobacteria. *Annu. Rev. Plant. Physiol.* **30:**27–40.
20. **Padan, E.** 1979. Impact of facultatively anaerobic photoautotrophic metabolism on ecology of cyano-

bacteria (blue-green algae). *Adv. Microb. Ecol.* **3**:1–48.
21. **Padan, E., and Y. Cohen.** 1982. Anoxygenic photosynthesis, p. 215–235. *In* N. C. Carr and B. A. Whitton (ed.), *The Biology of Cyanobacteria.* Blackwell Scientific Publications, Ltd., Oxford.
22. **Poindexter, J. S.** 1981. The caulobacters: ubiquitous unusual bacteria. *Microbiol. Rev.* **45**:123–179.
23. **Potts, M.** 1979. Nitrogen fixation (acetylene reduction) associated with communities of heterocystous and nonheterocystous blue-green algae in mangrove forests of Sinai. *Oecologia* **39**:359–373.
24. **Revsbech, N. P., B. B. Jørgensen, T. H. Blackburn, and Y. Cohen.** 1983. Microelectrode studies of the photosynthesis and O_2, H_2S, and pH profiles of a microbial mat. *Limnol. Oceanogr.* **28**:1062–1074.
25. **Rosenberg, E., N. Kaplan, O. Pines, M. Rosenberg, and D. Gutnick.** 1983. Capsular polysaccharides interfere with adherence of *Acinetobacter calcoaceticus* to hydrocarbons. *FEMS Microbiol. Lett.* **17**:157–160.
26. **Rosenberg, M., D. Gutnick, and E. Rosenberg.** 1980. Adherence of bacteria to hydrocarbons: a simple method for measuring cell-surface hydrophobicity. *FEMS Microbiol. Lett.* **9**:29–33.
27. **Sar, N., and E. Rosenberg.** 1987. Fish skin bacteria: colonial and cellular hydrophobicity. *Microb. Ecol.* **13**:193–202.
28. **Shilo, M.** 1982. Photosynthetic microbial communities in aquatic ecosystems. *Phil. Trans. R. Soc. London Ser. B* **297**:565–574.
29. **Van Rijn, J., and M. Shilo.** 1985. Carbohydrate fluctuations, gas vacuolation, and vertical migration of sum-forming cyanobacteria in fishponds. *Limnol. Oceanogr.* **30**:1219–1228.
30. **Walsby, T.** 1975. Gas vesicles. *Annu. Rev. Plant Physiol.* **36**:427–439.
31. **Waterbury, J. B., and R. Y. Stanier.** 1978. Patterns of growth and development in pleurocapsalean cyanobacteria. *Microbiol. Rev.* **42**:2–44.
32. **Zuckerberg, A., A. Diver, Z. Peeri, D. L. Gutnick, and E. Rosenberg.** 1979. Emulsifier of *Arthrobacter* RAG-1: chemical and physical properties. *Appl. Environ. Microbiol.* **37**:414–420.

Chapter 19

Mechanisms for Release of the Benthic Cyanobacterium *Phormidium* Strain J-1 to the Water Column

Y. Bar-Or, M. Kessel, and M. Shilo

INTRODUCTION

Detachment of microorganisms adhering to interfaces, such as mat-forming benthic cyanobacteria, is essential for their dispersal and colonization of new surfaces. All benthic cyanobacteria tested were found to possess hydrophobic cell surface characteristics (7), and hydrophobic interactions were thus proposed to be the mode of attachment of these organisms to their substratum surfaces. In addition, several benthic cyanobacterial species were found to possess acidic polysaccharides, which are associated with the cell wall and which bind and adsorb clay particles (3, 3a). In this chapter we describe various ways by which attachment of the benthic cyanobacterium *Phormidium* strain J-1 to the sediment is modulated, enabling its release to the water column.

Y. Bar-Or and M. Shilo • Division of Microbial and Molecular Ecology, Institute of Life Sciences, The Hebrew University of Jerusalem, Jerusalem 91904, Israel. *M. Kessel* • Department of Membrane and Ultrastructure Research, The Hebrew University-Hadassah Medical School, Jerusalem 91904, Israel.

EXPERIMENTAL TECHNIQUES

Hydrophobicity was measured by partitioning in a biphasic system of hexadecane and water by the method of Rosenberg et al. (18). Cell-bound and extracellular flocculants were assayed by measuring the rate of sedimentation of a standard bentonite suspension in water (8). Cells were prepared for electron microscopy as previously described (2) by using antibody stabilization of capsular polymers and staining with ruthenium red.

MODULATION OF CELL SURFACE HYDROPHOBICITY

Inhibition of protein synthesis in *Phormidium* strain J-1 by incubation with chloramphenicol (10 μg/ml) led to a sharp decrease in cell surface hydrophobicity (2). A similar effect was achieved after treatment with sodium dodecyl sulfate (0.25%) followed by digestion with pronase (200 μg/ml). These results suggested the involvement of a protein(s) or protein complexes in the surface hydrophobicity of *Phormidium* strain J-1. Transmission electron microscopy revealed the presence of a minicapsule enveloping the cells. In the presence of chloramphenicol or after treatment with protease there was extensive damage to the outer layer of the minicapsule (2), which was presumably reflected in a decrease in hydrophobicity. The shift to relative hydrophilicity in the presence of chloramphenicol may simulate the response of the organism to starvation conditions in nature, such as during a decrease in light intensity or a decrease in the abundance of nutrients. Under such conditions the hydrophobic proteins are probably not synthesized, and this allows the cells to detach from their substrata and migrate to environments with more favorable conditions. Under the favorable conditions the hydrophobic proteins are resynthesized and enable the organism to reattach to the benthos interface.

Phormidium strain J-1 produces considerable amounts of an amphiphilic polymer termed emulcyan, which has emulsifying (9) and surface-active (N. Hershkovitz and Y. Elkana, unpublished results) properties. Most of the emulcyan produced by young cultures is cell bound, whereas in old cultures it is excreted to the growth medium (9). We have found that cells from old cultures (14 to 20 days) were typically embedded in a thick slime layer (Fig. 1). After preparation for electron microscopy, including a dehydration step, the slime resembled a network of fibrous bundles (Fig. 2), which could be visualized only after pretreatment with whole-cell antiserum and staining with the cationic dye ruthenium red. Old cultures (14 to 20 days) of *Phormidium* strain J-1 exhibited a marked decrease in cell surface hydrophobicity, and emulcyan has been postulated to play a role in masking the surface hydrophobicity by virtue of its amphiphilic properties (9).

Old cells of *Phormidium* strain J-1 are therefore released to the water column by two processes: (i) masking of surface hydrophobicity by emulcyan and (ii) accumulation

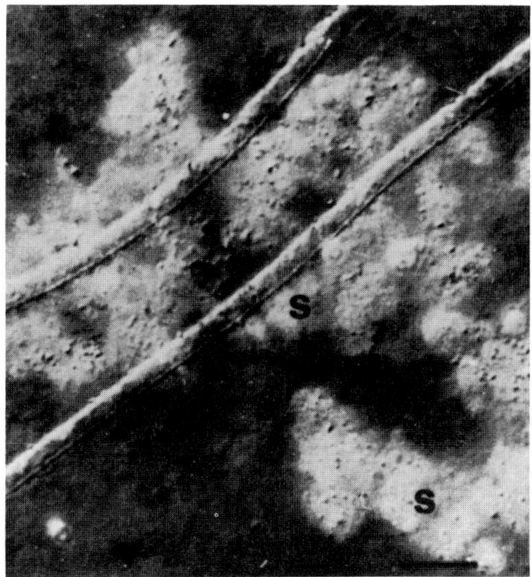

Figure 1. Nomarski interference micrograph, showing slime patches (S) adjacent to and surrounding filaments of *Phormidium* strain J-1. The cells were treated with specific antiserum. Bar, 5 μm.

of a slime layer which physically separates the cells from the benthos. These two processes may be related, since emulcyan may be a component of the slime layer. Washing old cells with water or fresh medium removed the slime layer and restored their initial surface hydrophobicity. This, again, may simulate natural conditions, under which detachment and reattachment are facilitated by production and subsequent removal of a slime layer.

MODULATION OF ADSORPTION TO CLAY PARTICLES

The attachment of *Phormidium* strain J-1 (and several other benthic cyanobacteria) to the benthos is enhanced by coflocculation with sedimentary clay particles (3a). Coflocculation is apparently due to a cell-associated flocculant (8) composed of an acidic polysaccharide backbone to which protein, fatty acids, and sulfate groups are linked (3). Divalent cations are required for cofloccula-

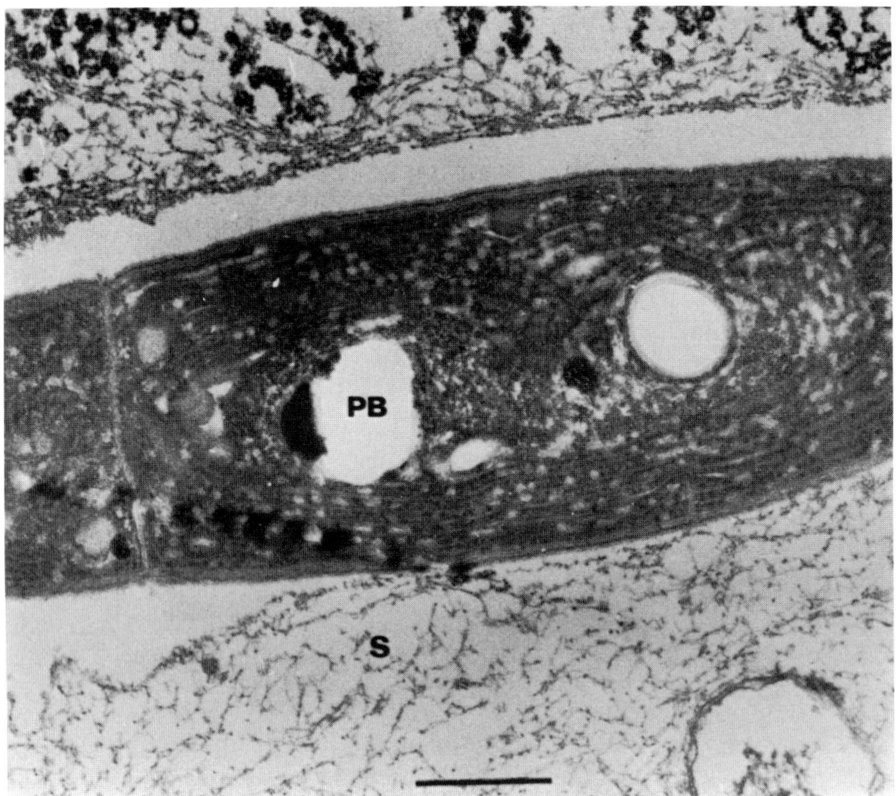

Figure 2. Transmission electron micrograph of the slime layer enveloping *Phormidium* strain J-1 cells. The cells embedded in slime (S) were treated with antiserum prior to fixation. The slime is bound loosely to the cell wall and is therefore sometimes detached during preparation for electron microscopy. Polyphosphate bodies (PB) were also lost, leaving empty spaces behind. Bar, 0.5 μm.

tion, as well as for flocculation. The cell-associated flocculant could be extracted from young cells (10 days) only by the relatively drastic treatment of preincubation with sodium dodecyl sulfate followed by protease. It could be extracted from older cells (14 to 20 days) with sodium dodecyl sulfate alone (2). In stationary-phase cultures most of the flocculant produced was found in the growth medium (8). In addition, stationary-phase cells were found to adsorb only about half of the amount of clay absorbed by young cells (3a).

It thus seems that detachment of *Phormidium* strain J-1 cells is made possible by several distinct processes: modulation of synthesis of hydrophobic surface proteins; masking of surface hydrophobicity by emulcyan; physical separation of the cells from the benthos as a result of the accumulated slime layer; and a decrease in the ability to bind to sedimentary clay particles owing to the gradual release of the cellular flocculant to the medium.

DISCUSSION AND CONCLUSIONS

Attachment of *Phormidium* strain J-1 to the benthos is apparently due to both hydrophobic interactions and binding to negatively charged clay particles via cation bridges. Detachment of the cells and their release to the water column thus depend on mechanisms which can overcome both types of interaction. Excretion of the amphiphilic emulcyan modulates the cell surface hydrophobicity, whereas movement of the cell-associated flocculant to the surrounding milieu releases the tightly bound clay particles. The slime layer which gradually covers old cells may physically separate them from both hydrophobic and hydrophilic, charged surfaces. *Phormidium* strain J-1 thus maintains two subpopulations: one that is attached to the benthos and one that is detached and facilitates dispersal.

In most reports on hydrophobic bacteria the surface hydrophobicity, determined by adhesion to hydrocarbons, did not reach 100%. In other words, not all cells adhere to the nonaqueous phase. Furthermore, different tests for hydrophobicity (adhesion to hydrocarbons, salting out, or contact angle between cells and water) do not always yield similar results for the same organisms (10, 16), suggesting a different degree of affinity toward different hydrophobic substrata. It can therefore be surmised that bacteria which are usually considered attached are, in fact, divided into more adhering and less adhering subpopulations. Variation in attachment to a certain substratum within the same cell population may be due to modulation of surface hydrophobicity. Such modulation is known to take place in many different microbial species and is facilitated by various genetic, environmental, and physiological mechanisms. The presence of R plasmids in *Escherichia coli* has been shown to influence its surface hydrophobicity (10). We have recently found that resistance of *Phormidium* strain J-1 mutants to cyanophages is accompanied by a significant decrease in hydrophobicity. Various bacteria were reported to exhibit different degrees of hydrophobicity depending on their growth conditions (15). The response to changes in the C-to-N ratio in the growth medium was species specific. *Leptospira biflexa* cells responded to starvation conditions by an increase in their attachment to solid surfaces, and this has been proposed to constitute a survival strategy, since fatty acids, which they require for growth and maintenance, concentrate at such surfaces (11).

A change in surface hydrophobicity is part of the developmental cycle of *Myxococcus xanthus*. A shift to relative hydrophilicity occurs prior to sporulation of the cells, and in sporulation-defective mutants there is also no shift to hydrophilicity (12). Interestingly, *Bacillus cereus* spores are more hydrophobic than the vegetative cells (6).

Reports on modulation of electrostatic interactions are very scarce. Polymeric bridging between attached microorganisms and their substratum surfaces is common (4, 14). Such polymers, which often accumulate and form a visible slimy layer, could be destroyed by various chemical and enzymatic treatments (5). However, it is not known whether such processes occur naturally. Various soil bacteria adsorb clay particles through surface charges (13), and certain strains adsorb less clay per unit cell area owing to variations in surface charged groups.

Separate mechanisms for attachment to hydrophilic and hydrophobic surfaces have been found for several microorganisms (1, 17). Modulation of these mechanisms in *Phormidium* strain J-1 is well coordinated, enabling the release of the cells from both hydrophobic benthic surfaces and sedimentary clay particles to the water column.

LITERATURE CITED

1. **Baker, J. H.** 1984. Factors affecting the bacterial colonization of various surfaces in a river. *Can. J. Microbiol.* **30**:511–515.

2. **Bar-Or, Y., M. Kessel, and M. Shilo.** 1985. Modulation of cell surface hydrophobicity in the benthic cyanobacterium *Phormidium* J-1. *Arch. Microbiol.* **142**:21–27.

3. **Bar-Or, Y., and M. Shilo.** 1987. Characterization of macromolecular flocculants produced by *Phormidium* J-1 and by *Anabaenopsis circularis* PCC 6720. *Appl. Environ. Microbiol.* **53**:2226–2230.

3a. **Bar-Or, Y., and M. Shilo.** 1988. The role of

cell-bound flocculants in coflocculation of benthic cyanobacteria with clay particles. *FEMS Microbiol. Ecol.* **53**:169–174.

4. **Corpe, W. A.** 1970. An acid polysaccharide produced by a primary film forming bacterium. *Dev. Ind. Microbiol.* 11:402–412.
5. **Corpe, W. A.** 1974. Detachment of marine periphytic bacteria from surfaces of glass slides. *Dev. Ind. Microbiol.* **15**:281–287.
6. **Doyle, R. J., F. Nedjat-Haiem, and J. S. Singh.** 1984. Hydrophobic characteristics of *Bacillus* spores. *Curr. Microbiol.* **10**:329–332.
7. **Fattom, A., and M. Shilo.** 1984. Hydrophobicity as an adhesion mechanism of benthic cyanobacteria. *Appl. Environ. Microbiol.* **47**:135–143.
8. **Fattom, A., and M. Shilo.** 1984. *Phormidium* J-1 bioflocculant: production and activity. *Arch. Microbiol.* **139**:421–426.
9. **Fattom, A., and M. Shilo.** 1985. Production of emulcyan by *Phormidium* J-1: its function and activity. *FEMS Microbiol. Ecol.* **31**:3–9.
10. **Ferreiros, C. M., and M. T. Criado.** 1984. Expression of surface hydrophobicity encoded by R-plasmids in *Escherichia coli* laboratory strains. *Arch. Microbiol.* **138**:191–194.
11. **Kefford, B., B. A. Humphrey, and K. C. Marshall.** 1986. Adhesion: a possible survival strategy for leptospires under starvation conditions. *Curr. Microbiol.* **13**:247–250.
12. **Kupfer, D., and D. R. Zusman.** 1984. Changes in cell surface hydrophobicity of *Myxococcus xanthus* are correlated with sporulation-related events in the developmental program. *J. Bacteriol.* **159**:776–779.
13. **Marshall, K. C.** 1968. Interaction between colloidal montmorillonite and cells of *Rhizobium* species with different ionogenic surfaces. *Biochim. Biophys. Acta* **156**:179–186.
14. **McCourtie, J., and L. J. Douglas.** 1985. Extracellular polymer of *Candida albicans*: isolation, analysis and role in adhesion. *J. Gen. Microbiol.* **131**:495–503.
15. **McEldowney, S., and M. Fletcher.** 1986. Effect of growth conditions and surface characteristics of aquatic bacteria on their attachment to solid surfaces. *J. Gen. Microbiol.* **132**:513–523.
16. **Onaolopo, J. A., and R. M. M. Klemperer.** 1986. Effect of R-plasmid on surface hydrophobicity of *Proteus mirabilis*. *J. Gen. Microbiol.* **132**:3303–3307.
17. **Paul, J. H., and W. M. Jeffrey.** 1985. Evidence for separate adhesion mechanisms for hydrophilic and hydrophobic surfaces in *Vibrio proteolytica*. *Appl. Environ. Microbiol.* **50**:431–437.
18. **Rosenberg, M., D. Gutnick, and E. Rosenberg.** 1980. Adherence of bacteria to hydrocarbons: a simple method for measuring cell-surface hydrophobicity. *FEMS Microbiol. Lett.* **9**:29–33.

Chapter 20

Adhesion and Desorption during the Growth of *Acinetobacter calcoaceticus* on Hydrocarbons

Eugene Rosenberg, Mel Rosenberg, Yuval Shoham, Nachum Kaplan, and Nechemia Sar

INTRODUCTION

The importance of cell adhesion in the growth physiology of many bacterial species is well documented (see, e.g., references 5, 17, 36, and 40). In cases in which the carbon or energy source is a water-insoluble material such as chitin or cellulose, cell adhesion to the substrate can facilitate growth (1, 18, 38), but cell contact is not an absolute requirement, because extracellular enzymes can degrade the polymers into water-soluble substrates. However, the growth of microorganisms on hydrocarbons presents a special problem, since not only are the hydrocarbons immiscible with water, but also their breakdown cannot occur extracellularly. The final step in aromatic (8) or aliphatic hydrocarbon degradation is the introduction of molecular oxygen into the molecules by membrane-associated enzymes (6).

It has been shown that to facilitate the hydrocarbon-cell interactions, the petroleum-degrading organism *Acinetobacter calcoaceticus* RAG-1 (i) adheres avidly to hydrocarbons and other surfaces possessing hydrophobic properties (27) and (ii) produces a potent extracellular emulsifying agent (25). The purified emulsifying agent, referred to as emulsan, is an anionic heteropolysaccharide (42) containing fatty acid side chains (3), with an average molecular weight of 9.9×10^5 and an axial ratio of 60. The remarkable surface properties of emulsan depend on both its amphipathic nature (41) and its high molecular weight (37).

Production of emulsifying agents is widespread in *A. calcoaceticus* strains (32). In strain BD4, emulsification requires the presence of both a rhamnose-containing polysaccharide and a protein (15). The chemical structure of the polysaccharide has been elucidated (14), and the protein has been partially purified (15). The diversity of mi-

Eugene Rosenberg, Yuval Shoham, Nachum Kaplan, and Nechemia Sar • Department of Microbiology, George S. Wise Faculty of Life Sciences, Tel Aviv University, Ramat Aviv 69978, Israel. ***Mel Rosenberg*** • Department of Human Microbiology, Sackler Faculty of Medicine, Tel Aviv University, Ramat Aviv 69978, Israel.

crobial surfactants has recently been reviewed (21).

By using information obtained from studies with strains RAG-1 and BD4, an attempt will be made in this chapter to derive a dynamic model for the growth of *A. calcoaceticus* on hydrocarbons. The requirement of adhesion for growth on hydrocarbons will be extended to other heterotrophic bacteria in their natural environments.

BACTERIAL STRAINS AND GROWTH CONDITIONS

A. calcoaceticus RAG-1 (ATCC 31012) was originally isolated from the Mediterranean coast after enrichment on crude oil (20). The nonadherent mutant MR-481 was derived from RAG-1 (29). The partial adhesion revertant strain RV23-4 was derived from MR-481 by selection for growth on hexadecane (29). Strain TR3 is an emulsan-deficient mutant of RAG-1 (19). *A. calcoaceticus* BD4, originally isolated by Taylor and Juni (39), was provided by K. Bryn. Strain BD413, a minicapsulated (12) tryptophan auxotroph (11, 34) derived from BD4, was a gift from E. Juni. Strain BD4-R7 is a capsule-deficient, bacteriophage-resistant mutant of BD4 (15).

Growth experiments were performed at 30°C either in 250-ml flasks containing 50 ml of medium or in 14-mm-diameter Klett tubes containing 4.2 ml of medium. The flasks were incubated in a model G-53 Gyrotory shaker (New Brunswick Scientific Co., Inc., Edison, N.J.) at 150 rpm, and the tubes were incubated in a New Brunswick model G-76 Gyrotory water bath shaker at 330 rpm. Turbidity was monitored with a Klett-Summerson colorimeter fitted with a green filter.

Brain heart infusion medium was a product of Difco Laboratories, Detroit, Mich. Ac medium consisted of 0.2% sodium acetate in PUM buffer (22.2 g of $K_2HPO_4 \cdot 3H_2O$, 7.26 g of KH_2PO_4, 1.8 g of urea, 0.2 g of $MgSO_4 \cdot 7H_2O$, and distilled water to 1,000 ml; final pH, 7.1). Hx medium consisted of PUM buffer supplemented with hexadecane (olefin-free, 99% purity; Fluka AG, Buchs SG, Switzerland). GM medium contained (per liter of deionized water) 5.0 g of glucose, 9.17 g of $K_2HPO_4 \cdot 3H_2O$, 3.0 g of KH_2PO_4, 4.0 g of $(NH_4)_2SO_4$, 0.2 g of $MgSO_4 \cdot 7H_2O$, and 100-μg/ml tryptophan (final pH, 7.0). EM medium contained (per liter of deionized water) 22.2 g of $K_2HPO_4 \cdot 3H_2O$, 7.26 g of KH_2PO_4, 4.0 g of $(NH_4)_2SO_4$, 0.2 g of $MgSO_4 \cdot 7H_2O$, and 5.0 ml of ethanol. Solid media were prepared by the addition of 2% agar (Difco) to liquid media.

MEASUREMENT OF BACTERIAL HYDROPHOBICITY

Techniques

BATH Method

The bacterial adhesion to hydrocarbon (BATH) technique for measuring hydrophobicity was described previously (27, 29). Cells were harvested by centrifugation, washed twice, and suspended in 1.2 ml of PUM buffer to an initial A_{400} of ca. 1.5 (model 240 spectrophotometer; Gilford Instrument Laboratories, Inc., Oberlin, Ohio; light path, 1 cm). Various volumes of *n*-hexadecane were added, and after 10 min of incubation at 30°C the contents of each tube were mixed uniformly in a Vortex mixer (The Vortex Manufacturing Co., Cleveland, Ohio) for 120 s. After allowing 15 min for the phases to separate, we measured the A_{400} of the lower aqueous phase and compared it with that of the bacterial suspension before the mixing procedure. The fraction of adherent cells was taken as the percent decrease in A_{400} of the aqueous phase after mixing and phase separation, as compared with that of the original suspension.

DOS Method

In the direction-of-spreading (DOS) method (31, 33), a small drop of water (5 μl) was placed at the interface between the bacterial lawn and each of the following three surfaces: (i) the indicated growth medium solidified with 2% agar, (ii) a microscope glass cover slip (20 by 20 mm; Chance Propper Ltd., Warley, England [the cover slips were washed successively with ethanol and distilled water and then dried at room temperature]), and (iii) sterile polystyrene cover slips (25-mm diameter, no. 142; Lux Scientific Corp., Newburg Park, Calif. [not treated for tissue culture]). The direction to which the water drop spread was noted. Figure 1 exemplifies the method. The scoring system is summarized in Fig. 2.

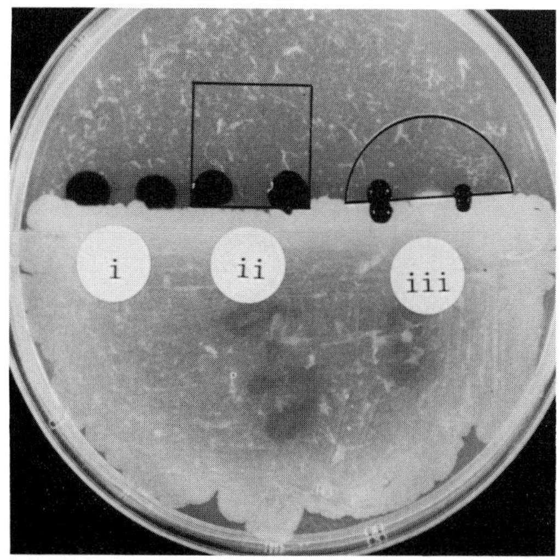

Figure 1. Measurement of the hydrophobicity of the colony surface of *A. calcoaceticus* BD4-R7 by the DOS method. After the bacterial lawn was allowed to develop for 48 h, 5-μl water drops were introduced at the border between the bacterial growth (the lower part) and one of the following surfaces: (i) 2% agar, (ii) a glass cover slip, and (iii) a polystyrene cover slip. The relative colony hydrophobicity was determined as described in the text.

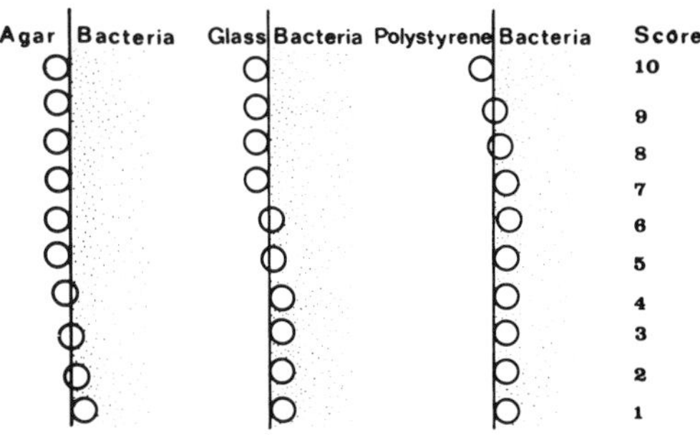

Figure 2. Scoring system for measuring the surface hydrophobicity of microbial lawns by the DOS method. A small drop of water was placed at the interface between a bacterial lawn and agar, a bacterial lawn and a glass cover slip, and a bacterial lawn and a polystyrene cover slip. For an example, see Fig. 1. The direction of spread of the drop determines the score, which is measured from 1 (the least hydrophobic bacteria) to 10 (the most hydrophobic bacteria).

Determination of BD4 and RAG-1 Capsular Polysaccharides

Culture samples were centrifuged in the cold at 12,000 × *g* for 20 min. The cell pellets were then suspended in distilled water for determination of the BD4 capsular polysaccharides by the sulfuric acid-cysteine procedure (4) with the pure capsule polysaccharide of BD4 as a standard (14), and for determination of the RAG-1 capsular polysaccharides by determining polymeric galactosamine with emulsan as the standard, as previously described (24).

MATERIALS

Emulsan, the polyanionic emulsifying agent of RAG-1, was purified from the cell-free supernatant fluid obtained from a culture grown on EM medium as described previously (25). The purified product had an emulsifying activity of 210 U/mg, an O-ester content of 0.6 μmol/mg, a reduced viscosity of 600 cm^3/g, and a residual protein content of less than 2%. The BD4 capsular polysaccharide was obtained from BD4 cells grown on GM medium. The polysaccharide was sheared off the washed cells and purified by acetone precipitation and differential centrifugation as described previously (14). The final product contained rhamnose, glucose, mannose, and glucuronic acid in the molar ratios of 4:1:1:1.

Emulsan depolymerase was obtained from YUV-1 cells as described previously (37). The BD4 capsule-degrading enzyme was obtained from lysates of the BD4 bacteriophage ϕSL-1 (15). The crude enzyme preparation was obtained by centrifugation of the lysate at 6,000 × *g* for 20 min at 4°C, followed by precipitation of the enzyme from the supernatant fluid with ammonium sulfate (50% saturation). After standing overnight in the cold, the mixture was centrifuged at 15,000 × *g* for 20 min. The pellet was dissolved in water and dialyzed against deionized water. The resulting salt-free solutions served as the enzyme preparation.

MECHANISMS OF ADHESION

Adhesion of *A. calcoaceticus* RAG-1 to Hexadecane

Microscopic observations of samples taken during the growth of RAG-1 on hydrocarbons indicated that during the exponential phase of growth, most of the RAG-1 cells were adhering to the hydrocarbon/water interface. To study this adhesion phenomenon we developed a simple adhesion test, now referred to as the BATH test. The test is not specific for hydrocarbon-degrading bacteria, but, rather, is a measure of cell surface hydrophobicity.

RAG-1 cells showed 95 to 100% adhesion to hexadecane (Table 1). By sampling the aqueous phase and repeating the adhesion assay, it was possible to obtain a mutant of RAG-1, referred to as MR-481, which failed to adhere to 0.05 to 0.2 ml of hexadecane.

Table 1.
Adhesion of *A. calcoaceticus* Strains to Hexadecane

Strain[a] and hexadecane concn (ml)	Adhesion[b] (%)
RAG-1	
0.05	100
0.1	98
0.2	98
MR-481	
0.05	0
0.1	0
0.2	0
RV23-4	
0.05	30
0.1	40
0.2	55

[a] Cells were grown to stationary phase in brain heart infusion broth.
[b] Adhesion was measured as a function of hydrocarbon volume.

Role of Adhesion in Growth of Strain RAG-1 on Hexadecane

Strains RAG-1 and MR-481 both grew well on acetate as a carbon source, reaching approximately 200 Klett units in 15 h (Table 2). However, the nonadhering strain failed to grow on hexadecane medium for at least 30 h, whereas RAG-1 grew well after a short lag, reaching 100 Klett units in 15 h. The latter experiment was conducted with low initial cell densities (ca. 5 Klett units) and low mechanical agitation to maximize the need for adhesion in the cell-hydrocarbon interaction. When the experiment was carried out in the presence of emulsan (to increase the surface area of the hexadecane) strain MR-481 grew well. Therefore, MR-481 had the potential to grow at the expense of hexadecane once it came into close contact with the substrate.

After a few days of incubation of MR-481 in hexadecane medium, cells appeared which were capable of growth on hexadecane under these restrictive conditions. One of these mutants, referred to as RV23-4, was cloned and studied. Strain RV23-4 was a partial revertant with regard to both adhe-

Table 2.
Growth of *A. calcoaceticus* Strains in Acetate- and Hexadecane-Containing Media[a]

Strain and incubation time (h)	Turbidity (Klett units) in:	
	Ac medium	Hx medium
RAG-1		
5	55	6
10		60
15	205	110
MR-481		
5	53	5
10		4
15	200	3
RV23-4		
5	53	5
10		10
15	205	40

[a] Washed cells were inoculated into acid-washed Klett tubes (diameter, 14 mm) containing 4.2 ml of Ac medium (PUM buffer supplemented with 0.2% sodium acetate) or Hx medium (4.2 PUM buffer plus 0.5 ml of hexadecane) to give an initial turbidity of 5 Klett units.

sion to hexadecane (Table 1) and growth on hexadecane (Table 2). Studies with a number of different adhesion mutants of RAG-1 have demonstrated that adhesion is a crucial factor in the growth of RAG-1 on water-insoluble hydrocarbons in the absence of emulsification of the substrate.

Mechanism of RAG-1 Adhesion to Hydrocarbons

Henrichsen and Blom (10) described two types of fimbriae, ca. 3 and 5 nm in diameter, present on various *A. calcoaceticus* strains. The thick fimbiae were correlated with twitching motility. No physiological role was proposed for the thin fimbriae. Thick fimbriae (6.5 nm) are present on RAG-1, MR-481, and RV23-4. Thin fimbriae (3.5 nm) are the major organelles responsible for adhesion of the cells to hydrophobic surfaces (26). Washed cells containing thin fimbriae adhere avidly to hydrocarbons, including nonutilizable aromatic hydrocarbons. Nonadhering mutants, such as MR-481, lack the thin fimbriae, fail to adhere to hydrocarbons, and grow poorly on water-insoluble hydrocarbon substrates. Revertants of these mutants, such as RV23-4, regain the ability to adhere to hydrocarbons and again show the presence of the characteristic thin fimbriae (26). Additional evidence for the role of thin fimbriae in adhesion was obtained by shearing the cells, with concomitant loss of the fimbriae and cell surface hydrophobicity.

Role of Emulsans

Although emulsans were initially isolated and characterized as extracellular emulsifying agents (13, 20, 25), it is now clear that both BD4 and RAG-1 emulsans have their origin in capsular polysaccharides that are released from the cell surface during the stationary phase or under conditions of unbalanced growth. With radioactive ethanol as a carbon source, pulse-chase experiments established that the RAG-1 emulsan polysaccharide, released during chloramphenicol treatment, was present on the cell surface before addition of the chloramphenicol (30). Further evidence for a cell-associated form of emulsan came from chemical analyses of washed cells (Table 3), electron microscopy (2), and immunochemical studies (9).

The demonstration that the emulsan

Table 3.
Release of Emulsan from the RAG-1 Cell Surface

Growth stage[a]	Emulsan (% of cell dry wt)	
	Capsular	Extracellular
Exponential phase	14	0.6
Plus chloramphenicol[b]	0.5	15
Stationary phase	2.0	22

[a] Cells were grown in EM medium (see the text).
[b] Exponential-phase cells were harvested and suspended in EM medium containing 25 μg of chloramphenicol per ml; the chloramphenicol-treated cells were incubated for 4 h at 30°C.

Table 4.
Inhibitory Role of Cell-Associated Capsules in Adhesion of *A. calcoaceticus* to Hexadecane

Strain[a]	Capsular polysaccharide (% of cell dry wt)	Adhesion (%)
BD4	27	0
BD413	11	70
BD4-R7	0	97
BD4[b]	3	98
RAG-1	17	61
RAG-1[b]	3	93
TR3	4	85

[a] Strains BD43, BD4-R7, and BD413 were grown in GM medium to stationary phase. Strains RAG-1 and TR3 were grown in EM medium and harvested during exponential phase.
[b] These strains were enzymatically decapsulated. Decapsulation was carried out with BD4 phage capsule depolymerase (for BD4) (15) and emulsan depolymerase (for RAG-1) (38).

polysaccharides were cell surface polymers suggested that they might be involved in adhesion to hydrocarbons. However, the evidence (Table 4) indicated that the emulsan capsules decreased the ability of BD4 and RAG-1 to adhere to hydrocarbons (24). Removal of the emulsan capsule, either by mutation (BD4-R7 and TR3) or by enzymatic degradation, significantly enhanced cell surface hydrophobicity. During the exponential growth phase of RAG-1 on hydrocarbons, the cells were observed to be tightly bound to the oil droplets. When growth slowed, emulsan was released into the medium and an increasing percentage of the cells were found free in the medium. This increasing number of unbound RAG-1 cells at a time when the cells were becoming more hydrophobic, owing to release of capsular emulsan (Table 4), can be understood if one realizes that the surface of the emulsified oil (hydrocarbon-water) has become hydrophilic. It is known that purified emulsan inhibits the adhesion of cells to hydrophobic surfaces (22). The inhibition of adhesion can be reversed by degrading the emulsan with emulsan depolymerase.

COMPARISON OF BATH AND DOS METHODS

The hydrophobicity of colonies or surface layers of bacteria is not necessarily correlated to the hydrophobicity of the dispersed cells making up these colonies (33). The capsule-deficient BD4-R7 was very hydrophobic by both methods, whereas RAG-1 and TR3 adhered avidly to hexadecane, but showed only intermediate colony surface hydrophobicity (Table 5). Although strain BD4 failed to adhere to hexadecane, it yielded as high a surface hydrophobicity as strains RAG-1 and TR3 did (Table 5).

DYNAMICS OF GROWTH OF *A. CALCOACETICUS* ON PETROLEUM

Studies on the growth, surface properties, and production of emulsans by *A. calcoaceticus* RAG-1 and BD4 have led to a model of growth on water-insoluble hydrocarbons that comprises the following steps.

(i) Individual cells adhere to the surfaces of large oil drops. Adhesion of RAG-1

Table 5.
Comparison between Colony Hydrophobicity (DOS method) and Cell Surface Hydrophobicity (BATH method)

Strain	Hydrophobicity	
	BATH method (% adhesion)[a]	DOS method[b]
RAG-1	100	5
MR-481	25	4
TR3	97	6
BD4	3	6
BD4-R7	97	9

[a] Strains RAG-1, MR-481, and TR3 were grown on brain heart infusion agar; strains BD4 and BD4-R7 were grown on GM agar. Cells were removed from the agar medium and suspended directly in PUM buffer. Each cell suspension was washed twice by centrifugation and resuspension before adhesion to 0.2 ml of hexadecane was measured as described in the text.
[b] The hydrophobicity of each colony surface was determined by the DOS method as described in the text and illustrated in Fig. 1. The scoring system for the DOS method is shown in Fig. 2.

is mediated by the presence of hydrophobic fimbriae which are distributed over the entire surface of the bacterium.

(ii) The bacteria grow as a biofilm at the hydrocarbon/water interface, extracting water-soluble minerals, nitrogen, and phosphate from the aqueous phase and oxidizing *n*-alkanes to carboxylic acids, which provide the carbon and energy source. Growth is restricted to the hydrocarbon/water interface.

(iii) During the growth phase, the cells accumulate emulsan on their surfaces in the form of a capsule. As the encapsulated cells cover the oil droplets, they reduce the interfacial tension between the hydrocarbon and the water. This results in the emulsification of the oil and an increase in the oil-water surface area available for growth. During the growth phase, the individual oil droplets that have been colonized by the bacteria become depleted of *n*-alkanes. These droplets become enriched in nonutilizable aromatic and cyclic hydrocarbons.

(iv) When the *n*-alkane content of the colonized droplets becomes depleted, the cells become starved. The starved cells release their emulsan capsules. The extracellular emulsan attaches avidly to the oil droplet, thereby displacing the cells to the aqueous phase. Each "used" droplet is then covered with a stable monomolecular film of emulsan. The fatty acid esters and amides of emulsan are oriented into the oil phase, whereas the hydrophilic carboxylic and hydroxyl groups of the polysaccharide face the aqueous phase. The hydrophilic outer surface of the emulsan-coated oil prevents reattachment of the free bacteria. The released capsule-deficient bacteria are extremely hydrophobic and readily adhere to fresh hydrocarbon substrate.

CELL SURFACE AND COLONY HYDROPHOBICITY

The BATH test measures the adhesion of individual suspended cells to hydrocarbons. For a cell to partition to the hydrocarbon/water interface, it must contain a hydrophobic region on at least part of its outer surface. Other measures of cell surface hydrophobicity include hydrophobic interaction chromatography and binding of radiolabeled hydrophobic compounds (28).

The DOS test measures the hydrophobicity of the upper layer of cell lawns or colonies. The direction of spreading of a water drop between two surfaces is, in fact, a satisfactory definition of hydrophobicity. In principle, the forces which govern the spreading of a water drop between two surfaces are the same as those measured by the contact-angle method (7).

Comparison between cell surface and colony hydrophobicity of a number of different bacteria resulted in a wide variety of relationships. In some cases both the individual cells and the colony surface showed similar hydrophobicity values, whereas in other cases there was a large difference between cell and colony hydrophobicity. The differences between cell and colony hydrophobicity values can be explained in at least two ways. (i) The colony can be differentiated so that the cells at the surface of the colony have an intrinsically different cell surface hydrophobicity from that of the interior cells. (ii) The cells produce extracellular slime, which concentrates at the surface of the colony and thereby imparts to it a different hydrophobicity from that of the individual cells. It may be assumed that part or all of this extracellular slime can be removed from the cells during suspension and washing procedures. In this regard, it is interesting that Shapiro (35) has observed, by using scanning electron microscopy, a thin layer of extracellular surface material covering the colony. Also, during the development of *Myxococcus xanthus*, a loosely bound extracellular material is produced which can change the cell surface hydrophobicity (16).

GENERAL REQUIREMENT FOR ADHESION AND COLONY FORMATION IN THE GROWTH OF HETEROTROPHIC BACTERIA IN NATURE

Shilo has outlined the general requirements for adhesion and growth of benthic cyanobacteria (chapter 18 of this volume). The data presented here demonstrate that adhesion and colony formation are also requirements for the growth of *A. calcoaceticus* RAG-1 on hydrocarbon substrates. However, it can be argued that hydrocarbons, although released into water and onto land by natural oil seeps, are unusual substrates from which to generalize on the growth patterns of bacteria. What, then, are the typical growth substrates for heterotrophic bacteria in natural water and soils?

The most abundant sources of organic material for microorganisms are probably particulate matter, complex polysaccharides, and structural proteins. For bacteria to utilize these carbon sources, the cells must produce a battery of extracellular enzymes. Consider, for example, an individual bacterium dispersed in a natural body of water containing a low concentration of dispersed particulate matter. The cell secretes its extracellular enzymes, which diffuse into the water. The overall concentration of enzymes produced from a single cell is very low and would not produce a high enough concentration of permeable substrate to allow growth. If, on the other hand, the cell was adhering to the substrate, then the local concentration of enzyme and resulting permeable substrate would be high enough to initiate growth. Furthermore, the resulting colony of biofilm would produce a high local concentration of enzyme that would ensure efficient continued growth. This type of cooperative growth kinetics has been demonstrated in the model case of *M. xanthus* growing on casein (23).

The above arguments lead to the conclusion that bacteria in nature typically adhere to surfaces and form colonies, mats, and biofilms.

LITERATURE CITED

1. **Bayer, E. A., R. Kenig, and R. Lamed.** 1983. Adherence of *Clostridium thermocellum* to cellulose. *J. Bacteriol.* **156**:818–827.
2. **Bayer, E. A., E. Rosenberg, and D. Gutnick.** 1981. The isolation of cell surface mutants of *Acinetobacter calcoaceticus* RAG-1. *J. Gen. Microbiol.* **127**:295–300.
3. **Belsky, I., D. L. Gutnick, and E. Rosenberg.** 1979. Emulsifier of *Arthrobacter* RAG-1: determination of emulsifier-bound fatty acids. *FEBS Lett.* **101**:175–178.
4. **Dische, Z., and L. B. Shettles.** 1948. A specific color reaction of methylpentoses and a spectrophotometric micromethod for their determination. *J. Biol. Chem.* **175**:595–603.
5. **Dworkin, M.** 1973. Cell-to-cell interactions in the myxobacteria. *Symp. Soc. Gen. Microbiol.* **23**:125–142.
6. **Foster, J. W.** 1962. Bacterial oxidation of hydrocarbons, p. 241–261. *In* O. Hayaishi (ed.), *Oxygenases*. Academic Press, Inc., New York.
7. **Gerson, D. F., and D. Sheer.** 1980. Cell surface energy, contact angles and phase partition. III. Adhesion of bacterial cells to hydrophobic surfaces. *Biochim. Biophys. Acta* **602**:506–510.
8. **Gibson, D. T.** 1971. The microbial oxidation of aromatic hydrocarbons. *Crit. Rev. Microbiol.* **1**:199–223.
9. **Goldman, S., Y. Shabtai, C. Rubinowitz, E. Rosenberg, and D. L. Gutnick.** 1982. Emulsan in *Acinetobacter calcoaceticus* RAG-1: distribution of cell-free and cell-associated cross-reacting material. *Appl. Environ. Microbiol.* **44**:165–170.
10. **Henrichsen, J., and J. Blom.** 1975. Correlation between twitching motility and possession of polar fimbriae in *Acinetobacter calcoaceticus*. *Acta Pathol. Microbiol. Scand. Sect. B* **83**:103–115.
11. **Juni, E.** 1972. Interspecies transformation of *Acinetobacter*: genetic evidence for ubiquitous genus. *J. Bacteriol.* **112**:917–931.
12. **Juni, E., and A. Janick.** 1969. Transformation of *Acinetobacter calcoaceticus* (*Bacterium anitratum*). *J. Bacteriol.* **98**:281–288.
13. **Kaplan, N., and E. Rosenberg.** 1982. Exopolysaccharide distribution of and bioemulsifier production by *Acinetobacter calcoaceticus* BD4 and BD413. *Appl. Environ. Microbiol.* **44**:1335–1341.
14. **Kaplan, N., E. Rosenberg, B. Jann, and K. Jann.** 1985. Structural studies of the capsular polysaccharide of *Acinetobacter calcoaceticus* BD4. *Eur. J. Biochem.* **152**:453–458.

15. **Kaplan, N., Z. Zosim, and E. Rosenberg.** 1987. Reconstitution of emulsifying activity of *Acinetobacter calcoaceticus* BD4 emulsan by using pure polysaccharide and protein. *Appl. Environ. Microbiol.* **53**:440–446.
16. **Kupfer, D., and D. Zusman.** 1984. Changes in cell surface hydrophobicity of *Myxococcus xanthus* are correlated with sporulation-related events in developmental program. *J. Bacteriol.* **159**:776–779.
17. **Marshall, K. C.** 1976. *Interfaces in Microbial Ecology*. Harvard University Press, Cambridge, Mass.
18. **Murray, W. D., L. C. Sowden, and J. R. Colvin.** 1986. Localization of the cellulose activity of *Bacterioides cellulosolvens*. *Lett. Appl. Microbiol.* **3**:69–72.
19. **Pines, O., and D. L. Gutnick.** 1981. Relationship between phage resistance-emulsan production and interaction of phages with the cell surface to *Acinetobacter calcoaceticus* RAG-1. *Arch. Microbiol.* **130**: 129–133.
20. **Reisfeld, A., E. Rosenberg, and D. Gutnick.** 1972. Microbial degradation of crude oil: factors affecting the dispersion in sea water by mixed and pure cultures. *Appl. Environ. Microbiol.* **24**:363–368.
21. **Rosenberg, E.** 1986. Microbial surfactants. *Crit. Rev. Biotechnol.* **3**:109–132.
22. **Rosenberg, E., A. Gottlieb, and M. Rosenberg.** 1983. Inhibition of bacterial adherence to hydrocarbons and epithelial cells by emulsan. *Infect. Immun.* **39**:1025–1028.
23. **Rosenberg, E., K. H. Keller, and M. Dworkin.** 1977. Cell density-dependent growth of *Myxococcus xanthus* on casein. *J. Bacteriol.* **129**:770–777.
24. **Rosenberg, E., N. Kaplan, O. Pines, M. Rosenberg, and D. Gutnick.** 1983. Capsular polysaccharides interfere with adherence of *Acinetobacter calcoaceticus* to hydrocarbon. *FEMS Microbiol. Lett.* **17**:157–160.
25. **Rosenberg, E., A. Zuckerberg, C. Rubinowitz, and D. L. Gutnick.** 1979. Emulsifier of *Arthrobacter* RAG-1: isolation and emulsifying properties. *Appl. Environ. Microbiol.* **37**:402–408.
26. **Rosenberg, M., E. A. Bayer, J. Delarea, and E. Rosenberg.** 1982. Role of thin fimbriae in adherence and growth of *Acinetobacter calcoaceticus* RAG-1 on hexadecane. *Appl. Environ. Microbiol.* **44**:929–937.
27. **Rosenberg, M., D. Gutnick, and E. Rosenberg.** 1980. Adherence of bacteria to hydrocarbons: a simple method for measuring cell-surface hydrophobicity. *FEMS Microbiol. Lett.* **9**:29–33.
28. **Rosenberg, M., and S. Kjelleberg.** 1986. Hydrophobic interactions: role in bacterial adhesion. *Adv. Microb. Ecol.* **9**:353–393.
29. **Rosenberg, M., and E. Rosenberg.** 1981. Role of adherence in growth of *Acinetobacter calcoaceticus* RAG-1 on hexadecane. *J. Bacteriol.* **148**:51–57.
30. **Rubinowitz, C., D. L. Gutnick, and E. Rosenberg.** 1982. Emulsan production by *Acinetobacter calcoaceticus* in the presence of chloramphenicol. *J. Bacteriol.* **152**:126–132.
31. **Sar, N.** 1987. Direction of spreading (DOS): a simple method for measuring the hydrophobicity of bacterial lawns. *J. Microbiol. Methods* **6**:211–219.
32. **Sar, N., and E. Rosenberg.** 1983. Emulsifier production by *Acinetobacter calcoaceticus* strains. *Curr. Microbiol.* **9**:309–314.
33. **Sar, N., and E. Rosenberg.** 1987. Fish skin bacteria: colonial and cellular hydrophobicity. *Microb. Ecol.* **13**:193–202.
34. **Sawula, R. V., and I. P. Crawford.** 1972. Mapping of the tryptophan genes of *Acinetobacter calcoaceticus* by transformation. *J. Bacteriol.* **112**:797–805.
35. **Shapiro, J. A.**1985. Scanning electron microscope study of *Pseudomonas putida* colonies. *J. Bacteriol.* **164**:1171–1181.
36. **Shapiro, L.** 1976. Differentiation in the caulobacter cell cycle. *Annu. Rev. Microbiol.* **30**:377–407.
37. **Shoham, Y., and E. Rosenberg.** 1983. Enzymatic depolymerization of emulsan. *J. Bacteriol.* **156**:161–167.
38. **Sundarrj, C. P., and J. V. Bhat.** 1972. Breakdown of chitin by *Cytophaga johnsoni*. *Arch. Mikrobiol.* **85**:159–167.
39. **Taylor, W. H., and E. Juni.** 1961. Pathways for biosynthesis of a bacterial polysaccharide. I. Characterization of the organism and polysaccharide. *J. Bacteriol.* **81**:688–693.
40. **Varon, M., and M. Shilo.** 1968. Interaction of *Bdellovibrio bacteriovorus* and host bacteria. I. Kinetic studies of attachment and invasion of *Escherichia coli* B by *Bdellovibrio bacteriovorus*. *J. Bacteriol.* **95**:744–753.
41. **Zosim, Z., D. Gutnick, and E. Rosenberg.** 1981. Properties of hydrocarbon-in-water emulsions stabilized by *Acinetobacter calcoaceticus* emulsan. *Biotechnol. Bioeng.* **24**:281–292.
42. **Zuckerberg, A., A. Diver, Z. Peeri, D. L. Gutnick, and E. Rosenberg.** 1979. Emulsifier of *Arthrobacter* RAG-1: chemical and physical properties. *Appl. Environ. Microbiol.* **37**:414–420.

Chapter 21

Regulation of External Polymer Production in Benthic Microbial Communities

Christopher S. F. Low and David C. White

INTRODUCTION

Attachment to an inanimate or animate surface is a common mechanism of microorganisms to increase survival. Organisms may form specific consortia with component organisms of different physiological properties to increase the metabolic versatility of the complex. The importance of attachment to specific surfaces is a key feature in the distribution of microorganisms in nature, and the mechanisms and consequences of attachment have been the subject of several intensive studies (14, 23–26, 31, 32, 38, 39). The consequences of specific attachments of microbial parasites to tissues have been recognized from the earliest studies of infectious diseases, and the enormous literature based on these studies is far too complex and extensive to be reviewed here.

In this chapter, only the relatively nonspecific attachment of bacteria to substrata will be considered. It is clear from many studies, for example, the extensive work of Fletcher and co-workers, that the physiological status and metabolic activities of attached and nonattached microorganisms in the same monoculture can be very different (12, 13); hence, making generalizations about the regulation of external polymer production is fraught with risk.

The ubiquity of cell adhesion to a wide variety of surfaces (including other cells) reinforces the fundamental concept that surface contact or proximity must occur prior to attachment. This may represent a phase of relatively nonspecific attachment. This is a prelude to biologically directed adhesion, which may induce a specific response for specific substrata (30) or may evoke a general response. These two stages represent the primary physical attraction, which is reversible, and a secondary, biologically directed adhesion stabilization reaction, which is irreversible (23).

The irreversible step of adhesion often involves the elaboration of extracellular surface structures or polymers. Elucidation of the controls in this complex process is compounded by the diversity of the surface substrates to which the organisms attach, the enormous matrix of organisms that can po-

Christopher S. F. Low and David C. White • Institute for Applied Microbiology, University of Tennessee, 10515 Research Drive, Suite 300, Knoxville, Tennessee 37932-2567.

tentially adhere in monoculture as well as in adherent consortia (19), and the multiple mechanisms that organisms use in the attachment process. Cell surface components and structures that are known to influence adhesiveness include thin fimbriae (34), cell surface antigens (34), secreted lipopolysaccharides, and extracellular polysaccharide (glycocalyx) polymers and were magnificently illustrated in the work of Costerton et al. (3).

The multitude of responses of different bacteria to changes in the physicochemical parameters of the substrata that influence the specific structure or to changes in the chemistry of extracellular processes responsible for irreversible adhesion strongly indicate that few generalizations are possible and that specific mechanisms for a number of microbes will have to be defined before the fundamental patterns are elucidated. Many of the factors that could influence the specific microbial response in the formation of the irreversible adhesion component will not be considered further. An apparent prerequisite for adhesion in marine systems is the presence in the substratum of a carbohydrate-enriched layer scavenged from the bulk liquid phase. Inorganic as well as organic substrates for adhesion are rapidly coated with a layer of adsorbed macromolecules and lipids (11, 14). This is possibly part of the mysterious process of surface conditioning that can either increase or decrease the adhesion of microorganisms to a substratum. The nature of the conditioning film for substrata exposed to the marine environment has so far eluded chemical definition (1). Bacteria also adsorb antibodies and macromolecules from animal hosts, which may or may not promote cell attachment competency (34). This mechanism will not be considered further.

This review will concentrate on methods for studying two responses of specific microorganisms to metal substrata in the marine environment. The two phenomena are the formation of hydrophobic protein fimbriae and the elaboration of acetic extracellular polymer polysaccharides as the biological manifestation of irreversible attachment.

EXOPOLYMER SECRETION

For our purposes adhesion will be defined as the formation of a single interface between two material phases so that mechanical work can be passed through the interface without loss of intimate contact. An adhesive joint is formed from at least two adhesion interfaces with an intervening third phase. By these definitions, an adhesive joint will have two strengths of adhesion and an adhesion interface will have one (15).

Destructive physical and quantitative chemical methods of assessing adhesion involve the examination of the finished bioinorganic interface integrity. Washing with solvents, chemical degradation, and physical stress are the major processes which are coupled to spectroscopic, chromatographic, or gravimetric techniques for quantitative assessment of what remains bound to the test surface. A combination of analytical techniques (gas chromatography [GC], high-pressure liquid chromatography, and gas-chromatographic mass spectrometry [GC-MS]) can be used to elucidate the structure and components forming the biofilm and surface-binding matrix. However, owing to the complexity and diversity of organisms, these techniques do not always provide the definitive results expected, nor do they provide dynamic in situ monitoring of film formation kinetics. Nevertheless, GC-MS techniques (33) have been used to determine the structures of mono-, di-, and polysaccharides, biologically important sugars (deoxyribose, ribose, aminosaccharides, uronic acid[s], phosphorylated sugars, and glycosides). Complex biological saccharides are usually totally or partially hydrolyzed as well as methylated to identify, sequence, and determine the linkages in the polymer (33). Supercritical fluid chromatography (SFC) can be used to separate the neutral, glycolipid, and polar lipid fractions from various environmental sources. SFC provides the advantage of separating high-molecular-weight, high-boiling-point, and thermolabile compounds at moderate temperatures. With SFC, the high temperature used for volatilization of high-molecular-weight compounds is replaced by fluid density manipulations (18, 35).

Extracellular polymeric substances participate in the formation of microbial aggregates. They are readily detected as capsules and slimes in fungi, bacteria, and algae. Secreted capsular material adheres to the cell wall, whereas slime exists free in the suspending medium. The gel form of the extracellular polymeric substances is hydrated (ca. 98% water); it is dehydrated in the fixation process for electron microscopy and can appear as fibrous strands between the organism and the substrata. Antibody to the extracellular polymeric substances has been used by Costerton et al. (3) before fixation to show the diffuse nature of exopolymer adherent materials. Flocs are larger mats (20 to 200 μm in diameter) of mucus- or slime-coated organic particles (16). One means of demonstrating exopolysaccharide function in adhesion has been to use sodium periodate treatment to denature polysaccharide by oxidation and cleavage of vicinal hydroxyl groups (24). This clearly documents the polysaccharide nature of extracellular polymeric substances. Other bacterial components possibly involved in attachments to the bacterial cell walls include peptidoglycan (minor fraction in gram-negative bacteria and up to 80% in gram-positive bacteria) or teichoic and teichuronic acids (polymers of glycerol and ribitol phosphate with sugar substituents and ester-linked alanine) linked to muramic acid, lipopolysaccharides, and outer membrane proteins (39).

The irreversible step in adhesion of a number of microorganisms involves the elaboration of an extracellular polymer. These substances are primarily polysaccharide with reactive carboxyl, amino, or carbonyl groups on the carbohydrate residues. In eubacteria the uronic acid group seems to provide the acidic properties that characterize these attachment polymers. From examination of compendia of structural analysis of exopolymers (7), it was clear that uronic acids were the most unique yet universal component found in polymers on the outside of the cytoplasmic membrane in bacteria. A quantitative assay based on the predominance of D-glucuronic, D-galacturonic, D-mannuronic, and L-gulonic acids in these polymers was developed (5). Since the presence of the carboxyl group stabilizes the glycosidic linkage (20), low yields result when the polymers are subjected to acid hydrolysis prior to separation and analysis. Once the uronic acids are released from the polymer, they are subjected to lactonization that is not reproducible. The solution to the quantitative assay problems involved activating the carbonyl group by esterification and then reducing the carbonyl group to an alcohol group with sodium borodeuteride. The uronic acid residues were then reduced to primary alcohols while they were still in place in the polymers but containing the deuterium tracer. The deuterium-containing sugars were readily hydrolyzed and separated by capillary GC after anomeric carbon reduction and peracetylation. The deuterium was detected after electron-impact mass spectrometry, and the proportion of uronic acids was estimated. Quantitative recovery of gum arabic polymer D-glucosamine and D-galactosamine confirmed the success of the procedure. Mixtures of polygalactosuronic acid were derivatized to yield dideuterated galactose, which was recovered and mixed with authentic galactose in various proportions. The percentages of added dideuterated galactose correlated with the results of the GC-MS analysis. With this technique, uronic acid-containing polymers were detected in marine pseudomonads, Maldanid worm tubes, ptychodera fecal mounds, biofouled titanium and aluminum, and estuarine sediments (7, 8). It was also possible to show that cultures of the marine bacterium *Pseudomonas atlantica* formed increased amounts of extracellular polysaccharide glycocalyx with increased proportions of uronic acids as the age of the culture increased (38). The age of monoculture was estimated from the growth curve and the adenylate energy charge. Maximal glycocalyx formation corresponded to conditions of nutritional stress. The faster the rate of exopolymer synthesis, the higher the uronic acid concentration. These data clearly showed that the composition of the exopolymer glycocalyx could change during the growth cycle. During a period of 8 days the total extractable phospholipid content increased 3-fold, the total polysaccharide polymer content increased 24-fold, and the total uronic acid content increased 83-fold, representing an increase in proportions from 8 to 26%. The total content of arabinose and xylose in the polymer remained relatively constant. The galactose/galacturonic acid ratio decreased from 9.3 to 1.8, compared with an increase from 2.4 to 3.4 in the glucose/glucosuronic acid ratio (38). In an unidentified marine bacterium (22), changes were observed in poly-β-hydroxybutyrate (PHB). When these cells were starved for 24 h, total fatty acid levels and the ratio of monounsaturated fatty acids to saturated fatty acids decreased, and the level of short-chain fatty acids increased. Initially, starving cells contained PHB, but after 3 h no PHB was detected. Addition of a phosphorus buffer permitted an initial increase in PHB and then a prolonged delay prior to the disappearance of PHB.

The chemical detection of the glycocalyx by derivatization, hydrolysis with GC-MS, and detection of uronic acid proportions was a long and ardous procedure. It provided compositional data only, with little insight into the detailed structure of the glycocalyx. The role of exopolymer structure in attachment requires more detailed study. The preliminary experiments with *P. atlantica* suggested strongly that the cellular attachment strength to metals in seawater was increased as the proportion of uronic acids was increased. Various components of the glycocalyx polysaccharides have been suggested to be critical in the attachment of the cells to substrata. In the polymer of *P. atlantica*, uronic acid(s), acetylated hydroxyl groups, and pyruvate acetals all could contribute charged or hydrophobic groups. Attempts to change the adhesive strength of the isolated polymer or the cells by chemical means, such as methylating the polymer uronic acids, did result in a much less adherent polymer, but fragmentation in the polymer, deacetylation, and methylation of the uronic acids resulted under even the gentlest conditions for ester formation. Chemical modifications of polymers appeared to be just too blunt an instrument with which to dissect the structure-adhesion relationships.

GENETIC ANALYSIS

The sharp instrument with which to study the effects of specific changes in the extracellular polymers that result in changes in the specificity and strength of adhesion is at hand: the creation of mutants with modifications in the structure of the exopolymers. A dramatic and uncontestable change related to adhesion was detected in *Vibrio parahaemolyticus*. The irreversible attachment to substrata coincided with the elaboration of thin lateral flagella by the bacteria. These lateral flagella were clearly distinct from the sheathed polar flagellum used for motility. Genetic analysis by Belas et al. (2) has shown that genes for production of lateral flagella are switched on about 30 min after the organism has made contact with a surface. Clearly, a surface-sensing event has taken place (19). By transposon genetic manipulation [mini-Mu(LacZ Tet^r)], Belas et al. (2) were able to isolate peritrichous tetracycline-resistant *V. parahaemolyticus* with defective lateral flagella (nonswarmers). Since *V. parahaemolyticus* is Lac negative, nonswarmers were tested for β-galactosidase activity as a function of induction of lateral flagella. In at least 40% of the isolates, a surface-dependent β-galactosidase activity on agar, nitrocellulose, cellulose acetate, and wetted polyvinylidene fluoride filters was demonstrated. The time course of appearance for surface-dependent β-galactosidase was the same as that observed for production of lateral flagella (2). Engebrecht and co-workers were able to isolate the genes for bioluminesence from *Vibrio fischeri* (6) and have used them in the same way as the β-galactosidase to detect the activity of the specific operon that contains the genes. Thus, the activity of specific genes involved in adhesion can be monitored in real time as the adhesion process continues. The elegant work has provided the tools required to modify the exopolymer or extracellular organelle involved in irreversible adhesion once the specific mechanism(s) is known.

METHODS OF MEASUREMENT OF ADHESION

The creation of mutants with different adhesive propensities provides the biological material for the analysis of the role of extracellular polymers in adhesion. Clearly, chemical analysis of polymers provides insight into the factors affecting formation; however, new methods, particularly if they can be rapid and nondestructive, provide a means of interpreting the molecular consequences of the mutations involving adhesive properties.

Shear Force

One method of measuring adhesion is to apply a fixed or variable force to adhered cells. Adhesion is then expressed as a percentage of the cells which withstand the applied disruptive force (e.g., gravity flow, gentle washing, sonication, vortexing, and gravity and resultant buoyant density). One of the most accurate methods of measuring adhesive strength is to use the disk shearing device (40, 41), in which a disk is spun at a known rate and distance from another disk with adhered cells. Transmitted shear stress is then dependent upon rotation rate, separation distance, fluid viscosity, and radial position. These methods measure the adhesiveness of the fully formed biological adhesion joint. Dynamic assessment of adhesion, however, sets up a fluid shear against which the cell must adhere. The cells are carried along in a cylinder of nutrient liquid between two parallel test surfaces separated by a distance h. Opposing forces (attraction to the test surface and fluid shear) act upon the cells, and there is a mean residence time during which the cells are in contact with the test surfaces. The time required for biological adhesion to occur is then dependent upon the biological adhesion potential of the cells (15). By using the cell adhesion module, the dynamic shear force required for a cell to become attached to a surface can be determined as a function of the radius, r. In the cell adhesion module there are two types of flow rates: (i) the linear (overall) flow rate, which is set to ensure laminar flow (Reynolds number $[R_n] < 2{,}000$) (Fig. 1), and (ii) the decreasing differential flow of nutrient medium between the plates of the cell adhesion module. The differential flow results from the ever-increasing volume (r is continuously increasing) which must be filled (volume $= \pi \times r^2 \times h$) as the front of the flowing medium moves from the center of the test surface to beyond its outer edge, where the flow again becomes linear. The cells which form the inner ring of the biofilm (Fig. 2) are the most strongly adherent cells, since they are attached to the test surface at a point of higher differential flow (higher shear force) than are the cells at the periphery. By measuring the radius formed by the most strongly adherent cells, the maximum dynamic shear force can then be calculated (15). A variety of techniques exist for mark-

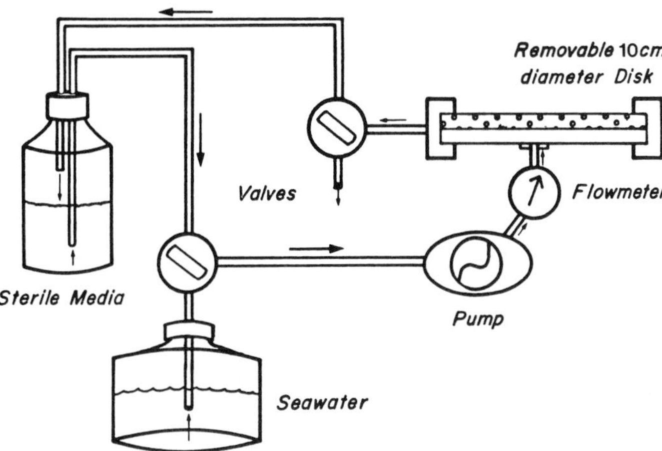

Figure 1. Cell adhesion module, used to monitor microbial attachment (adhesive strength) in a controlled shear gradient. Bacterial monoculture is pumped through the cell in a continuous-flow system as depicted or in a closed batch system (not shown). The test surface here is a glass disk (diameter, 10 cm) and is removable. Laminar flow rates are used (Reynolds number, <2,000), and the shear force, τ, is calculated from $\tau = 3Q\mu/\pi rh^2$, where Q is total volumetric flow rate (in cubic centimeters per second), μ is the dynamic viscosity of the medium (in centipoise), r is the radius (in centimeters), and h is the vertical distance between the plates (in centimeters).

ing the inner boundary of the biofilm with attached cells (measurement of diffuse reflectance by Fourier transforming-infrared spectroscopy [FT-IR] [28], biological staining and epifluorescence microscopy, tritium exchange and autoradiography, and destructive extraction followed by chromatography for signature compounds [36, 37, 43]).

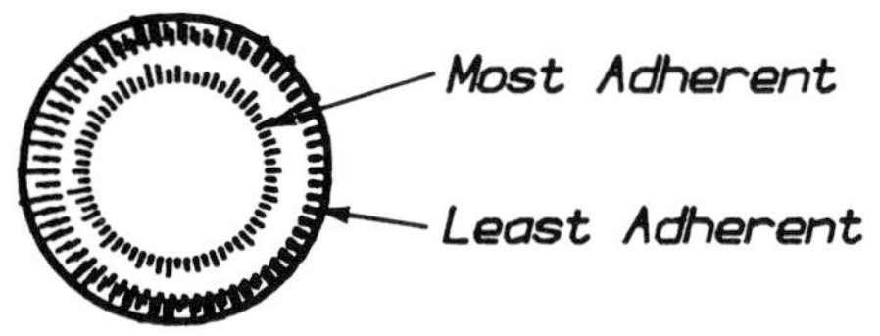

Figure 2. Cell adhesion module disk. The disk from the cell adhesion module (diameter, 10 cm), with attached microbial film and bacteria, is shown. The inner ring contains the most strongly adherent bacteria, since they have become attached in a region of high shear force compared with the least strongly adherent cells (attached under lower shear force). The cells become attached to the removable disk, forming a clear zone around the center inlet; the outer radius of the clear zone is measured and used to calculate the maximum dynamic shear force for the given bacterial strain. The mean least radius (maximum shear force) of attachment is monitored by a variety of methods (see the text).

Nondestructive Methods

AC Impedance

The basic concept of impedance measurement methods lies in the development of an electrical field structure at the electrode interface and a different electrical field structure in the bulk medium. The electrical interface (electrode/bulk medium) forms a compromise structure. To measure microbial growth, the impedance or the capacitance of the bulk medium is measured (32); however, the effects of adhesion require a surface to which the organisms can attach. Therefore, the methodology must allow for removal of the effects of bulk phase changes (e.g., bacterial growth, pH, charged species, and solvent dipoles) and analyze only the electrode interface changes. This region of the electrode/electrolyte interface is electrically neutral, but has a potential difference across the interface (the electrical double layer) (10). Therefore, the electrobacteriological experimental setup consists of at least two electrodes placed in the growth medium chambers which have been inoculated with the appropriate cultures. At each electrode there is an electrical double layer with its own direct current (DC) boundary potential. When an alternating current (AC) signal is impressed across these electrodes, the DC boundary potential becomes modulated by the AC signal and the resulting impedance can be calculated. When the appropriate experimental protocol (dual-chamber cells separated by a membrane filter and multiple electrodes) (21) is followed, the resultant impedance provides a calculated measure of the events at the test surface interface electrode. Practical applications of AC impedance methods (4) have shown that it is possible to measure microbially facilitated corrosion rates repeatedly in the same biofilm. This allows the detection of the consequences of succession in the biofilms. The measurements have been shown to correlate with the classical potentiometric DC measurements of corrosion rates (5). The DC methods unfortunately impress such large voltages on the surface that the measurements are destructive to the biofilm. AC impedance measurements offer the potential of nondestructive monitoring of biofilm chemistry, particularly when correlated with other nondestructive methods.

FT-IR

FT-IR provides spectral advantages sufficient to allow the examination of biofilms (27, 43). When FT-IR is used with the attenuated total reflectance (ATR) cell, living biofilms can be detected. With ATR, the incident light passes through the crystal a given number of times that are dependent upon the angle at which the crystal face was cut. The attenuated emitted light is then detected by the FT-IR spectrometer. The ATR system involves a crystal of germanium or zinc selenide shaped in such a way that the IR spectrometer sees an evanescent wave extending about 0.5 μm into the test system (Fig. 3). The IR spectra of whatever becomes attached to the crystal surface to the edge of the evanescent wave can be detected continuously and nondestructively. If a cell is created in which seawater flows over the cell (Fig. 3), the formation of biofilms can be detected. The formation of the conditioning film on surfaces exposed to seawater was readily monitored, and the steady accumulation of carbohydrate was monitored as an increase in the ether stretch absorbance (27). Geesey et al. (17) used FT-IR with ATR to directly demonstrate the corrosion of copper in seawater by the deposition of acidic bacterial polymers on the surface of the copper. They sputter-coated the ATR crystal with a thin film of copper, placed the crystal in a cell, and compared the effects of seawater with those of seawater plus attached bacterial exopolymer. They detected corrosion by measuring the accumulation of copper in the seawater and by directly watching the increase in water absorption at 1,640 cm^{-1} as the copper was removed from the film. Recent experiments in our laboratory by Nivens (unpublished) have shown that the attachment of *Caulobacter* species to the ATR crystal is correlated with the appearance of bacterial proteins and polysaccharides in the IR spectra. By using a crystal sputter-coated with a thin film of metal such as stainless steel, it will be possible to simultaneously monitor the attachment of bacteria to the coated crystal by FT-IR and AC impedance techniques. The effects of changing the electrical potential of the metals on adhesion chemistry of the bacteria can then be observed.

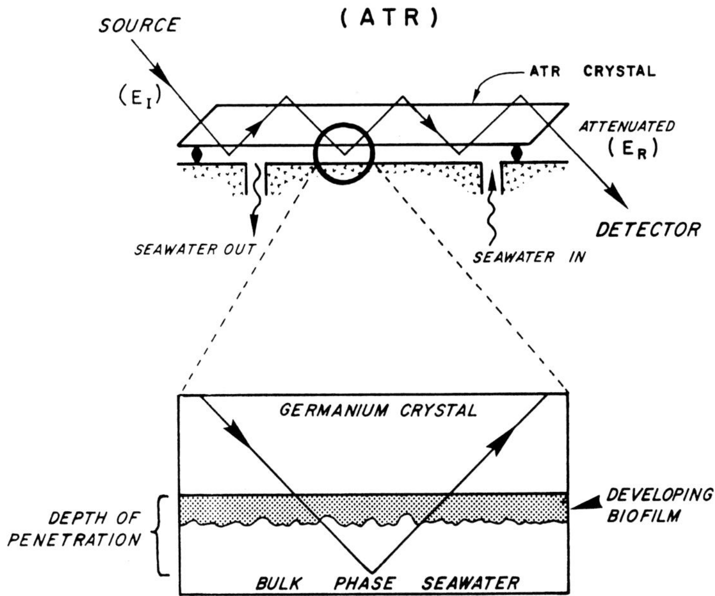

Figure 3. ATR schematic. The incident source of radiant energy impinges upon and is reflected through the ATR crystal. The angle of reflection and the number of times the light beam is reflected before it is detected are dependent upon the angle at which the crystal face was cut. Background spectra (without any attached biofilm) can be subtracted from the spectra taken as the biofilm (and as the bacteria) attaches, and wavelengths characteristic for the molecules of interest can be examined.

Surface-Enhanced Raman Spectroscopy

Surface-enhanced Raman spectroscopy is a nondestructive technique that complements FT-IR. In this analysis the molecules of material are activated by exposure to visible light from a laser. The Raman scattering of light from molecules results in a color shift in a portion of the scattered light by conversion of quanta to the vibrational energy of molecules. This is detected in a different region of visible light, with the difference representing the vibrational spectra of the activated molecules. Because the exciting and emitted light are both in the visible portion of the electromagnetic spectrum, the problems with solvents that are associated with the FT-IR are obviated. The absorption pattern of water, carbon dioxide, glass, etc., is not a problem in Raman spectroscopy, since any optically clear component is clear to the Raman spectrometer. In the past, the problem with Raman spectroscopy has been lack of sensitivity. Generally, Raman scattering is much less intense than Rayleigh scattering. Recently it has proved possible to enhance the intensity of Raman scattered light by a factor of 10^6 when the molecules of interest form a monolayer on a metallic surface. The enhancement of polarizability of molecules in the electron plasmon of silver has proved the most effective method of increasing the sensitivity of Raman spectroscopy to date (9). These developments mean that an apparatus like that illustrated in Fig. 4 could be developed. If the fiber-optic probe is used, it should be possible to detect and identify molecules attaching to the silver granules in specific parts of the biofilm.

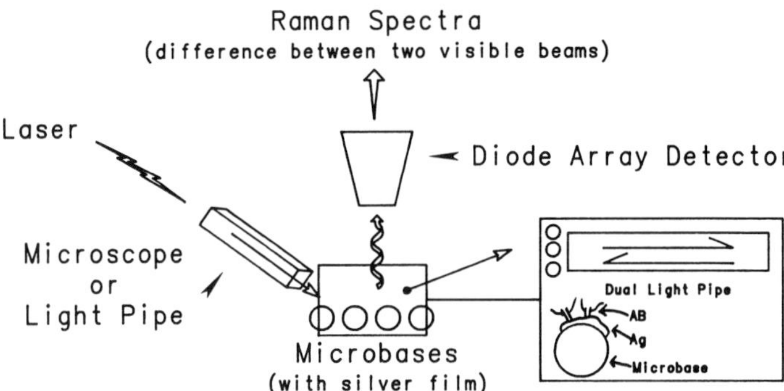

Figure 4. Surface-enhanced Raman spectroscopy schematic. The incident light from the laser can be focused through a microscope or fiber-optic tube on the microbases on which the silver surface is deposited. The plasmon of the silver on the microbases enhances the Raman scattered light from the molecules deposited on the microbases, and the scattered light is detected by a diode array detector. The surface-enhanced Raman spectroscopy system can be made specific (inset) by attaching antibodies or enzymes to the microbases and analyzing the changes in conformation when the molecule reacts with its antigen or substrate.

Ultrasensitive Destructive Measurement Techniques

Examination of biofilm development in monoculture or use of microbial consortia very often is limited by the sensitivity of the instruments used. To enhance the sensitivity and accuracy of these measurements (femto- and attomolar ranges), high-purity gases and solvents, efficient extractions, and ultrasensitive instruments are used. The methods developed for the extraction and concentration of cellular components, particularly of the membrane phospholipids (42), will be used prior to purification by high-pressure liquid chromatography and SFC. These membrane - lipopolysaccharide - exopolymer components are further separated by SFC or capillary GC for analysis by MS. If electron-withdrawing derivatives and soft chemical-ionization MS with detection of negative ions (29) are used, the sensitivity can be increased significantly. Direct analysis of biofilms or isolated cellular components has recently been shown by using fast-atom bombardment MS (see chapter 39 of this volume). This can increase the utility significantly and may be used to show whether specific microorganisms are located in specific parts of a biofilm. All of these analytical methods provide quantitative structural data for biofilm adhesive joints that are finished.

CONCLUSIONS

Applications of current techniques, when combined with mutants selected for their differences in adhesive properties, can yield considerable insight into the factors that control the elaboration of exopolymers. The questions of the mechanisms by which bacteria adhere to surfaces, the role(s) of extracellular polymer in attachment and adhesion, the strain or species divergent evolutionary mechanism(s) of attachment, and the signaling events which result in detectable physiological changes all deal with small numbers of cells (fewer than 10^4 cells per ml), and the limits of instrumental detection are easily reached. In some cases, observation and analysis of the attachment of one bacterium are required. The development of new nondestructive methodologies with severalfold increases in sensitivity will herald the next stage of understanding of this essential feature of the microbial mat.

Acknowledgments. This research was supported by grants N00014-83-K-0056 and N00014-87-K-0012 from the Department of the Navy, Office of Naval Research. The Nicolet 60SX FT-IR was purchased with grant N00014-83-G-0166 from the Department of Defense University Instrumentation Program through the Office of Naval Research. The VG-Trio-3 tandem MS was purchased with funds from the University of Tennessee and grants ARO 24187-LS-RI from the Department of Defense University Instrumentation Program through the Army Research Office, DEG-Lab Equipment 2-4-01018 from the Department of Education for laboratory equipment (to G. Sayler), and DE-F605-87ER75379 from the Department of Energy University Research Instrumentation Program. The Department of Energy grant will allow the construction of a surface-enhanced Raman spectrometer.

LITERATURE CITED

1. **Baier, R. E.** 1984. Initial events in microbial film formation, p. 57–62. *In* J. D. Costlow, and R. C. Tipper (ed.), *Marine Biodeterioration: an Interdisciplinary Study.* Naval Institute Press, Annapolis, Md.
2. **Belas, R., L. McCarter, M. Simon, and M. Silverman.** 1983. Genetic basis of bacterial surface recognition, p. 12–13. *In Proceedings of the Conference of Microbial Adhesion and Corrosion in the Marine Environment.* The Agouron Institute, La Jolla, Calif.
3. **Costerton, J. W., T. J. Marrie, and K.-J. Cheng.** 1985. Phenomena of bacterial adhesion, p. 3–43. *In* D. C. Savage and M. Fletcher (ed.), *Bacterial Adhesion.* Plenum Publishing Corp., New York.
4. **Dowling, N. J. E., J. Guezennec, and D. C. White.** 1987. Facilitation of corrosion of stainless steel exposed to aerobic seawater by microbial biofilms containing both facultative and absolute anaerobes, p. 27–38. *In* E. C. Hill, J. L. Sherman, and R. J. Watkinson (ed.), *Microbial Problems in the Offshore Oil Industry.* John Wiley & Sons, Ltd., Chichester, England.
5. **Dudman, W. F.** 1977. The role of surface polysaccharides in natural environments, p. 357–414. *In* I. Sutherland (ed.), *Surface Carbohydrates of the Prokaryotic Cell.* Academic Press, Inc., New York.
6. **Engebrecht, J., K. Nealson, and M. Silverman.** 1983. Bacterial bioluminescence: isolation and genetic analysis of functions from *Vibrio fischeri. Cell* 32:773–781.
7. **Fazio, S. A., D. J. Uhlinger, J. H. Parker, and D. C. White.** 1982. Estimations of uronic acids as quantitative measures of extracellular and cell wall polysaccharide polymers from environmental samples. *Appl. Environ. Microbiol.* **43**:1151–1159.
8. **Fazio, S. D., W. R. Mayberry, and D. C. White.** 1979. Muramic acid assay in sediments. *Appl. Environ. Microbiol.* **38**:349–350.
9. **Ferrell, T. L., T. A. Callcott, and R. J. Warmack.** 1985. Plasmons and surfaces. *Am. Sci.* **73**:344–353.
10. **Firstenberg-Eden, R., and G. Eden.** 1984. *Impedance Microbiology,* p. 7–34, 41–70, and 73–134. John Wiley & Sons, Inc., New York.
11. **Fletcher, M.** 1976. The effects of proteins of bacterial attachment to polystyrene. *J. Gen. Microbiol.* **94**:400–404.
12. **Fletcher, M.** 1986. Effect of solid surfaces on the activity of attached bacteria, p. 339–400. *In* D. C. Savage and M. Fletcher (ed.), *Bacterial Adhesion.* Plenum Publishing Corp., New York.
13. **Fletcher, M., and G. D. Floodgate.** 1976. The adhesion of bacteria to solid surfaces, p. 101–107. *In* R. Fuller and D. W. Lovelock (ed.), *Microbial Ultrastructure.* Academic Press, Inc. (London), Ltd., London.
14. **Fletcher, M., and K. C. Marshall.** 1982. Bubble contact angle method for evaluating substratum interfacial characteristics and its relevance to bacterial attachment. *Appl. Environ. Microbiol.* **44**:184–192.
15. **Fowler, H. W., and A. J. McKay.** 1980. The measurement of microbial adhesion, p. 48–66. *In* R. C. W. Berkeley, J. M. Lynch, J. Melling, P. R. Rutter, and B. Vincent (ed.), *Microbial Adhesion to Surfaces.* Ellis Horwood Ltd., Chichester, England.
16. **Geesey, G. G.** 1982. Microbial exopolymers: ecological and economic considerations. *ASM News* **48**:9–14.
17. **Geesey, G. G., M. W. Mittleman, T. Iwaoka, and P. R. Griffiths.** 1986. Role of bacterial exopolymers in their deterioration of metallic copper surfaces. *Mater. Perf.* **25**:37–40.
18. **Hedrick, D. B., and D. C. White.** 1987. Rapid extraction and fractionation of particulate BW agents for detection of signature biomarkers by mass spectrometry, p. 1–134. *In* M. D. Rouser (ed.), *Proceedings of the U.S. Army Chemical Defense Research and Engineering Center Conference, Edgewood Area.* Code CRDEC-SP-87008. U.S. Army Armaments, Munitions, and Chemical Command, Aberdeen Proving Ground.
19. **Lewin, R.** 1984. Microbial adhesion is a sticky problem. *Science* **224**:375–377.
20. **Lindberg, B., J. Longren, and S. Svensson.** 1975.

Specific degradation of polysaccharides. *Adv. Carbohydr. Chem. Biochem.* **31**:185–239.

21. **Little, B. J., P. Wagner, S. M. Gerchakov, M. Walch, and R. Mitchell.** 1986. The involvement of a thermophilic bacterium in corrosion processes. *Corrosion* **42**:533–536.
22. **Malmcrona-Friberg, K., A. Tunlid, P. Marden, S. Kjelleberg, and G. Odham.** 1986. Chemical changes in cell envelope and poly-beta-hydroxybutyrate during short term starvation of a marine bacterial isolate. *Arch. Microbiol.* **144**:340–345.
23. **Marshall, K. C., R. Stout, and R. Mitchell.** 1971. Selective sorption of bacteria from seawater. *Can. J. Microbiol.* **17**:1413–1416.
24. **McEldowney, S., and M. Fletcher.** 1986. Effect of growth conditions and surface characteristics of aquatic bacteria on their attachment to solid surfaces. *J. Gen. Microbiol.* **132**:513–523.
25. **Mirelman, D., and I. Ofek.** 1986. Introduction to microbial lectins and agglutinins, p. 1–19. *In* D. Mirelman (ed.), *Microbial Lectins and Agglutinins.* John Wiley & Sons, Inc., New York.
26. **Mitchell, R., and D. Kirchman.** 1984. The microbial ecology of marine surfaces, p. 49–56. *In* J. Costlow and R. C. Tipper (ed.), *Marine Biodeterioration: an Interdisciplinary Study.* Naval Institute Press, Annapolis, Md.
27. **Nichols, P. D., J. M. Henson, J. B. Guckert, D. E. Nivens, and D. C. White.** 1985. Fourier transform-infrared spectroscopic methods for microbial ecology: analysis of bacteria, bacteria-polymer mixtures and biofilms. *J. Microbiol. Methods* **4**:79–94.
28. **Nivens, D. E., P. D. Nichols, J. M. Henson, G. G. Geesey, and D. C. White.** 1986. Reversible acceleration of corrosion of stainless steel exposed to seawater induced by the extracellular secretions of the marine vibrio *V. natriegens. Corrosion* **41**: 201–210.
29. **Odham, G., A. Tunlid, G. Westerdahl, L. Larsson, J. B. Guckert, and D. C. White.** 1985. Determination of microbial fatty acid profiles at femtomolar levels in human urine and the initial marine microfouling community by capillary gas chromatography-chemical ionization mass spectrometry with negative ion detection. *J. Microbiol. Methods* **3**:331–344.
30. **Pethica, B. A.** 1980. Microbial and cell adhesion, p. 19–45. *In* R. C. W. Berkeley, J. M. Lynch, J. Melling, P. R. Rutter, and B. Vincent (ed.), *Microbial Adhesion to Surfaces.* Ellis Horwood Ltd., Chichester, England.
31. **Platt, R. M., G. G. Geesey, J. D. Davis, and D. C. White.** 1986. Isolation and partial chemical analysis of firmly bound exopolysaccharide from adherent cells of a freshwater sediment bacterium. *Can. J. Microbiol.* **31**:675–680.
32. **Pringle, J. H., M. Fletcher, and D. C. Ellwood.** 1983. Selection of attachment mutants during the continuous culture of *Pseudomonas fluorescens* and the relationship between attachment ability and surface composition. *J. Gen. Microbiol.* **129**:2557–2569.
33. **Radford, T., and D. C. DeJongh.** 1972. Carbohydrates, p. 313–349. *In* G. R. Waller (ed.), *Biochemical Applications of Mass Spectrometry.* John Wiley & Sons, Inc., New York.
34. **Rosenberg, M., and S. Kjelleberg.** 1986. Hydrophobic interactions: role in bacterial adhesion. *Adv. Microb. Ecol.* **9**:353–393.
35. **Schwartz, H. E., P. J. Barthel, S. E. Moring, and H. H. Lauer.** 1987. Packed vs. capillary column for supercritical fluid chromatography. *LC-GC.* **5**:490–498.
36. **Tunlid, A., B. H. Baird, M. B. Trexler, S. Olsson, R. H. Findlay, G. Odham, and D. C. White.** 1985. Determination of phospholipid ester-linked fatty acids and poly beta hydroxybutyrate for the estimation of bacterial biomass and activity in the rhizosphere of the rape plant *Brassica napus* (L.). *Can. J. Microbiol.* **31**:1113–1119.
37. **Tunlid, A., G. Odham, R. H. Findlay, and D. C. White.** 1985. Precision and sensitivity in the measurement of ^{15}N enrichment in D-alanine from bacterial cell walls using positive/negative ion mass spectrometry. *J. Microbiol. Methods* **3**:237–245.
38. **Uhlinger, D. J., and D. C. White.** 1983. Relationship between physiological status and formation of extracellular polysaccharide glycocalyx in *Pseudomonas atlantica. Appl. Environ. Microbiol.* **45**:64–70.
39. **Ward, J. B., and R. C. W. Berkeley.** 1980. The microbial cell surface and adhesion, p. 48–66. *In* R. C. W. Berkeley, J. M. Lynch, J. Melling, P. R. Rutter, and B. Vincent (ed.), *Microbial Adhesion to Surfaces.* Ellis Horwood Ltd., Chichester, England.
40. **Weiss, L.** 1961. The measurement of cell adhesion. *Exp. Cell Res. Suppl.* **8**:141–153.
41. **Weiss, L.** 1961. Studies on cellular adhesion in tissue culture. IV. The alteration of substrata by cell surfaces. *Exp. Cell Res.* **25**:504–517.
42. **White, D. C.** 1986. Environmental effects testing with quantitative microbial analysis: chemical signatures correlated with *in situ* biofilm analysis by FT/IR. *Toxicity Assessment* **1**:315–338.
43. **White, D. C., D. E. Nivens, P. D. Nichols, A. T. Mikell, B. D. Kerger, J. M. Henson, G. G. Geesey, and C. K. Clarke.** 1986. Role of aerobic bacteria and their extracellular polymers in the facilitation of corrosion: use of Fourier transforming infrared spectroscopy and "signature" phospholipid fatty acid analysis, p. 233–243. *In* S. C. Dexter (ed.), *Biologically Induced Corrosion.* National Association of Corrosion Engineers reference book 8. National Association of Corrosion Engineers, Houston, Tex.

Chapter 22

Cyanobacterial-Heterotrophic Bacterial Interaction

K. C. Marshall

INTRODUCTION

Microbial mats are areas of intense primary productivity, especially by a variety of oxygenic and, in some circumstances, anoxygenic cyanobacteria (3, 11). The availability of both dissolved and particulate organic carbon as a result of excretion from and lysis of these cyanobacteria results in the active growth of heterotrophic bacteria in the mats. As a consequence, one can expect that the cyanobacteria and the heterotrophic bacteria will undergo interactions ranging from beneficial (commensalism, protocooperation, and mutualism) to harmful (competition, amensalism, and parasitism). In this chapter I will briefly review some of the published work on beneficial and harmful interactions between cyanobacteria and heterotrophic bacteria and discuss studies on the specific adhesion of heterotrophic bacteria to cyanobacterial heterocysts.

K. C. Marshall • School of Microbiology, University of New South Wales, P.O. Box 1, Kensington, New South Wales 2033, Australia.

HARMFUL INTERACTIONS

Gliding Bacteria

Stewart and Brown (42) reported that *Cytophaga* strain N-5, isolated from a waste-stabilization pond in Texas, formed plaques on *Nostoc muscorum* and *Plectonema boryanum* and was also capable of killing or lysing gram-positive and gram-negative bacteria. More recent reports of lysis of cyanobacteria by gliding bacteria with low G+C contents involve *Flexibacter chinenses*, with a G+C content of 34.2 mol% (24, 25), and *Coleomitus rectisporus*, with a G+C content of 33.9 mol% (23, 24).

Lysis of cyanobacteria by gliding bacteria with high G+C contents has been reported by many workers (5, 12–14, 22, 24, 41). Shilo (41) described a nonfruiting myxobacter (now termed *Lysobacter* sp. [10]) that is capable of lysing many species of unicellular and filamentous cyanobacteria, as well as some heterotrophic bacteria. Direct microscopic observation of the lysis of a *Nostoc* sp. revealed that lysis occurred only when there was direct contact between the polar tip of the myxobacter and the cyanobacterium. It could be that one (or both) pole of this gliding bacterium is relatively hydrophobic, as described by Marshall and Cruickshank (28) for a *Flexibacter* isolate, thereby ensuring the correct orientation of the cell at the cyanobacterial surface.

Daft and Stewart (13, 14) reported the isolation of myxobacterial cultures capable of lysing over 40 cyanobacterial strains, including bloom-forming species. They observed that lysis of vegetative cells of the cyanobacteria depended on the presence of growing bacteria, which attached in a polar manner. Heterocysts were not lysed. The initial effect of isolate CP-1 was to cause lysis of the L2 layer of the cyanobacterium followed by rupture or disintegration of the other cell wall layers. In completely lysed cells there was a loss of polyphosphate and other granules, gas vacuoles, and most other cellular inclusions, although the thylakoid membranes remained intact. Daft and Stewart (14) proposed the use of bacterium CP-1 for preparing isolated cyanobacterial membranes and gas vesicles. In a more detailed ecological study of fresh waters, Daft et al. (12) found large numbers of nonfruiting myxobacteria that were able to lyse bloom-forming cyanobacteria, including *Anabaena*, *Aphanizomenon*, *Gloeotrichia*, *Microcystis*, and *Oscillatoria* spp. Burnham et al. (5) have described the entrapment and lysis of *Phormidium luridum* by an isolate of *Myxococcus xanthus*, whereby irregular clumps of the bacterium and cyanobacterium develop to the point of forming spherules consisting of green cyanobacterial colonies in various stages of degradation surrounded by a tightly woven mass of myxobacterial cells.

Why are gliding bacteria so predominant as cyanobacterial lytic agents? Rosenberg and Varon (39) have emphasized the wide range of lytic properties available to the myxobacteria. These include (i) antibiotics, many of which interfere with cell wall synthesis; (ii) lytic enzymes, including cell-wall-cleaving and proteolytic enzymes, nucleases, and lipases; (iii) helper mechanisms, such as the excretion of fatty acids that convert lysis-resistant organisms into a susceptible state; and (iv) bacteriocins.

Other Bacteria

Granhall and Berg (19) described the effect of heat-resistant, low-molecular-weight antibioticlike substances from *Cellvibrio* strains that lysed the vegetative cells of *Anabaena inaequalis*, but did not affect the viability of heterocysts and akinetes. Lysis of *Plectonema boryanum* by an atypical *Bacillus brevis* was reported by Reim et al. (37), whereas Burnham and Sun (6) found that the parasitic bacterium *Bdellovibrio bacteriovorus* inhibited photosynthesis and stimulated an autolytic process in *Phormidium luridum* and a *Synechococcus* sp.

BENEFICIAL INTERACTIONS

Benefit to the Heterotrophic Bacteria

Various studies have reported the benefit to heterotrophic bacteria of dissolved organic carbon excreted by cyanobacteria in microbial mats and blooms and of its role in the establishment of food chains (7, 20, 21, 29, 32, 33, 35). Gallucci (quoted in reference 33) demonstrated that bacteria epiphytic on cyanobacteria are attracted to the cyanobacteria through a positive chemotactic response. These epiphytic bacteria exhibit enhanced uptake of organic substrates, phosphorus, and thymidine (32) and reduction of tetrazolium salts (35), especially when compared with free-living bacteria (7, 35). The etiological agent of Legionnaires disease, *Legionella pneumophila*, has been isolated from microbial mats and shown to be stimulated by extracellular products of the cyanobacterium *Fischerella* sp. (43). In addition, *L. pneumophila* aerosols were found to be more stable when the bacterium was suspended in an extract of *Fischerella* sp. than in tryptose-saline solution (4). A relatively low-molecular-weight substance must have been involved in this stabilization, since the protective effect was lost when the extract was dialyzed.

Benefit to the Cyanobacteria

Oxygen levels in microbial mats may vary from supersaturation to depletion within the space of a few millimeters (33, 38). Supersaturation of oxygen leads to CO_2 depletion and sensitization of laboratory and naturally occurring cyanobacteria to photooxidative death (1, 15). Under these conditions, intense heterotrophic bacterial activity results in the induction of microenvironments around the cyanobacteria where oxygen becomes depleted and CO_2 is replenished, thereby protecting the cyanobacteria from photooxidative effects (35). The close association of heterotrophic bacteria with cyanobacterial heterocysts also results in more effective N_2 fixation at higher oxygen levels (26, 35) (see below).

Other benefits to cyanobacteria from associated heterotrophic bacteria include the exchange of vitamins and other growth factors leading to enhanced cyanobacterial growth (36). Escher and Characklis (16) have provided a model, based on the control of glyoxylate excretion by the activity of the ribulose biphosphate carboxylase-oxygenase system, to explain the interactions between algae and heterotrophic bacteria.

SPECIFIC ADHESION OF BACTERIA TO CYANOBACTERIAL HETEROCYSTS

Bacterial Associations with Heterocysts

The attachment of bacteria to the terminal heterocysts of several *Cylindrospermum* spp. has been noted (18, 40). Paerl (30) proposed a specific association between bacteria and the heterocysts of *Anabaena* and *Aphanizomenon* species on the basis that the bacteria were located mainly on the heterocysts, although the heterocysts constituted only 5 to 10% of the cyanobacterial filaments (31). Lupton and Marshall (26) reported the isolation of two bacteria, *Pseudomonas* strain SL10 and *Zoogloea* strain SL20, that adhered specifically to the heterocysts of *Anabaena* species, whereas a third isolate, *Flavobacterium* strain SL13, adhered randomly along the cyanobacterial filaments.

Light- and electron-microscopic examination revealed that *Pseudomonas* strain SL10 adhered perpendicularly to the surface of the heterocyst (26), probably indicating that one pole of the cell was more hydrophobic than the remainder of the rod-shaped cell (28). The polar end of the cell attached to the outer fibrous layer of the heterocyst (Fig. 1A). A small bulge consistently appeared at

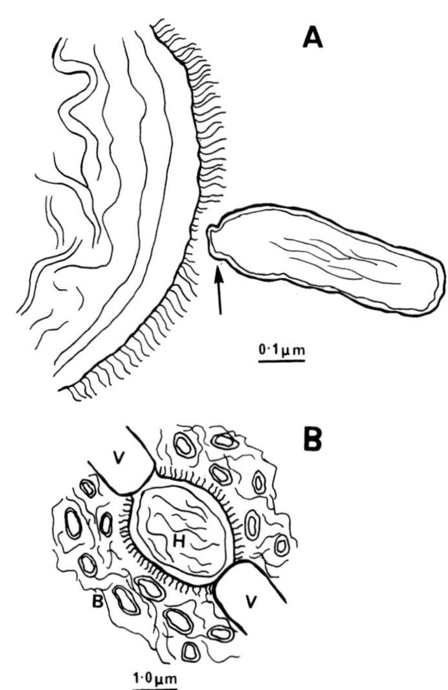

Figure 1. Modes of adhesion of bacteria to *Anabaena* heterocysts. (A) Polar attachment of *Pseudomonas* strain SL10 to the heterocyst envelope. Note the small bulge at the pole of the bacterial cell (arrow). (B) Multilayer adhesion of *Zoogloea* strain SL20 at the heterocyst surface. Abbreviations: B, bacteria; H, heterocyst; V, vegetative cells.

Table 1.
Numbers of *Pseudomonas* Strain SL10 and *Zoogloea* Strain SL20 Cells Adhering to Heterocysts of Various *Anabaena* Species[a]

Anabaena spp.	No. of bacteria/mg (dry wt) of heterocysts	
	Pseudomonas strain SL10	*Zoogloea* strain SL20
A. cylindrica	4.1×10^7	7.6×10^8
A. oscillarioides	2.5×10^7	5.8×10^8
A. flos-aquae	10.5×10^7	2.0×10^8
A. azollae	9.8×10^7	1.2×10^8

[a] Adapted from reference 26.

this pole of the cell (26), and in this respect, *Pseudomonas* strain SL10 resembled the organisms described by Paerl (32). Cells of *Zoogloea* strain SL20 were orientated in a random manner and tended to form aggregates around the heterocysts (Fig. 1B). The number of *Zoogloea* strain SL20 cells attached to heterocysts was approximately 10-fold greater than that of *Pseudomonas* strain SL10 cells (Table 1). Adhesion of strain SL10 to heterocysts of *Anabaena flos-aquae* and *Anabaena azollae* was significantly greater than to those of *Anabaena cylindrica* and *Anabaena oscillarioides*, whereas strain SL20 may adhere in greater numbers to heterocysts of *A. cylindrica* than to those of the other *Anabaena* species. These results reflect the fact that *A. azollae* and *A. flos-aquae* are morphologically and antigenically similar and are very distinct from *A. cylindrica* and *A. oscillarioides* (17).

Adhesion Isotherms

The adhesion of *Pseudomonas* strain SL10 and *Zoogloea* strain SL20 to heterocysts of *A. cylindrica* and *A. azollae* was assayed using ^{14}C-labeled bacteria to derive the appropriate adsorption isotherms (26, 27). The adhesion of strain SL10 followed a curvilinear relationship (type I or Langmuir adsorption isotherm), indicating saturation of the heterocyst surface by a monolayer of adhering bacteria at high concentrations of added strain SL10 (26). The adhesion of strain SL20 to heterocysts gave a sigmoid (type II) adsorption isotherm, indicating two steps in the process: first, a specific interaction between the bacteria and the heterocyst surface, resulting in monolayer adhesion; and, second, aggregation between bacterial cells at the surface, resulting in multilayer adhesion (26).

A comparison of monolayer adhesion of the two bacteria was made by using the mathematical expression for a type I isotherm: $q = CN/(C + K_d)$, where C is cell concentration at equilibrium (cells per milliliter), q is the number of cells adhering per unit weight of heterocysts, N is the adhesion capacity of the heterocysts (binding sites per milligram of heterocysts), and K_d is the equilibrium constant (cells per milliliter). A linearization of this isotherm is $C/q = (C/N) + (K_d/N)$, and a plot of C/q versus C results in a straight line with x and y intercepts of $-K_d$ and K_d/N, respectively. The reciprocal of K_d is the affinity constant, K_a. The product of K_a and N has been used to give an index of adhesion, since it can be shown that $K_aN = q([1/C] + K_a)$, and hence K_aN is a linear function of q and is independent of N, regardless of the range of cell concentrations used (2, 26).

Lupton and Marshall (27) determined that there are more potential binding sites for *Pseudomonas* strain SL10 on *A. azollae* heterocysts, but the affinity (K_aN) for these sites is lower than that for sites on *A. cylindrica* (Table 2). No differences were apparent between the number of binding sites or affinities for the adhesion of *Zoogloea* strain SL20 cells to *A. azollae* and *A. cylindrica* heterocysts. Strain SL10 exhibited a greater affinity for its binding sites than did strain SL20.

Basis of Adhesion Specificity

The bacteria and *A. cylindrica* heterocysts were treated separately with trypsin,

Table 2.
Available Binding Sites and Adhesion Index for Bacteria on *Anabaena* Heterocysts[a]

Bacterium and *Anabaena* species	Available binding sites/mg of heterocysts (N)	Adhesion index (K_aN)
Pseudomonas strain SL10		
A. cylindrica	2.10×10^7	0.42
A. azollae	7.87×10^7	0.24
Zoogloea strain SL20		
A. cylindrica	3.42×10^7	0.10
A. azollae	3.51×10^7	0.09

[a] Adapted from reference 27.

pronase, and periodate and by methylation to determine their effects on the specific adhesion of the bacteria to the heterocysts (27). In addition, the interaction between the bacteria and the heterocysts was determined in the presence of potential inhibitors such as EDTA, Tween 80, borate, and a range of lectins (27). Adhesion using treated strain SL10 or *A. cylindrica* heterocysts was as follows (expressed as percentage of untreated control adhesion). Trypsin, pronase, periodate, and methylation produced 79, 35, 119, and 23% adhesion for strain SL10 and 142, 170, 28, and 9% adhesion for the heterocysts. The inhibitors used were EDTA (62% of control adhesion), borate (73%), concanavalin A (50%), peanut agglutinin (97%), soybean agglutinin (78%), and garden pea agglutinin (76%). Adhesion using treated strain SL20 or *A. cylindrica* was as follows. Trypsin, pronase, and periodate produced 32, 15, and 58% adhesion for strain SL20 and 80, 70, and 8% adhesion for the heterocysts. Methylation produced 76% adhesion for the heterocysts. The inhibitors used were EDTA (57% adhesion), Tween 80 (30%), borate (41%), and concanavalin A (122%).

It was concluded that the initial reversible adhesion of strain SL10 to heterocysts of *A. cylindrica* involves triplet ion formation between divalent cations and carboxyl groups both on the pole of the bacterial cell and on the heterocyst. The specific adhesin of strain SL10 appears to be a pronase-sensitive, carbohydrate-binding protein, whereas the heterocyst adhesin is probably a carbohydrate with terminal glucosyl residues (27). Glucose is more common as a terminal sugar residue in heterocyst polysaccharides than in vegetative cell sheath polysaccharides (8, 9). Differences in the adhesion of strain SL10 to heterocysts of different *Anabaena* species may result from variations in the distribution and abundance of a common adhesin on the various heterocysts. Adhesion to *A. azollae* heterocysts also involved polar attachment and showed similar responses to treatments and inhibitors as did adhesion to *A. cylindrica*. The specific adhesin of strain SL20 involved extracellular fibrils (26) which are removed by both proteolytic enzyme and periodate treatments. Further work is required to establish the exact nature of the adhesin in this bacterium. The heterocyst adhesin is almost certainly a carbohydrate, but its properties are different from those of the adhesin interacting with strain SL10.

Ecological Significance

Paerl and Keller (30, 34, 35) have proposed that the specific association of bacteria with heterocysts of *Anabaena* species promotes nitrogen fixation by creating a more reduced microenvironment around the heterocyst, thereby protecting the oxygen-sensitive nitrogenase from the high oxygen levels found in blooms and algal mats. Considering the possibility that the bacteria may nonspecifically enhance cyanobacterial growth, Lupton and Marshall (26) showed that a nonadhering *Enterobacter aerogenes* strain stimulated the growth of *A. cylindrica* as effectively as the adhering *Pseudomonas* strain SL10 and *Zoogloea* strain SL20 did. Under normal conditions, strain SL10 and *E. aerogenes* did not affect the rate of acetylene reduction by the cyanobacteria, whereas strain SL20 gave a significant increase in acetylene reduction. When the cultures were oxygenated there was a dramatic decrease in acetylene reduction in all culture combinations. Acetylene reduction by cyanobacteria appeared to be promoted in the presence of both strain SL10 and strain SL20, but owing to the large variance observed in the results, the apparent stimulation was not statistically significant (26). These results suggest that specific attachment of bacteria to cyanobacterial heterocysts may enhance nitrogen fixation by the cyanobacteria, particularly under conditions where oxygen levels substantially exceed normal.

ACKNOWLEDGMENTS. I acknowledge the work of F. S. Lupton reported in the section on specific adhesion of bacteria to cyanobacterial heterocysts.

I acknowledge the support of the Bat Sheva de Rothschild Foundation in attending this seminar. Original studies reported were supported by a grant from the Australian Research Grants Committee.

LITERATURE CITED

1. **Abeliovich, A., and M. Shilo.** 1972. Photooxidative death in blue-green algae. *J. Bacteriol.* **111:** 682–689.
2. **Applebaum, B., E. Golub, S. E. Holt, and B. Rosan.** 1979. In vitro studies of dental plaque formation: adsorption of oral streptococci to hydroxyapatite. *Infect. Immun.* **25:**717–728.
3. **Bauld, J., R. V. Burne, L. A. Chambers, J. Ferguson, and G. W. Skyring.** 1980. Sedimentological and geobiological studies of intertidal cyanobacterial mats in North-Eastern Spencer Gulf, South Australia, p. 157–166. *In* P. A. Trudinger, M. R. Walter, and B. J. Ralph (ed.), *Biogeochemistry of Ancient and Modern Environments.* Australian Academy of Science, Canberra.
4. **Berendt, R. F.** 1981. Influence of blue-green algae (cyanobacteria) on survival of *Legionella pneumophila* in aerosols. *Infect. Immun.* **32:**690–692.
5. **Burnham, J. C., S. A. Collart, and B. W. Highison.** 1981. Entrapment and lysis of the cyanobacterium *Phormidium luridum* by aqueous colonies of *Myxococcus xanthus* PCO2. *Arch. Microbiol.* **129:** 285–294.
6. **Burnham, J. C., and D. Sun.** 1977. Electron microscope observations on the interaction of *Bdellovibrio bacteriovorus* with *Phormidium luridum* and *Synechococcus* sp. (Cyanophyceae). *J. Phycol.* **13:**203–208.
7. **Caldwell, D. E., and S. J. Caldwell.** 1978. A *Zoogloea* sp. associated with blooms of *Anabaena flos-aquae. Can. J. Microbiol.* **24:**922–931.
8. **Cardemil, L., and C. P. Wolk.** 1976. The polysaccharides from heterocyst and spore envelopes of a blue-green alga: methylation analysis and structure of the backbones. *J. Biol. Chem.* **251:**2967–2975.
9. **Cardemil, L., and C. P. Wolk.** 1979. The polysaccharides from heterocysts and spore envelopes of a blue-green alga: structure of basic repeating unit. *J. Biol. Chem.* **254:**736–741.
10. **Christensen, P., and F. D. Cook.** 1978. *Lysobacter*: a new genus of nonfruiting gliding bacteria with a high base ratio. *Int. J. Syst. Bacteriol.* **28:**367–393.
11. **Cohen, Y., Z. Aizenshtat, A. Stoler, and B. B. Jørgensen.** 1980. The microbial geochemistry of Solar Lake, Sinai, p. 167–172. *In* P. A. Trudinger, M. R. Walter, and B. J. Ralph (ed.), *Biogeochemistry of Ancient and Modern Environments.* Australian Academy of Science, Canberra.
12. **Daft, M. J., S. B. McCord, and W. D. P. Stewart.** 1975. Ecological studies on algal-lysing bacteria in fresh waters. *Freshwater Biol.* **5:**577–596.
13. **Daft, M. J., and W. D. P. Stewart.** 1971. Bacterial pathogens of freshwater blue-green algae. *New Phytol.* **70:**819–829.
14. **Daft, M. J., and W. D. P. Stewart.** 1973. Light and electron microscope observations on algal lysis by bacterium CP-1. *New Phytol.* **72:**799–808.
15. **Eloff, J. N., Y. Steinityz, and M. Shilo.** 1976. Photooxidation of cyanobacteria in natural conditions. *Appl. Environ. Microbiol.* **31:**119–126.
16. **Escher, A., and W. G. Characklis.** 1982. Algal-bacterial interactions within aggregates. *Biotechnol. Bioeng.* **24:**2283–2290.
17. **Gates, J. E., R. W. Fisher, T. W. Goggin, and N. L. Azrolan.** 1980. Antigenic differences between *Anabaena azollae* fresh from the *Azolla* fern leaf cavity and free-living cyanobacteria. *Arch. Microbiol.* **128:**126–129.
18. **Geitler, L.** 1932. *Cyanophyceae*, vol. 14. *In* L. Rabenhorst (ed.), *Rabenhorsts Kryptogamenflora.* Akademische Verlagsgesellschaft, Leipzig, German Democratic Republic.
19. **Granhall, U., and B. Berg.** 1972. Antimicrobial effects of *Cellvibrio* on blue-green algae. Arch. Mikrobiol. **84:**234–242.
20. **Harrison, P. G., and B. J. Harrison.** 1980. Interaction of bacteria, microalgae and copepods in a detritus microcosm through a flask darkly, p. 373–386. *In* K. R. Tenore and B. C. Coull (ed.), *Marine Benthic Dynamics.* University of South Carolina Press, Columbia.
21. **Herbst, V., and J. Overbeck.** 1978. Metabolic

coupling between the alga *Oscillatoria redekei* and accompanying bacteria. *Naturwissenschaften* **65**: 598–599.

22. **Li, Q., and S. Li.** 1981. Bacteria that lyse nitrogen-fixing blue-green algae. *Acta Hydrobiol. Sin.* 7:377–384.
23. **Li, Q., and S. Li.** 1983. A new species of *Coleomitus*. *Acta Microbiol. Sin.* **23**:185–192.
24. **Li, Q., and S. Li.** 1984. Ecological factors restricting blue-green algae population. *Acta Ecol. Sin.* 4: 310–315.
25. **Li, Q., S. Li, and D. Wang.** 1984. Isolation and characterization of *Flexibacter chinenses* sp. nov. *Acta Microbiol. Sin.* **24**:7–13.
26. **Lupton, F. S., and K. C. Marshall.** 1981. Specific adhesion of bacteria to heterocysts of *Anabaena* spp. and its ecological significance. *Appl. Environ. Microbiol.* **42**:1085–1092.
27. **Lupton, F. S., and K. C. Marshall.** 1984. Mechanisms of specific bacterial adhesion to cyanobacterial heterocysts, p. 144–150. *In* M. J. Klug and C. A. Reddy (ed.), *Current Perspectives in Microbial Ecology*. American Society for Microbiology, Washington, D.C.
28. **Marshall, K. C., and R. H. Cruickshank.** 1973. Cell surface hydrophobicity and the orientation of certain bacteria at interfaces. *Arch. Mikrobiol.* **91**: 29–40.
29. **Menge, B. A.** 1976. Organization of the New England rocky intertidal community: role of predation, competition and environmental heterogeneity. *Ecol. Monogr.* **46**:355–381.
30. **Paerl, H. W.** 1976. Specific associations of the blue-green algae *Anabaena* and *Aphanizomenon* with bacteria in freshwater blooms. *J. Phycol.* **12**:431–435.
31. **Paerl, H. W.** 1978. Role of heterotrophic bacteria promoting N_2 fixation by *Anabaena* in aquatic habitats. *Microb. Ecol.* 4:215–231.
32. **Paerl, H. W.** 1980. Attachment of microorganisms to living and detrital surfaces in freshwater systems, p. 375–402. *In* G. Bitton and K. C. Marshall (ed.), *Adsorption of Microorganisms to Surfaces*. John Wiley & Sons, Inc., New York.
33. **Paerl, H. W.** 1985. Influence of attachment on microbial metabolism and growth in aquatic ecosystems, p. 363–400. *In* D. C. Savage and M. Fletcher (ed.), *Bacterial Adhesion. Mechanisms and Physiological Significance*. Plenum Publishing Corp., New York.
34. **Paerl, H. W., and P. E. Keller.** 1978. Optimization of N_2-fixation in O_2-rich waters, p. 68–75. *In* M. W. Loutit and J. A. R. Miles (ed.), *Microbial Ecology*. Springer-Verlag KG, Berlin.
35. **Paerl, H. W., and P. E. Keller.** 1978. Significance of bacterial *Anabaena* (Cyanophyceae) associations with respect to N_2 fixation in aquatic habitats. *J. Phycol.* 14:254–260.
36. **Provasoli, L., and A. F. Carlucci.** 1974. Vitamins and growth regulators, p. 741–787. *In* W. D. P. Stewart (ed.), *Algal Physiology and Biochemistry*. University of California Press, Berkeley.
37. **Reim, R. L., M. S. Shane, and R. E. Cannon.** 1974. The characterization of a *Bacillus* capable of blue-green bactericidal activity. *Can. J. Microbiol.* **20**:981–986.
38. **Revsbech, N. P., and B. B. Jørgensen.** 1986. Microelectrodes: their use in microbial ecology. *Adv. Microb. Ecol.* **9**:293–352.
39. **Rosenberg, E., and M. Varon.** 1984. Antibiotics and lytic enzymes, p. 109–125. *In* E. Rosenberg (ed.), *Myxobacteria: Development and Cell Interactions*. Springer-Verlag, New York.
40. **Schwabe, G. H., and R. Mollenhauer.** 1967. Uber den Einfluss der Begleitbakterien auf das Lagerbild von *Nostoc sphaericum*. *Nova Hedwigia* 13:77–80.
41. **Shilo, M.** 1970. Lysis of blue-green algae by myxobacter. *J. Bacteriol.* **104**:453–461.
42. **Stewart, J. R., and R. M. Brown.** 1969. Cytophaga that kills or lyses algae. *Science* **164**:1523–1524.
43. **Tison, D. L., D. H. Pope, W. B. Cherry, and C. B. Fliermans.** 1980. Growth of *Legionella pneumophila* in association with blue-green algae (cyanobacteria). *Appl. Environ. Microbiol.* **39**:456–459.

Chapter 23

Microbial Films in the Mouth: Some Ecologically Relevant Observations

Mel Rosenberg

INTRODUCTION

One need not travel great distances to investigate an exciting, naturally occurring macroscopic biofilm—a look in the mirror is quite sufficient. Dental plaque, the macroscopic accumulation of bacteria on tooth surfaces, is a convenient, close-at-hand biofilm of great importance to humans. Dental caries and periodontal disease, the two most widespread diseases of modern humans, are linked to the activity of microorganisms on tooth surfaces (7, 12, 17). The soft tissues in the mouth are also colonized by biofilms, which are generally sparser than dental plaque, presumably owing to the shedding (desquamation) of the outermost epithelial cell layers. Finally, dental restorations, dentures, and devices are similarly prone to rapid biofouling in the oral cavity.

Microbial accumulations in the oral cavity appear to share several characteristics with biofilms in the open environment, as follows. (i) Similar to microbial films which form on surfaces in contact with water flow (pipes, rocks and sediments, submerged aquatic life, etc.), oral microorganisms must adhere to remain in their ecological niche and to survive. (ii) Similar to many environmental biofilms, bacterial accumulation on oral surfaces entails a complex physical and physiological coexistence involving a wide variety of bacterial species (7, 12). (iii) Maturing dental plaque results in a succession of anaerobic and acid-tolerant populations in close proximity to the tooth surface (7, 12). (iv) With prolonged passage of time, dental plaque accumulations become increasingly recalcitrant and calcified (dental calculus) and may be removed only by strong mechanical action (e.g., sonic oscillation). (v) Analogous to the chemical and structural changes in the underlying substratum as a result of many biofilms (e.g., corrosion of metals), tooth decay (dental caries) results from destruction of the protective enamel layer, primarily by acid accumulation (7, 12, 17). Finally, similar to the odorous, volatile gases produced by microbial mats and other biofilms, volatile gas elaborated by microbial accumulations in the mouth itself is the primary cause of bad breath (halitosis) (33, 36).

In this chapter I discuss some "ecological" aspects of oral biofilms which have attracted the interest of investigators in this

Mel Rosenberg • The Maurice and Gabriela Goldschleger School of Dental Medicine and the Department of Human Microbiology, Sackler Faculty of Medicine, Tel Aviv University, Ramat Aviv 69978, Israel.

and other laboratories: (i) adhesion, colonization, and desorption of oral bacteria; (ii) sampling and localization of adherent microbial populations on tooth surfaces; and (iii) measurement and monitoring of volatile, sulfur-containing gases in the oral cavity.

ADHESION, COLONIZATION, AND DESORPTION

Adhesion and Colonization

Cleaned surfaces inserted into the oral cavity or, alternatively, immersed in saliva become conditioned with a coat of salivary components prior to microbial adhesion (7, 12). This layer, termed the acquired pellicle, may differ to some extent in composition, distribution, or juxtaposition of adsorbed salivary molecules, depending on the substratum characteristics; it is likely, however, that many salivary components adsorb to a wide variety of oral surfaces (12).

Researchers studying the oral cavity are in general agreement that microbial adhesion is a crucial initial process in the colonization of oral surfaces; however, the actual mechanisms which promote oral adhesion have been a subject of debate for years (7, 12, 15–17, 30, 31). Similarly, the subsequent role of adhesion in biofilm buildup on oral surfaces is unclear: dental plaque accumulation appears to be due primarily to outgrowth of already attached bacteria, rather than to adsorption of free cells from saliva (12).

One striking observation of oral biofilms is the heterogeneity of their microbial composition. For example, high levels of *Streptococcus salivarius* are found attached to the tongue dorsum and buccal surfaces, but very low levels are observed in dental plaque. *Streptococcus mutans*, on the other hand, colonizes tooth surfaces but not soft oral tissues and is thus absent from mouths that are free of teeth or artificial hard surfaces (7). Striking differences in microbiota are also observed microscopically: whereas coccal forms are predominant on buccal epithelial cells, noncoccal forms, such as rods, filamentous bacteria, and fusiform bacteria, constitute a large proportion of dental plaque (7). Gibbons et al. have postulated that these differences reflect specific adhesion mechanisms (7, 10). Thus, according to this hypothesis, oral bacteria adhere with high affinity to surfaces on which they are able to proliferate, but adhere poorly to surfaces which they do not colonize.

During recent years, there has been an increasing number of reports that hydrophobic interactions play a significant role in bacterial adhesion in the open environment. In 1982, Weiss et al. (40) and Nesbitt et al. (21) reported findings suggesting that bacterial hydrophobicity may be important for oral adhesion as well. During the past 5 years, this possibility has been explored extensively in several laboratories.

The following lines of evidence support a major role for bacterial hydrophobicity in oral adhesion. (i) Many oral bacteria exhibit pronounced cell surface hydrophobicity, as measured by adhesion to hydrocarbons; they include fresh oral isolates (40), caries-associated and periodontal disease-associated laboratory strains (5, 8, 14, 19, 21, 23, 42), dental plaque itself (29), and bacteria obtained directly from saliva (15, 16). (ii) In several cases, clear-cut correlations were observed between adhesion to saliva-coated hydroxyapatite (SHA) beads (an in vitro tooth surface model system) and cell surface hydrophobicity in various oral strains (8, 19, 23) and following various treatments (11, 22); moreover, adhesion to SHA can be blocked by hydrophobic-bond-breaking agents, e.g., thiocyanate and tetramethyl urea (21, 38). (iii) Nonhydrophobic mutants of hydrophobic strains are deficient in adhering to tooth surfaces (39) or SHA (6, 9, 19). (iv) In an in vivo study, Svanberg et al. showed that a hydrophobic fresh isolate of *S. mutans* is retained in the human oral cavity, whereas its nonhydrophobic variant is not

(34). (v) Emulsan, a bacterial amphiphile which blocks adhesion to hydrophobic surfaces, similarly inhibits the adhesion of oral and nonoral strains to buccal epithelial cells (26).

If hydrophobic interactions constitute a major mechanism for oral adhesion, one would expect that adhesion to various oral surfaces would not be as specific as previously suggested (31). Indeed, although filamentous plaque bacteria do not adhere significantly to buccal epithelial cells in situ, adhesion readily takes place in a saliva-free in vitro system (35). Moreover, this binding is inhibited by amphipathic agents and saliva (M. Rosenberg and S. Guendelman, unpublished data). Thus, the heterogeneity of microbial populations on various tooth surfaces may not be a reflection of their adhesion capabilities, but rather their relative ability to colonize and proliferate on various oral substrata (31).

Several recent investigations have challenged the role of hydrophobicity in oral adhesion. Van der Mei et al. (37), using six different methods for studying hydrophobicity, were unable to demonstrate a clear correlation between hydrophobicity and adhesion to SHA. Similarly, Wyatt et al. (43) reported that although all strains of *Streptococcus sanguis* tested adhered to hexadecane, large differences in adhesion to SHA were observed. Knox et al. (13), using continuous-culture-grown cells, found no correlation between adhesion to hexadecane and to SHA. Clark et al. (2) found a strong correlation between the relative hydrophobicity of 42 oral *Actinomyces* strains and adhesion to SHA, but were unable to observe inhibition of adhesion of SHA by the surfactant Tween 80.

These apparent disagreements can be resolved only through further research and improved methodology. It is likely that SHA may be an interesting model system with which to study adhesion, but that it does not clearly mimic the tooth surface itself; moreover, this technique is used with various significant alterations in different laboratories. Similarly, improved bacterial hydrophobicity tests could be of great assistance (30). In any case, it is worth emphasizing that the current hydrophobicity tests do not measure the same characteristics (32) and that no single test is definitive in measuring these properties (4, 20). This point is well illustrated by the recent reports of Weerkamp et al. (38) and van der Mei et al. (37). These investigators did find overall correlations between the adhesion of wild-type *S. salivarius* and various mutants to hexadecane and SHA. One mutant (HB-7), however, was exceptional in that it adhered to hexadecane but not to SHA. The investigators inferred from this observation that adhesion to SHA cannot be accounted for by bacterial hydrophobicity. However, this mutant adheres poorly to the hydrophobic plastic polymethylmethacrylate (37). Opposite observations have been made with *Acinetobacter calcoaceticus* BD4, which adheres to polystyrene but not hexadecane (27, 30). Finally, it should be emphasized that adhesion characteristics, including cell surface hydrophobicity, may be lost upon subculturing (41).

To summarize, the extent to which hydrophobic interactions mediate adhesion within the oral cavity remains unresolved at present. Some investigators have proposed a major role for cell surface hydrophobicity in mediating oral adhesion (16, 23, 31, 38); others have urged fellow researchers to desist this line of study altogether (25); still others have proposed that hydrophobic interactions act in concert with additional mechanisms (3, 5, 21, 38). Indeed, the findings that the lactose-sensitive coaggregation of oral bacteria may have hydrophobic components (18) suggest that a precise line between stereochemically specific and nonspecific adhesion mechanisms may be hard to draw. Finally, bacteria may adhere by one mechanism to other bacteria, which are themselves bound by another mechanism to the tooth surface (1).

Bacterial Desorption from Oral Surfaces

Among the treatments to eliminate or reduce microbial buildup on surfaces, desorption of attached cells is a logical choice, since inhibition of the initial adhesion processes is often unfeasible. Desorption itself is a more complicated process than inhibition of adhesion, since a bacterium may adhere via a multitude of molecular interactions, which are difficult to break simultaneously; moreover, bacterial elaboration of extracellular polysaccharides may help cement already adherent cells.

The observation that a large proportion of oral bacteria adhere to oil droplets prompted us to propose swishing with an oil-water rinse as a technique of desorbing adsorbed bacteria from oral surfaces (E. Weiss, M. Rosenberg, and H. Judes, U.S. patent 4,525,342, June, 1985). In such a case, the actual adhesion mechanisms are irrelevant as long as the affinity of the cells for the oil droplets can overcome the adhesive forces. In addition to their ability to desorb oral bacteria, oils have been shown to be capable of desorbing bacteria from polystyrene (28) and fish skin (32). During our studies, the possibility arose that oil droplets could also bind amphipathic odor-causing molecules (Weiss et al., patent) and perhaps serve to reduce halitosis levels. This, in turn, prompted us to look for a simple means of studying halitosis, which is discussed later on.

SAMPLING OF TOOTH SURFACES FOR SPECIFIC ADHERENT MICROORGANISMS

Several convenient methods (Rodac plates, contact dip slides, adhesive tape, etc.) exist for sampling microorganisms from smooth surfaces. However, the localization and sampling of specific microorganisms adhering to rough surfaces, e.g., tooth surfaces, present a problem. Recent studies have shown that early dental caries formation is linked to the presence of *S. mutans* (17). The levels of desorbed *S. mutans* cells in saliva are currently used to assess dental caries levels and susceptibility.

We were interested in trying to develop a simple technique to pinpoint the growth of *S. mutans* on the tooth surfaces, rather than measure the levels of desorbed cells in saliva. To this end, we obtained impressions of the tooth surface by using a tacky, malleable material. Cells not adhering to the impression material were washed off in tap water. Liquid agar-containing media were cooled to 45°C and poured over the impressions, but growth was difficult to discern. To simplify the process, we decided to attempt to incubate the impression in a liquid medium. To confine outgrowth to the solid/liquid interface, the major carbon and energy source (sucrose) was impregnated in the impression material but was absent from the liquid medium. By incorporating selective, chromogenic components in the liquid outgrowth medium, specific growth of *S. mutans* on the impression surfaces was clearly observable (31a). This technique may have clinical and ecological implications in mapping the presence of this microorganism on tooth surfaces and, perhaps, in providing an indication of incipient caries formation. Minor modifications of the method may also facilitate sampling and concomitant localization of adherent populations on rough surfaces in the environment.

HALITOSIS

As stated above, the production of foul oral odors is analogous to putrid-gas elaboration by microbial mats and biofilms in the open environment. However, halitosis is the cause of substantial human discomfort; unfortunately, relatively little research has been directed toward this problem, which is said to affect the majority of adults in the Western world (36). About 85 to 90% of halitosis cases are reported to be oral in origin: the direct consequence of metabolic activity of bacterial accumulations on various oral surfaces (36). Volatile sulfur-containing gases such as hydrogen sulfide, methyl and ethyl mercaptan, and dimethyl sulfide are a major factor in oral malodor. Most halitosis cases can be alleviated by proper oral hygiene and dental techniques (36). We have recently suggested that monitoring the levels of halitosis may aid the dental practitioner in treating this common ailment. Previously, halitosis measurements have been made by gas chromatography (33, 36), which is too complicated and time-consuming for routine clinical testing, or by an organoleptic panel of volunteers, which is similarly unsuitable. We have recently tested a portable, simple-to-use industrial sulfide monitor, which can measure sulfide concentrations below 0.1 mg/liter; initial results suggest that such testing can provide relevant quantitative data (24). The use of such instruments for field measurement of volatile sulfur produced by microbial mats and other biofilms may also be of interest.

ACKNOWLEDGMENTS. I am grateful to Eugene Rosenberg, Ervin Weiss, Ilana Eli, Ron Doyle, and Jacob Gabbay for many discussions and joint research efforts related to the subjects discussed here. I am also grateful to Yardena Mazor for excellent technical assistance and to Rita Lazar for editorial help in preparing this manuscript.

LITERATURE CITED

1. **Ciardi, J. E., G. F. A. McCray, P. E. Kolenbrander, and A. Lau.** 1987. Cell-to-cell interaction of *Streptococcus sanguis* and *Propionibacterium acnes* on saliva-coated hydroxyapatite. *Infect. Immun.* **55**:1441–1446.
2. **Clark, W. B., M. D. Lane, E. Beem, S. L. Bragg, and T. T. Wheeler.** 1985. Relative hydrophobicities of *Actinomyces viscosus* and *Actinomyces naeslundii* strains and their adsorption to saliva-treated hydroxyapatite. *Infect. Immun.* **47**:730–736.
3. **Cowan, M. M., K. G. Taylor, and R. J. Doyle.** 1987. Energetics of the initial phase of adhesion of *Streptococcus sanguis* to hydroxyapatite. *J. Bacteriol.* **169**:2995–3000.
4. **Dillon, J. K., J. A. Fuerst, A. C. Hayward, and G. H. G. Davis.** 1986. A comparison of five methods for assaying bacterial hydrophobicity. *J. Microbiol. Methods* **6**:13–19.
5. **Doyle, R. J., W. E. Nesbitt, and K. G. Taylor.** 1982. On the mechanism of adherence of *Streptococcus sanguis* to hydroxylapatite. *FEMS Microbiol. Lett.* **15**:1–5.
6. **Fives-Taylor, P., and D. W. Thompson.** 1985. Surface properties of *Streptococcus sanguis* FW213 mutants nonadherent to saliva-coated hydroxyapatite. *Infect. Immun.* **47**:752–759.
7. **Gibbons, R. J.** 1980. Adhesion of bacteria to surfaces of the mouth, p. 351–388. *In* R. C. W. Berkeley, J. M. Lynch, J. Melling, P. R. Rutter, and B. Vincent (ed.), *Microbial Adhesion to Surfaces*. Ellis Horwood Ltd., Chichester, England.
8. **Gibbons, R. J., and I. Etherden.** 1983. Comparative hydrophobicities of oral bacteria and their adherence to salivary pellicles. *Infect. Immun.* **41**:1190–1196.
9. **Gibbons, R. J., I. Etherden, and Z. Skobe.** 1983. Association of fimbriae with the hydrophobicity of *Streptococcus sanguis* FC-1 and adherence to salivary pellicles. *Infect. Immun.* **41**:414–417.
10. **Gibbons, R. J., D. M. Spinell, and Z. Skobe.** 1975. Selective adherence as a determinant of the host tropisms of certain indigenous and pathogenic bacteria. *Infect. Immun.* **13**:238–246.
11. **Hesketh, L. M., J. E. Wyatt, and P. S. Handley.** 1987. Effect of protease on cell surface structure, hydrophobicity and adhesion of tufted strains of *Streptococcus sanguis* biotypes I and II. *Microbios* **50**:131–145.
12. **Kleinberg, I.** 1982. Dynamics of the oral ecosystem, p. 229–244. *In* W. A. Nolte (ed.), *Oral Microbiology*. C. V. Mosby Co., St. Louis.
13. **Knox, K. W., L. N. Hardy, L. J. Markevics, J. D. Evans, and A. J. Wicken.** 1985. Comparative studies on the effect of growth conditions on adhesion, hydrophobicity, and extracellular protein profile of *Streptococcus sanguis* G9B. *Infect. Immun.* **50**:545–554.
14. **Kozlovsky, A., Z. Metzger, and I. Eli.** 1987. Cell surface hydrophobicity of *Actinobacillus actinomycetemcomitans* Y_4. *J. Clin. Periodontol.* **14**:370–372.
15. **Leach, S. A.** 1980. A biophysical approach to interactions associated with the formation of the matrix of dental plaque, p. 159–183. *In* S. A. Leach (ed.), *Dental Plaque and Surface Interactions in the Oral Cavity*. IRL Press, Oxford.
16. **Leach, S. A., and E. A. Agalamanyi.** 1984. Hydrophobic interactions that may be involved in the formation of dental plaque, p. 43–50. *In* J. M. ten Cate, S. A. Leach, and J. Arends (ed.), *Bacterial*

Adhesion and Preventive Dentistry. IRL Press, Oxford.

17. **Loesche, W. J.** 1986. Role of *Streptococcus mutans* in human dental decay. *Microbiol. Rev.* 50:353–380.
18. **McIntire, F. C., L. K. Crosby, and A. E. Vatter.** 1982. Inhibitors of coaggregation between *Actinomyces viscosus* T14V and *Streptococcus sanguis* 34: β-galactosides, related sugars, and anionic amphipathic compounds. *Infect. Immun.* 36:371–378.
19. **Morris, E. J., N. Ganeshkumar, and B. C. McBride.** 1985. Cell surface components of *Streptococcus sanguis*: relationship to aggregation, adherence, and hydrophobicity. *J. Bacteriol.* 164:255–262.
20. **Mozes, N., and P. G. Rouxhet.** 1987. Methods for measuring hydrophobicity of microorganisms. *J. Microbiol. Methods* 6:99–112.
21. **Nesbitt, W. E., R. J. Doyle, and K. G. Taylor.** 1982. Hydrophobic interactions and the adherence of *Streptococcus sanguis* to hydroxylapatite. *Infect. Immun.* 38:637–644.
22. **Oakley, J. D., K. G. Taylor, and R. J. Doyle.** 1985. Trypsin-susceptible cell surface characteristics of *Streptococcus sanguis*. *Can. J. Microbiol.* 31:1103–1107.
23. **Olsson, J., and G. Westergren.** 1982. Hydrophobic surface properties of oral streptococci. *FEMS Microbiol. Lett.* 15:319–323.
24. **Pshigorski, I., J. Gabbay, I. Eli, S. Brenner, R. Bar-Ness, and M. Rosenberg.** 1987. A novel approach to halitosis measurement. *J. Dent. Res.* 66:911.
25. **Rosan, B., R. Wifert, and E. Golub.** 1985. Bacterial surfaces, salivary pellicles, and plaque formation, p. 69–76. *In* S. E. Mergenhagen and B. Rosan (ed.), *Molecular Basis of Oral Microbial Adhesion*. American Society for Microbiology, Washington, D.C.
26. **Rosenberg, E., A. Gottlieb, and M. Rosenberg.** 1983. Inhibition of bacterial adherence to hydrocarbons and epithelial cells by emulsan. *Infect. Immun.* 39:1024–1028.
27. **Rosenberg, E., N. Kaplan, O. Pines, M. Rosenberg, and D. Gutnick.** 1983. Capsular polysaccharides interfere with adherence of *Acinetobacter calcoaceticus* to hydrocarbons. *FEMS Microbiol. Lett.* 17:157–160.
28. **Rosenberg, M., H. Judes, and E. Weiss.** 1983. Desorption of adherent bacteria from a solid hydrophobic surface by oil. *J. Microbiol. Methods* 1:239–244.
29. **Rosenberg, M., H. Judes, and E. Weiss.** 1983. Cell surface hydrophobicity of dental plaque microorganisms in situ. *Infect. Immun.* 42:831–834.
30. **Rosenberg, M., and S. Kjelleberg.** 1986. Hydrophobic interactions: role in bacterial adhesion. *Adv. Microb. Ecol.* 9:353–393.
31. **Rosenberg, M., E. Rosenberg, H. Judes, and E. Weiss.** 1983. Bacterial adherence to hydrocarbons and to surfaces in the oral cavity. *FEMS Microbiol. Lett.* 20:1–5.

31a. **Rosenberg, M., E. Weiss, and I. Eli.** 1988. A novel replica technique for localization of caries-associated bacteria on tooth surfaces: development and initial experience. *Caries Res.* 22:42–44.

32. **Sar, N., and E. Rosenberg.** 1987. Fish skin bacteria: colonial and cellular hydrophobicity. *Microb. Ecol.* 13:193–202.
33. **Solis-Gaffar, M. C., H. P. Niles, W. C. Rainieri, and C. Kestenbaum.** 1975. Instrumental evaluation of mouth odor in a human clinical study. *J. Dent. Res.* 54:351–357.
34. **Svanberg, M., G. Westergren, and J. Olsson.** 1984. Oral implantation in humans of *Streptococcus mutans* strains with different degrees of hydrophobicity. *Infect. Immun.* 43:817–821.
35. **Tal, M., E. Weiss, H. Judes, and M. Rosenberg.** 1984. Adherence of dispersed plaque bacteria to buccal epithelial cells. *J. Dent. Res.* 63:547.
36. **Tonzetich, J.** 1977. Production and origin of oral malodor: a review of mechanisms and methods of analysis. *J. Periodontol.* 48:13–20.
37. **van der Mei, H. C., A. H. Weerkamp, and H. J. Busscher.** 1987. A comparison of various methods to determine hydrophobic properties of streptococcal cell surfaces. *J. Microbiol. Methods* 6:277–287.

37a. **van der Mei, H. C., A. H. Weerkamp, and H. J. Busscher.** 1987. Physico-chemical surface characteristics and adhesive properties of *Streptococcus salivarius* strains with defined cell surface structures. *FEMS Microbiol. Lett* 40:15–19.

38. **Weerkamp, A. H., H. C. van der Mei, and J. W. Slot.** 1987. Relationship of cell surface morphology and composition of *Streptococcus salivarius* K+ to adherence and hydrophobicity. *Infect. Immun.* 55:438–445.
39. **Weiss, E., H. Judes, and M. Rosenberg.** 1985. Adherence of a non-oral hydrophobic bacterium to the human tooth surface. *Dent. Med.* 3:11–13.
40. **Weiss, E., M. Rosenberg, H. Judes, and E. Rosenberg.** 1982. Cell surface hydrophobicity of adherent oral bacteria. *Curr. Microbiol.* 7:125–128.
41. **Westergren, G., and J. Olsson.** 1983. Hydrophobicity and adherence of oral streptococci after repeated subculture in vitro. *Infect. Immun.* 40:432–435.
42. **Wilson, P. A. D., W. M. Edgar, and S. A. Leach.** 1984. Some physical properties of oral streptococci that might have a role in their adhesion to oral surfaces, p. 99–112. *In* J. M. ten Cate, S. A. Leach, and J. Arends (ed.), *Bacterial Adhesion and Preventive Dentistry*. IRL Press, Oxford.
43. **Wyatt, J. E., L. M. Hesketh, and P. S. Handley.** 1987. Lack of correlation between fibrils, hydrophobicity and adhesion for strains of *Streptococcus sanguis* biotypes I and II. *Microbios* 50:7–15.

IV. Physiology of Major Mat-Building Microorganisms

Chapter 24

Aerobic-Anaerobic Metabolism in the Cyanobacterium *Oscillatoria limosa*

Lucas J. Stal, Heike Heyer, Susanne Bekker, Marlies Villbrandt, and Wolfgang E. Krumbein

INTRODUCTION

Intertidal sediments of the southern North Sea are colonized by dense populations of cyanobacteria (60). Cyanobacterial mats are characterized by steep gradients of oxygen, sulfide, pH, and light (44; chapter 11 of this volume), as well as diurnal shifts from aerobic during daytime to anaerobic during the night (25, 44, 60). Population dynamics studies of North Sea microbial mats have shown that among others, two species of cyanobacteria were of paramount importance. The first organism that colonizes the sediments is *Oscillatoria limosa.* Once established, cyanobacterial mats are dominated by the cyanobacterium *Microcoleus chthonoplastes* (60). Both field studies and pure-culture studies on nitrogen fixation have shown that *O. limosa*, a nonheterocystous, filamentous cyanobacterium, can fix nitrogen aerobically and therefore is able to colonize this low-nutrient environment (53, 55).

Cyanobacteria are predominantly oxygenic photoautotrophic organisms (61). They carry out a plantlike photosynthesis, using two photosystems (photosystems I and II) in series. Water is used as the electron donor, and the photosynthesis results in the liberation of oxygen. Some cyanobacteria are able to switch to an anoxygenic mode of photosynthesis in which only photosystem I is involved and H_2S is used as the electron donor (11, 15; chapter 25 of this volume). Other cyanobacteria have been shown to carry out oxygenic and anoxygenic photosynthesis simultaneously (9; chapter 29 of this volume). Depending on the species, elemental sulfur (11) or thiosulfate (chapter 29 of this volume) is produced during anoxygenic photosynthesis. Photosynthetically derived energy and reduction equivalents are used to fix CO_2 in the reductive pentose phosphate cycle. Part of the CO_2 fixed is diverted into the synthesis of reserve material.

The maximum period during which light intensity is sufficient to support photosynthesis in North Sea microbial mats is

Lucas J. Stal • Laboratorium voor Microbiologie, Universiteit van Amsterdam, Nieuwe Achtergracht 127, 1018 WS Amsterdam, The Netherlands. ***Heike Heyer, Susanne Bekker, Marlies Villbrandt, and Wolfgang E. Krumbein*** • Geomicrobiology Division, University of Oldenburg, P.O. Box 2503, D-2900 Oldenburg, Federal Republic of Germany.

about 12 h. However, photosynthesis can usually take place only during much shorter periods. Cyanobacteria should also be able to generate energy during the dark period. During the dark period energy is necessary not only for maintenance but also for growth of the organism, which may continue during the dark period at the same rate as in the light, for at least a limited period (65). During the dark period, cyanobacteria degrade storage compounds which were accumulated during the light period. In microbial mats, chemoheterotrophic metabolism of external carbon and energy sources is very unlikely, because of the extremely low affinities of cyanobacteria for organic substrates as compared with those of the chemoorganotrophic bacteria (52). This type of metabolism is therefore not discussed here (52). The chapter will focus on the metabolic strategies and flexibility of the microbial mat cyanobacterium *O. limosa.*

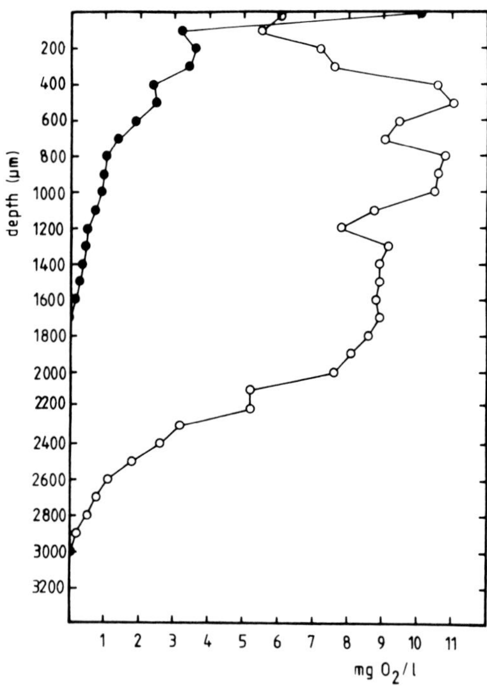

Figure 1. Oxygen profiles measured by microelectrodes in a core of a freshly colonized sandy sediment. The core was placed in seawater, but the sediment surface was exposed to the air. The sediment was illuminated by a slide projector (incident light intensity, 60 klux) (○) or incubated in the dark for 6 h (●). The layer of cyanobacteria was about 1 mm and was covered by approximately 0.5 mm of sand. Adapted from L. J. Stal and M. Villbrandt, manuscript in preparation.

NORTH SEA MICROBIAL MATS

O. limosa was isolated from microbial mats of the west coast of the North Sea island of Mellum, Federal Republic of Germany (55, 56). The western part of the island is an extended flat consisting mainly of fine sand. A considerable area of the flat belongs to the uppermost eulittoral and is flooded by the sea only irregularly during spring tides. This area between the upper eulittoral and the lower supralittoral is colonized by cyanobacteria which eventually give rise to the formation of tough coherent microbial mats (60). The pioneer cyanobacterium that colonizes the sediment is *O. limosa.* The established coherent mats consist mainly of *M. chthonoplastes*, but *O. limosa* is also common in such mats. The presence of *O. limosa* strongly correlates with nitrogenase activity in the mat (53).

Freshly colonized sediments consist of a layer of cyanobacteria loosely attached to the sand grains. The biomass is low, and there are no other dense populations of microorganisms. The cyanobacterial mat is usually hidden under a thin layer (0.5 mm) of sand, which protects the organisms against direct sunlight. The sediment is also oxidized in the deeper layers, and sulfate reduction does not occur, as was evident from the absence of ferrous sulfide (FeS). The loose sediment allows air to diffuse well into the sediment. The discrete layer of cyanobacteria was about 1 mm thick. Photosynthesis was low and could not be measured by the microelectrode technique (44), but nevertheless the sediment was supersaturated with oxygen (Fig. 1). Respiration rates were also low, and therefore the sediment remained oxygenated even after prolonged incubation in the dark.

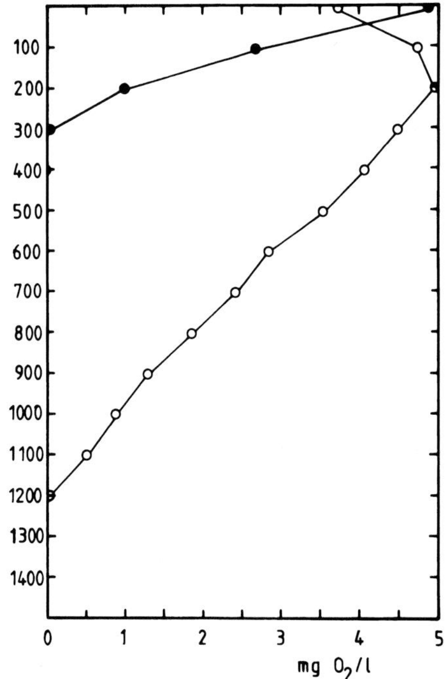

Figure 2. Oxygen profiles in an established *M. chthonoplastes* mat. The mat was about 1 mm thick and was not covered by sand. For details, see legend to Fig. 1.

Established mats are tough, coherent structures. The cyanobacterial mat contains almost no sediment and is almost entirely organic. The high standing crop of cyanobacteria gives rise to other dense populations of microorganisms such as sulfate-reducing bacteria and purple sulfur bacteria (mainly *Thiocapsa* spp.). These populations of microorganisms are vertically stratified and give rise to laminated structures. Between the cyanobacteria on top (1 mm) and the layer of purple sulfur bacteria below (0.5 mm), a rusty layer 0.5 mm thick is often seen. Microscopic analysis revealed that, if any, only cyanobacteria were present in this layer. We postulate that this layer consists of oxidized iron. Photosynthesis is very high in the cyanobacterial layer, but so is respiration. Therefore, the cyanobacterial mat is only moderately supersaturated with oxygen in the light and tends to be anaerobic during dark periods (Fig. 2). Only the upper 200 μm may be oxygenated. If the sediment is not submersed in water, air will diffuse into it. The upper layer of the cyanobacterial mat may consist of inactive or dead specimens, which protect the underlying cyanobacteria against photooxidation (sunglasses effect).

STORAGE COMPOUNDS

Definition and Significance

Cyanobacteria are known to produce a number of storage compounds (Table 1). According to Dawes and Senior (12), a compound is regarded as an energy storage compound if it meets the following criteria. (i) The compound is accumulated intracellularly if the exogenous source of energy is in excess of that required for growth and related processes. (ii) The compound is utilized if the exogenous source of energy becomes depleted and is no longer sufficient for growth or for maintenance or related

Table 1.
Storage Compounds in Cyanobacteria

Compound	Maximum amt (% [dry wt])	Function
Glycogen	10–20 (up to 60 if N starved)	Energy, carbon
Cyanophycin	8–16	Nitrogen, carbon, energy (?)
PHB	6–10	Carbon (?)
Phycobiliprotein	25	Nitrogen, carbon (?)
Polyphosphate		Phosphate, energy

Table 2.
Low-Molecular-Mass Carbohydrates in Cyanobacteria from North Sea Microbial Mats[a]

Carbohydrate	Chemical name	Occurrence
Sucrose	*O*-β-D-Fructofuranosyl-(2→1)-α-D-glucopyranoside	*Nostoc* and *Anabaena* spp.
Trehalose	*O*-α-D-Glucopyranosyl-(1→1)-α-D-glucopyranoside	*Oscillatoria* and *Calothrix* spp.
Glucosylglycerol	*O*-α-D-Glucopyranosyl-(1→2)-glycerol	LPP,[b] *Gloeocapsa*, *Synechocystis*, and *Spirulina* spp.

[a] Adapted from reference 59b.
[b] LPP, *Lyngbya*, *Phormidium*, *Plectonema* group.

processes. (iii) The compound is metabolized to produce energy in a form that the cell can utilize and to give the cell an ecological advantage over organisms that do not have these storage compounds. Such criteria also apply to storage compounds other than energy reserves.

The compounds that can be regarded as reserve material usually are polymers. The cellular content of a reserve polymer depends on the environmental conditions. The total amount of a storage compound accumulated in the cell can be considerable. However, because of the high molecular weight of these compounds, the effect on the cellular osmotic pressure will be negligible.

Marine cyanobacteria accumulate low-molecular-weight carbohydrates to adjust the osmotic pressure of the cytoplasm to the saline conditions of the environment (42, 59b) (Table 2). These carbohydrates cannot be regarded as storage compounds per se, because they accumulate in response to the concentration of salt in the environment (43). However, on the other hand, the concentration of these low-molecular-weight carbohydrates in the cytoplasm can be considerable, and therefore, in fact, they present a potential source of energy.

Glycogen

Cyanobacteria accumulate an (α1-4)-polyglucose with (α1-6) branches. This glycogen is accumulated and deposited as granules between the thylakoids (26). Exponentially photoautotrophically growing cyanobacteria may contain glycogen up to 10 to 20% of the cell dry weight (3, 32, 65). Nitrogen-starved cultures may contain up to 60% (dry weight) (3, 32). Glycogen represents a carbon and energy reserve.

PHB

Poly-β-hydroxybutyrate (PHB) is a common reserve material in bacteria (12). It serves as a carbon and energy reserve. It has so far been detected in a few cyanobacterial strains only. Thus, it has been observed in *Chlorogloeopsis* spp. (*Chlorogloea fritschii*) grown in the presence of acetate (8, 24), in *Gloeothece* spp. (R. Rippka, personal communication), and in *Spirulina platensis* (7). Its role in cyanobacteria remains obscure. Presumably, it cannot be used as a source of energy, since a complete tricarboxylic acid cycle is absent in cyanobacteria. Thus, acetate from PHB would serve as a carbon source only for biosynthetic reactions. PHB was also found in *O. limosa* (H. Heyer and L. J. Stal, submitted for publication). Synthesis of PHB in this organism could not be stimulated by acetate in the growth medium as was the case in *Chlorogloeopsis* spp. (8, 24). Preliminary results have shown that PHB accumulates to considerable amounts (±10% [dry weight]) in the late-exponential- and stationary-phase cultures of *O. limosa*. Van Gemerden (64) reported the conversion of glycogen in PHB in *Chromatium* spp. under anaerobic conditions in the dark, with elemental sulfur as an electron acceptor. Such a pathway was not found in *O.*

limosa. The amount of PHB in this organism, incubated in the dark under aerobic or anaerobic conditions, even in the presence of elemental sulfur, was virtually unchanged. PHB therefore seems not to serve as an energy store in *O. limosa*.

Polyphosphate

Many bacteria, including cyanobacteria, store phosphate as polyphosphate in granules, sometimes referred to as volutine. Polyphosphate may accumulate in stationary-phase cultures or in P-starved organisms if they are provided with phosphate (30, 62). P-starved cyanobacteria show increased uptake rates of phosphate (45).

Polyphosphate may serve as a phosphate store, e.g., for the synthesis of phospholipids or nucleic acids. Polyphosphate also may serve as an energy source (12, 33). *O. limosa* contains considerable amounts of polyphosphate, but the role of this polyphosphate in energy metabolism is uncertain.

Cyanophycin

Cyanophycin is a storage compound unique to cyanobacteria (31). It is found in electron-dense granules and is a polypeptide composed of arginine and aspartic acid in a 1:1 ratio. Cyanophycin is the trivial name for multi-L-arginine poly-L-aspartic acid. It consists of a poly-L-aspartic acid core to which L-arginyl residues are attached by peptide links (51). Synthesis of cyanophycin is not inhibited by chloramphenicol, which means that ribosomes are not involved (31, 49, 50). Evidence has been obtained that cyanophycin is produced as a result of protein catabolism in the cell and not from immediate products of CO_2 fixation (2). Cyanophycin usually is accumulated during the late exponential or stationary phase of growth and may account for 8 to 16% (dry weight) (2, 48). If cyanobacteria are grown under nitrogen limitation, the cyanophycin level is very low (1). If cultures are starved for nitrogen, cyanophycin disappears immediately (1, 17, 63). This, and the high N content of cyanophycin, strongly suggest that this polymer plays a role mainly as an N storage compound. However, it may also provide the cell with carbon and energy. It has been proposed that cyanobacteria may degrade arginine to ornithine via the dihydrolase route, which would allow the production of ATP by substrate-level phosphorylation, even under anaerobic conditions in the dark (52, 61).

O. limosa grown in the presence of nitrate contains about 2% cyanophycin (dry weight). In line with the role of cyanophycin as an N reserve, N-limited, N-fixing cultures contain only traces of cyanophycin (H. Heyer, unpublished results). If cyanophycin also played a role as an energy reserve, this would result in the production of ammonia. Indeed, ammonia was liberated by cultures of *O. limosa* incubated anaerobically in the dark. However, only a small amount of cyanophycin was degraded under such conditions (H. Heyer, unpublished results). Therefore, it was concluded that cyanophycin plays at most only a very minor role as an energy storage compound in *O. limosa*.

Phycobiliproteins

Cyanobacteria contain phycobiliproteins as light-harvesting pigments for photosynthesis (16). If cultures are starved for nitrogen, the phycobiliprotein content decreases rapidly (1, 3). The degradation of phycobiliprotein as a result of N starvation, however, always occurs after the degradation of cyanophycin (1, 2). The role of phycobiliprotein as a storage compound for N is uncertain. The decrease in phycobiliprotein can also be interpreted as a lower energy demand as a result of cessation of growth owing to nutrient limitation. Nevertheless, in non-N-limited cultures, phycobiliprotein might be present in excess of the amount needed for light harvesting. This surplus of phycobiliprotein might function as N storage or as protection against high light intensities (66). In *O. limosa*, the phycobiliprotein content of the cells was not affected by incubation of the cells in the dark. The phycobiliprotein content of *O. limosa* was not very high in any case and was a little lower under N-fixing conditions (58).

NITROGEN FIXATION

O. limosa is a cyanobacterium which is able to fix nitrogen aerobically without developing specialized cells (heterocysts) (55). The presence of the extremely oxygen-sensitive nitrogenase in oxygenic phototrophic cyanobacteria seems contradictory. Two

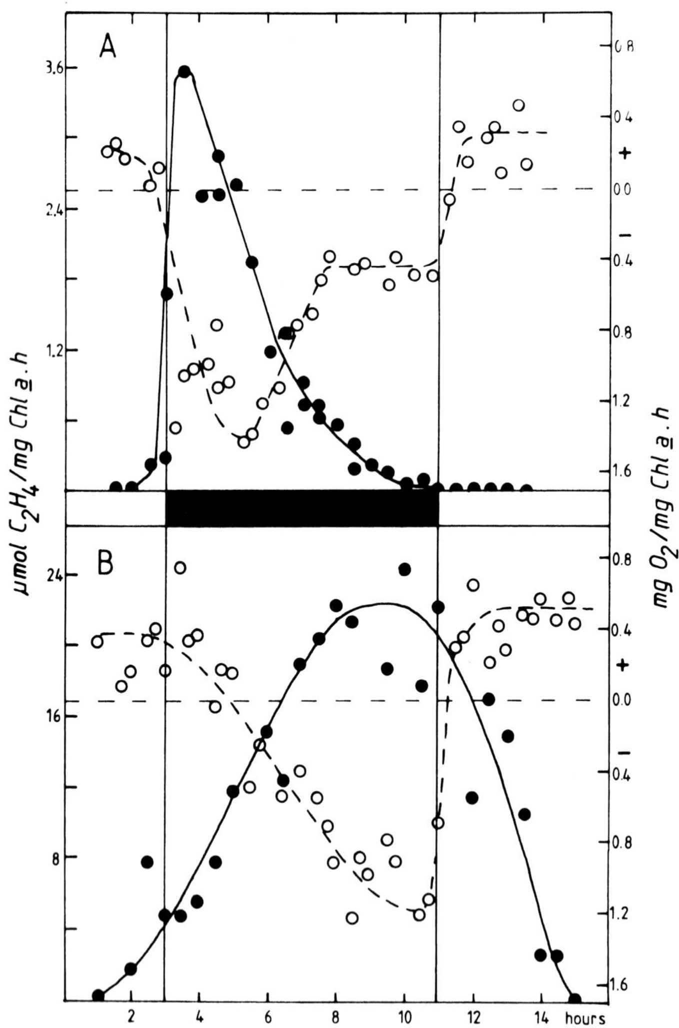

Figure 3. Nitrogenase activity (●) and photosynthesis (+) and respiration (−) (○) in cultures of *O. limosa* grown in a 16-h light, 8-h dark cycle. (A) Activities were measured in the light or in the dark according to their light-dark rhythm and as indicated by the black and white bars. (B) Cultures were transferred from a light-dark cycle to continuous light and measured in the light only. The original dark period is indicated by the black bar. Adapted from reference 59a.

strategies have been developed by nitrogen-fixing representatives of this group of organisms to cope with this problem. The first of these is a spatial separation of oxygenic photosynthesis and nitrogen fixation (27, 34, 59a). Heterocystous cyanobacteria fix nitrogen in the heterocysts. These cells do not contain photosystem II and consequently are not able to split H_2O and do not produce oxygen (14). Photosynthesis is carried out by the vegetative cells of these organisms. The other strategy is a temporal separation of nitrogen fixation and oxygenic photosynthesis (59a). Temporal separation of the processes was shown for several unicellular cyanobacteria (22, 34, 35), for the filamentous cyanobacterium *M. chthonoplastes* (38), and for *O. limosa* (57, 59a), when grown in light-dark cycles. For *O. limosa* it has been shown that even in continuous light there is a temporal separation, resulting in high respiration in the light, when the organism has induced nitrogenase (Fig. 3) (59a).

By using a light-dark cycle with anaerobic conditions during the dark period, Stal and Heyer (54) were able to induce a diurnal pattern of nitrogenase activity with two maxima in *O. limosa* (Fig. 4). Stal et al. (53) have shown that field samples which were incubated under anaerobic conditions in the presence of 3-(3,4-dichlorophenyl)-1,1-dimethylurea (DCMU) may show a diurnal pattern of nitrogenase activity which is quite similar to that shown in Fig. 4 (Fig. 5). However, it is stressed that the field experiments described in reference 53 were carried out on samples of the upper 1.3 mm of

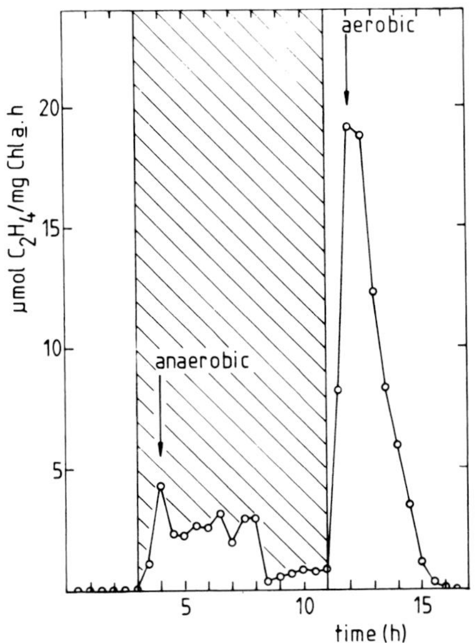

Figure 4. Nitrogenase activity in cultures of *O. limosa* grown in a 16-h light, 8-h dark aerobic-anaerobic cycle. One hour after the light was turned off, the cultures were flushed with oxygen-free N_2 (arrow, "anaerobic"). One hour after the light was turned on again, the cultures were transferred back to air (arrow, "aerobic"). Nitrogenase activity was measured under the same conditions. The shaded area indicates the dark period. Adapted from reference 54.

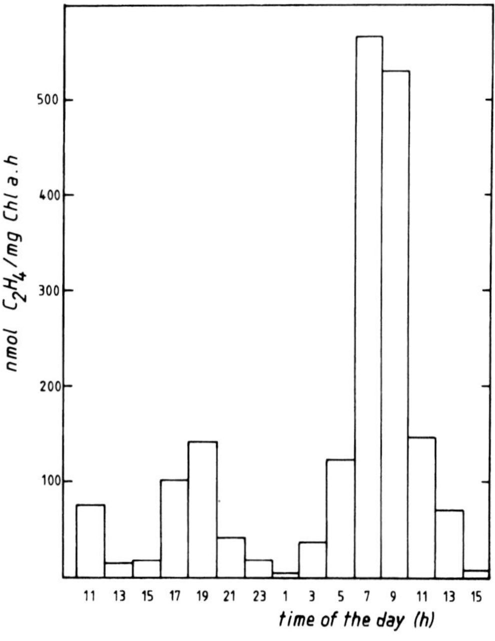

Figure 5. Diurnal nitrogenase activity in an established North Sea microbial mat. The upper 1.3 mm of the sediment was sampled. Assays were carried out on 100-mm^3 sediment samples submersed in seawater. The samples were incubated in natural light under a helium atmosphere in the presence of 10^{-5} M DCMU. Adapted from reference 53.

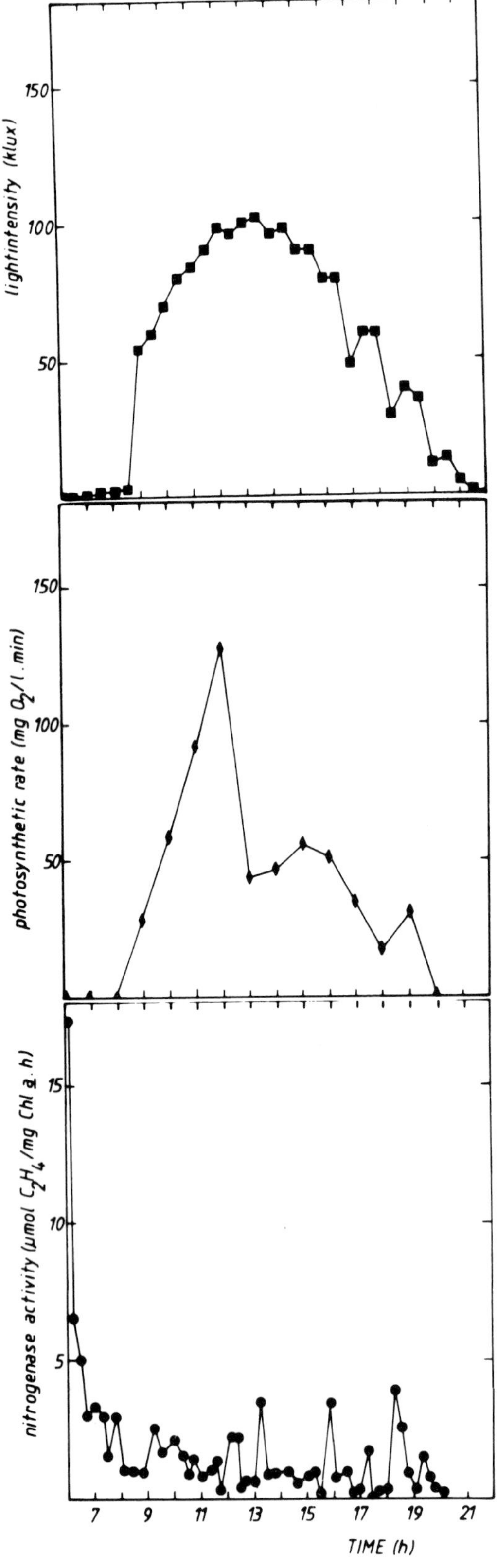

the mat, which were incubated under anaerobic conditions, submersed in seawater.

The field situation, however, is much more complicated, exhibiting more states than only aerobic conditions during the day and anaerobic conditions during the night. Pioneer populations of cyanobacteria on the sediment do not become anaerobic during the night, owing to air diffusion into the loose sediment. Only if the sediment is occasionally covered by water (high tide) do anaerobic conditions occur. In contrast, well-established mats become anaerobic during the night, and only the top 200 μm also contains some oxygen as a result of diffusion of air into the mat. Also, in this case, anaerobic conditions may persist up to the very surface of the mat, if it is covered by water. However, water covers the North Sea microbial mats quite rarely and then only for short periods (ca. 1 h).

Measurements of nitrogenase by the bell jar technique (52a) showed several differences from results of earlier experiments. Specific activities were much higher. The peak activity occurred, as expected, in the early morning, before photosynthesis was detected (Fig. 6). However, during the day a rather high but fluctuating activity was still observed. With the bell jar technique, presumably the deeper layers of the mat also contribute to the observed acetylene reduction. Stal et al. (53) have shown that the deeper layers of the mat show activities (Fig. 7) higher than those used for the experiments shown in Fig. 5. This was explained by the fact that only light of longer wavelengths (≥700 nm) penetrates the microbial mat relatively well (60; chapter 11 of this volume). This light will allow the generation of

Figure 6. Light intensity, photosynthesis, and nitrogenase activity measured in an established mat. Photosynthesis was measured by the oxygen microelectrode technique. The nitrogenase activity was measured in situ by the bell jar technique. Adapted from Stal and Villbrandt (in preparation).

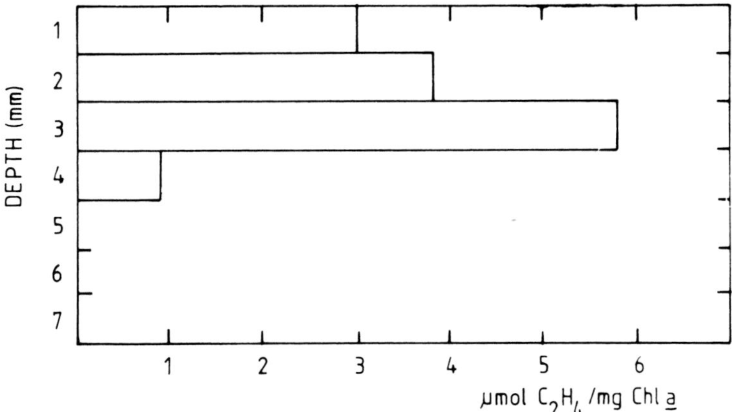

Figure 7. Depth profile of nitrogenase activity in a North Sea microbial mat. A 7-mm long, 10-mm core was sectioned into 1-mm slices and tested for acetylene reduction under anaerobic conditions in the light. Adapted from reference 53.

energy by cyclic electron transport, but not oxygenic photosynthesis. As a result, the oxygen concentrations are low in these layers, which promotes nitrogen fixation.

METABOLISM IN THE DARK

Aerobic Conditions

O. limosa grown in a light-dark cycle accumulated glycogen in the light to about 10% (dry weight) when nitrate was used as source of nitrogen and up to 30 to 40% if grown with N_2. When a nitrate-grown culture of *O. limosa* was incubated aerobically in the dark, its glycogen content did not decrease significantly (Fig. 8), even after prolonged incubation in the dark. In contrast, N_2-fixing *O. limosa*, grown under a 16-h light and 8-h dark cycle, accumulated glycogen during the light period up to 15 to 20 mg of glycogen per mg of chlorophyll *a*. During

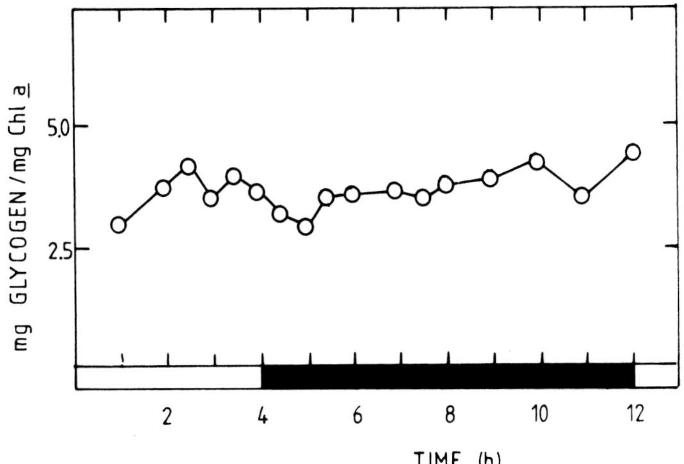

Figure 8. Glycogen content of nitrate-grown *O. limosa* in the light and in the dark.

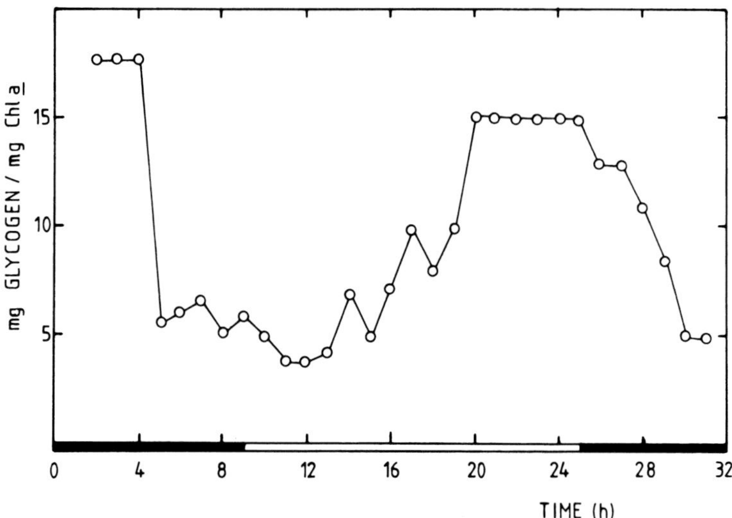

Figure 9. Glycogen content of N_2-grown *O. limosa* in the light and in the dark.

the subsequent dark period, this glycogen was degraded (Fig. 9). Degradation of glycogen, however, stopped when about 4 to 5 mg of glycogen per mg of chlorophyll *a* was left over, which was about the same content as was found in nitrate-grown cultures. Even an extended dark period of up to 48 h (Fig. 10) did not result in a further degradation of glycogen. This remaining material was isolated and enzymatically determined to be (α1-4),(α1-6)polyglucose. Therefore, we could not regard this remaining polyglucose as structural polysaccharide. In addition, under anaerobic conditions in the dark, some, but not all, of this glycogen is degraded (see below). Thus far, we do not have a satisfac-

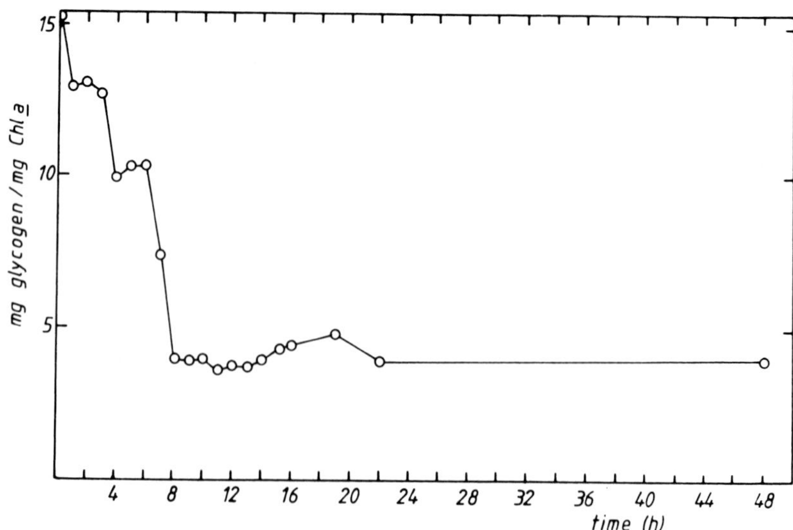

Figure 10. Decrease in glycogen in N_2-grown *O. limosa* during 48 h of incubation in the dark under aerobic conditions.

Table 3.
Products of Glycogen Fermentation in *O. limosa*[a]

Culture conditions	Amt (μmol/mg of chlorophyll *a*) of:			
	Glycogen decrease (glucose units)	Lactate	Ethanol	CO_2
N_2 grown	43	43	31	31
NO_3^- grown	25	28	28	23
N_2 grown with sulfur	38	39	41	ND[b]
NO_3^- grown with sulfur	33	24	28	ND

[a] Adapted from Heyer and Stal (submitted).
[b] ND, Not determined.

tory explanation for the fact that glycogen apparently cannot be completely degraded. It also remains obscure why more glycogen is degraded anaerobically, although its level does not decrease to zero. Thus, the question remained of which additional compound supplies the energy in nitrate-grown *O. limosa* incubated in the dark. Apparently, none of the other storage compounds was involved. Preliminary experiments have shown that the disaccharide trehalose, which occurs in *O. limosa* in considerable amounts as an osmoticum (59b), is also degraded. Experiments are in progress to resolve the role of trehalose in aerobic metabolism in the dark.

Anaerobic Conditions

O. limosa in the microbial mat has to cope with anaerobic conditions during the night (60). Also, the fact that this organism shows high nitrogenase activity anaerobically in the dark (54, 55) clearly stresses the point that *O. limosa* must possess very efficient ways to generate energy under such conditions. Indeed, it was found that glycogen was degraded anaerobically in the dark (59). Glycogen degradation continued for 12 to 24 h and was accompanied by an acidification of the medium. Heyer and Stal (submitted) found that glycogen was fermented in a heterofermentative way, producing 1 mol each of lactate, ethanol, and CO_2 per mol of glucose fermented (Table 3). The difference in glycogen content of cells grown on N_2 or nitrate was clearly reflected in the amounts of fermentation products per unit of biomass in both types of cultures. N_2-fixing cultures of *O. limosa* are N limited and therefore accumulate more glycogen. This higher amount of glycogen eventually resulted in correspondingly higher amounts of fermentation products.

Virtually nothing is known about anaerobic fermentative metabolism in cyanobacteria in the dark. Oren and Shilo (37) have reported fermentation in the cyanobacterium *Oscillatoria limnetica*. This organism apparently fermented glycogen in a homofermentative way, producing 1.4 to 1.8 mol of lactate per mol of glucose fermented. The pathway by which *O. limnetica* ferments its polyglucose is not known. Degradation via the oxidative pentose phosphate pathway does not seem likely in this case, because only a maximum of 1 mol of lactate per mol of glucose would be expected. Enzymes for homolactic fermentation have been detected in another cyanobacterium (47). On the other hand, a heterolactic acid fermentation in *O. limnetica* cannot be excluded completely, because no search for other fermentation products was conducted. The obligate chemolithotrophic *Thiobacillus neapolitanus* was shown to ferment endogenous glycogen in a heterolactic fermentation (5). Organisms that fix CO_2 through the Calvin cycle usually

Table 4.
Products of Trehalose Fermentation in *O. limosa*[a]

Culture conditions	Amt (μmol/mg of chlorophyll *a*) of:				
	Trehalose decrease (glucose units)	Acetate	H_2	H_2S	CO_2
N_2 grown	10.2	31	25.0	ND[b]	31
NO_3^- grown	10.8	28	10.5	ND	23
N_2 grown with sulfur	ND	36	14.5	1.6	ND
NO_3^- grown with sulfur	ND	36	2.5	27.1	ND

[a] Adapted from Heyer and Stal (submitted).
[b] ND, Not determined.

degrade glycogen via the oxidative pentose phosphate pathway. A heterolactic acid fermentation in such organisms seems useful because this pathway has several enzymes in common with the oxidative pentose phosphate pathway.

Under anaerobic conditions in the dark, *O. limnetica* also reduced elemental sulfur to sulfide. Elemental sulfur is produced by this organism in the light through anoxygenic photosynthesis, with sulfide as the electron donor (36). In the presence of elemental sulfur, endogenous polyglucose was oxidized mainly to CO_2, presumably via the oxidative pentose phosphate pathway. It was suggested (37) that sulfur served as terminal electron acceptor in an electron transport chain (sulfur respiration). However, even in the presence of elemental sulfur, lactate was still produced in considerable amounts (37). This might have been due to a limited availability of the sulfur, which is present only extracellularly. Energetically, no significant difference is expected between lactate fermentation and sulfur respiration. Homolactic fermentation, via glycolysis, will yield 3 mol of ATP per mol of glucose. Sulfur respiration, assuming that glucose is degraded via the oxidative pentose phosphate pathway (52), yields about 3.5 mol of ATP per mol of glucose (41).

The addition of elemental sulfur to cultures of *O. limosa* also resulted in the production of sulfide (59). In contrast to an earlier observation (59), the addition of sulfur did not significantly affect heterolactic acid fermentation (Table 3) (Heyer and Stal, submitted). Unexpectedly, analysis of the media of cultures of *O. limosa* incubated anaerobically in the dark in the presence or absence of elemental sulfur showed that considerable amounts of acetate were produced. No significant difference in acetate production was found between N_2-fixing and nitrate-grown cultures (Heyer and Stal, submitted). Strong evidence was obtained that the acetate originated from trehalose degradation. Trehalose is an osmoticum in *O. limosa* (59b). Although no further investigations of this phenomenon have been carried out, we suggest that *O. limosa* will adjust the osmotic value of the cytoplasm with inorganic ions or perhaps even with another, so far undetected, compatible solute. Elevated levels of trehalose as a result of incubation under higher salt concentrations resulted in the production of elevated amounts of acetate. Per mol of trehalose, 5 to 6 mol of acetate was produced (Table 4). *O. limosa* also produces hydrogen. Anaerobically in the dark, H_2 might be produced by nitrogenase and/or by a reversible hydrogenase (see below) in nitrogen-fixing cells, but only by hydrogenase in nitrate-grown cells, since such cells do not contain nitrogenase. The addition of elemental sulfur prevented H_2 production by the reversible hydrogenase almost completely in nitrate-grown cul-

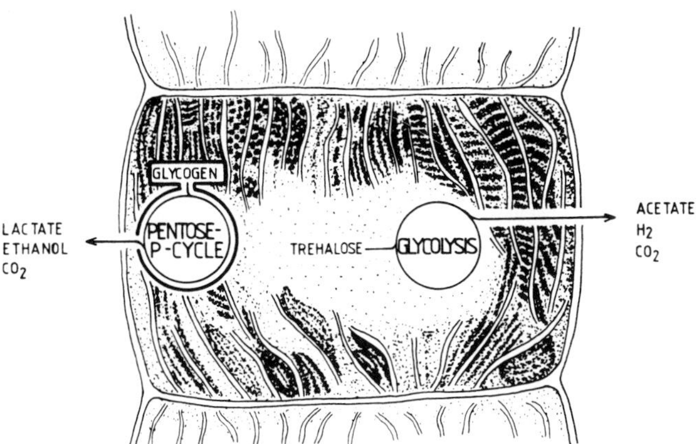

Figure 11. Proposed compartmentation of the two fermentative pathways in *O. limosa.*

tures (Table 4). Most probably, the hydrogen was used to reduce elemental sulfur. In N_2-fixing cultures only small amounts of sulfide were formed, as was reported earlier (59). This is still not understood. The H_2 observed in this case is most probably produced by nitrogenase. Preliminary investigations have suggested that N_2-fixing *O. limosa* cells can precipitate S^{2-} as FeS. The sheath of the organism might be enriched in Fe. This hypothesis is currently being investigated in our laboratory.

The pathway of acetate fermentation in *O. limosa* is not known. However, the production of 5 to 6 mol of acetate per mol of trehalose pointed strongly to a pathway in which glucose is degraded via glycolysis. This would imply that *O. limosa* possesses two independent fermentation pathways. Heyer and Stal (submitted) stressed that this can be explained only if a strict compartmentation occurs (Fig. 11). Glycogen is deposited as granules between the thylakoids (26, 46). Thylakoids are the sites of photosynthesis and of CO_2 fixation. Therefore, heterolactic fermentation might well occur at the same place. The osmoticum trehalose occurs free, dissolved in the cytoplasm, and the degradation of this compound therefore might follow a different pathway. This hypothesis is currently under investigation.

The presence of lactate fermentation has now been established in two cyanobacteria: *O. limnetica* (homolactic fermentation) and *O. limosa* (heterolactic fermentation). It should be stressed that although both organisms are assigned to the genus *Oscillatoria*, they are morphologically very different. Moreover, *O. limnetica* is a typical planktonic cyanobacterium. The organisms have in common, however, the fact that their respective habitats frequently become anaerobic (28). A survey of several cyanobacteria isolated from North Sea microbial mats revealed that no other strain possessed the capacity for lactic acid fermentation. Therefore, this type of metabolism seems to be quite rare among cyanobacteria. Sulfur reduction, on the other hand, seems to be much more common. Oren and Shilo (37) have reported that as well as *O. limnetica*, *Aphanothece halophytica* reduced elemental sulfur. Stal and Krumbein (59) reported sulfur reduction in *M. chthonoplastes*. In fact, most of the cyanobacteria isolated from North Sea microbial mats reduce elemental sulfur. However, in none of these strains is sulfur reduction correlated with a decrease in glycogen levels (L. J. Stal, unpublished results). Recently, we have obtained evidence for the involvement of organic osmotica in anaerobic metabolism of *M. chthonoplastes* in the dark.

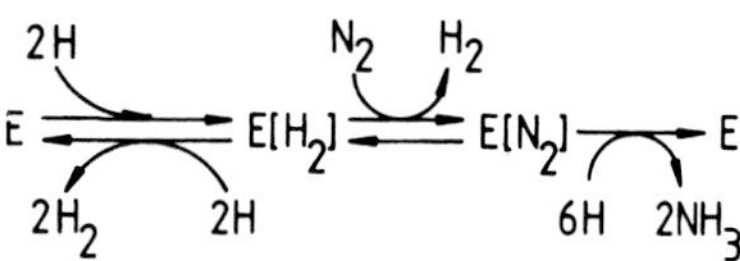

Figure 12. Model of nitrogenase-catalyzed H_2 evolution as suggested by Cleland (see reference 23).

HYDROGENASE

Three different types of hydrogenases have been detected in cyanobacteria (Table 5) (19). Nitrogenase also possesses hydrogenase activity. The reduction of N_2 to NH_3 is generally believed to involve eight electrons. Two electrons apparently are diverted for the reduction of H^+ to H_2. Cleland (see reference 23) proposed a mechanism for H_2 production by nitrogenase (Fig. 12). In the absence of a nitrogenase-reducible substrate, e.g., in an atmosphere of helium or argon, nitrogenase exclusively produces H_2, as long as energy and a low-redox-potential electron donor (reduced ferredoxin) are available. Recently, it has been proposed that nitrogenase may also function as an uptake hydrogenase (10). These observations, however, must be confirmed by more detailed investigations. H_2 production in some cases has been used to measure nitrogenase activity. This, however, is possible only if no uptake hydrogenase is present. Most N_2-fixing cyanobacteria (29) and *Anacystis nidulans* (39, 40) possess uptake hydrogenase. All uptake hydrogenases found in cyanobacteria thus far are membrane bound. They provide a reductant both to photosynthetic and to respiratory electron transport chains. The uptake of H_2 in cyanobacteria implies the presence of a suitable electron acceptor. In the well-known Knallgas reaction, electrons from H_2 are transferred to oxygen via the respiratory electron transport chain. Photosynthetic electron transport chains may transfer electrons to low-redox-potential electron carriers, which are used for CO_2 fixation and other reductant-requiring processes. The role of uptake hydrogenase in N_2-fixing cyanobacteria most probably is in the recycling of electrons which otherwise would have been lost by the action of nitrogenase (6). In addition, if H_2 is taken up in the Knallgas reaction, uptake hydrogenase might help in the protection of nitrogenase for oxygen and at the same time provide limited additional energy. Nothing is known about the role of uptake hydrogenase in the non-N_2-fixing cyanobacterium *A. nidulans* in nature. The anaerobic N_2-fixing cyanobacterium *Plectonema boryanum* does not possess uptake hydrogenase (29). This supports the hypothesis that uptake hydrogenase might play a role in oxygen protection of nitrogenase.

In the presence of acetylene, nitrogenase reduces acetylene to ethylene without producing H_2. Therefore, H_2 production and C_2H_2 reduction should be equal under conditions where no uptake of H_2 can occur, or in cyanobacteria that do not possess up-

Table 5.
Hydrogenases in Cyanobacteria

Enzyme	Localization	Reaction
Nitrogenase	Soluble	Produces ATP-dependent H_2 evolution
Uptake hydrogenase (aerobic enzyme)	Membrane bound	Provides reductant to photosynthetic and respiratory electron transport chains
Reversible hydrogenase (anaerobic enzyme)	Soluble	Produces and takes up H_2 at identical rates

take hydrogenase and if N_2 is excluded from the assay vial. Figure 13 shows the patterns of acetylene reduction and hydrogen production in several N_2-fixing cyanobacteria. In *Anabaena* strain 7120 and *P. boryanum* 73110, C_2H_2 reduction and H_2 production follow very similar patterns. The lag in the appearance of activity in *P. boryanum* was due to the time needed to induce nitrogenase in this organism. For *O. limosa* and *Anabaena* strain 1403-2, significant differences in C_2H_2 reduction and H_2 production were observed. The low H_2 evolution in *Anabaena* strain 1403-2 could not be explained. In *O. limosa* under such conditions, H_2 is very actively taken up (Table 6). The electron acceptor in this case is not known. If an atmosphere of N_2 was used, no H_2 evolution was observed. Theoretically, the rate of H_2 production under N_2 is only 1/4 of the rate of C_2H_2 reduction. The H_2 produced may be taken up and recycled for N_2 fixation.

The third type of hydrogenase is the so-called reversible hydrogenase. This soluble, anaerobic enzyme produces and takes up H_2 at identical rates (19–21). Although several reports on the existence of a reversible hydrogenase in cyanobacteria have been published (19), the role of such an enzyme remained obscure (13). Indeed, it was difficult to understand why this typical anaerobic enzyme should play a role in photoautotrophic oxygenic cyanobacteria. An exception is the photosynthetic H_2 production from H_2S during anoxygenic photosynthesis (4).

Recently, we have found a reversible hydrogenase activity in *O. limosa*; this activity plays a role in the fermentative metabolism of this organism (see above). H_2 production under anaerobic conditions in the dark occurs both in N_2-fixing and in nitrate-grown cultures (Fig. 14). When N_2-fixing cultures were tested in an argon atmosphere, high H_2 production was observed. This H_2 production, of course, was the result of both nitrogenase activity and the reversible hydrogenase. If H_2 production by nitrogenase was reduced by the addition of C_2H_2 or N_2, considerable H_2 production still remained. H_2 production under anaerobic conditions in the dark may last for 3 days (Fig. 15). The addition of elemental sulfur clearly abolished H_2 production by the reversible hydrogenase almost completely. *O. limosa* produced H_2 under anaerobic conditions in the dark. In the light, this organism takes up H_2, under both aerobic and anaerobic conditions. Also, in the dark, H_2 is taken up if a suitable electron acceptor is present. Under aerobic conditions a Knallgas reaction takes place. Interestingly, in the presence of elemental sulfur, H_2 is also taken up anaerobically in the dark with the formation of sulfide (L. J. Stal, F. Meier, and S. Bekker, manuscript in preparation). Indeed, this is to be expected if sulfur reduction in *O. limosa* is a dissimilatory process (Fig. 16).

CONCLUDING REMARKS

The cyanobacterium *O. limosa* is highly adapted to life in microbial mats. Microbial mats are systems with steep gradients of oxygen and other physicochemical parameters. Such gradients are not stable, but shift continuously. These shifts occur during the diurnal cycle, but, also, actual concentrations of oxygen vary instantaneously with incident light intensity. This is the case, e.g., on sunny but cloudy days. Therefore, the situation at any point in the cyanobacterial mat at any time may be light or dark, aerobic or anaerobic. The research carried out thus far has shown that *O. limosa* has a metabolic response to all of these situations (Fig. 17). In general, the metabolic responses are immediate; i.e., the metabolic potentials are constitutive (except perhaps anoxygenic photosynthesis). The fact that the enzymes are constitutive meets the demands of coping with rapidly changing situations in the field.

The production of lactate, ethanol, acetate, H_2, and CO_2 by *O. limosa* is of great

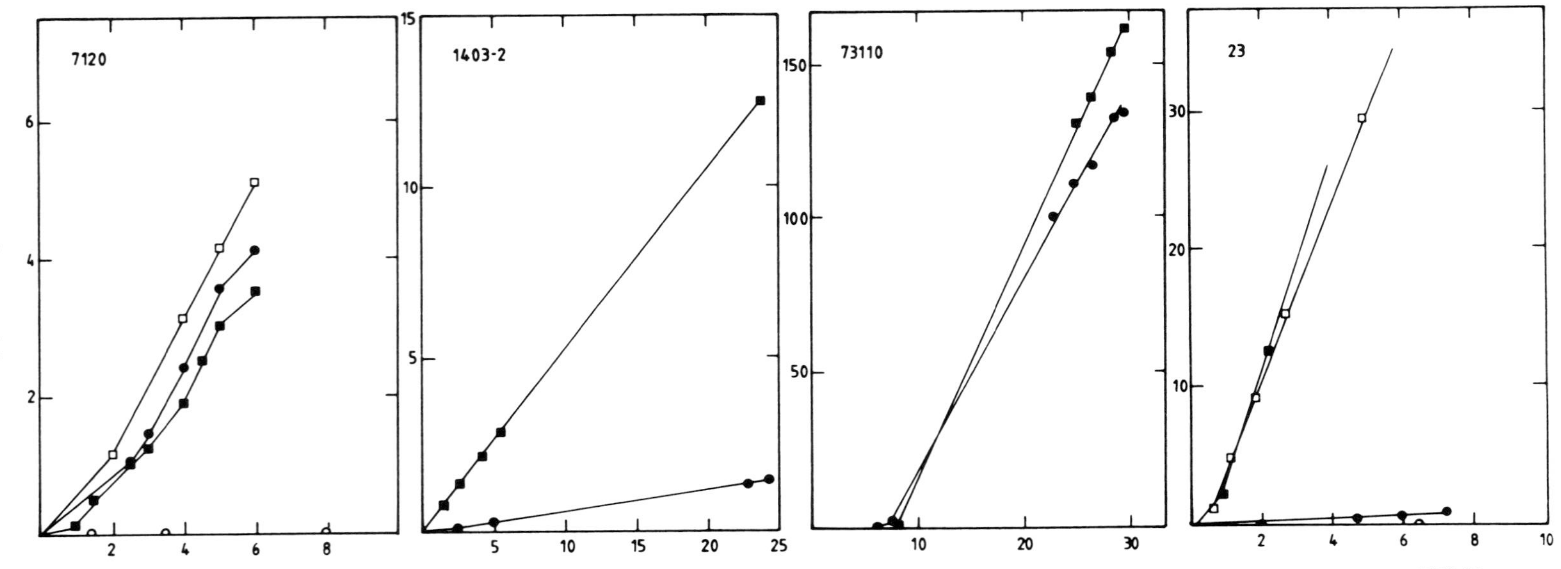

Figure 13. Nitrogenase-catalyzed H_2 evolution (●,○) and acetylene reduction (■,□) under anaerobic conditions in the light. Open symbols indicate experiments carried out under a N_2 atmosphere. Solid symbols indicate an argon atmosphere. The cultures tested were *Anabaena* strain 7120, *Anabaena cylindrica* 1403-2, *P. boryanum* 73110, and *O. limosa* 23. For culture collection numbers, refer to Table 6, footnote *c*. Adapted from L. J. Stal and S. Bekker (manuscript in preparation).

Table 6.
Acetylene Reduction, Hydrogen Production, and Hydrogen Uptake in Nitrogen-Fixing Cyanobacteria[a]

Organism[b]	Rate (μmol/mg of chlorophyll *a* per h) of:		
	C_2H_2 reduction	H_2 production	H_2 uptake[c]
O. limosa OL 23	13.30	0.16	1.82
Anabaena strain PCC 7120	0.69	0.94	0
Anabaena strain 1403-2	0.69	0.06	0
P. boryanum PCC 73110	8.07	5.97	0

[a] Adapted from Stal and Bekker (in preparation). Measurements were made under anaerobic conditions (argon atmosphere) in the light in the presence of 10 μM DCMU.
[b] Strain numbers refer to the Oldenburg collection (OL), Pasteur Culture Collection (PCC), and Sammlung von Algenkulturen of the Pflanzenphysiologisches Institut, University of Göttingen, Göttingen, Federal Republic of Germany.
[c] Uptake rate was measured after the addition of 11% H_2.

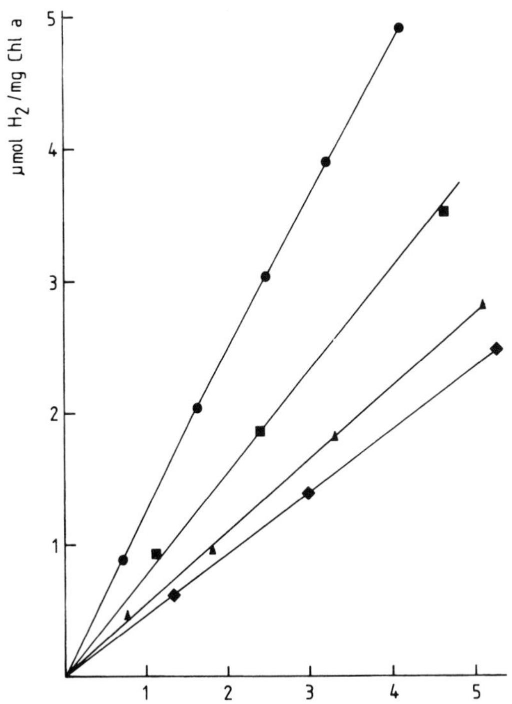

Figure 14. Anaerobic hydrogen evolution by *O. limosa* in the dark. Nitrogen-fixing cultures were incubated under an argon atmosphere (●), under N_2 (♦), or under argon with 14% acetylene (■). Nitrate-grown cultures were incubated under argon (▲). Chl a, Chlorophyll *a*. Adapted from Stal and Bekker (in preparation).

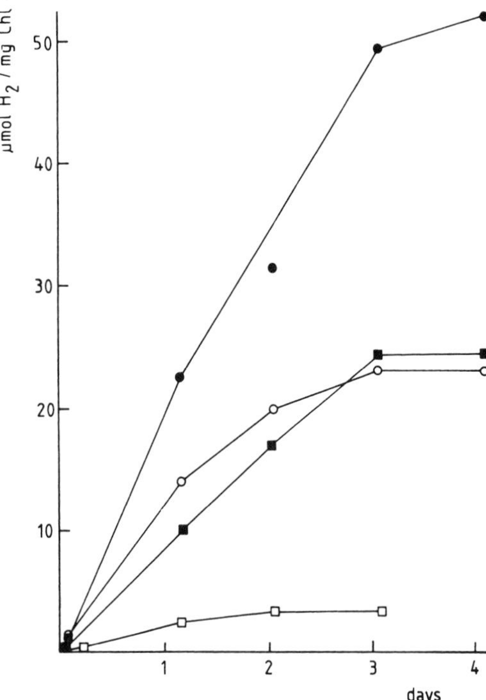

Figure 15. Effect of sulfur on anaerobic hydrogen evolution by *O. limosa* in the dark. All experiments were carried out under an argon atmosphere. Open symbols indicate the presence of sulfur; solid symbols indicate the absence of sulfur. Symbols: ●,○, experiments with nitrogen-fixing *O. limosa*; ■,□, H_2 evolution in nitrate-grown cells. Chl a, Chlorophyll *a*. Adapted from Stal and Bekker (in preparation).

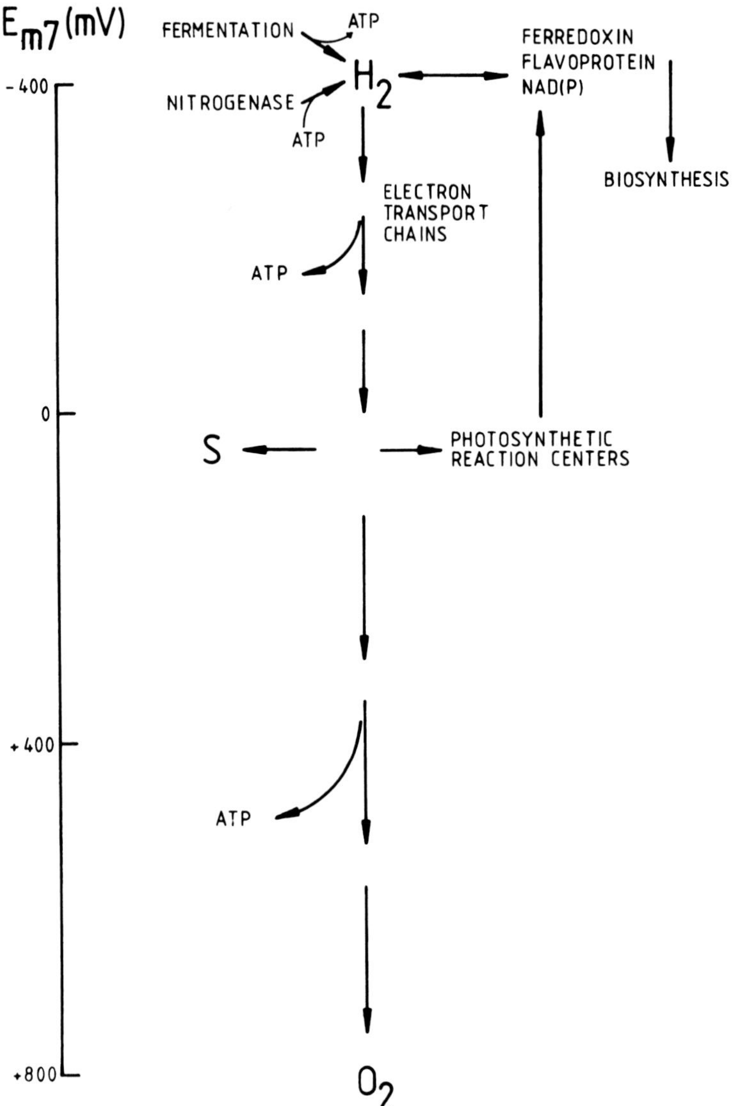

Figure 16. Hydrogen in cyanobacterial metabolism. H_2 can be produced by fermentation, photosynthesis, or nitrogenase. It can be taken up by photosynthetic and respiratory electron chains. Oxygen or sulfur may serve as a terminal electron acceptor. Adapted from reference 19.

ecological interest. Any of these products can be a substrate for sulfate-reducing bacteria. The work of Cohen (chapter 3 of this volume) has shown that sulfate reduction is spatially coupled to the cyanobacterial mat. Counts of sulfate-reducing bacteria in North Sea siliciclastic microbial mats revealed that the highest numbers were in the cyanobacterial layer (unpublished results). The ecological relationships between cyanobacteria and sulfate-reducing bacteria in microbial mats are certainly worthy of further study.

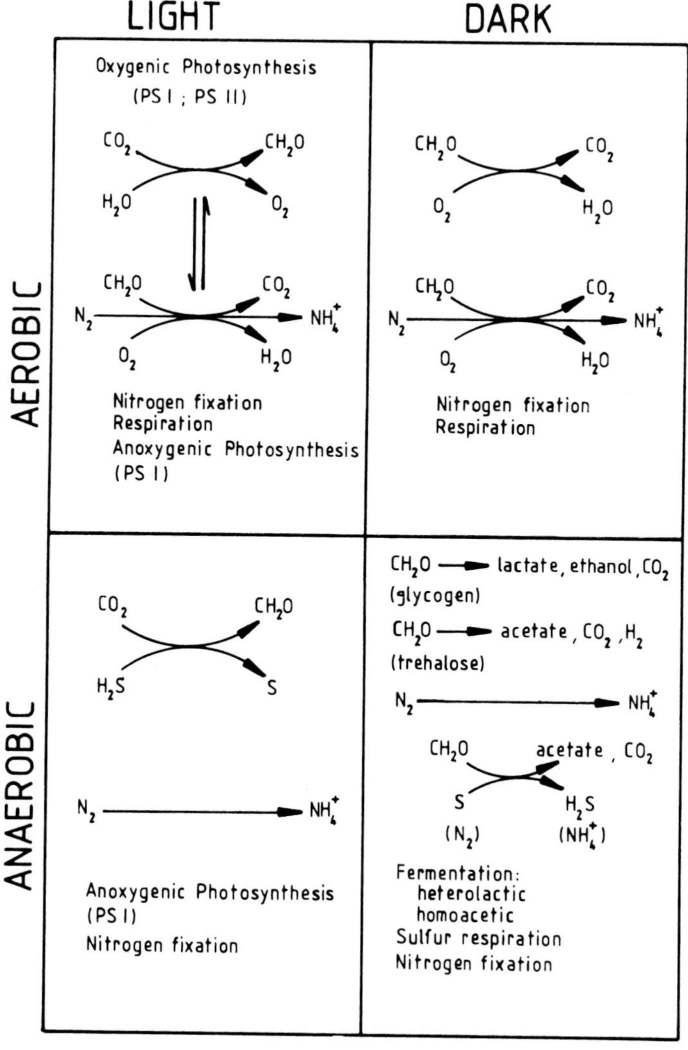

Figure 17. Summary of metabolic flexibility of *O. limosa*. Adapted from M. Shilo, A. Oren, and D. Friedberg, *Abstr. 3rd Int. Symp. Photosynth. Prokaryotes*, which contains a similar picture for *O. limnetica*.

Acknowledgments. The work on microbial mats in our laboratory has been generously supported by several grants from the Deutsche Forschungsgemeinschaft and Volkswagen Vorab of Niedersachsen.

LITERATURE CITED

1. **Allen, M. M., and F. Hutchison.** 1980. Nitrogen limitation and recovery in the cyanobacterium *Aphanocapsa* 6308. *Arch. Microbiol.* **128:**1–7.

2. **Allen, M. M., F. Hutchison, and P. J. Weathers.** 1980. Cyanophycin granule, polypeptide formation and degradation in the cyanobacterium *Aphanocapsa* 6308. *J. Bacteriol.* 141:687–693.

3. **Allen, M. M., and A. J. Smith.** 1969. Nitrogen chlorosis in blue-green algae. *Arch. Mikrobiol.* **69:** 114–120.

4. **Belkin, S., and E. Padan.** 1983. Low redox potential promotes sulphide- and light-dependent hydrogen evolution in *Oscillatoria limnetica*. *J. Gen. Microbiol.* **129:**3091–3098.

5. **Beudeker, R. F., W. de Boer, and J. G. Kuenen.** 1981. Heterolactic fermentation of intracellular

polyglucose by the obligate chemolithotroph *Thiobacillus neapolitanus* under anaerobic conditions. *FEMS Microbiol. Lett.* **12:**337–342.

6. **Bothe, H., J. Tennigkeit, G. Eisbrenner, and M. G. Yates.** 1977. The hydrogenase-nitrogenase relationship in the blue-green alga *Anabaena cylindrica. Planta* **133:**237–242.
7. **Campbell, J., III, S. E. Stevens, Jr., and D. L. Balkwill.** 1982. Accumulation of poly-β-hydroxybutyrate in *Spirulina platensis. J. Bacteriol.* **149:** 361–363.
8. **Carr, N. G.** 1966. The occurrence of poly-β-hydroxybutyrate in the blue-green alga *Chlorogloea fritschii. Biochim. Biophys. Acta* **120:**308–310.
9. **Castenholz, R. W., and H. C. Utkilen.** 1984. Physiology of sulfide tolerance in a thermophilic *Oscillatoria. Arch. Microbiol.* **138:**299–305.
10. **Chen, P.-C., H. Almon, and P. Böger.** 1986. Evidence for nitrogenase-catalyzed hydrogen uptake in nitrogen-fixing filamentous blue-green algae. *FEMS Microbiol. Lett.* **37:**45–49.
11. **Cohen, Y., E. Padan, and M. Shilo.** 1975. Facultative anoxygenic photosynthesis in the cyanobacterium *Oscillatoria limnetica. J. Bacteriol.* **123:**855–861.
12. **Dawes, E. A., and P. J. Senior.** 1973. The role and regulation of energy reserve polymers in microorganisms. *Adv. Microb. Physiol.* **10:**135–266.
13. **Eisbrenner, G., P. Roos, and H. Bothe.** 1981. The number of hydrogenases in cyanobacteria. *J. Gen. Microbiol.* **125:**383–390.
14. **Fay, P., W. D. P. Stewart, A. E. Walsby, and G. E. Fogg.** 1968. Is the heterocyst the site of nitrogen fixation in blue-green algae? *Nature* (London) **220:**810–812.
15. **Garlick, S., A. Oren, and E. Padan.** 1977. Occurrence of facultative anoxygenic photosynthesis among filamentous and unicellular cyanobacteria. *J. Bacteriol.* **129:**623–629.
16. **Glazer, A. N.** 1981. Photosynthetic accessory proteins with bilin prosthetic groups, p. 51–96. *In* M. D. Hatch and N. K. Boardman (ed.), *Biochemistry of Plants*, vol. VIII: *Photosynthesis.* Academic Press, Inc., New York.
17. **Gupta, R. S., and N. G. Carr.** 1981. Enzyme activities related to cyanophycin metabolism in heterocysts and vegetative cells of *Anabaena* spp. *J. Gen. Microbiol.* **125:**17–23.
18. **Hallenbeck, P. C., and J. R. Benemann.** 1978. Characterization and purification of the reversible hydrogenase of *Anabaena cylindrica. FEBS Lett.* **94:**261–264.
19. **Houchins, J. P.** 1984. The physiology and biochemistry of hydrogen metabolism in cyanobacteria. *Biochim. Biophys. Acta* **768:**227–255.
20. **Houchins, J. P., and R. H. Burris.** 1981. Occurrence and localization of two distinct hydrogenases in the heterocystous cyanobacterium *Anabaena* sp. strain 7120. *J. Bacteriol.* **146:**209–214.
21. **Houchins, J. P., and R. H. Burris.** 1981. Comparative characterization of two distinct hydrogenases from *Anabaena* sp. strain 7120. *J. Bacteriol.* **146:**215–221.
22. **Huang, T.-C., and T.-J. Chow.** 1986. New type of N_2-fixing unicellular cyanobacterium (blue-green alga). *FEMS Microbiol. Lett.* **36:**109–110.
23. **Jensen, B. B., and R. H. Burris.** 1985. Effect of high pN_2 and high pD_2 on NH_3 production, H_2 evolution, and HD formation by nitrogenases. *Biochemistry* **24:**1141–1147.
24. **Jensen, T. E., and L. M. Sicko.** 1971. Fine structure of poly-β-hydroxybutyric acid granules in a blue-green alga, *Chlorogloea fritschii. J. Bacteriol.* **106:**683–686.
25. **Jørgensen, B. B., N. P. Revsbech, and Y. Cohen.** 1983. Photosynthesis and structure of benthic microbial mats: microelectrode and SEM studies of four cyanobacterial communities. *Limnol. Oceanogr.* **28:**1075–1093.
26. **Jost, M.** 1965. Die Ultrastruktur von *Oscillatoria rubescens* D.C. *Arch. Mikrobiol.* **50:**211–245.
27. **Kallas, T., R. Rippka, T. Coursin, M. C. Rebière, N. Tandeau de Marsac, and G. Cohen-Bazire.** 1983. Aerobic nitrogen fixation by nonheterocystous cyanobacteria, p. 281–302. *In* G. C. Papageorgiou and L. Packer (ed.), *Photosynthetic Prokaryotes: Cell Differentiation and Function.* Elsevier Biomedical Press, Amsterdam.
28. **Krumbein, W. E., Y. Cohen, and M. Shilo.** 1977. Solar Lake (Sinai). 4. Stromatolitic cyanobacterial mat. *Limnol. Oceanogr.* **22:**635–656.
29. **Lambert, G. R., and G. D. Smith.** 1980. Hydrogen evolution by photobleached *Anabaena cylindrica. Planta* **153:**312–316.
30. **Lawry, N. H., and T. E. Jensen.** 1979. Deposition of condensed phosphate as an effect of varying sulfur deficiency in the cyanobacterium *Synechococcus* sp. (*Anacystis nidulans*). *Arch. Microbiol.* **120:** 1–7.
31. **Lawry, N. H., and R. D. Simon.** 1982. The normal and induced occurrence of cyanophycin inclusion bodies in several blue-green algae. *J. Phycol.* **18:**391–399.
32. **Lehmann, M., and G. Wöber.** 1976. Accumulation, mobilization and turnover of glycogen in the blue-green bacterium *Anacystis nidulans. Arch. Microbiol.* **111:**93–97.
33. **Merrick, J. M.** 1979. Metabolism of reserve materials, p. 199–219. *In* R. K. Clayton and W. R. Sistrom (ed.), *The Photosynthetic Bacteria.* Plenum Publishing Corp., New York.
34. **Mitsui, A., S. Kumazawa, A. Takahashi, H. Ikemoto, S. Cao, and T. Arai.** 1986. Strategy by which nitrogen-fixing unicellular cyanobacteria

grow photoautotrophically. *Nature* (London) **323:** 720–722.

35. **Mullineaux, P. M., J. R. Gallon, and A. E. Chaplin.** 1981. Acetylene reduction (nitrogen fixation) by cyanobacteria grown under alternating light-dark cycles. *FEMS Microbiol. Lett.* **10:**245–247.
36. **Oren, A., and E. Padan.** 1978. Induction of anaerobic, photoautotrophic growth in the cyanobacterium *Oscillatoria limnetica*. *J. Bacteriol.* **133:**558–563.
37. **Oren, A., and M. Shilo.** 1979. Anaerobic heterotrophic dark metabolism in the cyanobacterium *Oscillatoria limnetica*: sulfur respiration and lactate fermentation. *Arch. Microbiol.* **122:**77–84.
38. **Pearson, H. W., G. Malin, and R. Howsley.** 1981. Physiological studies on *in vitro* nitrogenase activity by axenic cultures of the blue-green alga *Microcoleus chthonoplastes*. *Br. Phycol. J.* **16:**139.
39. **Peschek, G. A.** 1979. Anaerobic hydrogenase activity in *Anacystis nidulans*: H_2-dependent photoreduction and related reactions. *Biochim. Biophys. Acta* **548:**187–202.
40. **Peschek, G. A.** 1979. Aerobic hydrogenase activity in *Anacystis nidulans*. The oxy-hydrogen reaction. *Biochim. Biophys. Acta* **548:**203–215.
41. **Pfennig, N., and H. Biebl.** 1976. *Desulfuromonas acetoxidans* gen. nov. and sp. nov., a new anaerobic, sulfur-reducing, acetate-oxidizing bacterium. *Arch. Microbiol.* **110:**3–12.
42. **Reed, R. H., L. J. Borowitzka, M. A. Mackay, J. A. Chudek, R. Foster, S. R. C. Warr, D. J. Moore, and W. D. P. Stewart.** 1986. Organic solute accumulation in osmotically stressed cyanobacteria. *FEMS Microbiol. Rev.* **39:**51–56.
43. **Reed, R. H., D. L. Richardson, S. R. C. Warr, and W. D. P. Stewart.** 1984. Carbohydrate accumulation and osmotic stress in cyanobacteria. *J. Gen. Microbiol.* **130:**1–4.
44. **Revsbech, N. P., B. B. Jørgensen, T. H. Blackburn, and Y. Cohen.** 1983. Microelectrode studies of the photosynthesis and O_2, H_2S, and pH profiles of a microbial mat. *Limnol. Oceanogr.* **28:**1062–1074.
45. **Riegman, R., and L. R. Mur.** 1985. Effects of photoperiodicity and light irradiance on phosphate-limited *Oscillatoria agardhii* in chemostat cultures. II. Phosphate uptake and growth. *Arch. Microbiol.* **142:**72–76.
46. **Ris, H., and R. N. Singh.** 1961. Electron microscope studies on blue-green algae. *J. Biophys. Biochem. Cytol.* **9:**63–80.
47. **Sanchez, J. L., N. J. Palleroni, and M. Doudoroff.** 1975. Lactate dehydrogenases in cyanobacteria. *Arch. Microbiol.* **104:**57–65.
48. **Simon, R. D.** 1973. Measurement of the cyanophycin granule polypeptide content in the blue-green alga *Anabaena cylindrica*. *J. Bacteriol.* **114:** 1213–1216.
49. **Simon, R. D.** 1973. The effect of chloramphenicol on the production of cyanophycin granule polypeptide in the blue-green alga *Anabaena cylindrica*. *Arch. Mikrobiol.* **92:**115–122.
50. **Simon, R. D.** 1976. The biosynthesis of multi-L-arginyl-poly(L-aspartic acid) in the filamentous cyanobacterium *Anabaena cylindrica*. *Biochim. Biophys. Acta* **422:**407–418.
51. **Simon, R. D., and P. Weathers.** 1976. Determination of the structure of the novel polypeptide containing aspartic acid and arginine which is found in cyanobacteria. *Biochim. Biophys. Acta* **420:** 165–176.
52. **Smith, A. J.** 1982. Modes of cyanobacterial carbon metabolism, p. 47–85. *In* N. G. Carr and B. A. Whitton (ed.), *The Biology of Cyanobacteria*. Blackwell Scientific Publications, Ltd., Oxford.

52a. **Stal, L. J.** 1988. Nitrogen fixation in cyanobacterial mats. *Methods Enzymol.* **167:**474–484.

53. **Stal, L. J., S. Grossberger, and W. E. Krumbein.** 1984. Nitrogen fixation associated with the cyanobacterial mat of a marine laminated microbial ecosystem. *Mar. Biol.* (Berlin) **82:**217–224.
54. **Stal, L. J., and H. Heyer.** 1987. Dark anaerobic nitrogen fixation (acetylene reduction) in the cyanobacterium *Oscillatoria* sp. *FEMS Microbiol. Ecol.* **45:**227–232.
55. **Stal, L. J., and W. E. Krumbein.** 1981. Aerobic nitrogen fixation in pure cultures of a benthic marine *Oscillatoria* (cyanobacteria). *FEMS Microbiol. Lett.* **11:**295–298.
56. **Stal, L. J., and W. E. Krumbein.** 1985. Isolation and characterization of cyanobacteria from a marine microbial mat. *Bot. Mar.* **28:**351–365.
57. **Stal, L. J., and W. E. Krumbein.** 1985. Nitrogenase activity in the non-heterocystous cyanobacterium *Oscillatoria* sp. grown under alternating light-dark cycles. *Arch. Microbiol.* **143:**67–71.
58. **Stal, L. J., and W. E. Krumbein.** 1985. Oxygen protection of nitrogenase in the aerobically nitrogen-fixing, non-heterocystous cyanobacterium *Oscillatoria* sp. *Arch. Microbiol.* **143:**72–76.
59. **Stal, L. J., and W. E. Krumbein.** 1986. Metabolism of cyanobacteria in anaerobic marine sediments, p. 301–309. *In GERBAM—2ᵉ Colloque International Bactériologie Marine, IFREMER, Actes de Colloques*, vol. 3. IFREMER, Brest, France.

59a. **Stal, L. J., and W. E. Krumbein.** 1987. Temporal separation of nitrogen fixation and photosynthesis in the filamentous, non-heterocystous cyanobacterium *Oscillatoria* sp. *Arch. Microbiol.* **149:**76–80.

59b. **Stal, L. J., and R. H. Reed.** 1987. Low molecular mass carbohydrate accumulation in cyanobacteria from a marine microbial mat in response to salt. *FEMS Microbiol. Ecol.* **45:**305–312.

60. **Stal, L. J., H. van Gemerden, and W. E. Krum-**

bein. 1985. Structure and development of a benthic marine microbial mat. *FEMS Microbiol. Ecol.* 31:111–125.

61. **Stanier, R. Y., and G. Cohen-Bazire.** 1977. Phototrophic prokaryotes: the cyanobacteria. *Annu. Rev. Microbiol.* 31:225–274.
62. **Stewart, W. D. P.** 1977. A botanical ramble among the blue-green algae. *Br. Phycol. J.* 12:89–115.
63. **Stewart, W. D. P., M. Pemble, and L. Al-Ugaily.** 1978. Nitrogen and phosphorus storage and utilization in blue-green algae. *Mitt. Int. Ver. Theor. Angew. Limnol.* 21:224–247.
64. **van Gemerden, H.** 1968. On the ATP generation by *Chromatium* in darkness. *Arch. Mikrobiol.* 64:118–124.
65. **van Liere, L., L. R. Mur, C. E. Gibson, and M. Herdman.** 1979. Growth and physiology of *Oscillatoria agardhii* Gomont cultivated in continuous culture with a light-dark cycle. *Arch. Microbiol.* 123:315–318.
66. **Wyman, M., R. P. F. Gregory, and N. G. Carr.** 1985. Novel role for phycoerythrin in a marine cyanobacterium, *Synechococcus* strain DC2. *Science* 230:818–820.

Chapter 25

Combined Molecular and Physiological Approach to Anoxygenic Photosynthesis of Cyanobacteria

Etana Padan

INTRODUCTION

The goal of this chapter is to summarize the recent studies on anoxygenic photosynthesis of cyanobacteria, with special emphasis on molecular and biochemical approaches. In view of these data I will reevaluate the suggestion that this type of photosynthesis bridges the gap between photosynthetic bacteria and plant-type photosynthetic organisms at the ecological, physiological, and biochemical levels (20–22).

ANOXYGENIC PHOTOAUTOTROPHIC PHYSIOLOGY AND ECOLOGY

The anoxygenically photosynthesizing cyanobacteria perform both plant-type oxygenic photosynthesis with photosystems I and II and bacterial-type anoxygenic photosynthesis solely with photosystem I. Thus, the latter occurs in the presence of H_2S, 3-(3,4-dichlorophenyl)-1,1-dimethylurea (DCMU), and 700-nm light (10) and does not show the enhancement phenomenon characteristic of oxygenic photosynthesis (18). This type of photosynthesis, first observed in *Oscillatoria limnetica*, is now known to be widespread among cyanobacteria (7, 9, 12, 13).

Sulfide oxidation in *O. limnetica* yields elemental sulfur with no further oxidation. The sulfur appears extracellularly as sulfur globules (8).

Thiosulfate has been shown to be a poor electron donor in *Anacystis nidulans* (27). Very recently, *Microcoleus chthonoplastes* has been found to efficiently oxidize sulfide to thiosulfate in a DCMU-insensitive reaction which leads to photoassimilation of CO_2 (12). It is very likely that further studies of sulfide oxidation in cyanobacteria will show as great a diversity of sulfide oxidation patterns as that observed in photosynthetic bacteria (26, 28, 29).

Depending on the reaction conditions, the electrons from sulfide in *O. limnetica* are channeled to CO_2 photoassimilation, N_2 fixation, or H_2 production (4a). Photoassimilation of CO_2 proceeds as long as H_2S (3.5

Etana Padan • Division of Molecular and Microbial Ecology, Institute of Life Sciences, The Hebrew University of Jerusalem, Jerusalem 91904, Israel.

mM) and CO_2 are supplied. Nitrogen fixation occurs in the absence of combined nitrogen (1). Hydrogen evolution occurs only when CO_2 is absent or when CO_2 fixation is otherwise inhibited (2–4). The ambient redox potential has a marked effect on the rate of hydrogen evolution, most probably owing to a direct effect on the hydrogenase (3).

The capacity to utilize sulfide in anoxygenic photosynthesis does not necessarily mean that the cyanobacteria grow anaerobically. *O. limnetica*, *O. amphigranulata*, and *Aphanothece halophytica* perform efficient anoxygenic photosynthesis in the presence of H_2S and DCMU (7, 17). *A. halophytica* does not grow at all under these conditions, yet the other two grow anaerobically: *O. limnetica* at a growth rate similar to the aerobic rate and *O. amphigranulata* at a lower rate. The lack of anaerobic growth capacity may stem from the need for O_2 in an essential metabolic reaction. Such a reaction is fatty acid oxidation in *A. halophytica* and other cyanobacteria (16). Similar to these cyanobacteria, *M. chthonoplastes* is capable of anoxygenic photosynthesis and yet cannot grow anaerobically in the presence of sulfide and DCMU (12). De Wit and van Gemerden recently succeeded in growing this cyanobacterium in the presence of sulfide and DCMU by bubbling air into the growth medium (R. de Wit and H. van Gemerden, personal communication). This interesting approach should be tried with *A. halophytica* and similar cyanobacteria.

The inability to grow anaerobically in the presence of sulfide and DCMU may also stem from the very low redox potential imposed on the cells and/or toxicity of sulfide to an essential enzyme(s) (21, 22). In fact, the very different apparent affinities of various cyanobacteria to sulfide may be caused in part by sulfide toxicity (28).

By measuring the induction of variable fluorescence of photosystem II in *O. limnetica* and *A. halophytica*, it has been shown that this photosystem is very sensitive to sulfide (19). A concentration of 0.1 mM sulfide, which is much lower than that required to operate photosystem I in anoxygenic photosynthesis (3.5 and 1.5 mM, respectively), completely blocks photosystem II. Such sensitivity has been shown to occur in other cyanobacteria, eucaryotic algae, and chloroplasts (19, 22). However, there are cyanobacteria that show marked resistance to sulfide and even perform oxygenic photosynthesis in the presence of up to 1.3 mM sulfide (7, 9, 12). These cyanobacteria include both strains that possess the capacity for facultative anoxygenic photosynthesis and those that obligately perform only oxygenic photosynthesis.

Taken together, the results for the cyanobacteria show a wide spectrum of adaptation patterns to sulfide, which may represent ancient steps in the evolution of photoautotrophic metabolism. One end of this spectrum, represented by the obligate oxygenic cyanobacteria, is shared by the plant-type photosynthetic organisms. They all possess the two photosystems, evolving oxygen while photoassimilating CO_2. They are very sensitive to sulfide and cannot utilize it as an electron donor. Furthermore, they are obligate aerobes which cannot endure a lack of oxygen and low redox potential. Thus, none of these organisms grows anaerobically even in the presence of cyclic photophosphorylation (in the presence of DCMU, organic substrates, and/or hydrogen) (21, 22). The other end of the spectrum is represented by cyanobacteria such as *O. limnetica*, as well as sulfur-oxidizing photosynthetic bacteria, which photosynthesize with one photosystem and can grow photosynthetically under totally anaerobic conditions.

BIOCHEMICAL AND MOLECULAR APPROACHES

To further understand anoxygenic photosynthesis, we studied the pattern of electron transport from sulfide in *O. limnetica* (2,

3, 4a, 5, 24a). Since sulfide-dependent CO_2 photoassimilation of CO_2 is DCMU insensitive and since it requires energy, we assumed that the cytochrome b_6f complex (cyt b_6f) must be directly or indirectly involved in the process. This complex is the main proton pump, which transduces light energy into ATP.

Two main patterns of sulfide oxidation were considered (Fig. 1). In the first pathway (Fig. 1A), sulfide is oxidized and NAD is reduced by a reversed electron transport, driven by the energy derived from the photosynthetic system. This electron transport is indirectly linked to the photosynthetic system. In the other pathway (Fig. 1B), sulfide donates its electrons directly to the photosynthetic system, at or before the cyt b_6f complex. In both pathways, sulfide oxidation should be inhibited by inhibitors of the cyt b_6f complex. Indeed, such an inhibition, both by 2,5-dibromo-3-methyl-6-isopropyl-*p*-benzoquinone (DBMIB) and the 2,4-dinitrophenyl ether of 2-iodo-4-nitrothymol (DNPINT) (2, 3, 4a) and by 2,3-dimercaptopropan-1-ol (BAL) (5), was found in intact cells of *O. limnetica*.

It is possible to choose between the two suggested pathways by using uncouplers such as carbonyl cyanide *p*-trifluoromethoxyphenylhydrazone (FCCP). In the first pathway an uncoupler should inhibit sulfide oxidation owing to deprivation of energy. In the second pathway the uncoupler should accelerate sulfide oxidation because of the uncoupling of electron transport. Since FCCP was shown to accelerate sulfide-dependent hydrogen evolution in intact cells, we have suggested that the donation site of sulfide electrons is directly at or before the cyt b_6f complex (2). Thus, sulfide electrons operate the main proton pump of the photosynthetic system.

To confirm the sulfide donation site and to further study this electron transport, it became essential to obtain a membrane preparation which performs anoxygenic photosynthetic electron transport and which does not leak protons. Such a preparation was obtained from *O. limnetica* by treating the cells with lysozyme in the presence of betaine and then bursting the spheroplasts by osmotic shock (24a). This membrane preparation showed sulfide-dependent electron transport to NAD which was dependent on light, membranes, and sulfide and was inhibited by DBMIB and DNPINT, as was the reaction in the intact cell.

To find whether this reaction is coupled to proton pumping, we monitored proton movement in the reaction system by using 9-aminoacridine, a fluorescent probe which distributes across the membrane according to the ΔpH (23, 24). A ΔpH of 2.5 units was formed across *O. limnetica* membranes in the presence of light, H_2, and NAD (24a). The ΔpH was sensitive to uncouplers and to the cyt b_6f complex inhibitors. We therefore concluded that anoxygenic electron transport in *O. limnetica* initiates electron transport, at or near the cyt b_6f complex, which is coupled to proton transport. In this way, energy is conserved for anoxygenic photoassimilation of CO_2.

The cyanobacterial electron transport, which is directly driven by photosystem I, is reminiscent of that of green bacteria such as *Chlorobium* spp. (14). Interestingly, the reaction center of the latter is claimed to be similar to photosystem I (6). On the other hand, photosystem II has recently been shown to be very similar and even homologous to the reaction center of the purple bacteria (11, 15, 25). In these bacteria, sulfide oxidation is indirectly driven by the photosynthetic system (14). Hence, it will be interesting to pursue the idea that *O. limnetica* and similar cyanobacteria are more related to the green bacteria.

The transition to anoxygenic photosynthesis involves an induction process, since it is inhibited by chloramphenicol (17). The redox potential has a marked effect on this process, which can be totally replaced by the addition of dithionite (4). To study the proteins which are synthesized during the in-

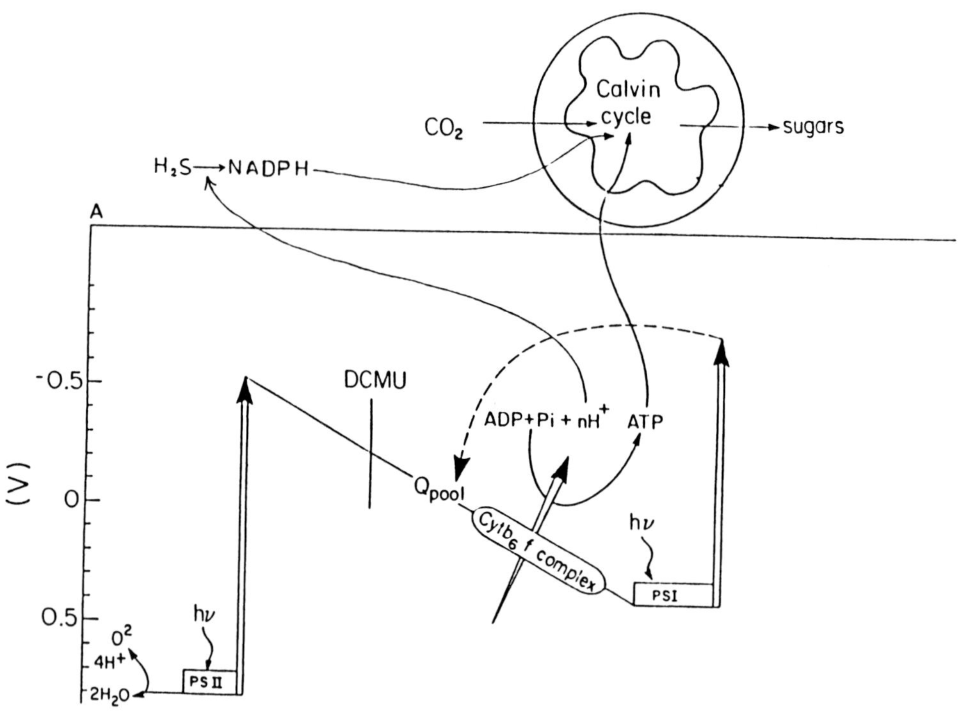
Calvin cycle
CO_2
sugars
H_2S → NADPH
A
-0.5
(V)
0
0.5
DCMU
ADP+Pi + nH^+
ATP
Q_{pool}
$Cytb_6$ f complex
hν
PSI
0^2
$4H^+$
$2H_2O$
PS II

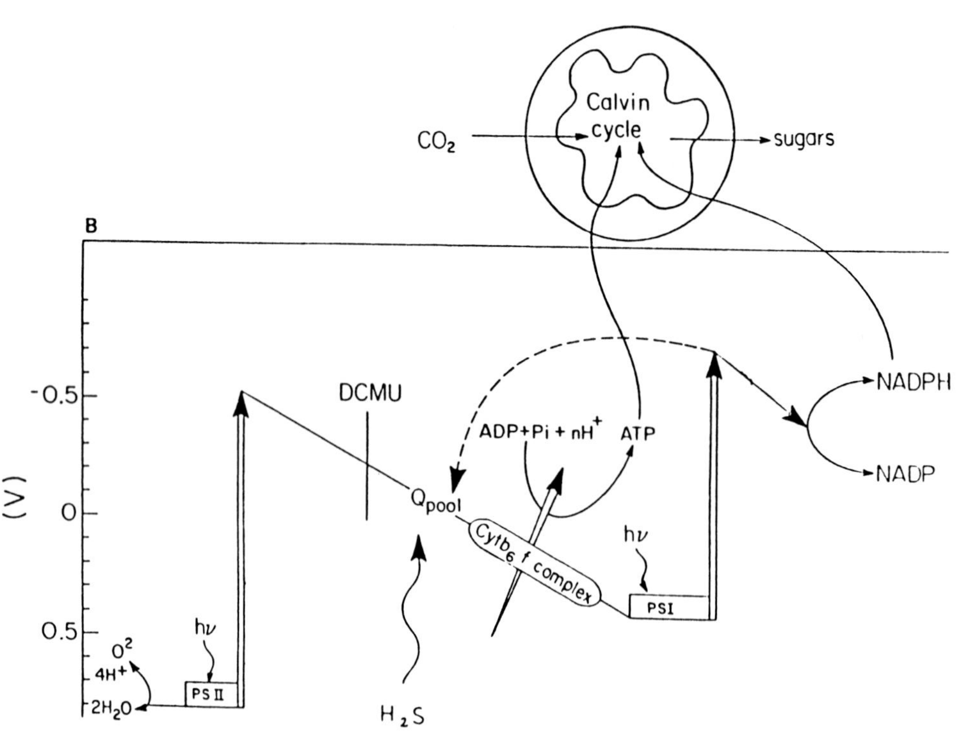
Calvin cycle
CO_2
sugars
B
-0.5
(V)
0
0.5
DCMU
ADP+Pi + nH^+
ATP
NADPH
NADP
Q_{pool}
$Cytb_6$ f complex
hν
PSI
0^2
$4H^+$
$2H_2O$
PS II
H_2S

Figure 1. Simplified scheme for alternative pathways for sulfide oxidation in phototrophic procaryotes. (A) Sulfide oxidation is indirectly linked to the photosynthetic system. (B) Sulfide donates its electrons directly to photosynthetic electron transport.

duction process, we have recently pulse-labeled the cells with [^{14}C]methionine for different times, resolved the proteins by polyacrylamide gel electrophoresis, and identified them by autoradiography (B. Arieli, B. Binder, Y. Shahak, and E. Padan, unpublished data). A major soluble protein of molecular weight 12,000 was found to be synthesized during the induction process. We are currently studying the location and function of this protein in the induced *O. limnetica* cells.

ACKNOWLEDGMENT. This work was supported by the National Council for Research and Development, Israel, and the Bundesministerium für Forschung und Technologie, Munich, Federal Republic of Germany.

LITERATURE CITED

1. **Belkin, S., B. Arieli, and E. Padan.** 1982. Sulfide-dependent electron transport in *Oscillatoria limnetica. Isr. J. Bot.* **31**:199–200.
2. **Belkin, S., and E. Padan.** 1978. Sulfide-dependent hydrogen evolution in the cyanobacterium *Oscillatoria limnetica. FEBS Lett.* **94**:291–294.
3. **Belkin, S., and E. Padan.** 1983. Low redox potential promotes sulfide- and light-dependent hydrogen evolution in *Oscillatoria limnetica. J. Gen. Microbiol.* **129**:3091–3098.
4. **Belkin, S., and E. Padan.** 1983. Na-dithionite promotes photosynthetic sulfide utilization by the cyanobacterium *Oscillatoria limnetica. Plant Physiol.* **72**:825–828.

4a. **Belkin, S., Y. Shahak, and E. Padan.** 1988. Anoxygenic photosynthetic electron transport. *Methods Enzymol.* **167**:380–386.

5. **Belkin, S., Y. Siderer, Y. Shahak, B. Arieli, and E. Padan.** 1984. 2,3-Dimercaptopropan-1-ol (BAL), an aerobic electron-transport inhibitor, but an anaerobic photosynthetic electron donor. *Biochim. Biophys. Acta* **766**:563–569.
6. **Blankenship, R. E.** 1985. Electron transport in green photosynthetic bacteria. *Photosynth. Res.* **6**:317–333.
7. **Castenholtz, R. W., and H. C. Utkilen.** 1984. Physiology of sulfide tolerance in a thermophilic *Oscillatoria. Arch. Microbiol.* **138**:299–305.
8. **Cohen, Y., B. B. Jørgensen, E. Padan, and M. Shilo.** 1975. Sulfide-dependent anoxygenic photosynthesis in the cyanobacterium *Oscillatoria limnetica. Nature* (London) **257**:489–492.
9. **Cohen, Y., B. B. Jørgensen, N. P. Revsbech, and R. Poplawski.** 1986. Adaptation to hydrogen sulfide of oxygenic and anoxygenic photosynthesis among cyanobacteria. *Appl. Environ. Microbiol.* **51**:398–407.
10. **Cohen, Y., E. Padan, and M. Shilo.** 1975. Facultative anoxygenic photosynthesis in the cyanobacterium *Oscillatoria limnetica. J. Bacteriol.* **123**: 855–861.
11. **Deisenhofer, J., O. Epp, K. Miki, R. Huber, and H. Michel.** 1985. Structure of the protein subunits in the photosynthetic reaction centre of *Rhodopseudomonas viridis* at 3Å resolution. *Nature* (London) **318**:618–624.
12. **de Wit, R., and H. van Gemerden.** 1987. Oxidation of sulfide to thiosulfate by *Microcoleus chthonoplastes. FEMS Microbiol. Ecol.* **45**:7–13.
13. **Garlick, S., A. Oren, and E. Padan.** 1977. Occurrence of facultative anoxygenic photosynthesis among filamentous unicellular cyanobacteria. *J. Bacteriol.* **129**:623–629.
14. **Knaff, D. B.** 1978. Reducing potentials and the pathway of NAD^+ reduction, p. 629–640. *In* R. K. Clayton and W. R. Sistrom (ed.), *The Photosynthetic Bacteria.* Plenum Publishing Corp., New York.
15. **Michel, H., and J. Deisenhofer.** 1986. X-ray diffraction studies on a crystalline bacterial photosynthetic reaction center: a progress report and conclusions on the structure of photosystem II reaction centers, p. 371–381. *In* L. A. Staehelin and C. J. Arntzen (ed.), *Encyclopedia of Plant Physiology*, vol. 19. Academic Press, Inc., New York.
16. **Oren, A., A. Fattom, E. Padan, and A. Tietz.** 1985. Unsaturated fatty acid composition and biosynthesis in *Oscillatoria limnetica. Arch. Microbiol.* **141**:138–142.
17. **Oren, A., and E. Padan.** 1978. Induction of anaerobic photoautotrophic growth in the cyanobacterium *Oscillatoria limnetica. J. Bacteriol.* **133**: 558–563.
18. **Oren, A., E. Padan, and M. Avron.** 1977. Quan-

tum yields for oxygenic and anoxygenic photosynthesis in the cyanobacterium *Oscillatoria limnetica. Proc. Natl. Acad. Sci. USA* 74:2152–2156.

19. **Oren, A., E. Padan, and S. Malkin.** 1978. Sulfide inhibition of photosystem II in cyanobacteria (blue-green algae) and tobacco chloroplasts. *Biochim. Biophys. Acta* **546**:270–279.

20. **Padan, E.** 1979. Facultative anoxygenic photosynthesis in cyanobacteria. *Annu. Rev. Plant Physiol.* **30**:27–40.

21. **Padan, E.** 1979. Impact of facultatively anaerobic photoautotrophic metabolism on ecology of cyanobacteria (blue-green algae). *Adv. Microb. Ecol.* **3**:1–48.

22. **Padan, E., and Y. Cohen.** 1982. Anoxygenic photosynthesis of cyanobacteria, p. 215–235. *In* N. C. Carr and B. A. Whitton (ed.), *The Biology of Cyanobacteria.* Blackwell Scientific Publications, Ltd., Oxford.

23. **Padan, E., and S. Schuldiner.** 1978. Energy transduction in the photosynthetic membranes of the cyanobacterium (blue-green alga) *Plectonema boryanum. J. Biol. Chem.* **253**:3281–3286.

24. **Schuldiner, S., H. Rottenberg, and M. Avron.** 1972. Determination of ΔpH in chloroplasts. 2. Fluorescent amines as a probe for the determination of ΔpH in chloroplasts. *Eur. J. Biochem.* **25**:64–70.

24a. **Shahak, Y., B. Arieli, B. Binder, and E. Padan.** 1987. Sulfide-dependent photosynthetic electron flow coupled to proton translocation in thylakoids of the cyanobacterium *Oscillatoria limnetica. Arch. Biochem. Biophys.* **259**:605–615.

25. **Trebst, A.** 1985. Topology of the plastoquinone and herbicide binding peptides of photosystem II in the thylakoid membrane. *Z. Naturforsch. Sect. C* **41**:240–245.

26. **Trüper, H. G.** 1984. Phototrophic bacteria and their sulfur metabolism, p. 357–382. *In* A. Muller and A. Krabs (ed.), *Sulfur, Its Significance for Chemistry, for the Geo-Bio and Cosmosphere and Technology*, vol. 5. Elsevier/North-Holland Publishing Co., Amsterdam.

27. **Utkilen, H. C.** 1976. Thiosulfate as electron donor in the blue-green algae *Anacystis nidulans. J. Gen. Microbiol.* **95**:177–180.

28. **van Gemerden, H.** 1984. The sulfide affinity of phototrophic bacteria in relation to the location of elemental sulfur. *Arch. Microbiol.* **139**:289–294.

29. **van Gemerden, H.** 1986. Production of elemental sulfur by green and purple sulfur bacteria. *Arch. Microbiol.* **146**:52–56.

Chapter 26

Ecology and General Physiology of Anoxygenic Phototrophic Bacteria in Benthic Environments

Pierre Caumette

INTRODUCTION

Anoxygenic phototrophic bacteria often develop as dense layers in a wide variety of anoxic, poorly illuminated environments found in metalimnia and hypolimnia of stratified lakes or lagoons. They also occur at the anoxic sediment surface of water bodies in the presence of sufficient light. Blooms of phototrophic bacteria have been observed as colored biomasses of mainly purple or green sulfur bacteria in such habitats (22, 23, 58, 74, 78). In addition to requiring anoxic conditions and photosynthetically active radiation, phototrophic purple and green sulfur bacteria need a suitable electron donor such as hydrogen sulfide. Most of the hydrogen sulfide which accumulates in anoxic layers is of biogenic origin, with the exception of that in sulfur springs and hydrothermal vents. In anoxic sediments hydrogen sulfide is derived mainly from bacterial breakdown of sulfur proteins via fermentation processes or from anaerobic respiration of sulfate or sulfur by sulfate- or sulfur-reducing bacteria (3, 84, 85). The second process can produce more than 95% of the biogenic sulfide found in anoxic layers of sulfate- or sulfur-rich habitats, which are encountered mainly in the marine environment, in karstic lakes (23), or in inland salt lakes (B. J. Tindall, Ph.D. thesis, University of Leicester, Leicester, England, 1980).

In anoxic sediments, little sulfide is oxidized. With exception of the anaerobic sulfur-oxidizing bacteria that use nitrate as the electron acceptor, sulfide can be oxidized only by phototrophic sulfur bacteria. This process takes place in the upper anoxic sediment reached by light (Fig. 1). Phototrophic bacteria oxidize the accumulated sulfide to intracellular sulfur (members of the family *Chromatiaceae*) or extracellular sulfur (members of the families *Ectothiorhodospiraceae* and *Chlorobiaceae*) and then to sulfate. Sulfur and sulfide can chemically react, leading to the formation of polysulfides (79). The polysulfides, as well as the intermediary oxidized products (sulfur, thiosulfate, and sulfite), can also be oxidized by phototrophic bacteria.

In addition to their presence on the sediment surface in shallow water bodies (41), phototrophic sulfur bacteria have been

Pierre Caumette • Laboratoire de Microbiologie, Faculté des Sciences de Saint-Jérôme, F-13397 Marseille Cedex 13, France.

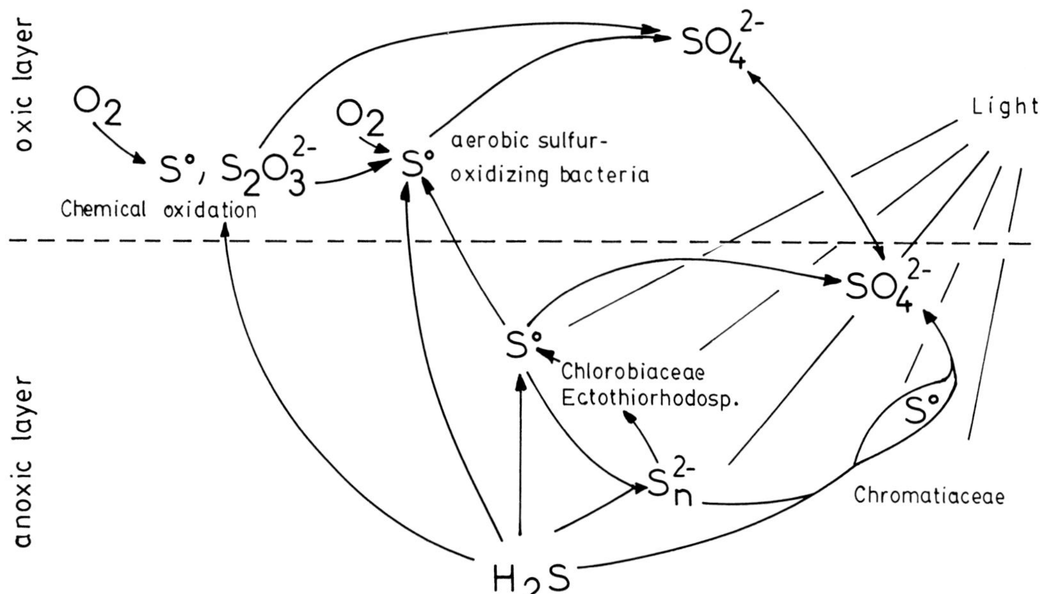

Figure 1. Oxidation pathways of hydrogen sulfide at the sediment surface of shallow water bodies.

observed in colored layers in laminated sediments underneath green layers of algae or cyanobacteria (6, 15, 27; B. K. Pierson, *V Int. Symp. Photosynth. Prokaryotes*, abstr. no. 10, 1985). They are usually found in the coastal marine environment (brackish coastal lagoons, hypersaline ponds, and salt marshes), in shallow freshwater ponds or lakes containing sulfate, or in sulfur hot springs. Their position at the surface of the anoxic layer on the sediment surface is comparable to the situation in stratified lakes, where planktonic phototrophic bacteria develop between the upper oxic and deeper anoxic waters (chapter 4 of this volume). In contrast, filamentous green bacteria often form mats at the sediment surfaces of different ecosystems such as hot springs and coastal marine waters (6; E. E. Mack and B. K. Pierson, *Workshop Green Photosynth. Bacteria*, abstr. no. 27, 1987). This chapter will be devoted to the selective advantages of benthic phototrophic bacteria in benthic ecosystems and the major role they play in such environments.

CHARACTERIZATION OF PURPLE AND GREEN SULFUR BACTERIA AND THEIR SYSTEMATIC POSITION AMONG THE ECOPHYSIOLOGICAL GROUP OF PHOTOTROPHIC BACTERIA

By definition, anoxygenic phototrophic bacteria are anaerobic photosynthetic procaryotes, and their photosynthesis does not lead to oxygen formation. Since they possess only photosystem I, they are not able to use water as an electron donor; they require more-reduced compounds such as hydrogen sulfide, sulfur, hydrogen, C_1 compounds, or organic compounds (24). In the light, two major metabolic pathways can be used by phototrophic bacteria: one involves the use of a mineral electron donor to reduce CO_2 into cell material (photolithotrophy), leading to the formation of elemental sulfur globules inside or outside the cells; the second involves photoassimilation of organic compounds (photoorganotrophy). By using such operational definitions, the phototrophic

Table 1.
Different Families or Groups of Anoxygenic Phototrophic Bacteria and Their Major Characteristics

Bacterial type	Family or group	Main electron donors	BChls	Sulfur globules
Purple bacteria	Purple nonsulfur bacteria	Organic compounds[a] (H_2S, $Na_2S_2O_3$, H_2)	BChl *a* or *b* and carotenoids	None
	Chromatiaceae[b]	H_2S, S^0, Na_2SO_3, $Na_2S_2O_3$, H_2 (organic compounds)[a,d]	BChl *a* or *b*[c] and carotenoids	Inside the cells
	Ectothiorhodospiraceae[b]	H_2S, S^0, $Na_2S_2O_3$, H_2 (organic compounds)[a,d]	BChl *a* or *b*[c]	Outside the cells
Green and brown bacteria	*Chlorobiaceae*[b]	H_2S, S^0, $Na_2S_2O_3$ (organic compounds[d] in the presence of CO_2)	BChl *c*, *d*, or *e*; small amount of BChl *a* and carotenoids	Outside the cells
	Chloroflexaceae	Organic compounds[d] (H_2S)	BChl *c* or *d*; small amount of BChl *a* and carotenoids	None or outside the filament
	Heliobacteria	Organic compounds[d]	BChl *g*; small amount of carotenoids	None

[a] Organic compounds can serve as electron donors, but are generally photoassimilated directly.
[b] The three families which form phototrophic sulfur bacteria.
[c] Only a few species.
[d] Organic compounds serve as the photosynthetic carbon source.

bacteria can be separated in two groups: those that grow mainly by photolithotrophy (phototrophic sulfur bacteria) and those that preferentially use organic compounds, including the purple nonsulfur bacteria and the filamentous gliding phototrophic bacteria.

From a systematic point of view, phototrophic bacteria are divided into purple and green bacteria according to their respective bacteriochlorophylls (BChls) and carotenoids (Table 1). The purple bacteria contain BChl *a* or *b* and numerous carotenoids (okenone, spirilloxanthin, rhodopinal, lycopenal, etc.) incorporated into a complex cell membrane system continuous with the photosynthetic membrane (56, 58, 67, 68). The specific amount of carotenoids yields colors ranging from yellow-brown to red to purple-violet. The green and brown bacteria contain BChl *c*, *d*, or *e* and carotenoids of the isorenieratene series (67) as light-harvesting pigments which are located in vesicles (chlorosomes) attached to the cell membrane at the intracytoplasmic periphery of the cells. They also contain a small amount of BChl *a* as a photosynthetic reaction center located in the membrane. In addition, other phototrophic bacteria contain β-carotene as their major carotenoid (6), whereas no photosynthetic apparatus membrane system could be found in the brown bacteria containing BChl *g*, suggesting that the pigments may be uniformly distributed in the cytoplasm (2, 17).

Phototrophic bacteria were characterized primarily on the basis of morphology, photosynthetic pigments, and nutritional requirements. However, recent investigations on the phylogenetic relatedness (69) among phototrophic purple and green bacteria have shown that their systematic differentiation must include a comparative analysis of the fine structure of certain macromolecules (59; J. Imhoff, *IV Int. Symp. Photosynth. Prokaryotes*, abstr. no. A 21, 1983; J. F. Imhoff, *V*

Table 2.
Genera of the Phototrophic Purple Bacteria[a]

Group and genus	Morphology	Division	Motility	Gas vacuoles
Purple nonsulfur bacteria				
Rhodopseudomonas	Rods	Budding	+	−
Rhodomicrobium	Ovoid cells	Budding	+	−
Rhodospirillum	Spirilloid cells	Binary	+	−
Rhodocyclus	Curved cells in circle	Binary	+ or −	−
Rhodopila	Spherical cells	Binary	+	−
Rhodobacter	Rods	Binary	+ or −	−
Purple sulfur bacteria				
Chromatiaceae[b]				
Chromatium	Rods to ovoid cells	Binary	+	−
Thiocystis	Spherical cells	Binary	+	−
Thiosarcina	Tetrads	Binary	−	−
Thiospirillum	Spirilloid cells	Binary	+	−
Thiocapsa	Spherical cells	Binary	−	−
Lamprocystis	Spherical cells	Binary	+	+
Lamprobacter	Rods	Binary	+	+
Thiodictyon	Rods	Binary	−	+
Thiopedia	Spherical cells in platelets	Binary	−	+
Amoebobacter	Spherical cells	Binary	−	+
Ectothiorhodospiraceae[c]				
Ectothiorodospira	Spirilloid to curved cells	Binary	+	+ or −

[a] Data from references 30, 35, and 76.
[b] Sulfur globules inside the cells.
[c] Sulfur globules outside the cells.

Int. Symp. Photosynth. Prokaryotes, abstr. no. 17, 1985). Both systems have led to a rearrangement of certain families and genera (30, 35), as well as of the higher taxa of phototrophic procaryotes which were recently reviewed by Trüper (73). Phototrophic sulfur bacteria are composed of three groups (Table 1): among the purple bacteria, the major phenotypic difference between members of the families *Chromatiaceae* and *Ectothiorhodospiraceae* is the ability to accumulate elemental sulfur as globules inside and outside the cells, respectively. The third family of sulfur bacteria (*Chlorobiaceae* or green sulfur bacteria) is made up of green bacteria that form sulfur globules outside their cells (Fig. 2). They include the genus *Chloroherpeton*, although members of this genus have gliding motility (see Table 3) (18).

The groups of phototrophic bacteria that grow by photoorganotrophy are listed in Tables 1 to 3. The purple nonsulfur bacteria, whose taxonomy has been recently reinves-

Figure 2. Some phototrophic bacteria often encountered in benthic environments. (a) *Rhodopseudomonas palustris*; (b) *Rhodobacter* sp.; (c) *Rhodospirillum* sp. (courtesy of Norbert Pfennig); (d) *Chromatium violascens*; (e) *Chromatium vinosum*; (f) *T. roseopersicina*; (g) *Chlorobium vibrioforme*; (h) *Chlorobium phaeobacteroides*. Bars, 10 μm.

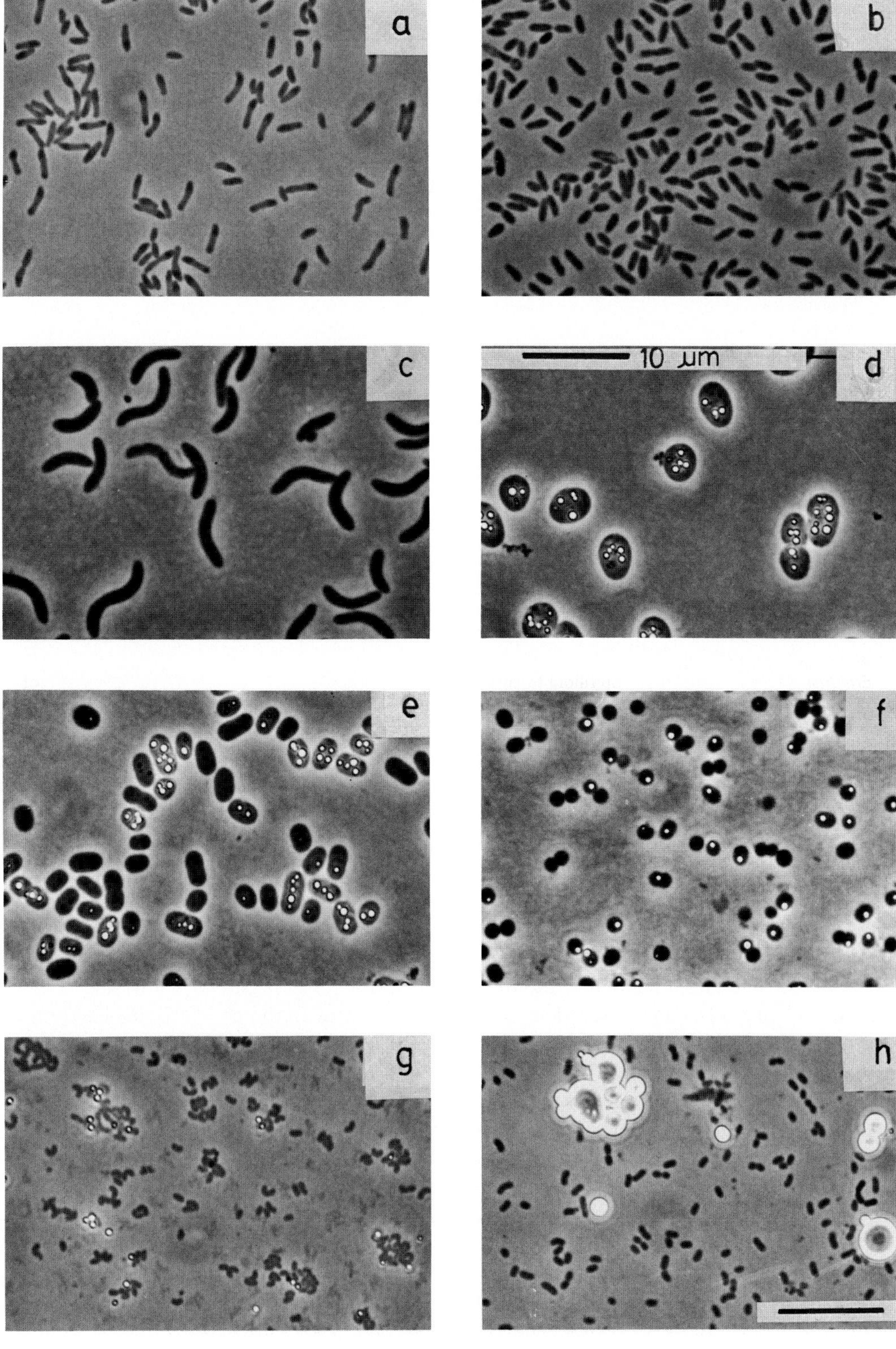
a
b
c
10 µm
d
e
f
g
h

Table 3.
Genera and Groups of Green, Brown, and Filamentous Phototrophic Bacteria[a]

Group and genus	Morphology	Motility	Gas vacuoles
Green sulfur bacteria[b]			
Chlorobiaceae			
Chlorobium	Straight or curved rods	−	−
Prosthecochloris	Irregular cells with appendages	−	−
Pelodictyon	Straight, curved, or ovoid cells	−	+
Ancalochloris	Spherical cells with appendages	−	+
Gliding green sulfur bacteria			
Chloroherpeton	Long flexing rods	Gliding	+
Filamentous green bacteria			
Chloroflexus	Filaments of 30 to 300 μm	Gliding	−
Chloronema	Filaments of 150 to 250 μm	Gliding	+
Oscillochloris	Filaments of few mm	Gliding	+
Heliothrix	Filaments	Gliding	−
Brown (Bchl *g*) phototrophic bacteria			
Heliobacterium	Long rods	Gliding	−
Heliobacillus	Long rods	+	−
Heliospirillum	Spirilloid rods	+	−

[a] Data from references 2, 7, 18, 73, and 76.
[b] Sulfur globules outside the cells.

tigated (35), are able to use a variety of simple organic compounds and fatty acids up to C_{20} (36), as well as methylamines (45), either as the sole photosynthetic carbon source or in conjunction with CO_2. Some species, isolated mainly from sulfide-rich marine environments, are also able to use reduced-sulfur compounds as electron donors (56, 80). The green filamentous gliding bacteria can form mats at the sediment surface in extreme environments. Species thus far isolated show a pronounced tendency to preferentially use organic compounds (7, 60; Mack and Pierson, *Workshop Green Photosynth. Bacteria*, 1987), although photolithotrophic growth was recently observed in *Chloroflexus* mats (19) with sulfide as the electron donor and excreted sulfur globules observed outside the filaments.

Recently, other genera and species of phototrophic bacteria have been described, including "heliobacteria," which contain BChl *g* (2, 17). For these strains, a new family (*Heliobacteriaceae*) has been suggested (2).

PHOTOTROPHIC BACTERIA IN BENTHIC ECOSYSTEMS

Almost all the phototrophic bacteria are encountered in benthic environments. Most of those found in shallow benthic environments are either purple or green sulfur bacteria. However, in some extreme environments, such as sulfur-containing hot springs, filamentous green bacteria are often the dominant organisms. In contrast, the purple nonsulfur bacteria can be isolated from any shallow benthic environment when they are enriched with suitable selective media, but they are generally found in small numbers

(78) and are considered ubiquitous organisms.

Characteristics of Benthic Ecosystems and Significance of Environmental Factors for Phototrophic Bacterial Growth

Oxygen, Sulfide, and Light Gradients

In most cases, organic matter-containing sediments in shallow water bodies are anoxic. The oxic/anoxic interface (chemocline or redoxcline) is generally found within the first millimeter or centimeter of sediment (9, 38, 63). The narrow interface between the oxygen and sulfide layers often reveals a transition zone of less than 1 mm, free of both compounds (40), in contrast to stratified lakes, where both compounds can be found in a large transition layer.

Oxygen residing in the overlying water column usually does not penetrate sediments deeper than 2 mm, although in sediments covered by cyanobacterial or algal mats it can be detected as deep as 10 mm (37, 38, 40). Below such depths, oxygen is depleted as a consequence of both chemical combination with sulfide and consumption by different heterotrophic and chemotrophic organisms, particularly the colorless sulfur-oxidizing bacteria (chapter 32 of this volume). In many sediments of shallow water bodies or tidal flats, adequate photosynthetically active radiation reaches depths of 2 to 8 mm (Fig. 3) (16, 39, 40). The blue and green parts of the light spectrum penetrate less deeply than does red and near-infrared light, which is used by phototrophic bacteria (see below). Light penetration into sediments depends on the overlying water depth; near-infrared light penetrates sediments under very shallow water bodies (less than 50 to 100 cm in depth) only. In water bodies deeper than 2 to 4 m, only wavelengths between 450 and 550 nm reach the sediment surface (Fig. 4); they can be used by phototrophic bacteria which have BChls and some specific carotenoids as light-harvesting pigments.

Effect of Temperature and pH

In temperate climates, the growth of phototrophic bacteria is affected by a strong seasonal temperature gradient. As such, this growth is frequently stimulated during spring and summer as the ambient temperature increases (9). Mesophilic phototrophic bacteria isolated from temperate benthic habitats exhibit large temperature ranges (R. Matheron, unpublished results). In contrast, mesophilic bacteria isolated from tropical areas have a narrow temperature range (P. Caumette, unpublished results). During the cold seasons in temperate climates, phototrophic bacteria remain active in the sediments. Several strains were isolated at low temperatures (5, 28, 52). Although they were active at temperatures near 5°C, their growth optima ranged between 25 and 30°C. Thus far, true psychrophilic phototrophic bacteria have not been isolated. In contrast, a few thermophilic phototrophic bacteria exist, and some were isolated from sulfur hot springs or hot hypersaline environments (Table 4) (32, 47, 60; Mack and Pierson, *Workshop Green Photosynth. Bacteria*, 1987).

Van Niel was the first to recognize the importance of pH in regulating the growth of purple and green bacteria (81). The majority of strains show optimal growth between pH 6.5 and 8. Usually, green bacteria can be selected for at pH 6.5 to 6.8, whereas purple bacteria can grow at pH values in excess of 7; most show optimal growth at pH 7.4 to 7.5 and can tolerate pH values up to pH 8.0. In most of the marine benthic environments, pH values range between 7 and 8, thus favoring purple bacteria. In addition, some halophilic purple sulfur bacteria are alkalophilic, with optimal growth between pH 8.0 and 9.0 (32; Tindall, Ph.D. thesis). So far, alkalophilic green sulfur bacteria have not been isolated, although they

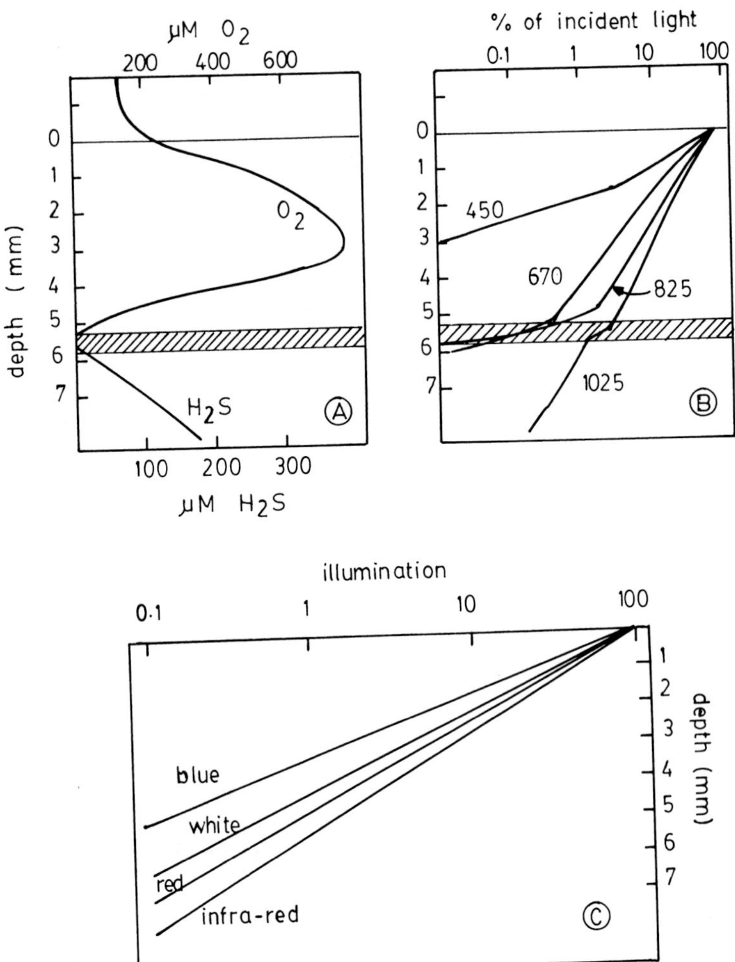

Figure 3. Typical oxygen, sulfide, and light gradients in the upper sediment of shallow water bodies. Panels A and B are adapted from reference 40; panel C is adapted from reference 16.

were observed in alkaline environments (72; Tindall, Ph.D. thesis).

Effect of Salinity

Most sediment sulfureta are located in shallow coastal marine environments, with salinity ranging from brackish to hypersaline. Some phototrophic bacteria isolated from these environments are halotolerant up to 2 to 4% NaCl, whereas strictly halophilic purple or green bacteria have frequently been isolated. These generally exhibit optimal growth at salinities between 2 and 5% NaCl (Table 5). In contrast, only a few purple bacteria have thus far been isolated from hypersaline habitats; some green sulfur bacteria have been observed (12) but not isolated. According to the classification of Trüper and Galinski (75), most of the purple bacteria isolated from hypersaline waters or sediments are moderate to halophilic bacteria sensu stricto, with optimal growth at salinities between 6 and 11% NaCl. Extremely halophilic bacteria have most commonly been isolated from alkaline brines in

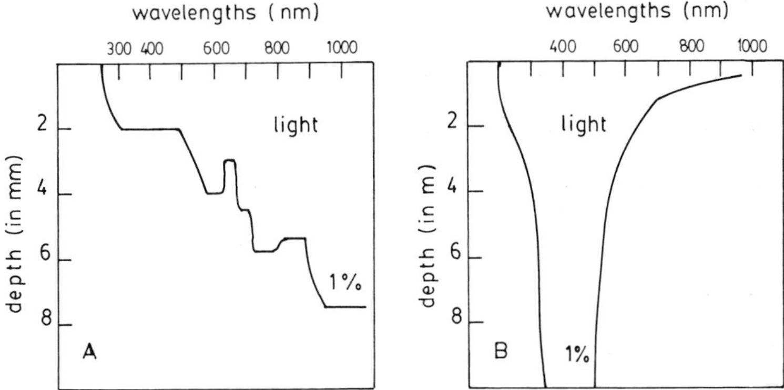

Figure 4. Depths to which 1% of incident light penetrates in natural habitats. (A) Upper sediment of shallow-water pond, covered by a cyanobacterial mat (adapted from reference 40); (B) deeper-water lake (adapted from reference 83).

athalassohaline environments such as desert lakes (32). They require about 20 to 25% NaCl for optimal growth.

Comparative Major Selective Advantages among Benthic Phototrophic Bacteria

Pigment Composition

In contrast to chlorophyll *a*, which absorbs wavelengths near 680 nm, BChl absorbs wavelengths above 700 nm. A large portion of the solar light spectrum between 370 and 1,100 nm can be used for photosynthesis in natural habitats (Fig. 5).

Jørgensen and Des Marais (40), as well as Stal et al. (70), found dense layers of *Chromatium* and *Thiocapsa* spp. underneath cyanobacterial mats, located at a sediment depth reached by wavelengths above 700 nm. Below this layer, wavelengths between 700 and 800 nm and above 900 nm were detected (40). At sufficient intensities, these spectral components can be used by green bacteria (700 to 800 nm) or purple bacteria containing BChl *b* such as *Thiocapsa pfennigii* (1,000 nm). Pierson (*V Int. Symp. Photosynth. Prokaryotes*, 1985) observed bacterial stratification in a laminated sediment: below a purple layer of *Chromatium* spp., a peach layer of *T. pfennigii* (at 0.3% of incident irradiance) and then a deeper green layer of *Prosthecochloris* spp. (0.1% of incident irradiance) were found.

Light intensity is an additional selective factor for the growth of purple and green bacteria. All the phototrophic bacteria can grow at photosynthetically active radiation intensities ranging between 20 and 40 microeinsteins · $m^{-2} \cdot s^{-1}$, or about 1 to 5% of incident light levels. Most of the pho-

Table 4.
Thermophilic Purple and Green Phototrophic Bacteria

Species	Growth temp (°C)		Reference
	Range	Optimum	
Ectothiorhodospira halochloris	33–50	45–48	32
Chromatium tepidum	34–57	48–50	47
Chloroflexus auriantiacus	≤80	55	60
Heliothrix oregonensis	≤56		61

Table 5.
Halophilic Phototrophic Bacteria Grouped According to Their Salt Requirements and Classification of Halophilic Organisms[a]

Bacterial type	Species	Reference	5 10 15 20 25 % NaCl
Marine to slightly halophilic (1.5 to 6% NaCl)	*Chromatium buderi*	57	
	Chloroherpeton thalassium	18	
	Ectothiorhodospira mobilis	57	
	Rhodobacter sulfidophilus	57	
	Pelodictyon phaeum	57	
	Rhodopseudomonas marina	35	
	Ectothiorhodospira vacuolata	31	
	Prosthecochloris phaeoasteroidea	57	
	Chlorobium chlorovibrioides	62	
	Chromatium purpuratum	33	
	Rhodobacter adriaticus	49	
	Prosthecochloris aestuarii	57	
	Chromatium vinosum HPC	1	
	Lamprobacter modestohalophilus	22	
Moderately halophilic (3 to 15% NaCl)	*Rhodospirillum mediosalinum*	46	
	Rhodospirillum salexigens	13	
	Chromatium salexigens	Caumette et al., in preparation	
Halophilic sensu stricto (9 to 24% NaCl)	*Ectothiorhodospira abdelmalekii*	34	
	Rhodospirillum salinarum	50	
Extremely halophilic (18 to 30% NaCl)	*Ectothiorhodospira halophila*	57	
	Ectothiorhodospira halochloris	32	

[a] ○, Optimum salinity.

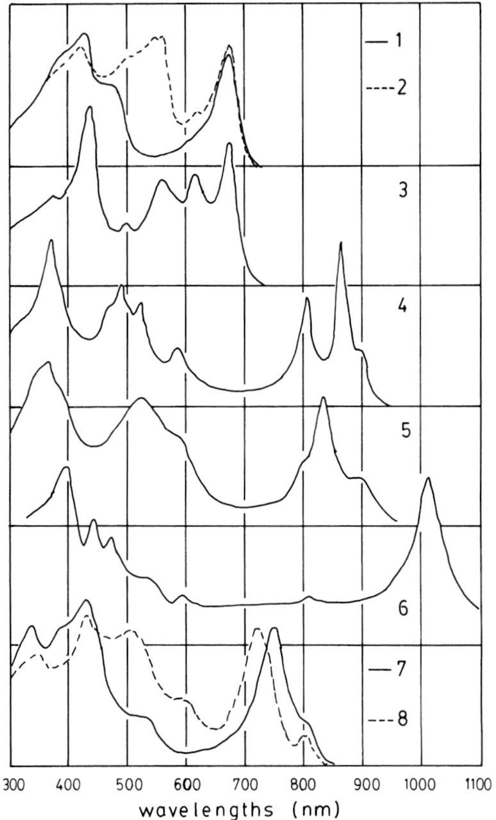

Figure 5. In vivo spectra of different phototrophic bacteria and algae. Spectra: 1, green algae; 2, red algae; 3, cyanobacteria; 4, purple bacteria containing BChl *a* plus spirilloxanthine; 5, purple bacteria containing BChl *a* plus okenone; 6, purple bacteria containing BChl *b*; 7, green bacteria containing BChl *c*; 8, green bacteria containing BChl *e*.

totrophic bacteria are able to grow at lower intensities, and some green and brown sulfur bacteria are efficient at only 0.01% of incident light levels. Green bacteria are usually dominant in sediments under deeper water bodies and hence reached by very low light intensities, since these bacteria use the wavelengths reaching the sediments (450 to 550 nm). In contrast, purple bacteria grow only at light intensities of about 0.5 to 1% of incident light levels. They are able to live in deeper water bodies at such light intensities by using wavelengths of 500 to 550 nm; such a bacterium is *Chromatium okenii*, which formed a purple layer at the sediment surface of a deep-water lake (K. Hanselmann, *V Int. Symp. Photosynth. Prokaryotes*, abstr. no. 12, 1985).

Metabolic Capacities

Sulfide tolerance. In sulfureta, phototrophic sulfur bacteria dominate the nonsulfur bacteria, including those using reduced-sulfur compounds as electron donors, owing to a higher sulfide tolerance (Table 6) (25, 26). Only few species of purple nonsulfur bacteria can tolerate low sulfide levels (0.4 to 2 mM) (25, 26, 44, 53). Members of the family *Chloroflexaceae* can also tolerate or use low sulfide concentrations (19). However, in the presence of sulfide, phototrophic sulfur bacteria are often selected for. Most of the green and brown sulfur bacteria can be enriched at high sulfide concentrations (4 to 8 mM), whereas the gas vacuolate green sulfur bacteria tolerate only low sulfide concentrations (0.4 to 2 mM) (54, 72). Although some large members of the family *Chromatiaceae* can be enriched at low sulfide concentrations (1 mM), most species in this group can tolerate 4 to 8 mM. A recent isolate grew in media containing 11 mM sulfide (P. Caumette, R. Baulaigue, and R. Matheron, manuscript in preparation).

Survival or growth in oxic environments. Diurnally benthic phototrophic bacteria are subjected to fluctuating oxic and anoxic conditions. Although they are not capable of phototrophic growth under aerobic conditions, they can survive when exposed to air. Some members of the family *Chromatiaceae* grew chemoorganotrophically or chemolithotrophically under semiaerobic conditions in the dark (42, 43), particularly some *Chromatium*, *Thiocystis*, and *Thiocapsa roseopersicina* strains (9, 43; chapter 28 of this volume). In contrast, the green sulfur bacteria were found to be more sensitive to oxygen. So far, no growth has been observed in any tested strain either under chemolithotrophic or chemoorganotrophic condi-

Table 6.
Major Ecophysiological Characteristics of the Main Phototrophic Bacteria Found in Benthic Environments

Bacterial type, family, or genus	Minimum light intensity required (microeinsteins · m^{-2} · s^{-1})	Light wavelength used (nm)	Chemotrophic growth[a]	Sulfide tolerance (mM)	Motility	Gas vacuole	Maintenance in fluctuating conditions
Purple nonsulfur bacteria							
Rhodopseudomonas	5–10	480–550 + 800–900	+	0.4–2	+	−	Good
Rhodospirillum	5–10		ND[b]	0.4–2	+	−	Good
Chromatiaceae							
Thiocapsa (BChl *a*)	2–10	480–550 + 800–900	+	1–8	−	−	Good
Thiocapsa (BChl *b*)	ND	480–550 + 1,000	−	1–8	−	−	Good
Chromatium	5–10	480–550 + 800–900	+	1–8	+	−	Good
Thiocystis	5–10		+		+	−	Good
Amoebobacter	ND		+/−		−	+	Low to good
Thiopedia	ND		−		−	+	Low to good
Ectothiorhodospiraceae							
Ectothiorhodospira	ND	480–550 + 800–900	−	1–8	+	−[c]	Good
Chlorobiaceae		450–550 + 705–755					
Chlorobium (green)	1–5		−	4–10	−	−	Low
Chlorobium (brown)	0.1–2		−	4–10	−	−	Low
Pelodictyon	0.05–2		−	0.4–2	−	+	Very low
Chloroflexaceae							
Chloroflexus			+/−	0.5	Gliding	−	Low to good
Chloroherpeton			ND		Gliding	+	

[a] Chemotrophic growth in the dark in the presence of oxygen.
[b] ND, Not determined.
[c] With the exception of one species (*Ectothiorhodospira vacuolata*).

tions (8, 42), although some green sulfur bacteria have been isolated from oxic environments (8). In stratified sediments, the green sulfur bacteria populate the most anoxic illuminated layers, underneath purple bacteria; at the oxic/anoxic interface they rarely form blooms comparable to those of purple bacteria (56). In contrast, the filamentous green bacteria are able to grow chemoorganotrophically when subjected to oxic fluctuating conditions at the sediment surface of natural environments (7).

Competition for organic substrates. In sulfide-rich sediments, purple and green sulfur bacteria can compete for organic substrates. Most members of the family *Chromatiaceae* can use different simple organic compounds as the photosynthetic carbon source (51, 55), particularly the small-celled *Chromatium*, *Amoebobacter*, and *Thiocapsa* strains (56, 76). Their affinity for organic substrates exceeds that of the green sulfur bacteria (29, 82). The latter group used relatively few organic compounds, including acetate and pyruvate, and can use them only in the presence of sulfide and CO_2 (56, 66). Thus, at the oxic/anoxic sediment interface, purple sulfur bacteria are favored by the presence of organic matter usually originating from decaying algae (9).

Examples of the Distribution and Diversity of Phototrophic Bacteria in Benthic Environments

Seasonal Distribution of Benthic Phototrophic Bacteria

The largest numbers of phototrophic bacteria were found in sediment densely populated by purple sulfur bacteria which formed visible colored layers; bacterial densities ranged between 10^7 and 10^8 CFU · ml^{-1} (27, 41), although they could be underestimated owing to the use of agar colony counts. In most of the sulfide-rich sediments under shallow water bodies, bacterial counts commonly range between 10^6 and 10^8 CFU · ml^{-1}. In temperate climates, the largest numbers are found during the summer: in a coastal lagoon, for instance, phototrophic bacterial counts at the sediment surface ranged between 10^4 CFU · ml^{-1} during the winter and 10^7 CFU · ml^{-1} in the warmest season (9).

Stratification and Vertical Colonization in Sediments

Some examples of stratification of different types of phototrophic sulfur bacteria in sediments are reported in Fig. 6. The different organisms stratify according to light penetration in sediments. Usually, in sediments covered by less than 1 m of water, purple to pink layers occur underneath cyanobacterial or algal mats (9, 40, 70). In the examples cited in Fig. 6, the dominant organisms in such layers are members of the genera *Chromatium* and *Thiocapsa*. *Thiocapsa* spp. were also found to be the dominant organisms in sediments on sheltered beaches in contact with air (27). In addition, Pierson (*V Int. Symp. Photosynth. Prokaryotes*, 1985) described a three-layered community of anoxygenic phototrophic bacteria underneath a cyanobacterial mat (Fig. 6); green sulfur bacteria were able to develop in the lowest layer where light was still available. In contrast, in sediments under deeper water bodies, purple or green sulfur bacteria can be favored according to the available light intensity (Fig. 6) rather than light quality. According to data in the literature, it can be stated that at the oxic/anoxic interface in sediments under deeper water bodies reached by light of wavelengths between 450 and 550 nm, purple bacteria can grow if photosynthetically active radiation intensities are higher than 0.5 to 1% of the incident irradiance; below such intensities, green bacteria are generally selected.

The position of the colored layer of phototrophic bacteria can be estimated by measuring phototrophic sulfide oxidation with a sulfide microelectrode (38, 40, 63). Experiments carried out on a marine coastal sediment covered by a *Beggiatoa* mat revealed a peak of phototrophic sulfide oxidation at the surface of the sulfide layer (P. Caumette and B. B. Jørgensen, unpublished results) (Fig. 7). In the dark, sulfide diffused up in the first 1 mm of water above the sediment (Fig. 7-1). Consequently, motile phototrophic bacteria of the *Chromatium* type were found in such a layer above the *Beggiatoa* mat. In this case, motile phototrophic bacteria were more efficient than *Beggiatoa* strains in oxidizing hydrogen sulfide by means of their chemotrophic metabolism used in the dark. When the sediment surface was subjected to infrared light, a peak of phototrophic sulfide oxidation was measured at the sediment surface (Fig. 7-2). This community remained stable as long as it was exposed to infrared light. When it was exposed to normal daylight, the oxic/anoxic interface migrated downward (4 to 6 mm depth) as a consequence of benthic oxygen

	Light reaching the layers		Species observed or isolated	Main light-harvesting pigments
	Intensity (%)	Quality (nm)		
cyanobact.				
purple layer	2.5	700	*Thiocapsa*	BChl *a*
orange layer	0.3	700–800 plus >1,000	*Thiocapsa pfennigii*	BChl *b*
green layer	0.1	700–800	*Prosthecochloris*	BChl *c*
12 meters of water depth + green algae				
purple layer	1–5	550	*Chromatium okenii*	Okenone
30 cm of water				
cyanobact.				
purple layer	0.3	>700	*Chromatium* spp.	BChl *a*
20 meters of water depth				
green bacteria	0.1	450–550	*Chlorobium* spp.	Series of isorenieratenes plus BChl *c*, *d*, *e*
air				
purple layer			*Thiocapsa*	BChl *a* plus spirilloxanthine series
sand				
50 cm water				
"pink" layer	5–20	>700	*Thiocapsa*	BChl *a*

Figure 6. Some examples of phototrophic bacterial layers at the surface or in laminated sediments. From top to bottom, the examples were originally discussed by Pierson (*V Int. Symp. Photosynth. Prokaryotes*, 1985), by Hanselmann (*V Int. Symp. Photosynth. Prokaryotes*, 1985), and in references 40, 48, 27, and 9, respectively.

production; this led to an increase of *Beggiatoa* activity, since these bacteria were more efficient in oxidizing hydrogen sulfide under these oxic conditions. Consequently, the phototrophic bacteria were detected deeper in the anoxic sediment in the dark.

In salinas, purple layers of phototrophic bacteria often occur below the crystal deposits, usually the gypsum crust. Such layers in the salinas in Giraud, France, were recently analyzed. They occurred underneath a green layer composed almost entirely of cyanobacteria, at a depth of 5 to 7 mm (Fig. 8) (12). The largest numbers of phototrophic bacteria were obtained from colony counts in 25% NaCl medium; however, few phototrophic

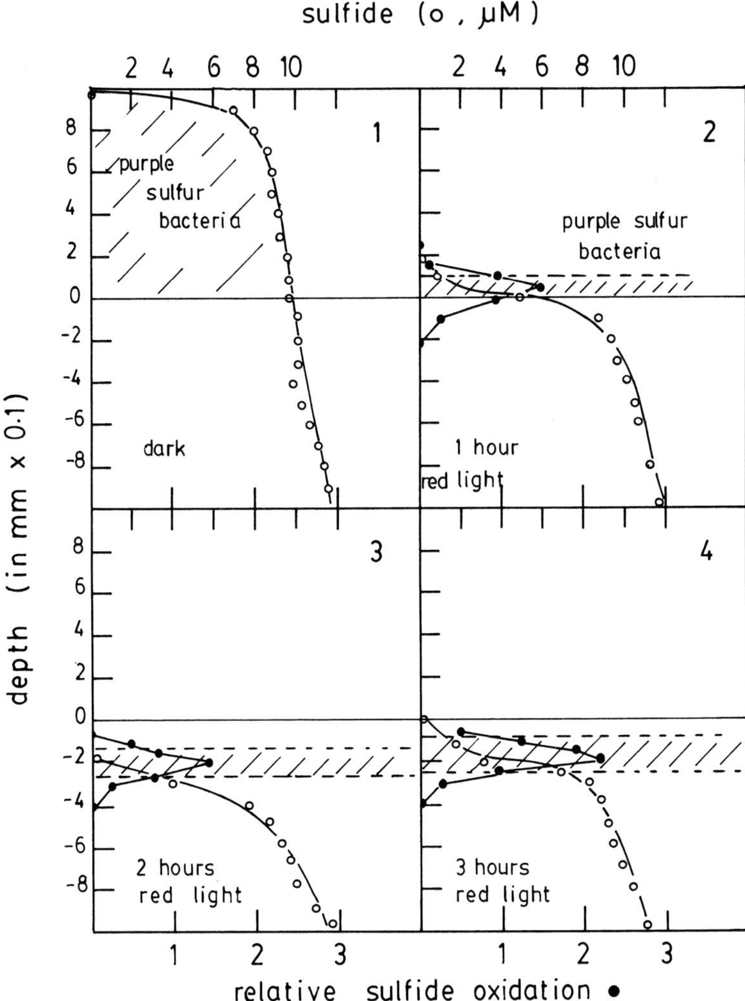

Figure 7. Position of the oxic/anoxic interface and phototrophic bacteria through light and dark exposures at the sediment surface in a shallow-water pond. Infrared light was used to prevent oxygenic photosynthesis. The sulfide oxidation was measured by microelectrodes.

bacteria also grew in NaCl-free media, since halotolerant organisms presumably originated from seawater and still lived in the concentrated brines. From the 25% NaCl medium, in addition to members of the family *Ectothiorhodospiraceae*, some members of the family *Chromatiaceae* were isolated (Caumette et al., in preparation). They were often observed in different hypersaline environments (4, 65).

Generic Distribution of Phototrophic Bacteria in Benthic Communities

In contrast to planktonic phototrophic bacteria, which are restricted to species able to float or swim, virtually all types of phototrophic bacteria can be found at the sediment surface (Table 7). Bacteria adapted to fluctuating physicochemical conditions are usually dominant. They migrate vertically in

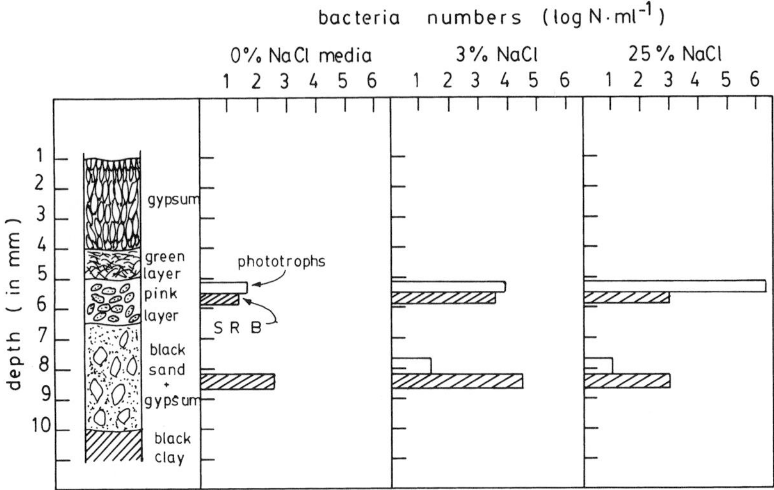

Figure 8. Numbers of phototrophic and sulfate-reducing bacteria in laminated sediments of salinas in Giraud, France. Numbers were estimated by agar counts in three different salted media. Adapted from reference 12.

accordance with sulfide and light gradients or are metabolically well adapted to fluctuating conditions (i.e., *Thiocapsa* spp.). A morphological adaptation to sediments is the gliding motility of some green bacteria, such as *Chloroflexus* spp., which densely populate the sediment surface of sulfur hot springs (60, 61; Mack and Pierson, *Workshop Green Photosynth. Bacteria*, 1987) or *Chloroherpeton* spp., which were isolated from marine coastal sediments (18).

T. roseopersicina and *Chromatium vinosum* were often found to be the dominant organisms in benthic environments and were sometimes accompanied by other species such as *Chromatium okenii* or *Amoebobacter* spp. (41; Hanselmann, *V. Int. Symp. Photosynth. Prokaryotes*, 1985), which are gas-vacuolated organisms. From an ecological point of view, the occurrence of vacuolated bacteria in benthic environments cannot be easily explained. However, *Amoebobacter* spp. possess slime capsules around the cells, which allow cell aggregation. They shares this property with *Thiocapsa* spp. The presence of slime capsules around the cells could have an important role in the adhesion of cells and cell aggregates to the sediment surface, as well as in the survival of such aggregates under oxic conditions.

In most of the colored layers of purple or green sulfur bacteria, the bacterial diversity is very low, often being restricted to only a few species. From an ecological point of view, such ecosystems are considered extreme habitats, as typified by hypersaline and hot environments; the extreme conditions lead to a low diversity. For instance, during the time when anoxic water is present in a shallow brackish coastal lagoon, the purple layer that extends to the overlying water is almost entirely composed of *T. roseopersicina*, which fills the disturbed ecosystem that is rich in hydrogen sulfide (9).

Most of the extremely halophilic anoxygenic phototrophic bacteria are members of the family *Ectothiorhodospiraceae*, which densely populates some alkaline hypersaline environments (32–34). In contrast, in marine hypersaline environments, the purple layers occurring at the sediment surface appear to be composed mostly of members of the family *Chromatiaceae* (65). In diverse salinas around the Mediterranean Sea, members of the family *Chromatiaceae* such as *Chromatium* spp. (65) or *Thiocapsa* spp. (4)

Table 7.
Main Genera or Species of Phototrophic Bacteria Observed or Isolated from Benthic Environments

Ecosystem	Phototrophic purple bacteria	Phototrophic green bacteria
Freshwater benthic	*Chromatium vinosum*	*Chlorobium limicola*
	Chromatium violascens	*Chlorobium vibrioforme*
	Chromatium minus	*Chlorobium phaeobacteroides*
	Thiocystis violacea	*Pelodictyon* sp.
	Thiocapsa roseopersicina	*Chloroherpeton*-like bacteria[a]
	Thiocapsa pfennigii	
	Thiosarcina sp.	
	Thiospirillum sp.	
	(*Thiopedia*, *Lamprocystis*, *Amoebobacter*, *Thiodictyon*)	
Hot freshwater benthic	*Chromatium tepidum*	*Chloroflexus* spp.
	Heliothrix sp.	
Marine to salt-water benthic	*Chromatium buderi*	*Chlorobium limicola*
	Chromatium vinosum	*Chlorobium vibrioforme*
	Chromatium gracile	*Chlorobium phaeovibrioides*
	Chromatium minus	*Prothecochloris* spp.
	Thiocystis violacea	*Pelodictyon* spp.
	Thiocystis roseopersicina	*Chloroherpeton thalassium*
	Thiocapsa sp. (purple-red)	*Chloroflexus*-like organisms
	Thiospirillum sp.	
	(*Thiopedia*, *Lamprocystis*, *Amoebobacter*, *Thiodictyon* spp.)	
	Rhodospirillum salinarum	
	Rhodospirillum mediosalinum	
	Rhodospirillum salexigens	
	Chromatium salexigens	
	Ectothiorhodospira spp.	

[a] From B. Eichler, *Workshop Green Photosynth. Bacteria*, abstr. no. 26, 1987.

were often observed in purple layers. However, so far only one strain has been isolated (*Chromatium salexigens* [Caumette et al., in preparation]). In these salinas, a decrease of the microbial diversity was observed as the salinity increased. Similar observations were made as the temperature increased in hot springs. Most of the thermophilic phototrophic bacteria populating hot spring sediments are related to the family *Chloroflexaceae*. However, members of the families *Chromatiaceae* and *Chlorobiaceae* were also often observed in hot springs containing significant levels of hydrogen sulfide (81); so far, one thermophilic species (*Chromatium tepidum*) has been isolated (47).

CONCLUSIONS

Benthic phototrophic bacteria develop by using sulfide as the main electron source. In purple layers formed by members of the family *Chromatiaceae*, sulfur is stored as an intermediary product inside the cells. Consequently, it is not available for other organisms unless the cells are lysed. In contrast, in colored layers formed by members of the family *Ectothiorhodospiraceae* or *Chlorobiaceae*, extracellular sulfur globules can by used by chemotrophic sulfur-oxidizing bacteria in the upper, semioxic layer and by sulfur-reducing bacteria in the deeper, anoxic layer. As a result, sulfur-reducing bacteria could

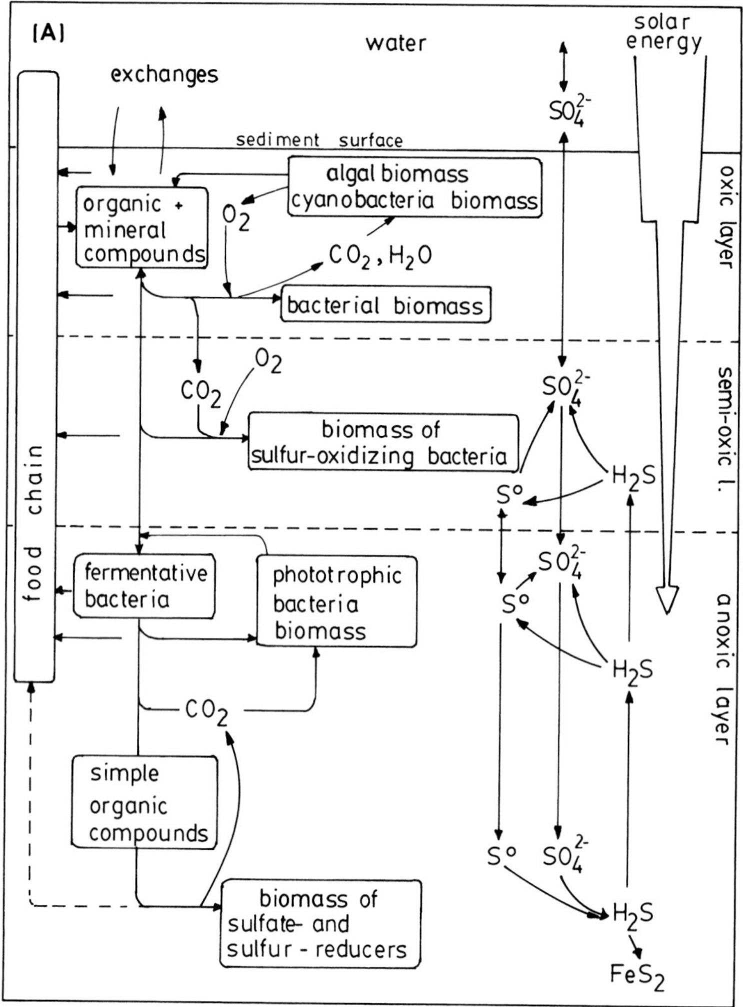

Figure 9. Main microbial activities contributing to both sulfur and carbon cycles in the upper layer of sediments in shallow water bodies. (A) Day; (B) night.

live in metabolically coupled relationships (syntrophisms) with such phototrophic bacteria in laminated sediments, as has been demonstrated in laboratory experiments (3).

During daylight hours, sulfide in the sediment surface is rapidly oxidized by both phototrophic bacteria and colorless sulfur-oxidizing bacteria, thereby preventing sulfide diffusion into the overlying oxic water. In contrast, during the night, chemotrophic sulfide oxidation is not so efficient, and, consequently, some sulfide diffuses up into the water column and often into the atmosphere (74).

When densely populating the sediment surface of certain shallow water bodies, phototrophic bacteria are an important source of biomass for higher organisms (10, 14, 15). Numerous ciliates can be found in the purple layers of phototrophic bacteria. Benthic copepods have also been observed above purple layers of phototrophic bacteria (9). Such copepods (*Tisbe* sp.) are highly resistant to anoxic conditions and are able to sink

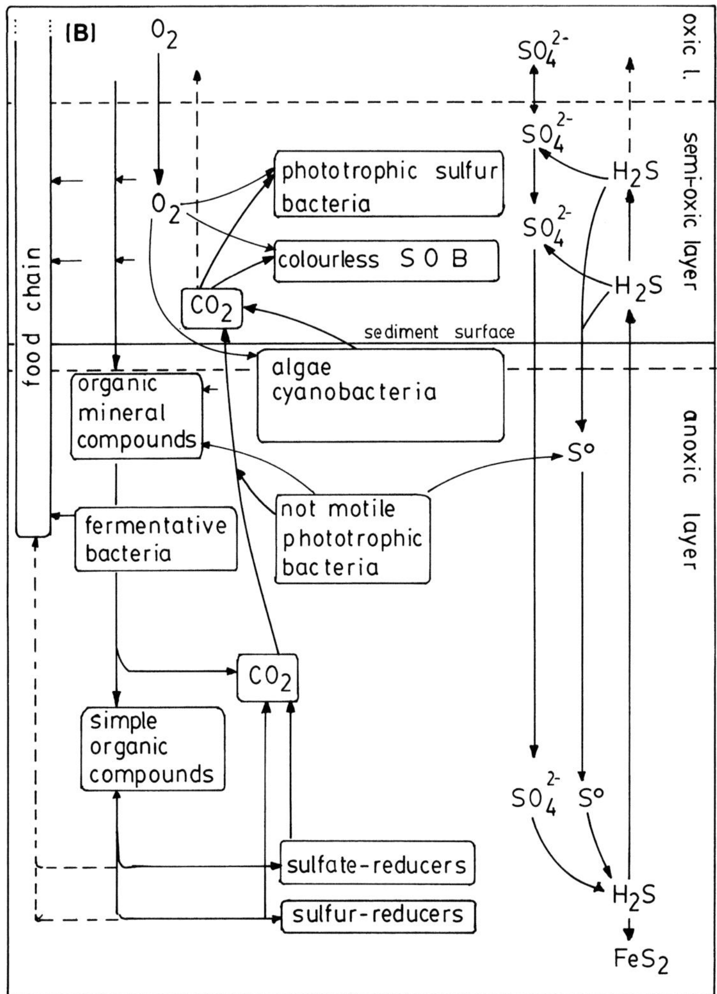

Figure 9. *Continued.*

into the anoxic layer for feeding purposes. Laboratory experiments showed direct trophic interactions between these copepods and phototrophic bacteria biomass (20, 21, 64).

Therefore, in the upper sediment of many shallow water bodies, mainly in the marine coastal environment, phototrophic bacteria contribute, as do other primary producers, to the first level of the food chain (Fig. 9). In laminated sediments, underneath cyanobacterial or algal layers, they contribute efficiently to the total photosynthetic production, and this production can be considered a complement of oxygenic primary production.

Phototrophic bacteria were initially described as secondary primary producers (56); at the sediment surface of shallow water bodies and in laminated sediments they may have an important role as summarized in Fig. 9, not only in the oxidation of reduced-sulfur compounds but also in the carbon budget and the food chain that develops in such habitats. Further observations of benthic ecosystems and both field and laboratory experiments are necessary for other nutrient cycling and trophic characteristics of phototrophic bacterial communities to be revealed, including their quantitative contribution to sulfide and other reduced-sulfur compound oxidation, as well as their importance in the carbon, nitrogen, and phosphate cycles which take place at the oxic/anoxic interface in benthic environments.

LITERATURE CITED

1. **Bauld, J., J. L. Favinger, M. T. Madigan, and H. Gest.** 1987. Obligately halophilic *Chromatium vinosum* from Hamelin Pool, Shark Bay, Australia. *Curr. Microbiol.* **14:**335–339.
2. **Beer-Romero, P., and H. Gest.** 1987. *Heliobacillus mobilis*, a peritrichously flagellated anoxyphototroph containing bacteriochlorophyll g. *FEMS Microbiol. Lett.* **41:**109–114.
3. **Biebl, H., and N. Pfennig.** 1978. Growth yields of green sulfur bacteria in mixed cultures with sulfur and sulfate reducing bacteria. *Arch. Microbiol.* **117:**9–16.
4. **Boon, J. J., J. W. de Leeuw, and W. E. Krumbein.** 1985. Biogeochemistry of Gavish Sabkha sediments. *Ecol. Stud. Anal. Synth.* **53.**
5. **Burke, M. E., E. Gorham, and D. C. Pratt.** 1974. Distribution of purple photosynthetic bacteria in wetland and woodland habitats of central and northern Minnesota. *J. Bacteriol.* **117:**826–833.
6. **Castenholz, R.** 1984. Composition of hot spring microbial mats: a summary, p. 101–119. *In* Y. Cohen, R. W. Castenholz, and H. O. Halvorson (ed.), *Microbial Mats: Stromatolites.* Alan R. Liss, Inc., New York.
7. **Castenholz, R. W., and B. K. Pierson.** 1981. Isolation of members of the family Chloroflexaceae, p. 290–298. *In* M. P. Starr, H. Stolp, H. G. Trüper, A. Balows, and H. G. Schlegel (ed.), *The Prokaryotes.* Springer-Verlag KG, Berlin.
8. **Caumette, P.** 1984. Distribution and characterization of phototrophic bacteria isolated from the water of Bietri Bay (Ivory Coast). *Can. J. Microbiol.* **30:**273–284.
9. **Caumette, P.** 1986. Phototrophic sulfur bacteria and sulfate reducing bacteria causing red waters in a shallow brackish lagoon (Prevost Lagoon, France). *FEMS Microbiol. Ecol.* **38:**113–124.
10. **Caumette, P., M. Pagano, and L. Saint-Jean.** 1983. Répartition verticale du phytoplancton, des bactéries et du zooplancton dans un milieu stratifié en baie de Biétri (Côte d'Ivoire). *Hydrobiologia* **106:**135–148.
11. **Cohen, Y., W. E. Krumbein, and M. Shilo.** 1977. Solar Lake (Sinai). 2. Distribution of photosynthetic microorganisms and primary production. *Limnol. Oceanogr.* **22:**609–620.
12. **Cornée, A.** 1983. Sur les bactéries des saumures et des sédiments des marais salants méditerranéens. *Documents du Greco*, vol. 52. Laboratoire de Géologie du Muséum, Paris.
13. **Drews, G.** 1981. *Rhodospirillum salexigens*, sp. nov., an obligatory halophilic phototrophic bacterium. *Arch. Microbiol.* **130:**325–327.
14. **Fenchel, T.** 1969. The ecology of marine microbenthos. *Ophelia* **6:**1–182.
15. **Fenchel, T., and B. B. Jørgensen.** 1977. Detritus food chain in aquatic ecosystems: the role of bacteria. *Adv. Microb. Ecol.* **1:**1–55.
16. **Fenchel, T., and B. T. Straarup.** 1971. Vertical distribution of photosynthetic pigments and the penetration of light in marine sediments. *Oikos* **22:**172–182.
17. **Gest, H., and J. L. Favinger.** 1983. *Heliobacterium chlorum*, an anoxygenic brownish green photosynthetic bacterium containing a new "form" of bacteriochlorophyll. *Arch. Microbiol.* **136:**11–16.
18. **Gibson, J., N. Pfennig, and J. B. Waterburry.** 1984. *Chloroherpeton thalassium* gen. nov., sp. nov., a non-filamentous, flexing and gliding green sulfur bacterium. *Arch. Microbiol.* **138:**96–101.
19. **Giovannoni, S. J., N. P. Revsbech, D. M. Ward, and R. W. Castenholz.** 1987. Obligately phototrophic *Chloroflexus*: primary production in anaerobic hot spring mats. *Arch. Microbiol.* **147:**80–87.
20. **Gophen, M.** 1977. Feeding of *Daphnia* on *Chlamydomonas* and *Chlorobium. Nature* (London) **265:** 271–273.
21. **Gophen, M., B. Z. Cavari, and T. Berman.** 1974. Zooplankton feeding on differentially labeled algae and bacteria. *Nature* (London) **247:**393–394.
22. **Gorlenko, V. M., G. A. Dubinina, and S. I. Kuznetsov.** 1983. The ecology of aquatic microorganisms. *Binnengewässer* **28.**
23. **Guerrero, R., C. Pedros-Alio, I. Esteve, and J. Mas.** 1987. Communities of phototrophic sulfur bacteria in lakes of the Spanish Mediterranean region. *Acta Acad. Abo.* **47:**125–151.
24. **Hansen, T. A.** 1983. Electron donor metabolism in phototrophic bacteria, p. 76–95. *In* J. G. Ormerod (ed.), *The Phototrophic Bacteria: Anaerobic Life in the Light.* Blackwell Scientific Publications, Ltd., Oxford.
25. **Hansen, T. A., A. B. J. Sepers, and H. van Gemerden.** 1975. A new purple bacterium that oxidizes sulfide to extracellular sulfur and sulfate. *Plant Soil* **43:**17–27.
26. **Hansen, T. A., and H. van Gemerden.** 1972. Sulfide utilization by purple non sulfur bacteria. *Arch. Microbiol.* **86:**49–56.
27. **Herbert, R. A.** 1985. Development of mass blooms of phototrophic bacteria on sheltered beaches in Scapa Flow, Orkney Island. *Proc. R. Soc. Edinb. Sect. B* **87:**15–25.
28. **Herbert, R. A., and A. C. Tanner.** 1977. The

isolation and some characteristics of phototrophic bacteria (Chromatiaceae and Chlorobiaceae) from antarctic marine sediments. *Appl. Bacteriol.* 43:437–445.

29. **Hofmann, P. A. G., M. J. W. Veldhuis, and H. van Gemerden.** 1985. Ecological significance of acetate assimilation by *Chlorobium phaeobacteroides*. *FEMS Microbiol. Ecol.* 31:271–278.
30. **Imhoff, J. F.** 1984. Reassignment of the genus *Ectothiorhodospira* Pelsh 1936 to a new family, *Ectothiorhodospiraceae*, fam. nov., and emended description of the *Chromatiaceae* Bavemdamm 1924. *Int. J. Syst. Bacteriol.* 34:338–339.
31. **Imhoff, J. F., B. J. Tindall, W. D. Grant, and H. G. Trüper.** 1981. *Ectothiorhodospira vacuolata*, sp. nov., a new phototrophic bacterium from soda lakes. *Arch. Microbiol.* 130:238–242.
32. **Imhoff, J. F., and H. G. Trüper.** 1977. *Ectothiorhodospira halochloris* sp. nov., a new extremely halophilic phototrophic bacterium containing bacteriochlorophyll b. *Arch. Microbiol.* 114:115–121.
33. **Imhoff, J. F., and H. G. Trüper.** 1980. *Chromatium purpuratum*, sp. nov., a new species of Chromatiaceae. *Zentralbl. Bakteriol. Mikrobiol. Hyg. Abt. 1 Orig. Reihe C*, p. 61–69.
34. **Imhoff, J. F., and H. G. Trüper.** 1981. *Ectothiorhodospira abdelmalekii*, sp. nov., a new halophilic and alkalophilic phototrophic bacterium. *Zentralbl. Bakteriol. Mikrobiol. Hyg. Abt. 1 Orig. Reihe C*, p. 228–234.
35. **Imhoff, J. F., H. G. Trüper, and N. Pfennig.** 1984. Rearrangement of the species and genera of the phototrophic purple nonsulfur bacteria. *Int. J. Syst. Bacteriol.* 34:340–343.
36. **Janssen, P. H., and C. H. Harfoot.** 1987. Phototrophic growth on n-fatty acids by members of the family Rhodospirillaceae. *Syst. Appl. Microbiol.* 9:9–11.
37. **Jørgensen, B. B.** 1977. The sulfur cycle of a coastal marine sediment. *Limnol. Oceanogr.* 22:814–832.
38. **Jørgensen, B. B.** 1982. Ecology of the bacteria of the sulphur cycle with special reference to anoxic-oxic interface environments. *Philos. Trans. R. Soc. London Ser. B* 298:543–561.
39. **Jørgensen, B. B., Y. Cohen, and D. Des Marais.** 1987. Photosynthetic action spectra and adaptation to spectral light distribution in a benthic cyanobacterial mat. *Appl. Environ. Microbiol.* 53:879–886.
40. **Jørgensen, B. B., and D. Des Marais.** 1986. Competition for sulfide among colorless and purple sulfur bacteria in cyanobacterial mats. *FEMS Microbiol. Ecol.* 38:179–186.
41. **Kaiser, P.** 1966. Ecologie des bactéries photosynthétiques. *Rev. Biol. Ecol. Sol* 3:409–472.
42. **Kämpf, C., and N. Pfennig.** 1980. Capacity of Chromatiaceae for chemotrophic growth. Specific respiration rates of *Thiocystis violacea* and *Chromatium vinosum*. *Arch. Microbiol.* 127:125–135.
43. **Kämpf, C., and N. Pfennig.** 1986. Isolation and characterization of some chemotrophic Chromatiaceae. *J. Basic Microbiol.* 26:507–515.
44. **Keppen, O. I., and V. M. Gorlenko.** 1975. A new species of purple budding bacteria containing bacteriochlorophyll b. *Mikrobiologiya* 44:258–264.
45. **Kobayashi, M.** 1976. Utilization and disposal of wastes by photosynthetic bacteria, p. 443–454. *In* H. G. Schlegel and J. Barnea (ed.), *Microbial Energy Conversion*. Unitar seminar. E. Goltze KG, Göttingen, Federal Republic of Germany.
46. **Kompantseva, E. I., and V. M. Gorlenko.** 1984. A new species of the temperate halophilic purple bacterium *Rhodospirillum mediosalinum*, sp. nov. *Microbiologiya* 53:954–961.
47. **Madigan, M. T.** 1986. *Chromatium tepidum*, sp. nov., a thermophilic photosynthetic bacterium of the family *Chromatiaceae*. *Int. J. Syst. Bacteriol.* 36:222–227.
48. **Matheron, R., and R. Baulaigue.** 1976. Sur l'écologie des Chromatiaceae et des Chlorobiaceae marines. *Ann. Inst. Pasteur Microbiol.* 127:515–520.
49. **Neutzling, O., J. F. Imhoff, and H. G. Trüper.** 1984. *Rhodopseudomonas adriatica*, sp. nov., a new species of the Rhodospirillaceae, dependent on reduced sulfur compounds. *Arch. Microbiol.* 137:256–261.
50. **Nissen, H., and I. D. Dundas.** 1984. *Rhodospirillum salinarum*, sp. nov., a halophilic photosynthetic bacterium isolated from a Portuguese saltern. *Arch. Microbiol.* 138:251–256.
51. **Ormerod, J. G., and R. Sirevag.** 1983. Essential aspect of carbon metabolism, p. 100–119. *In* J. G. Ormerod (ed.), *The Phototrophic Bacteria: Anaerobic Life in the Light*. Blackwell Scientific Publications, Ltd., Oxford.
52. **Osnitskaya, L. K., and V. I. Chudina.** 1978. Photosynthetic bacteria from lake Vanda (Antarctica). *Microbiology* (Eng. Transl. *Mikrobiologiya*) 47:131–137.
53. **Pfennig, N.** 1967. Photosynthetic bacteria. *Annu. Rev. Microbiol.* 21:285–324.
54. **Pfennig, N.** 1975. The phototrophic bacteria and their role in the sulfur cycle. *Plant Soil* 43:1–16.
55. **Pfennig, N.** 1977. Phototrophic green and purple bacteria: a comparative systematic survey. *Annu. Rev. Microbiol.* 31:275–290.
56. **Pfennig, N.** 1978. General physiology and ecology of photosynthetic bacteria, p. 3–18. *In* R. K. Clayton and W. R. Sistrom (ed.), *The Photosynthetic Bacteria*. Plenum Publishing Corp., New York.
57. **Pfennig, N., and H. G. Trüper.** 1974. The phototrophic bacteria, p. 24–64. *In* R. E. Buchanan and N. E. Gibbons (ed.), *Bergey's Manual of Determinative Bacteriology*, 8th ed. The Williams & Wilkins Co., Baltimore.
58. **Pfennig, N., and H. G. Trüper.** 1981. Isolation

of members of the families Chromatiaceae and Chlorobiaceae, p. 279–289. *In* M. P. Starr, H. Stolp, H. G. Trüper, A. Balows, and H. G. Schlegel (ed.), *The Prokaryotes*. Springer-Verlag KG, Berlin.

59. **Pfennig, N., and H. G. Trüper.** 1983. Taxonomy of phototrophic green and purple bacteria: a review. *Ann. Microbiol.* **134**:9–20.
60. **Pierson, B. K., and R. W. Castenholz.** 1974. A phototrophic gliding filamentous bacterium of hot springs, *Chloroflexus auriantiacus*, gen. nov., sp. nov. *Arch. Microbiol.* **100**:5–24.
61. **Pierson, B. K., S. J. Giovannoni, D. A. Stahl, and R. W. Castenholz.** 1985. *Heliothrix oregonensis*, gen. nov, sp. nov., a filamentous phototrophic gliding bacterium containing bacteriochlorophyll a. *Arch. Microbiol.* **142**:164–167.
62. **Pushkova, N. N., and V. M. Gorlenko.** 1982. A new green sulfur bacterium, *Chlorobium chlorovibrioides* nov. sp. *Mikrobiologiya* **51**:118–124.
63. **Revsbech, N. P., and B. B. Jørgensen.** 1986. Microelectrodes: their use in microbial ecology. *Adv. Microb. Ecol.* **9**:293–351.
64. **Rieper, M.** 1982. Relationships between bacteria and marine copepods, p. 169–172. *In* A. Bianchi (ed), *Bactériologie Marine*. Editions du Centre National de la Recherche Scientifique, Paris.
65. **Rodriguez-Valera, F., A. Ventosa, G. Juez, and J. F. Imhoff.** 1985. Variation of environmental features and microbial populations with salt concentrations in a multi-pond saltern. *Microbiol. Ecol.* **11**:107–115.
66. **Sadler, W. R., and R. Y. Stanier.** 1960. The function of acetate in photosynthesis by green bacteria. *Proc. Natl. Acad. Sci. USA* **46**:1328–1334.
67. **Schmidt, K.** 1978. Biosynthesis of carotenoids, p. 729–750. *In* R. K. Clayton and W. R. Sistrom (ed.), *The Photosynthetic Bacteria*. Plenum Publishing Corp., New York.
68. **Schmidt, K., N. Pfennig, and S. LiaaenJensen.** 1965. Carotenoids of Thiorhodaceae. *Arch. Mikrobiol.* **52**:132–146.
69. **Stackebrandt, E., and C. R. Woese.** 1981. The evolution of prokaryotes, p. 1–31. *In* M. J. Carlile et al. (ed.), *Molecular and Cellular Aspects of Microbial Evolution*. Cambridge University Press, Cambridge.
70. **Stal, L. J., H. van Gemerden, and W. E. Krumbein.** 1985. Structure and development of a benthic marine microbial mat. *FEMS Microbiol. Ecol.* **31**:111–125.
71. **Stanier, R. Y., N. Pfennig, and H. G. Trüper.** 1981. Introduction to the phototrophic prokaryotes, p. 197–211. *In* M. P. Starr, H. Stolp, H. G. Trüper, A. Balows, and H. G. Schlegel (ed.), *The Prokaryotes*. Springer-Verlag KG, Berlin.
72. **Tindall, B. J., and W. D. Grant.** 1983. The anoxygenic phototrophic bacteria, p. 115–155. *In* E. M. Barnes and G. C. Mead (ed.), *Anaerobic Bacteria in Habitats Other than Man*. Blackwell Scientific Publications, Ltd., Oxford.
73. **Trüper, H. G.** 1987. Phototrophic bacteria (an incoherent group of prokaryotes). A taxonomic versus phylogenetic survey. *Microbiologia SEM* **3**: 71–89.
74. **Trüper, H. G.** 1980. Distribution and activity of phototrophic bacteria at the marine water-sediment interface, p. 275–282. *In* A. Bianchi (ed.), *Biogéochimie de la Matière Organique à l'interface Eau-Sediment Marin*. Centre National de la Recherche Scientifique, Paris.
75. **Trüper, H. G., and E. A. Galinski.** 1986. Concentrated brines as habitats for micro-organisms. *Experientia* **42**:1182–1187.
76. **Trüper, H. G., and N. Pfennig.** 1978. Taxonomy of the Rhodospirillales, p. 19–30. *In* R. K. Clayton and W. R. Sistrom (ed.), *The Photosynthetic Bacteria*. Plenum Publishing Corp., New York.
77. **Trüper, H. G., and N. Pfennig.** 1981. Characterization and identification of the anoxygenic phototrophic bacteria, p. 299–312. *In* M. P. Starr, H. Stolp, H. G. Trüper, A. Balows, and H. G. Schlegel (ed.), *The Prokaryotes*. Springer-Verlag KG, Berlin.
78. **van Gemerden, H.** 1983. Physiological ecology of purple and green bacteria. *Ann. Inst. Pasteur Microbiol.* **134**:73–92.
79. **van Gemerden, H.** 1987. Competition between sulfur purple bacteria and green sulfur bacteria: role of sulfide, sulfur and polysulfides. *Acta Acad. Abo.* **47**:13–27.
80. **van Gemerden, H., and H. H. Beeftink.** 1983. Ecology of phototrophic bacteria, p. 148–185. *In* J. G. Ormerod (ed.), *The Phototrophic Bacteria: Anaerobic Life in the Light*. Blackwell Scientific Publications, Ltd., Oxford.
81. **Van Niel, C. B.** 1931. On the morphology and physiology of the purple and green sulphur bacteria. *Arch. Mikrobiol.* **3**:1–112.
82. **Veldhuis, M. J. W., and H. van Gemerden.** 1986. Competition between purple and brown phototrophic bacteria in stratified lakes: sulfide, acetate and light as limiting factors. *FEMS Microbiol. Ecol.* **38**:31–38.
83. **Wetzel, R. G.** 1975. *Limnology*. The W. B. Saunders Co., Philadelphia.
84. **Widdel, F.** 1988. Microbiology and ecology of sulfate and sulfur reducing bacteria, p. 469–584. *In* A. J. B. Zehnder (ed.), *Environmental Physiology of Anaerobes*. John Wiley & Sons, Inc., New York.
85. **Widdel, F., and N. Pfennig.** 1984. Dissimilatory sulfate or sulfur reducing bacteria, p. 663–679. *In* N. R. Krieg and J. G. Holt (ed.), *Bergey's Manual of Systematic Bacteriology*, vol. 1. The Williams & Wilkins Co., Baltimore.

Chapter 27

Chromatic Regulation of Photosynthesis in Cyanobacteria

Anton F. Post, Gabriel Zwart, Jean-Pierre Sweers, Arnold Veen, Dave Rensman, Andien van der Heuvel, and Luuc R. Mur

INTRODUCTION

Adaptation to variations in the ambient light regime is a feature of the photosynthetic apparatus in most, if not all, eucaryotic algae and procaryotic cyanobacteria. Reduction of the light flux in terms of intensity and/or duration of illumination triggers an increase of the cellular pigment content in a large number of cyanobacteria (6, 7, 13, 23, 24, 28). Such an increase is normally found for both chlorophyll *a* (Chl *a*) and the phycobiliproteins. The changes in pigment content reflect the major processes occurring during photoadaptation which affect the structure and organization of the photosynthetic apparatus. First, an increase in phycobiliproteins serves to enlarge the phycobilisome (PBS) antenna. The PBS transfers the harvested energy to photosystem II (PS II) exclusively (17, 22), so that an increase in its size results in a larger absorption cross-section of PS II (see, e.g., reference 15). Second, cyanobacteria react to low-light conditions by raising their Chl *a* contents, indicated by an increasing PS I/PS II ratio (13). Since the PS I Chl *a* antenna contains approximately three times more Chl *a* than its counterpart in PS II (16, 17, 20, 21) it follows that the majority of the Chl *a* pool in the cells will be associated with PS I.

The combination of the processes described above may allow cells to perform photosynthesis with a balanced distribution of excitation energy over both photosystems in polychromatic "white" light. However, in most habitats favoring cyanobacterial dominance, variations in the ambient light regime are not just limited to changes in intensity and periodicity. Cyanobacteria occurring in aquatic ecosystems with high biomass levels (6), in stratifying multispecies layers (2), and in dense benthic microbial mats (11) exhibit strong vertical shifts in the spectral distribution of the light impinging on the cells. The shift in spectral quality can be either a static feature, when cells are fixed in the light gradient, or a dynamic one, when cells are transported through the vertical gradient.

Anton F. Post, Gabriel Zwart, Jean-Pierre Sweers, Arnold Veen, Dave Rensman, Andien van der Heuvel, and Luuc R. Mur • Laboratory of Microbiology, University of Amsterdam, Nieuwe Achtergracht 127, 1018 WS Amsterdam, The Netherlands.

Both ways of experiencing changes in spectral distribution require a specific response of the cyanobacterial species involved. Being fixed in a given spectral band will force cells to adjust their photosynthetic pigment systems to optimize their light-harvesting potential, a process called chromatic adaptation. Three types of chromatic adaptation are described (27) for cyanobacteria growing in green light as compared to red light: first, an increase in all phycobiliproteins at a constant molar ratio; second, an increase in phycoerythrin (CPE) levels while phycocyanin (CPC) remains constant; finally, an increase in CPE levels while CPC levels decrease. The third of these is called complementary chromatic adaptation and is thought to involve complete turnover of PBS (1). Apparently, the chromatic effect was attributed for a long time to processes concerning optimal operation of PS II. More recently, Fujita et al. (8) showed that PS I/PS II ratios of cyanobacteria vary with the spectral quality of the light, yielding a higher ratio in green light.

Here we present a study on the effect of light quality on chromatic adaptation as expressed by photosynthetic performance and flexibility of the photosynthetic apparatus to respond to changes in light quality. Results of this study may be useful for the understanding of the basic principles underlying studies on in situ measurements of vertical O_2 profiles, photosynthetic activities, or action spectra determined on mats consisting of cyanobacteria (see, e.g., references 2, 11, and 12).

EXPERIMENTAL PROCEDURES

The cyanobacterium *Fremyella diplosiphon* 7601 from the Pasteur Culture Collection was grown in a mineral medium described previously (28). Continuous cultures were maintained in 300-ml culture vessels of cylindrical shape with their flat side facing the light source. The culture vessels were temperature controlled at 21°C by circulating water from a thermostatted water bath through a water jacket. Culture pH was constant in all experiments at 8.0 ± 0.1. Monochromatic illumination was obtained from a 2,000-W Osram 50250 light source by passing the collimated light beam through Schott double-band interference filters. The filters had half-band widths of less than 8 nm. Transmission maxima for the filters used in this study were 560, 630, 680, and 704 nm. In the cultures cells experienced a light intensity of 40 μmol quanta $\cdot$ m^{-2} $\cdot$ s^{-1}, allowing for maximal cell division rates.

Chl *a* was determined on 90% acetone extracts of cells pelleted on Whatman GF/F filters, and its concentration was calculated by the method of Jeffrey and Humphrey (10). After subsequent centrifugation, at 10,000 rpm for 10 min (twice) and at 36,000 rpm for 60 min, of samples sonicated in 0.05 M KPO_4 buffer (pH 7.6), supernatants were scanned spectrophotometrically and their concentrations were calculated using the equations of Bennett and Bogorad (1). Dry weight was determined on lyophilized samples that were further dried for 24 h in an exsiccator. Photosynthesis-light relationships were determined in a system described by Dubinsky et al. (3), using Schott double-band interference filters as described above. Photosynthetic unit sizes were determined in the same system by using an EG&G FX-279 xenon flashlamp (flash duration, 6 μs) connected to an EG&G PS-302 power supply. Flash frequency-dependent O_2 evolution was determined with a continuous far-red background light to saturate PS I activity and to offset the Kok effect (22). In vivo absorption spectra were determined on a Zeiss M4QIII spectrophotometer equipped with an integrating sphere. In vivo absorption coefficients were determined from the spectra (9) and normalized to Chl *a* for calculations of the absorption cross-sections (4). PS II fluorescence yields were measured on whole cells both with and without DCMU [3-(3,4)-dichlorophenyl)-1,1-

dimethylurea], using the method described by Post et al. (25). All measurements were performed at least in triplicate, and the data presented below are average values with standard deviations of less than 10%.

EFFECTS OF LIGHT QUALITY ON CHROMATIC ADAPTATION

Steady-state cultures of *F. diplosiphon* 7601 were obtained within 8 days of growth in monochromatic light with an intensity allowing for maximal cell division rates (0.035 h^{-1}). Monochromatic light was chosen at 560, 630, 680, and 704 nm, so that cells were exposed to wavelengths at the absorption maximum of a different pigment (system) in each experiment: respectively, CPC, CPE, Chl *a*, and the long-wavelength-absorbing Chl *a* in PS I. Cellular pigment contents changed markedly with culture conditions (Table 1). The complementary nature of chromatic adaptation showed clearly from the CPE and CPC contents of cells at the different spectral qualities. In light absorbed by Chl *a*, cells contained higher levels of allophycocyanin, suggesting increased numbers of PBS per cell. Chl *a* contents declined at wavelengths allowing for enhanced light harvesting by this pigment. Since the major changes occurred in the PBS antenna, serving as the light-harvesting antenna for PS II (17, 22), this

Table 1.
Chl *a* and Phycobiliprotein Contents for *F. diplosiphon* 7601[a]

λ (nm)	Content ($\mu g \cdot \mu g$ dry wt^{-1})			
	APC	CPC	CPE	Chl *a*
560	9 (1)	20 (2.1)	44 (3.8)	12
630	8 (1)	54 (7)	4 (0.5)	10
680	14 (1)	69 (4.5)	5.5 (0.3)	10
704	14 (1)	50 (3.3)	5 (0.3)	8

[a] Organism was grown at 40 $\mu E \cdot m^{-2} \cdot s^{-1}$ of light preferentially absorbed by phycobiliproteins (560 or 630 nm) or Chl *a* (680 or 704 nm). Parentheses indicate monomer molar ratios relative to allophycocyanin (APC).

finding suggests to some extent that chromatic adaptation might be limited to adjustment of the PS II absorption cross-sections only. However, since the PS I antenna consists of Chl *a* only, chromatically adapted cells may have to deal with severe imbalances of light harvesting by PS I and PS II at some wavelengths. Resulting imbalances in PS I and PS II activities would lead to failing in ATP generation during linear photosynthetic electron transport and in forming reductant, both needed in the Calvin-Benson cycle for CO_2 fixation. Such situations may lead to increased energy loss by (mainly PS II) fluorescence and to increased quantum requirements for photosynthetic O_2 evolution.

To find whether cells are provided with a mechanism of both short-term (seconds to minutes) and long-term (hours to days) balancing of PS II and PS I activities, we studied photosynthetic O_2 evolution in monochromatic light in relation to absorption cross-sections, PS I and PS II antenna sizes, and energy distribution between the two photosystems. The basic cellular properties, measured as the basis for the data given below and for calculations and interpretations, are given schematically in Fig. 1.

From the photosynthesis-light (P versus I) relationship (Fig. 1a), two parameters were derived. First, the parameter describing light-limited photosynthesis is the linear slope of the P-versus-I curve, α, which gives the efficiency with which incident light is utilized for O_2 evolution. Light-limited photosynthesis is determined by two components, as follows:

$$\alpha = k_{c\lambda} * \phi$$

where $\alpha = \mu mol\ O_2 \cdot \mu mol\ Chl\ a^{-1} \cdot min^{-1} / \mu mol\ quanta \cdot m^{-2} \cdot s^{-1}$
$k_c = m^{-2} \cdot \mu mol\ Chl\ a^{-1}$
$\phi = mol\ O_2 \cdot mol\ quanta^{-1}$

$k_{c\lambda}$, the absorption cross-section at the wavelength at which photosynthesis is mea-

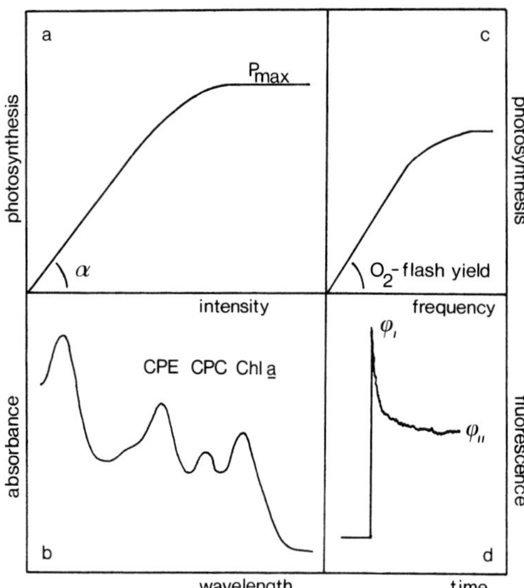

Figure 1. Schematic representation of measurements of cyanobacterial photosynthetic properties in chromatic adaptation. (a) Photosynthesis-light relationship to obtain light utilization efficiency, α, and photosynthetic capacity, P_{max}. (b) In vivo absorption spectra to determine wavelength-related absorption cross-sections at the absorption maxima for CPE (560 nm), CPC (630 nm), and Chl *a* (680 and 704 nm) after normalization to Chl *a* concentration. (c) Frequency-dependent O_2 production giving O_2 flash yields which are related to Chl *a* concentration to give PSU. (d) Determination of PS II fluorescence following a far-red preillumination to determine fluorescence yields for states 1 and 2, ϕ_I and ϕ_{II}. ϕ_I in the presence of DCMU equals F_{max}. Fluorescence was determined with 630 nm excitation and 685 nm emission and normalized to Ch1 *a* concentration.

sured, was derived from in vivo absorption spectra of whole cells, determined in an integrating sphere to correct for light scattering as a result of the particle nature of cells (9). $k_{c\lambda}$ gives the efficiency by which the incident light is absorbed by the pigment bed of the cells. From the measurements of α and $k_{c\lambda}$ we directly derive the efficiency with which absorbed quanta are used in O_2 evolution, and its reciprocal value, $1/\phi$, is the quantum requirement for photosynthetic O_2 evolution, with a minimum value of 8 quanta per O_2 produced as predicted by the Z scheme of photosynthesis. Cells incubated at the wavelength of growth showed light utilization and light-harvesting efficiencies that were very similar (Table 2), and the resulting quantum requirements approached the predicted minimum very closely. However, when green (560-nm) light-grown cells were incubated in orange (630 nm) or red (680 nm) light, $1/\phi$ increased about 50%. The reverse situation was observed in red light-grown cells: $1/\phi$ increased when CPE or CPC had to act as the main light-harvesting pigment in the green and orange light, respectively. Thus, apparently cells were adapted optimally to the wavelength of cultivation, whereas transfer to other wavelengths, although not blocking photosynthesis, led to an increase in $1/\phi$. Similar observations were found in O_2-flash yield studies with the non-CPE-containing cyanobacterium *Anacystis nidulans* (21, 22). Cells grown in orange light appeared not to be affected by wavelength changes in the 560- to 680-nm band.

The second parameter under study in P versus I relationships is the maximal rate of photosynthesis, P_{max}. P_{max} can also be divided into two components. First, since P_{max} is normalized to Chl *a*, whereas it reflects just the activity of reaction center II (RC II), it incorporates the photosynthetic unit (PSU) given as the number of micromoles of Chl *a* per mole e^- (1 RC II) or per mole O_2 (4 RC II) (see references 4 and 15 for more detailed discussion). PSU sizes were determined from O_2 evolution in saturating light flashes (Fig. 1d) short enough to allow every RC II trap to be closed and reopened only once during each flash. From the measurements of P_{max} and PSU we derived a second parameter, τ, a time constant:

$$\tau = \frac{60{,}000}{P_{max}{}^{*}\mathrm{PSU}} \quad \text{(ms)}$$

$$\text{where } P_{max} = \mu\text{mol } O_2 \cdot \mu\text{mol Chl } a^{-1} \cdot \min^{-1}$$

$$\mathrm{PSU} = \text{mol Chl } a \cdot \text{mol } O_2^{-1}$$

Table 2.
Light Utilization Efficiencies, In Vivo Absorption Coefficients, and Minimal Quantum Requirements for *F. diplosiphon* 7601[a]

λ (nm)		α	$k_{c\lambda}$	1/φ
Culture	Photosynthesis	(nmol O_2 · μmol Chl a^{-1} · min^{-1}/μE · m^{-2} · s^{-1})	(m^2 · mmol Chl a^{-1})	(mol quanta · mol O_2^{-1})
560	560	1.5	12.3	8.2
	630	0.8	8.1	10
	680	0.9	10.9	13
630	560	0.9	7.5	8.3
	630	1.6	13.5	8.3
	680	1.5	12.1	8.1
680	560	0.9	11.7	13
	630	1.7	21	12.4
	680	1.7	15	8.9
704	560	0.9	9.9	11
	630	1.4	18.8	13.4
	680	1.4	11.7	8.2

[a] Light utilization efficiencies (α), in vivo absorption coefficients ($k_{c\lambda}$), and minimal quantum requirements (1/φ) for *F. diplosiphon* 7601 grown at 40 μE · m^{-2} · s^{-1} of light preferentially absorbed by phycobiliproteins (560 or 630 nm) or by Chl *a* (680 or 704 nm) and incubated to study P-versus-I relationships at 560, 630, and 680 nm.

τ reflects here the minimal time needed for 4 e^- to be transferred from the water-splitting system to the reductant for PS I, NADP, during steady-state photosynthesis in a continuous light flux (4). As a result of chromatic adaptation, P_{max} increased at longer wavelengths due to decreasing PSU sizes (Table 3). τ proved to be a constant for all cultures at a value of 9 to 10 ms, which is relatively long as compared with green algae and diatoms (4, 5, 19) but among the shortest time constants found for cyanobacteria (22). Continuing within the concept of the PSU, which is, on a statistical basis, the number of pigments organized to produce 1 O_2 following the simultaneous closing of all RC II traps, absorption cross-sections of the PS I and the PS II pigment systems operating within a PSU were calculated for each culture. These cross-sections are compared with the number of chromophores assigned to either PS II or I (Table 3). Calculation of the absorption cross-sections was performed by the method of Dubinsky et al. (4) and was

Table 3.
P_{max} versus PSU for *F. diplosiphon* 7601[a]

λ for culture (nm)	P_{max} (μmol O_2 · μmol Chl a^{-1} · min^{-1})	PSU (mol Chl *a* · mol O_2^{-1})	RC I/RC II	n_{II}	n_I	σ_{II} (nm^2)	σ_I (nm^2)
560	4.1	1,660	2.4	1,313	1,460	36.1	0.9
630	4.9	1,360	1.7	1,130	1,154	27.5	3
680	6.4	1,010	1.4	816	810	14.6	10.5
704	7.4	922	1.2	686	722	8.6	9.4

[a] Maximal photosynthetic capacity (P_{max}), photosynthetic unit size (PSU), ratio of reaction centers of PS I versus PS II (RC I/RC II), number of chromophores (n_I and n_{II}), and absorption cross-sections (σ_I and σ_{II}) of the PS II and PS I population of a PSU for *F. diplosiphon* grown at 40 μE · m^{-2} · s^{-1} of light preferentially absorbed by the phycobiliproteins (560 or 630 nm) and Chl *a* (680 or 704 nm). τ (minimal turnover time) was 9 ms for each wavelength.

Table 4.
Fluorescence and Energy Spillover for *F. diplosiphon* 7601[a]

λ (nm)	+DCMU Fluorescence (mV · μg Chl a^{-1}) ϕ_I	$\phi_{II}\phi_I$	s	−DCMU Fluorescence (mV · μg Chl a^{-1}) ϕ_I	ϕ_{II}	s
560	30	16	0.48	17	12	0.29
630	1,050	704	0.33	746	609	0.18
680	2,360	2,172	0.08	1,581	1,498	0.06
704	2,685	2,642	0.02	1,916	1,954	0

[a] Maximal fluorescence yield (ϕ_I), steady-state fluorescence (ϕ_{II}), and fraction of energy spillover (s), determined with and without DCMU for *F. diplosiphon* 7601 grown at 40 μE · m^{-2} · s^{-1} of light preferentially absorbed by phycobiliproteins (560 or 630 nm) and Chl *a* (680 or 704 nm).

based on three cyanobacterial properties not determined in this study: (i) the PBS is structurally connected to PS II (26), and hence energy transfer is to PS II exclusively (17); (ii) the antenna of PS II consists of 50 Chl *a* per RC II (20) and the PS I antenna contains 150 Chl *a* per P_{700} (16); and (iii) the allophycocyanin core of the PBS is made of 33 monomers. The numbers of chromophores assigned to either PS I or PS II within a PSU were strikingly equal for all cultures (Table 3). This resulted in absorption cross-sections for PS I and PS II being apart as far as 3 and 97%, respectively, for green light-grown cells. Only in the red and far-red light were PS II and PS I absorption cross-sections close enough to allow balanced photosynthesis under culture conditions with quantum requirements approaching the theoretical minimum. However, it is obvious that the difference in PS I and PS II absorption cross-sections in green light-grown cells can by no means explain the minimal quantum requirements found. For every 32 quanta arriving in PS II, only 1 quantum can be harvested in PS I, which would predict a minimal quantum requirement of 128 quanta per O_2.

To study this discrepancy, we undertook to determine the capacity for spillover of energy harvested in PS II and subsequently transferred to PS I. As for higher plants and green algae, energy spillover in cyanobacteria can be determined in vivo from PS II fluorescence studies (25), and changes in the amount of spillover are related to the redox state of the photosynthetic electron transport chain (18). As was observed for *Synechococcus* strain 6301 (18), it was found for *F. diplosiphon* 7601 that DCMU allowed for complete state 1→2 transitions (Fig. 1d). Since DCMU closes all RC II traps, changes in fluorescence yield can only be explained from changes in energy spillover, assuming that energy dissipation in the form of heat is negligible. In other words, a decrease in fluorescence going from the high-fluorescent state 1 to state 2 means that an increased fraction of total excitation energy is made available for PS I activity, although it was originally harvested by the PS II antenna. In the presence of DCMU, maximal fluorescence yields were low for green light-grown cells and increased for cells grown at longer wavelengths (Table 4). Steady state 2 fluorescence was approximately 50% lower than F_{max} for green light-grown cells, whereas red light-grown cells yielded values hardly different from F_{max}. From the differences in state 1 and 2 fluorescence yields, the capacity for energy spillover, s, could be calculated as 0.48 for green light-grown cells and as low as 0.02 for far-red-light-grown cells. Although not as pronounced, a similar situation was observed in the measurements in which DCMU was omitted (Table 4).

The overall conclusion from these experiments is that in red light-grown cells, PS I and PS II absorption cross-sections are approximately equal and the slight imbalances can be offset by energy spillover with low capacity. In green and orange light-grown cells, the differences in PS I and PS II absorption cross-sections are large. However, even in the case of green light-grown cells this difference can be fully offset by the large capacity for energy spillover, which allows for up to 50% of the energy harvested in PS II to be transferred to PS I. The mechanism for regulation of the capacity of energy spillover is possibly founded in the adjustment of the PS I/PS II ratio to culture wavelength. Such an adjustment was found to be common to cyanobacteria when transferred from Chl *a* light to PBS light (8). Increasing PS I/PS II ratios may increase both the statistical chance of energy spillover and the physical proximity between PS I and PS II in the thylakoid membranes.

Looked at in a more ecological context, a well-developed ability in cyanobacteria to adapt to green light has two major advantages over other photosynthetic organisms. First, fully adapted cyanobacterial species capable of harvesting light in a wavelength band poorly absorbed by green algae and diatoms (see, e.g., reference 11) and utilized with low quantum requirements may yield a population which is successfully competitive in its habitat. Second, a high capacity for energy spillover equips species exposed to spectral gradients (due to vertical transport in the epilimnion or vertical migration in a bacterial mat) with a mechanism of short-term regulation of PS I and PS II activities (25) and, thus, balanced photosynthesis.

LITERATURE CITED

1. **Bennett, A., and L. Bogorad.** 1973. Complementary chromatic adaptation in a filamentous blue green alga. *J. Cell Biol.* **58:**419–435.
2. **Cohen, Y., B. B. Jørgensen, N. P. Revsbech, and R. Poplawski.** 1986. Adaptation to hydrogen sulfide of oxygenic and anoxygenic photosynthesis among cyanobacteria. *Appl. Environ. Microbiol.* **51:** 398–407.
3. **Dubinsky, Z., P. G. Falkowski, A. F. Post, and U. M. Van Hes.** 1987. A system for measuring phytoplankton photosynthesis in a defined light field with an oxygen electrode. *J. Plankton Res.* **9:**607–612.
4. **Dubinsky, Z., P. G. Falkowski, and K. Wyman.** 1986. Light harvesting and utilization in phytoplankton. *Plant Cell Physiol.* **27:**1235–1250.
5. **Falkowski, P. G., Z. Dubinsky, and K. Wyman.** 1985. Growth-irradiance relationships in phytoplankton. *Limnol. Oceanogr.* **30:**311–321.
6. **Foy, R. H., and C. E. Gibson.** 1982. Photosynthetic characteristics of planktonic blue-green algae: changes in photosynthetic capacity and pigmentation of *Oscillatoria redekei* Van Goor under high and low light. *Br. Phycol. J.* **17:**183–193.
7. **Foy, R. H., C. E. Gibson, and R. V. Smith.** 1976. The influence of daylength, light intensity and temperature on the growth rates of planktonic blue-green algae. *Br. Phycol. J.* **11:**151–163.
8. **Fujita, Y., K. Ohki, and A. Murakami.** 1985. Chromatic regulation of photosystem composition in the photosynthetic system of red and blue-green algae. *Plant Cell Physiol.* **26:**1541–1548.
9. **Gobel, F.** 1978. Direct measurement of pure absorbance spectra of living phototrophic microorganisms. *Biochim. Biophys. Acta* **538:**593–602.
10. **Jeffrey, S. W., and G. F. Humphrey.** 1975. New spectrophotometric equations for determining chlorophylls a, b, c1, c2 in higher plants, algae and natural phytoplankton. *Biochem. Physiol. Pflanz. (BPP)* **167:**191–194.
11. **Jørgensen, B. B., Y. Cohen, and D. J. Des Marais.** 1987. Photosynthetic action spectra and adaptation to spectral light distribution in a benthic cyanobacterial mat. *Appl. Environ. Microbiol.* **53:** 879–886.
12. **Jørgensen, B. B., Y. Cohen, and N. P. Revsbech.** 1986. Transition from anoxygenic to oxygenic photosynthesis in a *Microcoleus chthonoplastes* cyanobacterial mat. *Appl. Environ. Microbiol.* **51:**408–417.
13. **Kawamura, M., M. Mimuro, and Y. Fujita.** 1979. Quantitative relationship between two reaction centers in the photosynthetic system of blue-green algae. *Plant Cell Physiol.* **20:**697–705.
14. **Ley, A. C.** 1984. Effective absorption cross-sections in *Porphyridium cruentum*. *Plant Physiol.* **74:** 451–454.
15. **Ley, A. C., and D. C. Mauzerall.** 1982. Absolute absorption cross-sections for photosystem II and the minimum quantum requirement for photosyn-

thesis in *Chlorella vulgaris*. *Biochim. Biophys. Acta* **680**:95–106.

16. **Lundell, D. J., A. N. Glazer, A. Melis, and R. Malkin.** 1985. Characterization of a cyanobacterial photosystem I complex. *J. Biol. Chem.* **260**:646–654.
17. **Mimuro, M., and Y. Fujita.** 1977. Estimation of chlorophyll a distribution in the photosynthetic pigment systems I and II of the blue-green alga *Anabaena variabilis*. *Biochim. Biophys. Acta* **503**: 343–361.
18. **Mullineaux, C. W., and J. F. Allen.** 1986. The state 2 transition in the cyanobacterium *Synechococcus* 6301 can be driven by respiratory electron flow into the plastoquinone pool. *FEBS Lett.* **205**:155–160.
19. **Myers, J., and J.-R. Graham.** 1971. The photosynthetic unit in *Chlorella* measured by repetitive short flashes. *Plant Physiol.* **48**:282–286.
20. **Myers, J., J.-R. Graham, and R. T. Wang.** 1978. On spectral control of pigmentation in *Anacystis nidulans* (Cyanophyceae). *J. Phycol.* **14**:513–518.
21. **Myers, J., J.-R. Graham, and R. T. Wang.** 1980. Light harvesting in *Anacystis nidulans* studied in pigment mutants. *Plant Physiol.* **66**:1144–1149.
22. **Myers, J., J.-R. Graham, and R. T. Wang.** 1983. On the oxygen flash yields of two cyanophytes. *Biochim. Biophys. Acta* **722**:281–290.
23. **Post, A. F.** 1986. Transient state characteristics of adaptation to changes in light conditions for the cyanobacterium *Oscillatoria agardhii*. I. Pigmentation and photosynthesis. *Arch. Microbiol.* **145**:353–357.
24. **Post, A. F., J. G. Loogman, and L. R. Mur.** 1986. Photosynthesis, carbon flows and photosynthesis by *Oscillatoria agardhii* grown with a light/dark cycle. *J. Gen. Microbiol.* **132**:2129–2136.
25. **Post, A. F., A. Veen, and L. R. Mur.** 1986. Regulation of cyanobacterial photosynthesis determined from variable fluorescence yields of photosystem II. *FEMS Microbiol. Lett.* **35**:129–133.
26. **Redlinger, T., and E. Gantt.** 1982. A Mr 95.000 polypeptide in *Porphyridium cruentum* phycobilisomes and thylakoids: possible function in linkage of phycobilisomes to thylakoids and in energy transfer. *Proc. Natl. Acad. Sci. USA* **79**:5542–5546.
27. **Tandeau de Marsac, N.** 1977. Occurrence and nature of chromatic adaptation in cyanobacteria. *J. Bacteriol.* **130**:82–91.
28. **Van Liere, L., and L. R. Mur.** 1978. Light limited cultures of the blue-green alga *Oscillatoria agardhii*. *Mitt. Internat. Ver. Theor. Angew. Limnol.* **21**:158–167.
29. **Zevenboom, W., and L. R. Mur.** 1984. Growth and photosynthetic response of the cyanobacterium *Microcystis aeruginosa* in relation to photoperiodicity and irradiance. *Arch. Microbiol.* **139**:232–239.

Chapter 28

Phototrophic and Chemotrophic Growth of the Purple Sulfur Bacterium *Thiocapsa roseopersicina*

Hans van Gemerden and Rutger de Wit

INTRODUCTION

In microbial mats, purple sulfur bacteria frequently occur as dense populations. Just underneath a lamina of oxygenic phototrophic organisms, *Thiocapsa roseopersicina*, an immotile representative of the family *Chromatiaceae*, very often is the dominant species. By using microelectrodes it has been demonstrated that during the day, oxygen produced by the oxygenic phototrophs diffuses into the layer of purple sulfur bacteria. At night, photosynthesis does not occur, in contrast to respiration and sulfate reduction. Consequently, purple sulfur bacteria in laminated microbial ecosystems are often exposed to aerobic conditions during the day and to anaerobic conditions during the night (8, 14). The purple sulfur bacterium *Chromatium vinosum* is a motile species able to actively adjust its depth horizon to the prevailing oxygen/sulfide interface in microbial mats (7). However, *T. roseopersicina* is nonmotile and, consequently, is exposed to more drastic environmental fluctuations in these ecosystems. It has been reported that in microbial mats occurring on sheltered beaches (Orkney Islands, United Kingdom), *T. roseopersicina* was exposed to 20 to 200 μM oxygen during the day (5a). Therefore, it is of great ecological interest to study the effect of oxygen on the growth and physiology of this purple sulfur bacterium.

Under aerobic conditions pigment synthesis is inhibited in *C. vinosum* (6, 10) and other purple sulfur bacteria (1, 9), and it was demonstrated that *C. vinosum* cells remained viable for at least 12 h (6). It has been demonstrated unequivocally that many species of purple sulfur bacteria, including *T. roseopersicina*, are capable of chemolithotrophic growth in the dark if oxygen and an appropriate electron donor are present (1–3, 9–11). However, these conditions are unlikely to occur in microbial mats.

This chapter describes experiments performed with *T. roseopersicina* cultivated in chemostats either in the continuous presence of oxygen or under different semiaerobic–anaerobic regimens. Special attention is paid to pigment synthesis and to whether the organisms are growing phototrophically or chemotrophically. Finally, growth and pig-

Hans van Gemerden and Rutger de Wit • Department of Microbiology, University of Groningen, Kerklaan 30, 9751 NN Haren, The Netherlands.

ment synthesis by *T. roseopersicina* were analyzed under simulated natural conditions which occur in the layer of purple sulfur bacteria in microbial mats.

PROCEDURES FOR GROWING *T. ROSEOPERSICINA*

All experiments were performed with *T. roseopersicina* M1, isolated from a laminated microbial ecosystem on the island of Mellum, Federal Republic of Germany.

The continuous-culture equipment and culture medium have been described previously (2). The limiting substrate was either thiosulfate or sulfide as described below. Illumination was provided by incandescent light sources, and the intensity applied was 110 microeinsteins $\cdot$ $m^{-2} \cdot s^{-1}$. The oxygen level was kept constant at 52 ± 8 μM (semiaerobic conditions) by a controlled pulsing of air. The oxygen concentration in thiosulfate- and sulfide-limited continuous cultures was monitored by using Yellow Springs Instrument Co. and custom-made cathode-type oxygen electrodes, respectively. The specific oxygen uptake rate in the culture was measured from the linear decrease of the oxygen concentration after such a pulse was supplied (2).

In continuous cultures a steady state is achieved when the specific growth rate of the organism equals the washout rate (dilution rate = D), and, consequently, the population density and the environmental parameters remain constant.

Cultures were subjected to regimens with a time span of 24 h as follows. One set of experiments involved cultivation in continuous light with a pulsed oxygen supply, resulting in a semiaerobic-anaerobic regimen. In another set of experiments both light and oxygen were supplied in a pulsed mode so that the presence of oxygen (semiaerobic conditions) coincided with the light period, whereas anaerobiosis occurred during the dark period. In both sets of experiments, different semiaerobic periods were tested. Steady states cannot be achieved when organisms are cultivated under fluctuating conditions. Nevertheless, a time-dependent dynamic equilibrium was observed after prolonged cultivation with all regimens.

Thiosulfate was determined colorimetrically (16). Sulfide was measured by the methylene blue method (15) or with the aid of ion-specific electrodes if the concentration was below 0.2 mM (18). Bacteriochlorophyll *a* (BChl *a*) and elemental sulfur in methanol extracts of whole cells were determined by the methods of Stal et al. (13) and Van Gemerden (17), respectively. Protein was determined by using the Folin phenol reagent (12) after extraction of elemental sulfur and BChl *a* and subsequent solubilization of the pellet (5). Total sugar concentrations were determined by using the anthrone reagent, and glycogen was calculated by the method of Van Gemerden and Beeftink (19).

INFLUENCE OF OXYGEN AND LIGHT ON GROWTH

To study the influence of the presence of oxygen on the growth of *T. roseopersicina* in the light, we shifted a thiosulfate-limited continuous culture at steady state (D = 0.019 h^{-1}) from anaerobic to semiaerobic conditions (52 μM oxygen). The time course of protein and BChl *a* levels and the specific oxygen uptake rate (qO_2) are shown in Fig. 1. The BChl *a* concentration, which was constant at 1.3 mg $\cdot$ $liter^{-1}$ during the preceding photolithotrophic steady state, decreased according to the theoretical washout rate and was virtually zero after five volume changes. Evidently, BChl *a* was neither synthesized nor actively degraded in the presence of oxygen. Nevertheless, growth still continued, the energy being provided by the aerobic respiration of thiosulfate (chemolithotrophic growth). The culture had a typical

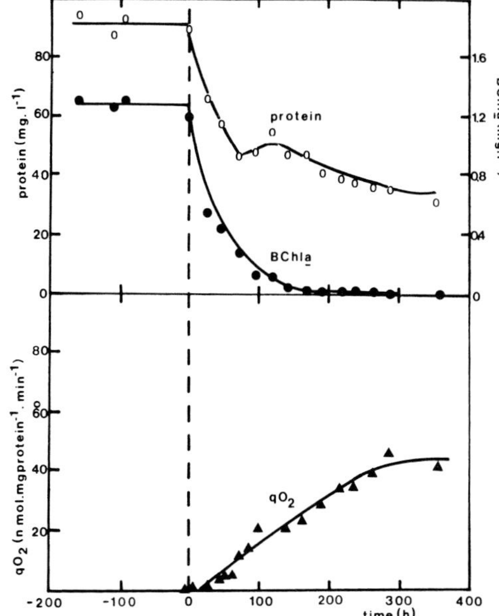

Figure 1. Time course of protein and BChl *a* concentrations and the specific oxygen uptake rate (qO_2) in an illuminated, thiosulfate-limited, continuous culture of *T. roseopersicina* M1 ($D = 0.019\ h^{-1}$) on being shifted from anaerobic to semiaerobic conditions at zero time. The curve shown for BChl *a* reflects the theoretical time course for a synthesis rate equal to zero after the shift (washout rate).

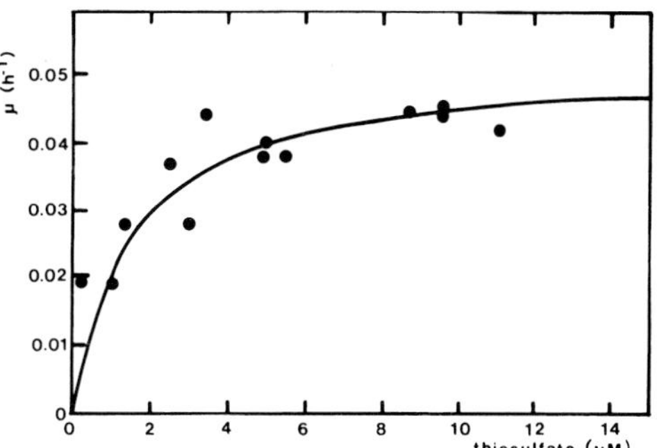

Figure 2. Steady-state data on the residual concentration of thiosulfate and the specific growth rate of *T. roseopersicina* M1 during chemolithotrophic growth in the dark. The curve reflects the relation calculated according to Monod kinetics.

milky white appearance due to the presence of intracellular sulfur. Eventually, a chemolithotrophic steady state was established in which the yield was approximately one-third of the yield obtained under photolithotrophic growth conditions. This observation can be explained by the fact that during phototrophic growth all electrons obtained by the oxidation of thiosulfate are used for the synthesis of cell material, whereas during chemolithotrophic growth part of the substrate is respired aerobically for energy generation (2). It was demonstrated earlier by Bogorov (1) that *T. roseopersicina* is able to grow chemolithotrophically on oxygen and thiosulfate similarly to colorless sulfur bacteria. Under these conditions *T. roseopersicina* can be properly described as a "colorless purple sulfur bacterium."

It is relevant to analyze whether growth occurred chemolithotrophically and/or photolithotrophically during the transient state. When exposed to oxygen, cells which still possess BChl *a* are capable of photosynthesis as well as respiration. It was observed that directly after the shift, respiration virtually did not occur. The specific respiration rate increased concomitantly with the decrease in the BChl *a* content. The organism gradually shifted to respiration when photosynthesis became inadequate as a result of the decreasing BChl *a* content. From these observations, it was concluded that cells preferentially photosynthesize rather than respire.

To judge the chemolithotrophic growth capacities of *T. roseopersicina*, we cultivated the organism under semiaerobic conditions in the dark with thiosulfate as the limiting substrate. The relation between the residual thiosulfate concentration and the specific growth rate after the establishment of steady states is shown in Fig. 2. The growth kinetics can be properly described by Monod kinetics. The maximum specific growth rate (μ_{max}) was $0.052\ h^{-1}$, and the saturation constant K_s was 1.5 μM thiosulfate. These data clearly demonstrate the potential of the organism to perform chemolithotrophic growth. The maximum specific growth rate was 68% of the attainable growth rate under phototrophic conditions. Obviously, *T. roseopersicina* is a versatile organism which is capable of both photolithotrophic and chemolithotrophic growth.

If oxygen were present continuously, *T. roseopersicina* would depend on its chemolithotrophic growth capacities when competing for substrates with colorless sulfur bacteria. The outcome of the competition for thiosulfate can be predicted from μ-*s* curves of the organisms involved, in which μ is the specific growth rate and *s* is the substrate concentration. The curve for *T. roseopersicina* is shown in Fig. 2, but reported maximum specific growth rates of specialized thiobacilli (4) exceed those of this purple sulfur bacterium by a factor of 7, and specialized thiobacilli appear to have higher growth affinities for thiosulfate as well. Although chemolithotrophic growth capacities of *T. roseopersicina* are noticeably good, this species evidently cannot compete successfully with specialized thiobacillus species in the continuous presence of oxygen. However, as mentioned above, such environmental conditions are unlikely to occur in microbial mats.

For *C. vinosum* it has been reported that BChl *a* synthesis becomes progressively less inhibited at decreasing oxygen concentrations (10). Consequently, *C. vinosum* possesses BChl *a* when the environmental oxygen concentration is below 30 μM, and the BChl *a* content increases with decreasing oxygen concentrations. A similar phenomenon was also observed in *T. roseopersicina* (C. S. Tughan, personal communication). Conceivably, at lower oxygen concentrations in the light, this would improve the competitive position of *T. roseopersicina* with respect to colorless sulfur bacteria.

However, continuous stable conditions are rare in microbial phototrophic communities, where light- and dark-induced changes occur on a diel basis. Particularly in microbial mats in which oxygenic phototrophs live in close proximity to purple sulfur bacteria, the latter will be confronted

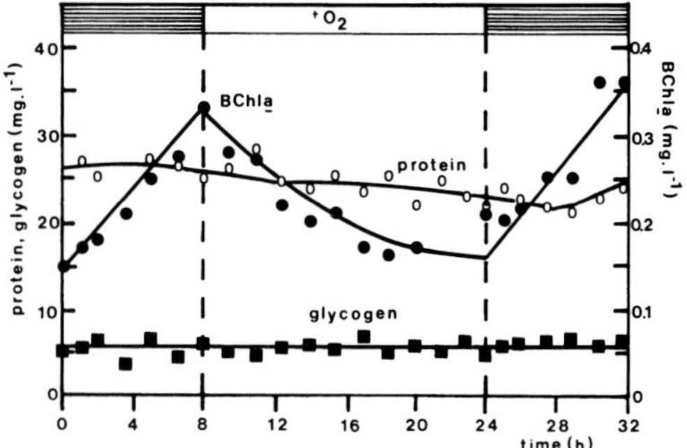

Figure 3. Time course of BChl *a*, protein, and glycogen concentrations in a sulfide-limited culture of *T. roseopersicina* M1 (D = 0.03 h^{-1}) after cultivation for five volume changes under a 16-h semiaerobic–8-h anaerobic regime in continuous light. The presence of oxygen is indicated at the top of the figure.

with oxygen during the day, whereas anaerobic conditions will prevail at night. Therefore, it is ecologically more relevant to study the influences of aerobic-anaerobic regimes than constant conditions. *T. roseopersicina* was cultivated in illuminated continuous culture, with sulfide as the limiting substrate (D = 0.03 h^{-1}), under different semiaerobic–anaerobic regimes, the total time span of a cycle being 24 h. After five volume changes, repeating patterns were observed. The time course of BChl *a*, protein, and glycogen concentrations over 32 h, after prolonged cultivation under a regimen of 16 h of semiaerobic and 8 h of anaerobic periods, is shown in Fig. 3. While protein and glycogen concentrations remained constant, the concentration of BChl *a* showed a fluctuating pattern, i.e., an increase during the anaerobic period and a decrease during the aerobic period. Again, the BChl *a* decrease in the presence of oxygen was due to washout only (compare Fig. 1). When the anaerobic period began, BChl *a* synthesis started immediately and continued at a rate exceeding the specific growth rate, thus compensating for the loss of BChl *a* during the semiaerobic period. Consequently, the specific contents of BChl *a*, which are expressed on a protein basis, were fluctuating, but always exceeded a value of 9 μg · mg^{-1}.

Table 1.
Concentrations of BChl *a* and Protein and the Ratio of BChl *a* to Protein in Sulfide-Limited Cultures of *T. roseopersicina* M1 under Various Aeration Regimens[a]

Aeration regime (h) ($+O_2/-O_2$)	Values at end of anaerobic period			Values at end of aerobic period		
	BChl *a* ($\mu g \cdot liter^{-1}$)	Protein ($mg \cdot liter^{-1}$)	BChl *a*/ protein ($\mu g \cdot mg^{-1}$)	BChl *a* ($\mu g \cdot liter^{-1}$)	Protein ($mg \cdot liter^{-1}$)	BChl *a*/ protein ($\mu g \cdot mg^{-1}$)
0/24	800	29.1	27	—[b]	—	—
4/20	880	24.7	36	770	23.5	33
7.5/16.5	695	27.3	26	600	34.0	18
12/12	829	28.4	29	711	33.2	21
16/8	325	24.8	13	205	22.0	9
21/3	184	26.9	7	129	27.9	5
24/0	—	—	—	0	9.6	0

[a] Concentrations were determined, after cultivation for five volume changes under different aeration regimes, at the ends of the anaerobic and aerobic periods. $D = 0.03\ h^{-1}$; reservoir concentration of sulfide, $S_{R'} = 1.8$ mM.
[b] —, Not applicable.

Similar patterns were also observed for cultures grown under other semiaerobic-anaerobic regimes. Table 1 lists the concentrations of BChl *a* and protein and the BChl *a* content at the end of the anaerobic and aerobic periods; maximal BChl *a* concentrations were always observed at the end of the anaerobic period. The level of BChl *a* clearly depended on the length of the aerobic period. When cultivated at regimens in which the aerobic period was no longer than 12 h, *T. roseopersicina* maintained its maximal BChl *a* content equal to that observed in steady states of continuously anaerobically grown cultures, whereas the minimum content was not lower than two-thirds of these values. However, when the organism was cultivated under progressively more aerobic regimes, both maximal and minimal BChl *a* contents decreased. As shown above, *T. roseopersicina* prefers photosynthesis over respiration, but the possession of BChl *a* is vital for photosynthesis. It is most remarkable that for all aerobic-anaerobic regimens, the yield of protein almost equalled the value for the anaerobic steady state, whereas it was only one-third of this value in the continuous presence of oxygen. The latter value represents the chemolithotrophic yield of cells that are devoid of BChl *a*. From these observations, it can be concluded that *T. roseopersicina* grew predominantly in a phototrophic mode during the semiaerobic–anaerobic regimens. Apparently, virtually no electrons were used for respiration. The increased rate of BChl *a* synthesis during the anaerobic periods and the low rate of respiration in the presence of oxygen are the decisive factors that enable the organism to grow predominantly as a phototrophic organism under these conditions. This also appears to be true for the much lower values of BChl *a* encountered for a 21-h semiaerobic–3-h anaerobic regimen (Table 1). Obviously, *T. roseopersicina* is very well adapted to changing aerobic-anaerobic conditions.

In microbial mats *T. roseopersicina* indeed encounters daily shifts from aerobic to anaerobic conditions. However, light is not continuously present in these ecosystems. Generally, aerobic periods coincide with daylight hours, whereas anaerobic conditions occur during the night in the natural environment. These natural conditions were simulated in the laboratory. Experiments were performed in sulfide-limited continuous cultures ($D = 0.03\ h^{-1}$) subjected to comparable semiaerobic-anaerobic fluctuations, but illuminated during the semiaerobic period only. The time course of protein, glycogen, and BChl *a* concentrations in a

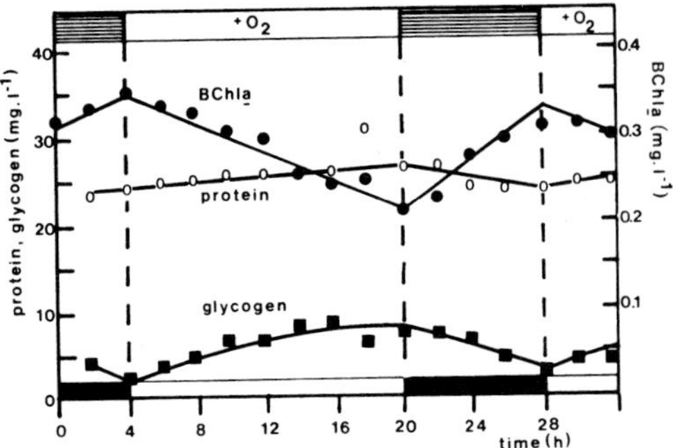

Figure 4. Time course of BChl *a*, protein, and glycogen concentrations in a sulfide-limited culture of *T. roseopersicina* M1 (D = 0.03 h^{-1}) after cultivation for five volume changes under a 16-h light, semiaerobic and 8-h dark, anaerobic regime. The presence of oxygen is indicated at the top, and dark periods (▬) are indicated at the bottom.

culture grown under a regimen of 16 h of semiaerobic light conditions and 8 h of anaerobic and dark conditions is shown in Fig. 4. Again, BChl *a* was washed out during the aerobic periods, but synthesis was observed during the anaerobic dark periods. The process of BChl *a* synthesis requires energy, which obviously was provided in the dark by the breakdown of glycogen.

For other regimens studied, similar patterns were observed. Table 2 lists the concentrations of BChl *a* and protein and the BChl *a* content for the different regimens at the end of the semiaerobic light period and the end of the anaerobic dark period. Again, the BChl *a* content clearly correlated with the length of the semiaerobic light period.

T. roseopersicina is noticeably well adapted to the environmental conditions encountered in microbial mats. Oxygen produced by its oxygenic neighbors inhibits its pigment synthesis. However, pigment synthesis occurs during the anaerobic night periods, and, consequently, phototrophic capacities can be expressed during the day even in the presence of oxygen. Although *T. roseopersicina* is frequently exposed to oxygen, the organism behaves predominantly as a phototroph. By possessing these physiological characteristics, *T. roseopersicina* elegantly circumvents problems caused by oxygen produced in oxygenic phototrophs. Apparently, these characteristics enable it to maintain its position in microbial mats, although it is unable to adjust its depth horizon to the oxygen/sulfide interface.

The present data demonstrate the high potential of the continuous cultivation technique for microbial ecology. This technique enables many environmental conditions to be accurately controlled. Once the restricting conditions are obeyed (e.g., the requirement of culture homogeneity does not easily facilitate the study of spatial relations), chemostats are excellent tools for studying physiological adaptations of microorganisms to fluctuating environmental conditions. The knowledge obtained will also facilitate a bet-

Table 2.
Concentrations of BChl *a* and Protein and the Ratio of BChl *a* to Protein in Sulfide-Limited Cultures of *T. roseopersicina* M1 under Aeration and Illumination Regimens[a]

Aeration and illumination regimen (h) (+O_2 [light]/ −O_2 [dark])	Values at end of anaerobic dark period			Values at end of aerobic light period		
	BChl *a* (μg · liter^{-1})	Protein (mg · liter^{-1})	BChl *a*/ protein (μg · mg^{-1})	BChl *a* (μg · liter^{-1})	Protein (mg · liter^{-1})	BChl *a*/ protein (μg · mg^{-1})
14/10	382	23.2	16	259	24.2	11
16/8	331	23.3	14	213	26.3	8
20/4	166	28.3	6	130	30.9	4
24/0	—[b]	—	—	0	9.6	0

[a] See Table 1, footnote *a*.
[b] See Table 1, footnote *b*.

ter interpretation of natural phenomena when these are occurring in highly structured laminated communities.

ACKNOWLEDGMENTS. The contribution of Geert Bos to the experimental work and the technical assistance of Marion Rademaker are highly appreciated.

This investigation was supported by a grant from the Foundation for Fundamental Biological Research, which is subsidized by the Netherlands Organization for the Advancement of Pure Science.

LITERATURE CITED

1. **Bogorov, L. V.** 1974. About the properties of *Thiocapsa roseopersicina* strain BBS, isolated from the estuary of the White Sea. *Microbiology* (Engl. Transl. *Mikrobiologiya*) 43:275–280.
2. **De Wit, R., and H. van Gemerden.** 1987. Chemolithotrophic growth of the phototrophic sulfur bacterium *Thiocapsa roseopersicina. FEMS Microbiol. Ecol.* 45:117–126.
3. **Gorlenko, V. M.** 1974. The oxidation of thiosulfate of *Amoebobacter roseus* in the dark under microaerophilic conditions. *Microbiology* (Engl. Transl. *Mikrobiologiya*) 43:624–625.
4. **Gottschal, J. C., S. De Vries, and J. G. Kuenen.** 1979. Competition between the facultatively chemolithotrophic *Thiobacillus A2*, and an obligately chemolithotrophic *Thiobacillus* and a heterotrophic *Spirillum* for inorganic and organic substrates. *Arch. Microbiol.* 121:241–249.
5. **Herbert, D., P. J. Phipps, and R. E. Strange.** 1971. Chemical analyses of microbial cells, p. 209–344. *In* J. R. Norris and B. W. Ribbons (ed.), *Methods in Microbiology*, vol. 5B. Academic Press, Inc. (London), Ltd., London.

5a. **Herbert, R. A., C. S. Tughan, R. de Wit, and H. van Gemerden.** 1988. Development of laminated mat microbial ecosystems in temperate marine environments, p. 105–111. *In* F. Megusar and M. Gantar (ed.), *Perspectives in Microbial Ecology: Proceedings of the 4th International Symposium on Microbiol Ecology, Ljubljana, 1986.* Slovene Society for Microbiology, Ljubljana, Yugoslavia.

6. **Hurlbert, R. E.** 1967. Effect of oxygen on viability and substrate utilization in *Chromatium. J. Bacteriol.* 93:1346–1352.
7. **Jørgensen, B. B.** 1982. Ecology of the bacteria of the sulphur cycle with special reference to anoxic-oxic interface environments. *Philos. Trans. R. Soc. London Ser. B* 298:543–561.
8. **Jørgensen, B. B., N. P. Revsbech, T. H. Blackburn, and Y. Cohen.** 1979. Diurnal cycle of oxygen and sulfide microgradients and microbial photosynthesis in a cyanobacterial mat sediment. *Appl. Environ. Microbiol.* 38:46–58.
9. **Kämpf, C., and N. Pfennig.** 1980. Capacity of Chromatiaceae for chemotrophic growth. Specific respiration rates of *Thiocystis violacea* and *Chromatium vinosum. Arch. Microbiol.* 127:125–135.
10. **Kämpf, C., and N. Pfennig.** 1986. Chemoautotrophic growth of *Thiocystis violacea, Chromatium gracile* and *C. vinosum* in the dark at various O_2 concentrations. *J. Basic Microbiol.* 26:517–531.
11. **Kondratieva, E. N., V. G. Zhukov, R. N. Ivanovsky, Y. P. Petushkova, and E. Z. Monosov.** 1976. The capacity of phototrophic sulfur bacterium *Thiocapsa roseopersicina* for chemosynthesis. *Arch. Microbiol.* 108:287–292.
12. **Lowry, O. H., N. J. Rosebrough, A. L. Farr, and R. J. Randall.** 1951. Protein measurement with the Folin phenol reagent. *J. Biol. Chem.* 193:265–275.
13. **Stal, L. J., H. van Gemerden, and W. E. Krumbein.** 1984. The simultaneous assay of chlorophyll and bacteriochlorophyll in natural microbial communities. *J. Microbiol. Methods* 2:295–306.
14. **Stal, L. J., H. van Gemerden, and W. E. Krumbein.** 1985. Structure and development of a benthic marine microbial mat. *FEMS Microbiol. Ecol.* 31:111–125.
15. **Trüper, H. G., and H. G. Schlegel.** 1964. Sulphur metabolism in Thiorhodaceae. I. Quantitative measurements on growing cells of *Chromatium okenii. Antonie van Leeuwenhoek J. Microbiol. Serol.* 30:225–238.
16. **Urban, P. J.** 1961. Colorimetry of sulphur anions. I. An improved colorimetric method for the determination of thiosulphate. *Z. Anal. Chem.* 179:415–422.
17. **Van Gemerden, H.** 1968. Growth measurements of *Chromatium* cultures. *Arch. Mikrobiol.* 64:103–110.
18. **Van Gemerden, H.** 1984. The sulfide affinity of phototrophic bacteria in relation to the location of elemental sulfur. *Arch. Microbiol.* 139:289–294.
19. **Van Gemerden, H., and H. H. Beeftink.** 1978. Specific rates of substrate oxidation and product formation in autotrophically growing *Chromatium vinosum* cultures. *Arch. Microbiol.* 119:135–143.

Chapter 29

Growth Responses of the Cyanobacterium *Microcoleus chthonoplastes* with Sulfide as an Electron Donor

Rutger de Wit and Hans van Gemerden

INTRODUCTION

The cosmopolitan cyanobacterial species *Microcoleus chthonoplastes* is very often the dominant organism in the green layers of microbial mats, since they occur in many marine and hypersaline environments. In these often complex microbial ecosystems, cyanobacteria, purple sulfur bacteria, and sulfate-reducing bacteria live in close proximity, and, consequently, the cyanobacteria are exposed to sulfide. Therefore it is of great ecological interest to analyze the impact of the presence of sulfide on *M. chthonoplastes*, since sulfide is known to be severely toxic to many aerobic as well as anaerobic organisms.

Several reports have demonstrated that many cyanobacteria are capable of performing anoxygenic photosynthesis with sulfide as the electron donor (3–8). In addition, it has been shown that photosystem II of several cyanobacteria is tolerant to sulfide (3–5, 7). It has been pointed out (4, 5) that the possession of one or both properties is the decisive factor in determining the tolerance of the species to sulfide. On the basis of these characteristics, the cyanobacteria were classified into four strategy groups (4, 5).

When *M. chthonoplastes* is exposed to sulfide, a 3-h lag period is observed before anoxygenic photosynthesis occurs, while oxygenic photosynthesis continues to operate. After the lag period both types of photosynthesis operate simultaneously (4, 5, 7). Apparently, *M. chthonoplastes* has both a sulfide-tolerant photosystem II and the ability to perform anoxygenic photosynthesis, but actual growth in the presence of sulfide has not been reported previously. In fact, growth exclusively at the expense of sulfide oxidation has been demonstrated only for *Oscillatoria limnetica* (11) and *Oscillatoria amphigranulata* (3).

The aim of this study was to elucidate the ability of *M. chthonoplastes* to grow in sulfide-containing media and to analyze the contribution of oxygenic and anoxygenic photosynthesis to growth in the presence of sulfide. Special attention was paid to the question of whether *M. chthonoplastes* is capable of growing on sulfide as the sole electron donor.

Rutger de Wit and Hans van Gemerden • Department of Microbiology, University of Groningen, Kerklaan 30, 9751 NN Haren, The Netherlands.

PROCEDURES FOR GROWING *M. CHTHONOPLASTES*

All experiments were carried out with *M. chthonoplastes* 11 isolated by L. J. Stal, University of Oldenburg, Oldenburg, Federal Republic of Germany, from a laminated microbial ecosystem on the island of Mellum, Federal Republic of Germany.

Single filaments of *M. chthonoplastes* were grown in liquid medium whose composition has been described previously (7). To ascertain whether the influence of contaminating bacteria in this monocyanobacterial culture was of minor importance, the chlorophyll *a*/protein ratio was determined and microscopic observations were made of the number of contaminating bacteria throughout the experiments. The growth experiments were performed with sulfide-pulsed batch cultures and continuous cultures operated as sulfidostats (see below). The temperature was 22 ± 1°C. Light was supplied by fluorescent and incandescent light sources. The illumination intensity was 35 to 45 microeinsteins · $m^{-2} \cdot s^{-1}$. In some experiments (see below), 3-(3,4-dichlorophenyl)-1,1-dimethylurea (DCMU) was added to a final concentration of 5 μM. Previously it had been demonstrated that this addition resulted in a complete inhibition of oxygen production (7). DCMU was used as solution of 1 mM in ethanol to ensure its rapid dispersion.

Oxygen-free culture medium was inoculated with cells precultivated aerobically in 500-ml conical flasks. Prior to inoculation, the inocula were aseptically flushed for 30 min with oxygen-free nitrogen.

Sulfide-pulsed batch cultures were grown in 6-liter bottles, with the headspace connected to a supply of oxygen-free nitrogen (10 kPa) to ensure that the volume sampled was replaced directly by nitrogen. To readjust the sulfide concentration to the desired level, 1 M Na_2S was frequently added aseptically to the cultures. The pH was readjusted daily to 8.0 ± 0.1 by the aseptic addition of 1 M HCl. Fluctuations in the pH were in the range of 7.9 to 8.3 throughout the course of the batch experiments. The effect of the simultaneous presence of oxygen, sulfide, and DCMU on growth was studied by using a DCMU-containing, sulfide-pulsed batch culture which was continuously flushed with air (1 to 3 liters · h^{-1}).

During the exponential growth phase in batch culture, the specific growth rate (μ) was calculated from the protein increase and the oxidation of sulfide. The assumption was made that the specific rate of sulfide consumption remained constant throughout the experiment.

Continuous cultures as described previously (2) were modified to sulfidostat cultures. To maintain the sulfide concentration at a desired level, a control device was installed that provided an intermittent flow of fresh medium (containing 1.8 mM sulfide) to the culture. The sulfide concentration in the culture was monitored with a sulfide-specific electrode and a double-junction reference electrode (with 1 M NaCl electrolyte solution in the outer jacket) connected with a millivolt meter of high impedance. The pH was kept constant at a value of 8.0 ± 0.1 by using a pH stat. The average dilution rate (*D*) was calculated from the outflow over 24 h.

To judge the contribution of oxygenic photosynthesis to the total photosynthesis, the time course of sulfide, thiosulfate, and oxygen concentrations was measured before and after the addition of DCMU. The original culture was grown at 0.35 mM sulfide. Subsamples were incubated in 600-ml serum flasks, with a headspace consisting of oxygen-free nitrogen, under the same light and temperature conditions as those used during precultivation. Oxygen concentrations in the subsample were measured polarographically with a gold-plated oxygen microelectrode (13).

Sulfide was determined by the methylene blue method (14). Thiosulfate was determined by the method of Urban (15).

After centrifugation and extraction of chlorophyll *a* and sulfur with methanol, the protein in solubilized pellets (1 M NaOH, 100°C) (9) was determined by using the Folin phenol reagent (10).

INFLUENCE OF SULFIDE ON GROWTH

To study growth responses and photosynthesis of *M. chthonoplastes* in the presence of sulfide, we cultivated the organism in sulfide-pulsed batch cultures (no DCMU added; see above). In cultures with average sulfide concentrations of 0.35 and 0.75 mM, the sulfide concentration fluctuated from 0.2 to 0.5 mM and from 0.6 to 0.9 mM, respectively. In these cultures, sulfide was clearly oxidized, and exponential growth occurred for at least six consecutive days. Eventually a stationary phase was reached. In cultures with sulfide concentration exceeding 1 mM, no growth was observed. Evidently, *M. chthonoplastes* is able to grow only in the presence of <1 mM sulfide.

Growth of *M. chthonoplastes* in the presence of sulfide is possible when this organism performs oxygenic or anoxygenic photosynthesis or both. It has been reported (4, 5, 7) that photosystem II is not completely inhibited by sulfide in this species. In addition, anoxygenic photosynthesis, which uses sulfide as the electron donor, is induced (5, 7, 8). For *M. chthonoplastes* 11, it has been reported that thiosulfate is the product of the biotic photooxidation of sulfide (7). To analyze the relative contribution of oxygenic and anoxygenic photosynthesis, we analyzed the effect of DCMU on the sulfide oxidation rate of a culture grown at an average sulfide concentration of 0.35 mM. The time course of sulfide and thiosulfate concentrations in a subsample taken from such a culture before and after the addition of DCMU is shown in Fig. 1. The concentration of oxygen was below 3 μM throughout the experiment. After the addition of DCMU, which was dissolved in ethanol, the rate of sulfide dis-

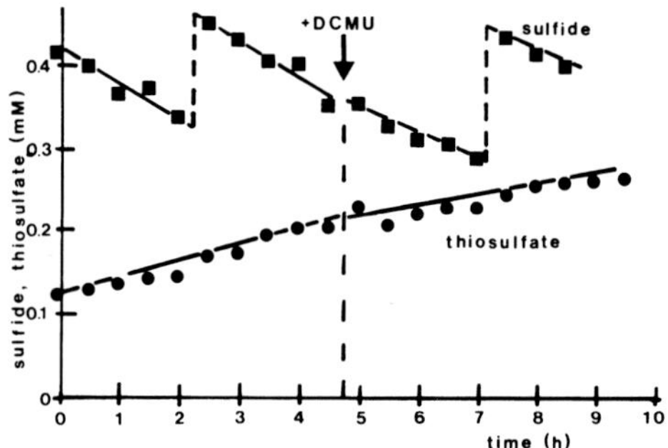

Figure 1. Time course of sulfide and thiosulfate levels in a subsample from a sulfide-pulsed batch culture of *M. chthonoplastes* 11 cultivated in the presence of sulfide (average concentration, 0.35 mM). During the experiment, DCMU (final concentration, 5 μM) was added as indicated. Throughout the experiment the concentration of oxygen was below 3 μM. The sulfide concentration was maintained above 0.3 mM by manual addition of Na_2S.

appearance decreased to 74% of its original value. An identical final concentration of ethanol (no DCMU) resulted in a decrease in the rate of sulfide disappearance to 78% of the original value. Evidently, the DCMU itself had little effect. The degree of inhibition of photosystem II increases with increasing sulfide concentration (5). It was observed in short-term $^{14}CO_2$ fixation experiments performed with *M. chthonoplastes* S.L. that at a sulfide concentration of 0.2 mM, the contribution of oxygenic photosynthesis to the total photosynthesis was no more than 5% (5). The present data are in agreement with these observations and showed that the growth of *M. chthonoplastes* at an average sulfide concentration of 0.35 mM was due predominantly to the utilization of sulfide as the electron donor in anoxygenic photosynthesis. Undoubtedly, this will also occur at higher sulfide concentrations. Therefore, it was of interest to determine the relation between the specific growth rate and the environmental sulfide concentration.

Sulfide-pulsed batch cultures were used to measure the specific growth rate at sulfide concentrations exceeding 0.35 mM, whereas sulfidostat continuous cultures were used to study lower concentrations (see above). In

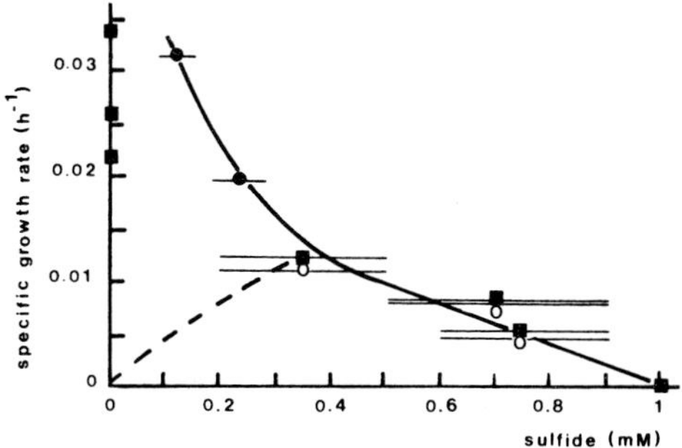

Figure 2. Relation between the specific growth rate (μ) of *M. chthonoplastes* 11 and the sulfide concentration as determined in sulfidostat continuous cultures (●) and in pulsed batch cultures. For the batch cultures the calculations were based on protein (■) and the total amount of sulfide being oxidized (○). The bars indicate the fluctuations of the sulfide concentration. ––––, Predicted relation when photosystem II is selectively inhibited and oxygen is externally supplied.

all cases, the constant chlorophyll *a*/protein ratio indicated the insignificant contribution of contaminants to the total protein concentration. The relation between the specific growth rate of *M. chthonoplastes* and the sulfide concentration (no DCMU added) is shown in Fig. 2. For sulfide concentrations exceeding 0.15 mM, the growth rate decreased with increasing sulfide concentration, until at sulfide concentrations exceeding 1 mM, growth did not occur. Evidently, sulfide is a toxic substance for *M. chthonoplastes*, as was observed for other organisms. However, the tolerance to sulfide is very species specific, a fact which has important ecological implications. Several cyanobacteria, for example, *Synechococcus* sp. and *Plectonema boryanum*, are unable to grow at sulfide concentrations exceeding 0.1 mM because photosystem II is completely inhibited and anoxygenic photosynthesis cannot be induced in these species (5). In contrast, it has been demonstrated that the cyanobacterium *O. limnetica* is able to grow at 2.5 mM sulfide (11), whereas the purple sulfur bacterium *Chromatium vinosum* grows even at 10 mM sulfide (16). Therefore, *M. chthonoplastes* can be considered to have a moderate tolerance to sulfide.

The growth kinetics of organisms which use sulfide as a substrate can normally be described by the Haldane equation (1), except for growth at very high sulfide concentrations (17). Accordingly, at low concentrations the specific growth rate will increase with increasing substrate concentrations. At higher concentrations, inhibition will occur and the growth rate will decline with increasing substrate concentrations. Consequently, as long as sulfide is the only electron donor, the relation between the specific growth rate and the substrate concentration is an optimum curve. However, the relation between the specific growth rate of *M. chthonoplastes* 11 and the sulfide concentration cannot be described by Haldane kinetics (Fig. 2). In particular, deviations occurred at low sulfide concentrations. This phenomenon can be explained by the fact that at low sulfide concentrations, photosystem II is still operative and therefore sulfide and water serve the same metabolic role (no DCMU). The contribution of oxygenic photosynthesis increased with decreasing sulfide concentrations, since the inhibition of photosystem II decreased (5). In the absence of sulfide, growth was due to oxygenic photosynthesis alone. At sulfide concentrations between 0.35 and 1 mM, growth occurred predominantly at the expense of sulfide oxidation. Only when the contribution of photosystem II was low could the growth kinetics be described by the Haldane equation.

The aforementioned data suggested that *M. chthonoplastes* should be able to grow with sulfide as the exclusive electron donor, when photosystem II is selectively inhibited by DCMU. However, we were unable to detect growth in anaerobic environments in the presence of DCMU. Although the contribution of photosystem II was low, its operation is apparently of paramount importance in anaerobic environments. Padan (12) postulated that oxygen is an essential nutrient for many cyanobacterial species, since it is required for the biosynthesis of unsaturated fatty acids. Fatty acid analyses of *M. chtho-*

noplastes 11 demonstrated that linoleate ($C_{18:2}$), linolenate ($C_{18:3}$), and tetradecadienoic acid ($C_{14:2}$) were present in appreciable amounts. If oxygen were indeed an essential nutrient for *M. chthonoplastes*, it is to be expected that the operation of photosystem II would be of paramount importance in anaerobic environments, because of the oxygen produced. It is anticipated that the required oxygen can be supplied by external sources as well. Consequently, *M. chthonoplastes* should be able to grow with sulfide as the exclusive electron donor in the presence of DCMU, provided that oxygen is supplied to the culture. This phenomenon was indeed observed in sulfide-pulsed batch cultures when air was continuously bubbled through.

Growth curves of *M. chthonoplastes* in sulfide-pulsed batch cultures in the absence of DCMU, under anaerobic conditions in the presence of DCMU, and with air supplied in the presence of DCMU are shown in Fig. 3. Again, the constant chlorophyll *a*/protein ratio proved that the protein was attributed almost exclusively to *M. chthonoplastes*. Growth occurred only if photosystem II was not inhibited by DCMU or when air was supplied to the culture. No growth of *M. chthonoplastes* on sulfide occurred under anaerobic conditions when photosystem II was inhibited by DCMU. These data clearly

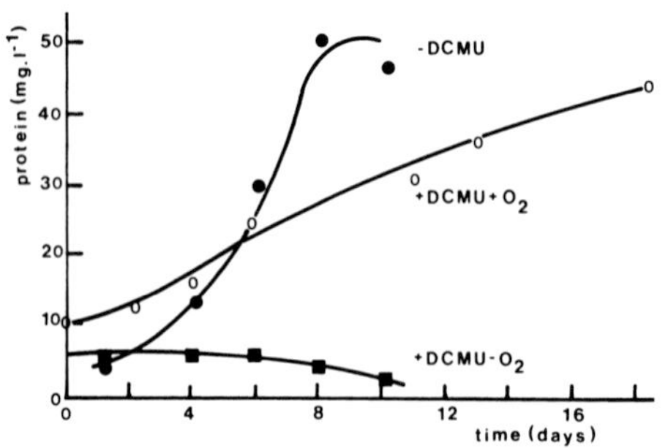

Figure 3. Time course of the concentration of protein in sulfide-pulsed batch cultures of *M. chthonoplastes* 11 in the absence of DCMU, under anaerobic conditions in the presence of DCMU, and with air supplied in the presence of DCMU. The average sulfide concentrations were 0.35 mM.

demonstrate that *M. chthonoplastes* is able to grow on sulfide as the exclusive electron donor and that oxygen is an essential nutrient. Evidently, photosystem II is able to provide the oxygen required by the organism, but oxygen can also be supplied from external sources. It is expected that the growth kinetics of *M. chthonoplastes* on sulfide, when photosystem II is selectively inhibited and oxygen is externally supplied, could be properly described by the Haldane equation. Assuming that all sulfide oxidized is used for growth, the relation of the specific growth rate of *M. chthonoplastes* and the sulfide concentration under these conditions can be predicted, by using the protein yield obtained on sulfide and the kinetic parameters determined in short-term uptake experiments. The yield of *M. chthonoplastes* on sulfide in terms of protein was 17.1 mg $\cdot$ mmol^{-1}. The short-term sulfide uptake experiments, which were performed under anaerobic conditions in the presence of DCMU, have been described previously. The kinetic parameters V_{max} and K_m were 2.6 μmol $\cdot$ mg of protein^{-1} $\cdot$ h^{-1} and 0.97 mM, respectively (7). The predicted relation between the specific growth rate of *M. chthonoplastes* and the concentration of sulfide as the electron donor, when photosystem II is selectively inhibited and oxygen is externally supplied, is indicated by the dashed line in Fig. 2. Obviously, oxygenic and anoxygenic photosynthesis both contribute significantly to growth at the lower sulfide concentrations.

M. chthonoplastes is a benthic species that thrives in microbial mats. This species is excluded from extremely sulfide-rich habitats because it cannot grow above 1 mM sulfide. In contrast, the cyanobacterium *O. limnetica* thrives in the anaerobic sulfide-rich layers of Solar Lake, Sinai, Egypt, during winter stratification (12). *O. limnetica* is indeed able to grow at higher sulfide concentrations (11). The sulfide concentrations in microbial mats in which daily transitions occur between aerobic and anaerobic conditions are normally below 1 mM. Consequently, *M. chthonoplastes* seems to be well adapted to the environmental conditions encountered in microbial mats.

It is noteworthy that *M. chthonoplastes* lives in close proximity to or may even coexist with purple sulfur bacteria in microbial mats. The purple sulfur bacteria are also able to oxidize sulfide phototrophically in the light or chemotrophically in the presence of oxygen. By doing so, they reduce the sulfide concentration in the environment of *M. chthonoplastes*, which results in less inhibition of its growth (Fig. 2). Finally, *M. chthonoplastes* does not have to compete with purple sulfur bacteria, since it shifts to oxygenic photosynthesis at low sulfide concentrations. These considerations show that the purple sulfur bacteria exert a positive effect on the growth of *M. chthonoplastes* in microbial mats.

ACKNOWLEDGMENTS. *M. chthonoplastes* 11 was kindly provided by Lucas J. Stal. The contribution of Wim van Boekel to the experimental work is highly appreciated.

This investigation was supported by a grant from the Foundation for the Fundamental Biological Research, which is subsidized by the Netherlands Organization for the Advancement of Pure Science.

LITERATURE CITED

1. **Andrews, J. F.** 1968. A mathematical model for the continuous culture of microorganisms utilizing inhibitory substrates. *Biotechnol. Bioeng.* **10:** 707–723.
2. **Beeftink, H. H., and H. van Gemerden.** 1979. Actual and potential rates of substrate oxidation and product formation in continuous cultures of *Chromatium vinosum*. *Arch. Microbiol.* **121:**161–167.
3. **Castenholz, R. W., and H. C. Utkilen.** 1984. Physiology of sulfide tolerance in a thermophilic *Oscillatoria*. *Arch. Microbiol.* **138:**299–305.
4. **Cohen, Y.** 1984. Oxygenic photosynthesis, anoxygenic photosynthesis, and sulfate reduction in cyanobacterial mats, p. 435–441. *In* M. J. Klug and C. A. Reddy (ed.), *Current Perspectives in Microbial Ecology*. American Society for Microbiology, Washington, D.C.
5. **Cohen, Y., B. B. Jørgensen, N. P. Revsbech, and R. Poplawski.** 1986. Adaptation to hydrogen sulfide of oxygenic and anoxygenic photosynthesis among cyanobacteria. *Appl. Environ. Microbiol.* **51:**398–407.
6. **Cohen, Y., E. Padan, and M. Shilo.** 1975. Facultative anoxygenic photosynthesis in the cyanobacterium *Oscillatoria limnetica*. *J. Bacteriol.* **123:** 855–861.
7. **De Wit, R., and H. van Gemerden.** 1987. Oxidation of sulfide to thiosulfate by *Microcoleus chthonoplastes*. *FEMS Microbiol. Ecol.* **45:**7–13.
8. **Garlick, S., A. Oren, and E. Padan.** 1977. Occurrence of facultative anoxygenic photosynthesis among filamentous and unicellular cyanobacteria. *J. Bacteriol.* **129:**623-629.
9. **Herbert, D., P. J. Phipps, and R. E. Strange.** 1971. Chemical analyses of microbial cells, p. 209–344. *In* J. R. Norris and B. W. Ribbons (ed.), *Methods in Microbiology*, vol. 5B. Academic Press, Inc. (London), Ltd., London.
10. **Lowry, O. H., N. J. Rosebrough, A. L. Farr, and R. J. Randall.** 1951. Protein measurement with the Folin phenol reagent. *J. Biol. Chem.* **193:** 265–275.
11. **Oren, A., and E. Padan.** 1978. Induction of anaerobic photoautotrophic growth in the cyanobacterium *Oscillatoria limnetica*. *J. Bacteriol.* **133:** 558–563.
12. **Padan, E.** 1979. Impact of facultatively anaerobic phototrophic metabolism on ecology of cyanobacteria (blue-green algae). *Adv. Microb. Ecol.* **3:**1–48.
13. **Revsbech, N. P., B. B. Jørgensen, T. H. Blackburn, and Y. Cohen.** 1983. Microelectrode studies of the photosynthesis and O_2, H_2S, and pH profiles of a microbial mat. *Limnol. Oceanogr.* **28:**1062–1074.
14. **Trüper, H. G., and H. G. Schlegel.** 1964. Sulphur metabolism in Thiorhodaceae. I. Quantitative measurements on growing cells of *Chromatium okenii*. *Antonie van Leeuwenhoek J. Microbiol. Serol.* **30:**225–238.
15. **Urban, P. J.** 1961. Colorimetry of sulphur anions. I. An improved colorimetric method for the determination of thiosulphate. *Z. Anal. Chem.* **179:** 415–422.
16. **Van Gemerden, H., and R. de Wit.** 1986. Strategies of phototrophic bacteria in sulphide-containing environments, p. 111–127. *In* R. A. Herbert and G. A. Codd (ed.), *Microbes in Extreme Environments*. Academic Press, Inc. (London), Ltd., London.
17. **Van Gemerden, H., and H. W. Jannasch.** 1971. Continuous culture of Thiorhodaceae. Sulfide and sulfur limited growth of *Chromatium vinosum*. *Arch. Mikrobiol.* **79:**345–353.

Chapter 30

Naturally Occurring Patterns of Oxygenic Photosynthesis and N_2 Fixation in a Marine Microbial Mat: Physiological and Ecological Ramifications

Hans W. Paerl, Brad M. Bebout, and Leslie E. Prufert

INTRODUCTION

Despite their widespread occurrence in nutrient-depleted, oligotrophic marine waters, microbial mats often exhibit remarkably high rates of primary production and biomass accumulation (18, 29). Localized productive mats are reminiscent of coral reefs, which also reside in nutrient-depleted waters (18, 29). Among potentially limiting nutrients, nitrogen is the most commonly cited as being chronically deficient (5). It follows that the availability of biologically utilizable nitrogen has been linked to levels of planktonic and benthic fertility in a wide range of marine environments (5, 12); microbial mat and reef communities are no exception (1, 8, 9, 21, 27). However, unlike overlying planktonic communities, in which nitrogen demands are met mainly by regeneration and physical inputs (upwelling, vertical and horizontal transport, and terrigenous runoff), mat and reef communities are sites of significant biological N_2 fixation (18, 21, 28). As such, N_2 fixation serves as an internal source of nitrogen for supporting primary production.

Laminated mat structures attributable to specific strains of photoautotrophic and chemotrophic microorganisms have been implicated in supporting ecologically significant rates of N_2 fixation (1, 2, 6, 7; H. W. Paerl, *in* R. S. Alberte and R. T. Barber, ed., *Photosynthesis in the Sea*, in press), thereby promoting localized enhancement of primary production. In microbial mats, production is often dominated by cyanobacteria, which are adept as initial (pioneer) colonizers and mat builders (7, 25, 29). Eucaryotic diatoms, anoxygenic phototrophic bacteria, and chemotrophic bacteria constitute important subdominant microbial groups once the structural integrity of the mats is established (7, 29; C. Polimeni, M.Sc. Thesis, University of North Carolina, Chapel Hill, 1976). Specific mat-building cyanobacterial, anoxyphotrophic, and chemotrophic bacterial genera are known to be capable of N_2 fixation

Hans W. Paerl, Brad M. Bebout, and Leslie E. Prufert • Institute of Marine Sciences, University of North Carolina at Chapel Hill, Morehead City, North Carolina 28557.

(6, 10, 18, 20, 21). Obviously, N_2 fixation offers physiological and ecological advantages among members of mat communities. However, mat N_2 fixation potentials are closely regulated by diverse environmental factors, including ambient O_2 concentration, inorganic and organic nutrient availability (3, 13, 17, 20, 30; Paerl, in press), and suppression by NH_3, the end product of N_2 fixation (20, 30). Inorganic nutrient (P, Fe, Mo, Co, and Cu) limitation of mat and detrital N_2 fixation has been discussed elsewhere (13, 15). In this chapter we will examine and discuss the often dominant role of dissolved O_2 in determining both spatial and temporal N_2 fixation patterns and potentials in a cyanobacteria-dominated mat located on a barrier island in coastal Atlantic Ocean waters bordering North Carolina.

We will focus on a paradoxical situation faced by dominant nonheterocystous filamentous and colonial cyanobacteria and associated heterotrophic eubacteria, thought to be the chief contributors to mat N_2 fixation rates. The paradox arises from the fact that resident microorganisms derive the bulk of their organic carbon requirements for growth and energy production from oxygenic photosynthetic primary production (4, 20, 30). However, associated O_2 evolution represents a potentially inhibitory step with respect to nitrogenase activity (NA) among these species (20, 30; Paerl, in press). In contrast to heterocystous cyanobacteria, which have some structural and physiological protection from O_2 inhibition of nitrogenase in the form of thick-walled, O_2-devoid heterocysts (8, 20), organisms of the dominant mat cyanobacterial genera, including *Lyngbya*, *Microcoleus*, *Oscillatoria*, and *Phormidium* spp., exhibit no such cellular differentiation. Accordingly, the question arises of how such nonheterocystous cyanobacteria and associated eubacteria contemporaneously and contiguously optimize oxygenic photosynthesis and O_2-sensitive N_2 fixation within the confines of a highly productive mat microenvironment (microzone). In addressing this paradox and its ecophysiological ramifications, we will examine data obtained from laboratory and field experiments in conjunction with diel (24-h) photosynthesis and N_2 fixation measurements of natural cyanobacterial-bacterial communities on Shackleford Banks, N.C.

LOCATION OF MATS

Intertidal microbial mats, composed primarily of filamentous nonheterocystous cyanobacteria (*Lyngbya aestuarii*, *Microcoleus chthonoplastes*, *Oscillatoria* spp., and *Phormidium* spp.), diatoms (*Amphora coffeiformis*, *Navicula complanata*, *Nitzschia closteria*, *Nitzschia filiformis*, and *Rhopaloidia* sp.), heterotrophic, photoautotrophic, and chemoautotrophic bacteria, protozoans, and invertebrates (nematodes, crustacean larvae, harpactacoid copepods, and snails), are a semipermanent feature of a lagoon formed by a sand spit at the western tip of Shackleford Banks (Fig. 1). Shackleford Banks is part of the Outer Banks system, separating the U.S. mainland from the Atlantic Ocean. Mats cover approximately 60 to 100 ha and are irregularly covered by high tides. This full-salinity habitat drains directly into the Atlantic Ocean via Beaufort Inlet (Fig. 1) on outgoing tides. The site has been the subject of numerous earlier studies (2, 3; Paerl, in press; Polimeni, M.Sc. thesis), which have demonstrated a strong inverse correlation of N_2 fixation potentials and O_2 tension (3; Paerl, in press) and a direct correlation of the potentials and organic matter availability. The mats are located approximately 5 km from the Institute of Marine Sciences, Morehead City, N.C.

PROCEDURES FOR DIEL STUDIES

A detailed description of methods used during the diel studies has been published elsewhere (3), but will be summarized here.

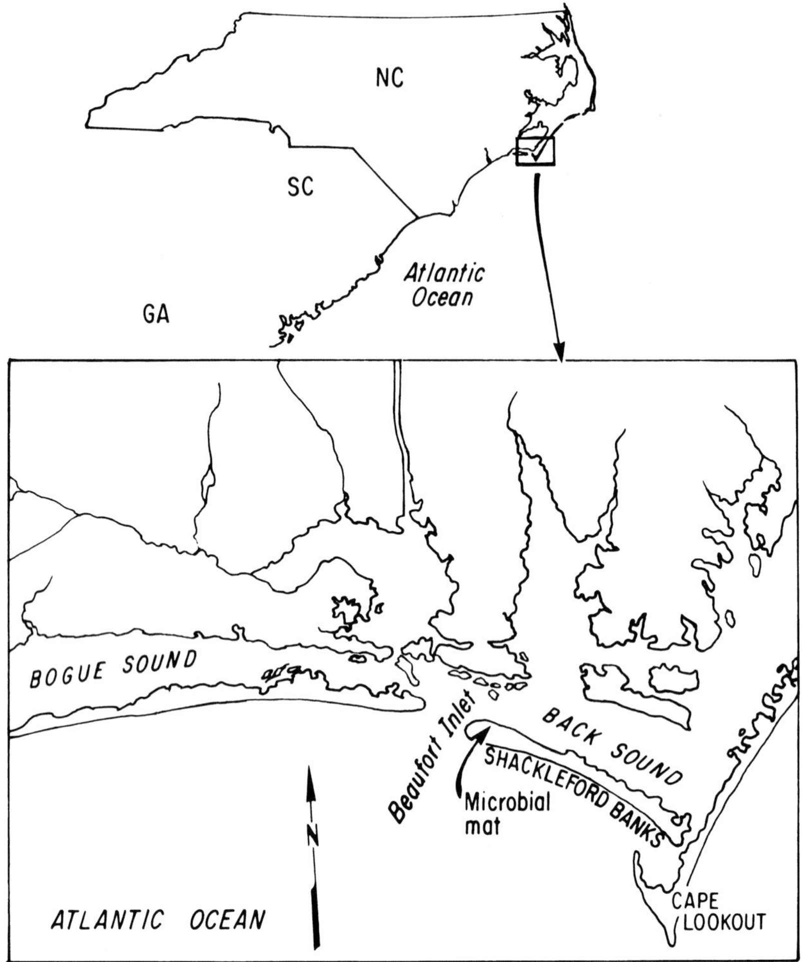

Figure 1. Location of the Shackleford Banks microbial mat examined in this study. The Institute of Marine Sciences is located on Bogue Sound approximately 6 km west-northwest of the mat. Reproduced from *Applied and Environmental Microbiology* (3).

Mat samples were collected for diel studies at the Shackleford Banks study site on three occasions: 2 June 1986 (day I), 18 June 1986 (day II), and 28 June 1986 (day III). A 1-m^2 section was cut from the surrounding mat with a scalpel and lifted from the underlying sand with a thin sheet of plywood. The mat was transported back to the laboratory, placed in a small wading pool, and covered with approximately 4 cm of circulating seawater. The mat was incubated outdoors under natural illumination and in situ temperature conditions. Measurements were made at 0400, 0800, 1200, 1600, 2000, and 0000. The mat section was subsampled for NA, primary productivity (CO_2 fixation), and dissolved O_2 levels.

A modification of the acetylene reduction (AR) technique (16, 26) was used to determine whole-mat NA. Twelve (six light and six foil-covered dark) 1-cm^2 pieces were placed into 25-ml Erlenmeyer flasks with 20 ml of seawater. The flasks were stoppered with rubber sleeve stoppers, resulting in a 10-ml headspace and a 20-ml aqueous phase. Purified acetylene was injected into the seawater phase of each flask through the stoppers, and the flasks were incubated in the wading pool for 3 h. They were then removed from the pool and shaken to liberate ethylene from the mat, and a 0.3-ml sample of headspace was analyzed by gas chromatography by using a Shimadzu GC9A apparatus with a flame ionization detector and a 2-m Porapak-T column at 80°C.

The $^{14}CO_2$ assimilation technique was used to determine mat primary productivity. Twelve (six light and six dark) replicate 1-cm^2 pieces were placed into 25-ml Erlenmeyer flasks, which were subsequently filled with seawater and stoppered with rubber sleeve stoppers. Sterile $NaH^{14}CO_3$ was injected through the stoppers, and the flasks were incubated in the pool for 3 h. At the end of the incubation period the mat pieces were removed from the flasks, washed twice with unlabeled seawater, air dried, powdered, fumed with HCl to remove inorganically precipitated CO_2, and counted in a Beckman LS-7000 liquid scintillation counter. Additional experiments have shown that mat $^{14}CO_2$ assimilation was linear for the entire 3-h incubation period during night and morning hours and for at least 1.5 h following isotope addition in the afternoon (3). Dissolved inorganic carbon concentrations in the mat were assumed to be in equilibrium with overlying seawater and were determined by infrared analysis with a Beckman model 864 infrared analyzer (14).

Photosynthetically active radiation (PAR) flux was monitored at 10-min intervals during diel studies by using a Li-Cor model 192 quantum sensor coupled to a model 550B printing integrator.

Measurements of dissolved O_2 concentrations within the mat were made by using Diamond Electrotech model 460 oxygen minielectrodes. A micromanipulator was used to position the electrode. A Keithley model 485 picoammeter was used to record the electrode output (11). These minielectrodes have sturdy steel casings which allowed repeated penetration into the sand mat, whereas more fragile glass microelectrodes proved highly susceptible to breakage. Increased O_2 consumption and decreased sensitivity due to the relatively large sensing tips of these electrodes permitted a spatial resolution of ca. 1 mm. Oxygen minielectrodes were calibrated by using air- and nitrogen-bubbled seawater standards. Absolute O_2 concentrations of the standards were determined by Winkler titrations.

PAR DEPRIVATION STUDIES

In 1985 and 1986, approximately 0.5-m^2 segments were cut from the mat, placed in large coolers, and immediately transported to the laboratory. These segments were cut into 100-cm^2 pieces, which were placed in sterile polyethylene beakers. A 200-ml portion of freshly collected seawater was then added to each beaker to cover each mat piece to a depth of 3 cm. Controls which were exposed to natural PAR (400 to 700 nm) cycles and left untreated, as well as samples incubated in the dark and/or nutrient amended, were set up in quadruplicate. All samples were kept at constant temperature in a small wading pool containing circulating seawater. Darkened samples were placed in a cooler fitted with light-blocking ports through which circulating seawater could pass. Several illuminated and darkened samples were enriched with 0.01 M (final concentration) reagent-grade maltose (Fisher Chemical Co.). Maltose was added both at the start of experimentation and after periods of PAR deprivation. Periodic dissolved O_2 measurements made at night in overlying waters indicated that O_2 saturation remained above 50%.

Both maltose-amended and unamended samples were deprived of PAR for periods ranging from 24 to 96 h. At 24-h intervals (sampled each night), 2-cm^2 pieces were cut from each sample, placed in 25-ml Erlenmeyer flasks, and assayed for NA by using the AR assay in a manner identical to that described for diel studies. AR assays were carried out for 2 h at each sampling interval. At each sampling interval, fresh seawater and maltose-amended seawater were provided to ensure inorganic nutrient sufficiency and continuous 0.01 M maltose concentrations in overlying seawater.

IMPACTS OF O_2 AND NH_3 ON N_2 AND CO_2 FIXATION

Freshly sampled mat material was exposed to an array of ambient O_2 tensions by introducing overlying seawater to either O_2-depleted (99% N_2–1% CO_2) or O_2-enriched (various O_2 concentrations in air) gas mixtures with a vacuum-purging manifold to which serum-stoppered 25-ml incubation flasks were attached. For all experiments 1-cm^2 mat pieces were placed in flasks, and 10 ml of freshly collected seawater was added to each flask. This left a gas headspace of approximately 15 ml. Following headspace purging and equilibration with newly introduced gas mixtures, dissolved O_2 concentrations were determined with either minielectrodes or a Yellow Springs Instruments BOD-type Clark electrode (YSI 5750) in overlying water. NA and $^{14}CO_2$ fixation rates were determined over periods ranging from 1 to 6 h after O_2 treatments by previously described (see diel studies) methodology. Incubations were conducted both under illumination (PAR, 200 microeinsteins $\cdot$ m^{-2} $\cdot$ s^{-1}) and in the dark.

Previous studies have shown 3-(3,4-dichlorophenyl)-1,1-dimethylurea (DCMU) to be an effective inhibitor of O_2 evolution in these mats (3, 15). In experiments discussed here, purified DCMU (Pfalz Chemical Co.) was added at 2×10^{-5} M (final concentration) 10 to 15 min before AR and $^{14}CO_2$ fixation assays to examine the impacts of this photosystem II inhibitor on mat NA and photosynthetic production potentials. As with the above-mentioned field and laboratory studies, DCMU was dispensed in 25-ml Erlenmeyer flasks containing 1-cm^2 mat pieces and 10 ml of seawater. Triplicate incubations were used for each treatment. Flasks were sealed with serum stoppers, mildly shaken for 30 s to ensure DCMU solubilization and diffusion into mat pieces, and incubated for 0.5 to 6 h under both illuminated (outdoor, natural illumination) and dark (foil-covered) conditions. All DCMU experiments were conducted in an outdoor pond containing circulating seawater. The impacts of DCMU on NA and $^{14}CO_2$ fixation were examined in the morning, at midday, and in the afternoon.

Ammonia is also known to be a potential inhibitor of NA (through end product suppression of nitrogenase). Recent analyses have shown NH_3 concentrations in mat pore water to be substantially higher than in ambient overlying waters (200 to 250 μg of N $\cdot$ $liter^{-1}$ in pore water versus <10 μg of N $\cdot$ $liter^{-1}$ in overlying water). Accordingly, we deemed it necessary to examine potential suppression of NA by NH_3. A range of NH_3 (as NH_4Cl) concentrations were added in triplicate to 1-cm^2 mat pieces placed in 25-ml Erlenmeyer flask with 10 ml of seawater. The flasks were then sealed, shaken, and placed in an outdoor circulating seawater pond for 4 h (midmorning) prior to NA determinations. Subsequent AR assays were conducted for 2 h under identical environmental conditions.

RESULTS OF DIEL STUDIES

Fortuitously, PAR conditions varied substantially among the three diel studies. Furthermore, since the diel studies were conducted within a 3-week period, microbial community composition and biomass varied little (Polimeni, Ph.D. thesis). We were thus able to examine primary production and N_2 fixation characteristics of essentially the same mat community during a heavily overcast day (day I), a cloudless bright day (day

II), and a partly cloudy day having lengthy, sunny intervals (day III). Daylight length varied little among the diel studies, with sunrise occurring near 0600 and sunset occurring near 2030 (Fig. 2). During day III significant PAR was not detected until 0730 because heavy overcast accompanied thunderstorms between dawn and 1000. While cloudy periods occurred throughout day I and later portions of day III, no rainfall coincided with these periods.

Not surprisingly, strong direct relationships were seen between PAR patterns (intensity and duration of clear periods) and depth-integrated CO_2 fixation rates (Fig. 2). During bright, cloudless days, maximum rates of CO_2 fixation occurred relatively early in the day (Fig. 2), as was the case during morning sampling intervals (centered around 0930) on days II and III. In contrast, day I revealed quite low rates of photosynthesis in response to heavy overcast (Fig. 2). Under continual overcast, photosynthetic production remained low, with no distinct peak either during morning hours or at midday. Not until late afternoon (1730) were maximum CO_2 fixation rates reported for this day. Because photosynthesis remained suppressed by cloudiness during day I, the maximum rate proved to be much lower than those reported for days II and III.

Days II and III showed relatively high photosynthetic rates during early to midmorning hours, evidently a response to bright conditions (Fig. 2). Following morning photosynthetic maxima, distinct midday suppression was evident on both days (Fig. 2). Although CO_2 fixation remained suppressed throughout the remainder of day II, a secondary late-afternoon optimum was seen during day III (Fig. 2). The different afternoon CO_2 fixation patterns between days II and III appeared closely linked to irradiance patterns; while day II had continually bright, cloudless conditions, the afternoon of day III included a lengthy period (1400 to 1830) of cloudiness. This cloudy period coincided with the secondary optimum of CO_2 fixation, which was most probably attributable to a cessation of photoinhibition, as witnessed 2 weeks earlier during day II.

These distinctly different diel periods reveal close regulation of photosynthetic production by both the intensity and the duration of PAR. Results suggest that periods of both suboptimal photosynthesis (day I) and photoinhibition (days II and III) are common features of the mat.

Vertical dissolved O_2 patterns in the mat and overlying water appear closely related to rates and daily histories of photosynthetic production (Fig. 2). Although one would a priori expect close spatial and temporal coupling of CO_2 fixation and O_2 evolution, it was possible to see, on a reasonably small scale, how diel photosynthetic periodicity is reflected as residual O_2 concentration patterns (Fig. 2). It must be stressed that such patterns reflect net values of O_2 production and consumption, the latter being mediated by photoautotrophs, heterotrophic bacteria, protozoans, and diverse invertebrates. It is possible that subtle population and activity shifts among the heterotrophic and autotrophic communities took place from one day to another, thereby affecting the net O_2 concentrations reported. However, general trends in O_2 concentration following variations in photosynthetic performance (CO_2 fixation) are clearly distinguishable for each day.

In general, maximal rates of photosynthetic production were correlated with vertical O_2 penetration into the mat (Fig. 2). Days II and III, which both exhibited distinct daytime photosynthetic maxima, also yielded maximal O_2 penetration into the mat. Oxygen maxima typically followed CO_2 fixation maxima on a temporal basis (Fig. 2). This sequence was not obvious during day III, when vigorous morning CO_2 fixation was followed by maximum vertical O_2 saturation by midafternoon. As mentioned above, since O_2 measurements represent a net balance among photosynthetic O_2 pro-

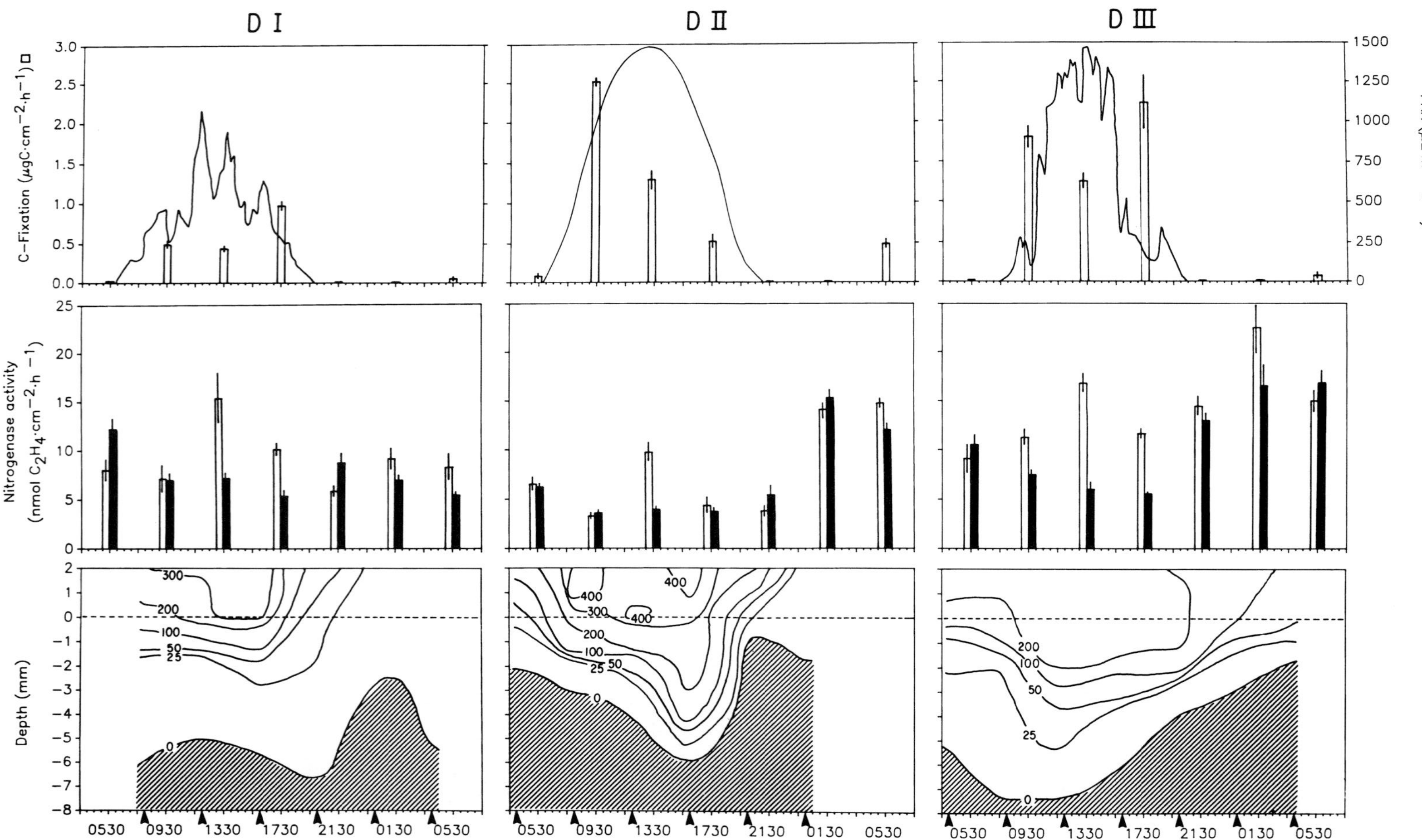

D I
D II
D III
C–Fixation (μgC·cm⁻²·h⁻¹) □
PAR (μE·m⁻²·s⁻¹)
Nitrogenase activity (nmol C2H4·cm⁻²·h⁻¹)
Depth (mm)
0530
0930
1330
1730
2130
0130
0530

duction, respiration, temperature-mediated solubility, and diffusive processes, it is difficult to pinpoint the precise nature of the observed lag between CO_2 fixation rates and O_2 concentration patterns in the mat.

AR measurements reveal several patterns, which are by and large distinct from PAR, CO_2 fixation, and O_2 saturation characteristics (Fig. 2). Also, AR patterns varied among diel studies. During day I, aside from a midday light-mediated maximum, AR remained remarkably constant between night and day, ranging from 5 to 12.2 nmol of $C_2H_4 \cdot cm^{-2} \cdot h^{-1}$. Rates of light- and dark-mediated AR were apparently different at some time points (Fig. 2). During daytime, these differences may have been due to enhancement of photoautotrophic (cyanobacterial and photosynthetic bacterial) processes; however, our understanding of community NA under light versus dark conditions is obscured by the fact that the variety of direct and indirect ways in which photogenerated reductant is proportionally produced and distributed throughout the mat (to N_2-fixing loci) remains ill understood. In a few cases, noticeable differences in AR under light and dark conditions were evident at night (Fig. 2). It is not clear why such differences occurred.

When comparing temporal AR patterns on day I with those on days II and III, we noticed some distinct differences. Most evident was the absence of a clear-cut day and night pattern on day I, whereas on days II and III there were large differences between daytime and nighttime AR rates (Fig. 2). Aside from a period of midday light-mediated stimulation (which was observed in all three diel studies), nighttime AR rates were on average 1.5 to 2 times higher than daytime rates during days II and III (Fig. 2). The enhancement of nighttime N_2 fixation appeared inversely related to dissolved O_2 concentrations; i.e., as O_2 saturation values decreased during evening hours, AR increased, and, conversely, as O_2 saturation increased in the early morning, AR abruptly (within the 4-h period between measurements) decreased. Interestingly, day I, which revealed relatively small changes in mat O_2 saturation, also showed very little temporal fluctuation in AR rates.

Although temporal fluctuations in AR appeared minimal during day I, the absolute rates of AR proved average when compared with days II and III. However, although strong daytime suppression of AR periodically led to submoderate rates during day II (and to a lesser extent day III), nighttime AR rates during days II and III substantially exceeded nighttime rates observed during day I.

PAR DEPRIVATION EFFECTS ON THE MAT

Deprivation of PAR led to either enhancement or reduction in NA, depending on the duration of deprivation and previous PAR history. During daylight hours, especially from late morning through afternoon, PAR deprivation generally led to immediate short-term (1- to 2-h) stimulation of NA (Fig. 3). Longer-term (2- to 4-h) PAR deprivation yielded either no significant impact or a slightly negative impact on NA (relative to

Figure 2. Summary of results of O_2 minielectrode measurements, made during the three diel studies. The times during which O_2 measurements were made are indicated by arrowheads on the *x* axis for each diel study. Dissolved O_2 concentrations were reported in micromoles of O_2 per liter and are plotted along isopleths. Anoxic conditions are indicated by shaded regions. Results of O_2 measurements in 2 mm of overlying seawater are included in the isopleth figures. The upper two frames of each diel study correspond, in time, to the lowest frame. The topmost frame includes traces of PAR flux during daylight hours. All histographs (bars) depicting photosynthetic CO_2 fixation (topmost frames) and light- (□) and dark- (■) mediated NA (middle frames) represent values midway between incubation periods. Error bars represent the variability among six replicates (bar, 1 standard error [SE]).

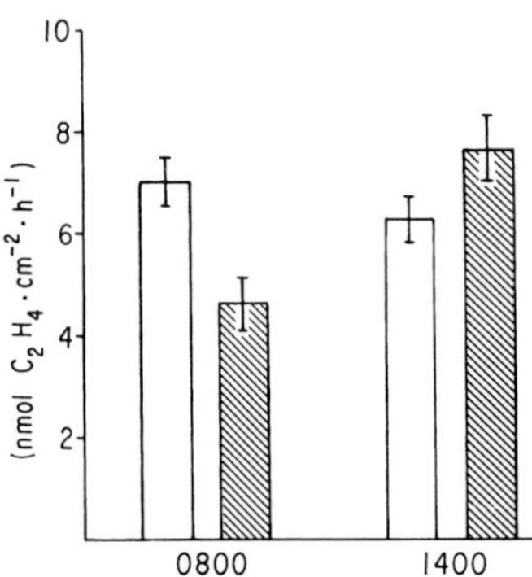

Figure 3. Results of short-term (1.5-h) PAR deprivation on mat NA. Relative impacts of morning (0800) and afternoon (1400) PAR deprivation are shown. Subsamples from a single mat piece previously exposed to natural irradiance were incubated either under illuminated (☐) or dark (▧) conditions. Variability among triplicate samples is shown (bar, 1 SE).

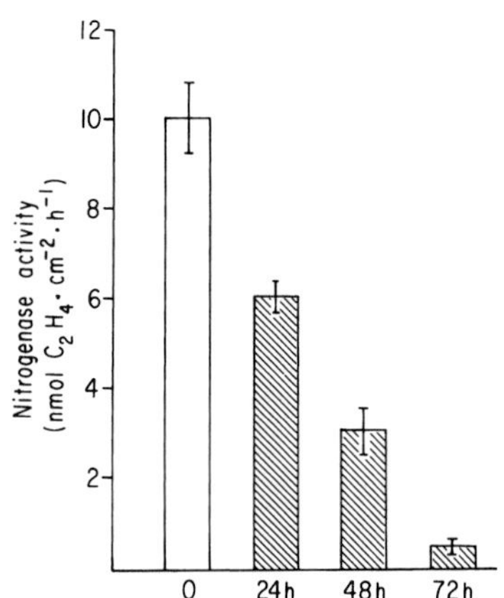

Figure 4. Effects of long-term PAR deprivation on mat NA. Initial illuminated conditions are shown as 0 on the time axis. AR assays (incubation period, 3 h) were conducted on PAR-deprived samples every 24 h. Variability among triplicates is shown (bar, 1 SE).

illuminated controls). During early-morning hours, either short- or long-term PAR deprivation led to significant decreases in NA relative to illuminated controls (Fig. 3). This differential response may be explained by the daylight buildup of inhibitory (to NA) O_2 levels and/or insufficient photoreductant (organic carbon pool). During early daylight hours both O_2 and organic carbon pool (both intra- and extracellular) levels would be expected to be low. It follows that the potential for O_2 inhibition of mat NA is relatively low.

Long-term PAR deprivation led to a gradual cessation of NA (Fig. 4). On average, 48 to 72 h of continued darkness arrested mat NA. Cessation of NA could have been due to either exhaustion of the organic carbon pool or the generation of inhibitory levels of NH_3 in the mat. To examine these alternatives, some mat pieces undergoing PAR deprivation were amended with 0.01 M maltose, while others were left unenriched. Parallel maltose-enriched and unenriched mat pieces were left under illuminated (naturally occurring PAR) conditions. All mat pieces were covered with seawater. NA was determined under the same conditions described for treatments. The results are shown in Fig. 5. Within 24 h of PAR deprivation, unenriched mat pieces showed a significant drop in NA when compared with illuminated controls. However, maltose-enriched samples incubated in the dark maintained significant rates of NA, which were similar in magnitude to those of illuminated controls (Fig. 5). After 48 h of PAR deprivation, results were even more clear-cut. Unenriched, PAR-deprived mat samples revealed minimal NA, whereas maltose-enriched, PAR-deprived samples exhibited substantially higher rates of NA (Fig. 5). Maximum NA was obtained under maltose-enriched, PAR-illuminated conditions (Fig. 5). For both unilluminated and illuminated samples, maltose enrichment led to rates of NA significantly higher than in illuminated controls, indicating that maltose may serve as an effective carbon and energy source for supporting mat N_2 fixation potentials.

Subsequent microautoradiographic examinations have shown maltose (as well as mannitol and glucose, which have also been used as carbon sources supporting mat N_2 fixation potentials) to be incorporated by both dominant cyanobacterial species (*L. aestuarii* and *M. chthonoplastes*), as well as by a wide variety of eubacteria closely associated

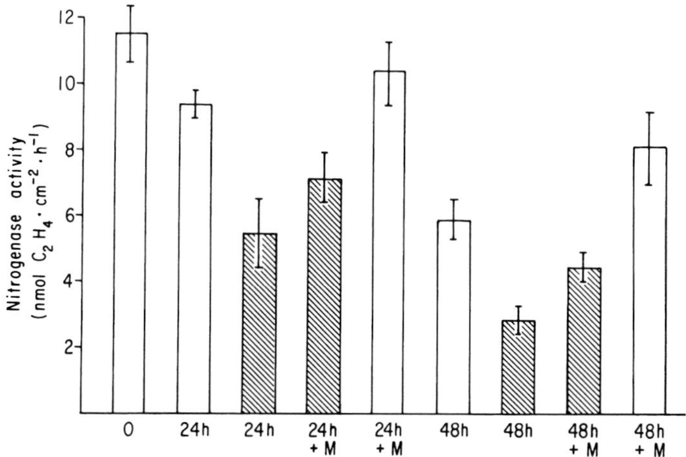

Figure 5. Relative impacts of 0.01 M maltose (+ M) enrichment on NA activity of PAR-deprived as well as illuminated (natural PAR cycle) mat samples. ▧, PAR deprivation, lasting either 24 or 28 h. Initial illuminated conditions are shown as 0 on the *x* axis. Variability among triplicates is shown (bar, 1 SE).

with either cyanobacteria, sand, or amorphous detritus under dark conditions (H. W. Paerl, manuscript in preparation). Under illuminated conditions, cyanobacterial maltose, mannitol, and glucose incorporation proved to be a relatively larger portion of total microbial (cyanobacterial plus eubacterial) community incorporation. Collectively, these results implicate both cyanobacteria and eubacteria as potential mat N_2 fixers; however, the relative importance of each microbial group as contributors to mat N_2 fixation potentials remains unclear. We are currently developing cell-specific immunological techniques (based on antibodies specific for the Fe protein subunit of nitrogenase) to directly identify and quantify specific microorganisms contributing to mat N_2 fixation potentials.

IMPACTS OF O_2 SATURATION ON MAT N_2 FIXATION POTENTIALS

Both laboratory and field experiments conducted throughout the past 3 years have repeatedly shown mat NA to be sensitive to changes in O_2 tension, under both dark and illuminated conditions (3, 15; Paerl, in press). Results from several relevant experiments are shown in Fig. 6. Subsaturated O_2 conditions led to immediate NA stimulation, whereas O_2 supersaturation strongly sup-

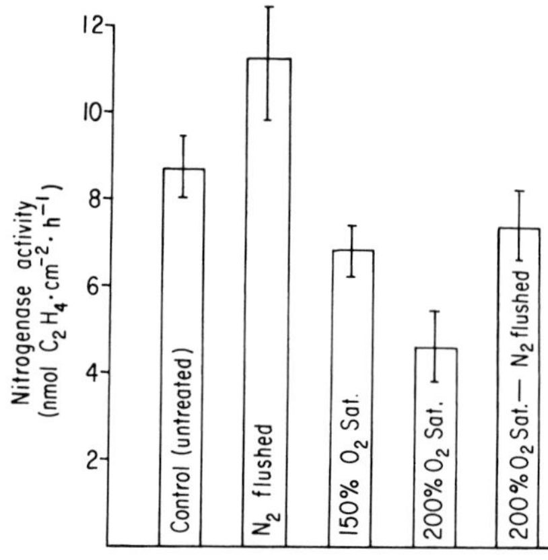

Figure 6. Relative impacts of O_2 subsaturation (N_2 flushed), O_2 supersaturation (150 and 200% O_2 saturation), and O_2 supersaturation followed by subsaturation on mat NA. AR assays following various treatments were conducted for 2 h under illuminated conditions. Variability among quadruplicates is shown (bar, 1 SE).

pressed NA relative to control conditions. Illuminated controls typically yielded higher O_2 saturation values than unilluminated controls did; this difference appears to be due to the presence of oxygenic photosynthesis in the former. The inhibitory or stimulatory impacts of differential O_2 saturation were detectable for up to 3 h in illuminated samples. Beyond 3 h, endogenous photosynthetically generated O_2 became the dominant force in dictating NA responses. In contrast, differential O_2 impacts on samples incubated in the dark were clearly detected for up to 6 h.

Interestingly, NA suppression brought on by O_2 enrichment could be readily reversed by subsequent flushing with an O_2-free gas mixture (Fig. 6). This indicates that mat NA can rapidly recover following periodic O_2 supersaturation. Diel experiments confirmed this ability. At present, it remains unclear whether recovery entails reactivation of previously suppressed NA or rapid de novo synthesis of nitrogenase.

Evidence is mounting that NA is confined to reduced (O_2-poor) microenvironments or microzones in both suspended detrital and mat environments. Both microscopic (microautoradiographic tetrazolium reduction observations) and microelectrode analyses have shown detritus and mats to contain patchy regions of microbial community activity and respiration, resulting in heterogeneous microzones with respect to internal O_2 gradients. Microzone-scale (10 to 1,000 μm) O_2 gradients exist around sand grains, within and around aggregated cyanobacterial and eubacterial populations (colonies), and within amorphous organic matrices constituting the adhesive framework holding detrital and mat materials together. Recent work in this laboratory has shown the presence and magnitude of NA to be closely linked to the extent of reduced microzone formation (13, 17).

EFFECTS OF DCMU ON MAT PHOTOSYNTHESIS AND N_2 FIXATION

The addition of 2×10^{-5} M DCMU led to the immediate cessation of O_2 evolution. Within 5 min of DCMU application, minielectrode analyses revealed distinct decreases in O_2 saturation at all depths in the mat previously exhibiting O_2 evolution (Fig. 7). As a result of the cessation of O_2 evolution, the oxic zone rapidly migrated upward so that within 45 min virtually no O_2 could be detected below a depth of 1.5 mm in the mat (Fig. 7). Parallel illuminated mat samples receiving no DCMU additions revealed detectable O_2 levels down to a depth of 5 mm, with noticeably higher O_2 saturation at all depths above 5 mm. The O_2 which remained in DCMU-treated samples was most probably due to downward diffusion of O_2 from overlying seawater, rather than O_2 evolution within the surface layers of the mat. We believe this to be the case, since photosynthetic $^{14}CO_2$ fixation, as well as O_2 evolution, was completely arrested within 5 min of DCMU addition in both intact mat communities and isolated dominant cyanobacterial (*L. aestuarii* and *M. chthonoplastes*) species.

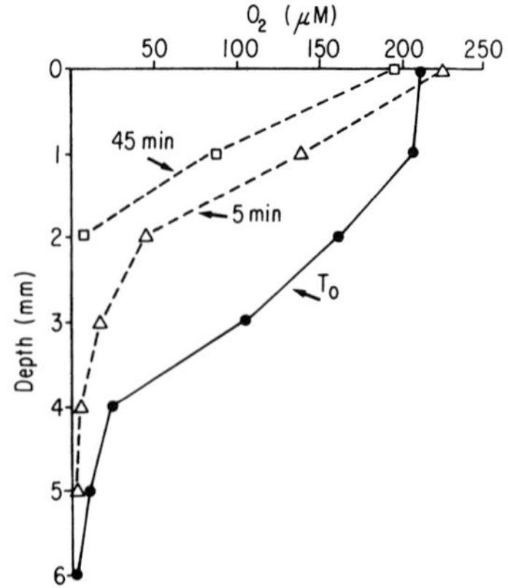

Figure 7. Effect of DCMU addition on the vertical distribution of O_2 concentrations in the mat (determined under illuminated conditions). Results are shown for conditions just before DCMU addition (T_0), as well as 5 and 45 min after DCMU addition.

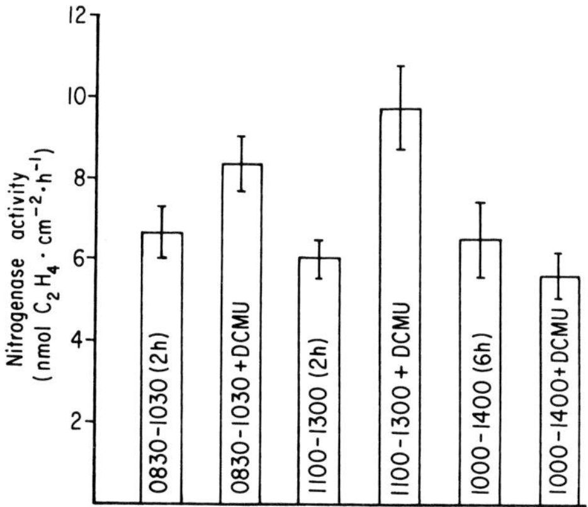

Figure 8. Impact of DCMU additions on mat NA determined at different times of the day. Results from both control (untreated, length of incubation period also shown) and DCMU-amended samples are presented. Variability among triplicates is shown (bar, 1 SE).

The impacts of DCMU on NA were striking and consistently positive among a set of at least four independent studies. As a rule, DCMU enhanced NA within the first 1 to 3 h following its addition (Fig. 8). The magnitude and duration of stimulation varied somewhat among experiments. Such variations could perhaps be due to temporal changes in algal and bacterial species composition and also to the time of day at which the experiments were conducted. When applied either very early or late in the day, DCMU led to minimal stimulation of NA (Fig. 8). During mid- to late-morning applications, DCMU led to maximal enhancement of NA (Fig. 8). This differential response may be linked to the fact that maximum photosynthetic activity, and hence the highest rates of O_2 evolution, took place during mid- to late morning. Since it could be demonstrated that increases in O_2 saturation consistently elicited the suppression of NA in mat samples (at any time of the day), it is concluded that suppression of O_2 evolution by DCMU during periods of vigorous photosynthesis contemporaneously reduced the impact of O_2 inhibition on NA, thus resulting in enhanced N_2-fixing activity within the mat.

Although it could be shown that in the short run (1 to 3 h), DCMU additions led to stimulation of NA, long-term impacts of DCMU proved to yield opposite effects. In samples incubated in the presence of DCMU (under illuminated conditions) for periods exceeding 3 h, mat NA rates gradually declined below those found under control (no DCMU additions) conditions (Fig. 8). Hence, although enhancing NA (presumably owing to the removal of O_2 inhibition) in the short run, the continued cessation of photosynthetic O_2 evolution and CO_2 fixation eventually led to declines in NA. At this point it could be argued that substantial periods of daylight anoxia following DCMU additions may have created toxic conditions for N_2-fixing cyanobacteria and eubacteria, thereby leading to observed declines in NA during periods exceeding 3 h. Several follow-up experiments were conducted in which the effects of DCMU were examined under maltose-amended versus unenriched conditions during illuminated as well as dark periods. These experiments conclusively show that the presence of maltose, a readily utilizable organic carbon source able to support growth and NA in both cyanobacteria (including *M. chthonoplastes*) and a variety of mat eubacteria (22, 30), leads to sustained rates of NA, often exceeding control rates, in the presence of DCMU.

IMPACTS OF NH_3 ON MAT N_2 FIXATION POTENTIALS

Ammonia enrichment up to 1,000 μg of $N \cdot liter^{-1}$ failed to suppress NA in the Shackleford mat (Fig. 9). Ammonia concentrations in the mat pore water ranged from 100 to 200 $\mu g \cdot liter^{-1}$; even under extensive periods of darkness, which ensure long-term (72-h) anoxia, pore-water NH_3 levels never exceeded 400 $\mu g \cdot liter^{-1}$. Direct experimental evidence, coupled with chemical analyses of pore water, strongly discounts internally generated NH_3 as being an impor-

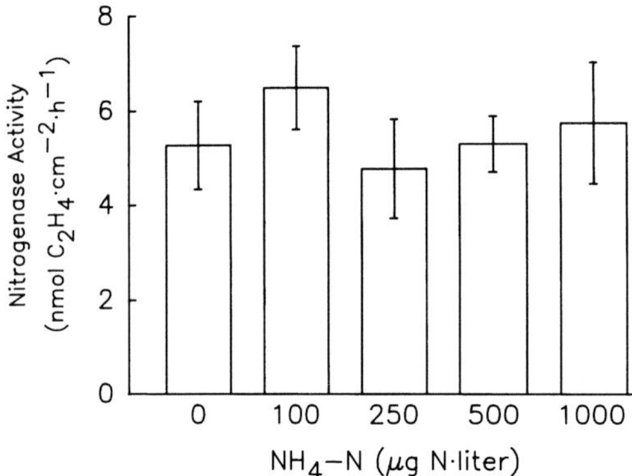

Figure 9. Impact of differential levels of ammonia enrichment on mat NA. Variability among triplicates is shown (bar, 1 SE).

tant regulator of NA observed in either diel or experimental studies discussed here. The possibility exists that NH_3 concentrations in O_2-depleted microzones bordering microbial aggregates exceed those measured in pore water. Hence, potential NH_3 suppression within mat microzones cannot be discounted at present. However, in terms of overall (gross) effects, NH_3 enrichment at levels greatly exceeding pore-water concentrations had little impact on mat NA, whereas even small changes in O_2 saturation led to immediate and substantial alterations of NA potentials.

CONCLUSIONS

Although previous studies have shown that availability of a number of inorganic nutrients known to be required for aquatic N_2 fixers (P, Fe, Mo, Co, and Cu) may regulate mat AR potentials (2, 15), diel and experimental results presented here strongly implicate O_2 tension and reductant (organic-matter) availability as key environmental factors controlling both spatial and temporal N_2 fixation characteristics. Our results are in accordance with previously published findings of Stal et al. (23, 24) that oxygenic photosynthesis and N_2 fixation in an *Oscillatoria* sp., originally isolated from a marine mat, are to a large extent incompatible. They found that although the *Oscillatoria* sp. was capable of N_2 fixation under subsaturated O_2 conditions, elevated O_2 levels (>0.15 atm [15.20 kPa]) proved inhibitory. Hence, periodic lowering of O_2 tension, accomplished by either lower PAR levels or under darkness, led to substantial enhancement of AR. We have observed an essentially similar response in an intact mat community.

Temporal optimization of CO_2 and N_2 fixation has also been reported for freshwater heterocystous cyanobacteria (16, 19). Among a variety of diel field studies in eutrophic lakes dominated by either *Anabaena* sp. or *Aphanizomenon* sp. blooms, mid- to late-morning CO_2 fixation (and O_2 evolution) maxima seldom coincided with N_2 fixation maxima (16). On clear days N_2 fixation peaks commonly occurred from 2 to 4 h after CO_2 fixation maxima, often coinciding with a drop in O_2 saturation values (16). These studies concluded that periodic O_2 supersaturation accompanying peak photosynthetic production led to inhibition of N_2 fixation in blooms. Although heterocystous species dominated such blooms, O_2 inhibition was still evident. It would thus appear that heterocysts offer only partial protection from O_2 inactivation of nitrogenase. Also, the possibility existed that some NA may have been present in vegetative cells, which are more susceptible than heterocysts to O_2 penetration.

Several ecophysiological mechanisms can operate to ensure spatial-temporal compatibility between oxygenic photosynthesis and N_2 fixation in marine mats. One mechanism, discussed in previous studies (3, 23, 24; Paerl, in press), is temporal separation of CO_2 and N_2 fixation optima, so that photosynthetic production of organic matter can precede N_2 fixation, which is dependent on organic matter as a source of reductant and carbon (acceptor) skeletons. For this strategy to be accomplished, a readily available organic carbon pool, periodically replenished

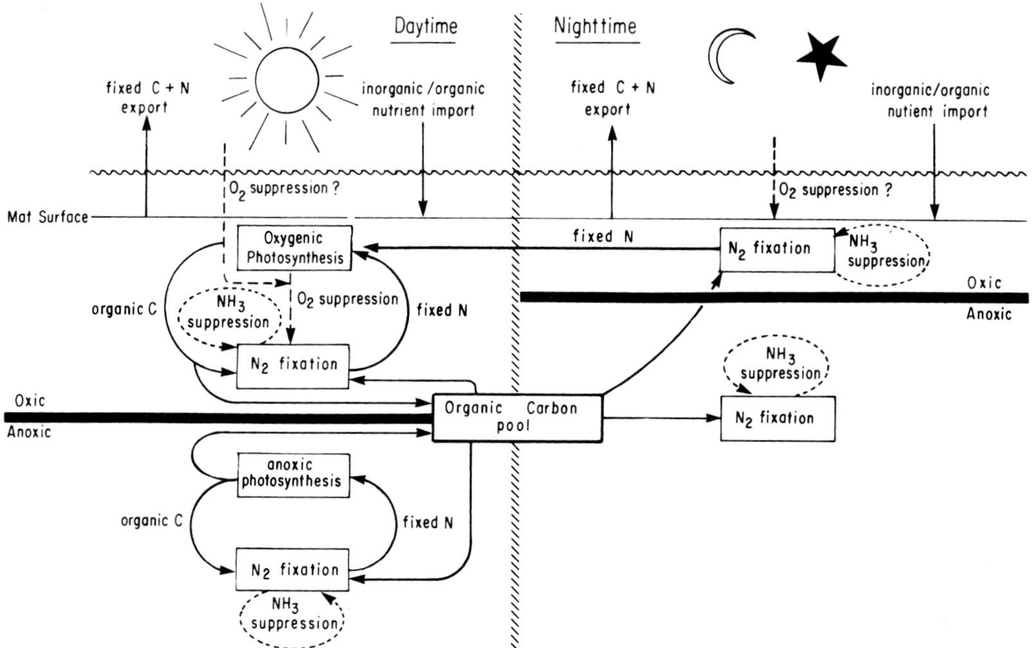

Figure 10. Schematic spatial and temporal representation of the interactions (both positive [———] and negative [----]) between oxygen photosynthesis and N_2 fixation in the mat microbial community during the day and night. Arrows signify major routes of either organic carbon or fixed nitrogen flux among diverse microbial primary producer and N_2-fixing populations. Note that the oxic/anoxic interface is deeper during the day, owing to photosynthetic O_2 production, than during the night, when net O_2 consumption forces this interface upward toward the mat surface. The centrally situated organic carbon pool represents both intra- and extracellular accumulations of fixed carbon dispersed throughout the mat.

by photoautotrophic (and perhaps chemoautotrophic) production, must be present. This pool would be composed of both intracellular (for N_2-fixing cyanobacteria and photosynthetic bacteria) and extracellular (for chemoautotrophic N_2 fixers reliant on exogenous organic carbon) compounds (Fig. 10). The scheme depicted in Fig. 10 implies that physical transport, either by chemical diffusion or by biotic migration of relevant N_2 fixers, ensures continued supplies of organic matter, which is especially crucial during nighttime, when synthesis of new organic matter is greatly restricted. Second, both structural and physiological protection from potential O_2 suppression of NA (Fig. 10) must be optimized. Structural protection during daylight high O_2 saturation periods can occur in localized O_2-poor microenvironments or microzones known to be present in the mat matrix (13; Paerl, in press). In addition, localized elevated respiration of organic matter by bacteria, associated with either cyanobacteria or eubacteria capable of N_2 fixation, provides protection by retarding the diffusion of O_2 into nitrogen-fixing cells (16, 18). Lastly, a host of intracellular O_2-scavenging mechanisms, all of which are reliant on adequate organic carbon supplies, have been shown to exist among diverse procaryotic N_2 fixers (4, 30). In consortium, these diverse O_2-protecting mechanisms are likely to play key roles in cooptimization of CO_2 and N_2 fixation.

At present the relative importance of cyanobacterial, chemotrophic, and anoxygenic phototrophic bacterial N_2 fixation in mat communities remains poorly understood. Although isolated cyanobacteria and eubacterial species can be shown to fix N_2 under defined laboratory conditions, it is difficult to qualitatively and quantitatively relate such findings to natural communities experiencing rapidly shifting PAR, photosynthetic, and O_2 regimes on relevant spatial and temporal scales illustrated here.

Because mat N_2 fixation is strongly regulated by photosynthetic histories, jointly reflected as O_2 saturation and organic matter accumulation, nighttime AR rates can commonly exceed daytime rates. While such observations are useful in identifying environmental controls of N_2 fixation, they are also of ecological relevance. Clearly, to accurately assess mat N_2 fixation (and resultant nitrogen inputs) in a meaningful manner, both nighttime and daytime measurements are essential. Previous studies, including our own, have routinely involved daytime AR measurements in estimating mat nitrogen inputs (2, 6, 10, 28, 30). Indications from this study are that diel N_2 fixation inputs based on daytime measurements may lead to underestimates on the order of 25 to 75%.

On the basis of these findings, we propose that marine mats such as those on Shackleford Banks may be much more important sources of fixed N than previously assumed. This conclusion in part helps to clarify how highly productive mat communities can proliferate and flourish in marine environments faced with chronic nitrogen depletion in surrounding waters.

Acknowledgments. We are grateful to J. Garner, V. Page, and H. Page for their assistance with manuscript preparation. K. Crocker, G. Suba, and J. Monnes assisted with field and laboratory work. We also appreciate the helpful review of this paper by L. Stal.

This work was supported by the National Science Foundation grant OCE 85-00740, North Carolina Sea Grant Project R/MER-5, and the North Carolina Biotechnology Center Project 86-G-01013. Travel to Israel (H.W.P.) was made possible in part through a Rothschild Foundation grant.

LITERATURE CITED

1. **Bauld, J.** 1984. Microbial mats in marginal marine environments: Shark Bay, Western Australia, and Spencer Gulf, South Australia, p. 39–58. *In* Y. Cohen, R. W. Castenholz, and H. O. Halvorson (ed.), *Microbial Mats: Stromatolites.* Alan R. Liss, Inc., New York.
2. **Bautista, M. F., and H. W. Paerl.** 1985. Diel N_2 fixation in an intertidal marine cyanobacterial mat community. *Mar. Chem.* **16:**369–377.
3. **Bebout, B. M., H. W. Paerl, K. M. Crocker, and L. E. Prufert.** 1987. Diel interactions of oxygenic photosynthesis and N_2 fixation (acetylene reduction) in a marine microbial mat community. *Appl. Environ. Microbiol.* **55:**369–384.
4. **Bothe, H.** 1982. Nitrogen fixation, p. 87–104. *In* N. G. Carr and B. A. Whitton (ed.), *The Biology of Cyanobacteria.* Blackwell Scientific Publications, Ltd., Oxford.
5. **Carpenter, E. J., and D. G. Capone.** 1983. *Nitrogen in the Marine Environment.* Academic Press, Inc., New York.
6. **Carpenter, E. J., C. D. Van Raalte, and I. Valiela.** 1978. Nitrogen fixation by algae in a Massachusetts salt marsh. *Limnol. Oceanogr.* **23:**318–327.
7. **Cohen, Y., R. W. Castenholz, and H. O. Halverson (ed.).** 1984. *Microbial Mats: Stromatolites.* Alan R. Liss, Inc., New York.
8. **Fogg, G. E., W. D. P. Stewart, P. Fay, and A. E. Walsby.** 1973. *The Blue-Green Algae.* Academic Press, Inc. (London), Ltd., London.
9. **Gotto, J. W., F. R. Tabita, and C. Van Baalen.** 1981. Nitrogen fixation in intertidal environments of the Texas Gulf Coast. *Estuarine Coastal Mar. Sci.* **12:**231–235.
10. **Jones, K.** 1974. Nitrogen fixation in a salt marsh. *J. Ecol.* **62:**553–565.
11. **Jørgensen, B. B., N. P. Revsbech, T. H. Blackburn, and Y. Cohen.** 1979. Diurnal cycle of oxygen and sulfide microgradients and microbial photosynthesis in a cyanobacterial mat sediment. *Appl. Environ. Microbiol.* **38:**46–58.
12. **McCarthy, J. J., and E. J. Carpenter.** 1983. Nitrogen cycling in near surface waters of the open ocean, p. 487–572. *In* E. J. Carpenter and D. G. Capone (ed.), *Nitrogen in the Marine Environment.* Academic Press, Inc., New York.
13. **Paerl, H. W.** 1985. Microzone formation: its role in the enhancement of aquatic N_2 fixation. *Limnol. Oceanogr.* **30:**1246–1252.

14. **Paerl, H. W.** 1987. *Dynamics of Blue-Green Algal Blooms in the Lower Neuse River, North Carolina: Causative Factors and Potential Controls.* Publication no. 229. University of North Carolina Water Resources Research Institute, Chapel Hill, N.C.

15. **Paerl, H. W., K. M. Crocker, and L. E. Prufert.** 1987. Limitation of N_2 fixation in coastal marine waters: relative importance of molybdenum, iron, phosphorus, and organic matter availability. *Limnol. Oceanogr.* **32**:525–535.

16. **Paerl, H. W., and P. E. Kellar.** 1979. Nitrogen-fixing *Anabaena*: physiological adaptations instrumental in maintaining surface blooms. *Science* **204**:620–622.

17. **Paerl, H. W., and L. E. Prufert.** 1987. Oxygen-poor microzones as potential sites of microbial N_2 fixation in nitrogen-depleted aerobic marine waters. *Appl. Environ. Microbiol.* **53**:1078–1087.

18. **Paerl, H. W., K. L. Webb, J. Baker, and W. J. Wiebe.** 1981. Nitrogen fixation in waters, p. 193–240. *In* W. J. Broughton (ed.), *Nitrogen Fixation*, vol. 1. *Ecology*. Claredon Press, Oxford.

19. **Peterson, R. B., E. E. Friberg, and R. H. Burris.** 1977. Diurnal variation in N_2 fixation and photosynthesis by aquatic blue-green algae. *Plant Physiol.* **59**:74–80.

20. **Postgate, F. R. S.** 1982. *The Fundamentals of Nitrogen Fixation.* Cambridge University Press, Cambridge.

21. **Potts, M., and B. A. Whitton.** 1977. Nitrogen fixation by blue-green algal communities in the intertidal zone of the Lagoon of Aldabra Atoll. *Oecologia* **27**:275–283.

22. **Smith, A. J.** 1982. Modes of cyanobacterial carbon metabolism. *Bot. Monogr.* (Oxford) **19**:47–85.

23. **Stal, L. J., S. Gossberger, and W. E. Krumbein.** 1984. Nitrogen fixation associated with the cyanobacterial mat of a marine laminated microbial ecosystem. *Mar. Biol.* (Berlin) **82**:217–224.

24. **Stal, L. J., and W. E. Krumbein.** 1985. Nitrogenase activity in the non-heterocystous cyanobacterium *Oscillatoria* sp. grown under alternating light and dark cycles. *Arch. Microbiol.* **143**:67–71.

25. **Stal, L. J., H. van Gemerden, and W. E. Krumbein.** 1985. Structure and development of a benthic marine microbial mat. *FEMS Microbiol. Lett.* **31**:111–125.

26. **Stewart, W. D. P., G. P. Fitzgerald, and R. H. Burris.** 1967. *In situ* studies on N_2 fixation using the acetylene reduction technique. *Proc. Natl. Acad. Sci. USA* **58**:2071–2078.

27. **Valiela, I.** 1983. Nitrogen in salt marsh ecosystems, p. 649–678. *In* E. J. Carpenter and D. G. Capone (ed.), *Nitrogen in the Marine Environment.* Academic Press, Inc., New York.

28. **Whitney, D. E., G. M. Woodwell, and R. W. Howarth.** 1975. Nitrogen fixation in Flax Pond: a Long Island salt marsh. *Limnol. Oceanogr.* **20**: 640–643.

29. **Whitton, B. A., and M. Potts.** 1982. Marine littoral, p. 515–542. *In* N. G. Carr and B. A. Whitton (ed.), *The Biology of Cyanobacteria.* Blackwell Scientific Publications, Ltd., Oxford.

30. **Yates, M. G.** 1977. Physiological aspects of nitrogen fixation, p. 219–270. *In* W. Newton, J. R. Postgate, and C. Rodriquez-Barruero (ed.), *Recent Developments in Nitrogen Fixation.* Academic Press, New York.

Chapter 31

Compatible Solutes in Halophilic Phototrophic Procaryotes

H. G. Trüper and E. A. Galinski

INTRODUCTION

The spectrum of life forms in extremely saline biotopes is, at least for eucaryotes, drastically limited. The only higher organisms are the brine shrimp (*Artemia salina*) and several species of brine flies (*Ephydra* spp.), so that these environments are mainly a domain of microorganisms with rather limited biogeochemical cycles of matter.

As primary producers, besides halophilic algae of the genera *Dunaliella* and *Asteromonas*, halophilic cyanobacteria and anoxygenic phototrophic bacteria are mainly found, whereas probably the best-known (because of their conspicuous red color) halobacteria belong to the heterotrophic inhabitants of this environment and possess complex nutrient requirements (9). A considerable number of other aerobic and anaerobic heterotrophic bacteria have been described recently and have amplified our knowledge of this special biotope (10). Thus in these ecosystems completely closed cycles of matter exist, which are catalyzed almost exclusively by microorganisms. These organisms have one problem in common: to overcome water stress.

H. G. Trüper and E. A. Galinski • Institut für Mikrobiologie der Rheinischen Friedrich-Wilhelms-Universität, Meckenheimer Allee 168, 5300 Bonn 1, Federal Republic of Germany.

STRATEGIES OF ADAPTATION

It may initially appear peculiar that a salt lake or a saltern is in principle a relatively dry environment. The high concentration of dissolved compounds (salts) is equivalent to low water concentration (a_w = 0.75 for a saturated salt solution), so that the organisms of this biotope are exposed to considerable water stress in addition to high ionic strength. A nonhalophilic organism that accidentally enters a salt lake is dehydrated very quickly and literally pickled. Adaptation to extremely saline environments therefore requires that an osmotic equilibrium between the cytoplasm and the surrounding medium is obtained and at the same time all physiological functions are maintained. Since the cytoplasmic membrane is obviously freely permeable to water and since a control by "water pumps" is rather unlikely, two basic strategies of adaptation in a saline medium of low water activity appear possible: (i) salt in the cytoplasm and (ii) organic osmolytes in the cytoplasm.

As regards the first strategy, the simplest solution would appear to be for the organism to tolerate in its cytoplasm salt concentrations as high as those in the surrounding medium. However, this mechanism of adaptation, which is used by members of the family *Halobacteriaceae*, requires a considerable number of physiological changes, mainly with respect to enzymes, to guarantee optimal functioning of metabolic and regulatory processes at high salinity (salt-adapted enzymes and cellular structures). The family *Halobacteriaceae*, perhaps a rather old group of organisms, belongs to the archaebacteria on the basis of several other properties and is thus phylogenetically distinct from all eubacteria (9).

With respect to the second strategy, most of the halophilic and halotolerant eubacteria, but also algae and fungi, possess normal, i.e., salt-sensitive, enzymes. They therefore have to avoid salts in the cytoplasm as far as possible. Instead, they accumulate organic osmolytes. These substances are found in the cytoplasm in molar concentrations and are absolutely necessary for the organisms to survive in highly saline environments. They may be taken up from the surrounding waters or brines or they may be entirely synthesized by the organisms themselves. The latter capability requires that the metabolism of the organism has high biosynthetic capacities. To describe these substances, the term "compatible solutes" was coined by Brown (1). An important property of these compounds is that they must be compatible with the cellular metabolism and must not interfere with physiological functions. In general, compatible solutes are organic osmolytes that do not negatively influence normal, i.e., salt-sensitive, enzymes even in high concentrations. The known compatible solutes belong to a few typical classes of compounds (see below).

Halophilic-Halotolerant Anoxyphotobacteria

The formation of hydrogen sulfide by bacterial sulfate reduction, characteristic of the sediment of saline environments, allows growth of anoxygenic phototrophic bacteria (purple bacteria), which are among the primary producers in these environments. Besides our own isolates from the Wadi Natrun, Egypt, which are extremely halophilic organisms (4), we have studied a larger number of phototrophic bacteria with more or less moderate salinity requirements. Most of them are typical marine organisms with a wide salt tolerance. Extremely halophilic phototrophic bacteria (optimal growth above 10% NaCl) so far seem to be limited to members of the genera *Ectothiorhodospira* and *Rhodospirillum* (Table 1).

Spectrum of Organic Substances

The compatible solutes so far found in phototrophic sulfur bacteria are (besides presently unidentified substances) mainly sucrose, trehalose, glycosylglycerol, and glycine betaine (Fig. 1). They seem to be typical of the saline environments under study. The trend appears to be as follows: marine and moderately halotolerant bacterial species accumulate mainly sugars and sugar derivatives (glycosylglycerol), whereas the really halophilic species depend primarily on glycine betaine as a compatible solute, as well as other components occurring in lower concentrations. Figure 2 shows the occurrence of three compatible solutes in *Ectothiorhodospira halochloris*.

Obviously, the capacity for synthesis or accumulation of glycine betaine (N-trimethylated glycine) is an important prerequisite for colonizing extremely saline environments. In a number of salt- and dehydration-resistant plants, mainly of the family Chenopodiaceae, the occurrence of betaine has been known for over a century (7), and the name betaine reflects its first isolation and description from *Beta vulgaris*, the original form of sugar beet. The accumulation of glycine betaine and other betaines has therefore frequently been viewed in connection with osmoadaptation and salt tolerance of

Table 1.
Marine and Halophilic Phototrophic Bacteria

Family	Marine and slightly halophilic species	Moderately and extremely halophilic species
Ectothiorhodospiraceae	*E. mobilis*	*E. halophila*
	E. shaposhnikovii	*E. halochloris*
	E. vacuolata	*E. abdelmalekii*
Chromatiaceae	*Chromatium buderi*	
	C. gracile	
	C. purpuratum	
	Thiocystis gelatinosa	
	Lamprobacter modestohalophilus	
Rhodospirillaceae	*Rhodopseudomonas marina*	*Rhodospirillum salinarum*
	Rhodobacter sulfidophilus	*R. salexigens*
	R. adriaticus	*R. mediosalinum*
Chlorobiaceae	*Prosthecochloris aestuarii*	

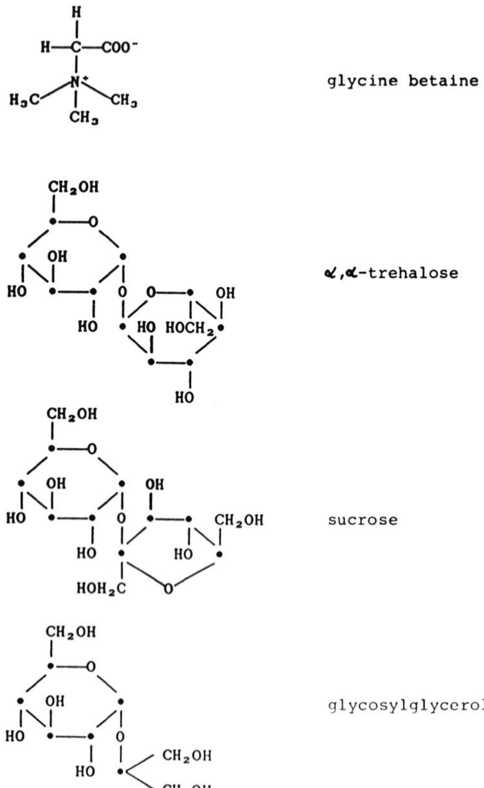

Figure 1. Compatible solutes of halophilic and halotolerant anoxyphotobacteria (for ectoine, see Fig. 5) of the genera *Chromatium*, *Ectothiorhodospira*, *Rhodobacter*, *Rhodopseudomonas*, and *Rhodospirillum*.

plants. It demonstrates a remarkably convergent evolution in adaptation to environments of low water activity. The description of the synthesis and accumulation of glycine betaine in *E. halochloris* (3) may, however, be considered to be the first report of the function of this compatible solute in bacteria. Subsequent descriptions of the occurrence of betaine in cyanobacteria and chemotrophic bacteria have shown that this substance is obviously far more widely distributed than was originally thought. It is remarkable that in halophilic and halotolerant cyanobacteria (5, 6) the same spectrum of solutes is found as in anaerobic phototrophic bacteria; also, the trend appears to be similar: the type of osmolyte used reflects, to a certain extent, the degree of halophily, and betaine is the predominant osmolyte in the hypersaline area.

E. HALOCHLORIS AS A MODEL ORGANISM

The special importance of glycine betaine for the functioning of metabolism can be demonstrated clearly by changing the physiological conditions of a suitable organ-

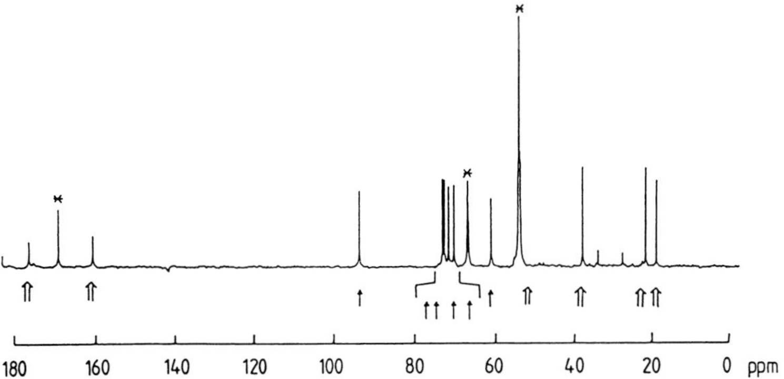

Figure 2. ^{13}C nuclear magnetic resonance spectrum of a halophilic microorganism (*E. halochloris*). All organic substances present at high concentration (osmolytes) are monitored. Symbols: *, glycine betaine; ↑, trehalose; ⇑, ectoine.

ism that is capable of synthesizing several different compatible solutes. Such a suitable model system is *E. halochloris*, an organism capable of synthesizing three compatible solutes, namely glycine betaine, trehalose, and ectoine (Fig. 2).

Trehalose and ectoine are metabolized as the carbon source becomes limited, so that cells in the stationary phase contain glycine betaine as the only significant cytoplasmic osmolyte (Fig. 3). In contrast, nitrogen limitation leads, as expected, to a relative increase in the level of the nitrogen-free osmolyte trehalose. The relative proportion of trehalose, however, never surpasses 20%. The cells seem to be unable to replace glycine betaine by trehalose (E. A. Galinski, doctoral dissertation, University of Bonn, Bonn, Federal Republic of Germany, 1986). This leads to the conclusion that it is critical which osmolyte is used and that some organic substances are better suited to function as compatible solutes than others.

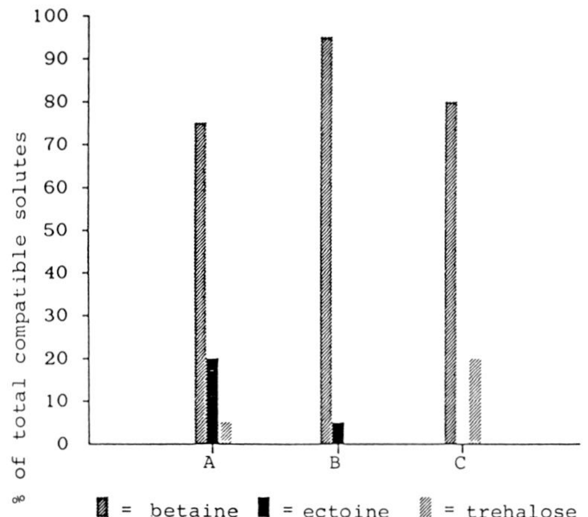

Figure 3. Percent ratios of compatible solutes in *E. halochloris* under different physiological conditions. A, Exponential growth phase, stressed cells, 15 mM NH_4Cl; B, stationary phase, standard medium; C, nitrogen-limited growth.

MOLECULAR PRINCIPLES OF COMPATIBLE SOLUTES

What are the differences between these compounds, which are so extremely important for the survival of organisms in saline environments, and what do their molecular structures have in common? The molecules are accumulated in high concentrations and therefore must be highly soluble. Therefore, low-molecular-weight substances with polar functional groups must be considered potential compatible solutes. It should further be postulated that these osmolytes are uncharged at normal cytoplasmic pH values, because high cytoplasmic ionic strength would lead to complete inhibition of enzymatic processes, as known from experiments

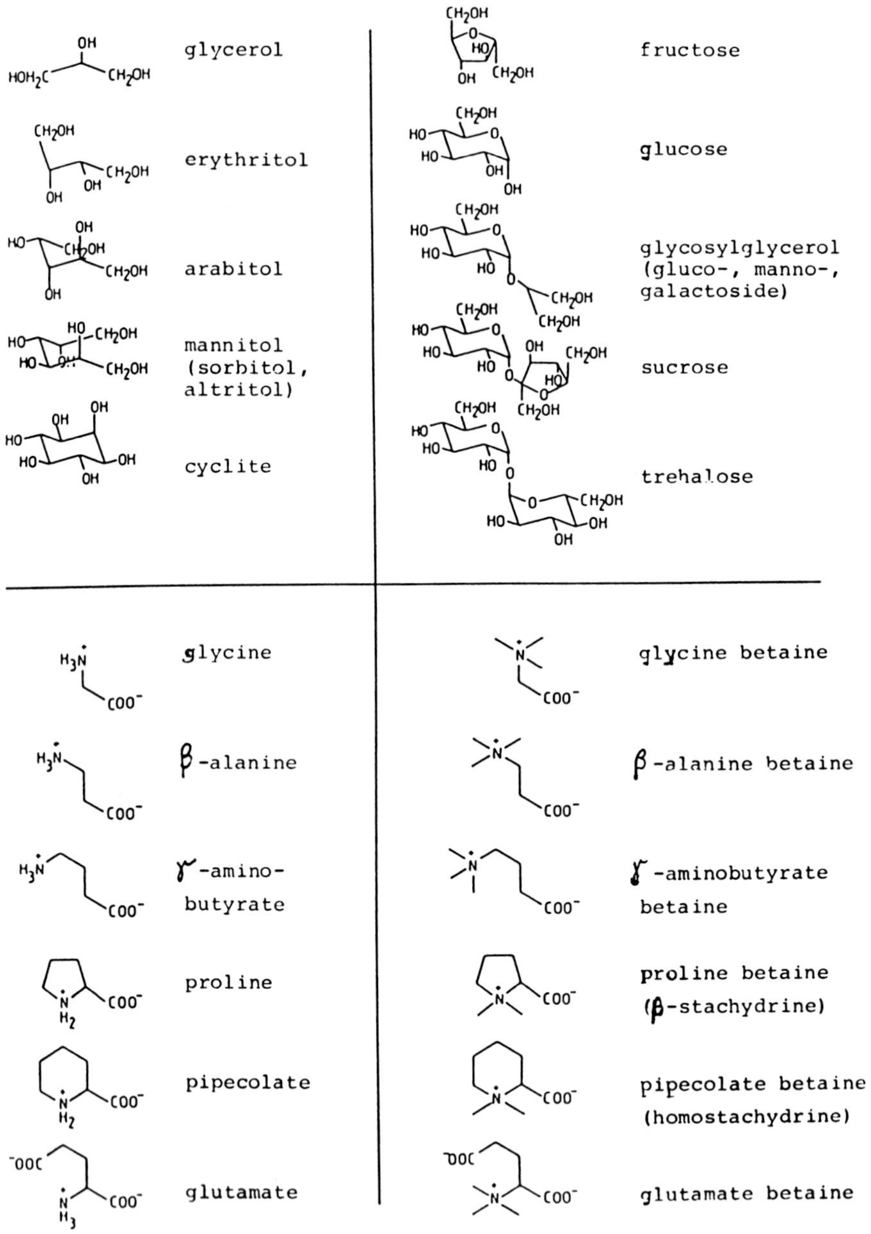

glycerol	fructose
erythritol	glucose
arabitol	glycosylglycerol (gluco-, manno-, galactoside)
mannitol (sorbitol, altritol)	sucrose
cyclite	trehalose
glycine	glycine betaine
β-alanine	β-alanine betaine
γ-amino-butyrate	γ-aminobutyrate betaine
proline	proline betaine (β-stachydrine)
pipecolate	pipecolate betaine (homostachydrine)
glutamate	glutamate betaine

Figure 4. Compatible solutes (polyols, sugars, amino acids, and betaines).

with salts. Potential candidates are therefore to be expected in the classes of sugars, polyols, amino acids, and betaines. Looking at a compilation of all compounds (Fig. 4) that so far have been connected in some kind with osmoadaptation (including reports about plants and lower animals with relatively low salt tolerances), it should be noted that typical examples of the classes mentioned above have been described again and again (most frequently glycerol, proline, and glycine betaine). Although the osmotic function has not been fully quantitatively explained in all other cases, it can be seen that small polyols (C_3 to C_6), including cyclitols and sugar-polyol derivatives (glycosylglycerol), are generally well suited as compatible solutes, whereas the utilization of sugars is apparently restricted mainly to sucrose and trehalose (and then only in relatively low cytoplasmic concentrations of up to 500 mM). Within narrow limits, osmoadaptation by amino acids is possible with glutamate (up to 100 mM). At higher osmolyte concentrations only neutral amino acids are suitable. Besides the cyclic forms (e.g., proline), α amino acids in their polar ionic forms are relatively insoluble. With increasing distance between the charged groups (e.g., in the β and γ amino acids) hydratation is favored, however, and the necessary solubility is obtained. In principle, a similar effect is seen in betaines, which represent a permanent dipolar ionic form of amino acids. In these compounds the decrease of charge density by methylation of the nitrogen atom has a great influence on hydratation and therefore on solubility (glycine, 3.3 molal; glycine betaine, 13.8 molal). Suitable compatible solutes of the dipolar ionic amino carbonyl compound type should therefore satisfy at least one of the following requirements: (i) cyclic molecular structure with hydrophilic and hydrophobic areas in the ring (e.g., proline), (ii) a relative distance between the charged groups of more than two bond lengths (e.g., β-alanine, γ-aminobutyrate), and (iii) low charge density by methylation of the amino group (e.g., betaine).

ECTOINE, A SO FAR UNKNOWN NATURAL COMPOUND

With respect to the calculations mentioned above, a so far unknown compound will be characterized in the following discussion. We isolated this compound from the extremely halophilic *E. halochloris*, in which it occurs in considerable concentrations (Fig. 3). We proposed the trivial name ectoine because we first found it in an *Ectothiorhodospira* sp. (2). The systematic name is 1,4,5,6-tetrahydro-2-methyl-4-pyrimidine carbonic acid (Fig. 5). As compared with the acetidine, pyrrolidine, and piperidine structures of 2-acetidine carbonic acid, proline, and pipecolinic acid, respectively, the hydrated pyrimidine ring system confers special properties to the molecule (8): (i) an additional polar group (N-H) in the ring increases the possibility of hydrogen bridge formation; (ii) the focus of the positive charge is farther from the carboxyl group because of resonance stabilization; (iii) the nitrogen-bound protons are hardly dissociable and result in a permanent dipolar ionic structure; and (iv) the delocalizing of the π bonds leads to a decreased charge density at the positively charged end of the molecule.

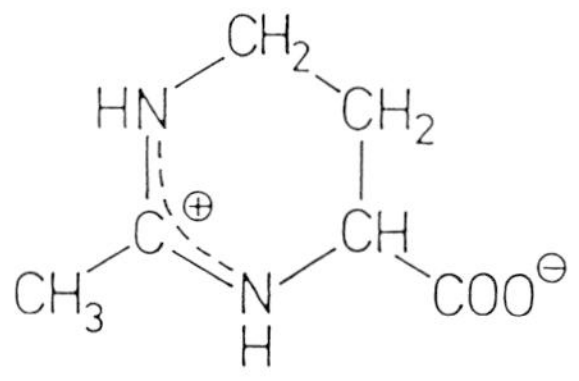

Figure 5. Molecular structure of ectoine (dipolar ionic resonance structure).

These structural specialties show that ectoine is formally similar to the known α imino acids, but structurally also to the β amino acids and betaines. This compound therefore fulfills in every respect the requirements for a good osmolyte. It is not surprising that ectoine obviously also functions as a compatible solute. We have evidence that this compound is not restricted to the genus *Ectothiorhodospira*, but occurs widely in halophilic and halotolerant organisms.

HOW DO COMPATIBLE SOLUTES FUNCTION?

The existing theories on the mode of action of compatible solutes are based almost exclusively on thermodynamic considerations and physicochemical observations of isolated systems. The studies done so far have arrived at some drastically contradictory conclusions, so that they can be used for model formation only to a limited extent. It is clear, however, that the mode of action of a compatible solute cannot be explained only by its osmotic function. Although the presence of organic osmolytes (instead of salt) in the cytoplasm avoids the damaging influence of high ionic strengths, it does not solve the problem of low water activity (water stress). At osmotic equilibrium the water concentrations in the medium and in the cytoplasm, i.e., on both sides of the cytoplasmic membrane, are equally low, so that for the organism the maintenance of hydratation of its proteins probably becomes a critical factor. It is not sufficient that compatible solutes do not interact with the proteins. They must (in contrast) specifically act toward a stabilization of the hydrate shells of the enzymes to maintain their physiological activity. At present the following action mechanisms of compatible solutes are being discussed.

The first is the replacement of water. This hypothesis demands that compatible solutes replace water molecules in the hydrate shells and thus avoid denaturation of proteins under water stress. This model is probably fulfilled for glycerol, which has been used in the laboratory as a stabilizer for enzymes and in the lyophilization of microorganisms.

The second mechanism is that of action as a minidetergent. This theory implies that hydrophobic areas of a protein molecule can be hydrophilized by detergents. This would increase the affinity of the protein for water, which would be equivalent to a stabilization of labile hydrate shells. Small molecules of hydrophilic and hydrophobic areas are being discussed as potential minidetergents. However, although spectroscopic indications already exist to support this theory, these observations are doubted. It remains to be determined whether this effect, which has been studied mainly with proline, can be proved beyond doubt.

The third mechanism is that of structure formation and breakage. In cases in which it could be shown that compatible solutes are excluded from the hydrate shells of proteins, it is postulated that these substances act on the structure of free water. Changes in the cluster structure are especially significant in this respect, and it may be expected that structure-breaking or structure-forming properties of the compatible solutes can indirectly influence the hydratation of proteins this way. This mechanism has been proposed for glycine betaine and is now under study in our laboratory.

Acknowledgment. This work is sponsored by a grant from the Deutsche Forschungsgemeinschaft.

LITERATURE CITED

1. **Brown, A. D.** 1976. Microbial water stress. *Bacteriol. Rev.* 40:803–846.
2. **Galinski, E. A., H. P. Pfeiffer, and H. G. Trüper.** 1985. 1,4,5,6-Tetrahydro-2-methyl-4-pyrimidinecarboxylic acid, a novel cyclic amino acid from halophilic phototrophic bacteria of the genus *Ectothiorhodospira. Eur. J. Biochem.* 149:135–139.
3. **Galinski, E. A., and H. G. Trüper.** 1982. Betaine, a compatible solute in the extremely halophilic phototrophic bacterium *Ectothiorhodospira halochloris. FEMS Microbiol. Lett.* 13:357–360.
4. **Imhoff, J. F., F. Hashwa, and H. G. Trüper.** 1978. Isolation of extremely halophilic phototrophic bacteria from the alkaline Wadi Natrun, Egypt. *Arch. Hydrobiol.* 84:381–388.
5. **MacKay, M. A., R. S. Norton, and L. J. Borowitzka.** 1984. Organic osmoregulatory solutes in cyanobacteria. *J. Gen. Microbiol.* 130:2177–2191.
6. **Reed, R. H., D. L. Richardson, S. R. C. Warr, and W. D. P. Stewart.** 1984. Carbohydrate accumulation and osmotic stress in cyanobacteria. *J. Gen. Microbiol.* 130:1–4.
7. **Scheibler, C.** 1869. Über das Betain, eine im Safte der Zuckerrübe (*Beta vulgaris*) vorkommende Pflanzenbase. *Ber. Dtsch. Chem. Ges.* 2:292–295.
8. **Schuh, W., H. Puff, E. A. Galinski, and H. G. Trüper.** 1985. Die Kristallstruktur des Ectoin, einer neuen osmoregulatorisch wirksamen Aminosäure. *Z. Naturforsch. Teil C* 40:780–784.
9. **Tindall, B. J., and H. G. Trüper.** 1986. The aerobic extremely halophilic archaebacteria. *Syst. Appl. Microbiol.* 7:202–212.
10. **Trüper, H. G., and E. A. Galinski.** 1986. Concentrated brines as habitats for microorganisms. *Experientia* 42:1182–1187.

Chapter 32

Comparative Ecophysiology of the Nonphototrophic Sulfide-Oxidizing Bacteria

J. Gijs Kuenen

INTRODUCTION

The diversity of the nonphototrophic sulfide-oxidizing or colorless sulfur bacteria is large. Sixteen genera are known already (Table 1), and almost every year new names can be added to the list. Not only have new species, and even new genera, been discovered (18), but also well-known bacterial species have now been found to be able to use reduced-sulfur compounds as a sole, or supplementary, source of energy, e.g., *Paracoccus denitrificans* (11), a *Hydrogenobacter* species (5), and, among the phototrophic bacteria, *Thiocapsa*, *Chromatium*, and *Ectothiorhodospira* species (21, 26).

Many of the colorless sulfur bacteria are taxonomically totally unrelated (27). They have in common only their ability to obtain metabolically useful energy from the oxidation of sulfur compounds. The colorless sulfur bacteria have not only diverse morphologies, but also widely different physiological types of metabolism and metabolic pathways. In fact, it is clear from the recent work on their phylogenetic characterization on the basis of RNA sequences (chapter 35 of this volume) that the group of sulfur oxidizers as we know it today must be the result of evolutionary convergence from very different roots (32). The spectacular forms of such organisms as the large *Beggiatoa* or *Achromatium* species drew the attention of early investigators such as Winogradsky. However, the main advance in our knowledge of the physiology of these organisms came, in the first place, from studying the smaller thiobacilli, since these were easiest to grow in pure culture. Only very recently have the techniques for growing pure cultures of autotrophic *Beggiatoa* species become available (38), and this has finally opened the way to a more detailed study of their physiology, biochemistry, and molecular biology.

The sulfur-oxidizing bacteria can be found whenever suitable sources of (in)organic sulfur compounds, in addition to a suitable electron acceptor, are available (24, 27). In most cases oxygen or nitrate is used. However, a *Desulfovibrio* species has recently been discovered (2) which can derive energy from a disproportionation reaction

J. Gijs Kuenen • Department of Microbiology and Enzymology, Delft University of Technology, Julianalaan 67, 2628 BC Delft, The Netherlands.

Table 1.
Genera of Colorless Sulfur Bacteria[a]

Genus	Affiliation	Genus	Affiliation
Thiobacillus	*Thiobacilliaceae*	*Beggiatoa*	*Beggiatoaceae*
Thiomicrospira	*Thiobacilliaceae*	*Thioploca*	*Beggiatoaceae*
Thiosphaera	*Thiobacilliaceae*	*Thiothrix*	*Leucotrichaceae*
Thiosphaera	*Thiobacilliaceae*	*Achromatium*	*Acromatiaceae*
Thiobacterium	Uncertain	*Thiodendron*	*Hyphomicrobia*
Macromonas	Uncertain	*Sulfolobus*	Archaebacteria
Thiospira	Uncertain	*Acidianus*	Archaebacteria
Thiovulum	Uncertain		
Thermothrix	Uncertain		

[a] For further taxonomic information, see J. G. Kuenen, *in* J. T. Staley, M. P. Bryant, and N. Pfennig, ed., *Bergey's Manual of Systematic Bacteriology*, vol. 3, in press.

when growing on thiosulfate or sulfite under anaerobic conditions; sulfide and sulfate are produced in this reaction.

The range of physical and chemical conditions required for growth of the sulfur oxidizers can be expected to be as wide as that for other procaryotes. Extreme thermophiles, growing at up to 95°C, as well as extreme acidophiles, growing at pH values down to 1.5, are known (18, 24, 27). However, further work is required to find whether nature harbors other types, e.g., obligately psychrophilic or extremely halophilic sulfide oxidizers.

The sources of reduced (in)organic sulfur compounds can be geological (volcanic or deposits), anthropogenic (industrial or agricultural), and, of course, biological, especially from dissimilatory sulfate reduction. In view of the recent findings that phytoplankton, as well as other marine algae, produces large amounts of dimethyl sulfide (DMS), the importance of organic sulfides in the global sulfur cycle is apparent (1). The discovery of bacteria that can effectively metabolize these compounds (48) has led to speculation that DMS also plays an important (global) role as a source of energy for the sulfur-oxidizing bacteria.

Given the wide distribution of sulfur-oxidizing capabilities over so many unrelated procaryotes, it is hardly surprising that the biochemistry of oxidation of sulfur compounds is also diverse. Recent work in several laboratories has shown that in the few organisms studied so far, there is very little unity in the biochemistry and localization of the oxidation pathway (23, 25; D. P. Kelly, *in* H. G. Schlegel and B. Bowien, ed., *Biology of Autotrophic Bacteria*, in press). For example, in *Thiobacillus versutus* the oxidation of thiosulfate probably proceeds entirely in the periplasm and results in the passage of electrons to cytochrome *c* (51). In *T. tepidarius* (53), *T. ferrooxidans* (17), and *T. denitrificans* (46) a substantial part of the oxidation occurs intracellularly. A few organisms possess, in addition, a substrate-level phosphorylation pathway (42). The cell yields (23, 34) of the different sulfur oxidizers, which range from 6 to 10 g of dry biomass per mol of thiosulfate (or sulfide) oxidized, reflect this. In two denitrifiers, *T. denitrificans* and *Thermothrix thiopara*, the yields are significantly higher. This is most probably related to a different electron transport chain in these organisms. For further details, see references 23, 34, 53, and 54.

The autotrophic pathway for CO_2 fixation in most of the sulfur oxidizers investigated is the Calvin cycle. Work with pure cultures of an autotrophic *Beggiatoa* species (38) has shown that this organism also possesses the Calvin cycle. The two key enzymes of this pathway, ribulose-bisphosphate carboxylase (RuBPcase) and ribulose-monophosphate kinase (RuMPkinase), are present in this organism at activities similar to those found in *Thiobacillus*, *Thiomicrospira*, and *Thiosphaera* species. In two thermophilic sulfur oxidizers, *Sulfolobus* (22) and *Hydrogenobacter* (47) species, a reverse citric acid cycle appears to operate. This implies that aerobic organisms are able to use a pathway which had previously been considered typical of anaerobic growth. In fact, the *Hydrogenobacter* species is a microaerophile, but to our knowledge the *Sulfolobus* species is not and grows under fully aerobic conditions. This biochemical diversity in both the sulfur and the carbon metabolism of these organisms emphasizes, once more, the fact that the sulfide oxidizers make up a group of unrelated bacteria. However, as a whole, these organisms are an ecologically important group and often share habitats.

For the aerobic oxidation of biologically produced sulfide, the colorless sulfur bacteria are limited primarily to aerobic/anaerobic interfaces, because oxygen and sulfide can coexist only at relatively low concentrations. The alternative electron acceptors, nitrate and nitrite, are produced by nitrification and are also primarily available at significant concentrations (i.e., at sufficiently high fluxes) at the same interface. Consequently, most of the colorless sulfur bacteria are typical gradient organisms, which are adapted to life at interfaces such as can be found, for example, in microbial mats.

In the following sections I will review and compare some of the general (eco)physiological properties of the (colorless) sulfur-oxidizing bacteria. This will be followed by a discussion of the physiological spectrum that can be found among these organisms. Finally I will deal with a few specific examples of adaptations to life in microbial films and at interfaces.

ECOPHYSIOLOGICAL ASPECTS OF AUTOTROPHIC AND HETEROTROPHIC METABOLISM

Many of the sulfide oxidizers are not obligately dependent on sulfide as an energy source or on CO_2 as the only carbon source. In fact, a spectrum of metabolic types exists (Table 2) (28). Apart from the highly specialized obligate chemolithoautotrophs, one can find versatile facultative chemolithoautotrophs and chemolithoheterotrophic sulfur oxidizers. In the group of chemolithoheterotrophs are provisionally included organisms such as some *Beggiatoa* strains (7) and *Thiospira* and *Macromonas* species (9, 10). These bacteria do not appear to derive metabolically useful energy from the oxidation, but do benefit from the oxidation of sulfide by detoxifying hydrogen peroxide formed during heterotrophic metabolism (24). Finally, Table 2 also lists the incidental

Table 2.
Physiological Types among the Nonphototrophic Sulfur-Oxidizing Bacteria

Physiological type	Energy source (electron donor)		Carbon source	
	Inorganic sulfur compound	Organic compound	CO_2	Organic compound
Obligate chemolithoautotroph	+	−	+	−
Facultative chemolithoautotroph	+	+	+	+
Chemolithoheterotroph	+	+	−	+
Chemoorganoheterotroph	−	+	−	+

sulfur oxidizers, which do not seem to benefit from the oxidations.

The physiology of the obligate and facultative chemolithotrophs has been studied in detail, and a number of reviews have extensively dealt with their comparative properties (3, 24, 28, 31, 35). The first group of these organisms comprises specialists with a metabolism that is totally geared to chemolithoautotrophic growth. These organisms are able to metabolize and grow on sulfur compounds at high rates. Some specialists are not limited to obtaining energy from sulfur compounds, but can also use other inorganic materials such as ferrous iron or hydrogen. Under growth-limiting conditions, the organisms may use these substances simultaneously (5, 17). This type of metabolism is called mixolithotrophy (5). The specialists studied so far have been shown to utilize organic compounds to a limited extent. Heterotrophic metabolism occurs primarily under starvation conditions at the expense of internal reserve materials, which can either be oxidized under aerobic conditions or even be fermented anaerobically (4, 36).

In contrast to the rigid metabolism of the specialist sulfur oxidizers, the facultative species appear to have metabolic machinery which is very versatile and can be accurately tuned to the required turnover of available (mixtures of) substrates. When provided with mixtures of organic and reduced-sulfur compounds during energy- and/or carbon-limited growth, these organisms can grow mixotrophically, whereby they display an efficient use of the available resources (13, 52). Additionally, they can grow effectively under short-term alternating autotrophic and heterotrophic growth conditions (15).

Similar studies on chemolithoheterotrophs have not been extensively done because of a lack of suitable test organisms. Recent work in our laboratory (12, 12a) has shown that one chemolithoheterotroph, *Thiobacillus* strain Q, can grow well on mixtures of, for example, acetate and thiosulfate or sulfide, provided that these compounds are growth limiting. In batch culture, when thiosulfate and acetate were both supplied, both compounds were completely metabolized, but the yields of such cultures were the same as those in batch cultures containing acetate alone. However, when the organism was grown in the chemostat under acetate limitation, the addition of thiosulfate caused a definite cell yield increase (Fig. 1), showing that thiosulfate was an effective supplementary energy source. Given the fact that this organism cannot grow autotrophically, it was hardly surprising to observe that acetate-grown cultures could induce only a limited capacity for thiosulfate oxidation. The maximum specific thiosulfate oxidation capacity (300 nmol/min per mg of protein) was never more than 30 to 40% of that found in facultative organisms and was less than 10% of that found in the specialists. A problem associated with the study of the chemolithoheterotrophs, and to a lesser extent with the mixotrophs, is sulfite toxicity. For this reason, potential mixotrophs and chemolithoheterotrophs should always be tested by using a high-quality thiosulfate and continuous cultures in which the concentrations of reduced-sulfur compounds can be kept limiting. If these precautions are not taken, it is possible that the lithotrophic potential of the organism will not be revealed.

Another example of mixotrophic growth by chemolithoheterotrophic organism is the growth of *Hyphomicrobium* strain EG, which was isolated by Suylen and Kuenen (48). This organism can be grown on either dimethyl sulfoxide (DMSO) or DMS as the sole carbon and energy source (Table 3) (49). DMS(O) is completely metabolized to carbon dioxide and sulfate. Assimilation of carbon from DMS(O) occurs via the serine pathway. Enzyme studies had shown that DMS(O) was metabolized to methylmercaptan. This, in turn, was further oxidized to formaldehyde, hydrogen peroxide,

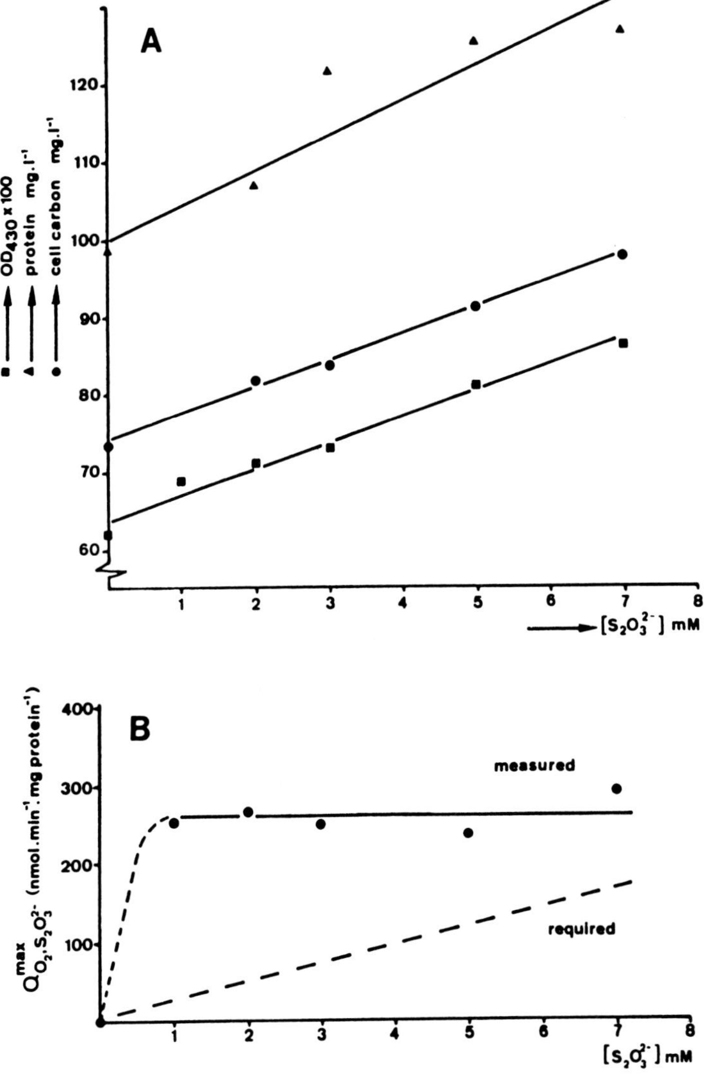

Figure 1. Biomass concentration (OD_{430}), protein, and cell carbon (A) and thiosulfate oxidation potential (B) of steady-state chemostat cultures grown under limitation by acetate (medium concentration, 10 mM) and with increasing amounts of thiosulfate in the medium (dilution rate $[D] = 0.09\ h^{-1}$, $T = 37°C$, pH = 7.5). The dotted line in panel B is the thiosulfate oxidation potential required to oxidize the thiosulfate present in the medium. Reproduced from reference 12a.

and hydrogen sulfide. The question arose of whether growth was truly mixotrophic in the sense that the sulfide formed in the pathway was also used as an energy source (49). This could be conveniently studied by taking advantage of the finding that *Hyphomicrobium* strain EG could also grow on a nonsulfur C_1 compound, methylamine. Methylamine-limited chemostat cultures showed a clear increase in yield when provided with sulfide (Table 3).

These two examples show that lithohet-

Table 3.
Dry Weight and Protein Content of *Hyphomicrobium* Strain EG Grown on Various Growth-Limiting (Mixtures of) Substrates in a Chemostat[a]

Concn of growth-limiting substrate in medium	Dry wt (mg/liter)	Protein content (mg/liter)
10 mM DMSO	167	69
10 mM DMS	191	79
10 mM MA[b]	108	47
9.5 mM MA + 10.4 mM Na_2S	212	77

[a] $D = 0.035\ h^{-1}$. Adapted from reference 49.
[b] MA, Methylamine.

erotrophic metabolism can be found among both heterotrophic and methylotrophic bacteria and further underline the importance of supplementary energy sources in this type of metabolism.

ECOPHYSIOLOGY AND ECOLOGICAL NICHES

At the sulfide/oxygen interface which occurs in many sediments, representatives of all metabolic types of sulfide oxidizers are often present, and it has been asked how such organisms might be able to coexist. In other words, what are their ecological niches? The key to the answer must lie in the nature of the interface, which is often highly dynamic because of, for example, diurnal or tidal changes. These changes lead to alterations in the fluxes of sulfide, oxygen, other nutrients such as organic compounds, and alternative electron acceptors (e.g., nitrate).

During the past 10 years, the role of organic compounds in the selection of thiobacilli and similar organisms has been studied in detail. This work has been reviewed (3, 28, 31), and therefore only a few key points will be summarized here. Pure-culture studies were done on bacteria that are representative of the different metabolic types. The experiments were done under laboratory conditions which could be considered relevant for the functioning of these types in the natural environment. This was achieved primarily by growing the organisms under nutrient-limiting conditions in the chemostat or in analogous continuous-flow systems. These experiments were complemented by mixed-culture studies and competition experiments involving the different types of organisms to obtain information about their competitiveness under the different growth conditions.

To study the flexibility and reactivity of the organisms, pure- and mixed-culture studies were also carried out under fluctuating conditions (for example, with an alternating supply of autotrophic [thiosulfate plus CO_2] and heterotrophic [acetate] substrates) and under alternating starvation and nutrient-limiting conditions for different periods. The outcome of these studies has allowed the construction of an ecophysiological profile for specialist and versatile types (Table 4). An important conclusion (3) of these studies was that not only is (metabolic) flexibility an important property for survival in nature, but also reactivity (i.e., the ability to respond rapidly to a changing environment) is a crucial factor for survival of a certain metabolic type.

A working model that was developed on the basis of these studies is shown in Fig. 2. It predicts that in a certain environment, depending on the relative turnover of the different organic and inorganic substrates, different types of sulfide oxidizers will come to the fore. Indeed, the different types of organisms could be enriched for in a chemostat culture fed with different ratios of the two substrates (14). For example, *Thiobacillus* strain Q, the chemolithoheterotroph discussed above, was the dominant organism (85% of the population) in a chemostat limited by 15 mM acetate and 10 mM thiosulfate. At this ratio of substrate concentrations, carbon dioxide fixation would not be energetically desirable. The results also showed that stable coexistence of different species, as indicated by the overlapping

Table 4.
Differences between a Metabolically Versatile and a Specialized Chemolithotrophic *Thiobacillus* sp.[a]

Item	Property	
	Specialist *Thiobacillus neapolitanus*	Versatile *Thiobacillus* strain A2
1	Few energy substrates utilized (S^{2-}, S^0, $S_2O_3^{2-}$, $S_4O_6^{2-}$, etc.)	Many different inorganic and organic energy substrates utilized
2	High specific growth rate on single substrate (μ_{max}[b] on thiosulfate, 0.35 h^{-1})	Low specific growth rate on single substrate (μ_{max} on thiosulfate, 0.10 h^{-1}); relatively high specific growth rate on mixed substrates (μ_{max} on thiosulfate plus acetate, 0.22 h^{-1})
3	High affinity for reduced-sulfur compounds	Relatively low affinity for reduced-sulfur compounds
4	High overcapacity of respiratory capacity during substrate-limited growth	Low overcapacity of respiratory capacity during substrate-limited growth
5	Low flexibility with respect to energy generation and organic-carbon assimilation; constitutive enzymes	High flexibility; inducible enzymes for energy generation and carbon assimilation
6	High reactivity toward few substrates	Low reactivity toward many substrates
7	Metabolic lesions	Many pathways, often overlapping
8	Low endogenous respiration	High endogenous respiration in autotrophically and heterophically grown cells
9	Very resistant to starvation	Less resistant to starvation
10	Ecological niche: environments with continuous or fluctuating supply of reduced sulfur compounds and a low turnover of organic compounds	Ecological niche: environments with simultaneous presence of both inorganic and organic substrates (mixotrophic conditions)

[a] Adapted from reference 3.
[b] μ_{max}, Maximum specific growth rate.

ranges in Fig. 2, was often possible. This was also indicated by mathematical modeling (16) of the competition experiment in the chemostat.

A FACULTATIVELY CHEMOLITHOAUTOTROPHIC DENITRIFIER FROM A WASTEWATER TREATMENT PLANT

Support for the model discussed above came from a study of a denitrifying, sulfide-oxidizing wastewater treatment system. This system consisted of a fluidized-bed reactor in which the biomass grew in a biofilm attached to sand particles. These particles were kept fluidized (floating) by the upflowing influent. This treatment plant was fed with organic acids (0.5 to 2 mmol/liter), sulfide (2 to 6 mmol/liter), and nonlimiting amounts of nitrate as the electron acceptor. Analysis of the microbial composition showed that facultative chemolithoautotrophs had become dominant (29). One representative species was selected for further study (45). This organism, *Thiosphaera pantotropha*, was a facultative chemolithoautotroph which could grow

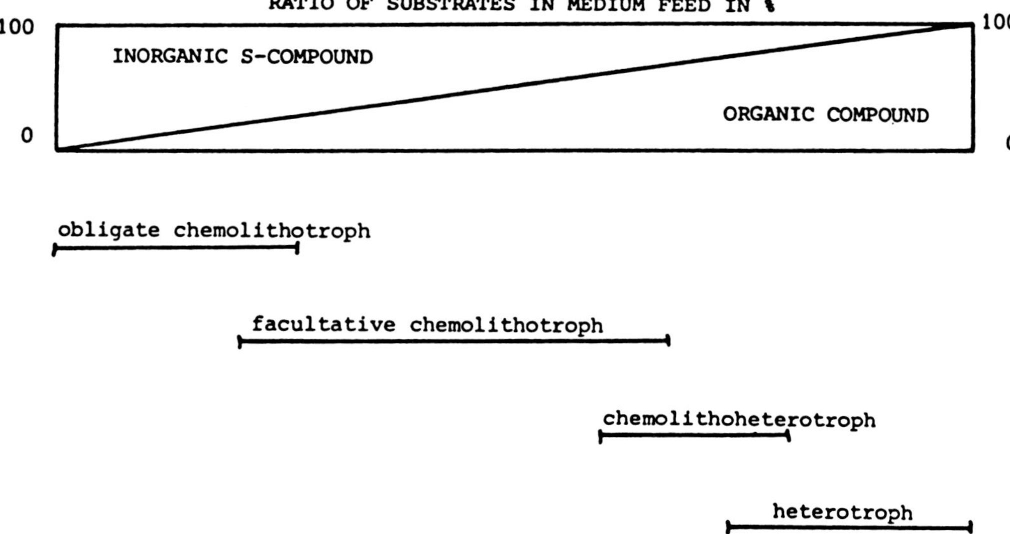

Figure 2. Model predicting the occurrence of sulfur-oxidizing bacteria as a function of the relative turnover rates of reduced inorganic sulfur compounds and organic substrates. Adapted from reference 28.

heterotrophically on a variety of organic compounds, autotrophically on sulfur compounds or hydrogen, and mixotrophically on mixtures of these compounds. As was expected, the organism could use oxygen or could denitrify. Unexpectedly, however, the organism turned out to be a constitutive denitrifier which continued to reduce nitrate or nitrite to nitrogen even under fully aerobic conditions. Figure 3 shows the corespiration by this organism of oxygen and nitrate in fully aerobic (>75% air saturation) cell suspensions which were well mixed and homogeneous, without clumps of cells (44). Details have been published elsewhere (43, 45).

The question was raised of why this type of organism might have become dominant in the biofilm of the fluidized-bed reactor. Closer inspection of the system showed that the influent contained not only nitrate as an electron acceptor, but also some oxygen. The oxygen became depleted in the lower part of the fluidized-bed column, and higher in the reactor only nitrate was available as an electron acceptor. It was also shown that the particles with the growing biofilm moved up and down within the fluidized bed of the column. Further experiments indicated that the cycling time of a particle moving up and down would be on the order of 4 to 24 h. This implies that the biomass particles would be exposed to alternating aerobic and anaerobic conditions. One current hypothesis is that *Thiosphaera pantotropha* (and other corespirers) would have a competitive advantage over organisms with a inducible denitrifying system, since such organisms would need to reinduce their denitrification capacity after an aerobic period. Repeated enrichment in laboratory-scale fluidized-bed reactors fed with a synthetic medium invariably yielded populations with a high percentage of constitutive denitrifiers which shared with *Thiosphaera pantotropha* the ability to denitrify under aerobic conditions (30). That the occurrence of aerobic denitrifiers is not unique for this particular type of system is becoming evident from the literature (30) and from a recent survey carried out in our laboratory. This has shown (Fig. 4) that there is a spectrum of aerobic denitrifiers with different threshold oxygen concentrations

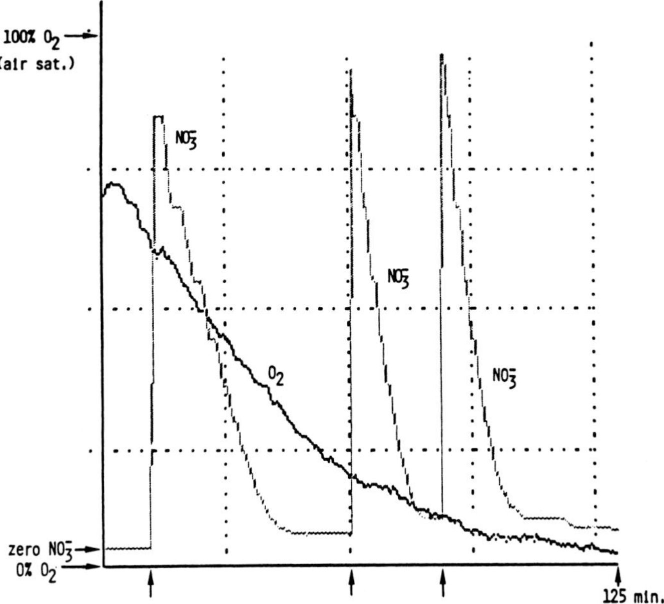

Figure 3. Simultaneous oxygen and nitrate consumption as a function of time in cell suspensions of *Thiosphaera pantotropha*. An aerobically grown culture was washed and suspended in growth medium with acetate as the carbon and energy source and ammonia as the nitrogen source. The suspension, which was completely homogeneous, was aerated and subsequently monitored with specific electrodes for dissolved oxygen and nitrate. At the times indicated by arrows, nitrate was injected to a final concentration of 10 μmol/liter. Adapted from reference 29.

above which aerobic denitrification is no longer possible (30).

The results obtained with *Thiosphaera pantotropha* indicate that its constitutive denitrification capacity may well be an adaption to life in a biofilm exposed to frequently changing aerobic and anaerobic conditions. The results also show that fluctuations in the availability of the electron acceptor can be an important selective force for organisms living at the aerobic/anaerobic interface.

A particularly puzzling point is that all aerobic denitrifiers so far investigated also appear to be heterotrophic nitrifiers. This implies that *Thiosphaera pantotropha*, grown aerobically in ammonium-containing minerals medium with acetate as the carbon and energy source, first nitrifies a part of the ammonium to nitrite and then immediately converts the nitrite to nitrogen gas. There is no intermediate accumulation of nitrite. In this way, up to 8 to 9 mmol of ammonium ions per liter may disappear from the culture under some conditions. If this also happened in nature, this short circuit in the nitrogen cycle might easily remain undetected. Further ecological implications have been discussed elsewhere (29, 30).

Before concluding this section on the ecological niches of the different metabolic types, it should be stressed that the role of organic compounds in the selection of different organisms is only one of the many selective forces that are exerted on the organisms living at the interface. It was mentioned above that even under the artificially strong selective forces in the chemostat, stable coexistence of different physiological types can be predicted and observed. Mathematical models further predict that the

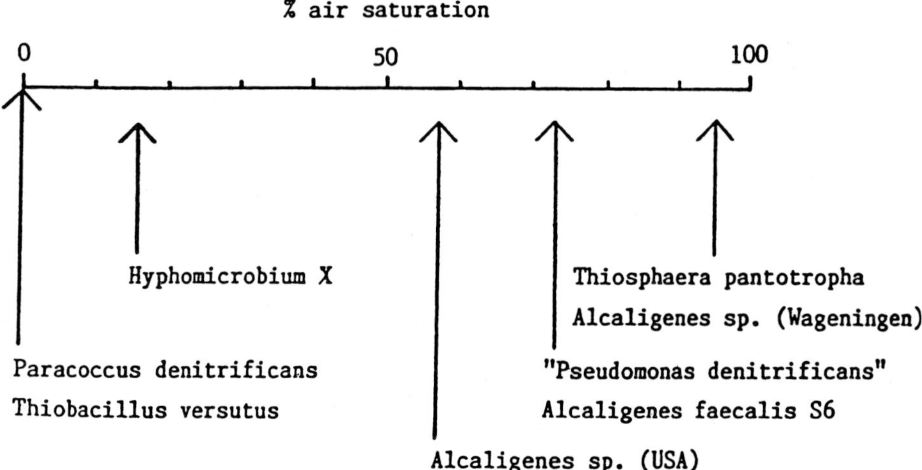

Figure 4. Sensitivities of the denitrification systems of various bacteria to dissolved oxygen concentration. The arrows indicate the concentration above which denitrification is inhibited by oxygen. Reproduced from reference 30.

number of coexisting species can never be more than the number of variables in a certain environment. It hardly needs to be emphasized that in most microbial biofilms or mats, the number of variables must be very large indeed, and, therefore, the number of coexisting species can also be large.

POPULATION EFFECTS

An interesting further dimension in the variables that can affect the outcome of competition is a population effect which is particularly relevant to the competition between the colorless sulfur bacteria and opportunistic sulfur oxidizers such as the phototrophic *Chromatium* or *Thiocapsa* species described in chapter 28 of this volume. These phototrophs are able to grow chemolithoautotrophically on sulfide or thiosulfate and oxygen in the dark. One might ask how effectively they might be able to compete with the specialized colorless sulfur bacteria, for example, with a (marine) specialist sulfur oxidizer such as *Thiobacillus thioparus*. It is known that the sulfide-oxidizing phototrophs often bloom in a dense layer at the aerobic/anaerobic interface, at densities on the order of 10^8 cells per g of sediment. During the day, *Chromatium* or *Thiocapsa* species can grow under anaerobic conditions and will be able to deplete the sulfide before it reaches the aerobic interface. During this period, these organisms can increase in biomass, whereas their potential competitor, *T. thioparus*, cannot, since it cannot obtain substrate. During the night, the sulfide moves upward and reaches the oxygen. If the phototroph is motile, as *Chromatium* species are, it might be able to follow the sulfide up to the interface. It must then compete with *T. thioparus*. *T. thioparus* has a Q_{max} (specific consumption rate) for sulfide (or thiosulfate) oxidation of 1,500 to 2,000 nmol/min per mg of protein. That of *Chromatium* species may be assumed to be similar to that found for *Thiocapsa roseopersicina*, i.e., 160 nmol of $S_2O_3^{2-}$ (or S^{2-}) per min per mg of protein (chapter 28 of this volume; H. van Gemerden and R. de Wit, personal communication). The K_s (Monod constant) of *T. thioparus* for $S_2O_3^{2-}$ (or S^{2-}) cannot be accurately measured (3), but is about 0.5 μmol/liter, and that of *Chromatium* and

Thiocapsa species is 1.5 μmol/liter. This implies that the affinity (Q_{max}/K_s) of *T. thioparus* for its substrate would be on the order of 30 times higher than that of *Chromatium* and *Thiocapsa* species. Thus, under sulfide limitation it would take 30 times more *Chromatium* biomass than *T. thioparus* biomass to convert the same amount of sulfide. Given 10^8 *Chromatium* cells per g, the equivalent activity in *T. thioparus* would require roughly 10^6 to 10^7 cells per g of this specialist. Such large numbers of cells are usually not found in these sediments. This calculation demonstrates that, potentially, during the night the motile *Chromatium* species (or the nonmotile *Thiocapsa* species in another situation) might be able to claim a major portion of the sulfide as a result of its successful growth during the day. This mass effect must obviously be a very important mechanism in the competition between species for limited resources. The same mass effect is also thought to play a role in the competition between autotrophic and heterotrophic nitrifiers, since the latter group can maintain a high population density at the expense of organic compounds. (The mathematical modeling of the mass effect can be found in reference 50.)

(ECO)PHYSIOLOGY AND ECOLOGY OF *BEGGIATOA* SP.

A colorless sulfur bacterium typical of microbial mats is *Beggiatoa* sp. The physiology and behavior of this organism are discussed some detail, since it is one of the few typical mat-building chemolithotrophs which have been well investigated. For analogous work on the mat-forming thermophile *Thermothrix thiopara*, see references 6 and 8. Information on the biology of other mat-forming colorless sulfur bacteria, *Thiothrix* and *Thioploca* species, can be found in a review by Larkin and Strohl (33).

The recent work by Nelson et al. (38–41) has given us more insight into the physiology of *Beggiatoa* species. These authors succeeded in growing a number of pure cultures of autotrophic marine strains in artificial sulfide-oxygen gradients (Fig. 5) (38). Test tubes with a sulfide-containing agar plug and a soft agar overlay were inoculated with *Beggiatoa* cells. The organisms rapidly moved to the sulfide/oxygen interface and started to multiply. Autotrophic growth could be proved from three pieces of evidence. First, experiments with radiolabeled carbon dioxide showed that over 95% of the *Beggiatoa* carbon originated from CO_2. Second, high levels of RuBPcase and RuMPkinase were found in these cultures (38). Third, the biomass yield calculated per sulfide ion oxidized to sulfate fell within the range of that normally found for a large variety of other autotrophic sulfide oxidizers (D. C. Nelson, *in* H. G. Schlegel and B. Bowien, ed., *Biology of Autotrophic Bacteria*, in press).

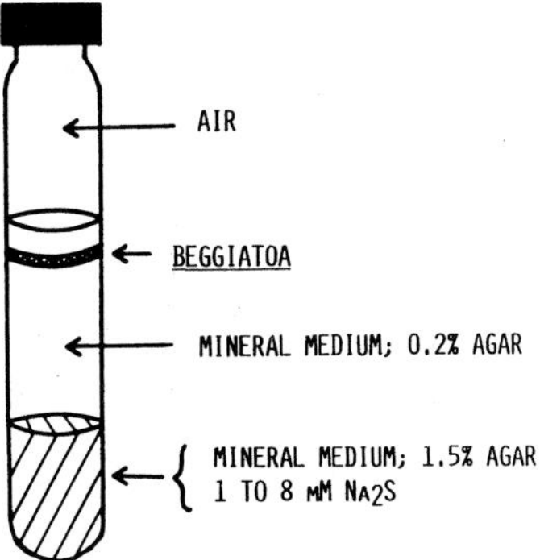

Figure 5. Sulfide gradient culture described in reference 38 and 40. The drawing was obtained from D. C. Nelson.

Nelson et al. (39, 40) carried out a detailed study of the sulfide and oxygen microprofiles in artificial gradients (Fig. 6). In the uninoculated control tube, the oxygen and sulfide coexisted at relatively high concentrations. Fluxes of oxygen and sulfide entering the interface could be calculated from the slopes of the gradient by using Fick's first diffusion law (39). The half-life of oxygen and sulfide in the control was more

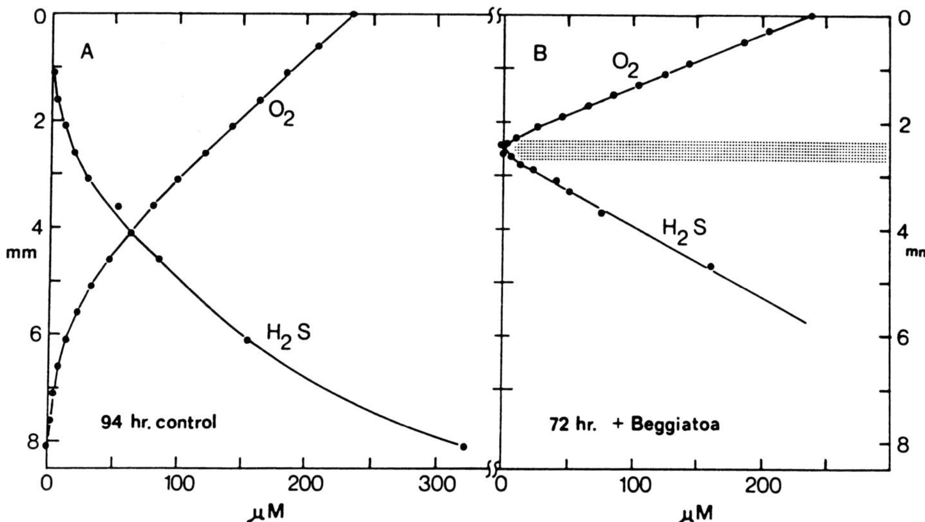

Figure 6. Oxygen and sulfide concentrations as a function of depth in artificial gradients prepared in tubes as shown in Fig. 5. Depth zero indicates the air/agar interface. (A) Uninoculated control at 94 h. (B) Profile in the presence of a *Beggiatoa* sp. growing at the shaded level, at 72 h. The data were obtained with microelectrodes for O_2 and S^{2-}. Reproduced from *Applied and Environmental Microbiology* (39).

than 50 min. In the *Beggiatoa* culture, the layer of oxygen-sulfide coexistence was less than 0.2 mm thick, and the half-life of oxygen and sulfide was only 3 to 5 s, about 1,000-fold less than in the control. Given the observation that the S^{2-}/O_2 consumption ratios of the chemical and biological reactions were the same, it could be concluded that in the presence of *Beggiatoa* species the oxidation of sulfide proceeded almost entirely biologically. Particularly important and relevant to life at the sulfide/oxygen interface is that *Beggiatoa* species are apparently able to outcompete the chemical oxidation by lowering the concentrations of the oxygen and sulfide. Indeed, from the fact that the initial growth of *Beggiatoa* species is exponential, it can be inferred that the K_s for sulfide and oxygen in this organism must be on the order of 1 µmol/liter or less. This is similar to the K_s values reported for thiobacilli grown under sulfide limitation in the chemostat (3). Thus, by their efficient sulfide oxidation at the sulfide/oxygen interface, the sulfur bacteria have the potential to protect the aerobic environment from toxic sulfide, and the anaerobic environment from toxic oxygen.

The *Beggiatoa* cell yield was calculated from the same gradient cultures. The protein increase was measured by daily harvesting of duplicates from a series of identical tubes. The sulfide consumption per day was calculated from the total sulfide profile in each tube at the time of harvesting. From these data, a yield of 8.6 g (dry weight) per mol of sulfide was estimated. This is well within the range found with other sulfur-oxidizing bacteria. The stoichiometry of the oxidation reaction during balanced growth was approximately $O_2/S^{2-} = 1.6$. This is the expected consumption if sulfide is completely oxidized to sulfate and about 20% of the electrons are used for the reduction of CO_2 to give the biomass found in the experiments (39). It was further shown that the growth of these *Beggiatoa* cultures was stimulated by the presence of acetate or propionate, demonstrating that they can grow mixotrophically (Nelson, in press).

In view of the positive results with the marine strains, Nelson et al. (40; Nelson, in press) also tested a number of freshwater strains for autotrophic growth in the gradient cultures, but all of the organisms studied so far appeared to be (litho)heterotrophs. The observation that strain 75-2a contained trace levels of RuBPcase which increased slightly at the end of the growth period was puzzling. However, it is not yet known whether, under special growth conditions, this strain is able to induce the enzyme to levels that would allow autotrophic growth. That autotrophic freshwater *Beggiatoa* strains are likely to exist was indicated by the fact that freshwater *Beggiatoa* blooms contained significant levels of RuBPcase (Nelson, in press).

The ecology of *Beggiatoa* species has been studied in quite some detail (19, 20, 37). This has been possible because the organism can be easily recognized under the microscope. Especially under conditions of continuous dark, the organisms may form large, dense mats on sulfide-rich sediments overlaid by aerobic water bodies. When these sediments are exposed to (day)light, a very complex diurnal pattern of competing and interacting sulfide-oxidizing bacteria may develop (19, 33). When the balance of sulfide and photosynthetic oxygen production in the top layer is such that the oxygen reaches below the photic zone, *Beggiatoa* species can grow at the sulfide/oxygen interface deep in the sediment during daylight hours. During the night, these organisms can move up to the top of the sediment and continue to oxidize the sulfur compounds. However, if light penetrates to the sulfide layer, a situation may develop whereby the phototrophs can use the sulfide both in the light and in the dark. This would be a situation similar to that described above for *Chromatium* species, as a result of which *Beggiatoa* species might be outcompeted (20). More complex situations may occur, depending on the dynamic balance between oxygen production and sulfide production. For example, Jørgensen (19) described a sulfide-rich sediment where *Beggiatoa* species can be found during the day under an oxygen-producing *Oscillatoria* bloom. Under the *Beggiatoa* layer he found a red layer of the motile *Chromatium* species. After a transient situation occurring after sunset, the sulfide production by this particular mat was such that the sulfide entered the first few centimeters of the water phase above the sediment. *Beggiatoa* species, being unable to move into the water, remained at the top of the sediment, but the motile *Chromatium* organisms were able to move into the water and take over the role of the *Beggiatoa* species as a chemolithotroph.

From these examples, it appears that *Beggiatoa* species not only must be able to grow at the interface, but also must be able to survive substantial anaerobic periods. The work of Nelson and Castenholz (36) has shown that intracellular sulfur formed during aerobic periods may serve as an electron acceptor for anaerobic heterotrophic metabolism in these organisms. In fact, one can imagine that by moving up and down through the oxygen/sulfide interface, the *Beggiatoa* species can regenerate its electron acceptor, sulfur, to be able to profit optimally from substrates in the anaerobic zone. Such a strategy might be particularly important for the many heterotrophic *Beggiatoa* strains. It also might serve for endogenous metabolism of the (facultatively) autotrophic strains.

Thus, laboratory experiments and field observations have shown that *Beggiatoa* species are typical interface organisms specialized on one hand to microaerophilic life in a solid substrate and on the other hand to life at a moving interface.

CONCLUDING REMARKS

The sulfide-oxidizing (nonphototrophic) bacteria are a very diverse group of organisms which share a common habitat where sulfide and either oxygen or nitrate (or nitrite) coexist. Such habitats can be found as interfaces in stratified lakes or microbial mats. The sulfide oxidizers are adapted to life at the interface in various ways, but they all share the high affinity for sulfide and oxygen which allows them to outcompete the spontaneous chemical oxidation of sulfide.

The highly dynamic nature of the interface, particularly when light is available or when tidal movements take place, requires organisms with a very adaptable, flexible metabolism, but also calls for a high reactivity to sudden changes in the environment. In line with these sometimes conflicting requirements, a spectrum of different metabolic types with widely different survival strategies can be found to coexist in nature.

Model competition experiments in the chemostat have demonstrated the advantages and disadvantages of the different types of physiology. Highly specialized organisms have an advantage when the turnover of sulfur compounds is high relative to that of organic compounds. They may also be able to dominate under conditions of intermittent sulfide supply, as can be expected during the 6- or 12-h tidal or diurnal changes at interfaces. The versatile, facultative sulfur oxidizers may come to the fore when mixtures of organic compounds, such as sugars or organic acids, and sulfide are available in similar quantities. The examples discussed, *T. versutus* (formerly *Thiobacillus* strain A2), *Thiosphaera pantotropha*, and the facultatively autotrophic *Beggiatoa* species, are typically versatile organisms. The heterotrophic sulfide oxidizers which are able to generate energy from sulfur compounds appear to be selected if the ratio of organic and sulfur compounds is such that an autotrophic potential is not required. These predictions are only trends, and it should be stressed that coexistence of different types is often observed.

In phototrophic microbial mats, colorless sulfur bacteria may have to compete with *Chromatium*, *Thiocapsa*, or *Ectothiorhodospira* species, which can also grow in the dark as chemolithoautotrophs. A simple calculation shows that when the phototrophs occur in blooms they may be able to contribute substantially to the total turnover of sulfur compounds in the dark, despite their relatively low affinity for sulfide. Thus, these phototrophs may represent significant competitors for the colorless sulfur bacteria.

Thiosphaera pantotropha is not only versatile in its ability to perform autotrophic or heterotrophic metabolism. It is also a typical interface organism in its ability to grow in biofilms and in its facultatively anaerobic character. The organism possesses a constitutive denitrification capacity which may be an adaption to fluctuating aerobic and anaerobic conditions in the presence of nitrate. The constitutivity of its denitrification system might confer a higher reactivity to the organism, allowing it to respond to a sudden onset of anaerobic, denitrifying conditions.

Beggiatoa species are typical mat-building colorless sulfur bacteria and are among the few colorless sulfur bacteria known to form blooms at the sulfide/oxygen interface. Most of the *Beggiatoa* species grown in pure culture are heterotrophs, a few may be chemolithoheterotrophs, and a number of recent isolates are facultative chemolithoautotrophs. All of them benefit from the presence of sulfide in some way. Their gliding motility allows them to move rapidly with the interface. This is very significant if it is realized that the sulfide/oxygen interface may move over several centimeters in a few hours. The movement of *Beggiatoa* organisms may not be a chemotactic, positive response, but rather a phobic response to light, oxygen, and sulfide (37). It is striking that both the autotrophic and heterotrophic strains are able to survive very well, or even grow, under anaerobic conditions. This shows that they have an effective anaerobic metabolism whereby, at least in some cases, intracellular sulfur can serve as the terminal electron acceptor.

The older literature listed *Beggiatoa* as a colorless blue-green alga. Indeed, this organism does share many properties with the cyanobacteria with respect to its adaption to life in microbial mats, including the capability to fix molecular nitrogen. As a matter of fact, an organism labeled as a "filamentous, gliding, sulfur-bearing, nitrogen-fixing, mat-building lithoautotroph" might be taken for an *Oscillatoria* species.

This last example shows that in the solutions that nature has found for survival in microbial mats, the metabolic strategies of phototrophs and nonphototrophs may show similar patterns.

ACKNOWLEDGMENTS. I thank L. A. Robertson and P. Bos for critical reading of the manuscript. I also thank D. C. Nelson for providing figures and information in press and for valuable discussions.

LITERATURE CITED

1. **Andreae, M. O., and H. Raemdonck.** 1983. Dimethyl-sulphide in the surface ocean and the marine atmosphere: a global view. *Science* **221**:744–747.
2. **Bak, F., and N. Pfennig.** 1987. Chemolithotrophic growth of *Desulfovibrio sulfodismutans* sp. nov. by disproportionation of inorganic sulfur compounds. *Arch. Microbiol.* **147**:184–189.
3. **Beudeker, R. F., J. C. Gottschal, and J. G. Kuenen.** 1982. Reactivity versus flexibility in thiobacilli. *Antonie van Leeuwenhoek* **48**:39–51.
4. **Beudeker, R. F., W. de Boer, and J. G. Kuenen.** 1981. Heterolactic fermentation of intracellular polyglucose by the obligate chemolithotroph *Thiobacillus neapolitanus* under anaerobic conditions. *FEMS Mirobiol. Lett.* **12**:337–342.
5. **Bonjour, F., and M. Aragno.** 1986. Growth of thermophilic, obligatorily chemolithoautotrophic hydrogen-oxidizing bacteria related to *Hydrogenobacter* with thiosulfate and elemental sulfur as electron and energy source. *FEMS Microbiol. Lett.* **35**:11–15.
6. **Brannan, D. K., and D. E. Caldwell.** 1980. *Thermothrix thiopara*: growth and metabolism of a newly isolated thermophile capable of oxidizing sulfur and sulfur compounds. *Appl. Environ. Microbiol.* **40**:211–216.
7. **Burton, S. D., and R. Y. Morita.** 1964. Effect of catalase and cultural conditions on growth of *Beggiatoa*. *J. Bacteriol.* **88**:1755–1761.
8. **Caldwell, D. E., T. L. Kieft, and D. K. Brannan.** 1983. Colonization of sulfide-oxygen interfaces on hot spring tufa by *Thermothrix thiopara*. *Geomicrobiol. J.* **3**:181–199.
9. **Dubinina, G. A., and M. Y. Grabovich.** 1983. Isolation of pure *Thiospira* cultures and investigation of their sulfur metabolism. *Mikrobiologiya* **52**:1–7. (In English.)
10. **Dubinina, G. A., and M. Y. Grabovich.** 1984. Isolation, cultivation, and characteristics of *Macromonas bipunctata*. *Mikrobiologiya* **53**:610–617. (In English.)
11. **Friedrich, C. G., and G. Mitringa.** 1981. Oxidation of thiosulphate by *Paracoccus denitrificans* and other hydrogen bacteria. *FEMS Microbiol. Lett.* **10**:209–212.
12. **Gommers, P. J. F., and J. G. Kuenen.** 1985. Physiological properties of a non-autotrophic sulphur-oxidizing bacterium. *Antonie van Leeuwenhoek* **51**:443–444.

12a. **Gommers, P. J. F., and J. G. Kuenen.** 1988. *Thiobacillus* strain Q, a chemolithoheterotrophic sulphur bacterium. *Arch. Microbiol.* **150**:117–125.

13. **Gottschal, J. C., and J. G. Kuenen.** 1980. Mixotrophic growth of *Thiobacillus* A2 on acetate and thiosulphate as growth limiting substrates in the chemostat. *Arch. Microbiol.* **126**:33–42.
14. **Gottschal, J. C., and J. G. Kuenen.** 1980. Selective enrichment of facultatively chemolithotrophic thiobacilli and related organisms in the chemostat. *FEMS Microbiol. Lett.* **7**:241–247.
15. **Gottschal, J. C., H. Nanninga, and J. G. Kuenen.** 1981. Growth of *Thiobacillus* A2 under alternating growth conditions in the chemostat. *J. Gen. Microbiol.* **126**:85–96.
16. **Gottschal, J. C., and T. F. Thingstad.** 1982. Mathematical description of competition between two and three bacterial species under dual substrate limitation in the chemostat: a comparison with experimental data. *Biotechnol. Bioeng.* **24**:1403–1418.
17. **Hazeu, W., W. Bijleveld, J. T. Grotenhuis, E. Kakes, and J. G. Kuenen.** 1986. Kinetics and energetics of reduced sulfur oxidation by chemostat cultures of *Thiobacillus ferrooxidans*. *Antonie van Leeuwenhoek* **52**:507–518.
18. **Huber, R., G. Huber, A. Segerer, J. Seger, and K. O. Stetter.** 1986. Aerobic and anaerobic extremely thermophilic autotrophs, p. 44–51. *In* H. W. van Verseveld and J. A. Duine (ed.), *Microbial Growth on C_1 Compounds*. Martinus Nijhoff, Dordrecht, The Netherlands.
19. **Jørgensen, B. B.** 1982. Ecology of the bacteria of the sulphur cycle with special reference to anoxic-oxic interface environments. *Philos. Trans. R. Soc. London Ser. B* **298**:543–561.

20. **Jørgensen, B. B., and D. J. Des Marais.** 1986. Competition for sulfide among colorless and purple sulfur bacteria in cyanobacterial mats. *FEMS Microbiol. Ecol.* **38:**179–186.
21. **Kämpf, C., and N. Pfennig.** 1980. Capacity of Chromatiaceae for chemotrophic growth. Specific respiration rates of *Thiocystis violacea* and *Chromatium vinosum*. *Arch. Microbiol.* **127:**125–137.
22. **Kandler, O., and K. O. Stetter.** 1981. Evidence for autotrophic CO_2 assimilation in *Sulfolobus brierleyi* via a reductive carboxylic acid pathway. *Zentralbl. Bakteriol. Mikrobiol. Hyg. I Abt. Orig. C* **2:**111–121.
23. **Kelly, D. P.** 1982. Biochemistry of the chemolithotrophic oxidation of inorganic sulphur. *Philos. Trans. R. Soc. London Ser. B* **298:**499–528.
24. **Kelly, D. P., and J. G. Kuenen.** 1984. Ecology of the colourless sulphur bacteria, p. 211–240. *In* G. A. Codd (ed.), *Aspects of Microbial Metabolism and Ecology*. Academic Press, Inc. (London), Ltd., London.
25. **Kelly, D. P., J. Mason, and A. P. Wood.** 1986. Energy metabolism in chemolithotrophs, p. 186–194. *In* H. W. van Verseveld and J. A. Duine (ed.), *Microbial Growth on C_1 Compounds*. Martinus Nijhoff, Dordrecht, The Netherlands.
26. **Kondratieva, E. N., V. G. Zhukov, R. N. Ivanovsky, Y. P. Petushkova, and E. Z. Monsonov.** 1966. The capacity of phototrophic sulphur bacterium *Thiocapsa roseopersicina* for chemosynthesis. *Arch. Microbiol.* **108:**287–292.
27. **Kuenen, J. G.** 1975. Colourless sulfur bacteria and their role in the sulfur cycle. *Plant Soil* 43:49–76.
28. **Kuenen, J. G., and R. F. Beudeker.** 1982. Microbiology of thiobacilli and other sulphur-oxidizing autotrophs, mixotrophs and heterotrophs. *Philos. Trans. R. Soc. London Ser. B* **298:**473–497.
29. **Kuenen, J. G., and L. A. Robertson.** 1986. Application of pure culture physiology concepts to mixed populations, p. 191–205. *In* V. Jensen, A. Kjøller, and L. H. Sørensen (ed.), *Microbial Communities in Soil*. Elsevier Applied Science Publishers Ltd., London.
30. **Kuenen, J. G., and L. A. Robertson.** 1987. Ecology of nitrification and denitrification, p. 161–218. *In* J. Cole and S. Ferguson (ed.), *Nitrogen and Sulphur Cycles*. Cambridge University Press, Cambridge.
31. **Kuenen, J. G., L. A. Robertson, and H. van Gemerden.** 1985. Microbial interactions among aerobic and anaerobic sulfur-oxidizing bacteria. *Adv. Microb. Ecol.* **8:**1–59.
32. **Lane, D. J., D. A. Stahl, G. J. Olsen, D. J. Heller, and N. R. Pace.** 1985. Phylogenetic analyses of the genera *Thiobacillus* and *Thiomicrospira* by 5S rRNA sequences. *J. Bacteriol.* **163:**75–81.
33. **Larkin, J. M., and W. R. Strohl.** 1983. *Beggiatoa, Thiothrix*, and *Thioploca*. *Annu. Rev. Microbiol.* **37:**341–367.
34. **Mason, J., D. P. Kelly, and A. P. Wood.** 1987. Chemolithotrophic and autotrophic growth of *Thermothrix thiopara* and some thiobacilli on thiosulphate and polythionates, and a reassessment of the growth yields of *Thx. thiopara* in chemostat culture. *J. Gen. Microbiol.* **133:**1249–1256.
35. **Matin, A.** 1978. Organic nutrition of chemolithotrophic bacteria. *Annu. Rev. Microbiol.* **32:**433–468.
36. **Nelson, D. C., and R. W. Castenholz.** 1981. Use of reduced sulfur compounds by *Beggiatoa* sp. *J. Bacteriol.* **147:**140–154.
37. **Nelson, D. C., and R. W. Castenholz.** 1982. Light responses of *Beggiatoa*. *Arch. Microbiol.* **131:**146–155.
38. **Nelson, D. C., and H. W. Jannasch.** 1983. Chemoautotrophic growth of a marine *Beggiatoa* in sulfide-gradient cultures. *Arch. Microbiol.* **136:**262–269.
39. **Nelson, D. C., B. B. Jørgensen, and N. P. Revsbech.** 1986. Growth pattern and yield of a chemoautotrophic *Beggiatoa* sp. in oxygen-sulfide microgradients. *Appl. Environ. Microbiol.* **52:**225–233.
40. **Nelson, D. C., N. P. Revsbech, and B. B. Jørgensen.** 1986. Microoxic-anoxic niche of *Beggiatoa* spp.: microelectrode survey of marine and freshwater strains. *Appl. Environ. Microbiol.* **52:**161–168.
41. **Nelson, D. C., J. B. Waterbury, and H. W. Jannasch.** 1982. Nitrogen fixation and nitrate utilization by marine and freshwater *Beggiatoa*. *Arch. Microbiol.* **133:**172–177.
42. **Peck, H. D.** 1960. Adenosine 5′ phosphosulfates as an intermediate in the oxidation of thiosulfate by *Thiobacillus thioparus*. *Proc. Natl. Acad. Sci. USA* **46:**1053–1057.
43. **Robertson, L. A., and J. G. Kuenen.** 1984. Aerobic denitrification—old wine in new bottles? *Antonie van Leeuwenhoek* **50:**525–544.
44. **Robertson, L. A., and J. G. Kuenen.** 1984. Aerobic denitrification: a controversy revived. *Arch. Microbiol.* **139:**351–354.
45. **Robertson, L. A., and J. G. Kuenen.** 1983. *Thiosphaera pantotropha* gen. nov. sp. nov., a facultatively anaerobic, facultatively autotrophic sulphur bacterium. *J. Gen. Microbiol.* **129:**2847–2855.
46. **Schedel, M., and H. G. Trüper.** 1980. Anaerobic oxidation of thiosulfate and elemental sulfur in *Thiobacillus denitrificans*. *Arch. Microbiol.* **124:**205–210.
47. **Shiba, H., T. Kawasumi, Y. Igarashi, T. Kodama, and Y. Minoda.** 1985. The CO_2 assimilation via the reductive tricarboxylic acid cycle in

an obligately autotrophic aerobic hydrogen-oxidizing bacterium, *Hydrogenobacter thermophilus*. *Arch. Microbiol.* 141:198–203.

48. **Suylen, G. M. H., and J. G. Kuenen.** 1986. Chemostat enrichment and isolation of *Hyphomicrobium* EG. *Antonie van Leeuwenhoek* 52:281–293.
49. **Suylen, G. M. H., G. C. Stefess, and J. G. Kuenen.** 1986. Chemolithotrophic potential of a *Hyphomicrobium* species, capable of growth on methylated sulphur compounds. *Arch. Microbiol.* 146:192–198.
50. **Tiedje, J. M., A. J. Sexstone, D. M. Myrold, and J. A. Robinson.** 1982. Denitrification: ecological niches, competition and survival. *Antonie van Leeuwenhoek* 48:569–583.
51. **Wei-Ping, L.** 1986. A periplasmic location for the thiosulphate-oxidizing multi-enzyme system from *Thiobacillus versutus*. *FEMS Microbiol. Lett.* 34: 313–317.
52. **Wood, A. P., and D. P. Kelly.** 1981. Mixotrophic growth of *Thiobacillus* A2 in chemostat culture on formate and glucose. *J. Gen. Microbiol.* 125:55–62.
53. **Wood, A. P., and D. P. Kelly.** 1986. Chemolithotrophic metabolism of the newly-isolated moderately thermophilic, obligately autotrophic *Thiobacillus tepidarius*. *Arch. Microbiol.* 144:71–77.
54. **Wood, A. P., D. P. Kelly, and P. R. Norris.** 1987. Autotrophic growth of four *Sulfolobus* strains on tetrathionate and the effect of organic nutrients. *Arch. Microbiol.* 146:382–389.

Chapter 33

Laboratory Model Systems for the Experimental Investigation of Gradient Communities

Julian Wimpenny

INTRODUCTION

Microbial mats have become, par excellence, classic examples of spatially heterogeneous microbial ecosystems. It would be unwise to forget, however, that the great majority of natural microbial habitats have at least some elements of spatial heterogeneity. The latter can range from spatially extensive shallow gradients in ocean ecosystems to the highly complex multiphasic physical and chemical gradients found in an average soil profile.

The intimate physiological behavior of microbes has been thoroughly explored in homogeneous laboratory growth systems. In particular the technique of continuous culture has played a vital part in unraveling some of these mysteries. At the other extreme, natural ecosystems must be investigated where they are found. There is no substitute for true in situ investigations. Techniques now available for such studies are becoming increasingly powerful and sophisticated.

Between these two extremes it seems clear that there is a role for simplified laboratory model systems which can replicate the behavior of pure cultures or simplified communities of organisms in spatially structured systems incorporating physicochemical gradients. A rather extreme example illustrates this case. The bacterium *Bacillus cereus*, grown in a gel-stabilized culture in counter gradients of glucose and oxygen, produces a sequence of sharply defined growth bands. This phenomenon (which is not restricted to this species) will be discussed in more detail later. The point is that it would be almost impossible to predict this behavior from a complete knowledge of the physiology of the organism gained from studies using homogeneous laboratory culture systems.

It is evident that scientific investigations can proceed in two entirely different directions at the same time. One approach we can call quasi-holistic; the other is superficially close to scientific reductionism. Holism suggests that the behavior of the entire system should be investigated with nothing altered or removed. Abstracting elements from the complete system invalidates any conclusions that might be drawn about the real phenomenon. Reductionists, on the other hand,

Julian Wimpenny • Department of Microbiology, University College, Newport Road, Cardiff Cf2 1TA, United Kingdom.

would like to predict the behavior of the system from a knowledge of its constituent parts. Most of us are neither true holists nor complete reductionists. What we can discern are tendencies to one approach or another. We ought not to regard this as anything but healthy for science as a whole. Both approaches should be applied simultaneously. They ought to lead to a more complete explanation of natural phenomena.

Laboratory investigations of natural microbial systems seem also to be dominated by these structural differences in scientific method. Thus we see **microcosms** and **models**. Microcosms are tools of the holist and owe their provenance to the natural system from which they directly derive. Virtually all of the complexity of the real habitat is retained. In the laboratory, conditions may be altered or held constant or manipulated in some way; however, the system is seldom dramatically simplified. On the other hand, a model is always some kind of abstraction. Elements of the real world are included in the model while all else is held constant or ignored. Models can evolve, of course. As more is found out about the behavior of a model, so it must be elaborated to include other elements closer to reality. Clearly, models appeal to reductionists!

My colleagues and I have taken the second approach. We believe that there is room for a family of model systems that fall between the homogeneous culture vessel and the reality of the in situ system. Our models, and others to be discussed here, are all spatially heterogeneous; in particular, they include the ability to generate solute diffusion gradients.

OPEN MODEL SYSTEMS

Multistage Continuous Culture Systems

Multistage continuous culture systems consist of more than one fermentor vessel linked together in a cascade. Traditionally, flow is in one direction only. The dilution rate in each vessel can vary since it depends on the volume of the vessel and on the possibility of solute additions at different points in the system.

The kinetics of multistage continuous culture systems has been discussed by Herbert (17), and Margalef (27) has compared multistage systems with natural ecological successions: "Any system of chemostats affords an ideal way for mapping time series into space; going down the row of flasks is equivalent to progressing along an ecological succession." For example, Slezak and Sikyta (39) set up a multistage continuous flow system to model a batch culture in which each of the growth stages was represented by a single vessel. The non-steady-state time series of a microbial batch culture was thus mapped onto a steady-state spatial array.

Multistage chemostats have been used to observe the sequential degradation of complex wastes such as coke oven effluents (1) and diesel oil (35) and, in addition, to investigate nitrification and denitrification in effluent streams (16).

A five-stage continuous flow system was used to investigate carbon flow in anaerobic microbial communities (41). Spatial separation of functional groups of bacteria associated with sulfate reduction, methanogenesis, and acidogenesis, characteristic of that seen in a natural sediment, took place in this system. Similar systems have been constructed to observe the degradation of paper and paper mill effluents (25, 33).

As Margalef stressed, the multistage continuous flow system provides a good example of time-dependent—in other words, successional—systems. That is, processes occurring in a single space as a function of time are modeled at a single time as a function of space. They are not good models of space-dependent systems unless the latter are predominantly unidirectional. For example, in a sediment ecosystem where food enters via a single surface only, it might seem appropriate to employ multistage systems. To do so, however, is to ignore the part played by diffusion in transferring highly significant solutes like sulfide, ammonia, and methane in the reverse direction. Multistage chemostats, in other words, only tell half of the story.

Bidirectionally Linked Multistage Systems (Gradostats)

Margalef also suggested growth systems in which "a certain amount of diffusion and contamination 'upstream' has been permitted as in natural systems" (27). Such ideas led in the end to the development of true bidirectionally linked multistage systems. Margalef also speculated on possible three-dimensional open systems, though he felt that they were too complex to establish practically. He imagined a 10-by-10-by-10 array of vessels, for example, all linked together with rubber tubing and pumps:

> The adequately programmed pumps could forward the liquid in either direction and by this mechanism, the effects of turbulent mixing could be simulated by pumping fluid from one vessel to another, stirring, pumping the same volume back again to the first and so on.

A simple bidirectionally linked system developed by Cooper and Copeland (10) was used to investigate estuarine ecology. It consisted of five 9-gallon plastic containers each connected to its neighbor by glass tubes through which solute transfer took place. Reservoirs and medium outlets were placed in each end of the end vessels. A salinity gradient was formed as fresh water flowed one way down the system while salt water was fed in from the second reservoir in the opposite direction. The system was colonized by the same microflora as was present at different positions in the estuary itself. The results were qualitatively but not quantitatively the same as patterns found in the estuary.

Lovitt and Wimpenny (23, 24) described the "gradostat." This was a bidirectionally linked system of small laboratory fermentors. Flow in one direction was by gravity over weirs, and that in the other direction was by tubing pumps. Reservoirs and overflows were located at each end of the system as in the Cooper-Copeland model (Fig. 1A).

The principle of bidirectionality seemed important, though not the way in which it was achieved. Our original gradostat we named the tubing-coupled or TC-gradostat, since each fermentor was connected to its neighbors by normal tubing lines. There are other ways of achieving bidirectional exchange. In a later, much more compact version, vessels consisted of glass tubing sections separated from one another by plates in which are located small exchange ports (holes!). Here exchange between vessels is by mechanically assisted diffusion and the device is referred to as the diffusion-coupled (DC) gradostat (Fig. 1B). The latest version is simpler still and consists of an accurately machined Delrin (acetal plastic) bar rotating in a glass tube. Open sections are turned out of the bar and form growth chambers. These are linked to each other only by the gap between the rotating bar and the inner glass wall of the vessel. Exchange rates depend on the rotation rate of the Delrin bar. This system has been called the shear-coupled or SC-gradostat (Fig. 1C) and because of its simplicity offers the possibility of higher spatial resolution by increasing the number of linked growth chambers. A 12-vessel version of the shear-coupled gradostat has been constructed and is undergoing trials at the moment.

Many of the properties of a gradostat have been elucidated by computer simulation (53). If a solute is placed in the first reservoir and water in the second, after a certain time solute distribution across the array of vessels will be a linear function of the original concentration. This will be true if all the vessel volumes and the flow rates in each direction are the same. The distribution of solutes we term rather loosely a "gra-

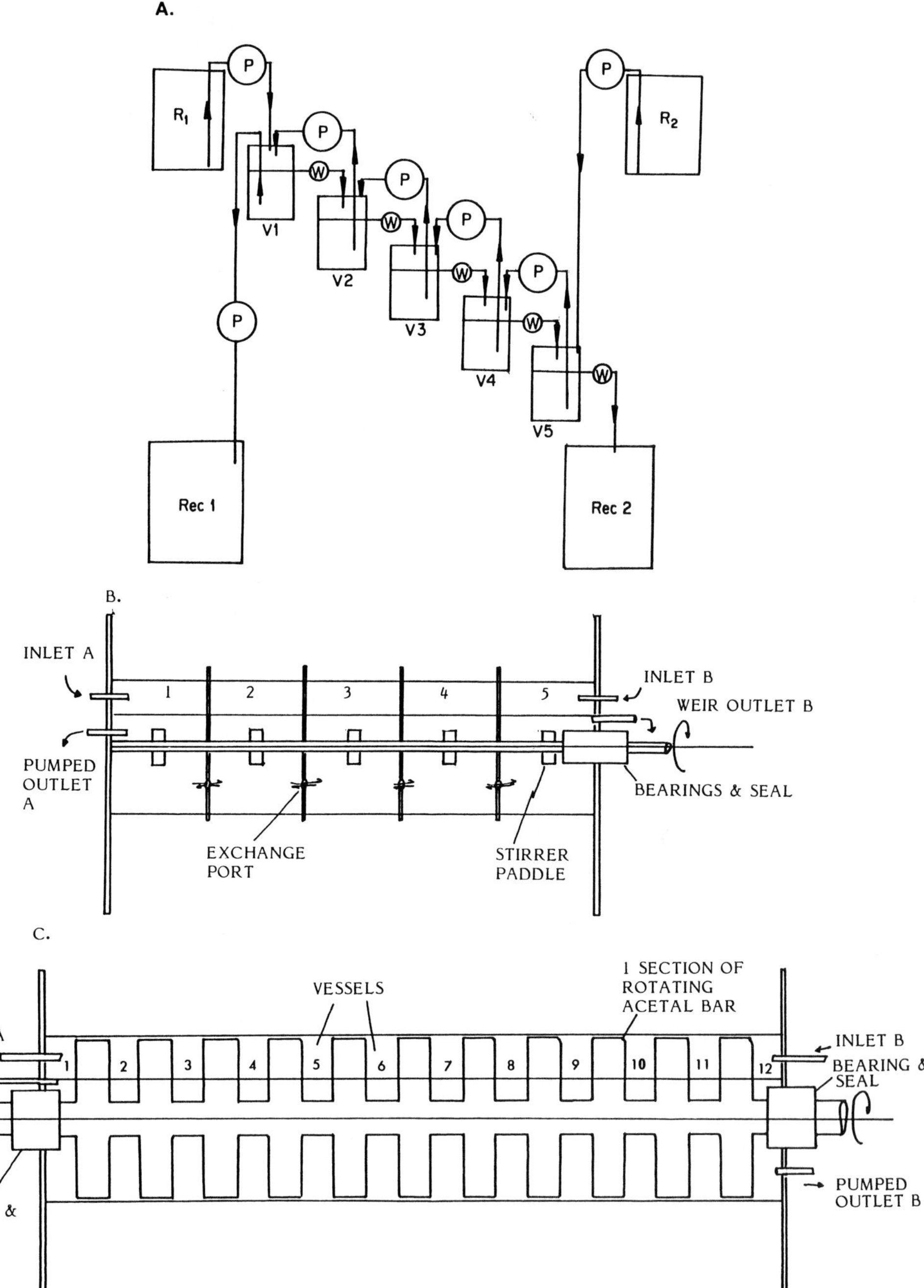

Figure 1. Diagrams of three practical gradostat designs. (A) The tubing-coupled gradostat. V1 through V5, Vessels; R1 and R2, medium reservoirs; Rec 1 and Rec 2, receivers; P, pumped lines; W, weir overflows. (B) The diffusion-coupled gradostat. Culture transfer is via exchange ports drilled into plates that separate each vessel. Vessels are numbered 1 through 5. Each vessel is stirred by paddles on a common shaft. (C) The shear-coupled gradostat. Vessels (numbered 1 to 12) consist of regions turned out of a common acetal plastic bar. The latter rotates inside a precision glass tube. Exchange between vessels is governed by the speed of rotation and the gap between the acetal bar and the wall of the vessel.

dient." This is, strictly speaking, a misnomer since concentration in each vessel is homogeneous. The stepped concentration distribution only approaches a true gradient as the number of linked vessels approaches infinity. The linear distribution from source to sink reflects the equilibrium distribution of solutes from sources to sinks in a classical diffusion system.

Retention time in the gradostat differs from that in both single- and multistage chemostats and is position dependent. Suppose a gradostat contains an initially homogeneously distributed concentration of a solute and that a flow of water is initiated from both reservoirs. Concentration falls most rapidly in the end vessels, through which fresh solvent enters and solution flows out. It has been shown by computer simulation techniques that organisms growing in opposing solute gradients can be retained in the center of the array growing at their μ_{max} even though they are being washed out of the system in the end vessels.

As in nature, solutes may be conservative if they are not altered in any way by the presence of cells, or they may be nonconservative if they are substrates or products of metabolism. Distribution of nonconservative solutes is controlled by the presence of cells that interact with them. If an organism is growing in the gradostat on stoichiometrically equivalent concentrations of two different substrates entering the array from different ends, then growth occurs in the center vessel. Growth here may be dual substrate limited. Now the center vessel becomes a sink for the two substrates and a source for cells. Under steady-state conditions, provided no other interactions occur, cells will be linearly distributed away from, and substrates towards, the center vessel.

Other simulations indicate that the actual position of maximum growth in a gradostat when essential substrates enter the system from opposing directions depends on relative concentrations and also on the growth yields of the cells for the substrate. In contrast, growth position is largely unaffected by either specific growth rate or substrate affinity.

The gradostat provides a number of separate habitats, one for each vessel, and ought therefore to allow the survival of different competing species of bacteria. The competitive exclusion principle suggests that where two organisms are competing for the same nutrient in a homogeneous environment, one or the other should be eliminated. This is normally borne out in actual continuous culture experiments. Jager and his colleagues have mathematically analyzed the behavior of the simplest two-vessel gradostat configuration and have proved that coexistence is possible if only under rather closely defined conditions (22, 40a). The same conclusions were reached by Fredrickson and Stephanopoulos (12), who analyzed a two-vessel multistage chemostat in which part of the output of the second vessel was fed back to the first.

Practical Experiments with Gradostats

Since the first experiments of Cooper and Copelands with bidirectional exchange in 1973, and since the description by Lovitt and Wimpenny of the gradostat in 1979, there have been a growing number of results obtained using such systems (Table 1). The main conclusions are that a wide range of experiments are possible using bidirectionally linked multistage culture systems. Using single species, physiological adaptation across spatial gradients is easy to investigate. Selection pressures can be designed into a gradostat profile, making this instrument a powerful tool for selection and enrichment. Interactions between organisms can readily be investigated in the gradostat. Thus the range of different ecological spaces possible can allow the coexistence of organisms with mutually exclusive habitat domains. As experiments with *Methylophilus methylotrophus* show, steady-state starvation and death can

Table 1.
Some Applications of Gradostats in Microbiology

Expt and organism(s)	Sources: Left hand	Sources: Right hand	Outcome	Reference
Pure-culture experiments				
Escherichia coli	Oxygen, nitrate	Glucose	Adaptive response to redox gradient	24
Paracoccus denitrificans	Nitrate	Succinate	Growth position correlated with predictive model	25
E. coli, then *Pseudomonas denitrificans*, then *Clostridium acetobutylicum*	Oxygen	Glucose	Facultative anaerobe replaced by both obligate species in turn	—[a]
Methylophilus methylotrophus	Methanol		Starvation led to death away from nutrient source	—[b]
Rhodopseudomonas capsulata and *Rhodopseudomonas marinalagilis*	NaCl		Species segregated according to salt sensitivity	—[c]
E. coli	Streptomycin		Selected streptomycin-resistant strains	—[b]
Desulfovibrio desulfuricans and *P. denitrificans*	Oxygen, sulfate	Lactate	Aerobe in vessel 1 and anaerobe in vessels 3 or 4, depending on sulfate concentration	—[d]
E. coli strains resistant to streptomycin or to tetracycline	Streptomycin	Tetracycline	Strong selection pressure for plasmid resistance transfer, which took place rapidly	—[e]
Natural culture experiments				
Estuarine community	Fresh water	NaCl	Estuarine community qualitatively but not quantitatively the same as natural system	10
Estuarine water	Fresh water	NaCl	Iron and phosphate behavior followed	4
Fresh- and saltwater inoculum	Fresh water	NaCl	Biodegradation of nitrilotriacetic acid followed	20
Oil storage tank inoculum	Nitrate	Lactate	Community selected containing SRB[f] and sulfide-oxidizing denitrifiers in different vessels	—[a]
Freshwater sediment inoculum	Oxygen	Glucose	SRB detected, also aerobic and facultative anaerobes	—[b]

[a] R. W. Lovitt, Ph.D. thesis, University of Wales, Cardiff, Wales, 1982.
[b] R. Earnshaw and J. W. T. Wimpenny, unpublished observations.
[c] J. W. T. Wimpenny, R. G. Earnshaw, H. Gest, J. M. Hayes, and J. L. Favinger, submitted for publication.
[d] H. Abdollahi, Ph.D. thesis, University of Wales, Cardiff, Wales, 1984.
[e] Z. Numan and J. W. T. Wimpenny, unpublished observations.
[f] SRB, Sulfate-reducing bacteria.

be observed in gradostats in vessels distant from nutrient sources.

Separate-Space Gradostats

In the standard gradostat configuration, cells and solutes are always inextricably mixed together. For some purposes, it would be interesting if different species could be restrained within each vessel while solutes were free to diffuse between them.

Such a system has been described by MacFarlane and Herbert (25). This system was constructed of three chambers separated by porous membranes, each chamber consisting of a chemostat with inputs and a simple overflow outlet. Solutes but not cells could diffuse between neighboring vessels. In one such system, *Clostridium butyricum*, *Desulfovibrio desulfuricans*, and *Chromatium vinosum* grew in vessels 1, 2, and 3, respectively. Glucose was fermented to a range of substrates, including ethanol, in the first vessel. The ethanol diffused into the center chamber and was oxidized to acetic acid. Sulfate was reduced to sulfide at the same time. Finally, in the third vessel the sulfide was photo-oxidized back to sulfate by the illuminated phototroph.

In this diffusion-linked system, interactions between pure cultures can be investigated without direct physical contact. In the gradostat, cells and solutes always transfer together. Neither model is a complete representation of a natural ecosystem, since conditions in the latter generally lie somewhere between the two extremes.

CLOSED GRADIENT MODEL SYSTEMS

The addition of gelling agents to culture media stabilizes them against bulk fluid movement, and movement of solute molecules is through molecular diffusion only. Cells in such media are generally immobilized, though this is not always true for motile organisms, especially with low concentrations of gelling agent. Such model systems allow the experimenter to examine growth pattern and position and interactions between organisms in spatially differentiated solute gradients.

Because cells and solutes are contained within the gels, the latter are closed systems behaving like batch rather than continuous cultures. Their virtues complement those of the gradostat. Thus they can have very high spatial resolution, whereas the gradostat lacks this. On the other hand, gel systems never reach a true steady state, unlike conditions in the gradostat.

Gel-diffusion models can be established in any suitable container; for example, large beakers, measuring cylinders, and test tubes have all been used in our laboratory. A simple example is a two-layer system, divided into a source zone containing the diffusible solute of interest and a growth region lacking the source component but containing a cell inoculum and all the other basal nutrients. The lower layer also contains full-strength agar, which forms a solid, resilient gel. Source and growth layers are normally but not necessarily equal in volume. The upper layer has the same basal nutrient profile as the source zone, but the agar concentration is reduced to interfere as little as possible with growth and solute diffusion. Semisolid gels can be formed with 0.4% (wt/vol) agar, for example.

Sampling the gels is easily accomplished using sterile cork borers. A gel slicer can then be used to produce sections down to about 1 mm thick. The slices may be used for chemical determinations or for the estimation of total and viable counts.

Growth pattern in the gels can be monitored by using a commercially available gel scanner. If scanners are to be used, there are advantages to growing the cultures in selected optically clear tubes. To follow growth, each tube should be marked so that the beam passes through the same part of the tube each time. In many modern computer-controlled gel scanners, base-line "noise" from the tube itself can be subtracted from the final sample.

Certain physicochemical parameters may be assessed if suitable electrodes are available. It is comparatively easy to buy or construct "needle" electrodes for pH, pO_2, redox potential, and sulfide activity. These are deployed using a micromanipulator.

Some Applications of Gel-Stabilized Gradient Systems

Growth in Opposing Gradients of Essential Solutes

Paracoccus denitrificans and *Pseudomonas pyocyanaea* were grown in gel-stabilized gradients at the expense of oxygen entering the system from the surface and of a range of substrates diffusing upwards from a source layer at the base of the tube (H. J. Ewers and J. W. T. Wimpenny, unpublished observations). Substrate concentration was varied in the source layer. Some of the results are shown in Fig. 2. The position of the growth band in this model depends on the inner substrate concentration; that is, the greater the concentration of substrate, the higher up the gel the band of growth appears. In each experiment the position of the band changes as might be predicted assuming a moving boundary solution to the reaction-diffusion equations governing growth.

In most of the experiments, single growth bands were observed, especially in the early phases of growth. In some cases, especially with *P. denitrificans*, multiple banding was seen. In almost all experiments, prolonged incubation led to multiple banding. The reason for multiple banding is not known at present. In old cultures especially, one should not ignore the possibility that cell lysis and regrowth occur. In addition, other physicochemical changes were probably taking place in the medium, in particular the accumulation of inhibitors. Banding phenomena have been reported before and will be discussed in a later section.

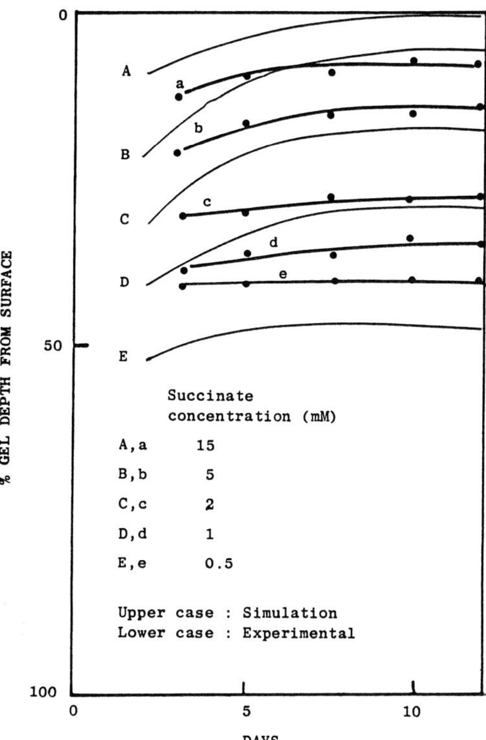

Figure 2. Growth of *Pseudomonas aeruginosa* in oxygen-succinate gradients, given as band position as a function of substrate concentration (experimental and computer simulation results). Total gel length, 8 cm; source zone, 4 cm; succinate concentration, from 0.5 to 15 mM. The following values were used in the simulation: $D_{\text{succinate}} = 0.029$ cm^2 h^{-1}; $D_{\text{oxygen}} = 0.072$ cm^2 h^{-1}; K_s succinate = 0.02 mM; K_s O_2 = 0.02 mM; $1/Y_{\text{succ}} = 0.0096$ mmol mg^{-1} dry cells; $1/Y_{O_2} = 0.03$ mmol mg^{-1} dry cells; $\mu_{\text{max}} = 0.43$ h^{-1}; initial cell concentration = 2 mg liter^{-1}; oxygen concentration at interface = 0.25 mM. Peak position was determined at different times up to 12 days.

The Oil Storage Tank Water Base

The storage of oil presents some problems directly attributable to microbial growth. There is almost always a water layer beneath the oil and adjacent to the steel base of the tank. Microbial growth at the expense of hydrocarbon fractions quickly removes much of the oxygen in this layer, allowing anaerobic bacteria, in particular sulfate-reducing species, to develop. These grow at the steel surface and cause sulfide corrosion which can, in bad cases, penetrate the steel tank bottom. The system was investigated in the laboratory, using a gel-stabilized model (50). The model was constructed in 250-ml beakers. The semisolid growth layer contained 0.4% (wt/vol) agar and mineral nutrients. An inoculum from the tank was also incorporated into this zone, which was poured over a steel plate placed in the base of the beaker. After this layer had set, a similar volume of gas oil was poured over it. Each beaker was incubated at room temperature in the dark for periods up to 90 days. The beakers were photographed, and pH, pO_2, and E_h gradients were measured with needle electrodes. Cores were removed for slicing followed by chemical analysis or viable-count determinations. Control gels lacked either cells or the steel plate. Only in the complete system were extensive growth and activity seen. Ferrous ions diffused up the gel and met oxygen diffusing downwards. The former were oxidized to a reddish-orange precipitate, forming a distinct band across the gel which separated aerobic from anaerobic zones. The anaerobic zone became black as sulfate-reducing bacteria oxidized the steel plate, producing iron sulfides. These bacteria appeared in highest numbers in the gel just below the iron oxide plate. Hexadecane-oxidizing bacteria were found in the upper aerobic zone. A large population of aerobic, facultative anaerobic, and anaerobic species were associated with the oxide plate. The system "organized" its available space in a beautiful series of patterns driven by the oxidation of hydrocarbons on the one hand and by anaerobes scavenging oxidation and metabolic products from the aerobes and hydrogen from the steel surface on the other.

Growth of Beggiatoa sp. in Counter-Gradients of Oxygen and Sulfide

Nelson and Jannasch (30) used gel-stabilized gradient systems to investigate the biology of several strains of *Beggiatoa* sp. The *Beggiatoa* group are seen by Nelson and his colleagues as "gradient" organisms that move by gliding motility to a point where opposing gradients of sulfide and oxygen meet (31). Here they proliferate as thin opaque disks or veils. Oxygen and sulfide form linear gradients which converge within the bacterial plate. The authors suggested that the species was microaerophilic and that the gel provided a microxic environment most suitable for its growth.

A Gel-Stabilized Sediment Model

R. A. Herbert and his colleagues have employed a gel-stabilized model to simulate a Tay estuary sediment ecosystem (26). Models were set up in sterile 1-liter Quickfit reaction vessels. A volume of 100 ml of sterile basal salts medium with added ammonium sulfate, cycloheximide, and agar was poured into each container. Once this had set, 100 ml of well-mixed material from the top 5 cm of a Kingoodie Bay sediment was added to the system. Finally, 600 ml of basal salts medium made semisolid with 0.3% agar was poured carefully over the sediment and allowed to set. The carbon source for these systems came from the sediment itself. pH, oxygen, E_h, NO_2^-, NO_3^-, and NH_4^+ profiles were measured at regular intervals, and samples were removed to determine population densities of nitrifying, nitrate-respiring, and sulfate-reducing bacteria.

Results of this experiment were in good agreement with actual measurements in Kingoodie Bay sediments themselves. The system was judged to be a useful laboratory model which gave reproducible results and seemed suitable for further development. MacFarlane et al. pointed out that the only major difference between the natural habitat and the laboratory model was in the lack of bioturbation and in the absence of gas or solute exchange in the closed model when compared with the real sediment (26).

Effects of Pollutants on Sediment Communities

Morgan and Watkinson (P. Morgan and J. Watkinson, 108th Meeting of the Soc. Gen. Microbiol., St. Andrews, Scotland, 1987) used a gel-stabilized model to examine the effects of 3-nitrophenol and 1:1 hexadecane-naphthalene on the biochemistry and microbiology of freshwater and marine sediment communities. Oxygen, pH, and ion gradients and populations of sulfate-reducing, nitrogen-recycling, xenobiotic-metabolizing, and heterotrophic bacteria were all assessed in these systems.

Periodic Growth Bands in Gel-Stabilized Systems

Williams (46–49) first reported multiple growth bands in gel-stabilized microbial cultures in 1938–1939. Such structures have also been noted in agar-stabilized streptomycete cultures by Nitsch and Kutzner (32) and in cultures of nitrogen-fixing azotobacters (42).

We first noticed periodic growth while developing gel systems as ecological models. A range of bacteria were examined, and it was concluded that facultatively anaerobic and certain obligately aerobic bacteria generated a pattern consisting of more than one growth band, but that most strict aerobes did not.

B. cereus grown in a basal Casamino Acids-containing medium in opposing glucose-oxygen counter gradients produced bands most prolifically, and work focused on this organism (9, 52, 54). Band formation was only apparent when oxygen was present and when there was an actual gradient of glucose in the gel. Motility was unimportant since band formation was still present at agar concentrations in the upper layer at which motility was almost certainly impossible. On incubation, growth in the lower parts of the gel led to major changes in pH, which quickly fell to around 5.1. Aerobic growth near the surface of the gel led to alkaline pH values around 7.8 to 8.0. It was assumed that this was due to the aerobic oxidation of amino acids leading to the liberation of free ammonia.

pH gradients were seen as one trigger mechanism for band formation. Thus, the higher the buffering capacity of the growth zone, the shallower the pH gradient, the more growth took place, and the deeper into the gel were the bands. Further evidence was provided in experiments in which the addition of an alkali-containing layer above the growth zone also led to band formation even when oxygen was absent.

Computer Simulations of Growth in Diffusion Systems

"Simple" Growth in Solute Gradients

Microbial growth in opposing essential solute gradients has been investigated by computer simulation based on a simple mathematical model which assumes that growth rates are determined by Monod saturation kinetics. It is also assumed that solutes are translocated by Fickian diffusion. Both the models and the computer programs used to solve them numerically were produced by S. Jaffe. The equations for the model are indicated below. In addition, the values of each parameter used in the model are described in Table 2.

For two substrate simulations:

$$\frac{\delta S_1}{\delta t} = D_{S_1}\frac{\delta^2 S_1}{\delta x^2} - \frac{1}{Y_{S_1}} \cdot C \cdot \mu_{max} \cdot F(S_1, S_2)$$

$$\frac{\delta S_2}{\delta t} = D_{S_2}\frac{\delta^2 S_2}{\delta x^2} - \frac{1}{Y_{S_2}} \cdot C \cdot \mu_{max} \cdot F(S_1, S_2)$$

$$\frac{\delta C}{\delta t} = D_C\frac{\delta^2 C}{\delta x^2} - C \cdot \mu_{max} \cdot F(S_1, S_2)$$

The growth function $F(S_1, S_2)$ is the product of two Monod equations:

$$F(S_1, S_2) = \frac{S_1}{[K_{s(S_1)} + S_1]} \cdot \frac{S_2}{[K_{s(S_2)} + S_2]}$$

Where a single substrate is also an inhibitor:

$$\frac{\delta S}{\delta t} = D_S\frac{\delta^2 S}{\delta x^2} - \frac{1}{Y_S} \cdot C \cdot \mu_{max} \cdot F(S_1)$$

Table 2.
Assumptions and Parameters of the Simple Growth in Solute Gradients Model

Parameter	Assumption or value
Boundary conditions.....	Left-hand boundary impermeable to solutes and cells Right-hand boundary impermeable to S_1 and cells Right-hand boundary concentration of S_2 fixed at 0.25 mM
System dimensions	
Source zone	30 mm
Growth zone	30 mm
Initial concentration	
Source zone	S_1 = 1 mM; S_2 = 0 mM; cells = 0 mg liter^{-1}
Growth zone	S_1 = 0 mM; S_2 = 0 mM; cells = 10 mg liter^{-1}
Diffusion coefficients	
Cells	0 cm^2 h^{-1}
S_1	0.024 cm^2 h^{-1}
S_2	0.072 cm^2 h^{-1}
Cell parameters	
μ_{max}	0.2 h^{-1}
$1/Y_{S_1}$	0.02 mmol mg^{-1} dry cells
$1/Y_{S_2}$	0.02 mmol mg^{-1} dry cells
$K_{s(S_1)}$	0.02 mM
$K_{s(S_2)}$	0.02 mM

The growth function is

$$F(S) = \frac{S}{(K_s + S)} \cdot \frac{K_i}{(K_i + S)}$$

In these models S_1, S_2, and S are substrate concentrations; C is the cell concentration; Y_{S_1}, Y_{S_2}, and Y_S are yield coefficients; $K_{s(S_1)}$, $K_{s(S_2)}$, and K_s are substrate affinity coefficients; K_i is an inhibition constant; μ_{max} is specific growth rate; and x and t represent distance and time, respectively.

Parameters to be varied concern, first, the cell (maximum specific growth rate, affinity for substrate, and cell yield) and, second, external variables, including diffusivity and solute concentration. It has been assumed that a substrate such as glucose is diffusing upwards from a source region while a second substrate, assumed to be oxygen, is diffusing from a constant source at the surface of the gel system.

Each of the key parameters has been varied, and results showing position of the growth peak at one particular time interval are plotted against values for the parameter in Fig. 3. Results show that critical cell parameters like substrate affinity and growth rate (not shown) are unimportant space determinants. Critical to growth position are diffusion coefficient, substrate concentrations, and yield coefficient. It may be surprising that yield coefficient is a key space determinant; however, there is a close relationship between **substrate concentration** and **cell yield.** Thus, if yield for one of the two limiting substrates is very low, balanced growth can only occur nearer its source, just as if under normal yield conditions the substrate were present in low concentrations. These simulations emphasize the importance of environmental factors in determining growth position, especially when the population is growing under diffusion-limited conditions.

Another major space determinant is the presence of a diffusible growth inhibitor. Figure 4 shows the results of a simulation in

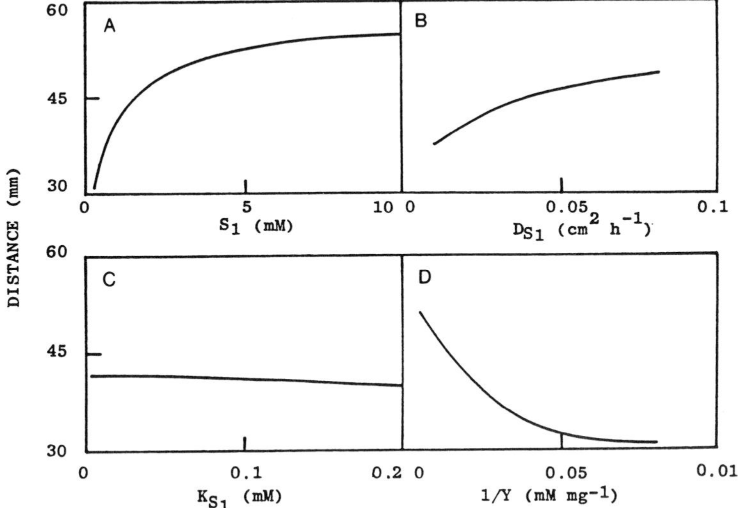

Figure 3. Simulation of microbial growth in opposing solute gradients. The following parameters were varied: (A) the concentration of internal substrate, S_1; (B) diffusion coefficient for S_1; (C) affinity constant of the cell for S_1; (D) reciprocal growth yield for the cell on S_1. The simulation was for 100 h of growth, and the line indicates the position of the growth peak. Other parameters as in Table 2.

which growth is at the expense of a single substrate diffusing from one direction only. It is assumed that the substrate is also a growth inhibitor. It is obvious that position in this case is determined by the inhibition constants for the substrate: the higher the latter, the more distant the peak of growth from the source.

So far these simulations consider only a single organism. A natural question concerns competition between bacteria. Do competition kinetics which apply in homogeneous environments also apply in spatially hetero-

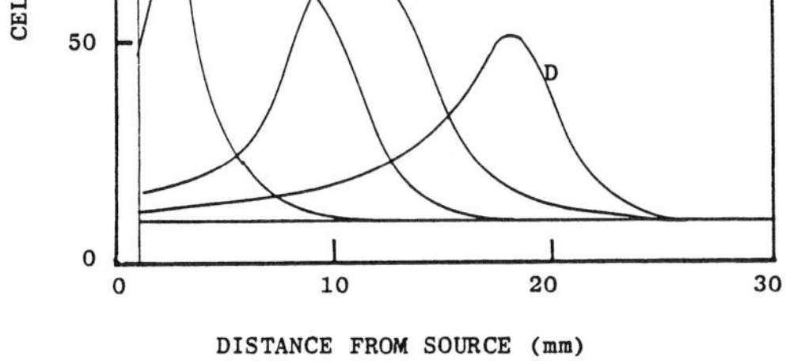

Figure 4. Inhibition as a determinant of position. In this simulation cell growth is dependent on a single solute diffusing from the left-hand boundary. This solute is also a noncompetitive growth inhibitor. In each simulation the affinity constant (K_s) is 0.02 mM. The inhibition constant (K_i) is varied. Other parameters as in Table 2.

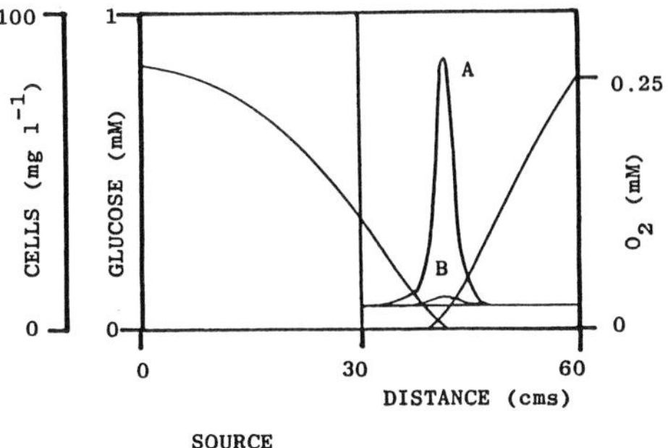

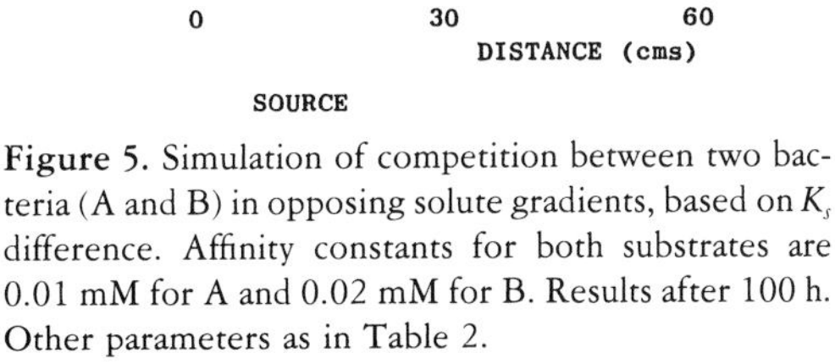

Figure 5. Simulation of competition between two bacteria (A and B) in opposing solute gradients, based on K_s difference. Affinity constants for both substrates are 0.01 mM for A and 0.02 mM for B. Results after 100 h. Other parameters as in Table 2.

geneous systems? Figure 5 confirms that they do. Organisms with lower substrate affinities are outcompeted at the growth position, as would be predicted.

Periodic Growth in Solute Gradients

The mechanism for band formation in cultures of *B. cereus* growing on an amino acids-containing medium in opposing oxygen-glucose gradients was investigated by S. Jaffe using the mathematical models. No combination of Monod growth kinetics on their own or coupled with inhibition, lag, death, or any other obvious growth function ever showed the sort of patterns seen in the experiments. Periodic behavior could be simulated as follows. Cells were assumed to exist in two states, active and inactive. Activation of inactive cells was assumed to be asymmetric with inactivation. Thus, growth took place until conditions altered far enough to lead to its cessation. On the other hand, growth could only start again at some different value for the critical parameter. The latter might be concentration of a critical solute, or it could be interpreted as time. In this case there would be an asymmetry between the time taken for growth to cease and the time it takes to start growing once more. This interpretation (52) is broadly in line with that suggested by Hoppensteadt et al. (18, 19), who investigated circular periodicities produced by *Escherichia coli* histidine auxotrophs if grown as surface cultures on a glucose-buffered salts medium when a small amount of concentrated histidine was allowed to diffuse outwards from the center of the plate.

Periodic structures like those discussed here add yet more complexity to manifestations of microbial growth in spatially heterogeneous systems. Such structures have been seen in marine sediments (34), so that they ought not to be overlooked when discussing the ecology of microorganisms.

MICROBIAL FILM FERMENTORS

Microbial film forms on any surface which is exposed to water and appropriate nutrients (7, 15). Laboratory model film fermentors have been used by workers in a number of different fields. The Robbins device (28) has been employed by different workers in aquatic environments and was discussed in detail by Hamilton (15). It consists of a spool tube section incorporated into a recirculation loop fed with nutrients and a bacterial inoculum. Film forms on replaceable studs or on removable sections of the tube. Costerton and Lashen (11) have used this device to investigate biocide effects primarily in oil field situations. Fundamental investigations into biofilm physiology by Hamilton and his colleagues have used modified forms of the Robbins device (e.g., reference 14).

Characklis and his colleagues used an annular reactor system incorporating removable glass sections. This device subjects microbial film to controlled high-shear fields and has been used to investigate the relationship between bacterial adsorption, polysaccharide production, and growth rate (29, 36). More recently, Characklis (W. G. Characklis, *in* R. Nagabhushanam and R. D. Turner, ed., *Marine Biodeterioration*, in press) has used rectangular capillary reaction vessels with microscopy and image analysis to monitor all stages of adsorption, growth, detachment, cell separation, and desorption during early film formation.

The Cardiff Constant-Depth Film Fermentor

Our aim was to develop an enclosed system capable of aseptic operation under any gassing or nutrient regime. The upper surface of the film was to be scraped to maintain the mature film at a constant depth so that (i) the system would operate under quasi-steady-state conditions, (ii) microelectrode experiments could be carried out with confidence because the film depth was known, and (iii) experiments would have a high degree of replicability. While such a model is an abstraction of real systems, it should be capable of generating almost any type of microbial film and retain at least some of the essential characters of "filmness."

The film fermentor was used initially to investigate the growth and structure of dental plaque (8). It was further developed by Peters and Wimpenny (A. C. Peters and J. W. T. Wimpenny, *Biotechnol. Bioeng.*, in press). Fifteen separate film pans are located in a PTFE (polytetrafluoroethylene) disk which rotates beneath a PTFE scraper bar. Each film pan is drilled with five or six holes in which can be located plugs on which the film is grown. Either 75 or 90 separate film plugs are therefore present in the fermentor at the start of each run. Substrate plugs are recessed into their holes with a steel template. Film grows within the space between the top of the plug and the scraper bar. Since each film plug is inserted separately, it follows that the use of a range of templates could allow films of different thicknesses to be formed in a single fermentor run. The film pans are irrigated with sterile medium, and there are inlets and outlets for medium and gas mixtures. The system can be sampled aseptically during a run by removing individual pans through the sampling port with a stainless-steel tool.

This version of the film fermentor was used to investigate film produced by a river water consortium. Under nutrient-limiting conditions the film doubled in mass every 60 h until a steady-state protein level was reached after about 20 days. The films could be removed from the fermentor, and oxygen gradients were measured within them by using microelectrodes with an appropriate positioning device. Films were also fixed and examined using scanning and transmission electron microscopy. This fermentor has also been used to grow dental plaque film at higher temperatures (37°C). A plaque community has been isolated and its growth rate has been monitored. In addition, other substratum surfaces including amalgam, gold, and acrylic, all relevant to the dental profession, have been used instead of PTFE.

MULTIDIMENSIONAL GRADIENT SYSTEMS

A knowledge of the relationship of an organism to all the physicochemical variables that affect its survival and growth is essential to an understanding of its place in microbial ecosystems. Any techniques that can help clarify these relationships deserve consideration. Among a number of possible candidates are the plate diffusion methods pioneered by Szybalski and Bryson (40) but applied later to the effects of pH (37, 38) and to metal ion activity (45).

Mapping microbial responses to two environmental variables at once was first reported by Baas-Becking and Ferguson-Wood (2, 3), who recorded responses of estuarine bacteria to pH and redox potential. A system was devised to grow photosynthetic algae in two dimensions of light and temperature (13, 43), and Caldwell et al. used steady-state two-dimensional gradient plates to investigate microbial growth (5, 6). We have developed two-dimensional gradient plates based on the Szybalski wedge plate technique.

Two-dimensional pH-versus-NaCl gradient plates were constructed in 10-cm-square disposable plastic petri dishes. After wedges for the pH gradient are poured, the plate is rotated through 90° and the second (salt) gradient is formed. After equilibration an inoculum is spread or poured as an agar overlay over the surface of the plate, and the plates are incubated appropriately until growth patterns are formed. Values for pH and NaCl gradients, temporal changes, reproducibility of the technique, and its application to a group of aerobic heterotrophic bacteria have been reported (43, 51, 55).

If a number of two-dimensional plates are incubated at different temperatures, the growth profiles for each plate can be stacked on top of one another to generate a solid object which represents the habitat domain for three factors. Similarly, two-dimensional plates can be constructed in which all layers contain a different uniform concentration of a solute. Both approaches could be taken together so that four factors could be varied simultaneously, including temperature and solute concentration as well as the two dimensions established in the original plates.

Three-dimensional experiments were carried out for a number of strains of two species of bacteria, *Serratia marcescens* and *Micrococcus luteus* (56). Both organisms were maintained on nutrient agar slopes and grown in pH-NaCl gradients in a Casamino Acids-containing medium as described earlier. In this experiment the third variable was incubation temperature, which ranged from 5 to 35°C.

Eight strains of *S. marcescens* and two of *M. luteus* were investigated in this way. The two micrococci were easy to distinguish by their sensitivity to NaCl. While the two micrococci showed similar temperature responses, these were quite different from those of the *Serratia* strains. What was clear was the significant difference among most of the strains. Such data have been confirmed in replicate experiments, suggesting that these multidimensional response systems can have taxonomic value.

A representative strain of *S. marcescens* was grown on two-dimensional salt-pH gradient plates containing a range of six different nitrate concentrations. Six identical sets of these plates were incubated at different temperatures. Nitrate concentration inhibits both growth of this organism and, to a lesser extent, its ability to produce prodigiosin.

The experiments outlined here suggest that gradient plates constitute a powerful tool in discriminating between closely related species or between strains of the same organism, and imply a value in identification and taxonomy. The gradient-plate technique has applications in microbial ecology. For instance, the microflora of one habitat can be easily compared with that from another on such plates. Competition studies using mixed populations are possible. Preliminary results indicate that certain pairs of bacteria can be differentiated on gradient plates; however, further, more detailed experimental work is needed here, especially in discriminating two organisms from confluent mixtures on the plates. There are applications on the theoretical side of microbial ecology, too. Hutchinson (21) has suggested that the niche of an organism consists of an *n*-dimensional hyperspace of environmental factors and cellular responses to such factors. It has not always been easy to visualize an *n*-dimensional hyperspace; however, the techniques outlined here suggest ways of realizing at least four dimensions in a simple practical experiment.

POSTSCRIPT

None of the systems described here has been applied to microbial mat ecosystems, and it is a trifle presumptuous to suggest experiments that could be done. However, of all the models mentioned it seems likely that gel-stabilized gradient systems could be helpful here. They offer the possibility of reconstructing a stable gradient system which could be illuminated by an appropriate light source following any preprogrammed light regime. Such systems could contain a small number of known pure cultures of organisms which might be predicted to constitute a minimal mat community. The establishment of a number of identical culture systems would allow them to be sacrificed at different times so that changes in physical and chemical conditions could be related to changes in the microbiology of the community. Knowing some of the kinetic parameters of the pure cultures could then allow numerical models of the mat community to be formulated and tested. The value of such models can only be determined by experiment!

Acknowledgments. I would like gratefully to acknowledge the collaboration and help given by all the following over the course of this work: Bob Lovitt, Philip Coombs, Steve Jaffe, Hamid Abdollahi, Paul Waters, Adrian Peters, and Richard Earnshaw.

LITERATURE CITED

1. **Abson, J. W., and K. H. Todhunter.** 1961. Plant for continuous biological treatment of carbonization effluents. *Soc. Chem. Ind. Monogr.* **12:**147–164.
2. **Baas-Becking, L. G. M., and E. J. Ferguson-Wood.** 1955. Biological processes in the estuarine environment. 1. Ecology of the sulphur cycle. *K. Ned. Akad. Wet. B* **58:**160–181.
3. **Baas-Becking, L. G. M., E. J. Ferguson-Wood, and I. R. Kaplan.** 1956. Biological processes in the estuarine environment. VIII. Iron bacteria as gradient organisms. *K. Ned. Akad. Wet. B* **59:** 398–407.
4. **Bale, A. J., and A. W. Morris.** 1981. Laboratory simulation of chemical processes induced by estuarine mixing: the behaviour of iron and phosphate in estuaries. *Estuarine Coastal Shelf Sci.* **13:**1–10.
5. **Caldwell, D. E., and P. Hirsch.** 1973. Growth of microorganisms in two-dimensional steady-state diffusion gradients. *Can. J. Microbiol.* **19:**53–58.
6. **Caldwell, D. E., S. H. Lai, and J. M. Tiedje.** 1973. A two-dimensional steady-state diffusion gradient for ecological studies. *Bull. Ecol. Res. Comm.* (Stockholm) **17:**151–158.
7. **Characklis, W. G.** 1981. Fouling biofilm development: a process analysis. *Biotechnol. Bioeng.* **23:** 1923–1960.
8. **Coombe, R. A., A. Tatevossian, and J. W. T. Wimpenny.** 1982. Bacterial thin films as *in vitro* models for dental plaque, p. 239–249. *In* R. M. Frank and S. A. Teach (ed.), *Surface and Colloid Phenomena in the Oral Cavity: Methodological Aspects.* IRL Press, London.
9. **Coombs, J. P., and J. W. T. Wimpenny.** 1982. Growth of *Bacillus cereus* in a gel-stabilised nutrient gradient system. *J. Gen. Microbiol.* **128:**3093–3101.
10. **Cooper, D. C., and B. J. Copeland.** 1973. Responses of continuous-series estuarine microecosystems to point-source input variations. *Ecol. Monogr.* **43:**213–236.
11. **Costerton, J. W., and E. S. Lashen.** 1984. The inherent biocide resistance of corrosion-causing biofilm bacteria. *Mater. Perform.* **23(2):**13–16.
12. **Fredrickson, A. G., and G. Stephanopoulos.** 1981. Microbial competition. *Science* **213:**972–984.
13. **Halldal, P., and C. S. French.** 1958. Algal growth in crossed gradients of light intensity and temperature. *Plant Physiol.* **33:**249–252.
14. **Hamilton, W. A.** 1985. Sulfate-reducing bacteria and anaerobic corrosion. *Annu. Rev. Microbiol.* **39:**195–217.
15. **Hamilton, W. A.** 1987. Biofilms: microbial interactions and metabolic activities. *Symp. Soc. Gen. Microbiol.* **41:**361–385.
16. **Hawkes, H. A.** 1977. Eutrophication of rivers, effects, causes and control, p. 159–192. *In* A. G. Callely, C. F. Forster, and D. A. Stafford (ed.), *Treatment of Industrial Effluents.* Hodder & Stoughton, London.
17. **Herbert, D.** 1961. A theoretical analysis of continuous culture systems. *Soc. Chem. Ind. Monogr.* **12:**21–53.
18. **Hoppensteadt, F. C., and W. Jager.** 1979. Pattern formation by bacteria. *Lect. Notes Biomath.* **38:**68–81.
19. **Hoppensteadt, F. C., W. Jager, and C. Poppe.** 1984. A hysteresis mode for bacterial growth patterns. *Lect. Notes Biomath.* **55:**123–134.
20. **Hunter, M., T. Stephenson, P. W. W. Kirk, R. Perry, and J. N. Lester.** 1986. Effect of salinity gradients and heterotrophic microbial activity on biodegradation of nitrilotriacetic acid in laboratory simulations of the estuarine environment. *Appl. Environ. Microbiol.* **51:**919–925.
21. **Hutchinson, G. E.** 1965. *The Ecological Theater and the Evolutionary Play*, p. 26–78. Yale University Press, New Haven, Conn.
22. **Jäger, W. H., J. W.-H. So, B. Tang, and P. Waltman.** 1987. Competition in the gradostat. *J. Math. Biol.* **25:**23–42.
23. **Lovitt, R. W., and J. W. T. Wimpenny.** 1981. Physiological behaviour of *Escherichia coli* grown in opposing gradients of glucose and oxygen plus nitrate in the gradostat. *J. Gen. Microbiol.* **127:**269–276.
24. **Lovitt, R. W., and J. W. T. Wimpenny.** 1981. The gradostat, a bidirectional compound chemostat, and its application in microbiological research. *J. Gen. Microbiol.* **127:**261–268.
25. **MacFarlane, G. T., and R. A. Herbert.** 1985. The use of compound continuous flow diffusion chemostats to study the interaction between nitrifying and nitrate-reducing bacteria. *FEMS Microbiol. Ecol.* **31:**249–254.

26. **MacFarlane, G. T., M. A. Russ, S. M. Keith, and R. A. Herbert.** 1984. Simulation of microbial processes in estuarine sediments using stabilised systems. *J. Gen. Microbiol.* **130**:2927–2933.
27. **Margalef, R.** 1967. Laboratory analogues of estuarine plankton systems, p. 515–524. *In* G. H. Lauf (ed.), *Estuaries: Ecology and Populations.* Horn-Shafer, Baltimore.
28. **McCoy, W. F., J. D. Bryers, J. Robbins, and J. W. Costerton.** 1981. Observations on fouling biofilm formation. *Can. J. Microbiol.* **27**:910–917.
29. **Nelson, C. H., J. A. Robinson, and W. G. Characklis.** 1985. Bacterial adsorption to smooth surfaces: rate, extent, and spatial pattern. *Biotechnol. Bioeng.* **27**:1662–1667.
30. **Nelson, D. C., and H. W. Jannasch.** 1983. Chemoautotrophic growth of marine *Beggiatoa* in sulphide-gradient cultures. *Arch. Microbiol.* **136**: 262–269.
31. **Nelson, D. C., B. B. Jørgensen, and N. P. Revsbech.** 1986. Growth pattern and yield of a chemoautotrophic *Beggiatoa* sp. in oxygen-sulfide microgradients. *Appl. Environ. Microbiol.* **52**:225–233.
32. **Nitsch, B., and H. J. Kutzner.** 1973. Wachstum von Streptomycetin in Schuttelagarkultur: eine neue Methode zur Festellung des c-Quellen-Spektrums. *Symp Tech. Mikrobiol.* (Berlin) **3**:481–486.
33. **Parkes, R. J.** 1980. Methods of enrichment isolation and analysis in laboratory systems, p. 173–189. *In* A. T. Bull and J. M. Slater (ed.), *Microbial Interactions and Communities.* Academic Press, Inc., New York.
34. **Perfil'ev, B. V., and D. R. Gabe.** 1969. *Capillary Methods of Investigating Microorganisms.* Oliver & Boyd, Edinburgh. (English translation.)
35. **Pritchard, P. M., R. M. Ventullo, and J. M. Suflita.** 1976. The microbial degradation of diesel oil in a multistage continuous culture system. *In* J. M. Sharply and A. M. Kaplan (ed.), *Proceedings of the 3rd International Biodegradation Symposium.* Applied Science Publishers, London.
36. **Robinson, J. A., M. G. Trulear, and W. G. Characklis.** 1984. Cellular reproduction and extracellular polymer formation by *Pseudomonas aeruginosa* in continuous culture. *Biotechnol. Bioeng.* **26**:1409–1417.
37. **Sacks, L. E.** 1956. A pH gradient plate. *Nature* (London) **178**:269.
38. **Sacks, L. E., A. D. King, Jr., and J. E. Schade.** 1986. A note on pH gradient plates for fungal growth studies. *J. Appl. Bacteriol.* **61**:235–238.
39. **Slezak, J., and B. Sikyta.** 1961. Continuous cultivation of microorganisms: a new approach and possibilities for its use. *J. Biochem. Technol. Eng.* **3**:357.
40. **Szybalski, W., and V. Bryson.** 1953. Genetic studies on microbial cross-resistance to toxic agents. I. Cross-resistance of *Escherichia coli* to fifteen antibiotics. *J. Bacteriol.* **64**:489–499.
40a. **Tang, B.** 1986. Mathematical investigation of growth of microorganisms in the gradostat. *J. Math. Biol.* **23**:319–339.
41. **Thompson, L. A., D. B. Nedwell, M. T. Balba, I. M. Banat, and E. Senior.** 1983. The use of multiple-vessel, open flow systems to investigate carbon flow in anaerobic microbial communities. *Microb. Ecol.* **9**:189–199.
42. **Tschapek, M., and N. Giambiagi.** 1954. The formation of Liesegang rings by *Azotobacter* under oxygen inhibition./Die bildung von Liesegang'schen Ringen durch Azotobakter bei O_2-Hemmung. *Kolloid Z.* **135**:47–48.
43. **Van Baalen, C., and P. Edwards.** 1973. Light-temperature gradient plate, p. 267–273. *In* J. Stern (ed.), *Handbook of Phycological Methods.* Cambridge University Press, Cambridge.
44. **Waters, P., and D. Lloyd.** 1985. Salt, pH and temperature dependencies of growth and bioluminescence of three species of luminous bacteria analysed on gradient plates. *J. Gen. Microbiol.* **131**:2865–2869.
45. **Weinberg, E. D.** 1957. Double-gradient agar plates. *Science* **125**:196.
46. **Williams, J. W.** 1938. Bacterial growth "spectrum" analysis. I. Methods and application. *Am. J. Med. Technol.* **4**:58–61.
47. **Williams, J. W.** 1938. Bacterial growth "spectrums." II. Their significance in pathology and bacteriology. *Am. J. Med. Technol.* **14**:642–645.
48. **Williams, J. W.** 1939. Growth of microorganisms in shake cultures under increased oxygen and carbon dioxide tensions. *Growth* **3**:21–33.
49. **Williams, J. W.** 1939. The nature of gel mediums as determined by various gas tensions and its importance in growth of microorganisms and cellular metabolism. *Growth* **3**:181–196.
50. **Wimpenny, J. W. T., J. P. Coombs, R. W. Lovitt, and S. G. Whittaker.** 1981. A gel-stabilised model ecosystem for investigating microbial growth in spatially ordered solute gradients. *J. Gen. Microbiol.* **127**:277–287.
51. **Wimpenny, J. W. T., H. Gest, and J. L. Favinger.** 1986. The use of two-dimensional gradient plates in determining the responses of non-sulphur purple bacteria to pH and NaCl concentration. *FEMS Microbiol. Lett.* **37**:367–371.
52. **Wimpenny, J. W. T., S. Jaffe, and J. P. Coombs.** 1984. Periodic growth phenomena in spatially organized microbial systems. *Lect. Notes Biomath.* **55**:388–405.
53. **Wimpenny, J. W. T., and R. W. Lovitt.** 1984. The investigation and analysis of heterogeneous environments using the gradostat, p. 295–312. *In*

J. M. Grainger and J. M. Lynch (ed.), *Microbiological Methods for Environmental Biotechnology.* SAB Technical Series no. 19. Academic Press, Inc. (Ltd.), London.

54. **Wimpenny, J. W. T., R. W. Lovitt, and J. P. Coombs.** 1983. Laboratory model systems for the investigation of spatially and temporally organised microbial ecosystems. *Symp. Soc. Gen. Microbiol.* **34**:67–117.

55. **Wimpenny, J. W. T., and P. Waters.** 1984. Growth of micro-organisms in gel-stabilized two-dimensional diffusion gradient systems. *J. Gen. Microbiol.* **130**:2921–2926.

56. **Wimpenny, J. W. T., and P. Waters.** 1987. The use of gel-stabilized gradient plates to map the responses of microorganisms to three or four environmental factors varied simultaneously. *FEMS Microbiol. Lett.* **40**:263–267.

V. Evolution of Mat-Forming Photosynthetic Procaryotes

Chapter 34

Efficiency of Biosolar Energy Conversion by Aquatic Photosynthetic Organisms

M. Avron

The process of photosynthesis, which determines the productivity of aquatic photosynthetic organisms, is intrinsically a highly efficient process. This was amply demonstrated by the observation that the photochemical reactions of photosynthesis can operate at close to the theoretical efficiency, delivering one electron for each quantum of light absorbed, and that the overall photosynthetic process also requires no more than 8 to 10 quanta per O_2 molecule evolved in the process (6). Nevertheless, it is the common experience of all investigators who have measured the energetic efficiency of photosynthesizing higher plants and algae in the field that no more than 1 to 3% of the energy available in the solar irradiation impinging upon the photosynthetic material can be recovered in the synthesized organic material (1, 3).

It is the purpose of this discussion to briefly outline the factors which account for the limitation of long-term energy productivity by intensive algal photosynthetic systems to no more than about 3% of the energy available in the impinging solar light. This translates to a productivity of about 25 g of organic matter per m^2 per day (4) (Table 1).

Light impinging upon an aquatic algal growth pond is only partially available for absorption. Around 10% of this light is reflected or scattered and thus does not reach the photosynthesizing organisms. Of the light which reaches the algae, only about 50%, at most, can be absorbed by them. The other 50% is present at wavelengths which are not within the absorption range of green photosynthetic organisms (i.e., not within the range of 400 to 700 nm). This already assumes that the pond absorbs all of the photosynthetically active radiation, even in wavelength regions (ca. 500 nm) at which absorption by green algae and plants is relatively low.

Photosynthetic efficiency considerations dictate that only about 25% of the absorbed light can be converted into energy stored within organic products. Thus, if eight quanta, with an average energy content of 55 kcal (230 kJ) per mole-quanta (at ca. 500 nm), are necessary to fix one molecule of CO_2 to the level of a carbohydrate, 440 kcal (1,841 kJ) of light energy is invested in this fixation. The energy available within one-sixth of a glucose molecule is about 110 kcal (460 kJ)/mol of C. Thus, the energy effi-

M. Avron • Department of Biochemistry, Weizmann Institute of Science, Rehovot 76100, Israel.

Table 1.
Factors Limiting Maximal Long-Term Productivity in Algal Ponds[a]

Factor	Single-factor effect (%)	Cumulative effect (%)
Solar light		100
Scattering and reflecting properties of surface	10	90
Absorption spectrum (depth of culture)	50	45
Photosynthetic efficiency (25%)	75	11.3
Light saturation (7–95%)	60	4.5
Respiration, photorespiration, excretion	5	4.3
Photoinhibition	10	3.8
Temperature	20	3.1

[a] Mean daily intensity (30° latitude), 4,000 kcal (16,736 kJ)/m^2 per day; energy productivity at 3% efficiency, 120 kcal (502 kJ)/m^2 per day; algal biomass productivity (5 kcal/g [21 kJ/g]), 24 g/m^2 per day.

ciency of the photosynthetic process cannot exceed 110/440, or 25%, of the energy of the absorbed quanta. This already ignores the energy requirements for nitrogen fixation and other essential biological elements of growth.

However, when dealing with photosynthetic productivity under natural outdoor conditions, we cannot stop here. Several major additional considerations contribute to the low energetic productivity of such systems. A major factor is the fact that such systems are, for a large fraction of the day, exposed to a light intensity which greatly exceeds the intensity required to saturate the photosynthetic apparatus. Thus, most photosynthetic organisms are light saturated at light intensities below 0.1 cal (0.42 J)/cm^2 per min, whereas solar energy exceeds 1.0 cal (4.2 J)/cm^2 per min for several hours per day on sunny days in temperate zones. It was estimated that light supersaturation leads to an average loss of about 60% of the absorbed light energy.

It has long been recognized that part of the energy stored by the photosynthetic process is reutilized for the maintenance of the organism. This is clearly obvious during dark periods, but of course occurs continuously. Furthermore, loss of energy occurs in the process of photorespiration (2), and a further loss is entailed when organic products are excreted to the medium and are not recovered in the harvested algae. These maintenance, photorespiratory, and excretion losses have been estimated to account for a loss of about 5% of the produced organic matter.

A factor which has recently been shown to be more significant than hitherto estimated is the loss due to photoinhibition (5). This process has long been known to markedly decrease photosynthetic efficiency, possibly owing to a slowly reversible destruction of an essential component in the photosynthetic electron flow path. It has been suggested that much of the benefit observed by researchers in several laboratories as a result of mixing of algal ponds may be partially or fully due to decreasing the inhibitory effect of photoinhibition. This is due both to the more efficient removal of the photosynthetically produced supersaturating oxygen from the medium and to the prevention of long exposure periods of individual algae to photoinhibitory light intensities. Recently, direct experiments on mass-cultured *Spirulina* spp. indeed supported the significant effect of photoinhibition on algal yield (A. Vonshak, personal communication). It was estimated that photoinhibition accounts for at least a 10% reduction in the photosynthetic energy conversion yield.

Finally, one has to consider the limitations imposed in real-life outdoor situations by temperature, which may be too low or too high to permit maximal photosynthesis. Both situations occur frequently in algal ponds: the former in the early morning during winter and the latter at midday in summer. It is estimated that temperature limitations account for a 20% loss in the photosynthetic productivity.

Table 1 is an attempt to consider all these factors and estimate their cumulative effect on the long-term productivity that can be expected in algal ponds; it shows that the cumulative effect of the enumerated factors leads to an overall energy storage of no more than about 3% of the available solar energy. When translated into organic matter (assuming that 1 g of organic matter releases about 5 kcal [21 kJ] in reconversion to CO_2 plus H_2O), this means that in an algal pond not more than about 25 g of organic matter can be produced per m^2 per day. This limitation is, of course, intrinsic and does not depend on the type of algae grown.

Such calculations lead one to view with caution any reported productivities which greatly exceed 25 g/m^2 per day. Reports of 40 and even 60 g/m^2 per day have appeared. Often they relate to rather short-term experiments, in which some of the factors enumerated above should not be considered.

Of course, many of the above estimates may be subject to error. Nevertheless, they do indicate an order of magnitude of intrinsic limitations on the maximal energy conversion efficiency which can be expected in real-life situations. Such considerations are sufficient to explain the apparent large discrepancy between the highly efficient photosynthetic process and the relatively poor energetic conversion efficiency (0.1 to 1.0%) observed in commercial agriculture.

LITERATURE CITED

1. **Boardman, N. K.** 1977. The energy budget in solar energy conversion in ecological and agricultural systems, p. 307–318. *In* R. Buvet et al. (ed.), *Living Systems as Energy Converters*. Elsevier/North-Holland Publishing Co., Amsterdam.
2. **Canvin, D. T.** 1979. Photorespiration, p. 368–398. *In* M. Gibbs and E. Latzko (ed.), *Encyclopedia of Plant Physiology*, vol. 6. Springer-Verlag KG, Heidelberg, Federal Republic of Germany.
3. **Goldman, J. C.** 1979. Outdoor algal mass cultures. I. Applications. *Water Res.* 13:1–19.
4. **Goldman, J. C.** 1979. Outdoor algal mass cultures. II. Photosynthetic yield limitations. *Water Res.* 13:119–136.
5. **Powles, S. B.** 1984. Photoinhibition of photosynthesis induced by visible light. *Annu. Rev. Plant Physiol.* 35:15–44.
6. **Radmer, R. J., and B. Kok.** 1977. Light conversion efficiency in photosynthesis, p. 125–135. *In* A. Trebst and M. Avron (ed.), *Encyclopedia of Plant Physiology*, vol. 5. Springer-Verlag KG, Heidelberg, Federal Republic of Germany.

Chapter 35

Phylogenetic Analysis of Microorganisms and Natural Populations by Using rRNA Sequences

Seán Turner, Edward F. DeLong, Stephen J. Giovannoni, Gary J. Olsen, and Norman R. Pace

INTRODUCTION

It is a common theme in microbial ecology that many components of natural microbial assemblages resist cultivation in the laboratory. Thus, they escape characterization or even detection. We are developing technical strategies based on the molecular phylogeny of rRNA sequences to define the component organisms of mixed, naturally occurring microbial populations (9, 12). These approaches sidestep the need for isolation or cultivation of individual population members. Since only the in situ biomass is required for these methods, the results project a relatively unbiased picture of the community. Because the analysis is phylogenetic in character, population constituents are related to known organisms in terms of their fundamental properties. A phylogenetic characterization of organisms is more than an exercise in taxonomy, since evolutionary relationships are established in a credible and quantitative way. Closely related organisms are expected to be similar in their general biochemical properties; the degree to which novel organisms resemble known ones will depend upon how close the evolutionary relationships are.

The basic premise of molecular phylogenetic analysis is that the amount of difference between homologous macromolecular sequences from different organisms reflects the extent to which the organisms have diverged from one another since their last common ancestor. Because they are present in all organisms, the RNA elements of the ribosome—the rRNAs—are well suited for phylogenetic analysis of microorganisms, among which great biochemical diversity must be expected. Of the three rRNAs (5S, 16S-like, and 23S-like), the 5S and 16S molecules have received the most attention, largely for historical and technical reasons. The 5S rRNA, because of its relatively small size (ca. 120 nucleotides), could be sequenced by the late 1960s. The 16S rRNA

Seán Turner, Edward F. DeLong, Stephen J. Giovannoni, Gary J. Olsen, and Norman R. Pace • Department of Biology and Institute for Molecular and Cellular Biology, Indiana University, Bloomington, Indiana 47405.

was too large for sequencing at that time, although it was subject to partial sequence analysis by the so-called oligonucleotide cataloging approach used by Woese and coworkers in their analysis of procaryote phylogeny (see reference 20 and references therein). More recently, complete 16S rRNA sequences have been inferred from cloned genes. The relatively large number of nucleotide characters available in complete 16S rRNA sequences makes it possible to establish an outline of the evolutionary relationships among all life-forms (11, 20).

In one approach toward evaluating organisms present in mixed microbial populations, we have directly isolated the rRNAs (or their genes) from naturally occurring biomass (see below). Sequence determinations and phylogenetic analyses with sequences from known organisms as references then reveal the natures of population constituents. However, the characterization of population members by sequencing is a laborious and expensive undertaking that is not readily incorporated into routine studies of microbial ecology. We therefore are developing methods for the more convenient identification of organisms. As outlined below, an approach based on phylogenetic group-specific hybridization probes is applicable even to the analysis of single cells. Both these experimental avenues rely upon a reference data base of rRNA sequences from known organisms, so that we have developed a protocol for rapidly analyzing 16S rRNA sequences from pure cultures. The procedure is sufficiently straightforward that phylogenetic analysis with 16S rRNA could be incorporated into routine characterizations of organisms. In this chapter we exemplify the use of that methodology with a broad phylogenetic survey of cyanobacteria, one of the dominant phototrophic components of microbial mats.

ANALYSIS OF NATURAL POPULATIONS BY USING rRNA SEQUENCES

The procedures that we have used for the rRNA-based identification of natural

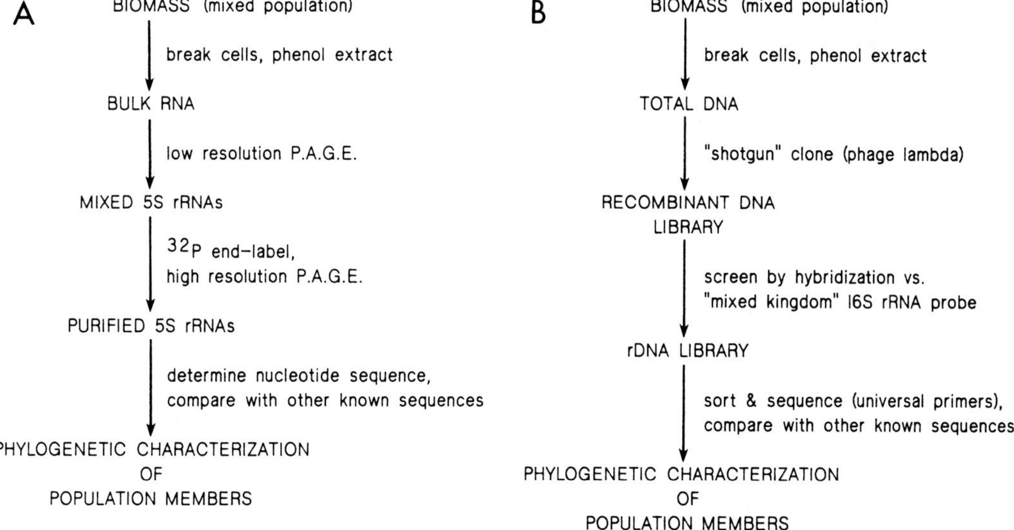

Figure 1. Flow charts for analyses of (A) 5S rRNAs and (B) 16S rRNA genes from natural populations. Reprinted from *Annual Review of Microbiology* (9) with permission of the publisher. P.A.G.E., Polyacrylamide gel electrophoresis.

population constituents are outlined in Fig. 1 (9). In one approach (Fig. 1A), 5S rRNA is isolated from the mixed population and the species-specific types are separated by high-resolution gel electrophoresis for subsequent sequencing. In a second, more general approach (Fig. 1B), 16S rRNA genes are shotgun cloned with DNA purified from natural samples. In the latter approach it does not matter that the original DNA was from a mixed population of organisms; individual rRNA genes are isolated in plaque-purified recombinant bacteriophages. The different types of cloned rRNA genes then are sorted in the laboratory and subjected to sequence analysis. With reference to the data base of available rRNA sequences, the relationships of the original population constituents to previously characterized organisms are established.

When we first undertook the characterization of mixed, natural microbial populations by using in situ samples, procedures for rapidly determining 16S rRNA gene sequences had not yet been developed. We therefore focused initially on 5S rRNA. So far, three different types of natural populations have been characterized by using 5S rRNA sequences: (i) the component organisms of symbioses, first discovered around submarine hydrothermal vents, in which procaryotic endosymbionts confer sulfur-based chemoautotrophy upon the invertebrates *Riftia pachyptila*, *Calyptogena magnifica*, and *Solemya velum* (17); (ii) the bacteria inhabiting the 91°C source pool of Octopus Spring in Yellowstone National Park, Wyoming (18); and (iii) the organisms in a leaching pond atop a copper recovery dump at the Chino Mine in southern New Mexico (7). The details and some results of these analyses have been reviewed (9, 12).

The communities mentioned above are composed of relatively few organisms, and so the mixtures of their 5S rRNAs were readily resolved into component species for sequencing. However, this approach is not always applicable. First, difficulties may arise in purifying individual forms of 5S rRNA if the community is too complex (more than ca. 5 to 10 organisms) or if the mixed RNA population contains 5S rRNAs with nearly identical electrophoretic properties. Moreover, minor population constituents might not be detected. Finally, because of its relatively small size, 5S rRNA comparisons permit phylogenetic assignments of only limited precision compared with those possible when using the much larger 16S rRNA (9, 12). We therefore have adopted the approach of directly cloning 16S rRNA genes from DNA purified from in situ samples (Fig. 1B). This method obviates each of the above-mentioned problems with 5S rRNA. Thus far, only the Octopus Spring community has been inspected using this methodology, with results generally consistent with those of the 5S rRNA analysis (12; D. J. Lane, unpublished results).

HYBRIDIZATION PROBES FOR MICROBIAL DETECTION AND PHYLOGENETIC IDENTIFICATION

Information derived from 5S or 16S rRNA sequences has potential for addressing ecologically relevant questions concerning the morphology, abundance, and population dynamics of individual community constituents. One approach involves the use of oligo- or polydeoxynucleotide hybridization probes that are complementary to specific rRNA sequences. rRNA is a particularly attractive hybridization target, since as many as 10^4 ribosomes are present in each cell. When these probes are used in conjunction with microautoradiography, detection and phylogenetic characterization of single cells are possible. Many sequences in the rRNAs are diagnostic of particular phylogenetic groups. Because of the presence of both variable and conserved regions in rRNA sequences, probes may be designed that are specific at various phylogenetic levels (kingdom, genus, etc.). Radioactive hybridization probes that are complementary to these group-specific signature sequences may be used to detect, identify, and quantify members of phylogenetic groups of interest in natural samples.

We have studied some examples of phylogenetic group-specific probes by using oligodeoxynucleotides that individually are complementary to one of the 16S rRNAs of the three primary lines of descent: the archaebacteria, the eubacteria, and the eucaryotes (2). The probe targets were chosen by examining the 16S rRNA sequences available from organisms representing each of the three kingdoms; regions invariant within a kingdom, but variable between kingdoms, were identified. Oligonucletides of 16 to 20 nucleotides in length and complementary to these kingdom-specific regions were then synthesized. A universal probe, complementary to the 16S rRNAs from all three kingdoms, served as a positive hybridization control. A probe with an rRNA-like sequence (as opposed to an rRNA-complementary sequence) was used as a negative control; it should not bind to the rRNA. Appropriate hybridization and wash temperatures were determined for annealing ^{32}P-labeled oligonucleotides to purified rRNA targets immobilized on nylon membranes, all as detailed by Giovannoni et al. (2). Probes were also hybridized to fixed whole cells attached to glass fiber filters, and the extent of binding was assessed by autoradiography and scintillation counting. In all cases, probes bound specifically to a variety of members of the targeted kingdom; little or no interkingdom binding was observed. Figure 2 shows some results of hybridizations of these kingdom-specific oligonucleotide probes to single cells.

Such phylogenetic group-specific probes should be useful for a variety of studies in microbial ecology. In principle, any phylogenetically coherent group is amenable to detection by this approach. Sequences that are specific for ecologically important groups, such as cyanobacteria, methanogens, or sulfate-reducing bacteria, are readily identified. Oligodeoxynucleotide probes could be used to estimate the relative amounts of specific organisms in bulk samples of a mixed population: the extent of hybridization of a phylogenetic group-specific probe to the sample, relative to the binding of a universal probe, would reflect the abundance of that group in the population. (Such estimates would be complicated by the fact that the rRNA content in most cells varies with the cell growth rate.) The relative distributions and population dynamics of organisms could also be monitored by using group-specific oligodeoxynucleotide probes. This approach is particularly relevant for the characterization of microorganisms that are difficult to grow or to detect by other means. In situ analyses should permit, for example, the enumeration and phylogenetic identification of microbial cells in thin sections of endosymbiont-containing organisms or in sections of stratified microbial communities.

Although rRNA sequence data bases are rapidly expanding and simplified protocols for rRNA sequencing are now available, these resources are not available in many laboratories. For that reason, we are developing methods for producing organism-specific probes that do not require the acquisition and analysis of large numbers of sequence data. In one method, radioactively labeled probe is synthesized by reverse transcription of 16S rRNA present in bulk cellular RNA preparations; 16S rRNA templates are selected by using 16S-specific oligonucleotide primers for reverse transcriptase. The technique yields polydeoxynucleotide probes that are heterogeneous in size, ranging from about 50 to several hundred nucleotides.

Shorter probes, which are more effective with less permeable fixed cells such as those of gram-positive bacteria, can be generated by including chain-terminating dideoxynucleotides in the reverse transcription reactions. Reverse transcripts have been successfully used in hybridization experiments

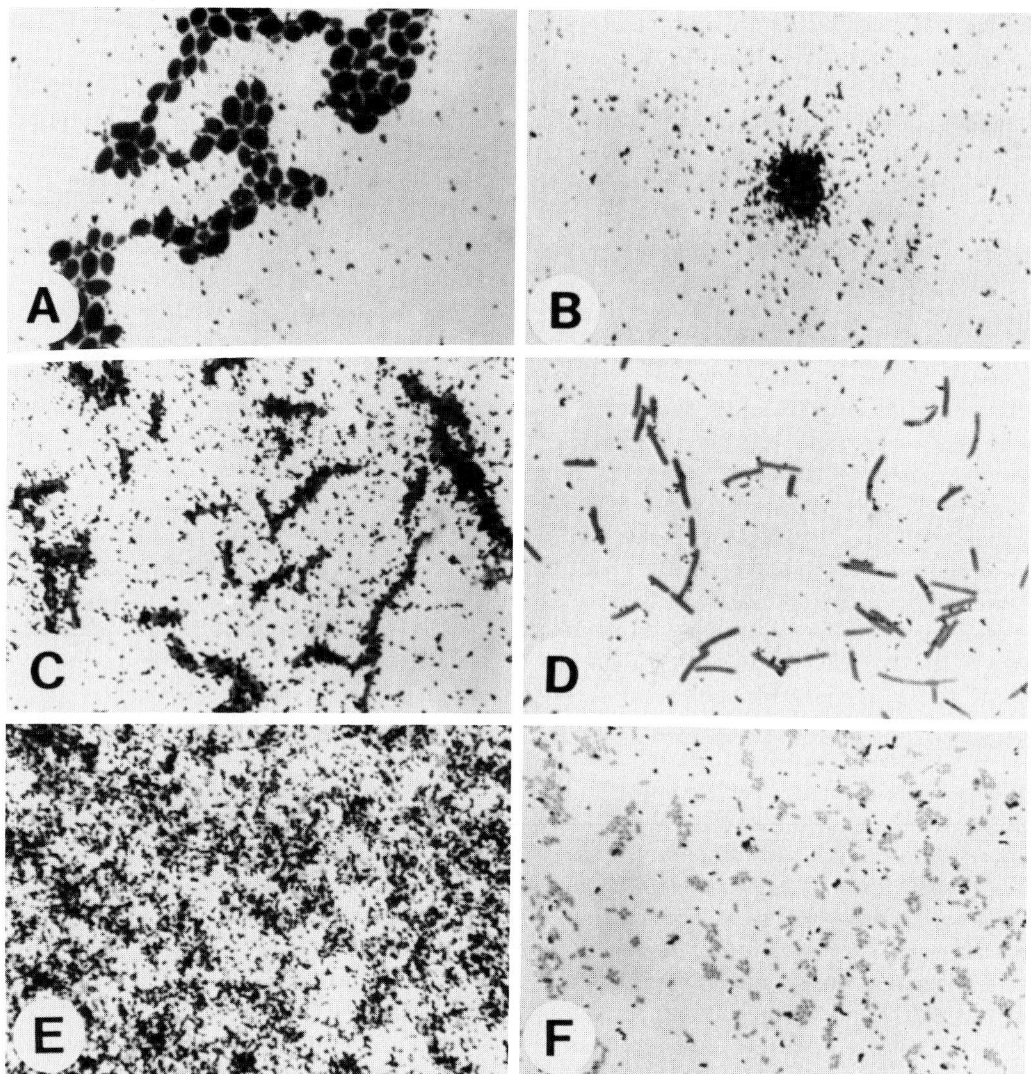

Figure 2. Microautoradiographs of single cells after hybridization to ^{35}S-labeled DNA probes. The indicated cells, fixed to glass slides, were hybridized with the indicated, ^{35}S-labeled, kingdom-specific oligodeoxynucleotide probes. After exposure and development, the cells were stained with Giemsa stain and visualized by bright-field microscopy. Panels: *Saccharomyces cerevisiae* (spores) hybridized with the eubacterial (A) and eucaryotic (B) probes; *Bacillus megaterium* hybridized with the eubacterial (C) and eucaryotic (D) probes; *Escherichia coli* hybridized with the eubacterial (E) and eucaryotic (F) probes. Reprinted from *Journal of Bacteriology* (2).

in the same manner as oligodeoxynucleotide probes. Specificity is controlled by adjusting the stringency of the hybridization conditions or posthybridization wash. Genus-specific probes are readily produced by this method (E. F. DeLong, unpublished results).

PHYLOGENY OF CYANOBACTERIA

Phylogenetic analyses of mat community organisms require a data base of reference sequences with which those of new organisms can be compared. Because cyanobacteria are conspicuous in most microbial mats, they are natural candidates for inclusion in such a data base. Using dideoxynucleotide-terminated sequencing with reverse transcriptase and synthetic oligodeoxynucleotide primers, we have determined partial 16S rRNA sequences of nearly 30 cyanobacterial strains grown in axenic culture (2a). The methodology has been published elsewhere (5a, 6) and hence is only briefly reviewed here.

High-molecular-weight RNA is obtained from lysed cells by a series of phenol extractions followed by ethanol and salt precipitations. The majority of this RNA is of ribosomal origin. 16S rRNA is targeted for primer extension sequencing by using synthetic DNA oligomers that anneal specifically to the 16S rRNA. Each oligonucleotide primer is complementary to a strategically located stretch of the 16S rRNA (Fig. 3). Because the primers are specific for 16S rRNA, the presence of contaminating RNAs is of little consequence. However, with some organisms it has been found necessary to purify the 16S rRNA by polyacrylamide gel electrophoresis prior to sequencing, to remove presumptive polysaccharides that interfere with the sequencing reactions. Sequencing is effected by reverse transcriptase extension of a primer with the four deoxynucleoside triphosphates, one of which is radiolabeled. Chain elongation is terminated at an A, T, G, or C residue by inclusion of the respective dideoxynucleotide in a reaction. Separation of the reaction products by polyacrylamide gel electrophoresis results in a sequencing ladder that corresponds to the complement of the 16S rRNA sequence (Fig. 4). The three primers in Fig. 3 collectively yield approximately 1,000 nucleotides

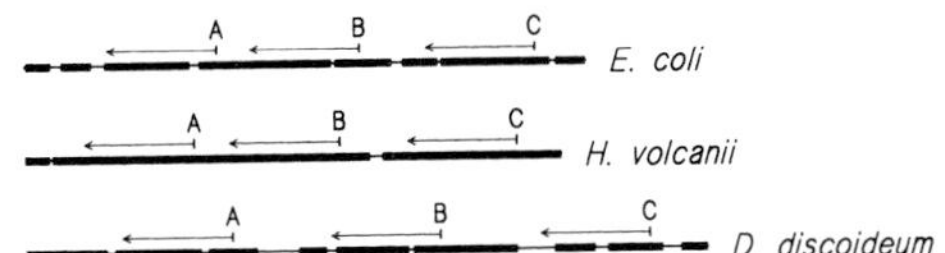

Figure 3. Hybridization sites of (A, B, and C) and approximate amounts of sequence accessible from (arrows) three primers specific for 16S-like rRNAs, shown on linear representations of the 16S-like rRNAs from *Escherichia coli* (a eubacterium), *Halobacterium volcanii* (an archaebacterium), and *Dictyostelium discoideum* (a eucaryote). Symbol: ▬, regions with sufficient intrakingdom homology to be generally useful in the inference of phylogenies. Reprinted from *Proceedings of the National Academy of Sciences of the United States of America* (6) with permission of the publisher.

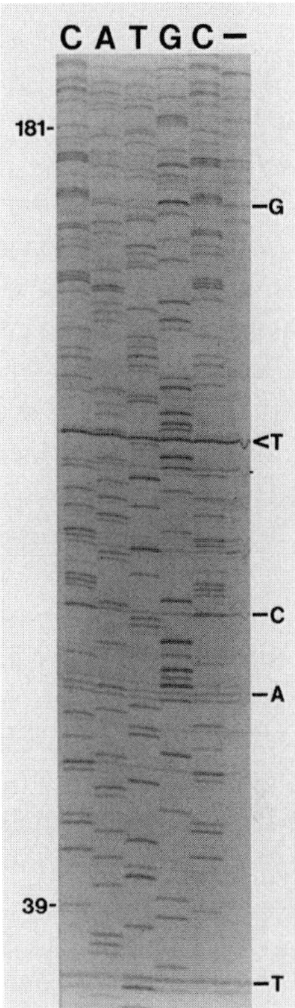

Figure 4. Autoradiograph of a 16S rRNA reverse transcriptase sequencing gel with bulk, high-molecular-weight *Halobacterium volcanii* RNA as template for the reverse transcriptase reactions, as outlined in the text. Distances from the 3′ terminus of the primer are indicated on the left. Terminations not mediated by dideoxynucleotide incorporations in which the correct nucleotide is evident (–) or not evident (<) are indicated on the right. These anomalies are discussed in detail elsewhere (6). Sequencing reactions in which ddCTP (C), ddATP (A), ddTTP (T), ddGTP (G), or no dideoxynucleotides (−) were used are indicated at the top. Reprinted from *Proceedings of the National Academy of Sciences of the United States of America* (6) with permission of the publisher.

of sequence; more complete sequence information can be obtained by the use of additional primers.

The sequences to be analyzed are aligned with each other on the basis of conserved sequence and secondary structural elements, and their pairwise similarities are determined. These similarities are then used to infer phylogenetic trees by using a least-squares distance matrix method (8b). Thorough discussions of sequence alignment and methods for constructing phylogenetic trees have been presented elsewhere (8b, 9, 12).

About 10 relatedness groups (phyla) of eubacteria have been defined on the basis of partial (RNase T_1-generated oligonucleotide catalogs) and complete 16S rRNA sequences (20). Together with the chloroplasts, the cyanobacteria make up one of these groups (Fig. 5). Previous studies had involved too few strains (eight) to develop a comprehensive overview of cyanobacterial diversity (1, and references therein). We used the taxonomic system of Rippka et al. (13) as a basis for selecting a broad diversity of cyanobacterial strains for sequence comparisons (2a). This taxonomy divides the cyanobacteria into five sections, based on morphology and mode of reproduction. Section I is composed of unicellular cyanobacteria that replicate by binary fission, including division by budding. Section II, the pleurocapsalean cyanobacteria, contains unicellular forms that replicate by multiple internal fissions to produce baeocytes. Section III includes filamentous cyanobacteria that do not produce heterocysts (cells specialized for aerobic dinitrogen fixation). The heterocyst-forming, uniseriate (Section IV) and branching (Section V) filamentous cyanobacteria constitute the remaining groups.

Figure 6 is a phylogenetic tree that contains members of Sections I, II, and III and a single representative (*Anabaena* sp.) of the heterocystous cyanobacteria. Inferred phylogenetic relationships among members of Sections IV and V are included in Fig. 7. It is particularly noteworthy that the deeper (i.e., earlier) branchings in the tree are separated by relatively short segments, giving the tree a fanlike appearance and indicating that modern cyanobacterial diversity has resulted from a rapid evolutionary radiation (Fig. 6). The relatively deep positions of many of the branch points in the tree make their branching orders uncertain, as discussed in detail elsewhere (2a). This uncertainty in the deeper branching orders is a consequence of the statistical uncertainty arising from the randomness of sequence changes and the imprecise knowledge of the number of multiple mutations at any given site (8b). Although exact branching orders near the roots of the trees cannot be determined, a considerable amount of useful information emerges from Fig. 6 and 7 and is discussed below.

The deepest identified branches in the cyanobacterial tree (Fig. 6) separate *Gloeobacter violaceus* (Section I), two closely related *Pseudanabaena* strains (Section III), and the remaining lineages into three distinct groups. *G. violaceus* is unique among the cyanobacteria in that it lacks thylakoids and phycobilisomes. Instead, its photosynthetic light-harvesting apparatus is part of the cytoplasmic membrane, with phycobiliproteins arranged in an underlying cortical layer (15). The early divergence of this organism from the main cyanobacterial lineage is consistent with this remarkably different light-harvesting apparatus. The result suggests that the common ancestor of the modern cyanobacteria may have lacked thylakoids. Unlike *G. violaceus*, the *Pseudanabaena* strains inspected have no obvious characteristics to suggest an early evolutionary divergence from the main cyanobacterial line. All known *Pseudanabaena* strains are aquatic, are characterized by deep constrictions between the cells of the filaments, and contain intracellular, polar gas vacuoles (13). Closer biochemical and ultrastructural inspection of these strains may reveal currently unrecognized features that will further distinguish these organisms from the other cyanobacteria and clarify the phylogenetic relationships between them.

The unicellular cyanobacteria in Section I and the nonheterocystous, filamentous forms in Section III are scattered throughout the tree (Fig. 6), indicating that organisms with these gross morphotypes are polyphyletic. In addition, orientation of cell division planes has been used as a partial basis for subdividing the members of Section I into several genera (13). An earlier phylogenetic study with RNase T_1-generated 16S rRNA oligonucleotide catalogs appeared to support this approach to the definition of natural relationships (1). However, the additional sequence data lead us to conclude that the various modes of cell division are not well correlated with phylogenetic groups. An example in Fig. 6 is the local cluster that contains *Gloeothece* strain PCC 6501 (cell division in one plane), *Synechocystis* strain PCC 6308 (cell division in two planes), and *Synechocystis* (*Eucapsis*) strain PCC 6906 and *Gloeocapsa* strain PCC 73106 (cell division in three planes). This cluster also contains the section III filamentous cyanobacterium *Spirulina* strain PCC 6313 (cell division in one plane). Thus, taxonomic classification based principally on morphology does not necessarily reflect phylogenetic relationships.

The pleurocapsalean organisms (section II) so far studied constitute a coherent, phylogenetic subgrouping, to the exclusion of other cyanobacteria (Fig. 6). Although only three of the six genera or groups of this section have been analyzed in this study (*Dermocarpa*, *Myxosarcina*, and *Pleurocapsa* spp.), all the strains in the section have very similar DNA base composition (4). Taken together with the fact that their mode of reproduction distinguishes these members from all other cyanobacteria (19), this suggests that reproduction by multiple fissions

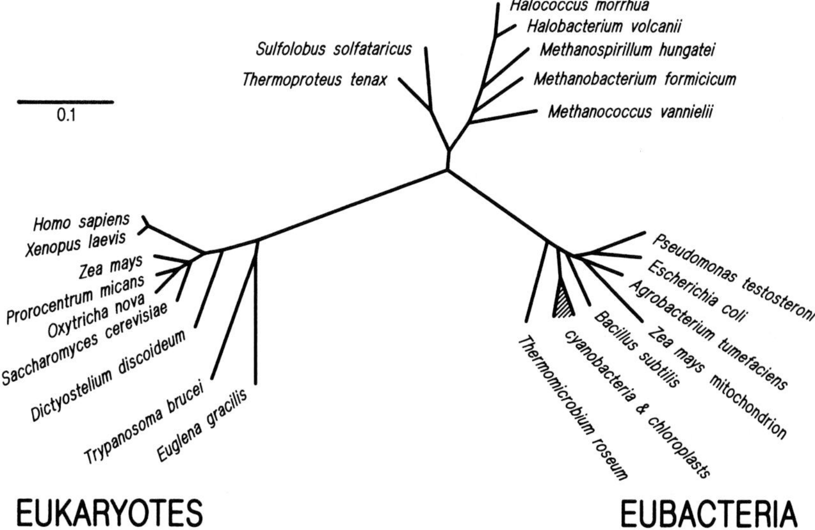

Figure 5. Unrooted phylogenetic tree, based on 16S-like rRNA sequences, illustrating the three primary lines of descent and some of their major subgroups. The hatched area depicts evolutionary diversity among oxygenic, phototrophic cells and organelles. Evolutionary distances are proportional to the segment lengths, which reflect the estimated number of fixed mutational events. The scale bar corresponds to an evolutionary separation of 0.1 accepted point mutation per sequence position. Adapted from reference 8a.

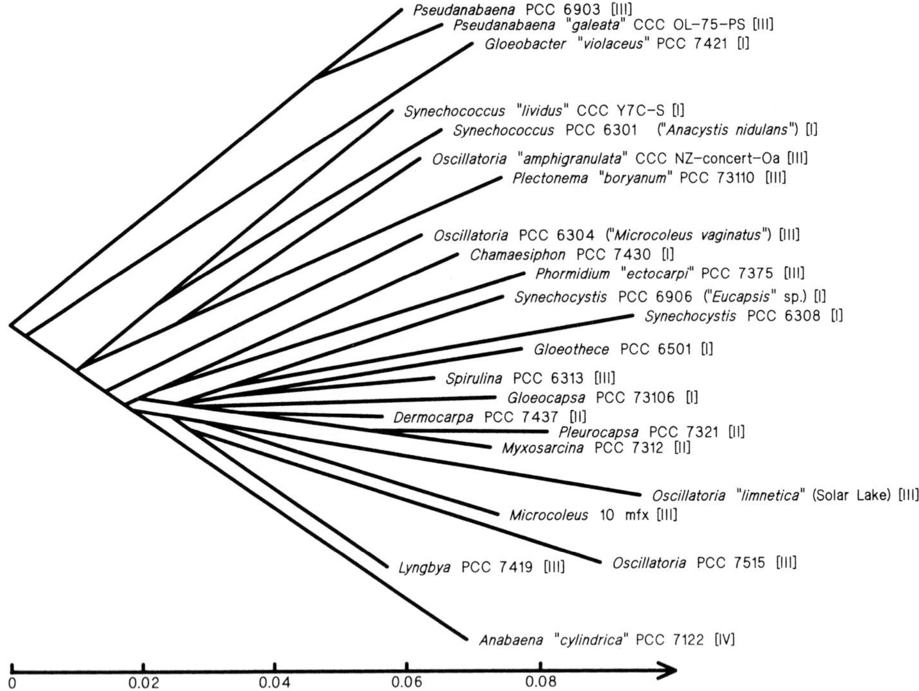

Figure 6. Rooted-tree topology illustrating evolutionary relationships among 16S-like rRNAs from cyanobacteria. *Agrobacterium tumefaciens*, *Bacillus subtilis*, and *Pseudomonas testosteroni* were used as outgroup organisms to locate the root. Roman numerals denote the taxonomic sections to which each organism has been assigned (13). The scale at the bottom represents the number of fixed point mutations per sequence position. This is a different tree depiction from that in Fig. 5; here, evolutionary distance is represented by only the horizontal component of segment lengths. Sequences from six heterocystous cyanobacteria were included in the tree analysis, but, because of space limitations, only one (*Anabaena* sp.) is shown. Adapted from Giovannoni et al. (2a).

(baeocyte formation) is of monophyletic origin.

Similarly, the members of Sections IV and V, the heterocystous, filamentous forms, cluster as a distinct phylogenetic subgroup, with the organisms in Section V arising from within the lineages of Section IV (Fig. 7). This confirms similar conclusions drawn from DNA-DNA hybridization data (5). The organisms in Sections IV and V are among the most morphologically diverse of the cyanobacteria. Aside from heterocysts, many strains also produce akinetes (resting cells that sometimes develop as a result of insufficient light) (14). It is noteworthy in Fig. 7 that strains that produce hormogonia as part of their developmental cycle (*Scytonema* strain PCC 7110, *Chlorogloeopsis* strain PCC 6718, *Fischerella* strain PCC 7414, and *Calothrix* strain PCC 7102) branch more deeply in the heterocyst-forming subline than do strains belonging to genera that do not produce hormogonia (*Anabaena* strain PCC 7122 and *Nodularia* strain PCC 73104). On this basis, one might predict that *Cylindrospermum*, another genus whose members do not form hormogonia, would be more closely related to *Anabaena* and *Nodularia* than to the other genera in Sections IV and V. Sequencing of the 16S rRNA of a *Cylindrospermum* strain is under way in our laboratory.

The ability to form heterocysts (i.e., to differentiate vegetative cells into ones spe-

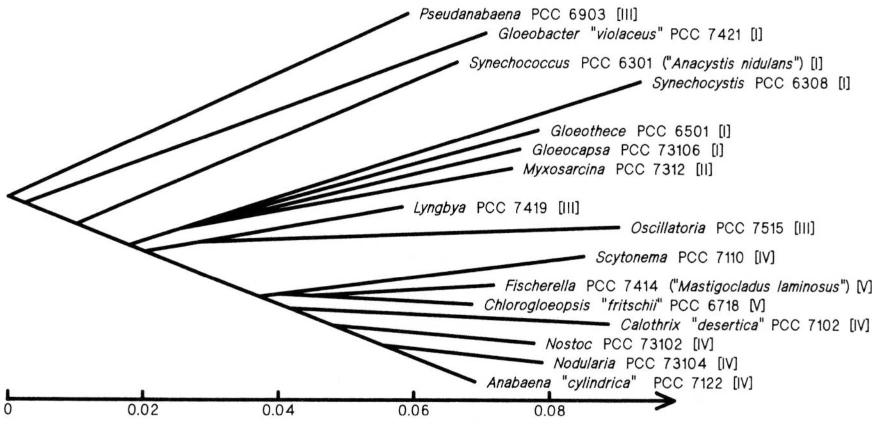

Figure 7. Rooted-tree topology illustrating evolutionary relationships among 16S-like rRNAs from heterocystous cyanobacteria. The presentation is as in Fig. 6. Adapted from Giovannoni et al. (2a).

cialized for aerobic dinitrogen fixation in the absence of a source of fixed nitrogen) apparently had a single origin. However, the formation of heterocysts is not obligatory for nitrogen fixation by cyanobacteria. Some unicellular forms (*Gloeothece* spp.) that branch more deeply in the tree than do the heterocystous cyanobacteria are also capable of aerobic N_2 fixation, but apparently without differentiating into specialized cells (13). In addition, some marine synechococci that also have this ability have been reported (8). Moreover, the evolution of dinitrogen fixation almost certainly predates that of the ability to form heterocysts: eubacteria from altogether different taxa manifest this metabolic trait.

As discussed in more detail elsewhere (2a), these sequence comparisons confirm the demonstration of Bonen et al. (1) that cyanobacteria and chloroplasts are closely related (Fig. 8). However, the present analysis permits more precise definition of the relationship: the green chloroplasts are seen

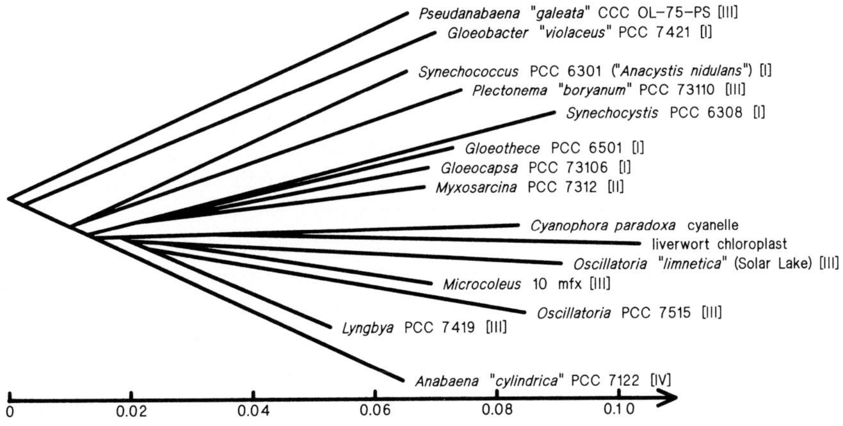

Figure 8. Rooted-tree topology illustrating evolutionary relationships among 16S-like rRNAs from selected cyanobacteria, a chloroplast, and the cyanelle of *Cyanophora paradoxa*. The presentation is as in Fig. 6. Adapted from Giovannoni et al. (2a).

as one of the cyanobacterial lineages. Although Fig. 8 includes only one example of a chloroplast rRNA, that of liverwort (*Marchantia polymorpha*), analyses that also include the chloroplast rRNAs of *Zea mays*, *Nicotiana tabacum*, *Chlamydomanas reinhardii*, and *Euglena gracilis* show that they form, with the liverwort chloroplast, a coherent subgroup of the cyanobacteria. Also associated with the chloroplast line is the photosynthetic organelle (cyanelle) of the flagellate *Cyanophora paradoxa*.

Because the 16S rRNAs of different organisms accumulate mutations at different rates, evolutionary distances cannot be accurately correlated with geological time. Hence, we have not attempted to infer the time that passed between the divergence of the major eubacterial phyla and the radiation of cyanobacterial diversity. It seems clear, though, that cyanobacterial diversification occurred within a relatively short span of evolutionary distance. The events responsible for this apparent burst of evolution in the cyanobacterial line of descent are uncertain. However, as the first organism to exploit water as an electron donor for photosynthesis, the common ancestor of the cyanobacteria had available a novel and profoundly fertile physiological niche. We suggest that rapid evolution of morphology (and behavior) ensued, leaving its record in the molecular relationships described.

The morphological similarities of microfossils in the earliest known stromatolites with some modern cyanobacteria have been cited as evidence for a particularly ancient origin of cyanobacteria (16). However, the high degree of sequence similarity of cyanobacterial rRNAs to each other and to those of other major eubacterial taxa indicate that this group did not arise as early as some other groups in the eubacterial line of descent, for example, members of the family *Chloroflexaceae*, which are represented by *Thermomicrobium roseum* in Fig. 5 (10). Some of the latter organisms are obligately anaerobic, phototrophic, filamentous forms that also form laminated microbial mats (3) and are morphologically similar to some stromatolitic microfossils. Thus, we do not believe that the occurrence of cyanobacterialike morphologies in fossil stromatolites necessarily constitutes evidence for the presence of cyanobacteria. On the other hand, if cyanobacteria were present in the 3.5-billion-year-old stromatolites, then the other major diversifications in the eubacterial line of descent must have occurred before that time.

CONCLUSIONS

Microorganisms that are observed in naturally occurring populations often elude cultivation and hence characterization. This chapter reviews some recently developed methods for using rRNA sequences to analyze phylogenetic aspects of organisms, including uncultivable ones. A phylogenetic characterization is more than an exercise in taxonomy, because if an uncharacterized organism can be related to a known one, the two are expected to share many cellular properties and their degree of resemblance will depend on how closely they are related. In one approach to the analysis of mixed, naturally occurring populations that consist of only a few organisms, 5S rRNAs are isolated from the mixed population, resolved into individual types by high-resolution gel electrophoresis, and sequenced. Comparison with previously determined sequences provides a phylogenetic characterization of the unknown organism with regard to known ones. In a second approach, suitable for complex populations, 16S rRNA genes from the mixed population are shotgun cloned for subsequent sequence and phylogenetic analyses. A third method for evaluating natural populations involves the use of oligodeoxynucleotide hybridization probes that bind specifically to the rRNAs of defined phylogenetic groups. Because each cell contains many ribosomes, binding of isotopically labeled probes to individual cells is detectable by using microautoradiography.

ACKNOWLEDGMENTS. This work was supported by Public Health Service grant GM34527 to N.R.P. from the National Institutes of Health and by Office of Naval Research grants N14-86-K-0268 to G.J.O. and N14-87-K-0813 to N.R.P.

We thank John Waterbury (Woods Hole Oceanographic Institution) for his interest and assistance in supplying cyanobacterial strains.

LITERATURE CITED

1. **Bonen, L., W. F. Doolittle, and G. E. Fox.** 1979. Cyanobacterial evolution: results of 16S ribosomal ribonucleic acid sequence analyses. *Can. J. Biochem.* 57:879–888.
2. **Giovannoni, S. J., E. F. DeLong, G. J. Olsen, and N. R. Pace.** 1988. Phylogenetic group-specific oligodeoxynucleotide probes for identification of single microbial cells. *J. Bacteriol.* 170:720–726.

2a. **Giovannoni, S. J., S. Turner, G. J. Olsen, S. Barns, D. J. Lane, and N. R. Pace.** 1988. Evolutionary relationships among cyanobacteria and green chloroplasts. *J. Bacteriol.* 170:3584–3592.

3. **Giovannoni, S. J., D. M. Ward, N. P. Revsbech, and R. W. Castenholz.** 1987. Obligately phototrophic *Chloroflexus*: primary production in anaerobic, hot spring microbial mats. *Arch. Microbiol.* 147:80–87.
4. **Herdman, M., M. Janvier, J. B. Waterbury, R. Rippka, R. Y. Stanier, and M. Mandel.** 1979. Deoxyribonucleic acid base composition of cyanobacteria. *J. Gen. Microbiol.* 111:63–71.
5. **Lachance, M.-A.** 1981. Genetic relatedness of heterocystous cyanobacteria by deoxyribonucleic acid-deoxyribonucleic acid reassociation. *Int. J. Syst. Bacteriol.* 31:139–147.

5a. **Lane, D. J., K. G. Field, G. J. Olsen, and N. R. Pace.** 1988. Reverse transcriptase sequencing of rRNA for phylogenetic analysis. *Methods Enzymol.* 167:138–144.

6. **Lane, D. J., B. Pace, G. J. Olsen, D. A. Stahl, M. L. Sogin, and N. R. Pace.** 1985. Rapid determination of 16S ribosomal RNA sequences for phylogenetic analyses. *Proc. Natl. Acad. Sci. USA* 82:6955–6959.
7. **Lane, D. J., D. A. Stahl, G. J. Olsen, D. J. Heller, and N. R. Pace.** 1985. Phylogenetic analysis of the genera *Thiobacillus* and *Thiomicrospira* by 5S rRNA sequences. *J. Bacteriol.* 163:75–81.
8. **León, C., S. Kumazawa, and A. Mitsui.** 1986. Cyclic appearance of aerobic nitrogenase activity during synchronous growth of unicellular cyanobacteria. *Curr. Microbiol.* 13:149–153.

8a. **Olsen, G. J.** 1987. The earliest phylogenetic branchings: comparing rRNA-based evolutionary trees inferred with various techniques. *Cold Spring Harbor Symp. Quant. Biol.* 52:829–837.

8b. **Olsen, G. J.** 1988. Phylogenetic analysis using ribosomal RNA. *Methods Enzymol.* 164:793–812.

9. **Olsen, G. J., D. J. Lane, S. J. Giovannoni, N. R. Pace, and D. A. Stahl.** 1986. Microbial ecology and evolution: a ribosomal RNA approach. *Annu. Rev. Microbiol.* 40:337–365.
10. **Oyaizu, H., B. Debrunner-Vossbrinck, L. Mandelco, J. A. Studier, and C. R. Woese.** 1987. The green non-sulfur bacteria: a deep branching in the eubacterial line of descent. *Syst. Appl. Microbiol.* 9:47–53.
11. **Pace, N. R., G. J. Olsen, and C. R. Woese.** 1986. Ribosomal RNA phylogeny and the primary lines of evolutionary descent. *Cell* 45:325–326.
12. **Pace, N. R., D. A. Stahl, D. J. Lane, and G. J. Olsen.** 1986. The analysis of natural microbial populations by ribosomal RNA sequences. *Adv. Microb. Ecol.* 9:1–55.
13. **Rippka, R., J. Deruelles, J. B. Waterbury, M. Herdman, and R. Y. Stainer.** 1979. Generic assignments, strain histories and properties of pure cultures of cyanobacteria. *J. Gen. Microbiol.* 111: 1–61.
14. **Rippka, R., and M. Herdman.** 1985. Division patterns and cellular differentiation in cyanobacteria. *Ann. Inst. Pasteur (Paris)* 136A:33–39.
15. **Rippka, R., J. B. Waterbury, and G. Cohen-Bazire.** 1974. A cyanobacterium which lacks thylakoids. *Arch. Microbiol.* 100:419–436.
16. **Schopf, J. W., and B. M. Packer.** 1987. Early Archean (3.3-billion to 3.5-billion-year-old) microfossils from Warrawoona Group, Australia. *Science* 237:70–73.
17. **Stahl, D. A., D. J. Lane, G. J. Olsen, and N. R. Pace.** 1984. Analysis of hydrothermal vent-associated symbionts by ribosomal RNA sequences. *Science* 224:409–411.
18. **Stahl, D. A., D. J. Lane, G. J. Olsen, and N. R. Pace.** 1985. Characterization of a Yellowstone hot spring microbial community by 5S rRNA sequences. *Appl. Environ. Microbiol.* 45:1379–1384.
19. **Waterbury, J. B., and R. Y. Stanier.** 1978. Patterns of growth and development in pleurocapsalean cyanobacteria. *Microbiol. Rev.* 42:2–44.
20. **Woese, C. R.** 1987. Bacterial evolution. *Microbiol. Rev.* 51:221–271.

Chapter 36

Evolution of Photosynthesis in Anoxygenic Photosynthetic Procaryotes

Beverly K. Pierson and John M. Olson

INTRODUCTION

We believe that the evolution of photosynthesis in the anoxygenic photosynthetic bacteria is deeply entrenched in the origin and early history of cellular life on Earth and that solar energy conversion mediated by porphyrins or their derivatives was among the earliest of bioenergetic processes known. Anoxygenic phototrophy seems to be very ancient and may well have been present in the earliest cells or even protocells. The most primitive photosystems were probably abiogenic in origin and hence variable and perhaps randomly associated with protocells. When true cellular life emerged with a nucleic acid-based genetic code and some translational program (with or without a transcriptional process) to produce polypeptides, so that the selection process leading to adaptive evolution could occur, the rapid evolution of early phototrophs could have proceeded.

This theory on the early origins of phototrophy is contrary to long-prevailing ideas on the early evolution of bioenergetics. We will first explain why we find the classical ideas inadequate and our ideas more appropriate interpretations of the true roots and ancestry of all photosynthesis. Then we will explore in depth our somewhat less controversial ideas on the rapid diversification and evolution of the ancestors of the many different anoxygenic phototrophic bacteria found on earth today.

Beverly K. Pierson • Biology Department, University of Puget Sound, Tacoma, Washington 98416. ***John M. Olson*** • Institute of Biochemistry, Odense University, DK-5230 Odense M, Denmark.

Early Bioenergetics

Fermentation Is neither Simple nor Primitive

Contemporary thoughts on the origins of bioenergetic systems have been heavily influenced by the idea that evolution of the earliest cells occurred in an organic aquatic environment with a ready supply of fermentable substrate that could sustain growth and evolution by substrate-level phosphorylation (12, 40). Even those who have championed a very early origin of photosynthesis (4, 45, 46) have been compelled to believe that some form of fermentation as in the Embden-Myerhof-Parnas (EMP) pathway was the primitive and hence the ancestral form of energy metabolism. Photosynthesis is seen by many as a far more complex process that by virtue of this complexity must have evolved much later, probably even after respiration.

The "facts" used to support the early origin of fermentation pathways are as follows. First, the EMP pathway of fermentation is claimed to be the simplest bioenergetic system known and, therefore, the most primitive. Second, it is claimed to be primitive because it could be sustained by the fermentable substrates present in the early environment. Third, it has been claimed to be a universal (or nearly so) metabolic pathway in extant organisms and therefore to be primitive (12).

In our view, the first of these assumptions is incorrect (38). The EMP pathway is not simple at all. It is a complex pathway requiring the sequential participation of approximately 10 large, well-functioning enzymes to extract a rather poor yield of metabolically useful energy from the very stable glucose molecule. It is indeed very difficult to imagine how this sort of system could have evolved gradually in an otherwise energy-insufficient cell. Since it is impossible to obtain any useful energy from glucose with only a partial system, it is equally hard to imagine how 10 big enzymes could have come together over a reasonably short period in a primitive cell to achieve this end. In addition, the EMP pathway has the added complication of requiring a phosphorylating system to operate at all, as well as pyridine nucleotide carriers to maintain redox balance. It is hard to imagine a simplified version of the contemporary pathway that could fulfill the energy needs of the cell. Hence, we do not believe that fermentation is a good candidate for a primitive bioenergetic system.

We regard the second assumption as equally controversial. It does not seem likely that the dilute organic hydrosphere that was the probable environment for the early evolution of life could sustain the growth and multiplication of many cells by using a substrate that would have become limiting very rapidly in the absence of a good source of its continuous production. Fermentation would have produced just too little energy from the amount of substrate available. Since the organic molecules available by the time "cells" had evolved would have been highly varied, the necessity of additional enzyme systems for bioenergetics would have had to be met. Contemporary fermenters must process a lot of substrate to grow. Since the primitive fermentation systems were hardly likely to be any more efficient than these, fermentation itself is a poor candidate for early metabolism.

We regard the basis of the third argument as more sound. It has been suggested that universality of a character among extant organisms is an argument for antiquity and that the presence of the EMP in nearly all organisms suggests that it is an ancient character. Although we agree that the principle relating universality with antiquity is a good one, the fact is that many new organisms have been described (e.g., among the archaebacteria) in which the EMP pathway as a bioenergetic system is lacking.

Photosynthesis Provides an Intriguing Alternative for Energy Conversion in Primitive Organisms

Although contemporary photosynthetic mechanisms are indeed complex and involve many proteins and other components, it is easy to imagine and even construct simpler systems that are likely candidates for primitive versions of the process. The origin of a simple photochemical system based on a porphyrin derivative has been explained elsewhere (38, 39). The prebiological (chemical) evolution of porphyrin-based energy conversion systems has also been discussed (19). In the context of the evolution of the photochemical reaction center, a porphyrin-derived system could be constructed that requires only minor participation of small polypeptides and perhaps one or two enzymes. Such a system could indeed be simple compared with the fermentation pathways and could potentially supply energy to the primitive cell in the form of a transmembrane electron transfer and/or ion gradient, thereby eliminating the initial requirement of a phosphorylating system. The gradient is always H^+ in contemporary photosynthetic organisms, but any electrochemical gradient is a potential source of energy for a cell. Such gradients could have provided energy directly under the most primitive conditions and could later have been coupled to ATP synthesis. As soon as cells were defined by membranes, the potential energy of gradients as a driving force for cellular activity was a likely possibility. The location of a primitive reaction center (RC) in such a membrane would have allowed for solar radiation to create a charge separation across the membrane. The evolution of such a simple system is discussed below.

This type of energy conversion system could have used the unlimited energy of the Sun, leaving most of the abiogenically produced organic compounds, limited in amount, for use as building blocks for cellular materials. Much more growth and multiplication of primitive cells could thus have occurred in phototrophic cells than in fermenting cells, before the available organic substrates were exhausted.

Photosynthesis, however, is not universal. It is found widely dispersed among the eubacteria, and its presence in so many unrelated groups is used as an argument for its antiquity in the evolution of eubacteria (63). Porphyrin-based photosynthesis is absent in the archaebacteria, however. It is clear that early in the evolution of bacteria, these two groups diverged. We cannot argue convincingly that photosynthesis did arise in the common ancestors to both groups, but if one looks for remnants of this possibility, one has to look at the essential components of the most primitive type of photosynthesis described below. Indeed, throughout the archaebacteria we find the presence of protoporphyrin IX, other related porphyrin derivatives, and even cytochromes. We therefore believe that there is adequate evidence to suggest that primitive photosystems based on protoporphyrin (not chlorophylls [Chls]) may have been operative in the common ancestors of both groups. We further believe that anaerobic respiration evolved from photosynthesis and that this process may have occurred before the two groups diverged.

Early Evolution of Photosynthesis

Most of the Evolution of Photosynthesis into Complex Contemporary Systems Probably Occurred in the Ancestral Anoxygenic Phototrophs

We believe that photosynthetic microorganisms were highly diverse and that most of the characteristics of contemporary anoxygenic phototrophs were probably well established and highly evolved before the oxygenic phototrophs became widespread between 3.0 and 2.5 billion years ago. We assume that oxygenic phototrophs were widespread at that time to account for the increase in the partial pressure of O_2 that occurred between 2.5 and 1.5 billion years ago (17). Phylogenetic evidence, geologic evidence, and biochemical and molecular evidence will be used to support this idea. The evolution of the diversity of pigment systems probably occurred in microbial mats, an argument that will be supported by recent ecological data. Evidence will be presented to support the notion that evolution of photosynthetic diversity in surficial or shallow aquatic mats was possible before the accumulation of oxygen in the atmosphere, despite a high UV flux in the Precambrian era.

Definitions

The contemporary photosynthetic apparatus is a complex assemblage of structural and molecular components that have different evolutionary histories. We will therefore consider each of them separately. The contemporary photosynthetic apparatus of the anoxygenic phototrophs includes an RC, a light-harvesting (LH) system, an electron transfer system (ETS), and a proton-driven ATP synthase enzyme (43). The RC includes all Chls (or other porphyrin derivatives) involved directly in photochemistry, their associated electron donors or acceptors, and their associated proteins. The ETS includes further redox centers involved in establishing electrochemical gradients across the photosynthetic membrane. The ATP synthase system refers to the membrane-bound reversible enzyme that catalyzes the synthesis of ATP from ADP and P_i with a flow of protons into the cell or the hydrolysis of ATP by using the resulting energy to drive a flow of protons out of the cell. The LH system refers to all Chls, their associated proteins, and other pigments that function to absorb light and transfer the excitation energy to the RC Chls.

We are defining photosynthesis as the conversion of solar radiation to useful biological energy in the form of ATP or, more simply, in the form of an electrochemical gradient. We are not including in our definition the synthesis of organic molecules from CO_2. The evolution of the biosynthetic process of CO_2 fixation has a history of its own and will be considered only insofar as it has an impact on the evolution of the photochemical energy-converting system. In this chapter we are discussing the evolution of Chl-based photosynthesis and not the evolution of the carotenoid-based bacteriorhodopsin photosynthesis in the halobacteria, which is an independent process that is likely to be of more recent origin (38, 52).

The Anoxygenic Phototrophs Comprise a Diverse Array of Procaryotes Found among Many Different Groups of Eubacteria

In our theory of evolution of photosynthesis we will try to account for the evolution of all groups of phototrophic procaryotes. The anoxygenic phototrophs include the well-known purple bacteria (sulfur and nonsulfur), the green sulfur bacteria, the green filamentous anoxygenic bacteria, the recently described *Heliobacterium* sp. and relatives, and certain members of the cyanobacteria. The group is very diverse and has an ancient history. The major characteristics of the contemporary members of these groups are listed in Table 1.

Our Theory

Photosynthesis as a source of cellular energy probably existed with the earliest

Table 1.
Diversity of Photosynthetic Systems in Anoxygenic Phototrophs

Group	RC	LH system	Reductants	Primary C source
C. aurantiacus	RC-2, Bchl *a*	Chlorosomes, Bchl *a* andBchl *c*	H_2, H_2S, organic	Organic (CO_2)[a]
Purple sulfur bacteria	RC-2, Bchl *a* or *b*	Membrane bound, Bchl *a* or *b*	H_2, H_2S	CO_2
Purple nonsulfur bacteria	RC-2, Bchl *a* or *b*	Membrane bound, Bchl *a* or *b*	H_2, H_2S, organic	Organic (CO_2)[a]
Green sulfur bacteria	RC-1, Bchl *a*	Chlorosomes, Bchl *a* and *c*, *d*, or *e*	H_2, H_2S	CO_2
H. chlorum	RC-1, Bchl *g*	Membrane bound, Bchl *g*	Organic	Organic
Cyanobacteria (anoxygenic)	RC-1, Chl *a*	Membrane bound, Chl *a*	H_2S, H_2, Fe^{2+}?	CO_2

[a] Although organic molecules are the primary carbon sources in these organisms, CO_2 fixation is also known to occur.

cells in the form of a simple RC based on protoporphyrin IX, perhaps loosely associated with short polypeptides in the primitive cell membrane (38, 39). The function of this RC was charge separation and the formation of a small ion gradient across the membrane (38, 39, 52). This earliest RC may have occurred in the common ancestor of both the eubacteria and the archaebacteria. The earliest evolutionary changes were in the RC, which quickly evolved to a Chl *a* RC-1 type (with a low-potential Fe-S center as the electron acceptor). Later phases of the evolution of the RC included elaboration of other Chls to compete for light, which was limiting in the crowded mat environment, and diversification resulting from competition for electron donors. RC-2 (with a higher-potential quinone as the electron acceptor) evolved from RC-1. An alternative view that RC-1 evolved from RC-2 and that bacteriochlorophyll *a* (Bchl *a*) preceded Chl *a* will be discussed briefly. Simultaneously, the ETS was evolving to include a larger number of carriers which functioned to increase the magnitude of the electrochemical gradient resulting from each quantum of light. The ETS was developed in at least a simple form very early, as was the enzyme ATP synthase. Respiration arose easily from minor changes in the photosynthetic electron transfer system. As the energy transformation system became more efficient, competition for light became stronger and accessory pigment systems developed and diversified to keep solar energy flowing to the RCs. This extensive competition for light, resulting in the elaboration of numerous pigment-protein complexes, occurred in thick mat communities. These early events preceded the evolution of the water-splitting photosystem and occurred under high UV flux in shallow mat communities. Protection of the mat organisms may have been provided by ferrous iron in the shallow water column and, more significantly, by ferric iron in the sediments.

GEOLOGIC RECORD

The major evidence that phototrophic cellular life was abundant very early in the history of the Earth comes from the presence of stromatolites, which are fossils of laminated microbial mat communities (2, 58). Early Precambrian stromatolites appear to have been formed by filamentous phototrophs growing in matlike communities in shallow sediments periodically experiencing surface exposure to the air (5, 58). The association of such macroscopic fossil evidence with microfossils preserved within the cherts would be convincing evidence for their procaryotic origin (49). Although there are not many examples of well-documented ancient microfossils (from ca. 3.5 billion years ago), those that have been well characterized are certainly morphologically similar to extant filamentous phototrophs belonging to either the family *Chloroflexaceae* or the cyanobacteria (38, 49, 57). Recent claims that some of these ancient microfossils suggest the presence of oxygen-evolving cyanobacteria as early as 3.5 billion years ago (48) are based on morphological criteria alone; there is no geochemical evidence to support them. Although the geochemical and fossil records (both stromatolites and microfossils) are certainly consistent with the idea that massive microbial communities composed of phototrophs dominated life on Earth 3.5 billion years ago, they do not prove it. The same data would also be consistent with the existence at that time of massive matlike communities of chemotrophs (both heterotrophic and autotrophic). It is interesting, however, that the oldest well-documented stromatolites appear to have been formed in shallow evaporitic environments. This is consistent with the idea that light was required for their formation (5, 58). Some authors have been careful to point out, however, that even the evidence suggesting the presence of phototaxis in stromatolites could be interpreted otherwise (58). Nevertheless, given the widespread distribution of microbial mats clearly dominated by phototrophs in several environments in the world today, the role of migrating filamentous phototrophs in these mats, and the similarities of these contemporary mats and microorganisms to the stromatolites and microfossils of the Archean age, it seems likely that diverse photosynthetic organisms were well established by 3.5 billion years ago.

RCs

Diversity of RCs

The essential part of the photosynthetic apparatus is the RC. The well-developed electron transfer chains and LH systems of contemporary photosynthetic organisms increase the transformation efficiency of solar to chemical energy by the RC but are not essential to the process. The earliest phototrophic organisms would have been distinguished from nonphototrophic organisms by the presence of multiple simple RCs. Most of the interesting stages in the evolution of photosynthesis are due to changes in the type of RC. It is likely that the ancestral phototrophs had very simple RCs that increased in complexity and gave rise to the five different types of RC seen in contemporary phototrophs (Fig. 1).

Although diversity exists among contemporary RCs, they all function in a similar manner. A photoreactive donor pigment (usually a special pair of Chl molecules) loses an electron, which passes through an initial electron acceptor and reduces a more stable acceptor. In this way the donor pigment molecules are photooxidized, an acceptor is photoreduced, and light energy has created a charge separation across a membrane. All RCs contain one of the following four Chls as the photoreactive pigment: Chl *a*, Bchl *a*, Bchl *b*, or Bchl *g*. All other Chls function as LH pigments and do not undergo photochemistry. Contemporary RCs can be di-

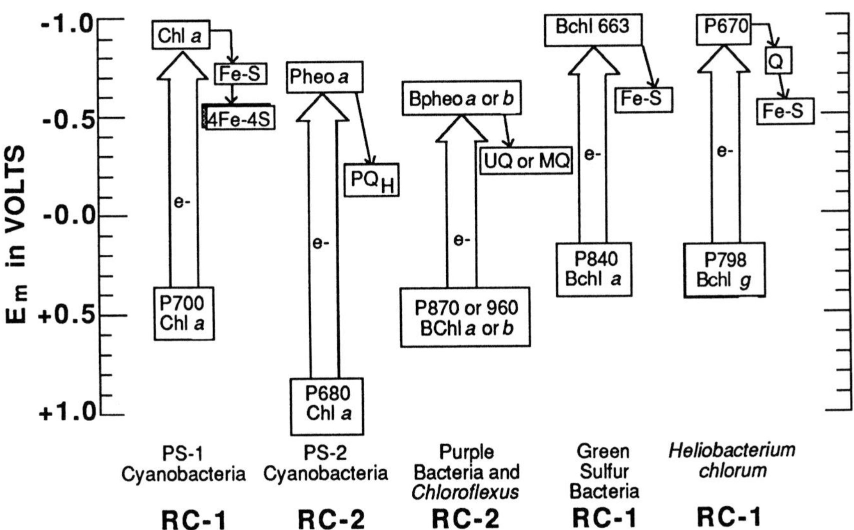

Figure 1. Contemporary RCs with electron acceptors positioned on a volt scale. The absorption maxima of the RC Chls (primary electron donors) are indicated. The initial electron acceptors are pheophytins in all RC-2s and Chls in the RC-1s. The secondary electron acceptors are quinones in all RC-2s and Fe-S centers in the RC-1s. The initial electron acceptor (P670) in *H. chlorum* (gram-positive line) is probably a Chl *a* or a Bchl *c*, and the secondary electron acceptor may involve both a quinone and an Fe-S center. Modified from reference 38.

vided into two types distinguished by the nature of the stable electron acceptor and its midpoint potential (E_m) (Fig. 1). RC-1s have a low-potential iron-sulfur (Fe-S) center (near −0.5 V) as an electron acceptor. RC-2s have a higher-potential quinone molecule (near 0 V) as an electron acceptor. Among contemporary phototrophs, only the cyanobacteria have both an RC-1 and an RC-2 type of RC. When the cyanobacteria perform anoxygenic photosynthesis, RC-2 is shut down and only RC-1 functions. Both of these contain Chl *a*, and when both are functioning the cyanobacteria are oxygenic. All the anoxygenic phototrophic bacteria have either an RC-1 or an RC-2. The purple bacteria have an RC-2 containing Bchl *a* or Bchl *b*, and the filamentous bacteria have an RC-2 containing Bchl *a*. The green sulfur bacteria have an RC-1 containing Bchl *a*, and *Heliobacterium chlorum* has an RC-1 containing Bchl *g*.

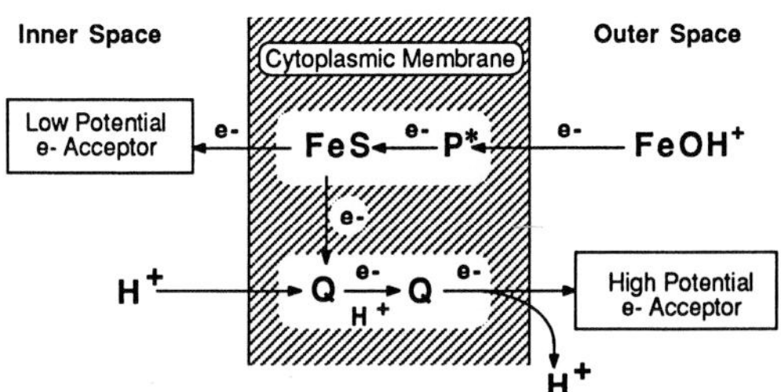

Figure 2. Primitive RC with an excited porphyrin (P*) that could be protoporphyrin IX. The excited porphyrin becomes photoreduced, accepting an electron from an external donor. The electron is transferred across the membrane through a primitive Fe-S center to a low-potential internal electron acceptor, resulting in a linear electron flow that causes an asymmetric charge distribution across the membrane. The introduction of a quinone (Q) permits linear transport of the electron back to the exterior of the cell, coupled with H^+ translocation to the outside. Reproduced from *Origins of Life* (39) with permission of the publisher.

Origin of RCs

Although the differences among contemporary RCs are significant, the overall similarity in the structure and function of RCs suggests a common origin. We believe that this origin coincided with the origin of cellular life itself and that the early evolution of the RC proceeded very rapidly and resulted in a diverse and well-established photobiota very early in the history of the Earth.

The precursors of the photoreactive Chls of contemporary RCs were probably porphyrins that were formed abiotically during chemical evolution and biogenically very early in biological evolution (19, 28, 32). The potential for abiotic formation of porphyrins from simple precursors has been recognized for some time (19), and the formation has been reproduced in the laboratory under anoxic conditions (50). Most of the precursor porphyrins formed under these conditions absorb only in the UV range and would not be particularly useful for photosynthesis. One of these early porphyrins (uroporphyrin) has been shown to form H_2 from organic molecules (EDTA) when excited with UV radiation (30, 31). To progress from the precursor uroporphyrin to the simplest of contemporary Chl molecules (Chl *a*), some enzymatic conversion is required (36). Very few steps are required, however, to go from uroporphyrin to protoporphyrin IX (the immediate precursor molecule to both hemes and Chls). We believe that the primitive ancestral RC contained protoporphyrin IX. In the stepwise progression to Chl, this porphyrin is the first that has visible absorption bands (38). The strongest absorption is at 404 nm, although there is a weak band at 633 nm. In photochemistry the excited protoporphyrin IX acts as an electron acceptor, becoming photoreduced (31, 38). The earliest RCs based on protoporphyrin IX may have provided a simple light-driven unidirectional flow of electrons across the primitive cell membrane from an external electron donor to an internal electron acceptor (Fig. 2). Such a flow could have provided energy in the form of asymmetric charge distribution across the membrane. Suitable early electron acceptors would have been a low-potential Fe-S center and perhaps ultimately a primitive ferredoxin. Involvement of a quinone with this primitive RC could have coupled electron transfer with proton translocation back across the membrane to the outside (Fig. 2), resulting in the primitive and simple creation of a proton gradient across the membrane. Association of protoporphyrin IX with simple polypeptides could have facilitated its insertion in the membrane. Porphyrins readily associate with amino acids, and the abiogenic formation of "proteinoids" from protoporphyrin IX and L-amino acids has been demonstrated (18). We believe that this simple photochemical RC functioned very early in the history of life, perhaps before the divergence of archaebacteria and eubacteria. Although no porphyrin-dependent photochemistry is known in the archaebacteria, protoporphyrin IX is synthesized in many members of this group, as well as in all phyla of eubacteria.

Each change that occurred in the stepwise transformation of protoporphyrin to Chl would have presented an advantage to the primitive phototroph. Protoporphyrin IX readily chelates metals, and the formation of such chelates with iron and magnesium probably occurred at the same time, resulting in the nearly simultaneous origin of heme and magnesium protoporphyrin. The presence of Mg in the porphyrin ring changes the photochemistry so that the excited pigment loses rather than gains an electron (38). Iron porphyrins do not exhibit photochemistry, but do function readily in chemical redox reactions. The appearance of magnesium protoporphyrins and heme groups simultaneously would have permitted a distinct improvement in the primitive RC (Fig. 2) by providing a photooxidizable pigment with a good electron transfer agent, a primitive cytochrome. Interaction of the heme group with suitable peptides readily occurs (18) and would permit the association of the resulting primitive cytochrome with the membrane, at least peripherally. Another change that would have had strong selective advantage would have been the conversion of magnesium protoporphyrin IX to protochlorophyll *a*. The significance of this change was to greatly increase the intensity of absorption in the red (622-nm) band (38, 47), providing for more efficient quantum capture and also equipping the primitive photosynthetic apparatus for efficient excitation transfer between pigment molecules, thereby leading to the development of LH systems. An even greater increase in intensity of absorption in the red band would have occurred when protochlorophyll *a* was replaced with Chl *a*. We assume that this replacement of primitive RC pigments by ever more suitable forms progressed very rapidly until Chl *a* was synthesized and incorporated into RCs. The rationale for this belief is that each intermediate requiring only one or very few synthetic steps from its precursor would have conferred a significant advantage directly on the energy-converting system of the RC and hence would have had a strong selective advantage for the organism.

The primitive RC depicted in Fig. 2 thus would have been rapidly replaced by the improved RC depicted in Fig. 3. This ancestral RC contained the essential components of all contemporary RCs. A true Chl molecule could be photooxidized and then reduced by a primitive cytochrome. Electrons could be pulled off to the interior of the cell via an Fe-S center to a low-potential internal electron acceptor (ferredoxin?) or could be cycled back to the peripheral cytochrome from the Fe-S center via quinones. When the electrons were cycled back via the quinones, protons could have been translocated to the exterior of the cell, creating an electrochemical gradient of potential energy. When the electrons were pulled off to the internal electron acceptor, a linear flow of electrons would have been established across the membrane and the primitive cytochrome would have had to be reduced by an external electron donor.

This RC (Fig. 3) might have been the

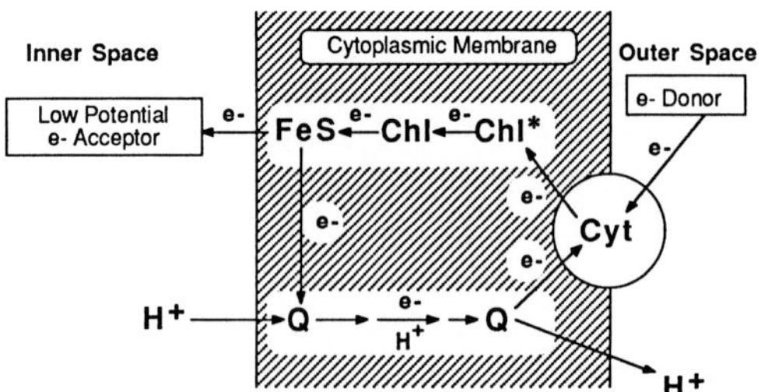

Figure 3. Improved RC containing all the essential components of a contemporary RC-1. This RC is thought to be the ancestral RC to all contemporary phototrophs. The RC pigment is a Chl which becomes oxidized when excited. The electron acceptor is an Fe-S center which can transfer electrons to a quinone (Q), which transfers them to a primitive cytochrome that functions as the electron donor to Chl^+, thus providing for cyclic electron flow coupled to H^+ translocation to the exterior of the cell. Alternatively, a linear flow of electrons across the membrane can be maintained in the presence of a suitable external electron donor. In this case the electrons are donated to an internal low-potential acceptor via the Fe-S center and can be used to sustain primitive autotrophy. Modified from reference 39.

ancestral RC for all anoxygenic phototrophs. The organisms containing it were probably ancestral to all eubacteria. We believe that the divergence of archaebacteria from eubacteria occurred after the formation of magnesium and iron protoporphyrins, but before the synthesis of Chl *a* in RCs.

The progression of RCs described above would have resulted in an RC of the RC-1 type containing Chl *a* very early in the evolution of eubacteria. We believe that Chl *a* may have been the first RC Chl in the ancestral phototrophs, because chlorophyllide *a* is synthesized first on the biosynthetic pathway to the other Chls (36). Three additional enzymatic steps are required to synthesize bacteriochlorophyllide *a* from chlorophyllide *a* (36).

Autotrophy, the Competition for Reductants, and the Origin and Evolution of RC-2

The earliest phototrophs were most probably photoheterotrophs. As the photochemical systems became more efficient, growth became more rapid and organic substrates would have been exhausted. Thus, autotrophy would have had to evolve. Although autotrophy may have begun with the primitive RC (Fig. 2), this RC would not have been able to provide both sufficient reductant for CO_2 fixation and an electrochemical energy gradient. Successful photoautotrophy would have been possible for the RC depicted in Fig. 3, which could create photochemical energy in the absence of an external reductant by cyclic electron flow coupled to H^+ translocation or could pull electrons from an external reductant to sustain CO_2 fixation. It is not our intent to discuss the evolution of autotrophy, but only to point out that as the RC evolved rapidly to the point shown in Fig. 3, autotrophy could have replaced heterotrophy as the supply of organic molecules in the environment was depleted. The result of autotrophy would have been a demand for reductants. Competition for limiting reductants could have been the impetus for the evolution of RC-2 (35). The first reductants used by RC-1 were probably low-potential reductants such as H_2 and H_2S. As these reductants were exhausted, photoautotrophs would have been forced to use more and more positive sources of reducing power until, of necessity, an RC with a more positive potential primary electron donor evolved. In addition to using H_2 and H_2S, RC-2 could have used much higher-potential reductants such as the abundant ferrous ion (36, 38). In the new organisms, RC-2 was probably linked to RC-1 from the very beginning, so that electrons from the higher-potential reductants (such as Fe^{2+}) could be delivered to $NAD(P)^+$ directly from RC-1 in a primitive Z scheme. If the new organisms found themselves in an environment containing H_2 or H_2S, they could shut down RC-2 and use only RC-1 with H_2 or H_2S as the electron donor. As the environment became more restricted (fewer varieties of reductants available), there may have been little advantage to having two different RCs, and different lines of descent would have retained either RC-1 or RC-2. The two together would have been advantageous only in Chl *a*-containing organisms in which linkage in series permitted the extraction of electrons from water. The lower-energy RCs containing Bchl *a*, *b*, or *g* rather than Chl *a*, however, would have been unable to extract electrons from water even if the RCs were linked. Hence, all the contemporary anoxygenic phototrophs descended from these lines contain either RC-1 or RC-2 but never both.

Origin of Diversity of RC Chls

Competition for light to sustain RC function would have resulted in the diversity of contemporary RC pigments. With the rapid proliferation of photoheterotrophs and photoautotrophs, the biomass in the primitive microbial environments would have increased considerably and the phototrophic organisms would have competed for light. Even before cells became highly pigmented following the evolution of LH systems, the diversity in RC pigments might have conferred an advantage to cells growing in dense communities.

Of the four different RC Chls (Table 1), Bchl *g* and Chl *a* are isomers (39). During biosynthesis at least three more enzymatic steps are required, however, to synthesize Bchl *a* and Bchl *b* than to synthesize Chl *a* (36, 47). Bchls *a*, *b*, and *g* have major absorption bands farther to the red and near infrared (NIR) than Chl *a* does, and this enables the Bchl *a*-, *b*-, and *g*-containing RCs to use light of different (lower-energy) wavelengths.

Alternative View of Origin and Early Evolution of RCs

Although we have tried to present a reasonable account of the origin and evolution of RCs, we recognize that this proposed sequence of evolutionary events occurring more than 3.5 billion years ago is purely hypothetical and is based on relatively little evidence. We believe that we should (at least briefly) present an alternative view that can just as easily be constructed from the minimal evidence. It may well be that the first primitive RC with protoporphyrin IX (Fig. 2) was an RC-2 rather than an RC-1. In this case, a quinone rather than an Fe-S center would have been the first electron acceptor transferring electrons to a high-potential electron acceptor inside the cell. The improved RC (Fig. 3) likewise would have been an RC-2 (with quinone as the first electron acceptor) that might have used reverse electron flow from the quinone pool with H_2S as the electron donor to support autotrophy. It is also possible that the improved RC-2 was based on Bchl *a* rather than Chl *a*, although this is difficult to reconcile with the known biosynthetic pathways of these pigments. If this were the case, however, then competition for reductants might have driven the evolution of RC-2 to more positive redox potentials of the electron donor, leading to the replacement of Bchl *a* with Chl *a*. As this RC-2 evolved to ever more positive redox potentials, there would have been an advantage to having an additional RC (RC-1) in series with RC-2 for the direct reduction of ferredoxin with electrons from $FeOH^+$. The evolution of RC-1 from RC-2 would have required the replacement of a quinone molecule in RC-2 by an Fe-S center in RC-1 as the first step.

Evidence for a Common Origin: the RC Polypeptides

We have previously discussed the evolutionary significance of sequence homologies in RC polypeptides (38). These homologies exist among the L, M, and H polypeptides from RC-2s of different purple bacteria containing Bchl *a* (*Rhodobacter capsulatus* and *Rhodobacter sphaeroides*) and Bchl *b* (*Rhodopseudomonas viridis*) (33, 62, 65). The homology extends to the D_1 and D_2

polypeptides of Chl *a* RC-2s from chloroplasts (14, 62). Particular regions appear to be more highly similar in amino acid sequence (14). These homologies suggest a common origin for the L and M subunits of the purple bacterial RC-2s (from both Bchl *a* and *b* RCs) and the D_1 and D_2 polypeptides of Chl *a* RC-2s. Data are not yet available on possible homologous relationships with the RC-2 polypeptides of *Chloroflexus aurantiacus* or with any of the RC-1 polypeptides.

ATP SYNTHASE AND ETSs

ATP Synthase

In contemporary phototrophs most ATP is made by ATPase (ATP synthase), the membrane-bound complex that couples the inward flow of protons to the phosphorylation of ADP. This enzyme can also use ATP hydrolysis to drive the outward flow of protons. The proton concentration difference across the membrane is established by the chemiosmotic mechanism as electrons are passed along the ETS. This mechanism of coupling electron flow to proton extrusion to build the electrochemical gradient for ATP synthesis appears to be ubiquitous. The high degree of homology, as well as similarity in structure and function, of ATPases argues strongly for a common ancestry of this enzyme among eubacteria and eucaryotic organelles (9, 13, 21, 56). Although a functionally analogous enzyme appears to be present in several archaebacteria, homology and common origins for archaebacterial and eubacterial ATPases have not yet been determined (20, 25). The significance of its role in the bioenergetics of some archaebacteria is not always clear (23).

The role of ETSs in providing an electrochemical gradient, whether created by a photosynthetic apparatus or a respiratory one, is also a universal phenomenon. The origin and evolution of this mechanism of ATP synthesis have been explored by Raven and Smith (45, 46). Assuming that fermentation was well established as the primitive means of generating energy in the form of ATP, these authors proposed that the ATPase enzyme originally functioned to extrude the accumulating waste product of protons from these cells at the cost of hydrolysis of some of the ATP produced by substrate-level phosphorylation. The enzyme originally functioned therefore solely as an ATPase. These authors also noted that the pressure would have been great to develop another, less costly mechanism to maintain intracellular pH and proposed the very early evolution of a simple cyclic photochemical (Chl-based) system which, in cycling two electrons from photochemistry through a nonheme iron, quinone, and primitive cytochrome, managed to transport two protons out of the cell. The first photosynthesis therefore would have functioned only indirectly as an energy source by sparing the use of ATP to drive the extrusion of protons (45, 46). Eventually, the two independent proton pumps, the ATPase and the Chl-based photosystem, would have become linked when an ATPase arose that could be reversed, hence coupling the inward flow of protons to the synthesis of ATP and becoming an ATP synthase (46). At this point photosynthesis first functioned as an energy-converting system. Raven and Smith (46) have also recognized the possibility that a simple prerespiratory system could have functioned instead of the photosystem to extrude protons.

Electron Transfer

We have previously discussed the origin and early evolution of the primitive RC and early electron donors and acceptors. After the establishment of the primitive photochemical system involving iron-sulfur proteins and quinones (Fig. 3) as electron acceptors, elaboration of the ETS might have functioned to create larger proton gradients and hence more energy stored per quantum.

All contemporary photosynthetic bacteria contain quinones and Fe-S centers in the ETS. Their universal presence in all photosystems and their simplicity as electron carriers support their inclusion as the earliest components of the ETS of ancestral anoxygenic phototrophs (Fig. 2).

To create a cyclic electron flow that pumped protons out of the cell, an electron carrier such as a cytochrome must have been included in the system. This system (Fig. 3), which we consider to be a more advanced or improved RC (39), is nearly identical to the system proposed by Raven and Smith (46) to function as a proton pump. In our scheme, however, this ETS evolved simply from the primitive RC containing protoporphyrin (Fig. 2) and was functioning from the beginning as a primary energy source rather than as a pH regulator. We originally proposed cytochrome *c* (cyt *c*) as the primitive cytochrome in the first cyclic ETS, but a *b*-type cytochrome could have functioned just as well in this capacity. The argument in favor of a *b*-type cytochrome is that the heme is more simply derived from the protoporphyrin IX precursor. This would be in agreement with the view (11) that cyt *b* is the more ancient of the two cytochromes. It is also easier to imagine the evolution of respiratory ETSs from such a scheme containing cyt *b* rather than cyt *c*, since most of the shorter systems associated with anaerobic respirations contain *b*-type cytochromes, whereas many lack *c*-type cytochromes and *a*-type cytochromes.

The argument in favor of cyt *c* is a compelling one, however. All known anoxygenic photosystems have *c*-type cytochromes as the electron donor to the oxidized RC Chl. The presence of cyt *b* has not been detected in *H. chlorum* (8), suggesting that only *c*-type cytochromes are essential to cyclic photosynthetic ETSs. Cytochromes of the *b* type can be extremely difficult to detect, however, and all other known phototrophs contain a cyt bc_1 complex with *b*- and *c*-type cytochromes and Fe-S centers (24, 29, 43). The presence of this type of complex or some remnant of it may yet be detected in *H. chlorum*. We are inclined to believe that the cyt bc_1 complex became established fairly early in the evolution of the ETS. It appears that photochemical energy conversion can function in some contemporary anoxygenic phototrophs in the absence of the direct participation of cyt c_2 with the RC as long as the cyt bc_1 complex is intact (7, 43). Resolution of whether cyt *b* or cyt *c* came first and of the origin of the cyt bc_1 complex is not yet possible.

Many bacteria including anoxygenic phototrophic bacteria can use respiration to meet cellular energy demands if provided with a suitable oxidant and reductant. To move from a simple photochemical ETS to a respiratory one requires the simple acquisition of a terminal oxidase and an initial reductase to remove electrons from the donating substrate. In some phototrophs and many other bacteria, the protoheme-containing cytochrome, cyt *o*, can function in the capacity of terminal oxidase. During evolution, the development of an *a*-type cytochrome could have allowed for the conservation of even more respiratory energy from an ETS when using O_2 as the terminal electron acceptor. In this way respiration could have evolved very quickly and easily from the relatively simple ETS of the primitive ancestral phototrophs.

LIGHT-HARVESTING SYSTEMS

Contemporary LH Pigments

All photosynthetic procaryotes have some sort of LH system that functions to absorb quanta and funnel excitation energy to the RCs. The LH pigments greatly increase the number of quanta that can be absorbed both by providing more absorbing molecules and by expanding the range of wavelengths that can be used beyond those that can be absorbed directly by the RC pigment-protein complexes themselves.

Table 2.
Major Pigments Present in the Photosystems of the Phototrophic Procaryotes

Group	RC pigments	LH pigments
Purple bacteria	Bchl *a*	Bchl *a* and carotenoids
	or	
	Bchl *b*	Bchl *b* and carotenoids
Green sulfur bacteria	Bchl *a*	Bchl *a* and Bchl *c*, *d*, or *e* and carotenoids
H. chlorum	Bchl *g*	Bchl *g*; few if any carotenoids
C. aurantiacus	Bchl *a*	Bch *a* and *c* and carotenoids
Cyanobacteria	Chl *a*	Chl *a* and phycobilins

The earliest LH pigments were probably the same as the RC Chls. Efficient transfer of energy among the pigment molecules would not have been possible until the major pigments present in the primitive phototrophs had evolved at least to the level of protochlorophyll. Only then was there sufficient intensity of the red absorption band for energy transfer to be efficient (38). Once this point had been reached, however, the presence of a large number of molecules identical to the RC Chl or protochlorophyll would have conferred a considerable advantage to an organism. The LH pigments would have had to be very close to the RC pigments, however, and association with small polypeptides would have permitted a means of organizing the LH pigments near the RC within the cell membrane.

The fact that most anoxygenic phototrophs have Chl molecules identical to the RC molecules functioning as their major LH pigments suggests that the LH system arose from the early RC pigments. Even phototrophs that have special accessory LH Chls (green sulfur bacteria and the *Chloroflexus* group of organisms) and those that have carotenoids or other special LH pigments (such as phycobilins in the cyanobacteria) still contain Chls identical to their RC Chls in the part of the LH system that is the immediate source of excitation transfer to the RC (66).

Whereas the RCs show considerable similarity among all phototrophs, much greater diversity exists in the LH systems. Table 2 summarizes the diversity of pigments in the LH systems of phototrophic procaryotes. All of the purple sulfur and nonsulfur bacteria contain only one type of LH Chl, Bchl *a* or *b*, depending upon which type of Chl is present in the RC. Likewise, the LH pigment of *H. chlorum* is Bchl *g*, as in the RC. Both the green sulfur bacteria and *C. aurantiacus* differ from these other anoxygenic phototrophs in having LH Chls (Bchl *c*, *d*, or *e*) that differ from the RC Bchl *a*. These Chls are found associated with a cytologically separate LH structure, the chlorosome, which is attached to but distinct from the cell membrane. Both groups, however, have Bchl *a* located at the attachment site of the chlorosome to the membrane, and this Bchl *a* funnels excitation energy absorbed by the accessory Chls in the chlorosome to the RC in the membrane (66). The cyanobacteria also have specialized structures, the phycobilisomes, which contain the phycobilins and transfer their absorbed energy to the RC in the thylakoid membranes via Chl *a* located in the membrane (66).

Evolution of Size and Diversity of LH Systems

The early origin of the limited diversity of the four RC pigments (Chl *a*, Bchl *g*, Bchl *a*, and Bchl *b*) as a result of competition for light was discussed above. We believe that simultaneously, or nearly so, there was an increase in the size of primitive LH systems composed of the same pigments as the RC and the LH systems became more organized to carry out more efficient energy transfer to the RC. Most of the LH pigments in extant organisms are associated with highly organized protein complexes composed of low-molecular-mass polypeptides (5 to 6 kilodaltons) to which the Chls are noncovalently bound. The LH systems range in size from 25 to 1,000 pigment molecules per RC (66).

The absorption characteristics of the Chls are greatly influenced by their organization in these protein complexes. Hence, organisms containing the same type of Chl organized with slightly different protein complexes can have very different absorption spectra associated with these complexes. For example, the light-harvesting Bchl *a* of *Chromatium tepidum* is organized in two different pigment-protein complexes, one with absorption maxima at 800 and 855 nm and one with a maximum at 918 nm (10), whereas the Bchl *a* protein complex from *Rhodospirillum rubrum* absorbs at 808 and 889 nm (54). Consequently, the evolution of diversity of LH systems in the anoxygenic phototrophic bacteria involves changes in two different parameters: diversity of pigment molecules themselves and slight changes in the polypeptides and resulting pigment-protein associations.

Table 3 lists the major absorption maxima characteristic of all the different LH Chls present in the photosynthetic membranes or whole cells of anoxygenic phototrophs. The diversity of the absorption characteristics is impressive, and when they are considered together with the absorption characteristics of Chl *a* and phycobilins of cyanobacteria and the multitude of carotenoids present in the photosynthetic procaryotes, it can be seen that these pigments absorb most of the visible and NIR portion of the solar spectrum (6). Photosynthesis has evolved to use nearly all of the available solar energy on the surface of the Earth, and most of the diversity in absorption of this radiation is found in the Chl pigments. Most of the carotenoid pigments absorb over a fairly narrow range (400 to 550 nm). The phycobilins also absorb over a fairly narrow range (550 to 650 nm). All the Chls taken together, however, have major absorption bands spanning the entire range of the solar spectrum from the near UV (325 nm) to the NIR (1,040 nm). There is, however, one notable gap in which no known pigments absorb radiation. This gap (930 to 1,010 nm) includes a region of the solar spectrum (930 to 970 nm) in which very little energy reaches the surface of the Earth (Fig. 4). Absorption by water vapor in the atmosphere causes this trough in the solar spectrum, and therefore one might not expect pigments to be absorbing over this range. Bchl *a* is known to absorb at 920 nm, on the blue side of this trough (10), and Bchl *b* absorbs on the red side (above 1,010 nm). It seems, however, that pigments that absorb from 970 to 1,010 nm on the long-wavelength side of the trough should also exist. If the absorption maxima of different Chls are examined (Table 3), it can be seen that in the near-UV and blue regions of the spectrum the intense absorption maxima possessed by all Chls are quite close to each other and overlap with the absorption maxima of other pigments as well (carotenoids, cytochromes, and other cellular constituents such as flavins and nonheme iron compounds). The real diversity in the absorption characteristics of the different Chls is in the wavelengths of

Table 3.
Absorption Maxima of LH Chls in Living Cells or Photosynthetic Membranes of Anoxygenic Phototrophic Bacteria

Chl	Major absorption maxima (nm)
Bchl *a*	375, 590, 790–810, 830–920
Bchl *b*	400, 600–610, 835–850, 1,015–1,040
Bchl *c*	325, 450–460, 740–755
Bchl *d*	325, 450, 725–745
Bchl *e*	345, 450–460, 710–725
Bchl *g*	420, 575, 670(?), 788
Chl *a*[a]	435, 670–680

[a] Chl *a* is included because it is the Chl pigment present in cyanobacteria, some of which can perform anoxygenic photosynthesis, and because we believe that it is ancestral to all other Chls.

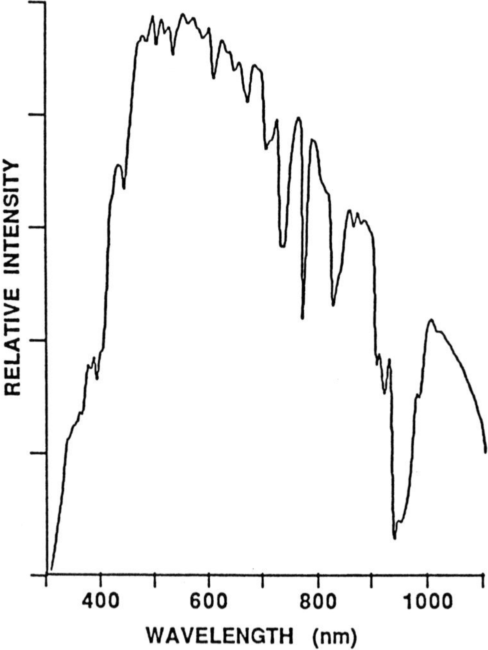

Figure 4. Incident solar spectrum at the surface of a microbial mat at Great Sippewissett Salt Marsh.

their red and NIR absorption bands, which range from 670 to 1,040 nm and do not overlap with absorption spectra of any other biological molecules. We believe that the great diversity of these LH Chl systems arose from the competition for light very early in shallow environments that were becoming very densely covered with phototrophic microorganisms accumulating in thick mats.

Evolution of Pigments in the Mat Environment

We believe that diversification of Chls in RCs and LH Chl-protein complexes probably occurred very early in mat environments where diverse absorbances in the red and NIR bands would be most useful to the organisms. Since the penetration of red and NIR radiation into water is much poorer than that of shorter-wavelength radiation, pigments absorbing in the red and NIR regions are useful only in shallow water or surficial environments. Penetration of wavelengths higher than 1,000 nm is less than 100 cm in clear water. Since no other biological molecules absorb at these wavelengths and since transfer of excitation energy occurs readily among molecules with strong absorption bands in the red and NIR regions, these diverse Chl molecules would have been ideally suited for LH in dense surface or shallow water mat communities.

Data collected on penetration of radiation in such mats today indeed reveal that red and NIR radiation penetrates deeply into mats (15, 16) and that selection for an attenuation of this radiation by different Chls at different depths in the mats allows for multilayering of phototrophic bacteria containing different Chl-protein complexes. Figures 5a to d show a series of spectra of incident radiation reaching different depths in the surficial sand mats at Great Sippewissett Salt Marsh, Mass. The spectra were recorded with a battery-operated portable spectroradiometer (no. 1800; LI-COR, Lincoln, Nebr.). By comparing Fig. 5a with an incident solar spectrum at the surface of the mat (Fig. 4), one can see that the first millimeter of the mat selectively reduces radiation from 300 to 700 nm but that most of the far-red and NIR radiation penetrates further. The first millimeter of this section of mat contained cyanobacteria and diatoms with Chl *a* and carotenoids (42a). The selective attenuation of red radiation around 670 nm absorbed by Chl *a* can be seen in Fig. 5a as the deep trough occurring from 660 to 700 nm. Below the rest of the dense cyanobacterial layer (at a depth of about 2 mm in the mat), all radiation below 500 nm and from 660 to 700 nm has been almost totally attenuated (Fig. 5b). Far-red (beyond 700 nm) and NIR radiation continue to penetrate. At 1 mm deeper in the mat, below the next layer, composed of purple sulfur bacteria containing Bchl *a* (42a), attenuation of radiation near 800 nm and from 830 to 890 nm becomes apparent (Fig. 5c) owing to

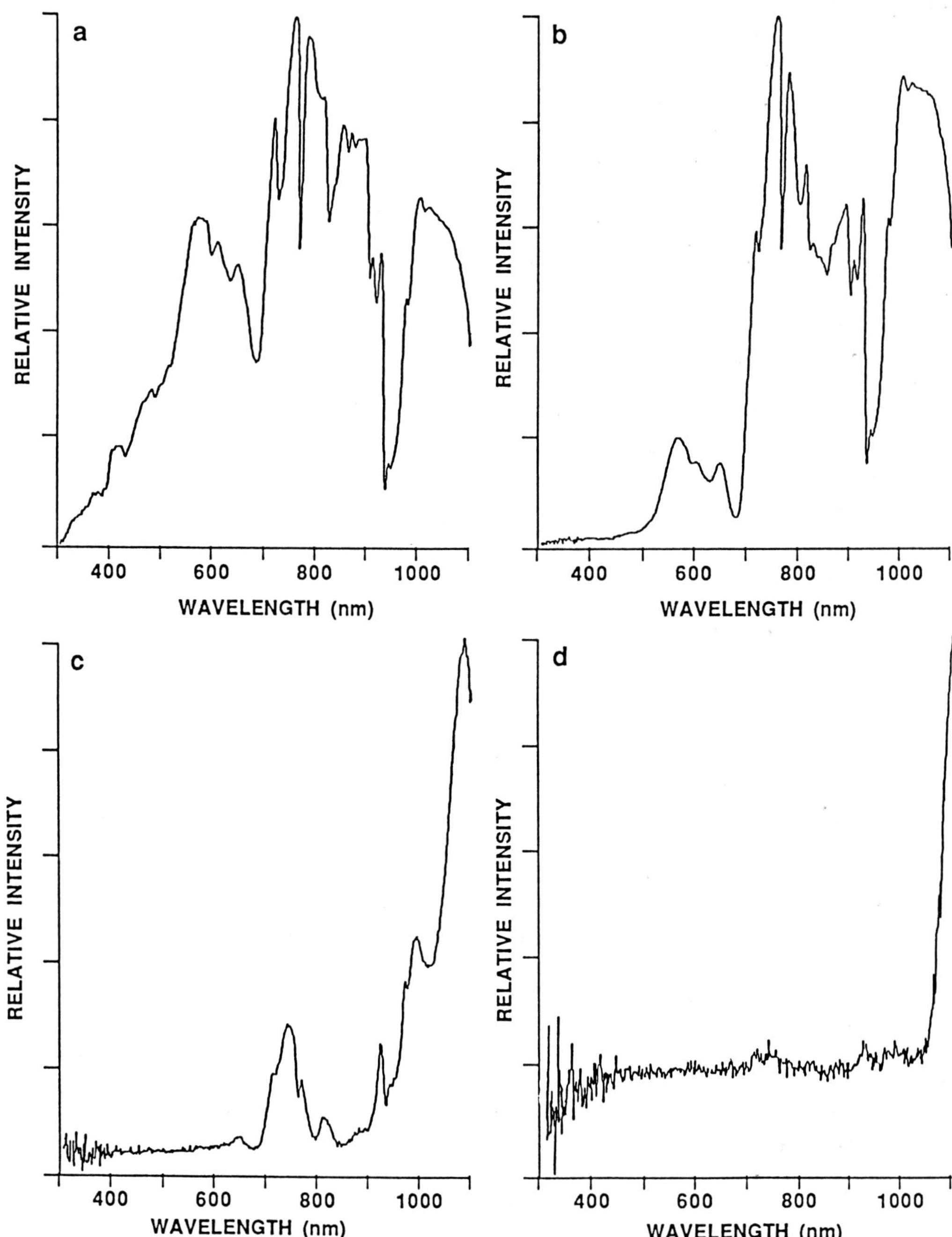

Figure 5. Spectra of radiation transmitted through the microbial mat layers at Great Sippewissett Salt Marsh. (a) Radiation transmitted through the surface gold layer (thickness, 1.0 mm) containing diatoms and cyanobacteria with carotenoids and Chl *a*. (b) Radiation transmitted through both the surface gold layer and the next 1 mm of mat containing cyanobacteria. (c) Radiation transmitted through the first two layers and the purple sulfur bacterial layer containing Bchl *a*. (d) Radiation reaching the bottom of the microbial mat.

absorption by the Bchl *a*-protein complexes of these bacteria. A significant trough is also seen at 1,020 nm owing to absorption by Bchl *b* present in this layer. It should be noted that radiation absorbed by Bchl *c* (740 nm) is still transmitted through this layer. In the next millimeter down, the remaining radiation is totally absorbed by Bchl *b* (*Thiocapsa pfennigii*) and Bchl *c* (*Prosthecochloris aestuarii*) (42a).

It is clear from these studies that thickly layered mats of phototrophs can develop when the organisms making up the different layers have different absorption characteristics in the red and NIR regions. We believe that an elaboration of diverse Chl-based LH systems must have occurred very early in the history of the Earth to permit the widespread development of thick microbial mats of phototrophic origin in the Precambrian era. It appears unlikely that dense, multilayered mat communities greater than 1 or 2 mm thick could have developed in the absence of diverse Chl LH systems.

The origin and evolution of the accessory Chls Bchl *c*, *d*, and *e* as LH systems for the green sulfur bacteria and members of the family *Chloroflexaceae* probably occurred after the diversification of the Chl *a* and Bchl *a*, *b*, and *g* RCs and associated LH systems. No RCs are based on Bchl *c*, *d*, or *e*, and these pigments are housed in chlorosomes, unique LH structures. The nearly exclusive association of these pigments with this specialized LH organelle suggests a more recent origin. The fact that the pigments, polypeptides, and resultant structure are unique to two groups of organisms that do not appear to be closely related to each other (63) suggests either a common origin for the structure accompanied by an early divergence or, possibly, a lateral gene transfer. The latter suggestion is an intriguing possibility. Although the chlorosomes of *C. aurantiacus* and the green sulfur bacteria have some significant differences in structure and composition (66), they are indeed very similar. The essential part of the photosynthetic apparatus, the RC, is membrane bound in both groups of organisms. The chlorosomes, however, vary in size and number with species and growth conditions. The synthesis of Bchl *a* and the LH Bchl *c* are under different control in *C. aurantiacus*, and in many ways the regulation of the chlorosome and its pigments is independent of that of the rest of the photosynthetic apparatus. If the genes for the chlorosome structural components and pigments are clustered together, one might imagine their being acquired as a single unit, perhaps via a plasmid. Without the chlorosome, the RC and membrane-bound LH Bchl *a* complexes in *C. aurantiacus* are quite similar to those of purple bacteria (66). The evolution of the chlorosome is an important question in the overall evolution of the anoxygenic photosynthetic bacteria, but we have little evidence with which to speculate, primarily owing to our lack of knowledge of chlorosome genetics.

Carotenoids

Carotenoids play a significant role as LH pigments in many planktonic phototrophic bacteria. Most of the LH Chl complexes of phototrophic bacteria contain carotenoids. However, in mat communities, carotenoids confer little advantage as LH pigments to organisms growing below the first 1 or 2 mm. Consequently, we doubt that the high diversity of carotenoids found in phototrophs today evolved under these circumstances. We believe that the initial function (and even today, the most important function) of carotenoids associated with the photosynthetic apparatus was as a protective agent against photooxidative damage. The development of the oxygen-evolving photosystem II would have initially created toxic conditions for early phototrophs, and protection by carotenoids probably evolved quickly after O_2 began to accumulate. All phototrophic bacteria today contain some carotenoids, although the heliobacteria group of organisms appear to have only very small amounts of these pigments and, interestingly, can grow only under strictly anoxic conditions (3). Carotenoid-deficient mutants of phototrophs such as *Rhodobacter sphaeroides* (51) and *C. aurantiacus* (42) form a functional photosynthetic apparatus and grow photosynthetically as long as oxygen is excluded from their environments. In addition to their protective function, carotenoids serve a significant LH function in deep aquatic environments where wavelengths in the 400- to 550-nm range penetrate and where the longer wavelengths absorbed by the far-red and NIR bands of BChls do not penetrate (27).

LH Polypeptides

Most of the LH Chls of anoxygenic phototrophic bacteria are associated with low-molecular-weight polypeptides. One of the most convincing lines of evidence for a common origin of photosynthesis is the high degree of homology that exists among these polypeptides from diverse groups of bacteria (66). The best characterizations of the LH polypeptides are those from the pigment-protein complexes of purple bacteria. The sequence homology and orientation of these complexes in membranes have been recently reviewed (66) in detail.

All of the purple photosynthetic bacteria examined to date have at least one LH complex associated with the RC which transfers excitation energy to the RC. This complex is termed the B870 or B890 complex in Bchl *a*-containing bacteria and the B1015 complex in Bchl *b*-containing bacteria. A similar complex (B808–866) in *C. aurantiacus* has recently been described (59). The complex is composed of two small polypeptides termed α and β, which have very low homology with each other even in the same organism. The α polypeptide isolated from the distantly related *C. aurantiacus*, however, has an impressive 27 to 39% homology with the α polypeptides isolated from the B890 (B870) and B1015 complexes of *Rhodospirillum rubrum*, *Rhodopseudomonas viridis*, *Rhodobacter sphaeroides*, and *Rhodobacter capsulatus* (60). Likewise, the β polypeptide from the B808–866 complex of *C. aurantiacus* has a high homology with the β polypeptides from the B890 (B1015) complexes of the same bacteria (59). Some purple bacteria have an additional LH complex (B800–850 in *Rhodobacter sphaeroides* and *Rhodobacter capsulatus*) which is also composed of α and β polypeptides. The homology of the *Chloroflexus* B806–866 α and β polypeptides with those of the B800–850 complexes is lower than with the B890 complexes (59). Of particular interest will be the much-needed comparison of sequences of Chl *a* LH polypeptides and Bchl *g* LH polypeptides with the above sequences.

All the Bchl *c* in the chlorosomes of *C. aurantiacus* is associated with a 5.6-kilodalton polypeptide (61). The Bchl *c* of the green sulfur bacterium *Chlorobium* sp. is associated with a 7.5-kilodalton polypeptide isolated from chlorosomes. Both proteins appear to bind six or seven Bchl *c* molecules. It may be significant that the Bchl *c*-binding protein in *Chloroflexus* chlorosomes contains only 51 amino acid residues (61), whereas the corresponding protein in *Chlorobium* chlorosomes contains 74 residues (10a). The two proteins show about 30% homology only over a short section of about 13 residues. A common origin is possible but by no means certain.

PHYLOGENY

Recent phylogenetic evidence based on molecular sequences of 16S rRNA clearly establishes photosynthesis as an ancient property in the eubacterial line of descent (53, 63, 64). There are two results of the phylogenetic analysis based on oligonucleotide catalogs and sequence data that support this view. First, photosynthesis based on Chl-containing RCs with similar ETSs is found in 5 of the approximately 10 eubacterial phyla (Fig. 6). The high degree of similarity of photosynthesis in these different phyla and its complexity argue for a common photosynthetic ancestor of all these phyla. It seems likely that this will also mean that there was a common ancestor of all 10 phyla and that it was a phototroph (63).

The second convincing result of such phylogenetic analysis is that one of the deeply branching phyla in the eubacterial line contains the phototrophic filamentous organism *C. aurantiacus* (41). That the organisms in this phylum truly represent a very deep branch in the kingdom rather than a fast clock has been established (41). Since photosynthesis is present in one of the ancient branches, it seems possible that the ancestral eubacterium was phototrophic.

It has been argued that these ancestral phototrophs were thermophiles, on the basis of the observations that thermophily is also widespread in the eubacterial phyla and is a characteristic of several of the organisms in the deepest branches including *C. aurantiacus* (1). It is clear from many recent observations (25a; E. D'Amelio and Y. Cohen, unpublished results), however, that *Chloroflexus*-like organisms are abundant in nonthermal environments, and it may indeed be that thermophily, although present in contemporary organisms from the ancient branches, is a derived characteristic from mesophilic ancestors.

There are no phylogenetic data to argue in favor of or against the possibility of a common photosynthetic ancestor for both eubacteria and archaebacteria. We believe that the possibility is worth considering seriously on biochemical grounds explained above. It is possible that a protoporphyrin-containing phototrophic ancestor led to two lines of descent: the archaebacterial line, containing cytochromes and other porphyrin-derived compounds, and the eubacterial line, containing cytochromes and also the photochemically active Chls.

An alternative phylogenetic scheme relating the halobacteria (bacteriorhodopsin-containing phototrophic archaebacteria) and

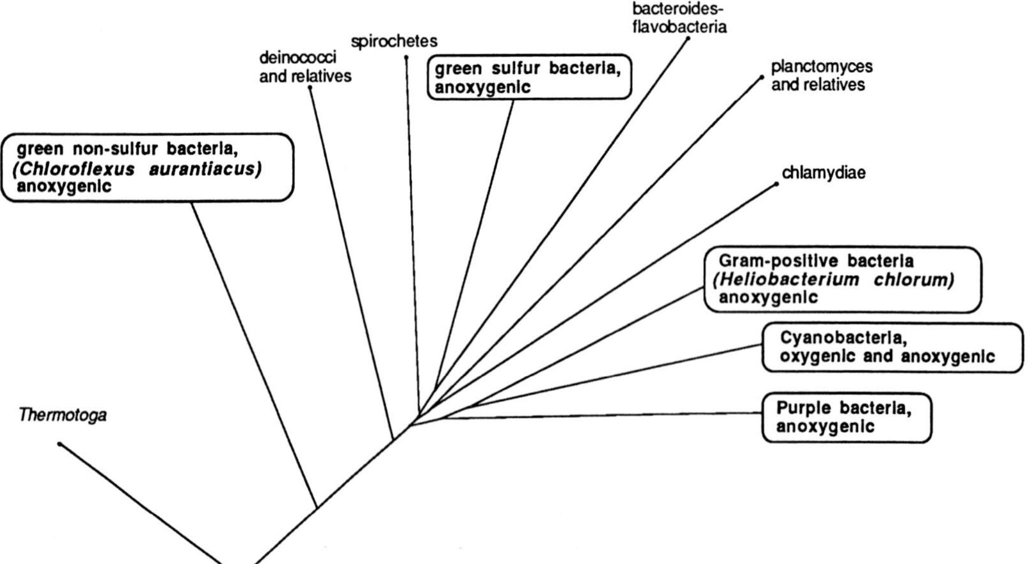

Figure 6. Eubacterial phylogenetic tree. Phyla containing phototrophs are set off in boxes. Modified from reference 63.

phototrophic eubacteria as being derived from a common phototrophic ancestor has been presented (22). This photocyte hypothesis is not acceptable to us (38) and has also been rejected by others (34). Halobacterial photosynthesis bears no relation to Chl-based photosynthesis and appears to be both a much more recently evolved system and an isolated phenomenon (38). There is also no strong phylogenetic evidence to support the photocyte hypothesis from 16S rRNA analyses (63).

EXPOSURE TO UV RADIATION DURING EARLY EVOLUTION

Early in the Precambrian era in the absence of O_2, there was no ozone in the primitive Earth atmosphere, and the surface of the Earth received a much higher flux of UV radiation than it does today. Although it is generally agreed that the early atmosphere on Earth was deficient in oxygen, it is not certain to what level. The flux of UV reaching the surface of the Earth is very dependent on the level of oxygen in the atmosphere (17). Under the worst possible conditions of less than 10^{-5} present atmospheric level of oxygen, there would have been nearly no attenuation of UV radiation by the atmosphere, and levels reaching 0.01 to 0.1 W m^{-2} nm^{-1} over the damaging range of 240 to 270 nm would have reached the surface (17). Under these conditions it seems doubtful at first that early organisms could have survived or grown.

Since it is clear that microorganisms evolved and microbial mats were well established early in the Precambrian era, some mechanism(s) must have been functioning to protect these organisms from the deleterious effects of the high UV flux. Several mechanisms that might have afforded protection have been discussed previously. These include avoidance behavior; the matting habit; absorbance of UV radiation by nitrates, nitrites, and organic substances in the water; and repair mechanisms under light and dark conditions (26, 44). These authors have suggested that in the mat environment the surface cell layers may have been killed by direct exposure to UV radiation, but may have continued to afford adequate protection to permit the growth of cells in lower layers.

Role of Iron in UV Protection

We agree with the conclusions of Rambler and Margulis (44) that exposure to high UV fluxes was not a serious deterrent to early microorganisms on Earth. It does not seem likely, however, that the thick, matting growth habit would have protected the earliest anoxygenic phototrophs, which initially may have formed very thin films among the sediment particles. It also does not seem likely that in the absence of oxygen the concentration of nitrates and nitrites would have been high enough to provide a UV screen. Likewise, we do not think that the concentration of organic molecules was likely to have been high enough to provide an effective screen. We have suggested iron as an abundant and potentially very effective UV screen (37). Both ferrous and ferric iron absorb strongly from 220 to 270 nm (37), and both were abundant in the Precambrian era (55). Ferrous iron (1 mM) in solution above pH 6.5 ($\varepsilon_{265} \approx 12$ cm^{-1} M^{-1}) could have afforded considerable protection ($A_{265} = 1.2$) to microorganisms growing in 1 m of water. Although ferric iron is insoluble in water, its presence in sediment particles could have afforded similar protection to organisms growing in surficial sediments in the absence of a covering water column. A 1-mm layer of sediment containing 0.1% Fe^{3+} would have had an A_{265} greater than 1.5 (37). It seems clear that in the presence of Fe^{2+} in the water and Fe^{3+} in the sediments, both shallow-water and surface mats of phototrophic organisms could have flourished despite the high UV flux characteristic of the Precambrian era.

SUMMARY AND CONCLUSIONS

The major events in our theoretical scheme for the evolution of anoxygenic photosynthesis are illustrated in Fig. 7. Included in this figure are the environmental factors that were probably most significant in the evolution of phototrophs and an estimation of when the major evolutionary events might have occurred. This figure summarizes our theory of the evolution of photosynthesis. We note here some of the most significant aspects of our theory and recognize some interesting alternatives.

We believe that the first primitive photosynthetic RC based on protoporphyrin IX functioned to convert solar energy to useful chemical energy in the form of transmembrane charge separation in the earliest cells on Earth. These cells were heterotrophs. Very rapidly, the RC pigment evolved to Chl *a* and the true ancestral RC of all contemporary phototrophs began functioning with Fe-S centers, quinones, and primitive cytochromes. The advent of heme occurred nearly simultaneously with Chl and permitted not only photochemical electron transport but also respiratory electron transport. Therefore, anaerobic respiration also evolved very early. Sometime during the development of the first true RC-1 containing Chl, the archaebacterial and eubacterial lines of descent diverged.

Competition for organic-carbon sources among primitive photoheterotrophs drove the selection for simple autotrophy. The presence of autotrophy resulted in an increased demand for reductants. As strong reductants were depleted, RC-2 evolved, permitting the use of a wider range of reductants. As growth became dense in shallow and surficial mats, competition for light led first to diversification of the RC pigments and second to an expansion and then diversification of LH pigments. The first LH pigments were identical to the RC pigments. The diversification of Chl pigments permitted growth in layered microbial mat communities where far-red and NIR radiation penetrated to greater depths.

At first these ancient phototrophic procaryotes differed little from each other except in the nature of their Chl pigments. It was only after this diversification of pigments around 3.5 billion years ago that the five lines of descent of contemporary anoxygenic phototrophs would have become distinct. The first line of descent to diverge was that of *Chloroflexus* spp., retaining heterotrophy, primitive autotrophy, and an RC-2 based on Bchl *a*, while losing RC-1. The second and third lines of descent, both losing RC-2, were those of the green sulfur bacteria and *Heliobacterium* spp. The green sulfur bacteria had Bchl *a*, and *Heliobacterium* spp. had Bchl *g*. The green sulfur bacteria had a more advanced autotrophy (the reductive tricarboxylic acid cycle), whereas *Heliobacterium* spp. remained photoheterotrophs. Following this divergence, advanced autotrophy developed in the main lineage, leading to the reductive pentose phosphate pathway, and the fourth line of descent diverged, leading to the purple bacteria containing RC-2 with either Bchl *a* or *b*. In the original primary line of descent containing Chl *a* in RC-1 and RC-2, the two RCs remained linked in series, permitting the use of water as a reductant and the release of oxygen. This splitting of water probably occurred by 3.0 billion years ago and produced the oxygen required for the subsequent development of aerobic respiration.

In consideration of our alternative view on the origin and evolution of RCs discussed above and recent phylogenetic evidence (Fig. 6) (63), we recognize an alternative sequence of events to that depicted in Fig. 7. If the ancestral RC were an RC-2, the common ancestor to all eubacteria might have been a phototroph containing Bchl *a* and RC-2. The phylum containing *Chloroflexus* spp. might have diverged from this ancestor and its remaining descendants before the evolution of RC-1 and Chl *a*. The common ancestor to the four remaining phyla con-

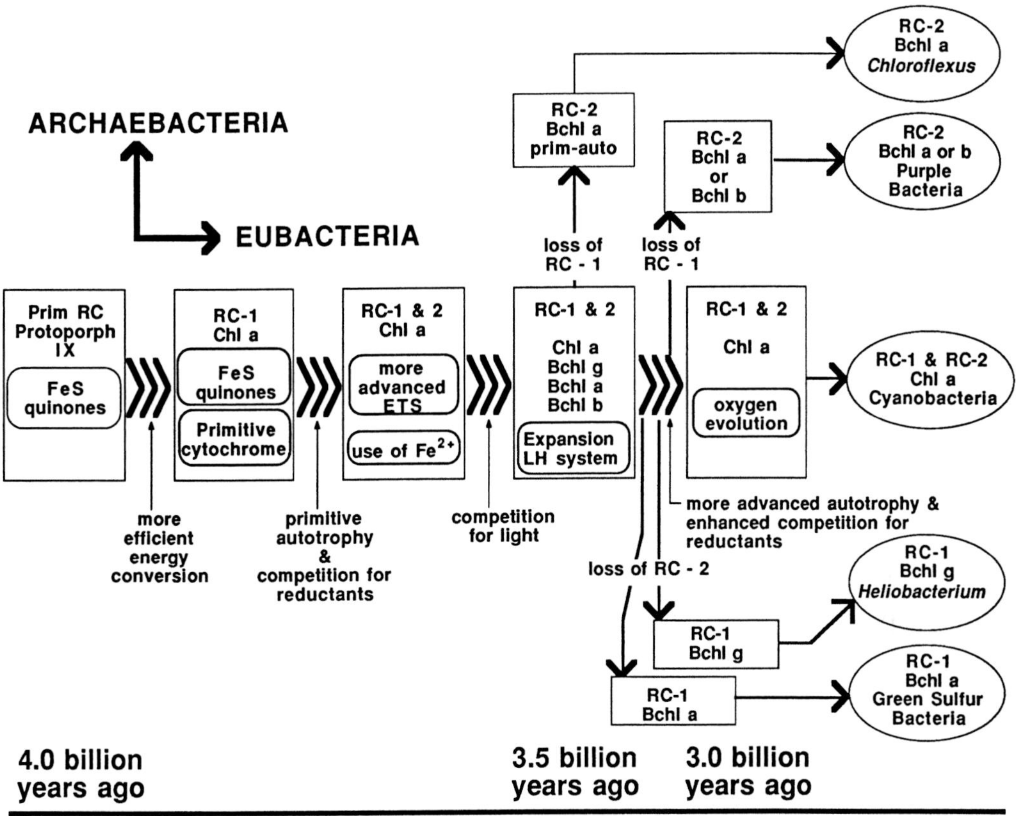

Figure 7. Diagram of the evolution of photosynthesis. See the text for a detailed explanation of this figure. The boxes represent major ancestral organisms (or groups of closely related organisms) in the early evolution of anoxygenic phototrophs. The ovals represent the five phyla of the contemporary phototrophic descendants from these ancestors. The specific Chls represent the RC Chls only. The vertical arrows indicate major events in the evolution of photosynthesis. The time line is not linear and is included to indicate reasonable guesses of when some of these major events might have occurred. The first box represents the evolution of the earliest cells with a primitive RC containing protoporphyrin IX and using Fe-S centers and quinones, around 4.0 billion years ago. During the stepwise progression to a more efficient photosystem containing Chl *a* and primitive cytochromes, the archaebacteria and eubacteria diverged. The second box and all the remaining events in this figure occur in the eubacterial line of descent. Divergence of the ancestors of the five major phototrophic phyla might have begun around 3.5 billion years ago with the divergence of the ancestors of the contemporary green, filamentous, nonsulfur bacteria. The other anoxygenic phototrophic ancestors diverged from each other more recently. The cyanobacteria were even more recent in origin, probably not becoming a distinct phylum until after the origin of oxygen evolution.

taining phototrophs would have been a photosynthetic bacterium containing RC-1 and RC-2, Bchl *a*, and perhaps Chl *a* as well. The presence of Bchl *a* in three of the five phyla containing phototrophs argues for its antiquity as an RC pigment. The known biosynthetic pathways of Chl *a* and Bchl *a*, however, support the evolutionary view presented in Fig. 7. It is worth noting that our knowledge of biosynthesis of Bchl *a* has been obtained from studies with the purple bacteria. Nothing is known of the biosynthetic pathway in *Chloroflexus* spp. If a different pathway were found that did not involve chlorophyllide *a*, this would be compelling evidence for the rejection of Chl *a* as the ancestral RC chlorophyll and would support the alternative view.

The following conclusions may be drawn from the above discussion. (i) Cyanobacteria are the culmination of the main line of descent from the primitive phototroph. They are in the most recent branch of the extant phototrophs and, as such, should exhibit the most modern features of photosynthesis, as well as retain evidence of the primitive features. (ii) *Chloroflexus* spp. are in the most ancient branch of the extant phototrophs. Therefore, features that this genus shares with the other branches must be primitive.

SOME PREDICTIONS

The major separation that occurred between the archaebacteria and eubacteria (nearly 4 billion years ago) and the diversification into several lines of descent among eubacteria (around 3.5 billion years ago) might have left behind some ancestral intermediates. If descendants of some of these "forgotten" early procaryotes exist today, then we might be able to find the following organisms: archaebacteria with a porphyrin-based photosystem, Chl *a*-containing procaryotes with RC-1 only, Chl *a*-containing procaryotes with RC-2 only, procaryotes with both Chl *a* and Bchl *a*, cyanobacteria (Chl *a* with RC-1 and RC-2, using Fe^{2+} as an electron donor to RC-2), and Bchl *a*-containing procaryotes with both RC-1 and RC-2. These are only a few examples of phototrophic missing links. Of particular significance to the verification of our theory would be the discovery of an archaebacterium with a primitive porphyrin-based photosystem.

Although it is possible that descendants of primitive Chl *a*-containing phototrophs with only one RC (similar to all the Bchl-containing phototrophs) could have survived to the present, it does not seem likely that such organisms would be abundant. Once water began to function as a reductant in the Chl *a*-containing phototrophs retaining both RC-1 and RC-2, the autotrophic descendants of these organisms (the contemporary cyanobacteria) had such a strong advantage over all other phototrophs that they literally covered the Earth and remain the most abundant and ubiquitous of all phototrophic procaryotes in the world today. Anoxygenic Chl *a*-containing phototrophs with only one RC would not be expected to compete well for the same wavelengths of light. All the known extant phototrophs with only one RC can survive in the shadow of Chl *a*-containing cyanobacteria because their Bchls absorb different wavelengths of light.

ACKNOWLEDGMENTS. We thank Scott Pierson for preparation of the figures.

Portions of this work were supported by an M. J. Murdock Charitable Trust Corporation Grant of Research to B.K.P. and by National Science Foundation grant BSR-8521724 to B.K.P.

LITERATURE CITED

1. **Achenbach-Richter, L., R. Gupta, K. O. Stetter, and C. R. Woese.** 1987. Were the original eubacteria thermophiles? *Syst. Appl. Microbiol.* **9**:34–39.
2. **Awramik, S. M.** 1984. Ancient stromatolites and microbial mats, p. 1–22. *In* Y. Cohen, R. W. Castenholz, and H. O. Halvorson (ed.), *Microbial Mats: Stromatolites*. Alan R. Liss, Inc., New York.
3. **Beer-Romero, P., and H. Gest.** 1987. *Heliobacillus mobilis*, a peritrichously flagellated anoxyphototroph containing bacteriochlorophyll *g*. *FEMS Microbiol. Lett.* **41**:109–114.
4. **Broda, E.** 1978. *The Evolution of the Bioenergetic Process*. Pergamon Press, Oxford.
5. **Byerly, G. R., D. R. Lowe, and M. M. Walsh.** 1986. Stromatolites from the 3,300-3,500-Myr Swaziland Supergroup, Barberton Mountain Land, South Africa. *Nature* (London) **319**:489–491.
6. **Clayton, R. K.** 1971. *Light and Living Matter*, vol. 2, p. 31–36. McGraw-Hill Book Co., New York.
7. **Daldal, F., S. Chena, J. Appelbaum, E. Davidson, and R. C. Prince.** 1986. Cytochrome c_2 is not essential for photosynthetic growth of *Rhodopseudomonas capsulata*. *Proc. Natl. Acad. Sci. USA* **83**:2012–2016.
8. **Fuller, R. C., S G. Sprague, H. Gest, and R. E. Blankenship.** 1985. A unique photosynthetic re-

action center from *Heliobacterium chlorum*. *FEBS Lett.* **182**:345–349.

9. **Futai, M., and H. Kanazawa.** 1983. Structure and function of proton-translocating adenosine triphosphatase (F_0F_1): biochemical and molecular biological approaches. *Microbiol. Rev.* 47:285–312.
10. **Garcia, D., P. Parot, A. Vermeglio, and M. T. Madigan.** 1986. The light-harvesting complexes of a thermophilic purple sulfur photosynthetic bacterium *Chromatium tepidum*. *Biochim. Biophys. Acta* **850**:390–395.

10a. **Gerola, P. D., P. Højrup, J. Knudsen, P. Roepstorff, and J. M. Olson.** 1988. The bacteriochlorophyll *c*-binding protein from chlorosomes of *Chlorobium limicola* f. *thiosulfatophilum*, p. 43–52. *In* J. M. Olson, J. G. Ormerod, J. Amesz, E. Stackebrandt, and H. G. Trüper (ed.), *Green Photosynthetic Bacteria*. Plenum Publishing Corp., New York.

11. **Gest, H.** 1981. Evolution of the citric acid cycle and respiratory energy conversion in prokaryotes. *FEMS Microbiol. Lett.* **12**:209–215.
12. **Gest, H., and J. W. Schopf.** 1983. Biochemical evolution of anaerobic energy conversion: the transition from fermentation to anoxygenic photosynthesis, p. 135–148. *In* J. W. Schopf (ed.), *Earth's Earliest Biosphere: Its Origin and Evolution*. Princeton University Press, Princeton, N.J.
13. **Godinot, C., and A. DiPietro.** 1986. Structure and function of the ATPase-ATPsynthase complex of mitochondria as compared to chloroplasts and bacteria. *Biochimie* **68**:367–374.
14. **Hearst, J. E.** 1986. Primary structure and function of the reaction center polypeptides of *Rhodopseudomonas capsulata*—the structural and functional analogies with photosystem II polypeptides of plants, p. 382–389. *In* L. A. Staehelin and C. J. Arntzen (ed.), *Encyclopedia of Plant Physiology, New Series, Photosynthesis III*, vol. 19. Springer-Verlag KG, Heidelberg.
15. **Jørgensen, B. B., Y. Cohen, and D. J. Des Marais.** 1987. Photosynthetic action spectra and adaptation to spectral light distribution in a benthic cyanobacterial mat. *Appl. Environ. Microbiol.* **53**: 879–886.
16. **Jørgensen, B. B., and D. J. Des Marais.** 1986. Competition for sulfide among colorless and purple sulfur bacteria in cyanobacterial mats. *FEMS Microbiol. Ecol.* **38**:179–186.
17. **Kasting, J. F.** 1987. Theoretical constraints on oxygen and carbon dioxide concentrations in the Precambrian atmosphere. *Precambrian Res.* **34**: 205–229.
18. **Kolesnikov, M. P., N. I. Voronova, and I. A. Egorov.** 1981. Molecular complexes of amino acids with porphyrins as possible precursors of pigment-protein systems. *Origins Life* **11**:223–231.
19. **Krasnovsky, A. A.** 1976. Chemical evolution of photosynthesis. *Origins Life* **7**:133–143.
20. **Kristjansson, H., M. H. Sadler, and L. I. Hochstein.** 1986. Halobacterial adenosine triphosphatases and the adenosine triphosphatase from *Halobacterium saccharovorum*. *FEMS Microbiol. Rev.* **39**:151–157.
21. **Kröger, A., J. Paulsen, and I. Schröder.** 1986. Phosphorylative electron transport chains lacking a cytochrome bc_1 complex. *J. Bioenerg. Biomembr.* **18**:225–234.
22. **Lake, J. A., M. W. Clark, E. Henderson, S. P. Fay, M. Oakes, A. Scheinman, J. P. Thornber, and R. A. Mah.** 1985. Eubacteria, halobacteria, and the origin of photosynthesis: the photocytes. *Proc. Natl. Acad. Sci. USA* **82**:3716–3720.
23. **Lancaster, J. R.** 1986. A unified scheme for carbon and electron flow coupled to ATP synthesis by substrate-level phosphorylation in the methanogenic bacteria. *FEBS Lett.* **199**:12–18.
24. **Ljungdahl, P. O., J. D. Pennoyer, D. E. Robertson, and B. L. Trumpower.** 1987. Purification of highly active cytochrome bc_1 complexes from phylogenetically diverse species by a single chromatographic procedure. *Biochim. Biophys. Acta* **891**: 227–241.
25. **Lübben, M., and G. Schafer.** 1987. A plasma-membrane associated ATPase from the thermoacidophilic archaebacterium *Sulfolobus acidocaldarius*. *Eur. J. Biochem.* **164**:533–540.

25a. **Mack, E. E., and B. K. Pierson.** 1988. Preliminary characterization of a temperate marine member of the *Chloroflexaceae*, p. 237–241. *In* J. M. Olson, J. G. Ormerod, J. Amesz, E. Stackebrandt, and H. G. Trüper (ed.), *Green Photosynthetic Bacteria*. Plenum Publishing Corp., New York.

26. **Margulis, L., J. C. G. Walker, and M. Rambler.** 1976. Reassessment of roles of oxygen and ultraviolet light in Precambrian evolution. *Nature* (London) **264**:620–624.
27. **Matheron, R., and R. Baulaigue.** 1977. Influence de la pénétration de la lumière solaire sur le développement des bactéries phototrophes sulfureuses dans les environnements marins. *Can. J. Microbiol.* **23**:267–270.
28. **Mauzerall, D.** 1976. Chlorophyll and photosynthesis. *Philos. Trans. R. Soc. London Ser. B* **273**:287–294.
29. **Melandri, B. A., and G. Venturoli.** 1984. Photosynthetic electron transport, p. 95–148. *In* L. Ernster (ed.), *Bioenergetics. New Comprehensive Biochemistry*, vol. 9. Elsevier Biomedical Press, Amsterdam.
30. **Mercer-Smith, J. A., and D. C. Mauzerall.** 1981. Molecular hydrogen production by uroporphyrin and coproporphyrin: a model of the origin of

photosynthetic function. *Photochem. Photobiol.* **34:** 407–410.

31. **Mercer-Smith, J. A., and D. C. Mauzerall.** 1984. Photochemistry of porphyrins: a model for the origin of photosynthesis. *Photochem. Photobiol.* **39:** 397–405.
32. **Mercer-Smith, J. A., A. Raudino, and D. C. Mauzerall.** 1985. A model for the origin of photosynthesis. III. The ultraviolet photochemistry of uroporphyrinogen. *Photochem. Photobiol.* **42:** 239–244.
33. **Michel, H., K. A. Weyer, H. Gruenberg, and F. Lottspeich.** 1985. The 'heavy' subunit of the photosynthetic reaction center from *Rhodopseudomonas viridis*: isolation of the gene, nucleotide and amino acid sequence. *EMBO J.* **4:**1667–1672.
34. **Olsen, G. J., and C. R. Woese.** 1986. Archaebacterial phylogeny: perspectives on the urkingdoms. *Syst. Appl. Microbiol.* **7:**161–177.
35. **Olson, J. M.** 1970. The evolution of photosynthesis. *Science* **168:**438–446.
36. **Olson, J. M.** 1981. Evolution of photosynthetic reaction centers. *BioSystems* **14:**89–94.
37. **Olson, J. M., and B. K. Pierson.** 1986. Photosynthesis 3.5 thousand million years ago. *Photosynth. Res.* **9:**251–259.
38. **Olson, J. M., and B. K. Pierson.** 1987. Evolution of reaction centers in photosynthetic prokaryotes. *Int. Rev. Cytol.* **108:**209–248.
39. **Olson, J. M., and B. K. Pierson.** 1987. Origin and evolution of photosynthetic reaction centers. *Origins Life* **17:**419–430.
40. **Oparin, A. I.** 1938. *The Origin of Life.* Macmillan, New York.
41. **Oyaizu, H., B. Debrunner-Vossbrinck, L. Mandelco, J. A. Studier, and C. R. Woese.** 1987. The green non-sulfur bacteria: a deep branching in the eubacterial line of descent. *Syst. Appl. Microbiol.* **9:**47–53.
42. **Pierson, B. K., L. M. Keith, and J. G. Leovy.** 1984. Isolation of pigmentation mutants of the green filamentous photosynthetic bacterium *Chloroflexus aurantiacus*. *J. Bacteriol.* **159:**222–227.

42a. **Pierson, B. K., A. Oesterle, and G. L. Murphy.** 1987. Pigments, light penetration, and photosynthetic activity on the multi-layered microbial mats of Great Sippewissett Salt Marsh, Massachusetts. *FEMS Microbiol. Ecol.* **45:**365–376.

43. **Pierson, B. K., and J. M. Olson.** 1987. Photosynthetic bacteria, p. 21–42. *In* J. Amesz (ed.), *Photosynthesis. New Comprehensive Biochemistry*, vol. 15. Elsevier Biomedical Press, Amsterdam.
44. **Rambler, M. B., and L. Margulis.** 1980. Bacterial resistance to ultraviolet irradiation under anaerobiosis: implications for pre-Phanerozoic evolution. *Science* **210:**638–640.
45. **Raven, J. A., and F. A. Smith.** 1976. The evolution of chemiosmotic energy coupling. *J. Theor. Biol.* **57:**301–312.
46. **Raven, J. A., and F. A. Smith.** 1981. H^+ transport in the evolution of photosynthesis. *BioSystems* **14:**95–111.
47. **Schiff, J. A.** 1981. Evolution of the control of pigment and plastid development in photosynthetic organisms. *BioSystems* **14:**123–147.
48. **Schopf, J. W., and B. M. Packer.** 1987. Early Archean (3.3-billion to 3.5-billion-year-old) microfossils from Warrawoona Group, Australia. *Science* **237:**70–73.
49. **Schopf, J. W., and M. R. Walter.** 1983. Archean microfossils: new evidence of ancient microbes, p. 214–239. *In* J. W. Schopf (ed.), *Earth's Earliest Biosphere: Its Origin and Evolution.* Princeton University Press, Princeton, N.J.
50. **Simionescu, C. I., R. Mora, and B. C. Simionescu.** 1978. Porphyrins and the evolution of biosystems. *Bioelectrochem. Bioenerg.* **5:**1–17.
51. **Sistrom, W. R., M. Griffiths, and R. Y. Stanier.** 1956. The biology of a photosynthetic bacterium which lacks colored carotenoids. *J. Cell. Comp. Physiol.* **48:**473–515.
52. **Skulachev, V. P.** 1976. A hypothesis of the evolution of biological energy transducers. *Origins Life* **7:**145–160.
53. **Stackebrandt, E., and C. R. Woese.** 1984. The phylogeny of prokaryotes. *Microbiol. Sci.* **1:**117–122.
54. **Trüper, H. G., and N. Pfennig.** 1981. Characterization and identification of the anoxygenic phototrophic bacteria, p. 299–312. *In* M. P. Starr, H. Stolp, H. G. Trüper, A. Balows, and H. G. Schlegel (ed.), *The Prokaryotes.* Springer-Verlag KG, Berlin.
55. **Walker, J. C. G., C. Klein, M. Schidlowski, J. W. Schopf, D. J. Stevenson, and M. R. Walter.** 1983. Environmental evolution of the Archaean-early Proterozoic earth, p. 260–290. *In* J. W. Schopf (ed.), *Earth's Earliest Biosphere: Its Origin and Evolution.* Princeton University Press, Princeton, N.J.
56. **Walker, J. E., I. M. Fearnley, N. J. Gay, B. W. Gibson, F. D. Northrop, S. J. Powell, M. J. Runswick, M. Saraste, and V. L. J. Tybulewicz.** 1985. Primary structure and subunit stoichiometry of F_1-ATPase from bovine mitochondria. *J. Mol. Biol.* **184:**677–701.
57. **Walsh, M. M., and D. R. Lowe.** 1985. Filamentous microfossils from the 3,500-Myr-old Onverwacht Group, Barberton Mountain Land, South Africa. *Nature* (London) **314:**530–532.
58. **Walter, M. R.** 1983. Archean stromatolites: evidence of the earth's earliest benthos, p. 187–213. *In* J. W. Schopf (ed.), *Earth's Earliest Biosphere: Its*

Origin and Evolution. Princeton University Press, Princeton, N.J.

59. **Wechsler, T. D., R. A. Brunisholz, G. Frank, F. Suter, and H. Zuber.** 1987. The complete amino acid sequence of the antenna polypeptide B806-866-β from the cytoplasmic membrane of the green bacterium *Chloroflexus aurantiacus*. *FEBS Lett.* 210:189–194.
60. **Wechsler, T. D., R. Brunisholz, F. Suter, R. C. Fuller, and H. Zuber.** 1985. The complete amino acid sequence of a bacteriochlorophyll *a* binding polypeptide isolated from the cytoplasmic membrane of the green photosynthetic bacterium *Chloroflexus aurantiacus*. *FEBS Lett.* **191**:34–38.
61. **Wechsler, T. D., F. Suter, R. C. Fuller, and H. Zuber.** 1985. The complete amino acid sequence of the bacteriochlorophyll *c* binding polypeptide from chlorosomes of the green photosynthetic bacterium *Chloroflexus aurantiacus*. *FEBS Lett.* **181**: 173–178.
62. **Williams, J. C., L. A. Steiner, G. Feher, and M. I. Simon.** 1984. Primary structure of the L subunit of the reaction center from *Rhodopseudomonas sphaeroides*. *Proc. Natl. Acad. Sci. USA* **81**:7303–7304.
63. **Woese, C. R.** 1987. Bacterial evolution. *Microbiol. Rev.* **51**:221–271.
64. **Woese, C. R., E. Stackebrandt, T. J. Macke, and G. E. Fox.** 1985. A phylogenetic definition of the major eubacterial taxa. *Syst. Appl. Microbiol.* **6**: 143–151.
65. **Youvan, D. C., E. J. Bylina, M. Alberti, H. Begusch, and J. E. Hearst.** 1984. Nucleotide and deduced polypeptide sequences of the photosynthetic reaction-center B870 antenna, and flanking polypeptides from *R. capsulata*. *Cell* **37**:949–957.
66. **Zuber, H., R. Brunisholz, and W. Sidler.** 1987. Structure and function of light-harvesting pigment-protein complexes, p. 233–271. *In* J. Amesz (ed.), *Photosynthesis. New Comprehensive Biochemistry*, vol. 15. Elsevier Biomedical Press, Amsterdam.

VI. Biogeochemistry of Microbial Mats

Chapter 37

Microbial Mats: Chemical Postmortem

Zeev Aizenshtat

INTRODUCTION

Recent cyanobacterial mats are morphologically similar to fossil stromatolites (see chapter 40 of this volume). Several questions arise when trying to relate the recent mats to their ancient fossil analogs. (i) Are there any means, namely, chemical, isotopic, or detailed morphological analyses, that will indicate the depositional environments controlling the formation of mats? (ii) Is it possible by examining a young cyanobacterial mat and its depositional environment to portray its geochemical record? (iii) How would such organic matter and mineral matrix look while aging?

Our previous work on the microbial mats of Solar Lake, Sinai (2) was carried out through an international collaborative research program, ORGAST, which is a cooperative research project run by geochemists from the Universities of Delft (J. W. de Leeuw and P. A. Schenck), San Francisco (J. J. Boon and A. L. Burlingame), Bristol (G. Eglinton), and Jerusalem (Z. Aizenshtat), as part of a more general effort to understand the chemical signature of the communal ecosystem. This research group has established that the morphological similarity of recent microbial mats to stromatolites may stem from the fact that some constituents of the microbial mat organisms are conserved. This unchanged fraction was related primarily to the calcified polysaccharide sheath material, which was a dominant feature in the benthic mode of growth and was well preserved in the fossil record.

To understand the chemical topology of these systems, various analytical methods have been used. In general they can be separated into methods for studying the bulk organic matter and methods for studying lipids, biomarkers, interstitial water, or mineralogical structure.

A good example of a useful bulk parameter is carbon isotope composition of microbial mats. Stable carbon isotope data can be determined for both ancient stromatolites and recent mats, yet one should define which fraction of the carbon should be analyzed. It is not yet known what characteristics of microbial mats determine their carbon isotops composition (2, 9). Stromatolites of Precambrian age have often shown $\delta^{13}C$ values of $-20‰$ (PDB), whereas recent microbial mats (e.g., Solar Lake, Laguna Figaroa, etc.) yielded $\delta^{13}C$ values of -6 to $-12‰$. However, one should not jump to conclusions, since the influence of the chemical postmortem (diagenesis or catagenesis) might be responsible for the difference in

Zeev Aizenshtat • Energy Research Center, Organic Chemistry, and Casali Institute, The Hebrew University of Jerusalem, Jerusalem 91904, Israel.

the isotopic signature between the ancient and the recent mats.

The sulfur cycle also affects the organic matter and mineralogy of the sediments (2, 4). However, only meager experimental data are available on the interrelation between the carbon and sulfur cycles. The bulk of the organic matter of the cyanobacterial mat is composed mostly of sugars and amino acids. Terrestrially derived organic matter, if deposited (allochthonous), will yield a characteristic distribution which is easy to differentiate from cyanobacterial lipids. The lipid fraction is much easier to study. The lipid fraction derived from various sections of the Solar Lake cyanobacterial mats was extensively analyzed by the ORGAST team (2, 5). Some additional work was reported by Boon et al. (6), and others have since tried to fingerprint the contributions of the various bacteria to the mat by using their molecular signatures. It has been suggested that part of the lipid fraction, mostly from the nonfunctional molecules, has a long-term survival potential. Therefore, these lipids may preserve the record of microorganisms present as early as the Precambrian era; however, other complications such as a harsh chemical environment might inflict major changes. (See also A. H. Knoll, this volume.)

The aim of the microbiologist is to study the influence of depositional environmental changes (pH, Eh, salinity, sulfate, phosphate, HCO_3^-, and nitrogen nutrients) on a microbial mat during its growth. I claim that these changes may not only determine the distribution of major microbial species, but also introduce considerable imprint on the diagenetic stage, hence causing a different maturation pathway.

In this overview I have selected some organic-matter parameters which are important contributors to the communal microbial mat. These parameters were also studied in recent sediments from different origins and environmental regimens. A major effort was directed at understanding the preservation and alteration of the bulk organic matter during its chemically controlled postmortem.

The microbial activities in the cyanobacterial mat decrease exponentially with depth. With burial and reduced microbial activities, the chemical transformation becomes less membrane controlled and more ecochemically controlled. As long as the membranes are preserved intact, some susceptible compounds such as carotenoids and unsaturated molecules are protected from the chemically reactive environment. With burial and degradation of many microbial communities, most of the cell contents are subjected to chemical transformations, the nature of which is determined by the chemical environment in the interstitial water of the sedimentary column.

The bulk of the organic matter produced by benthic microbial mats is different from the organic matter sedimented from a planktonic source. The major differences I would like to focus on are as follows. (i) In microbial mats most of the accumulated carbon is in the form of sheath matter, whereas in plankton-derived sediment the bulk of the carbon is in humic-fulvic substances. (ii) The relative amount of extractable lipid in recent sediments of planktonic origin is higher than that found in microbial mats of benthic nature. (iii) The thermal behavior, namely, pyrolitical profiles and pyroproducts, of the young organic matter originating from the two sources is different.

The use of axenic cultures of various bacteria was found helpful in studying the chemical topology of certain biological markers. However, during benthic communal growth many of these microorganisms produce protective shields, namely, glycocalyxes, capsular substances, or mucilage envelopes, some of which are characterized as lipopolysaccharides and others as aminopolysaccharides or even agarous or chitinic matter.

OBSERVATIONS

In Solar Lake, most of the residual carbon below the photosynthetic active layers is sheath. This insoluble sedimentary matter is also called protokerogen.

Oxygen-Containing Lipids

In this overview, lipids are defined as the soluble matter in organic solvents. Subfractionation of the lipid fraction is discussed below.

In most cases the lipid fraction contains straight-chain fatty acids with an even number of carbon atoms, dominated by the C_{16} and C_{18} fraction. The top 3-mm sections of the Solar Lake mat are rich in unsaturated acids, whereas at a depth of 10 mm very few of these survive. At 10 cm and deeper no unsaturated lipids can be found in the Solar Lake sedimentary column (10). Polyunsaturated di- and trimethyl ketones with long carbon chains have been found at moderate to high concentrations in marine sediments (8). Some alkenones were also reported by Brassell and Eglinton (7) and were suggested to be climatic indicators. However, these have been linked to possible terrestrial input.

Only small amounts of normal and branched alcohols are extracted with the lipids; the dominating alcohols in young sediments are sterols and hopanols. It is difficult to attribute the distribution of sterols to particular bacterial species of the cyanobacterial mat communities. No clear correlation could be derived between the bacterial composition and the sterol distribution. However, the existence of large quantities of dinosterols (4-methyl stanols) was correlated to the presence of dinoflagellates. Surprisingly, 4-methyl sterols (dimasterol and 4-methyl gorgosterols) have been found by Boon et al. (6) in the Solar Lake sediments at depths below 3 mm. This is an apparent contradiction to the absence of dinoflagellates in the water column of Solar Lake or in the benthic cyanobacterial mat. The use of these as geological markers depends on their transformation to steranes and hopanes. At this stage the prevailing chemical conditions are of prime importance, since chemical transformations may erase the origin imprint.

Phytols were reported only as a hydrolysis products of the living organisms at the top of the mat with other even-numbered alcohols and, of course, large amounts of glycerol.

Hydrocarbons

The Solar Lake microbial mats contain only a small fraction of hydrocarbons. In the top 1 to 2.5 mm the range is C_{15} to C_{20}, with some unsaturated compounds as well as measurable amounts of squalene (no C_{40} polyhydrogenated carotene) (2, 5). At a depth of 0 to 3 mm phyt-1-ene and squalene were present at 128 and 15.3 ppm, respectively, whereas below 1 cm they were present at 0.5 and 1.4 ppm, respectively. Figure 1 shows the hydrocarbon fraction in the same mat from the center of the lake, where for 9 months of the year it experienced a temperature of 65°C.

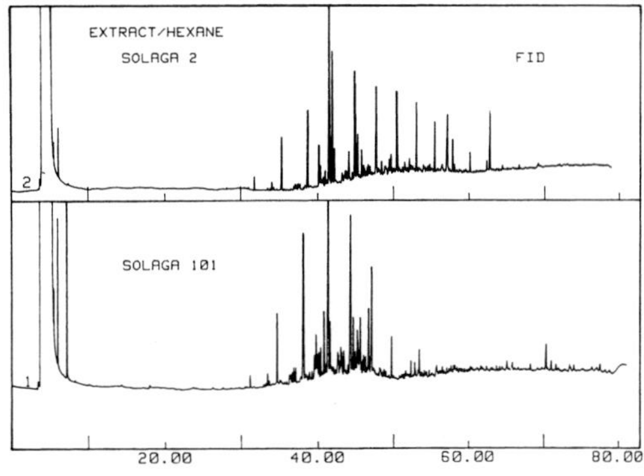

Figure 1. Gas chromatography profiles of the hydrocarbon fraction extracted from the Solar Lake cyanobacterial mats. Top, Solaga 2 (depth 2 to 4 cm) from the Shallow Flat mat; bottom, Solaga 101 (core of top 2 cm of sediment) from the bottom of the lake under 5 m of water.

Almost no aromatic fraction could be detected, and it is important to note that despite the reducing environment of deposition, no perylene could be detected (1, 4).

Pigments

Chlorines derived from chlorophyll and carotenoids dominate the top of the cyanobacterial mat. At lower layers the carotenoids decrease from 840 ppm (at a depth of 0 to 3 mm) to 32 ppm (at 10 to 20 mm). In some cases (see chapter 12 of this volume) carotenoids are well protected by the intracellular structure. However, carotenoids are susceptible both to removal by saturation to form C_{40} polyhydrogenated carotenoids and to reactions with active reduced-sulfur species, of which the most reactive at low temperature is polysulfide (4).

Metal transformation in and out of the lipid fraction could be controlled by phase transfer catalysis or carrier (PTC), as was demonstrated by Lipiner et al. (G. Lipiner, I. Willner, and Z. Aizenshtat, *Proc. 13th Org. Geochem. Meet.*, abstr. no. 67, 1987). Some of the chlorines react or are physically attached to the protokerogen to be later released as porphyrins, in most cases as nickel or vanadyl (VO) metallocomplexes. The highly reducing environment will preselect vanadyl metal complexation if vanadium is available (Lipiner et al., *Proc. 13th Org. Geochem. Meet.*, abstr. no. 67, 1987).

Humic Substances and Protokerogen

We were unable to isolate any humic substances (basic extractable acid-precipitable organic matter) from the Solar Lake sediments, even though many other marine-derived sediments of reducing or anoxic environments contain large amounts of humic matter. Other microbial mats that may contain transported terrestrial organic matter do show evidence of humic and fulvic acids (base-extractable and acid-soluble organic matter) and in some cases even some lignite when input from higher plants is evident. Partial oxidation and long-term exposure of the cyanobacterial mats lacking humic substances to atmospheric oxygen proved to form humic matter (2).

Our previous studies on conditions of melanoidin formation and their thermal behavior (3) led us to the conclusion that melanoidins of the humic type are not formed under the sedimentation conditions prevailing in the Solar Lake cyanobacterial mats. It is also evident that the protokerogen produced in Solar Lake is different from the protokerogens studied at other lagoonal sites. The protokerogen in the Solar Lake sedimentary column exhibits thermal similarity to the sheath matter with possible structural resemblance to the lipopolysaccharide (see chapter 39 of this volume).

The requirement for a large pool of monosugars with reductive sites was proven by us for the formation of sugar-based melanoidins. Polymerization of sugars without amino acids is possible under slightly basic conditions (3). The Maillard reaction shows that sugars and amino compounds polymerize to yield melanoidin. To simulate special conditions similar to the Solar Lake upper sediment interstitial waters, monosaccharides, amino acids, Na_2CO_3 or K_2CO_3, and polysulfides were reacted (3). Melanoidins were formed with a high free-radical concentration. This was surprising, since we thought that at certain stages the free radicals would be quenched by the polysulfides. The melanoidins formed under these conditions are rich in sulfur, part of which is thermally released (3). The production of melanoidins is hindered or almost stopped if $CaCO_3$ is used instead of Na_2CO_3 or K_2CO_3 in a pH 8 solution. In many of the high-salinity depositional environments $CaCO_3$ is also precipitating and therefore may be one of the factors in inhibiting humic substances (12). Moreover, it has been demonstrated that most free sugars in microbial mats are immediately reutilized by one or another process (sugar samples from deeper than the

3-mm monosugars, 5 mg/liter). Cohen (chapter 3 of this volume) shows that a coat of $CaCO_3$ is deposited on the sheath matter at an early stage of deposition.

The current definition of kerogen as the organic matter in sediments that is not soluble in organic solvents basic or acidic solutions excludes an unidentified group of usually high-molecular-weight polymeric matter. Protokerogen is the marine equivalent of terrestrial humus; it is young and resembles the biopolymeric systems. It has been suggested by Boon et al. (6) that in Solar Lake this protokerogen is constructed mostly of the sheath matter. Methods used to identify solids showed that the protokerogen still contained a large amount of O, N, and, at the more advanced stages of burial, S (2, 4).

Pyrolysis methods used by the groups at Delft University and Hebrew University, although different, showed that most fragments formed during pyrolysis derive from saccharides and amino-acids, as well as some sulfur-containing moieties. The Curie point analysis (Delft group) showed a large peak for CH_3SH and some CH_3—S—CH_3, whereas longer thermal treatments (Fig. 2 and 3) showed higher-molecular-weight sulfur-containing compounds and the release of elemental sulfur from the sulfur-rich protokerogen from depths of 30 to 90 cm (4). Figures 1 to 3 must be examined in the context of the data discussed previously in references 2 and 5 and other chapters in that volume. If one compares the Solaga 101 hydrocarbons with ZN_1 (the top 1 mm of the upper mat), they show almost the same distribution, but most of the unsaturated hydrocarbons have disappeared.

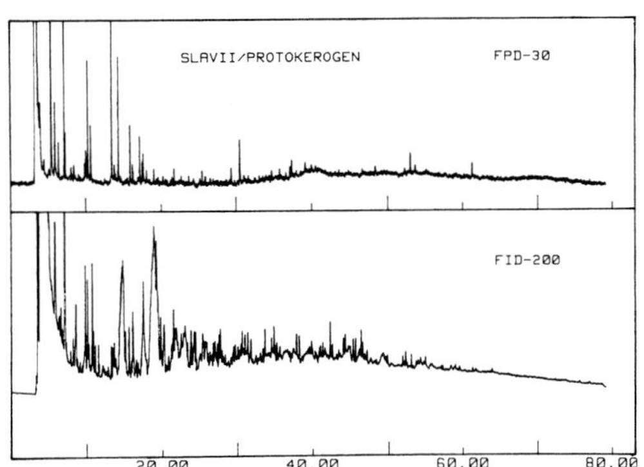

Figure 2. Pyrolysis-gas chromatography profiles of SLA VII protokerogen (depth, 2 to 4 cm) sulfur products: (top) FPD profile; (bottom) FID profile. The products are presently under gas chromatography-mass spectrometry examination, and a preliminary study has revealed that although the SLA VII protokerogen yields mostly oxygen-containing compounds, the SLA 101 sample (from the bottom of the lake) produces unsaturated and mono- and bicyclic isoprenoids up to C_{40}. This difference must be examined in view of other parameters, taking into account the special thermal setting of Solar Lake.

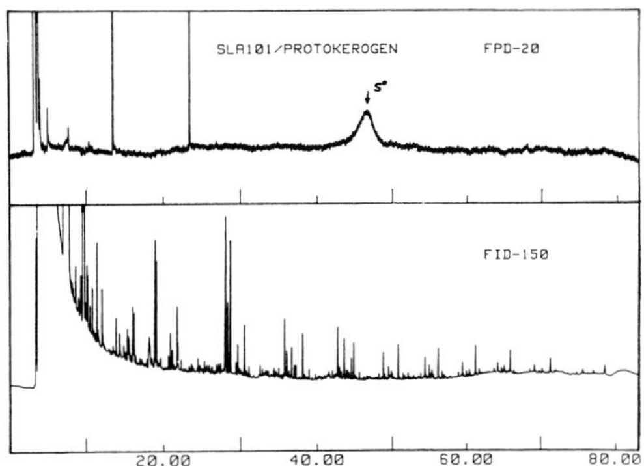

Figure 3. Pyrolysis-gas chromatography profile of SLA 101 protokerogen (equivalent sample to Solaga 101 presented in Fig. 1. Elemental S^0. (Top) FPD profile; (bottom) FID profile.

Most of the sheath material is nonhydrolyzable, and so is the protokerogen. However, if polysulfides enrich it with sulfur, where does this sulfur react? In the case of the small amount of lipopolysaccharide-like structures which have been found by Fredrickson et al. (chapter 39 of this volume), some of the α- or β-hydroxy acids may be enriched by addition of sulfur to the double bond or substitution of −OH with −SH. Also, other unsaturated sites may be attacked (Fig. 4). The very schematic pathway proposed in Fig. 4 demonstrates the facile attachment of large amounts of S_x (x = 8, 6, or 3) to double bonds, stabilization via ring closure, and low-temperature $H_2S + S^0$

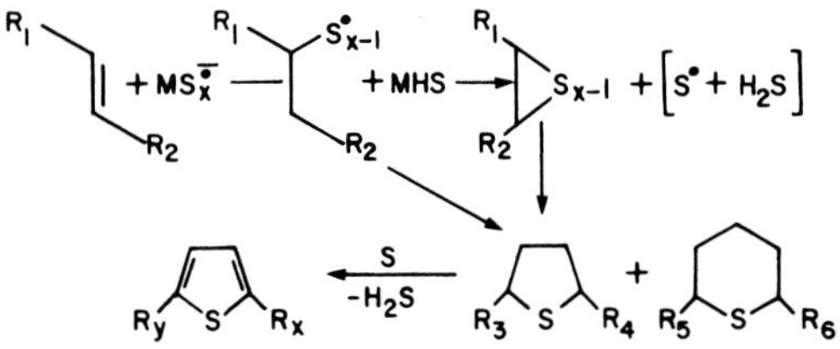

Figure 4. Schematic pathway of sulfur attachment to protokerogen, where x (in S_x) = 8, 6, or 3.

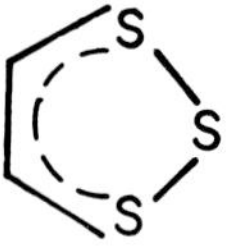

Figure 6. Structure of a sulfur-containing product at low temperatures.

release during steps of further stabilization and aromatization. The addition of sulfur or polysulfides to double bonds could also follow the mechanism of β-carbon attack (Fig. 5). In both mechanistic approaches, the first stabilized product is the saturated sulfur-containing five-member ring (dialkylthiacyclopentane). However, it should be noted that at low temperatures, S_x (x = 6, 8, or 3 atoms of sulfur) could catenate, and structures such as that shown in Fig. 6 could be produced. This somewhat speculative explanation was demonstrated by polysulfide reactions with poly- and monounsaturated hydrocarbons (in a phase transfer catalysis system). Elemental sulfur can react at much higher temperatures with saturated hydrocarbons (much easier with branched isoprenoids) (J. C. Schmidt, J. Connan, and P. Albrecht, Letter, *Nature* [London] **329:** 54–56, 1987; J. C. Schmidt, N. Perakis, J. Connan, and P. Albrecht, *Proc. 13th Org. Geochem. Meet.* abstr. no. 111, 1987), releasing H_2S, and with high-sulfur-containing polymers.

The possibility exists that if carotenes are exposed to polysulfides under phase transfer catalysis conditions, they may react to form sulfur-rich polymers. When heated to 200°C, low-temperature products [40°C NH_4^+ and $N^+(R_4)$ phase transfer carrier] release elemental sulfur which sublimes. The same phenomenon was observed for the 6 to 8% sulfur protokerogen (4). Another reaction in which sulfur can react with the sheath matter has been shown for agarous matter (Fig. 7).

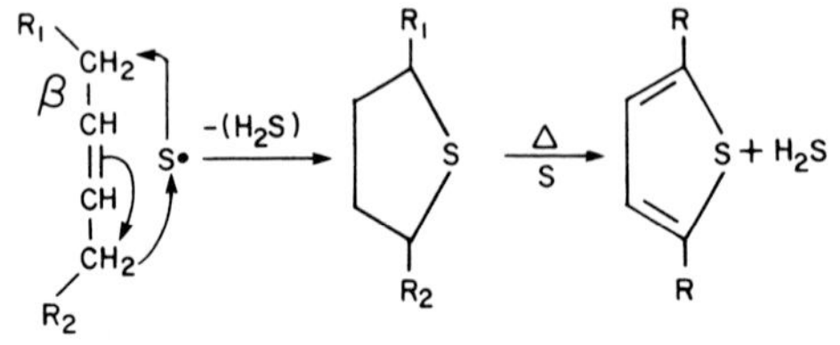

Figure 5. Mechanism of β-carbon attack for addition of sulfur or polysulfide to double bonds.

The possible formation of alkyl thiols may lead to saccharide sulfur-containing residues. The reaction of saccharides and polysaccharides with reduced sulfur, mostly H_2S, in the presence of hydrogen donors at somewhat elevated temperatures (100 to 200°C) and under pressure was demonstrated by Mango (11). However, it is important to note that all experiments were run under highly basic (0.1 N NaOH) conditions which could lead to hydrolysis of the polysugars and hence to formation of sugar melanoidins or transition molecules susceptible to attack by H_2S + S^0 (formed under the reaction conditions). Much milder conditions were used by Moers et al. (M. E. C. Moers, J. W. de Leeuw, and P. A. Schenck, *Proc. 13th Org. Geochem. Meet.*, abstr. no. 21–31, 1987) (15 to 50°C), showing the interaction of glucose with sulfur to form formyl triophenes and other substituted thiophenes under pyrolysis conditions. These works, coupled with our experiments, show that in general two types of sulfur-containing structures could be obtained.

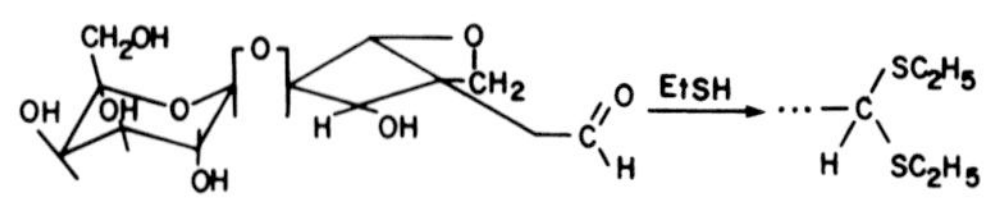

Figure 7. Reaction of sulfur with sheath material.

One is most probably S—S-catenated and is thermally susceptible, and the other type contains C—S—C stable bonds and produces, in the presence of excess S^0, the thiophene aromatic ring. All of these reactions may also have some bearing on the fact that biomarkers of kerogens or bituminous materials (asphalts as well) mature faster or at lower temperatures if they are sulfur rich than indicated by the so-called normal kinetics.

It is important in the present context to remember that large amounts of available iron will result in the formation of pyrite rather than in the secondary enrichment of the organic matter with sulfur. As long as the microbial mats are controlled by the microorganisms, one should consider the chemical influence to be secondary. If the cell membranes are still intact, transformation of reactants will be selected by the membrane. This is not the case for planktonic deposition of organic matter owing to the presence of different species and, more importantly, to transportation through the water column. Plankton-derived organic matter undergoes much more alteration before it reaches the safety of the anoxic or reducing environment. Also, in most cases, planktonic matter rides down on large inorganic allochthonous matter, most of which may be terrestrially derived. The lipids of higher plants are generally more resistant to degradation. Therefore, most depositional environments rich in plankton-derived matter show both high levels of humic matter and indications of allochthonous matter, which in most cases introduces enough iron to form large quantities of pyrite. In contrast, in microbial mats, despite the fast turnover at the top of the mat, the organic matter produced is chemically altered in an anoxic and generally sulfur-reducing, carbonate-rich environment.

The interdisciplinary approach, incorporating analytical and microbiological understanding of the difference between planktonic and benthic input to organic-rich sediments, complemented by the elucidation of the chemical processes following the domination of the microbial environment, will yield a better model of the link between the carbon and sulfur cycles.

ACKNOWLEDGMENTS. I thank Irena Miloslavski for pyrolysis work, gas chromatography, and gas chromatography-mass spectrometry of the Solar Lake protokerogens isolated by Achikam Stoler.

LITERATURE CITED

1. **Aizenshtat, Z.** 1973. Perylene and its geochemical significance. *Geochim. Cosmochim. Acta* 37:559–567.
2. **Aizenshtat, Z., G. Lipiner, and Y. Cohen.** 1984. Biogeochemistry of carbon and sulfur cycle in the microbial mats of the Solar Lake (Sinai), p. 281–312. *In* Y. Cohen, R. W. Castenholz, and H. O. Halvorson (ed.), *Microbial Mats: Stromatolites.* Alan R. Liss, Inc., New York.
3. **Aizenshtat, Z., Y. Rubinsztain, P. Ioselis, I. Miloslavski, and R. Ikan.** 1987. Long-living free radicals study of stepwise pyrolyzed melanoidins and humic substances. *Org. Geochem.* 11:65–71.
4. **Aizenshtat, Z., A. Stoler, Y. Cohen, and H. Nielsen.** 1983. The geochemical sulfur enrichment of recent organic matter by polysulfides in the Solar Lake, p. 207–227. *In* M. Bjorøy, P. Albrecht, C. Cornford, K. de Groot, G. Eglinton, E. Galimov, D. Leythaeuser, R. Pelet, J. Rullkotter, and G. Speers (ed.), *Advances in Organic Geochemistry 1981.* John Wiley & Sons, Ltd., Chichester, England.
5. **Boon, J. J.** 1984. Tracing the origin of chemical fossils in microbial mats: biogeochemical investigation of Solar Lake cyanobacterial mats using analytical pyrolysis methods, p. 313–342. *In* Y. Cohen, R. W. Castenholtz, and H. O. Halvorson (ed.), *Microbial Mats: Stromatolites.* Alan R. Liss, Inc., New York.
6. **Boon, J. J., H. Hines, A. L. Burlingame, J. Klok, W. I. C. Rijpstra, J. W. de Leeuw, K. E. Edmunds, and G. Eglinton.** 1983. Organic geochemical studies of Solar Lake laminated cyanobacterial mats, p. 207–227. *In* M. Bjorøy, P. Albrecht, C. Cornford, K. de Groot, G. Eglinton, E. Galimov, D. Leythaeuser, R. Pelet, J. Rullkotter, and G. Speers (ed.), *Advances in Organic Geochemistry 1981.* John Wiley & Sons, Ltd., Chichester, England.

7. **Brassell, S. C., and G. Eglinton.** 1986. Molecular geochemical indicators in sediments. *ACS Symp. Ser.* **305**:10–32.
8. **DeLeeuw, J. W.** 1986. Sedimentary lipids and polysaccharides, as indicators for source of input, microbial activity and short-term diagenesis. *ACS Symp. Ser.* **305**:33–61.
9. **Des Marais, D. J.** 1984. Evolutionary aspects of microbial mats and their possible global impact—discussion, p. 343–389. *In* Y. Cohen, R. W. Castenholz, and H. O. Halvorson (ed.), *Microbial Mats: Stromatolites.* Alan R. Liss, Inc., New York.
10. **Edmunds, K. L. H., and G. Eglinton.** 1984. Microbial lipids and carotenoids and their early diagenesis in the Solar Lake laminated microbial mat sequence, p. 343–389. *In* Y. Cohen, R. W. Castenholz, and H. O. Halvorson (ed.), *Microbial Mats: Stromatolites.* Alan, R. Liss, Inc., New York.
11. **Mango, F. D.** 1983. The diagenesis of carbohydrates by hydrogen sulfide. *Geochim. Cosmochim. Acta* **47**:1433–1441.
12. **Taguchi, K., and Y. Sampei.** 1986. The formation and clay mineral and $CaCO_3$ association reaction of melanoidins. *Org. Geochem.* **10**:1081–1089.

Chapter 38

Lipid Biochemical Markers and the Composition of Microbial Mats

David M. Ward, Jentaie Shiea, Y. Bin Zeng, Gary Dobson, Simon Brassell, and Geoffrey Eglinton

INTRODUCTION

Morphologic remains of microorganisms are well preserved as microfossils within stromatolites (2, 76), but shape by itself is a poor criterion for identification of most microorganisms. The long-term survival of lipids likely to be syngenetic with a few Precambrian sediments (1, 3, 23, 27, 40, 43, 48, 79) suggests that chemical fossils might provide independent and more detailed information about the types of microorganisms present during the early evolution of microbial life. The need to study appropriate samples (biological as well as geological) to better define which chemicals mark major physiologic innovations (such as oxygenic photosynthesis) and phylogenetic innovations (such as the eucaryotic cell) has been recognized (59).

David M. Ward • Department of Microbiology, Montana State University, Bozeman, Montana 59717. *Jentaie Shiea* • Department of Chemistry, Montana State University, Bozeman, Montana 59717. *Y. Bin Zeng, Gary Dobson, and Geoffrey Eglinton* • Organic Geochemistry Unit, School of Chemistry, University of Bristol, Cantock's Close, Bristol BS8 1TS, England. *Simon Brassell* • Department of Geology, Stanford University, Stanford, California 94305-2115.

Numerous geochemical studies have been carried out on modern microbial mats (see below) found in lagoonal settings similar to those in which ancient mats formed (85). There is a striking similarity in the microbial communities which inhabit these mats, probably reflecting the common environmental features of such habitats on the modern Earth. Although all are considered cyanobacterial mats, many are known to contain diatoms and photosynthetic bacteria, and thus they harbor organisms with the full variety of photosynthetic physiologies. The presence of diatoms and, in many cases, meiofauna means that both eucaryotic and eubacterial phylogenies are represented. The complexity of the communities makes it difficult to assign the microbial origins of specific compounds found in the mats (22).

It can be speculated that mats formed by distinctive types of microorganisms might have existed at different times during evolutionary history (75). Thus, mats which formed before the evolution of eucaryotes should be totally procaryotic, and mats which formed before the evolution of cyanobacteria might be formed by some other physiologic group, such as anoxygenic photosynthetic bacteria. Mats of these types exist now only in specialized environments in which extreme conditions of high temperature, low pH, and elevated H_2S levels provide suitable environmental features (e.g.,anoxic conditions on an otherwise oxic Earth) and refuge from competing organisms which might have evolved more recently (16; chapter 1 of this volume). We are studying microbial mats from different hot springs which could be considered analogs of stromatolites formed by (i) anoxygenic photosynthetic bacteria only (New Pit Spring), (ii) cyanobacteria with photosynthetic bacteria (Octopus Spring), and (iii) eucaryotic algae only (Nymph Creek). One aim is to correlate the various classes of lipids of the Octopus Spring mat with the lipids of its known or possible microbial inhabitants. A second aim is to observe differences in the lipid contents of the various mat types which might reflect the distribution of the different mat-building phototrophs. In this chapter we restrict the results and discussion of our study to the major lipids found in various analyses. Complete results and discussion have been (21a, 86, 87) or will be detailed elsewhere.

SAMPLES AND SAMPLE COLLECTION

Ward et al. (chapter 1 of this volume) have given complete descriptions of the locations and microbiology of the three mats sampled in this study. The Nymph Creek mat (47°C, pH 3) is built by the eucaryotic alga *Cyanidium caldarium* and lacks cyanobacteria and photosynthetic bacteria, which cannot grow at such a low pH. The New Pit Spring mat (55°C, pH 6.3, 34 μM sulfide) is built by the anoxygenic photosynthetic bacterium *Chloroflexus* sp. and lacks cyanobacteria and eucaryotic microorganisms, which are unable to grow under these conditions (11, 16). The Octopus Spring mat (53 to 55°C, pH 8) is built by the cyanobacterium *Synechococcus lividus* and the photosynthetic bacterium *Chloroflexus aurantiacus* and lacks eucaryotic microorganisms.

All mats were sampled at comparable temperatures above the upper temperature limits of metazoans known to inhabit these hot spring microbial mats (88). Samples were collected with a spatula or stainless steel coring tube which was prerinsed in methanol or CH_2Cl_2, transferred to prerinsed aluminum foil, and immediately frozen on dry ice. They were stored in the laboratory in darkness at −70°C. Before lipid analysis, samples were lyophilized, separated if necessary from extraneous pine needles, and ground with a prerinsed mortar and pestle to a fine powder.

LIPID ANALYSIS

Only ultrapure (EM Science Omnisolve grade) or redistilled solvents were used, and all materials which contacted samples were prerinsed in these solvents.

Total lipids were extracted by a modified Bligh and Dyer method (7; CH_2Cl_2 instead of chloroform) and were separated into neutral, glycolipid-containing, and phospholipid-containing fractions by column chromatography using techniques similar to those of Langworthy et al. (51). Methanolysis of phospholipids was carried out in concentrated HCl-methanol (5:95 (vol/vol) at 90°C for ca. 8 h). Glycerol diether core lipids were isolated by thin-layer chromatography from the neutral products of sequential acid and alkaline hydrolysis of the total extract. All fractions were treated with *N,O*-bis(trimethylsilyl)trifluoroacetamide (BSTFA) to form trimethylsilyl (TMS) ethers of free hydroxy groups before analysis by gas chromatography-mass spectrometry (GC-MS) (21a).

Free lipids were extracted by cold sonication with CH_2Cl_2; separated into hydrocarbon, wax ester, alcohol, and fatty acid fractions; derivatized; and analyzed by GC-

MS as described by Dobson et al. (21a). Elemental sulfur was removed from the hydrocarbon-containing fraction by passage of the fraction over copper filings preactivated with concentrated HCl. GC-MS analysis of free lipids of the New Pit and Nymph Creek mats was done by using a Varian 3700 GC instrument coupled to a VG-MM16 mass spectrometer; data were processed by a VG-2035 data system and PDP 8-A computer. Components were separated by using a DB-5 coated fused silica column (30 m by 0.25 mm [internal diameter]). After injection (split-splitless) at 35°C, the temperature was increased at 20°C/min to 80°C and then at 4°C/min to 300°C and was then held isothermal for 20 min. Mass spectra were collected from 33 to 550 amu at a scan rate of 0.7 s per scan (interscan delay, 0.3 s). The electron energy of the MS was 70 eV, or 40 eV when enhancement of alkane $M^{+\cdot}$ ion was required; the electron beam current was 200 mA, and the accelerating potential was 4.0 kV.

Compounds were identified on the basis of relative retention time and interpretation of their mass spectra. The peak area was corrected to concentration by using as internal standards *n*-tetracosane (for hydrocarbons, wax esters, and fatty acid methyl esters), TMS-derivatized *n*-nonanol (for nonsteroidal alcohols) or 3β-hydroxy-24-ethyl-$\Delta^{5,22}$-cholestadienol (for sterols). Replicate injection of standards gave a relative standard error of ca. 25%. After volume correction, concentrations were normalized to the dry weight of lyophilized sample extracted to determine the number of micrograms per gram (dry weight).

RESULTS OF LIPID ANALYSIS

The abundances of lipids of various classes in the three mats are shown in Table 1. The neutral lipid-containing fractions were of similar abundance to the glycolipid- and phospholipid-containing fractions in both the Octopus Spring and New Pit Spring mats. GC-MS analysis of the neutral lipid fractions showed that wax esters were the predominant neutral lipid components of these two mats (data not shown, but see below). The wax ester-containing fractions were the major free lipid fractions in these mats; however, no wax esters were found in the Nymph Creek mat.

The New Pit and Nymph Creek mats contained significant amounts of elemental sulfur, which was removed before GC-MS analysis of hydrocarbon components.

Table 1.
Abundances of Lipid Fractions Recovered from Hot-Spring Microbial Mats

Compound class	Fraction (μg/g [dry wt])[a] in:		
	Octopus Spring	New Pit Spring	Nymph Creek
Total lipid extract			
Glycolipid fraction	6,100	1,100	
Phospholipid fraction	6,200	600	
Neutral lipid fraction	4,000	1,600	
Free-lipid extract			
S^0	ND[b]	4,400	40
Hydrocarbons	72	235	50
Wax esters	2,405	1,050	ND
Alcohols	401	58	165
Fatty acids	65	12	35

[a] Fraction and S^0 weights are gravimetric; free-lipid class weights were determined by summing GC peaks.
[b] ND, Not detected.

Octopus Spring Polar Lipid Components

Methanolysis of the phospholipid-containing fraction released predominantly *n*-hexadecanoic, *n*-octadecanoic, and mono-octadecenoic acids (Fig. 1), with smaller amounts of hexadecenoic and various *n*-, iso- and anteiso-alkanoic acids. Also present in this fraction were alkan-1,2-diols, the most abundant being the brC_{20} homolog, as shown. 1-*O*-Alkylglycerol monoethers, most of which contained either a branched C_{17} or *n*-C_{18} alkyl group, were also present, as well

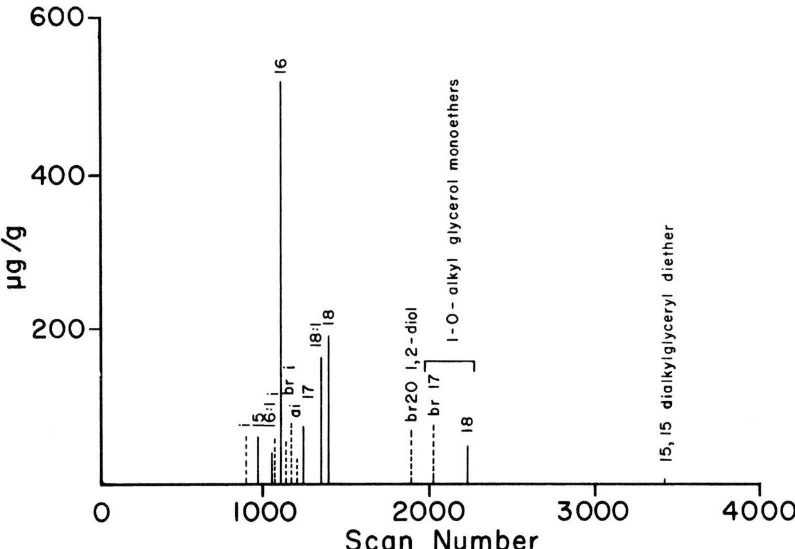

Figure 1. Methanolysis products of the phospholipid-containing fraction of the Octopus Spring cyanobacterial mat. Numbers refer to the carbon number of a compound (*n* unless specified; e.g. :*x* qualifies a preceding number, indicating that the compound of that chain length contains *x* double bonds). -------, Branched hydrocarbons of C number corresponding to the next following *n* peak (i, iso; ai, anteiso; br, branching in unknown position).

as a trace of C_{15},C_{15}-dialkylglyceryl diether. In separate studies of diether core lipids, traces of phytanyl, phytanyl diethers, and diethers with C_{16} and C_{17} alkyl groups have been identified.

Comparative Analysis of Free Lipids

Figures 2 and 3 present data on the free lipids of the New Pit Spring and Nymph Creek mats. The predominant hydrocarbons of the New Pit mat were di- and triunsaturated hydrocarbons ranging from C_{30} to C_{33}. C_{19}, C_{20}, and C_{21} monounsaturated alkenes were also present in smaller amounts. The Nymph Creek mat contained predominantly *n*-C_{17}, with smaller amounts of other *n*-alkanes ranging from C_{18} to C_{31}, and the chains with odd numbers of carbon atoms predominated.

The New Pit Spring mat contained a suite of wax esters ranging from C_{29} to C_{36} (maximum, C_{32}) and including both normal and branched components. No unsaturated wax esters were detected.

Phytol predominated over *n*- and branched alcohols in the Nymph Creek mat, but was in lower relative abundance in the New Pit Spring mat, where *n*-octadecanol, *n*-heptadecanol, *n*-hexadecanol, and an octadecenol were more abundant. Sterols were major components of the Nymph Creek mat alcohol fraction, whereas only a few sterols were detected in low or trace concentrations in the New Pit Spring mat. The variety and abundances of the sterols found in all mats are summarized in Table 2. The alcohol-containing fraction of the Nymph Creek mat also contained a series of monoglycerides and α- and β-tocopherols (Fig. 3).

Fatty acids, minor components in both mats, were similar; hexadecanoic and octadecanoic acids predominated over other *n*-, iso-, and anteiso-fatty acids. Unsaturated fatty acids were present only in the Nymph Creek mat.

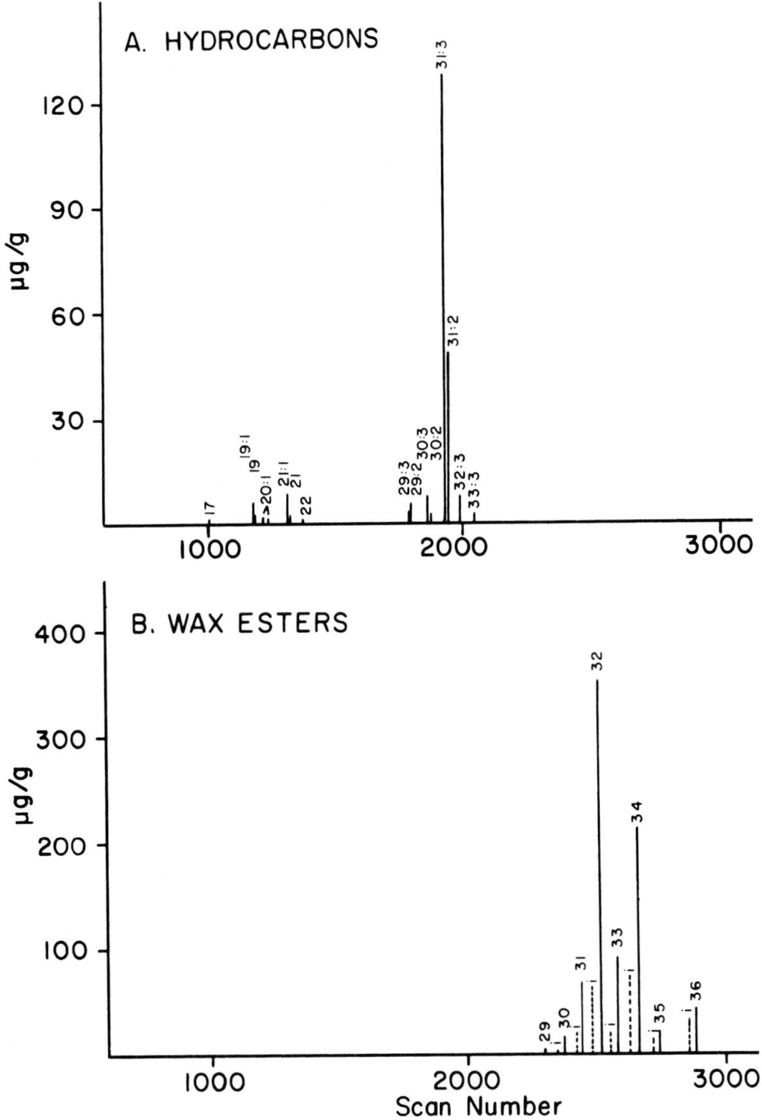

Figure 2. Free lipids of the New Pit Spring *Chloroflexus* mat. See the legend to Fig. 1 for a definition of symbols. Letters designate sterols described in Table 2. Abbreviation: Ph, phytol (position and configuration of double bonds in long-chain polyunsaturates have not yet been determined).

POLAR LIPID COMPONENTS AND THE OCTOPUS SPRING MAT COMMUNITY

It is interesting to compare the lipids released by methanolysis of the phospholipid-containing fraction of the Octopus Spring mat with the components of thermophilic microorganisms which have been isolated either from the mat or from thermal effluents and which might be mat inhabitants (Table 3 and references therein). Such a comparison would not reveal much

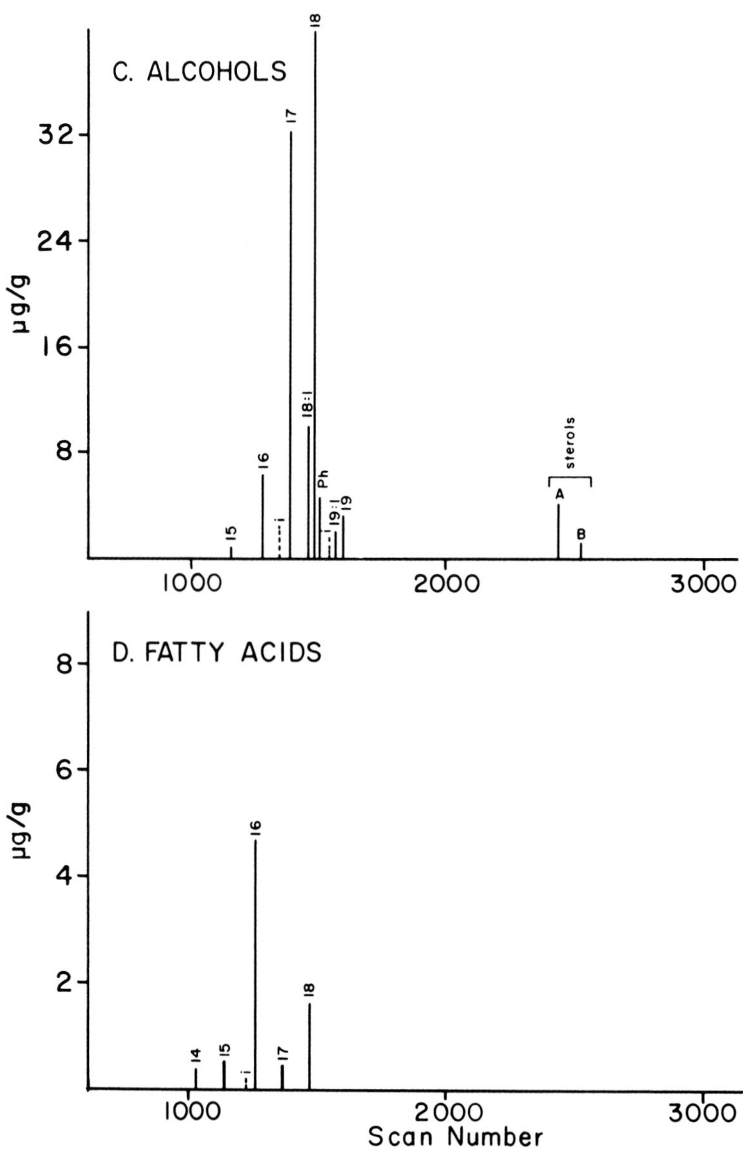

Figure 2. *Continued.*

about most natural systems owing to the complexity of the community and to the lack of uniqueness of the biomarkers therein. However, in this extreme environment the lipids of members of distinct physiological groups appear to be more unique and thus may be more informative biomarkers than is usually the case. This, combined with a simpler community structure, may permit recognition of how distinctive lipid signatures reflect community structure.

The most abundant methanolysis products were C_{16}, $C_{16:1}$, C_{18}, and $C_{18:1}$ fatty acids. These are known to be the major fatty acids of the two known mat phototrophs, *S. lividus* and *C. aurantiacus*. In addition, the major free lipids in the mat are characteristic of these phototrophs (see below). Two of the aerobic heterotrophic isolates could also be a source of C_{18} and $C_{18:1}$ fatty acids.

Several different phospholipid components may reflect the inputs of aerobic heterotrophic community members. The C_{20} alkan-1,2-diol (as well as other diols not reported here) is a major lipid of an aerobic heterotrophic thermophile, *Thermomicrobium roseum*, which was isolated from thermal effluent rather than from a hot-spring cyanobacterial mat, but which might be an important community member of the mat. Fatty acids (especially iso- and anteiso-branched) which appear to be characteristic of other aerobic or facultative heterotrophic thermophiles of this mat, such as *Thermus* or *Bacillus* species, were also among the major lipids found in the mat.

Monoethers are known components of both the clostridia which have been isolated from the mat and the novel sulfate reducer in the mat, *Thermodesulfobacterium commune.* Since the major diether lipids of *T. commune* (Table 3) were in far lower abundance, clostridia seem a more likely source of the monoethers. Phytanyl diethers (and acyclic biphytanyl tetraethers [86]) characteristic of the other known mat inhabitant which terminates anaerobic food chains, *Methanobacterium thermoautotrophicum*, were also present at low concentration. The predominant diether (C_{15},C_{15}) cannot be assigned to a known mat inhabitant and thus may suggest the presence of another, as yet unknown, community member which forms glyceryl ethers.

It is interesting that the relative abundance of lipids in the mat roughly corresponds to the expected trophic structure of the mat. The mat-building phototrophs are responsible for the initial incorporation of energy into the mat; lipids likely to have been contributed by these organisms are most abundant. During subsequent aerobic and anaerobic decomposition of phototroph fixed carbon (and energy), less energy should be available to aerobic and anaerobic heterotrophic microorganisms because of inefficiency in energy transfer from producers to consumers. Lipids characteristic of heterotrophic isolates of this mat constituted a significant proportion of the mat lipids, but were less abundant than those characteristic of phototrophs. Microorganisms which terminate anaerobic food chains should receive the least energy and thus should be the least abundant community members. Only traces of lipids characteristic of the sulfate reducer or methanogen of this mat were found.

FREE LIPIDS AND THE COMMUNITY STRUCTURE OF DIFFERENT MAT TYPES

Free-lipid data for the mats studied here and the Octopus Spring mat (21a) also suggest that lipid distribution reflects the distribution of the major microorganisms which form the mats. Among the hydrocarbons, 7-methylheptadecane was found only in the Octopus Spring mat, the only mat which contains cyanobacteria. There is good evidence that hydrocarbons with mid-chain branching are produced by cyanobacteria (25, 28, 36, 37, 58), although other sources of such compounds cannot be ruled out. *Thermomicrobium roseum*, for example, produces both alcohols and fatty acids with mid-chain branching (70). The long-chain polyunsaturated hydrocarbons predominating in the New Pit Spring *Chloroflexus* mat (also present in the Octopus Spring mat) have been found to be major hydrocarbons produced by *C. aurantiacus* (J. Shiea and D. M. Ward, unpublished results).

Wax esters characteristic of *C. aurantiacus* were found only in the Octopus Spring and New Pit Spring mats, where this organism is a major photosynthetic community member and the dominant member, respectively. No wax esters were found in the

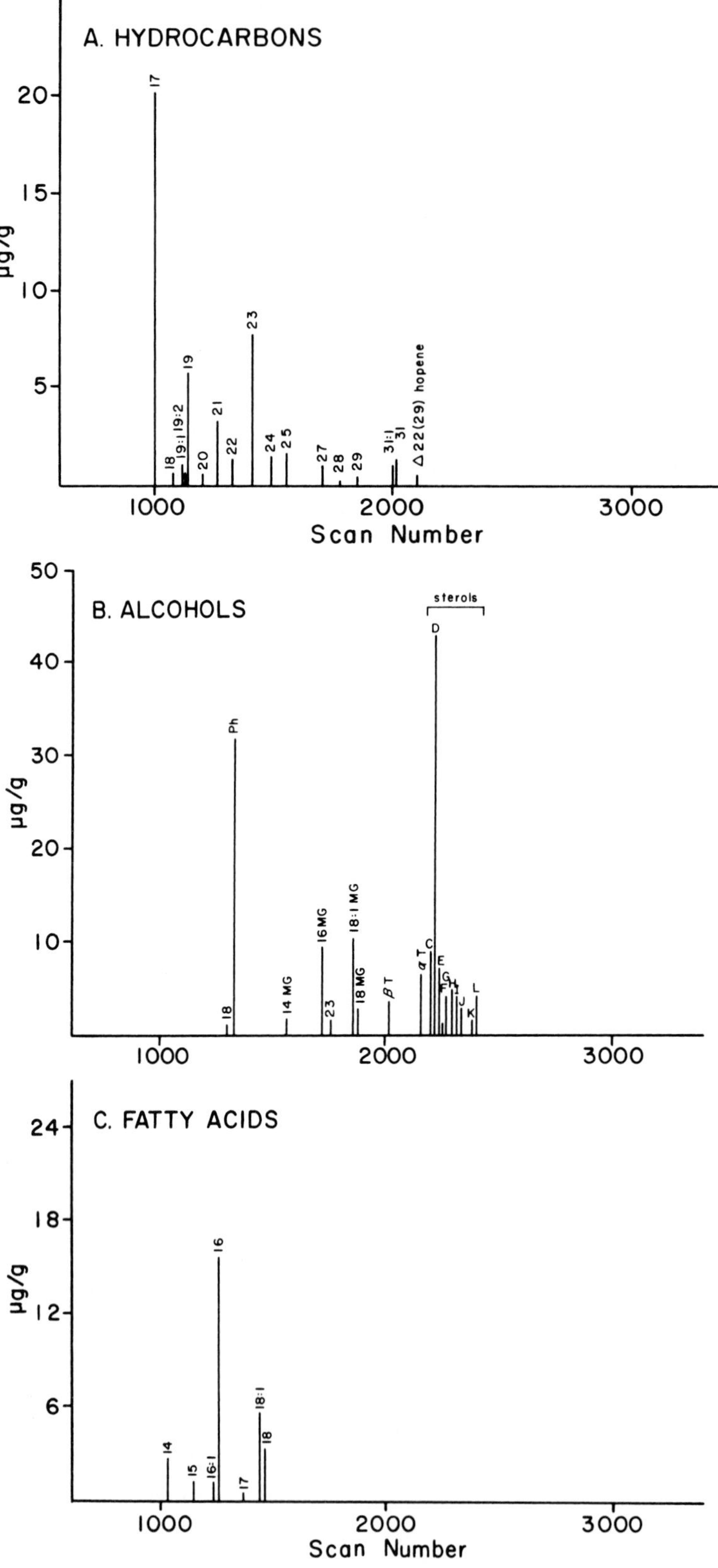
A. HYDROCARBONS
µg/g
Scan Number
17
18
19:1
19:2
19
20
21
22
23
24
25
27
28
29
31:1
31
Δ22(29) hopene
B. ALCOHOLS
sterols
Ph
18
14 MG
16 MG
23
18:1 MG
18 MG
βT
αT
C
D
E
F
G
H
I
J
K
L
C. FATTY ACIDS
14
15
16:1
16
17
18:1
18

Table 2.
Sterol Abundances in Hot-Spring Microbial Mats

Peak[a]	No. of C atoms	Sterol	Relative amt[b] in mat from: Octopus Spring[c]	New Pit Spring	Nymph Creek
A	C_{27}	Cholest-5-en-3β-ol	+	+	
		Unknown stanol	tr		
C	C_{28}	24-Methylcholesta-5,7,9(11),22-tetraen-3β-ol			++
D		24-Methylcholesta-5,7,22-trien-3β-ol			+++
E		24-Methylcholesta-7,22-dien-3β-ol			++
F		24-Methylcholest-5,7-dien-3β-ol			+
G		24-Methylcholest-7-en-3β-ol			+
		24-Methylcholest-5-en-3β-ol		tr	
		Unknown steradienol	tr		
		Unknown stenol	tr		
	C_{29}	4α,24-Dimethylcholesta-5,22-dien-3β-ol			tr
		24-Ethylcholesta-5,7,22-trien-3β-ol			tr
		24-Ethylcholesta-5,7-dien-3β-ol		tr	
H		24-Ethylcholesta-5,22-dien-3β-ol	tr		+
J		24-Ethylcholest-7-en-3β-ol			+
B		24-Ethylcholest-5-en-3β-ol	+	+	
I		4α,24-Dimethylcholest-8(14)-en-3β-ol			+
L		4α,24-Dimethylcholesta-5,7-dien-3β-ol			+
		24-Ethyl-5α(H)-cholestan-3β-ol	tr		
		24-Ethyl-5β(H)-cholestan-3β-ol	tr		
		Unknown Δ^5 stenol			tr
	C_{30}	Unknown Δ^5 stenol		tr	
		Unknown Δ^{22} stenol			tr
K		Unknown 4α-stenol			+
		Mixture of two unknown stenols		tr	
		Unknown stenol	tr		

[a] Corresponds to assignments in Fig. 2 and 3.
[b] tr, <1 μg/g (dry weight); +, 1 to 5 μg/g (dry weight); ++, 5 to 10 μg/g (dry weight); +++, >10 μg/g (dry weight).
[c] Data from Dobson et al. (21a).

Nymph Creek mat, which does not contain *C. aurantiacus.*

Similarly, the major nonsteroidal alcohol in each of the three mats seems to reflect the alcohol esterified to the predominant chlorophyll of the community. Thus, in the cyanobacterial (Octopus Spring) and eucaryotic algal (Nymph Creek) mats (chlorophyll *a* predominating) phytol is most abundant, whereas in the photosynthetic bacterial mat (New Pit Spring) octadecanol, the major alcohol esterified to bacteriochlorophyll *c* (34), predominates. Sterols previously reported in the eucaryotic alga *Cyanidium caldarium* (notably C_{28} components, e.g., compounds D, E, and G in Fig. 3B and Table 2 [77]) were abundant only in the Nymph Creek mat.

Figure 3. Free lipids of the Nymph Creek *Cyanidium caldarium* mat. See the legends to Fig. 1 and 2 for definitions of symbols. Abbreviations: MG, monoglyceride; βT, beta tocopherol; αT, alpha tocopherol.

Table 3.
Major Lipids of Known or Possible Octopus Spring Mat Isolates

Organism	Polar-lipid alkyl groups[a]					Free lipids and pigments[a]	Reference
	Fatty acid	Diol	Ether				
			Mono	Di	Tetra		
Phototrophs							
Synechococcus lividus	16, 18 16:1, 18:1					Chl *a*-O-phytol; phycocyanin; carotenoids	17, 26
Chloroflexus aurantiacus	16–18 18:1 (16:1, 17:1)[b]					Bchl *c*-O-octadecanol; β-, γ-, glycosidic-carotenoids, *n*-, monounsaturated 28–38 wax esters	5, 34, 35, 47, 49[c]
Aerobic and facultative heterotrophs							
Isosphaera pallida	18:1, 18						32
Thermomicrobium roseum	18, 12Me18	*n*- and mid-br 18–28					69, 70
Thermus aquaticus[d]	i15-17 ai15,ai17					Phytoene; Δ-carotene; polar carotenoids	38, 39, 44, 61–64, 66, 67, 71, 72
Bacillus stearothermophilus	i15-17 ai15,ai17						12, 13, 18, 19, 60, 63, 78, 90
Anaerobic fermentors							
Clostridium thermosulfurogenes[f]	i15,i17[e] 30diacid		i17, i15				53
Terminal anaerobes							
Thermodesulfobacterium commune			ai17, ai19, i16–19, 16, 18	ai17, ai18, i16–18, 16, 18			51
Methanobacterium thermoautotrophicum				Phytanyl	Acyclic biphytanyl		41, 55, 74, 81–83

[a] See figure legends for definitions of symbols. Me, methyl; the preceding number refers to the position of the methyl group, and the following number indicates the C number of the alkyl group which is methylated (e.g., 12Me18 is a C_{17} chain with methylation at C-12; thus, the compound contains 18 C atoms). mid, Mid-chain; -O-alkyl refers to the alcohol esterifiying a designated chlorophyll (Chl) or bacteriochlorophyll (Bchl).
[b] Includes fatty acids from wax esters.
[c] K. L. H. Edmunds, Ph.D. thesis, University of Bristol, Bristol, England, 1982.
[d] Also other *Thermus* or *Thermus*-like species.
[e] Also aldehydes as components of plasmalogens.
[f] Also *Clostridium thermohydrosulphuricum*.

Table 4.
Correlations between Distributions of Biomarkers and Predominant Phototrophic Microorganisms in Hot Spring and Lagoonal Microbial Mats

Mat	Phototroph composition			Mid-chain branched alkanes[a]	Predominant nonsteroidal alcohol[b]	Sterols		Reference
	Eucaryotes[c]	Cyanobacteria	Photosynthetic bacteria			No.	ug/g (dry wt)[b]	
Hot spring mats								
Octopus Spring		+	+	7Me17	Phytol	2[d]	9	21a
New Pit Spring			+		Octadecanol	2[d]	6	This study
Nymph Creek	+				Phytol	14	80	This study
Lagoonal mats								
Solar Lake	+	+	+	8Me16	Phytol	14–26	60	8, 22, 31, 50
Laguna Mormona	+	+	+	7-,8Me17	Phytol	>15	33.3	14
Laguna Guerrero Negro	+	+	+	8Me16, 7-,8Me17	Phytol[e]	>11[e]	NS	45, 46, 68
Gavish Sabkha	+	+	+	8Me15,8Me16, 4-6Me17, 6-9 Me19, 6-7Me20	Phytol	18	NR	20, 24, 29–31
Abu Dhabi		+	+	7-,8Me17	Phytol	>15	24	15
Hao Atoll		+	+	7Me17, 4-,7Me18	NS	21	>200	9, 10
Baffin Bay		+	+	7-,8Me17	NS	21	NR	4, 42, 89

[a] See Table 3 for a definition of symbols.
[b] NR, Not reported; NS, not studied.
[c] Diatoms or obvious animal or plant contamination.
[d] Traces of other components were also present (see Table 2).
[e] S. C. Brassell and R. P. Philp, unpublished results.

COMPARISONS WITH LAGOONAL MICROBIAL MATS

A major objective of studies of modern mats is to improve our ability to interpret the types of microorganisms present in ancient sediments by examining the types of biochemical markers recovered. As mentioned above, lagoonal microbial mats are either known or likely to contain microorganisms representing the complete spectrum of photosynthetic physiologies and phylogenies which have evolved, thus complicating the association between lipids in the mats and their sources. The value of hot-spring microbial mats lies in the simplicity and possible evolutionary relevance of the microbial communities they contain. Individual mat communities are produced by distinct types of phototrophic microorganisms, which might have predominated at different times during the changing environmental conditions of an evolving Earth. They provide an experimental means of sorting out the biomarkers which may distinguish each community. The distributions of potentially useful biomarkers in various lagoonal mats, as well as in the mats we have studied, are compared in Table 4 (see also references therein). We restrict our discussion to hydrocarbons and alcohols (such as phytol and sterols) which might be converted during diagenesis to geolipids (e.g., phytane and steranes) which have been recovered from Precambrian sediments.

Hydrocarbons with mid-chain branching are found in all of the lagoonal mats, consistent with the predominance of cyanobacteria in all of these mats. The range of specific molecules of this type varies, but a similar variation has been reported among the many cyanobacteria whose hydrocarbons have been studied (28, 37, 58). Our results are consistent with the assignment of such hydrocarbons as cyanobacterial markers. However, such hydrocarbons found in geologic samples may not originate solely from cyanobacteria (see above and reference 79).

In our study phytol was the major nonsteroidal alcohol in the two mats in which chlorophyll *a*-containing phototrophs were predominant, whereas octadecanol predominated in the mat containing the green photosynthetic bacterium *Chloroflexus* sp. as the major phototroph. All lagoonal mats for which nonsteroidal alcohols are reported contain phytol as the predominant alcohol. Chlorophyll *a* from oxygenic phototrophs (e.g., cyanobacteria) is presumably the most abundant source of phytol. However, purple photosynthetic bacteria, common to all lagoonal mats, possess mainly bacteriochlorophylls *a* and *b*, which also esterify phytol (33).

All of the lagoonal mats contain a variety of sterols (>11 to 26 sterols). In cases for which concentrations were reported, sterol concentrations ranged from approximately 24 μg/g to more than 200 μg/g. Among hot-spring mats, sterols were found in such abundance and variety only in the Nymph Creek mat, which is known to contain a eucaryotic alga as well as a fungus (11). Significant amounts of only two sterols (Table 2) (plus additional trace sterols) were found in the two mats which lack eucaryotic community members. Since the anoxygenic photosynthetic bacterial mats do not provide the oxic environment necessary for sterol biosynthesis (chapter 1 of this volume), the sterols found in these mats are likely to be extraneous and are probably not produced by organisms within the mats. A similar source may explain the presence of sterols found at similar concentrations in the Octopus Spring mat.

The possible contribution of sterols by the abundant cyanobacteria of lagoonal mats has been discussed in light of the reported presence of sterols in cyanobacteria (56, 84). However, in many such studies, evidence for the production of sterols is complicated by (i) reporting of *Cyanidium caldarium*, a eucaryotic alga, as a cyanobacterium (84); (ii) the use of unialgal rather than axenic cultures (57, 65); (iii) reporting of relative rather than absolute sterol abundances (21, 73); (iv) the lack of evidence for de novo biosynthesis (54, 73); and (v) the relatively simple nature of sterols reported to occur in cyanobacteria (84; D. M. Ward and S. Brassell, unpublished results). It seems far more likely that the sterols of lagoonal microbial mats originate from eucaryotic organisms such as diatoms and meiofauna, which are known to inhabit many of these mats. Those mats for which eucaryotic inputs have not been described occur in habitats similar to those known to contain eucaryotes. Seasonal reductions in salinity are known to permit diatoms and meiofauna to intermittently inhabit some lagoonal mats (29–31, 50). Thus, on ecological grounds one might expect the presence of similar eucaryotes which may have been missed in microbiological investigations. Although some bacteria cannot be excluded as possible sources of sterols (6), the contrast between totally procaryotic mat communities (of physiologically distinct type) and mat communities known to contain eucaryotes suggests that an abundant and diverse sterol assemblage is characteristically eucaryotic.

Hopanols, hopanoic acids, and/or other hopanoid compounds were present in all lagoonal mats in which they were sought (8–10, 14, 15, 20, 22, 68). In our study hopanols were found only in the Octopus Spring cyanobacterial mat (21a). We were surprised that none were found in the Nymph Creek mat, since a possible mat inhabitant, *Bacillus acidocaldarius*, is known to produce them, but our methods were not designed to release the hopane polyols characteristic of this organism (52). Hopenes were found in the Octopus Spring and Nymph Creek mats. Taylor (80) has made the interesting suggestion that hopanoid compounds are produced only by aerobic and/or facultatively aerobic bacteria. If this is correct, geolipids derived from hopanoid compounds might serve as biochemical markers for the presence of oxygen in the atmosphere of the Earth. Further studies of hopanoids in hot-spring mats may reveal whether they are restricted to oxic environments.

Analysis of hot-spring microbial mats has improved our understanding of the structure of these simple yet diverse communities. Future studies should be directed at confirming the suggested relationship between abundances of source microorganisms and their lipid biomarkers. Studies of these well-defined, physiologically and phylogenetically distinct communities have also helped us resolve the sources of specific types of biomarkers which are found in more complex modern mats. Although the sources suggested for some biomarkers seem likely (e.g., mid-chain methylated hydrocarbons from cyanobacteria), those suggested for other biomarkers can be disputed (e.g., sterols). Further work should be directed at proving or disproving whether the traces of sterols found in hot-spring cyanobacterial mats are actually biosynthesized by the cyanobacteria producing such mats.

ACKNOWLEDGMENTS. We thank National Science Foundation (grant BSR-8506602) for supporting sample collection and travel, the National Park Service for granting permission to collect samples in Yellowstone National Park, the United Kingdom Natural Environment Research Council for supporting the GC-MS facility (grants GR3/2951 and GR3/3758), the Natural Environment Research Council (grant GR3/6103) and the National Aeronautics and Space Administration (grant NGL 05-003-003) for supporting research on microbial polar lipids, and SEDC (China) for providing a studentship (Y.B.Z.).

LITERATURE CITED

1. **Arefev, O. A., M. N. Zabrodina, V. M. Makushina, and A. A. Petrov.** 1980. Relic tetra and pentacyclic hydrocarbons in the old oils of the Siberian platform. *Izv. Akad. Nauk SSSR Ser. Geol.* **1980:**135–140.
2. **Awramik, S. M.** 1984. Ancient stromatolites and microbial mats, p. 1–22. *In* Y. Cohen, R. W. Castenholz, and H. O. Halvorson (ed.), *Microbial Mats: Stromatolites.* Alan R. Liss, Inc., New York.
3. **Barghoorn, E. S., W. G. Meinschein, and J. W. Schopf.** 1965. Paleobiology of a Precambrian shale. *Science* **148:**461–472.
4. **Behrens, E. W., and S. A. Frishman.** 1971. Stable carbon isotopes in blue-green algal mats. *J. Geol.* **79:**94–100.
5. **Beyer, P., H. Falk, and H. Kleinig.** 1983. Particulate fractions from *Chloroflexus aurantiacus* and distribution of lipids and polyprenoid forming activities. *Arch. Microbiol.* **134:**60–63.
6. **Bird, C. W., J. M. Lynch, F. J. Pirt, W. W. Reid, C. J. W. Brooks, and B. S. Middleditch.** 1971. Steroids and squalene in *Methylococcus capsulatus* grown on methane. *Nature* (London) **230:**473–474.
7. **Bligh, E. G., and W. J. Dyer.** 1959. A rapid method of total lipid extraction and purification. *Can. J. Biochem. Physiol.* **37:**911–917.
8. **Boon, J. J., H. Hines, A. L. Burlingame, J. Klok, W. I. C. Rijpstra, J. W. de Leeuw, K. E. Edmunds, and G. Eglinton.** 1983. Organic geochemical studies of Solar Lake laminated cyanobacterial mats, p. 207–227. *In* M. Bjorøy, P. Albrecht, C. Cornford, K. de Groot, G. Eglinton, E. Galimov, D. Leythaeuser, R. Pelet, J. Rullkotter, and G. Speers (ed.), *Advances in Organic Geochemistry 1981.* John Wiley & Sons, Ltd., Chichester, United Kingdom.
9. **Boudou, J. P., J. Trichet, N. Robinson, and S. C. Brassell.** 1986. Profile of aliphatic hydrocarbons in a recent Polynesian microbial mat. *Int. J. Environ. Anal. Chem.* **26:**137–155.
10. **Boudou, J. P., J. Trichet, N. Robinson, and S. C. Brassell.** 1986. Lipid composition of a recent Polynesian microbial mat sequence. *Org. Geochem.* **10:** 705–709.
11. **Brock, T. D.** 1978. *Thermophilic Microorganisms and Life at High Temperatures.* Springer-Verlag, New York.
12. **Card, G. L.** 1973. Metabolism of phosphatidylglycerol, phosphatidylethanolamine, and cardiolipin of *Bacillus stearothermophilus. J. Bacteriol.* **114:**1125–1137.

13. **Card, G. L., C. E. Georgi, and W. E. Militzer.** 1969. Phospholipids from *Bacillus stearothermophilus*. *J. Bacteriol.* **97**:186–192.

14. **Cardoso, J., P. W. Brooks, G. Eglinton, R. Goodfellow, J. R. Maxwell, and R. P. Philp.** 1976. Lipids of recently-deposited algal mats at Laguna Mormona, Baja California, p. 149–174. *In* J. O. Nriagu (ed.), *Environmental Biogeochemistry*, vol. 1. Ann Arbor Science Publishers, Inc., Ann Arbor, Mich.

15. **Cardoso, J. N., C. D. Watts, J. R. Maxwell, R. Goodfellow, G. Eglinton, and S. Golubic.** 1978. A biogeochemical study of the Abu Dhabi algal mats: a simplified ecosystem. *Chem. Geol.* **23**:273–291.

16. **Castenholz, R. W.** 1977. The effect of sulfide on the blue-green algae of hot springs. II. Yellowstone National Park. *Microb. Ecol.* **3**:79–105.

17. **Castenholz, R. W.** 1984. Composition of hot spring microbial mats: a summary, p. 101–119. *In* Y. Cohen, R. W. Castenholz, and H. O. Halvorson (ed.), *Microbial Mats: Stromatolites.* Alan R. Liss, Inc., New York.

18. **Cho, K. Y., and M. R. J. Salton.** 1966. Fatty acid composition of bacterial membrane and wall lipids. *Biochim. Biophys. Acta* **116**:73–79.

19. **Daron, H. H.** 1970. Fatty acid composition of lipid extracts of a thermophilic *Bacillus* species. *J. Bacteriol.* **101**:145–151.

20. **de Leeuw, J. W., J. S. Sinninghe Damsté, J. Klok, P. A. Schenck, and J. J. Boon.** 1985. Biogeochemistry of Gavish Sabkha sediments. I. Studies on neutral reducing sugars and lipid moieties by gas chromatography-mass spectrometry, p. 350–367. *In* G. M. Friedman and W. E. Krumbein (ed.), *Hypersaline Ecosystems. The Gavish Sabkha.* Springer-Verlag KG, Heidelberg, Federal Republic of Germany.

21. **de Souza, N. J., and W. R. Nes.** 1968. Sterols: isolation from a blue-green alga. *Science* **162**:363.

21a. **Dobson, G., D. M. Ward, N. Robinson, and G. Eglinton.** 1988. Biogeochemistry of hot spring environments: extractable lipids of a cyanobacterial mat. *Chem. Geol.* **68**:155–179.

22. **Edmunds, K. L. H., and G. Eglinton.** 1984. Microbial lipids and carotenoids and their early diagenesis in the Solar Lake laminated microbial mat sequence, p. 343–389. *In* Y. Cohen, R. W. Castenholz, and H. O. Halvorson (ed.), *Microbial Mats: Stromatolites.* Alan R. Liss, Inc., New York.

23. **Eglinton, G., P. M. Scott, T. Belsky, A. L. Burlingame, and M. R. Calvin.** 1964. Hydrocarbons of biological origin from a one-billion-year-old sediment. *Science* **145**:263–264.

24. **Ehrlich, A., and I. Dor.** 1985. Photosynthetic microorganisms of the Gavish Sabkha, p. 296–321. *In* G. M. Friedman and W. E. Krumbein (ed.), *Hypersaline Ecosystems. The Gavish Sabkha.* Springer-Verlag KG, Heidelberg, Federal Republic of Germany.

25. **Fehler, S. W. G., and R. J. Light.** 1970. Biosynthesis of hydrocarbons in *Anabaena variabilis*. Incorporation of [methyl-^{14}C]- and [methyl-$^{2}H_3$]-methionine into 7- and 8-methylheptadecanes. *Biochemistry* **9**:418–422.

26. **Fork, D. C., N. Murata, and N. Sato.** 1979. Effect of growth temperature on the lipid and fatty acid composition, and the dependence on temperature of light-induced redox reactions of cytochrome *f* and of light energy redistribution in the thermophilic blue-green alga *Synechococcus lividus*. *Plant Physiol.* **63**:524–530.

27. **Fowler, M. G., and A. G. Douglas.** 1987. Saturated hydrocarbon biomarkers in oils of late Precambrian age from Eastern Siberia. *Org. Geochem.* **11**:201–213.

28. **Gelpi, E., H. Schneider, J. Mann, and J. Oro.** 1970. Hydrocarbons of geochemical significance in microscopic algae. *Phytochemistry* **9**:603–612.

29. **Gerdes, G., and W. E. Krumbein.** 1984. Animal communities in recent potential stromatolites of hypersaline origin, p. 59–83. *In* Y. Cohen, R. W. Castenholz, and H. O. Halvorson (ed.), *Microbial Mats: Stromatolites.* Alan. R. Liss, Inc., New York.

30. **Gerdes, G., W. E. Krumbein, and E. Holtkamp.** 1985. Salinity and water activity related to zonation of microbial communities and potential stromatolites of the Gavish Sabkha, p. 238–266. *In* G. M. Friedman and W. E. Krumbein (ed.), *Hypersaline Ecosystems. The Gavish Sabkha.* Springer-Verlag KG, Heidelberg, Federal Republic of Germany.

31. **Gerdes, G., J. Spira, and C. Dimentman.** 1985. The fauna of the Gavish Sabkha and the Solar Lake—a comparative study, p. 322–345. *In* G. M. Friedman and W. E. Krumbein (ed.), *Hypersaline Ecosystems. The Gavish Sabkha.* Springer-Verlag KG, Heidelberg, Federal Republic of Germany.

32. **Giovannoni, S. J., W. Godchaux III, E. Schabtach, and R. W. Castenholz.** 1987. Cell wall and lipid composition of *Isosphaera pallida*, a budding eubacterium from hot springs. *J. Bacteriol.* **169**:2702–2707.

33. **Gloe, A., N. Pfennig, H. Brockmann, Jr., and W. Trowitzsch.** 1975. A new bacteriochlorophyll from brown-colored Chlorobiaceae. *Arch. Microbiol.* **102**:103–109.

34. **Gloe, A., and N. Risch.** 1978. Bacteriochlorophyll c_s, a new bacteriochlorophyll from *Chloroflexus aurantiacus*. *Arch. Microbiol.* **118**:153–156.

35. **Halfen, L. N., B. K. Pierson, and G. W. Francis.** 1972. Carotenoids of a gliding organism containing bacteriochlorophylls. *Arch. Mikrobiol.* **82**:240–246.

36. **Han, J., H. W.-S. Chan, and M. Calvin.** 1969.

Biosynthesis of alkanes in *Nostoc muscorum*. *J. Am. Chem. Soc.* **91**:5156–5159.

37. **Han, J., E. D. McCarthy, and M. Calvin.** 1968. Hydrocarbon constituents of the blue-green algae *Nostoc muscorum*, *Anacystis nidulans*, *Phormidium luridum* and *Chlorogloea fritschii*. *J. Chem. Soc. Sect. C* **1968**:2785–2791.
38. **Heinen, W., H. P. Klein, and C. M. Volkmann.** 1970. Fatty acid composition of *Thermus aquaticus* at different growth temperatures. *Arch. Mikrobiol.* **72**:199–202.
39. **Hensel, R., W. Demharter, O. Kandler, R. M. Kroppenstedt, and E. Stackebrandt.** 1986. Chemotaxonomic and molecular-genetic studies of the genus *Thermus*: evidence for a phylogenetic relationship of *Thermus aquaticus* and *Thermus ruber* to the genus *Deinococcus*. *Int. J. Syst. Bacteriol.* **36**:444–453.
40. **Hoering, T. C.** 1978. Molecular fossils from the Precambrian Nonesuch shale, p. 243–255. *In* C. Ponnamperuma (ed.), *Comparative Planetology*. Academic Press, Inc., New York.
41. **Holzer, G., J. Oro, and T. G. Tornabene.** 1979. Gas chromatographic-mass spectrometric analysis of neutral lipids from methanogenic and thermoacidophilic bacteria. *J. Chromatogr.* **186**:795–809.
42. **Huang, W., and W. G. Meinschein.** 1978. Sterols in sediments from Baffin Bay, Texas. *Geochim. Cosmochim. Acta* **42**:1392–1396.
43. **Jackson, M. J., T. G. Powell, R. E. Summons, and I. P. Sweet.** 1986. Hydrocarbon shows and petroleum source rocks in sediments as old as 1.7 $\times 10^9$ years. *Nature* (London) **322**:727–729.
44. **Jackson, T. J., R. F. Ramaley, and W. G. Meinschein.** 1973. Fatty acids of a non-pigmented thermophilic bacterium similar to *Thermus aquaticus*. *Arch. Mikrobiol.* **88**:127–133.
45. **Javor, B. J., and R. W. Castenholz.** 1984. Invertebrate grazers of microbial mats, Laguna Guerrero Negro, Mexico, p. 85–94. *In* Y. Cohen, R. W. Castenholz, and H. O. Halvorson (ed.), *Microbial Mats: Stromatolites*. Alan R. Liss, Inc., New York.
46. **Javor, B. J., and R. W. Castenholz.** 1981. Laminated microbial mats, Laguna Guerrero Negro, Mexico. *Geomicrobiol. J.* **2**:237–273.
47. **Kenyon, C. N., and A. M. Gray.** 1974. Preliminary analysis of lipids and fatty acids of green bacteria and *Chloroflexus aurantiacus*. *J. Bacteriol.* **120**:131–138.
48. **Klomp, U. C.** 1986. The chemical structure of a pronounced series of iso-alkanes in South Oman crudes. *Org. Geochem.* **10**:807–814.
49. **Knudsen, E., E. Jantzen, K. Bryn, J. G. Ormerod, and R. Sirevag.** 1982. Quantitative and structural characteristics of lipids in *Chlorobium* and *Chloroflexus*. *Arch. Microbiol.* **132**:149–154.
50. **Krumbein, W. E., Y. Cohen, and M. Shilo.** 1977. Solar Lake (Sinai). 4. Stromatolitic cyanobacterial mats. *Limnol. Oceanogr.* **22**:635–656.
51. **Langworthy, T. A., G. Holzer, J. G. Zeikus, and T. G. Tornabene.** 1983. Iso- and anteiso-branched glycerol diethers of the thermophilic anaerobe *Thermodesulfotobacterium commune*. *Syst. Appl. Microbiol.* 4:1–17.
52. **Langworthy, T. A., and W. R. Mayberry.** 1976. A 1,2,3,4-tetrahydroxy pentane-substituted pentacyclic triterpene from *Bacillus acidocaldarius*. *Biochim. Biophys. Acta* **431**:570–577.
53. **Langworthy, T. A., and J. L. Pond.** 1986. Membranes and lipids of thermophiles, p. 107–135. *In* T. D. Brock (ed.), *Thermophiles: General, Molecular and Applied Microbiology*. John Wiley & Sons, Inc., New York.
54. **Levin, E. Y., and K. Bloch.** 1964. Absence of sterols in blue-green algae. *Nature* (London) **202**:90–91.
55. **Makula, R. A., and M. E. Singer.** 1978. Ether-containing lipids of methanogenic bacteria. *Biochem. Biophys. Res. Commun.* **82**:716–722.
56. **Nes, W. R., and M. L. McKean.** 1977. *Biochemistry of Steroids and Other Isopentenoids*, p. 458–460. University Park Press, Baltimore.
57. **Nishimura, M., and T. Koyama.** 1977. The occurrence of stanols in various living organisms and the behavior of sterols in contemporary sediments. *Geochim. Cosmochim. Acta* 41:379–385.
58. **Oehler, J. H.** 1976. Experimental studies in Precambrian paleontology: structural and chemical changes in blue-green algae during simulated fossilization in synthetic chert. *Geol. Soc. Am. Bull.* **87**:117–129.
59. **Oehler, J. H., J.-D. Arneth, G. Eglinton, S. Golubic, J. H. Hahn, J. M. Hayes, J. W. Hoefs, A. Hollerbach, C. E. Junge, W. E. Krumbein, D. M. McKirdy, M. Schidlowski, and J. W. Schopf.** 1982. Reduced carbon compounds in sediments: state of the art report, p. 289–306. *In* H. D. Holland and M. Schidlowski (ed.), *Mineral Deposits and the Evolution of the Biosphere*. Springer-Verlag KG, Berlin.
60. **Oo, K. C., and K. L. Lee.** 1971. The lipid content of *Bacillus stearothermophilus* at 37° and at 55°. *J. Gen. Microbiol.* **69**:287–289.
61. **Oshima, M.** 1978. Structure and function of membrane lipids in thermophilic bacteria, p. 1–10. *In* S. M. Friedman (ed.), *Biochemistry of Thermophily*. Academic Press, Inc., New York.
62. **Oshima, M., and T. Ariga.** 1976. Analysis of the anomeric configuration of a galactofuranose containing glycolipid from an extreme thermophile. *FEBS Lett.* **64**:440–442.
63. **Oshima, M., and A. Miyagawa.** 1974. Comparative studies on the fatty acid composition of mod-

erately and extremely thermophilic bacteria. *Lipids* **9**:476–480.

64. **Oshima, M., and T. Yamakawa.** 1974. Chemical structure of a novel glycolipid from an extreme thermophile, *Flavobacterium thermophilum. Biochemistry* **13**:1140–1146.
65. **Paoletti, C., B. Pushparaj, G. Florenzano, P. Capella, and G. Lercker.** 1976. Unsaponifiable matter of green and blue-green algal lipids as a factor of biochemical differentiation of their biomasses. II. Terpenic alcohol and sterol fractions. *Lipids* **11**:266–271.
66. **Pask-Hughes, R. A., H. Mozaffary, and N. Shaw.** 1977. Glycolipids in prokaryotic cells. *Biochem. Soc. Trans.* **5**:1675–1677.
67. **Pask-Hughes, R. A., and N. Shaw.** 1982. Glycolipids from some extreme thermophilic bacteria belonging to the genus *Thermus. J. Bacteriol.* **149**: 54–58.
68. **Philp, R. P., S. Brown, M. Calvin, S. Brassell, and G. Eglinton.** 1978. Hydrocarbon and fatty acid distributions in recently deposited algal mats at Laguna Guerrero, Baja California, p. 255–270. *In* W. E. Krumbein (ed.), *Environmental Biogeochemistry and Geomicrobiology.* Ann Arbor Science Publishers, Inc., Ann Arbor, Mich.
69. **Pond, J. L., and T. A. Langworthy.** 1987. Effect of growth temperature on the long-chain diols and fatty acids of *Thermomicrobium roseum. J. Bacteriol.* **169**:1328–1330.
70. **Pond, J. L., T. A. Langworthy, and G. Holzer.** 1986. Long-chain diols: a new class of membrane lipids from a thermophilic bacterium. *Science* **231**:1134–1136.
71. **Ray, P. H., D. C. White, and T. D. Brock.** 1971. Effect of temperature on the fatty acid composition of *Thermus aquaticus. J. Bacteriol.* **106**:25–30.
72. **Ray, P. H., D. C. White, and T. D. Brock.** 1971. Effect of growth temperature on the lipid composition of *Thermus aquaticus. J. Bacteriol.* **108**: 227–235.
73. **Reitz, R. C., and J. G. Hamilton.** 1968. The isolation and identification of two sterols from two species of blue-green algae. *Comp. Biochem. Physiol.* **25**:401–416.
74. **Ross, H. N. M., W. D. Grant, and J. E. Harris.** 1985. Lipids in archaebacterial taxonomy, p. 289–300. *In* M. Goodfellow and D. E. Minnikin (ed.), *Chemical Methods in Bacterial Systematics.* Academic Press, Inc. (London), Ltd., London.
75. **Schopf, J. W., J. M. Hayes, and M. R. Walter.** 1983. Evolution of earth's earliest ecosystems: recent progress and unsolved problems, p. 361–384. *In* J. W. Schopf (ed.), *Earth's Earliest Biosphere, Its Origin and Evolution.* Princeton University Press, Princeton, N.J.
76. **Schopf, J. W., and M. R. Walter.** 1983. Archaen microfossils: new evidence of ancient microbes, p. 214–239. *In* J. W. Schopf (ed.), *Earth's Earliest Biosphere, Its Origin and Evolution.* Princeton University Press, Princeton, N.J.
77. **Seckbach, J., and R. Ikan.** 1972. Sterols and chloroplast structure of *Cyanidium caldarium. Plant Physiol.* **49**:457–459.
78. **Shen, P. Y., E. Coles, J. L. Foote, and J. Stenesh.** 1970. Fatty acid distribution in mesophilic and thermophilic strains of the genus *Bacillus. J. Bacteriol.* **103**:479–481.
79. **Summons, R. E.** 1987. Branched alkanes from ancient and modern sediments: isomer discrimination by GC/MS with multiple reaction monitoring. *Org. Geochem.* **11**:281–289.
80. **Taylor, R. F.** 1984. Bacterial triterpenoids. *Microbiol. Rev.* **48**:181–198.
81. **Tornabene, T. G., and T. A. Langworthy.** 1979. Diphytanyl and dibiphytanyl glycerol ether lipids of methanogenic archaebacteria. *Science* **203**:51–53.
82. **Tornabene, T. G., T. A. Langworthy, G. Holzer, and J. Oro.** 1979. Squalenes, phytanes and other isoprenoids as major neutral lipids of methanogenic and thermoacidophilic "archaebacteria." *J. Mol. Evol.* **13**:73–83.
83. **Tornabene, T. G., R. S. Wolfe, W. E. Balch, G. Holzer, G. E. Fox, and J. Oro.** 1978. Phytanylglycerol ethers and squalenes in the archaebacterium *Methanobacterium thermoautotrophicum. J. Mol. Evol.* **11**:259–266.
84. **Volkman, J. K.** 1986. A review of sterol markers for marine and terrigenous organic matter. *Org. Geochem.* **9**:83–99.
85. **Walter, M. R.** 1977. Interpreting stromatolites. *Am. Sci.* **65**:562–571.
86. **Ward, D. M., S. C. Brassell, and G. Eglinton.** 1985. Archaebacterial lipids in hot-spring microbial mats. *Nature* (London) **318**:656–659.
87. **Ward, D. M., T. A. Tayne, K. L. Anderson, and M. M. Bateson.** 1987. Community structure and interactions among community members in hot spring cyanobacterial mats. *Symp. Soc. Gen. Microbiol.* **41**:179–210.
88. **Wiegert, R. G., and R. Mitchell.** 1973. Ecology of Yellowstone thermal effluent systems: intersects of blue-green algae, grazing flies (*Paraceonia*, *Ephydridae*) and water mites (*Partununiella*, *Hydrachnellae*). *Hydrobiologia* **41**:251–271.
89. **Winter, K., P. L. Parker, and C. van Baalen.** 1969. Hydrocarbons of blue-green algae: geochemical significance. *Science* **163**:467–468.
90. **Yao, M., H. W. Walker, and D. A. Lillard.** 1970. Fatty acids from vegetative cells and spores of *Bacillus stearothermophilus. J. Bacteriol.* **102**:877–878.

Chapter 39

Chemical Characterizations of Benthic Microbial Assemblages

H. L. Fredrickson, W. I. C. Rijpstra, A. C. Tas, J. van der Greef, G. F. LaVos, and J. W. de Leeuw

INTRODUCTION

The lipid content of microorganisms serves as a basis for chemotaxonomy (15), as a tool for the study of microbial ecology (20), and as a means of deciphering the organic geochemical record in recent sediments (12). Lipids with limited phylogenetic distributions are used as a criterion for differentiating microbial taxa in all these endeavors.

The quality of chemical information used for these purposes generally increases with increasing lipid molecular complexity. Microorganisms synthesize unique complex lipids from common building blocks (i.e., fatty acids, glycerol, phosphate, and saccharides), which, in general, are distributed widely through the phylogeny. Although some building blocks are sufficiently unique in themselves to provide useful information, differentiation of microbial taxa is less ambiguous at the macromolecular level of lipid organization than at the building-block level of organization.

Complex lipids can be analyzed by isolating the intact complex lipid, selectively degrading the molecule with reagents, and analyzing the components. For example, gram-negative bacteria possess a lipopolysaccharide (LPS) (Fig. 1) coat which is biochemically unique. LPS can be selectively extracted, and characteristic components (i.e., β-hydroxy fatty acids) can be analyzed by capillary gas chromatography-mass spectrometry (MS) after selective chemical hydrolysis and appropriate derivatization (10, 16).

Complex lipids with masses less than 2,000 to 3,000 daltons also can be analyzed in toto. Many taxonomically significant polar membrane lipids (15, 17) can be analyzed by soft-ionization MS methods such as field desorption (21) and fast-atom bombardment (FAB) (8, 11), which do not require thermal volatilization.

We have used these two approaches to

H. L. Fredrickson • Limnological Institute, "Vijverhof" Laboratory, Rijksstraatweg 6, 3631 AC Nieuwersluis, The Netherlands. *W. I. C. Rijpstra and J. W. de Leeuw* • Department of Chemistry and Chemical Engineering, Organic Geochemistry Unit, Delft University of Technology, De Vries van Heystplantsoen 2, 2628 RZ Delft, The Netherlands. *A. C. Tas, J. van der Greef, and G. F. LaVos* • Toegepast-Natuurwetenschappelijk Onderzoek–Centraal Instituut Voedingsmiddelen Onderzoek Institutes, Post Box 360, 3700 AJ Zeist, The Netherlands.

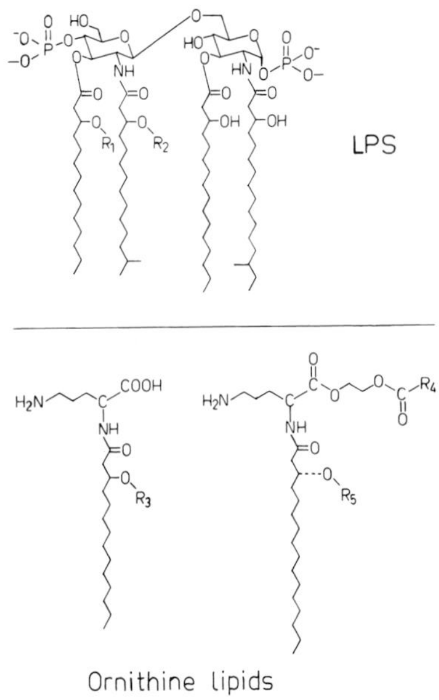

Figure 1. Partial structure of a bacterial LPS and of ornithine lipids.

study microbial lipids associated with one lacustrine and two hypersaline ecosystems. A selective hydrolytic procedure was used to measure amide-linked fatty acids derived from LPSs of gram-negative bacteria and ornithine lipids of both gram-negative and gram-positive bacteria (Fig. 1). These acids were selectively extracted from sediments associated with cyanobacterial mats of the Gavish Sabkha and Solar Lake. FAB-MS and FAB-tandem MS (FAB-MS/MS) were used to analyze the polar ether lipids of the extremely halophilic square archaebacterium originally isolated from the Gavish Sabkha and the polar lipids extracted from a methane-producing lacustrine sediment from Lake Vechten in The Netherlands.

SEDIMENT SAMPLES

Collection

Details of the collection of sediment samples from the Gavish Sabkha (5) and Solar Lake (2) and of related geochemical studies have been published. Briefly, cores were obtained from cyanobacterial mats which were permanently covered with water by pushing a Plexiglas tube into the sediment. The cores were extruded and sectioned with a knife, and the samples thus obtained were lyophilized. Samples from the mats in Solar Lake (depth, 0 to 3 mm) and from the Gavish Sabkha (2 to 7 cm), as well as a sample from a layer of gypsum (50 to 55 cm) below the mat in the Gavish Sabkha, were exhaustively extracted with organic solvents.

An anaerobic sediment sample from Lake Vechten (18) was obtained with a modified Jenkin mud sampler. Upon arrival at the laboratory (a journey of 45 min), undisturbed sediment cores were subcored with a Plexiglas tube, sliced into 2-cm sections, and immediately extracted as described below.

LPS and Ornithine Lipid Analysis

The LPSs which remained in the residue of the hypersaline sediments after solvent extraction were subjected to a selective sequential hydrolysis (10). First, the ester-linked fatty acids were removed by refluxing the residue in 1 M KOH in 96% methanol and extracting the acids with CH_2Cl_2. The residue remaining after saponification was refluxed in 4 N HCl, and the amide-linked fatty acids thus liberated were recovered by extraction with CH_2Cl_2. The resulting extract was methylated with diazomethane and silylated with BSTFA (E. Merck AG). Gas chromatography was performed on a Carlo Erba 4160 instrument equipped with an on-column injector and a 25-m CP Sil-5 (inner diameter, 0.32 mm; Chrompack) fused silica capillary column. The oven was programmed for a ballistic temperature increase from the injection temperature (100°C) to 130°C followed by an increase at 4°C/min until it reached 300°C. Gas chromatography-MS was performed on a Varian 3700 gas chromatograph connected to a Varian MAT 44 quadrupole mass spectrometer. Samples were ionized by electron impact (80 eV), and mass spectra were obtained from *m/z* 50 to 500 with a cycle time of 1.5 s. Derivatized fatty acids could be detected at nanogram concentrations with the flame ionization detector and at picogram concentrations with the mass spectrometer in the selective ion mode.

BACTERIAL SAMPLES

Culture

The square archaebacterium was obtained from Yehuda Cohen. *Halobacterium cutirubrum* was obtained from the Delft culture collection (number 81.95). Pure cell cultures were grown aerobically in the light. The medium contained NaCl (250 g), oxoidal peptone (10 g), $MgSO_4$ (9.8 g), sodium citrate (3.0 g), KCl (2.0 g), and $FeSO_4$ (50 mg) per liter of distilled water. The pH was adjusted to 6.5 with NaOH. Cells were harvested in the middle of the logarithmic phase of growth by centrifugation (10,000 × *g*).

Archaebacterial Polar Ether Lipid Analyses

Lipids were extracted by suspending the wet sample (1 g of cells or 10 g of Lake Vechten sediment) in chloroform-methanol-water (1:2:0.8 by volume) and treating the suspension for 10 min with an ultrasonic probe. The extraction mixture was centrifuged, the pellet was extracted three more times, and the supernatants were pooled. The ratio of the chloroform-methanol-water extraction mixture was adjusted to 1:1:0.95 (by volume), and the phases were separated. Lipids recovered in the chloroform phase were condensed into a volume of 1 ml by using a rotary evaporator.

Lipids were fractionated into polarity classes by column chromatography on activated (2 h at 100°C) silica gel (1 g of Kieselgel 60; Merck). Sequential elution with 20 ml each of chloroform, acetone, and methanol yielded three classes of lipids with increasing polarity. The least polar lipids, which eluted from the column with chloroform, were not analyzed. The acetone eluate, containing the majority of glycosyl lipids, was referred to as the total glycolipid eluate (TGL). The most polar lipids were eluted from the column with methanol. Most of the phosphorus-containing lipids were present in this eluate, which was referred to as the total phospholipid eluate (TPL).

FAB-MS experiments were carried out on a Finnigan MAT HSQ 30 high-resolution tandem mass spectrometer. The instrument was calibrated to standard mass values in the electron impact mode with Ultramark 3200 (Specialty Chemical) and in the FAB mode with KI and CsI. The FAB (Xe) gun was operated at 8 kV. Lipid samples were mixed into a matrix, and 5 μl was placed on a conventional probe and inserted into the ion source, which was maintained at room temperature (20°C). Glycerol or a mixture (1:2 by volume) of 1,1,3,3-tetramethylurea and triethanolamine (1) was used as the matrix. When operated in the MS mode the analyzer of the mass spectrometer was usually scanned from *m/z* 200 to 1,500 at a rate of 50 s/decade. Data were handled by a PDP 11/73 data system. Discrepancies between the measured mass values and the theoretical mass values were usually small (±0.3 daltons). A detection limit of 10 ng has been reported for purified fatty acids (11), but the large amount of Na in samples from hypersaline ecosystems raised the useful working concentration of polar lipid to microgram levels and made quantitative interpretation of these data tenuous.

MS/MS experiments were performed by using the Finnigan instrument as described above. Selected ions were collided (45 V) with argon (10^{-3} torr [0.13 Pa]) in a collision quadrupole, and daughter ions were separated by a second quadrupole. Data were handled by a PDP 11/73 data system. Discrepancies between the measured values and theoretical values were $\pm$ 0.5. All mass measurements by MS/MS are expressed as nominal mass values.

LPS ANALYSES

The lipids which were released only after sequential basic and acidic hydrolysis of samples from the cyanobacterial mats of Solar Lake and the Gavish Sabkha and from a gypsum layer of the Gavish Sabkha were analyzed after derivatization by capillary gas chromatography (Fig. 2). Components were identified (Table 1) by comparing retention times and mass fragmentation patterns with those of standards. Chromatograms of samples collected from the two cyanobacterial mats were comparable in that β-hydroxy fatty acids (β-OHFAs) were more abundant than non-OHFAs and that the patterns of the relative abundances of the major fatty acids, *n*-$C_{14:0}$-β-OHFA, *n*-$C_{16:0}$-β-OHFA, iso-$C_{17:0}$-β-OHFA, and *n*-$C_{18:0}$-β-OHFA, were similar. These OHFAs are thought to be derived primarily from the LPSs of gram-negative bacteria. It has been shown that *n*-$C_{16:0}$-β-OHFA is the predominant amide-linked LPS fatty acid in at least one cyanobacterium and that there is great diversity in LPS fatty acid components among gram-negative bacteria (10). These data support microscopic studies of the two mats (2, 9) which suggest that the microbial communities making up the mats were somewhat similar.

Although the amide-linked β-OHFA profiles of the two mat samples were grossly similar, some differences were apparent. Iso-$C_{15:0}$-β-OHFA, anteiso-$C_{15:0}$-β-OHFA, and iso-$C_{16:0}$-β-OHFA were relatively more abundant in the Gavish Sabkha mat sample than in the Solar Lake sample. *n*-$C_{18:0}$-α-OHFA was present only in the Solar Lake sample. The origin of this α-OHFA was not clear. Amide-linked α-OHFAs have not been previously reported as components of bacterial LPS, but are known to occur in the sphingolipids of higher organisms and a bacterium (20). The patterns of the relative abundances of the amide-linked non-OHFAs observed in the mat samples were less similar than the β-OHFA patterns. The non-OHFAs might reflect ornithine lipids of both gram-negative and gram-positive bacteria (15).

In contrast to the Solar Lake and Gavish Sabkha cyanobacterial mat samples, the gypsum layer sample from the Gavish Sabkha sediment core contained virtually no β-OHFAs (Fig. 2, lowest trace). This indicates that bacterial LPSs were absent from this sediment layer. The major component in the gas chromatography trace of the sample from the gypsum layer has been tentatively identified from its mass fragmentation pattern and appears to contain an alkyl chain ether-linked to a glycerol molecule. The origin of this compound is unknown.

This layer of gypsum is believed to have been deposited under highly evaporitic environmental conditions, which would favor the growth of extremely halophilic archaebacteria (9). Archaebacterial lipids contain ether-linked isoprenoid alkyl chains and lack ester- and amide-linked fatty acids (6). Ether linkages are stable to both basic and acidic hydrolyses, and so the analytical procedure used in this study would not reveal the presence of the known major archaebacterial lipids or, in consequence, the contribution of these microorganisms to the organic matter in the gypsum layer of the sediment.

Figure 2. Capillary gas chromatograms of lipids extractable only after acid hydrolysis of samples collected from the cyanobacterial mats of Solar Lake (top) and the Gavish Sabkha (middle) and a sample of the gypsum layer from the Gavish Sabkha (bottom). For an explanation of the peak numbers, see Table 1.

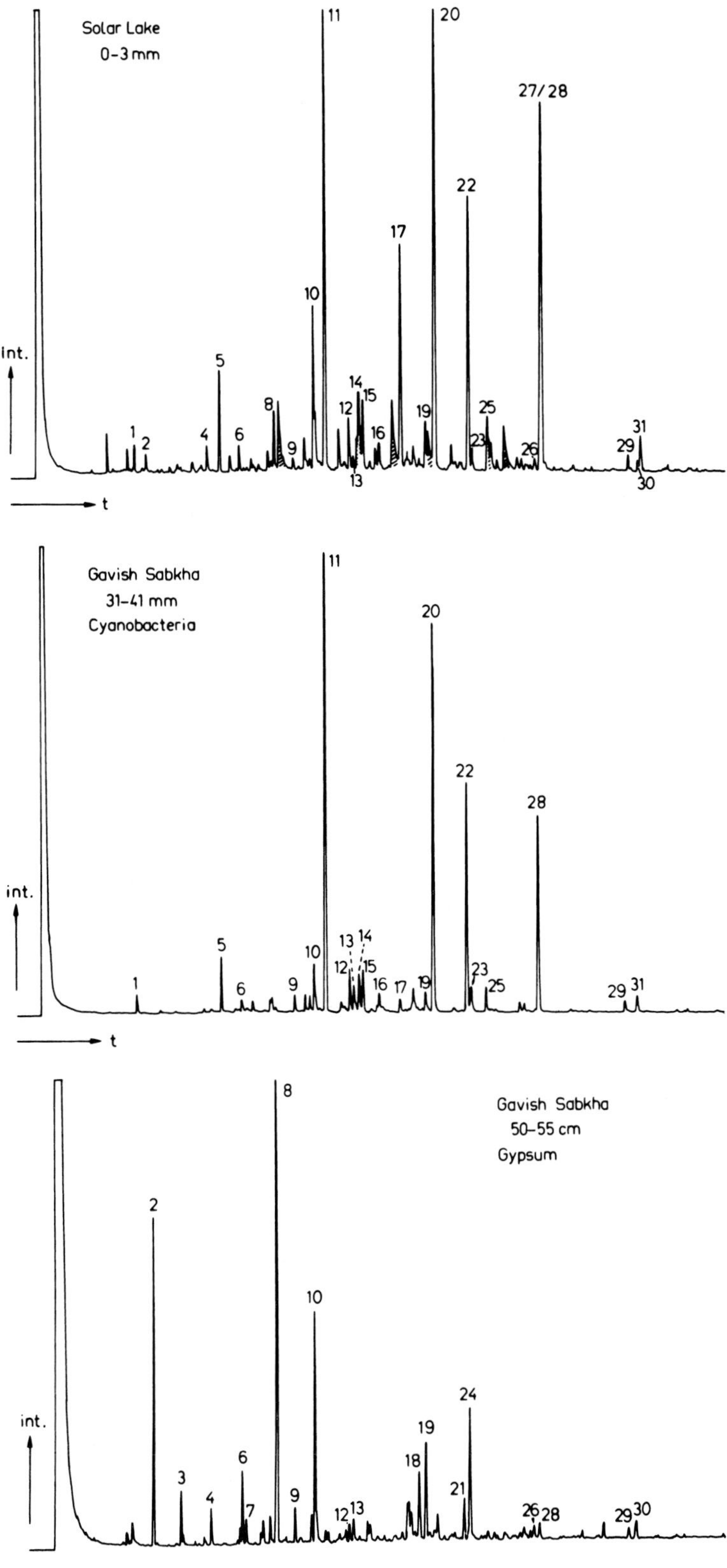

Solar Lake
0-3 mm
int.
t
1
2
4
5
6
8
9
10
11
12
13
14
15
16
17
19
20
22
23
25
26
27/28
29
30
31
Gavish Sabkha
31-41 mm
Cyanobacteria
int.
t
1
5
6
9
10
11
12
13
14
15
16
17
19
20
22
23
25
28
29
31
Gavish Sabkha
50-55 cm
Gypsum
int.
t
2
3
4
6
7
8
9
10
12
13
18
19
21
24
26
28
29
30

Table 1.
Identifications of the Major Compounds in the Cyanobacterial Mats

Peak	Compound[a]	Peak	Compound[a]
1	n-$C_{10:0}$-β-OHFA	17	i-$C_{16:0}$-β-OHFA
2	$C_{12:0}$-OH	18	Unknown
3	CH_2OMe-CH_2OMe-$CH_2OC_{11}H_{23}$?	19	n-$C_{18:0}$-FA
4	n-$C_{14:0}$-FA	20	n-$C_{16:0}$-β-OHFA
5	n-$C_{12:0}$-β-OHFA	21	isop-$C_{20:1}$-OH (phytol)
6	i-$C_{15:0}$-FA	22	i-$C_{17:0}$-β-OHFA
7	ai-$C_{15:0}$-FA	23	ai-$C_{17:0}$-β-OHFA
8	CH_2OMe-CHOMe-$CH_2OC_{13}H_{27}$?	24	Cyclopropyl-C_{19}-FA
9	i-$C_{16:0}$-FA	25	n-$C_{17:0}$-β-OHFA
10	n-$C_{16:0}$-FA	26	n-$C_{20:0}$-FA
11	n-$C_{14:0}$-β-OHFA	27	n-$C_{18:0}$-α-OHFA
12	i-$C_{17:0}$-FA	28	n-$C_{18:0}$-β-OHFA
13	ai-$C_{17:0}$-FA	29	n-$C_{20:1}$-β-OHFA
14	i-$C_{15:0}$-β-OHFA	30	n-$C_{22:0}$-FA
15	ai-$C_{15:0}$-β-OHFA	31	n-$C_{20:0}$-β-OHFA
16	n-$C_{15:0}$-β-OHFA		

[a] Abbreviations: n, normal; i, iso; ai, anteiso; isop, isoprenoid; $C_{n:x}$, n = total number of carbon atoms, x = number of double bonds; OH, alcohol; FA, fatty acid; α-OHFA, α-hydroxy fatty acid; β-OHFA = β-hydroxy fatty acid.

COMPLEX POLAR ETHER LIPID ANALYSES

To use the highly unique polar ether lipids of archaebacteria as markers for the contribution of archaebacteria to the organic matter of sediments, another method was required. Almost all studies on the molecular structures of archaebacterial ether lipids have involved an experimental approach which entailed hydrolysis of the polar head groups from the intact lipid, selective cleavage of ether linkages, identification of the pieces thus generated, and logical reconstruction of the original molecule (6, 13). Chappe and Albrecht (3) used this method and indirectly showed the presence of resistant parts of archaebacterial ether lipids in ancient sediments and petroleums.

It is also possible to analyze the archaebacterial ether lipids after the polar head group has been removed, but with the ether linkages still intact. Surprisingly, we found only one such report in the literature (4), even though this study revealed structural features about these types of ether lipids which would have been impossible to determine by using methods which entailed cleavage of the ether bond.

In this study we have chosen to investigate the ether lipids of archaebacteria in toto, with polar head groups and ether linkages intact, by using FAB/MS and FAB-MS/MS. This was done to circumvent some of the potential pitfalls inherent in molecular structural determinations which entail selective chemical degradation, analysis of degradation products, and logical reconstruction of the lipid molecule.

Lipids were extracted from the square archaebacterium and separated into polarity classes by silica gel column chromatography, and the polar lipid eluates from the column were analyzed by positive- or negative-ion FAB-MS. The positive-ion FAB mass spectrum of the mixture of polar lipids extracted from the square archaebacterium which eluted from the silica gel column with methanol (TPL eluate) is shown in Fig. 3. The most abundant ions in the positive-ion spectrum of the TPL eluate, m/z 115, 137, 185, and 207, resulted from the complexes formed from the glycerol matrix and Na.

In the positive-ion mode, lipids were

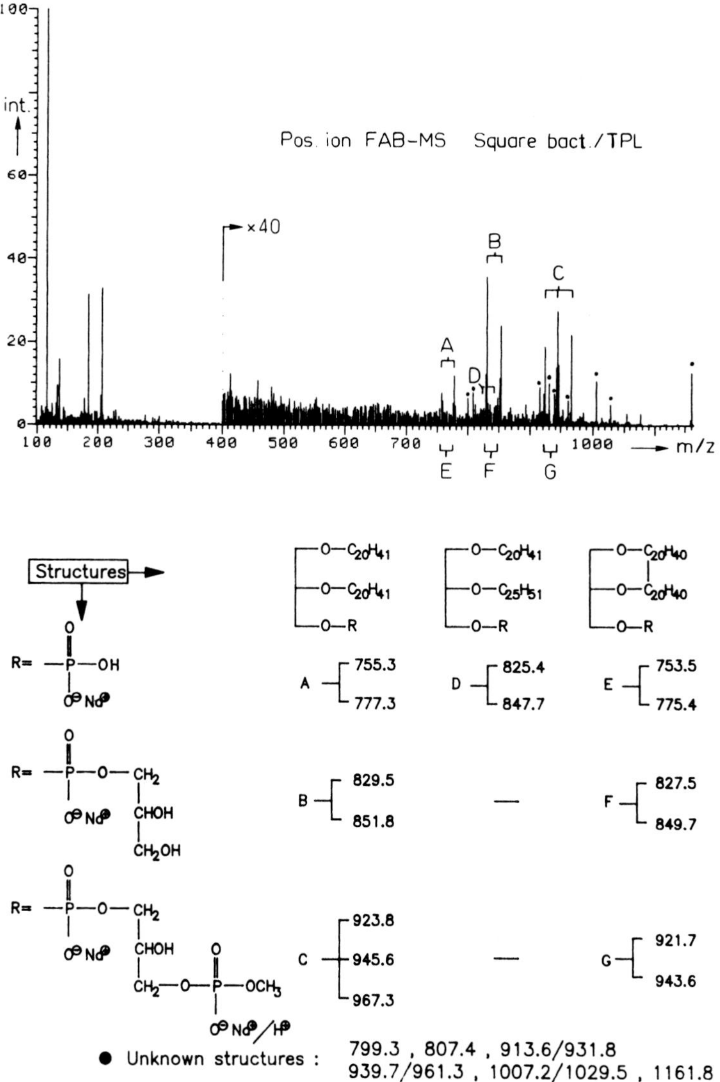

Figure 3. Positive-ion FAB mass spectrum of the lipids of the square archaebacterium which eluted from a silica gel column with methanol (TPL) and tentative assignments of the major ions. Peaks A to G are identified in the text.

revealed by their protonated or cationized (Na) derivatives. The diphytanyl diether analogs of phosphatidic acid (Fig. 3, structure A) and phosphatidylglycerol (structure B) were tentatively identified. The analogs of phosphatidic acid (structures A, D, and E) may have formed from fragmentation of a more complex lipid under Xe atom bombardment or may have already been present in the TPL eluate or both. The presence of the diphytanyl diether analog of phosphatidylglycerol (structure B) in the square archaebacterium extended the range of occurrence of this lipid. It has been found in all other extremely halophilic archaebacteria thus far analyzed (13). The positive-ion FAB-MS analysis revealed the presence of the diphytanyl diether analog of phosphatidylglycerophosphate as its methyl ester analog (structure C). We have observed this methylated lipid in the FAB mass spectra of other extremely halophilic archaebacteria, but have not yet determined whether the methylation resulted from the analytical procedure.

In addition to the diphytanyl diether analogs, the presence of other isoprenoid chain analogs was indicated by the positive-ion FAB-MS data. The phytanyl-sesterpanyl diether analog of phosphatidic acid (structure D) corresponded to ions at *m/z* 825 and 847. Phytanyl-sesterpanyl analogs in haloalkaliphilic archaebacteria have been reported (19). Neither unsaturated nor cyclopropyl isoprenoid chains have been reported to occur in the polar lipids of archaebacteria. Five-membered rings have been shown to occur in the C_{40} isoprenoid chains which cross-link tetraethers of extremely thermophilic archaebacteria (6), but are unknown in isoprenoid chains of diethers. Therefore, the presence of ions 2 daltons less than those expected for the phytanyl diether analogs of some polar lipids suggested that some of the isoprenoid chains may contain cyclic structures. Macrocyclic diether analogs of phosphatidic acid (structure E), phosphatidylglycerol (structure F), and *O*-methyl-phosphatidylglycerophosphate (structure G) were thought to be the most plausible structures of these ions. In the only other study which examined ether lipids without cleaving the ether linkages, Comita et al. (4) reported that the major ether lipid in the thermophilic methane-producing archaebacterium *Methanococcus jannaschii* was macrocyclic, but failed to find this lipid in a limited number of other methane-producing archaebacteria. However, no extreme halophiles were included in their study. The present results suggest the presence of macrocyclic ethers in at least one extreme halophile.

No other analyses of the lipids present in the TPL eluate from the square archaebacterium have been performed, and the tentative structures assigned here are based on circumstantial evidence. In separate studies we have determined the polar lipid content of a number of halophilic archaebacteria by FAB-MS/MS (H. L. Fredrickson, A. C. Tas, J. van der Greef, G. F. LaVos, J. J. Boon, and J. W. de Leeuw, *Biomed. Environ. Mass Spectrom.*, in press) and shown that the results were in general agreement with the results of degradation studies of lipids from related organisms (4, 6, 13). However, the ions of *m/z* 799, 807, 913 and 931, 939 and 961, 1,007 and 1,029, and 1,161, which appear in the spectrum of the square archaebacterium, did not correlate with structures mentioned in the literature.

The negative-ion FAB mass spectrum of the polar lipids extracted from the square archaebacterium which eluted from the silica gel column with acetone contained predominantly glycosyl lipids (Fig. 4). The ions of *m/z* 295, 297, 447, 592, and 595 were derived from the matrix. Polar lipids attributable to the square archaebacterium were revealed as deprotonated molecules ([M − H]). Diphytanyl diether analogs of sulfatidic acid (Fig. 4, structure H), *O*-methyl phosphatidylglycerophosphate (structure C), a disaccharide (structure I), a trisaccharide (structure J), and a sulfated trisaccharide (structure K) were tentatively identified. Structure H has not previously been found in archaebacteria; however, its mass (ion *m/z* 731) also could be explained by a deprotonated phosphatidic acid (Fig. 3, structure A without Na). Furthermore, this ambiguity between phosphate and sulfate also exists in other proposed structures (e.g., structures K and M), but can be quickly resolved by MS/MS. Glycosyl lipids with masses equivalent to structure I (14), structure J (7), and structure K (13) have been reported to be present among the polar lipids of other extremely halophilic archaebacteria.

Ions corresponding to the macrocyclic analogs of the trisaccharide (structure L) and the sulfated trisaccharide (structure M) were also present. Ions corresponding to the macrocyclic analogs of other components in the

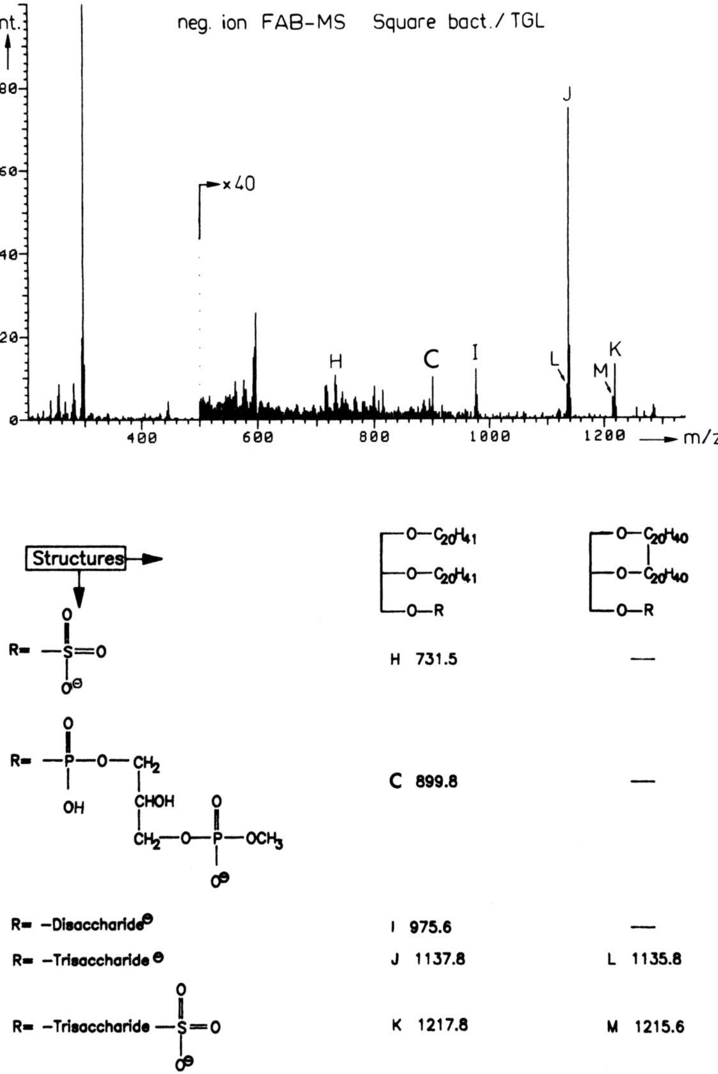

Figure 4. Negative-ion FAB mass spectrum of the lipids of the square archaebacterium which eluted from a silica gel column with acetone (TGL) and tentative assignments of the major ions. Peaks C, H, and I to M are identified in the text.

TGL eluate may have also been present, but their intensities were too close to the noise level. In a previous study of macrocyclic archaebacterial ether lipids (4), the polar head groups were removed before MS analysis, and so it was not possible to determine the structure of the intact lipid which contained the macrocyclic ether.

To obtain additional structural information, the ion of *m/z* 1,137 was discretely selected with the first mass spectrometer and collided with Ar atoms to induce molecular fragmentation. The negative-ion FAB-MS/MS spectrum and the proposed fragmentation mechanism are shown in Fig. 5. The precise structure of the trisaccharide chain was not determined in this study, and the structure shown assumed that this square archaebacterial lipid is the same as that reported for *Halobacterium cutirubrum* (13). The most intense daughter ions were explicable by cleavage of the glycosidic linkages of the carbohydrate chain between the first and second and between the second and third saccharide units. The initial fragments thus generated underwent subsequent loss of the phytanyl chain or of water and ketene as indicated in Fig. 5.

To determine whether it was possible to detect intact polar ether lipids of archaebacteria in sediments, we analyzed a sample from a methane-producing lacustrine sediment (Lake Vechten) by using the same analytical method as that used for the square archaebacterium. Extremely halophilic archaebacteria do not exist in this sediment, but some members of another group of archaebacteria, the methanogens, are known to contain similar polar ether lipids (13). The negative-ion FAB mass spectrum of the TPL eluate (Fig. 6) was complex because it probably contained phospholipids derived from a

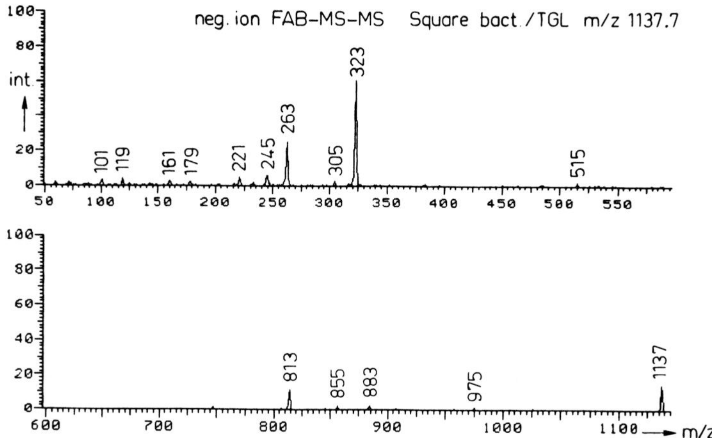

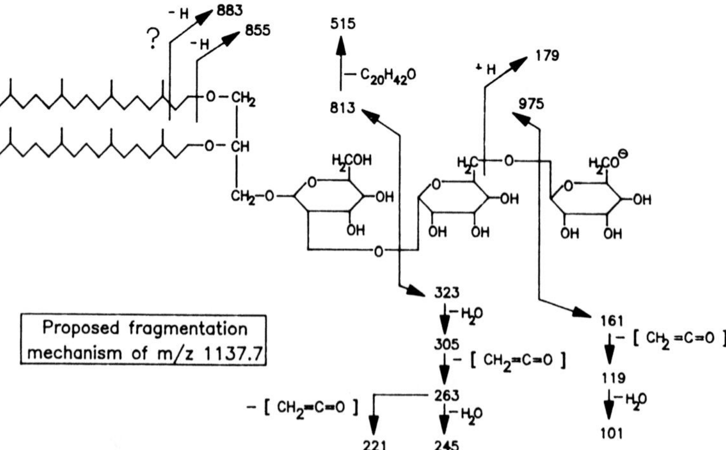

Figure 5. Negative-ion FAB mass spectrum of the ion of *m/z* 1,137.7 and a proposed fragmentation mechanism in which masses are indicated as nominal masses.

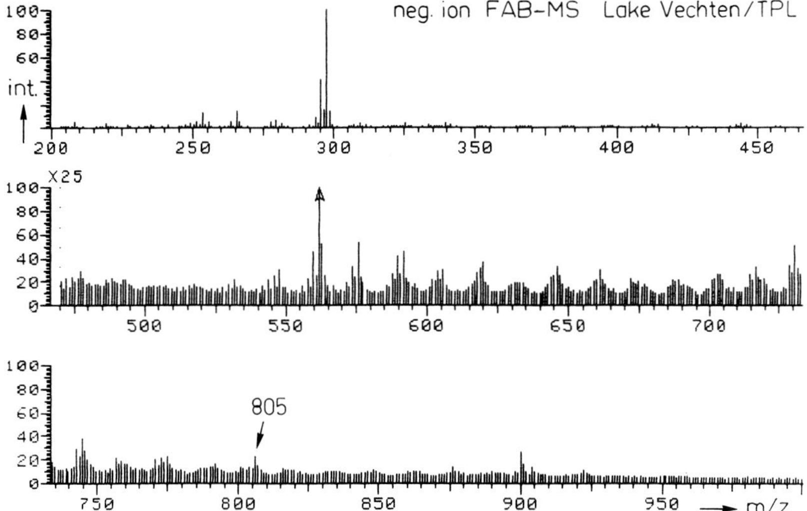

Figure 6. Negative-ion FAB mass spectrum of the sedimentary lipids of Lake Vechten which eluted from a silica gel column with methanol (TPL).

number of sources (e.g., archaebacteria, eubacteria, algae, and higher plants). Despite the complexity of this spectrum, the masses of some abundant ions coincided with the masses of polar lipids which are commonly found in archaebacteria. The ion of *m/z* 731 corresponded to the anion of the diphytanyl ether analog of phosphatidic acid or sulfatidic acid (Fig. 3, structure A without Na^+, and Fig. 4, structure H, respectively). The ion of *m/z* 805 suggested the presence of the diphytanyl diether analog of phosphatidylglycerol (Fig. 3, structure B without Na^+). The ions of *m/z* 899 and 921 could be explained by the presence of the anions which resulted from deprotonation of the diphytanyl diether analog of *O*-methyl-phosphatidylglycerophosphate (Fig. 3, structure G) and its monosodium salt, respectively.

Comparison of negative-ion FAB-MS/MS data of the ion of *m/z* 805 from the TPL eluate of the Lake Vechten sample with those of the ion of *m/z* 805 from the TPL eluate of *H. cutirubrum* (Fig. 7) showed that the lipid which produced this ion probably had an archaebacterial origin. The spectra are quite simple and are similar. Cleavage of the polar head group from the glycerol backbone resulted in a fragment ion (*m/z* 171) with low intensity which lost water and produced an ion of *m/z* 153 (Fig. 7). The decreased intensity of the ion of *m/z* 805 in the Lake Vechten spectrum was due to an increase in pressure in the collision chamber, which caused more collisions and hence more numerous but less intense fragment ions. The ion of *m/z* 385 was a background peak.

The preceding analyses of microbial macromolecular lipids exemplify the types of information on lipid molecular structure which can be obtained from two independent approaches. Each approach has advantages and disadvantages. By using the specific degradation and analysis approach, quantitative and detailed qualitative information on the individual lipid building blocks (i.e., fatty acids) can be determined precisely. However, logical reconstruction of the original molecule from the degradation products often remains ambiguous. For example, the structure of the macrocyclic isoprenoidyl diether of *M. jannaschii* (4) could not be proven from data which were obtained by cleavage of ether bonds, because the degradation product, a C_{40} iso-

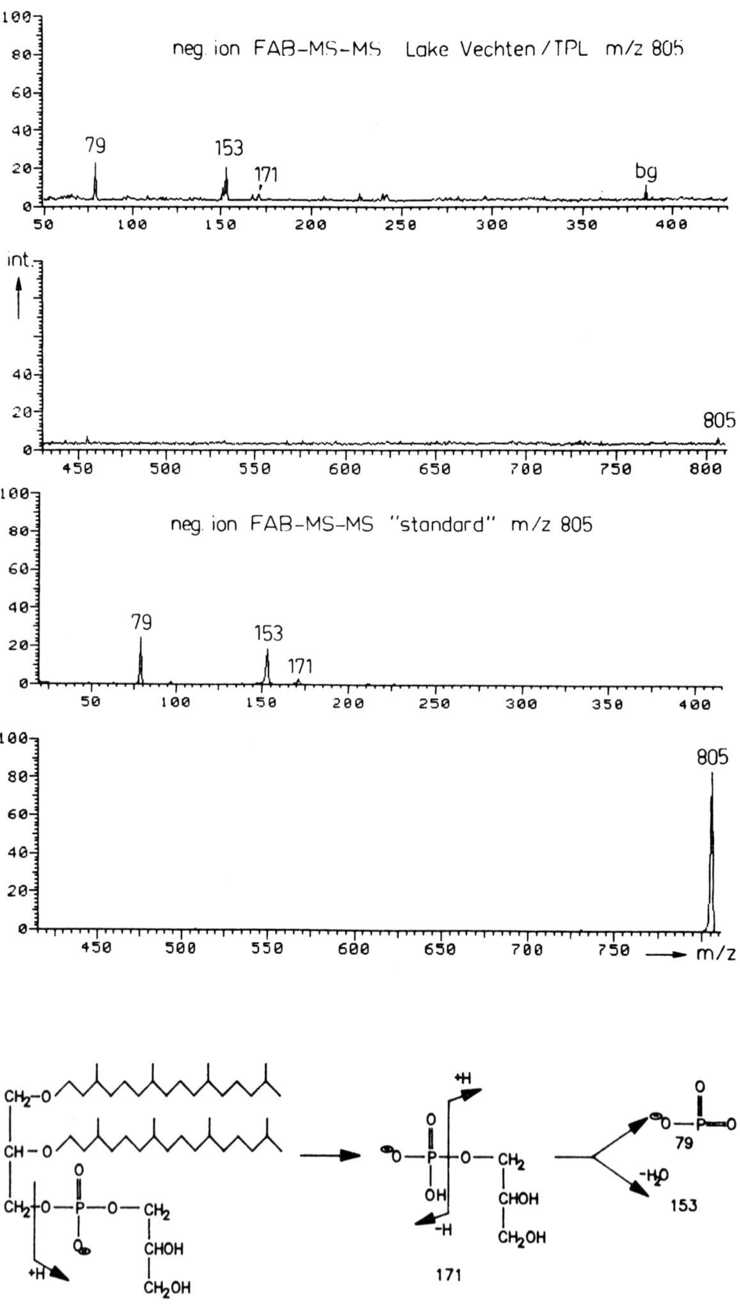

Figure 7. Comparison of the MS/MS data on the ion of *m/z* 805 from the TPL eluates of lipids extracted from Lake Vechten sediment and *H. cutirubrum* and (below) a proposed fragmentation mechanism in which masses are indicated as nominal masses.

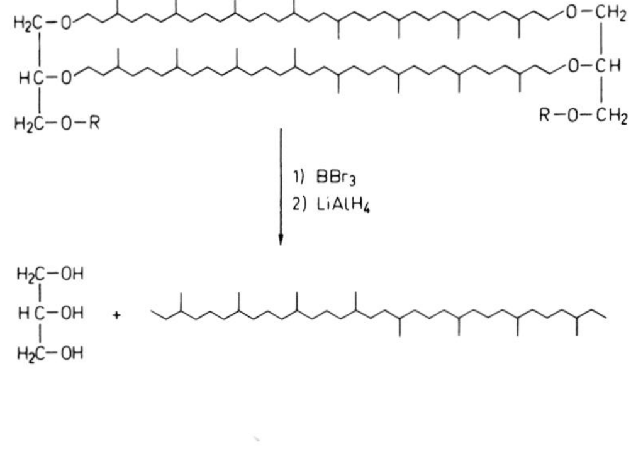

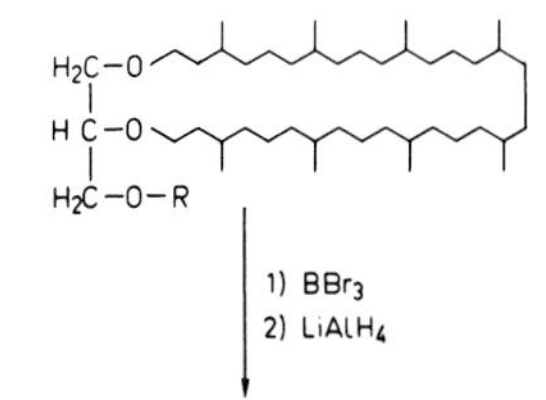

Figure 8. Example of the ambiguity which can arise in the logical reconstruction of complex lipids from degradation products.

prenoid alkane, could also have been derived from a bis-(diphytanyl)diglycerol tetraether (Fig. 8).

On the other hand, analysis of the complex lipids in toto can provide information about the arrangement of the individual lipid building blocks including the polar head groups and subtle features such as O-methylation. However, thorough interpretation of data generated by this analytical technique is currently limited by the lack of a theoretical explanation of molecular behavior under the influence of this analysis. As a result, the detailed structures of the lipid building blocks often remain obscure. Obviously, the most comprehensive determination of molecular structures of complex lipids is obtained by the use of both approaches in a complementary manner.

Acknowledgments. We thank Yehuda Cohen for supplying the square archaebacterium and Jaap Boon for transporting it to our laboratory.

LITERATURE CITED

1. **Arita, M., M. Iwamori, T. Higuchi, and Y. J. Nagai.** 1983. 1,1,3,3-Tetramethylurea and triethanolamine as a useful matrix for fast atom bombardment mass spectrometry of gangliosides and neutral glycosphingolipids. *Biochemistry* 93:319–322.
2. **Boon, J. J., H. Hines, A. L. Burlingame, J. Klock, W. I. C. Rijpstra, J. W. de Leeuw, K. E. Edmunds, and G. Eglinton.** 1983. Organic geochemical studies of Solar Lake laminated cyanobacterial mats, p. 207–227. *In* M. Bjorøy, P. Albrecht, C. Cornford, K. de Groot, G. Eglinton, E. Galimov, D. Leythaeuser, R. Pelet, J. Rullkotter, and G. Speers (ed.), *Advances in Organic Geochemistry 1981.* John Wiley & Sons, Inc., New York.
3. **Chappe, B., and P. Albrecht.** 1982. Polar lipids of archaebacteria in sediments and petroleums. *Science* 217:65–66.
4. **Comita, P. B., R. B. Gagosian, H. Pang, and C. E. Costello.** 1984. Structural elucidation of a unique macrocyclic membrane lipid from a new extremely thermophilic, deep sea hydrothermal vent archaebacterium, *Methanococcus jannaschii. J. Biol. Chem.* 259:15234–15241.
5. **de Leeuw, J. W., J. S. Sinninghe Damsté, J. Klok, P. A. Schenck, and J. J. Boon.** 1985. Biogeochemistry of Gavish Sabkha sediments. I. Studies on neutral reducing sugars and lipid moieties by gas chromatography-mass spectrometry, p. 350–367. *In* G. M. Friedman and W. E. Krumbein (ed.), *Hypersaline Ecosystems, Biogeochemistry of Gavish Sabkha Sediments.* Springer-Verlag KG, Berlin.
6. **De Rosa, M., A. Gambacorta, and A. Gliozzi.** 1986. Structure, biosynthesis, and physiological properties of archaebacterial lipids. *Microbiol. Rev.* 50:70–80.
7. **Evans, R. W., S. C. Kushwaha and M. Kates.** 1980. The lipids of *Halobacterium marismortui*, an extremely halophilic bacterium in the Dead Sea. *Biochim. Biophys. Acta* 619:533–544.
8. **Fenwick, G. R., J. Eagles, and R. Self.** 1983. Fast atom bombardment of intact phospholipids and related compounds. *Biomed. Mass Spectrom.* 10:382–386.
9. **Gerdes, G., W. E. Krumbein, and E. Holtkamp.** 1985. Salinity and water activity related to zonation of microbial communities and potential stromatolites of the Gavish Sabkha, p. 238–266. *In* G. M. Friedman and W. E. Krumbein (ed.), *Hypersaline Ecosystems, Biogeochemistry of Gavish Subkha Sediments.* Springer-Verlag KG, Berlin.

10. **Goossens, H., W. I. C. Rijpstra, R. R. Düren, J. W. de Leeuw, and P. A. Schenck.** 1985. Bacterial contribution to sedimentary organic matter: a comparative study of lipid moieties in bacteria and recent sediments. *Adv. Org. Geochem.* **10**:683–696.
11. **Jensen, N. J., K. B. Tomer, and M. L. Gross.** 1987. FAB MS/MS for phosphatidylinositol, -glycerol, -ethanolamine and other complex phospholipids. *Lipids* **22**:480–489.
12. **Johns, R. B.** 1986. *Biological Markers in the Sedimentary Record. Methods in Geochemistry and Geophysics*, vol. 24. Elsevier Biomedical Press, Amsterdam.
13. **Kates, M.** 1984. Adventures in membraneland. *J. Am. Oil Chem. Soc.* **61**:1826–1834.
14. **Kushwaha, S. C., M. Kates, G. Juez, F. Rodriguez-Valera, and D. J. Kushner.** 1982. Polar lipids of an extremely halophilic bacterial strain (R-4) isolated from salt ponds in Spain. *Biochim. Biophys. Acta* **711**:19–25.
15. **Lechevalier, M.** 1977. Lipids in bacterial taxonomy. A taxonomist's view. *Crit. Rev. Microbiol.* 7:109–210.
16. **Parker, J. P., G. A. Smith, H. L. Fredrickson, J. R. Vestals, and D. C. White.** 1982. Sensitive assay, based on hydroxy fatty acids from lipopolysaccharide lipid A, for gram-negative bacteria in sediments. *Appl. Environ. Microbiol.* **44**:27–39.
17. **Singer, S. J., and G. L. Nicolson.** 1972. The fluid mosaic model of the structure of membranes. *Science* **175**:720–731.
18. **Steenbergen, C. L. M., and H. Verdouw.** 1982. Lake Vechten: aspects of its morphology, climate, hydrology, and physicochemical characteristics. *Hydrobiologia* **95**:11–25.
19. **Tindall, B. J.** 1985. Qualitative and quantitative distribution of diether lipids in haloalkaliphilic archaebacteria. *Syst. Appl. Microbiol.* **6**:243–246.
20. **White, D. C.** 1983. Analysis of microorganisms in terms of quantity and activity in natural environments. *Symp. Soc. Gen. Microbiol.* **34**:37–66.
21. **Wood, W. W., P.-Y. Lau, and G. N. S. Rao.** 1976. Field desorption mass spectrometry of phospholipids. *Biomed. Mass Spectrom.* **3**:172–176.

Chapter 40

The Paleomicrobiological Information in Proterozoic Rocks

Andrew H. Knoll

INTRODUCTION

Microbiologists studying microbial mats and paleontologists interested in Precambrian evolution have traditionally drawn much support from one another. Paleontologists have used microbiological data to constrain interpretations of the Archean and Proterozoic records, while microbiologists have used geology to justify their interpretation of mats as primitive ecosystems characteristic of the young Earth. These interactions have been good and fruitful, but, with a few notable exceptions, they have remained at a general analogical level. As research on Proterozoic sedimentary rocks and their contained biotas grows increasingly refined, it has become clear that a much more specific interaction between neontologist and paleontologist is possible. Detailed taxonomic, ecological, and geochemical comparisons between modern microbial ecosystems and fossiliferous Proterozoic rocks from comparable environments have the potential to increase the effectiveness of paleobiological interpretation, thus strengthening the geological basis upon which the evolution of microbial mat communities is interpreted.

In this chapter I will review briefly the types and quality of paleomicrobiological information available in Proterozoic rocks and suggest questions that observations of ancient successions can generate about the microbiology, sedimentology, and geochemistry of modern microbial mats and associated ecosystems. Examples from fossiliferous Upper Proterozoic carbonates of Greenland and Svalbard and environmentally homologous sediments from the present-day Bahama Banks illustrate this approach in which, pace Hutton, the past provides a key to understanding the present.

SOURCES OF PALEOMICROBIOLOGICAL INFORMATION

Environment

Paleobiological analyses of Proterozoic sedimentary rocks begin with the determination of the stratigraphic and paleoenvironmental setting. Microbial communities vary with environment today, and it is clear that they also did so throughout the Proterozoic era (18, 23, 25). Although geologists cannot

Andrew H. Knoll • Botanical Museum, Harvard University, Cambridge, Massachusetts 02138.

determine past environmental conditions with the precision available in studying modern systems, sedimentological data place important constraints on the paleoenvironmental interpretation of ancient sediments, and many of the features used in paleoenvironmental determinations can be used to delimit specific modern environmental homologs.

Environmental parameters important to microbial community development include salinity; pO_2 and vertical oxygen gradient; the frequency, duration, and predictability of desiccation; the frequency, intensity, and predictability of wave and current energy and clastic particle influx; nutrient availability, and the presence or absence of animals or macroalgae. Within limits, many of these features can be constrained for ancient sedimentary rocks. For example, a given stratum can generally be interpreted as nonmarine, supratidal, low to high intertidal, subtidal subject to intermittent to persistent wave or current action, or subtidal below normal or storm wave base. Qualitative though these divisions may be, they do correspond to biologically significant thresholds in the onshore-offshore continuum.

A sedimentary facies is defined as a package of rocks deposited under an essentially uniform set of conditions. Facies recognition is based on the chemical composition of the sediments; their texture (grain size, shape, and size frequency distribution); their structural features such as bedding morphology, geometry, and sedimentary structures; and their lateral and vertical relationships to other conformable strata. For the purposes of the present argument, fossils are omitted from consideration because we wish to ascertain the distribution of fossils among environments determined by independent means. The facies concept can be illustrated by brief consideration of fossiliferous sediments from the 800- to 700-million-year-old Draken Conglomerate and Backlundtoppen formations, Spitsbergen, Norway (24). The Draken Conglomerate Formation is a 125- to 280-m-thick package of rocks whose lateral extent exceeds 600 km (when correlative units from East Greenland are included [Fig. 1 and 2]). The formation consists almost entirely of dolomite, with minor intercalations of quartzitic sandstone in its lower part (43). Fossiliferous beds are 5 to 20 cm thick, often thinning to zero over a lateral distance of a few tens of meters (Fig. 3A). Fossils occur in diagenetically silicified portions of intraformational conglomerates or gravel stones consisting of elongate, often sharp-edged microbial mat and mud clasts set in a matrix of carbonate and subordinate quartz sand (Fig. 3B and C). These rocks are interbedded with subequal thicknesses of laterally persistent carbonate mudstone and occasional oolites. Cross-bedding and clast imbrication are evident in the conglomerates, whereas interbedded carbonate muds exhibit millimeter-scale planar laminations. Evidence for evaporite deposition is limited to a thin but laterally persistent horizon near the base of the formation; desiccation cracks occur sporadically throughout the unit.

These features collectively indicate that Draken sediments accumulated on a broad carbonate tidal flat of negligible relief, save for wide, shallow channels. Muds and mats accumulated in quiet peritidal environments well away from strong waves or currents and were regularly ripped up and redeposited as sand to flake-shaped clasts during storms. Although salinities of ambient waters may have been somewhat elevated (normal marine plankton fossils in these rocks are rare and apparently washed in), rainfall must have been sufficiently high or circulation sufficiently strong to limit evaporite precipitation.

In contrast, endolithic microbial assemblages of the overlying Backlundtoppen Formation and its equivalent in East Greenland occur in laterally extensive 0.1- to 0.5-m-thick cross-bedded sets of oolites and pisolites often showing megarippled upper surfaces (Fig. 4A and B) (K. Swett and A. H.

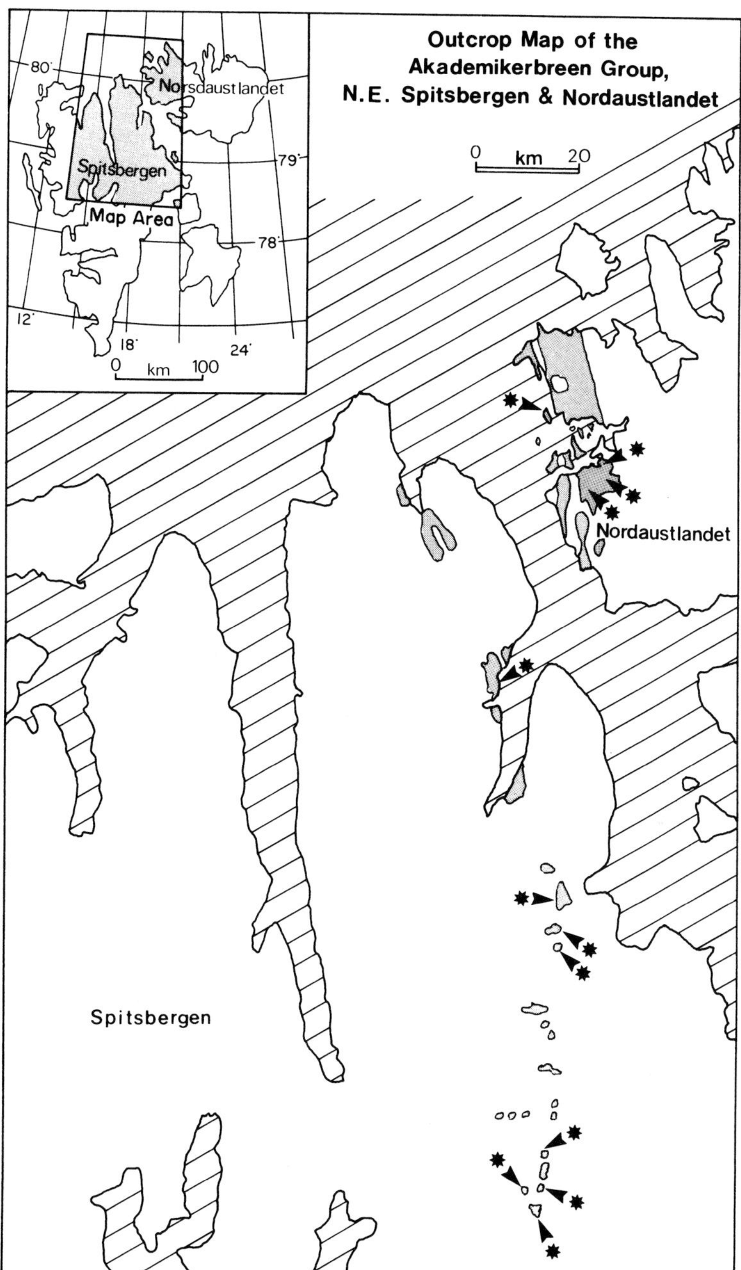

Figure 1. Map showing the outcrop distribution of the Upper Proterozoic Akademikerbreen Group and its equivalents in northeastern Spitsbergen and Nordaustlandet. Fossiliferous localities in the Draken Conglomerate and Backlundtoppen formations are denoted by asterisks.

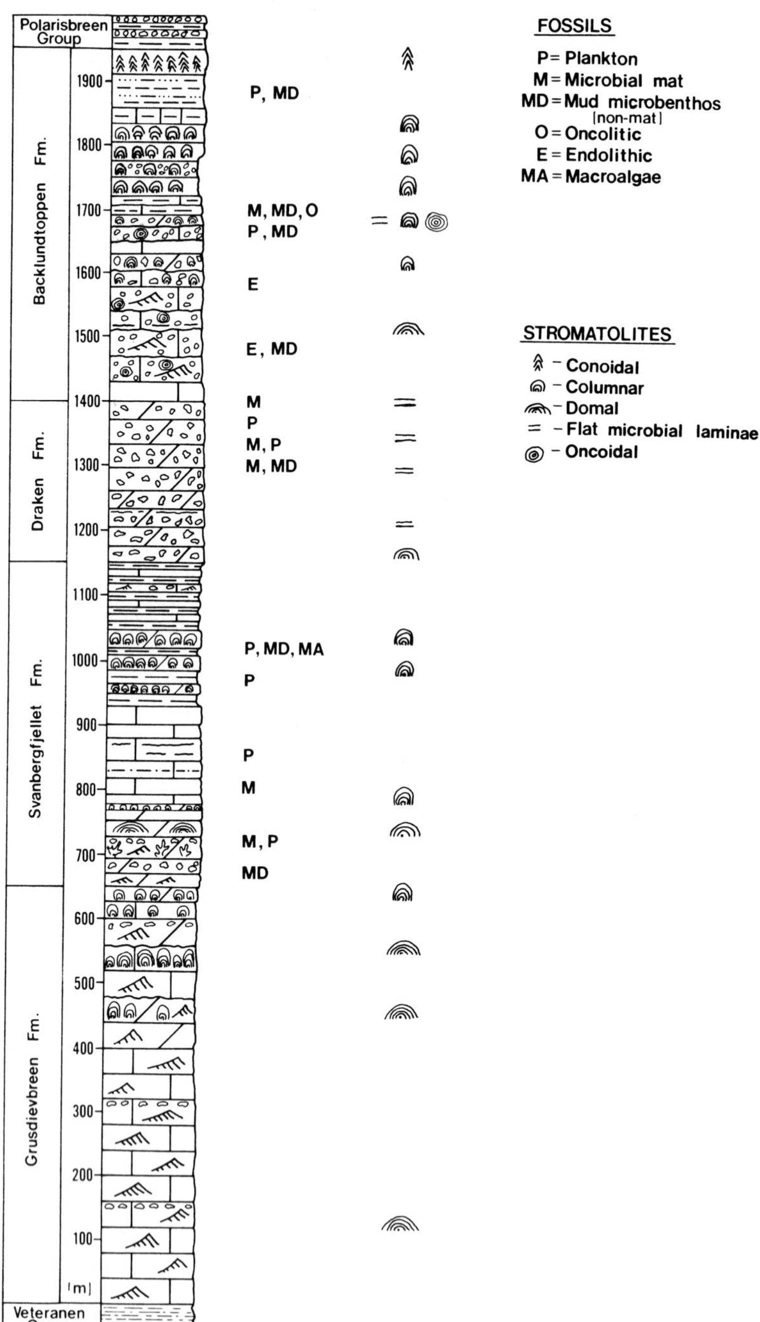

Figure 2. Generalized stratigraphic column of the Akademikerbreen Group in northeastern Spitsbergen, showing lithologies and the stratigraphic position of microfossil assemblages and microbialites (conoidal, columnar, and domal stromatolites; cryptalgalaminates; oncolites). Stratigraphic data were based on field measurements and reference 43.

Figure 3. (A) Field photograph showing characteristic bedding features of the Draken Conglomerate Formation. Note person at left for scale. (B) Close-up photograph of rocks illustrated in panel A, showing bedding and lithology. Note especially the dark clasts, which are fossil-bearing silicified microbial mat and mud clasts. The scale in the photograph is 15 cm long. (C) Photomicrograph of a silicified clast as shown in panel B. Note the abundance and diversity of organically preserved microfossils. Bar in lower right, 50 μm. See text for discussion.

Knoll, *Sedimentology*, in press). Cross-bedded sets occur in cosets of several units interbedded with thin mudstone units or, less frequently, low domal to columnar stromatolites. Upper surfaces of mudstones often show truncation and scouring by overlying oolitic beds. Most units in this member are limestone, although early diagenetic silicification and later dolomitization are common. Unlike the Draken sediments, these rocks accumulated in a shallow subtidal zone of persistent current activity and nearly normal salinity.

Both of these environments have close counterparts in the present-day Bahama Banks. Leeward of islands such as Andros, mats and muds accumulate in quiet peritidal environments and form the source materials for clasts deposited inter- or supratidally during storms (14). Oolites form in cross-bedded sets along platform margins subject to persistent currents (see, e.g., reference 15). The salient point is that the sedimentary characteristics of the two fossiliferous Proterozoic units permit them to be compared to reasonably similar environments observable today. This, in turn, makes possible environment-specific comparisons of ancient and modern microbial associations, stromatolite fabrics, and geochemical indicators. Paleoenvironmental information is critical to paleobiological interpretations of Proterozoic microfossils because it provides a crude indication of the physiological tolerances and relative competitive abilities of taxa distributed among various facies. In the absence of paleoenvironmental data, few evolutionary inferences can be gleaned from Proterozic rocks (23).

A relevant aside should be mentioned. In general, it appears that the peritidal, often restricted environments in which many of the best-studied modern microbial mats occur are more widely represented in Proterozoic carbonates than they are today. To a certain extent, this observation may be accurate, reflecting the fact that prior to the evolution of skeletonized organisms, most carbonate was precipitated near the edges of the oceans on broad, shallow platforms.

Morphology

The most obvious sources of paleomicrobiological information in Proterozoic rocks are the morphological remains of ancient organisms. Microfossil populations can be characterized well in terms of their morphology, morphometry (size frequency distribution), and behavior (division patterns, orientation, and population distribution within and among facies [25; J. Green, A. H. Knoll, and K. Swett, *J. Paleontol.*, in press]), and all of these features can be determined for extant populations living in comparable environments. Two principal uncertainties limit the interpretation of fossil morphologies. The first concerns the limited correlation between morphology and physiology among procaryotes. It is well known that strikingly similar morphologies can characterize populations that are metabolically dissimilar, for example, oscillatorian cyanobacteria and *Beggiatoa* species. Three sets of observations available in modern mats and associated sediments can be used to constrain the metabolic interpretation of fossil populations. The first is simply distribu-

Figure 4. (A) Field photograph of Backlundtoppen Formation pisolites, showing bedding surface defined by megaripples. The ovoid feature near the center is a chert nodule. (B) Close-up photograph of silicified pisoids and ooids within a cross-bedded stratum. See text for discussion. (C) Organically preserved endolithic cyanobacterium within a silicified pisoid; the grain surface is toward the top. (D) Modern cyanobacterial endolith *Hyella gigas* shown in etched ooid from the Bahama Banks; compare with panel C. (E and F) Large problematic microfossils within silicified clasts of the Draken Conglomerate Formation. The bar in panel C is 33 mm for panel B, 100 μm for panel C, 150 μm for panel D, and 15 μm for panels E and F.

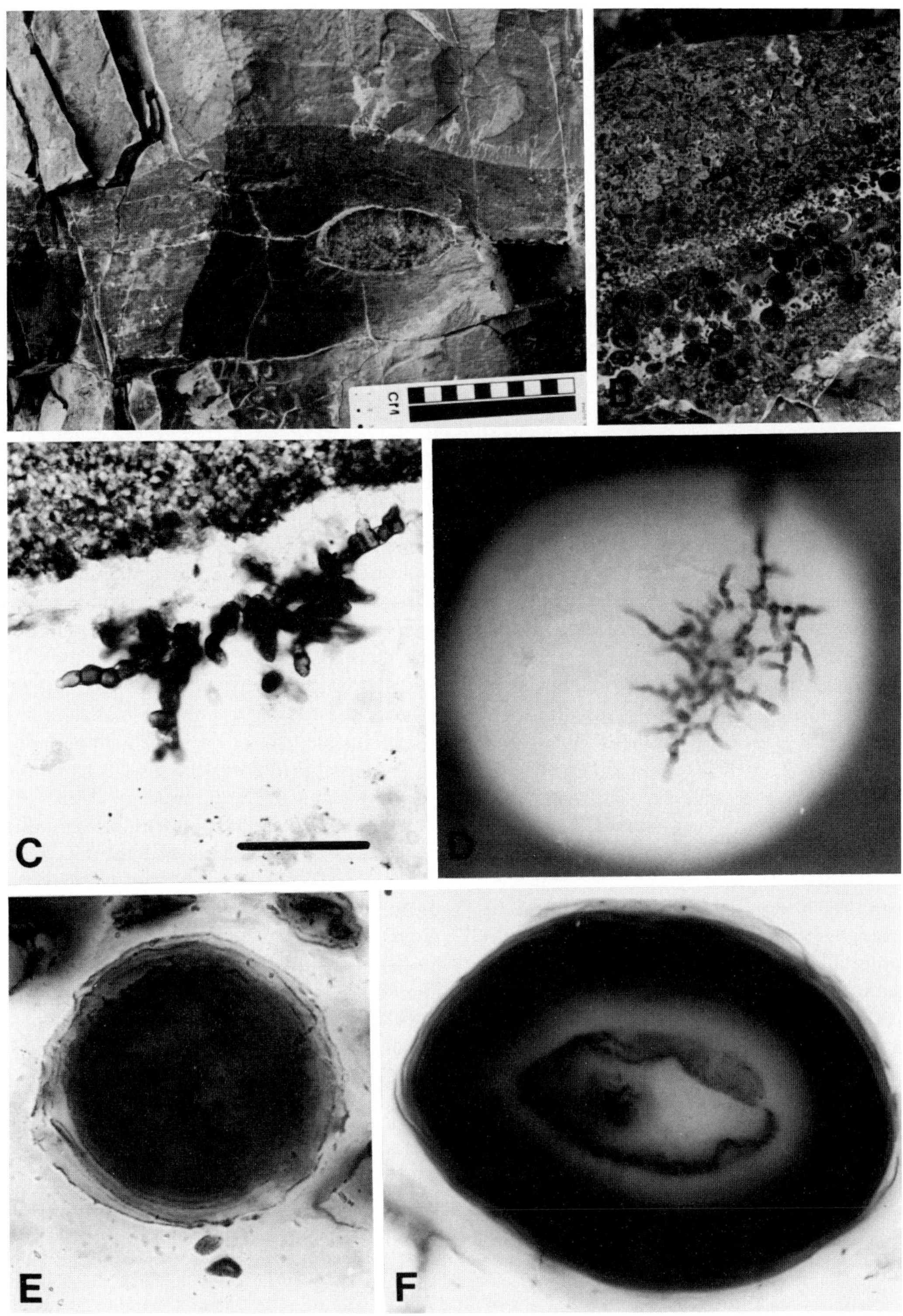
C
E
F

tional. Of the extant organisms known to compare morphologically to a given fossil taxon, which occur in environments similar to those in which the fossils are preserved? The second observation is taphonomic. Of the potential modern counterparts of a particular fossil species, which are likely to survive postmorten degradation under the sedimentary conditions inferred for deposition of the fossils? Do modern populations show taxon-specific patterns of degradation that might be applied to the interpretation of fossil remains? A number of ancient-modern taphonomic studies have demonstrated the utility of this approach (1, 7, 9, 11, 18, 20, 26, 27), but there remains a great need for additional research in the context of strict environmental control. The third set of observations concerns behavior (Green et al., in press). The distribution and orientation of microfossils within Proterozoic sediments are often distinctive and may be an important clue to interpretation. For example, in the Draken and Backlundtoppen formations and their East Greenland equivalents, nonseptate filamentous microfossils having a diameter of 2 to 3 μm are abundant. These are generally interpreted as cyanobacterial sheaths and are assigned to the species *Eomycetopsis robusta* Schopf emend. Knoll and Golubic (27), but consideration of microfossil orientations suggests that several biologically differentiable populations can be distinguished. Classical *E. robusta* sheaths commonly occur as densely interwoven populations that define laminae within Proterozoic mat-derived sediments; in the absence of early diagenetic compaction, alternating vertical and horizontal alignments of filaments can sometimes be discerned (Fig. 5A and B) (23, 44). This indicates that the constituent microorganisms were phototactic and responded behaviorally to repeated changes in local environments. By comparison with modern populations found in similar physical settings (10, 33), these fossils are interpreted as mat-building cyanobacteria showing noctidiurnal, seasonal, or aperiodic changes in orientation. In the Backlundtoppen Formation, Spitsbergen, morphologically identical populations of filaments occur in small oncolites and in thin, flat mats with which the oncolites are intimately associated (Fig. 5C) (A. H. Knoll, K. Swett, and E. Burkhardt, *J. Paleontol.*, in press); these may well represent a biological population similar to those of the aforementioned mat builders.

In contrast, *E. robusta*-like filaments found locally in the upper Draken Formation characteristically occur in bundles of a dozen or more filaments that are themselves interwoven into mats (Fig. 5D). It is likely that these populations were genetically different from those that produced the true *E. robusta* sheaths. Yet another behaviorally distinctive filament of the same general morphology occurs in ooids and pisoids from the Backlundtoppen Formation (Fig. 5E and F). Two features of this population, i.e., the tortuousity of individuals and the fact that filaments cut across concentric laminae of carbonate accretion, permit comparison with the modern endolithic cyanobacterium "*Plectonema terebrans*" found in present-day Bahamian ooids.

The second limitation of micropaleontological studies concerns the relationship between genetics and morphology. Labora-

Figure 5. (A) Microbial mat-building population of *E. robusta*, showing alternating horizontal and vertical orientation of individuals. (B) Higher-magnification photomicrograph of population shown in panel A. (C) Silicified oncoid showing dense tangle of *E. robusta* filaments forming a crescentic layer on the upper grain surface. (D) Mat-building population of *Eomycetopsis*-like filaments showing characteristic filament bundles. (E) Filamentous euendolith, showing characteristic tortuousity. The grain surface is marked by dolomite rhombs above the fossil. (F) Close-up photograph of panel E. The bar in panel E is 10 μm for panel A, 20 μm for panel B, 175 μm for panel C, 60 μm for panel D, 100 μm for panel E, and 40 μm for panel F.

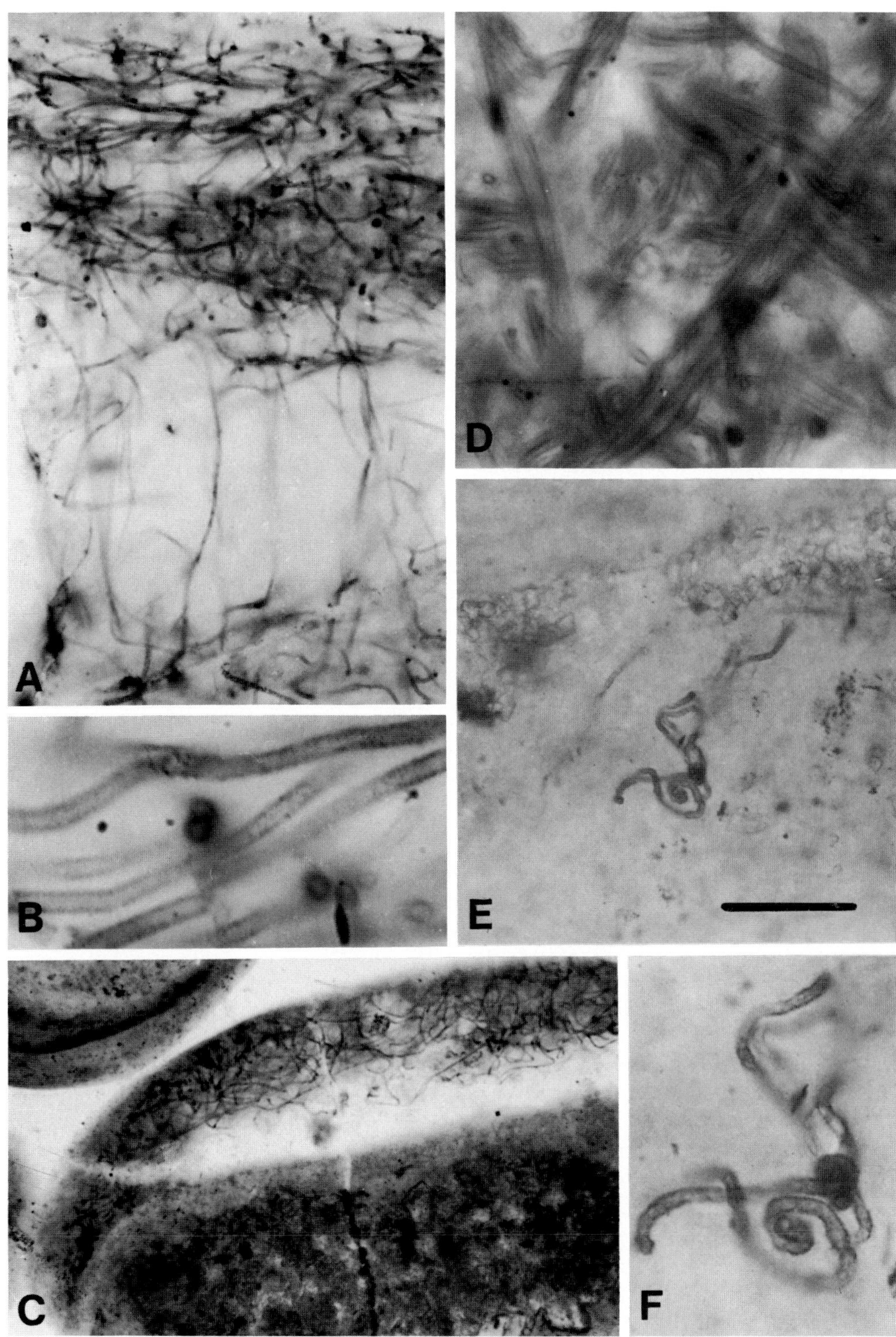
A
B
C
D
E
F

tory studies of cyanobacteria have shown that morphology within a clone can vary markedly as a function of culture conditions (5, 35, 42), but few data provide an indication of how these in vitro studies relate to natural populations distributed among environments recognizable in the geological record. In axenic cultures in which competition is absent, unusual morphs may persist under conditions where they would not occur in nature. In a recent study by Stal and Krumbein (39) on microbial mats from the North Sea, systematic classification based on morphology was found to correlate reasonably well with that based on physiology, although in culture some clones displayed characters not observed in nature. Cyanobacterial taxonomy remains in a state of ferment, but it does appear clear that under a given set of environmental conditions, a given genotype will display a predictable and limited set of phenotypes. This observation is a useful one for paleontologists, who will have to continue to delineate taxa on the basis of morphological discontinuities, size frequency distributions, discernible divisional patterns, behavioral features, and other characteristics of populations preserved in ancient sediments. It suggests that paleoenvironmental determination is important even for systematic comparisons of ancient and modern microorganisms.

Silicified clasts of peritidal carbonates from the Draken Formation and correlative beds in East Greenland contain at least seven paleontologically distinct microbial mat assemblages (13, 24; unpublished data). Principal mat-building populations (as inferred by density and orientation) occur in recurrent association with particular subordinate mat builders and mat dwellers, suggesting a structured pattern of community composition. Similar associations of taxa occur in modern Bahamian environments. For example, a distinctive surface-encrusting cyanobacterial population that is abundant in tidal flat carbonate clasts from the Proterozoic succession of East Greenland has a close morphological, developmental, and behavioral counterpart living today in physically similar Bahamian environments (Fig. 6) (13). Similarly, the assemblage of endolithic microfossil populations found in Proterozoic ooids and pisoids from the Backlundtoppen Formation and its equivalents in East Greenland compares closely to the community of cyanobacterial endoliths that bore into shallow subtidal ooids from Cat and Joulters cays today (Fig. 4C and D) (28; Green et al., in press).

The general point is that in the comparison of Late Proterozoic microfossil assemblages from Svalbard and East Greenland with modern microbial communities inhabiting shallow benthic habitats in the Bahamas, one can move beyond general comparisons to specific environment-by-environment considerations of assemblage composition and the morphology, development, reproduction, behavior, and taphonomy of constituent populations. As we learn more about Proterozoic microfossils, we need more specific data on the characteristics of microorganisms found in homologous modern environments. This is especially true of the various protists that occur as component species or allochthonous elements in modern mats and muds (see, e.g., reference 31). A better knowledge of these will aid immeasurably in understanding the many problematic fossils in Proterozoic sediments (Fig. 4E and F).

Microbial Trace Fossils

Much has been written about the stromatolites and oncolites that are conspicuous elements of many Proterozoic carbonates (see, e.g., reference 40). Their interpretation as microbially built structures seems secure in most cases, but much remains to be learned about the biological and sedimentological processes responsible for their formation and the extent to which paleobiological inferences about microbial community composition and distribution can be drawn from

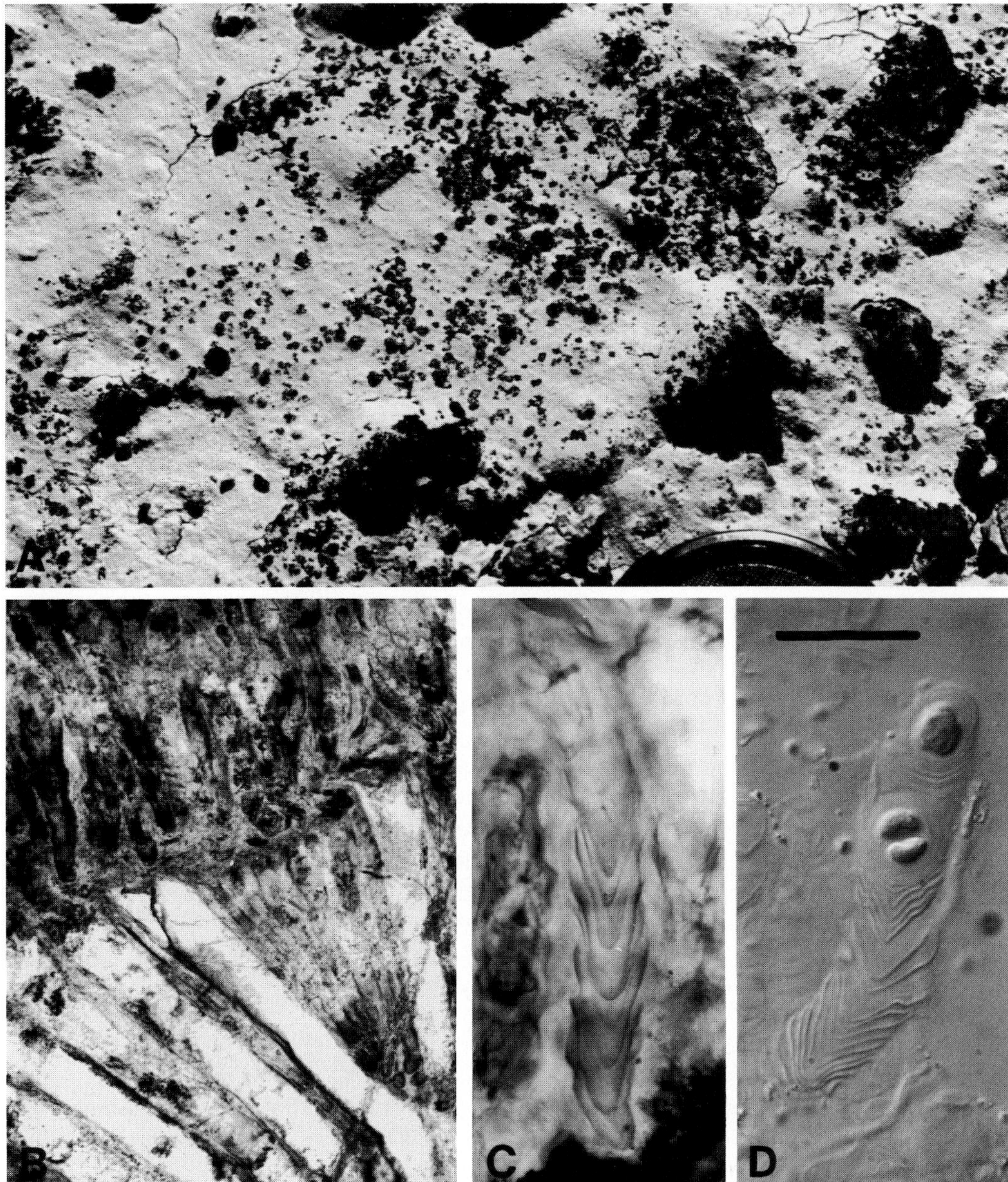

Figure 6. (A) Dark surface-encrusting populations of *Cyanostylon*-like cyanobacteria from Andros Island, Bahamas. Note the camera lens cap for scale. (B) Silicified fossils from Bed 18 of the Limestone-Dolomite "Series," central East Greenland (lateral equivalent of the Backlundtoppen Formation), showing successive surface-encrusting populations of *Polybessurus bipartitus*. (C) *P. bipartitus*, showing characteristic structure of stalklike envelopes. (D) Modern *Cyanostylon*-like cyanobacterium from the surface encrustations shown in panel A. The bar in panel D is 130 μm for panel B, 70 μm for panel C, and 20 μm for panel D.

their macromorphologies, microstructures, and/or petrographic textures. Unlike microbial mat procaryotes, most Proterozoic stromatolites cannot be compared directly with modern counterparts, because many of the physical environments colonized in the Proterozoic era by microbial mats now support large populations of seaweeds, calcareous algae, sessile invertebrates, and vagile mat-grazing animals. Nonetheless, neontological studies of the relationships among biota, environment, and stromatolite morphology and fabric provide analogies that can maximize paleobiological interpretation, just as for microfossils (4, 6, 19). The relationship between sedimentary environment and stromatolite macrostructure is often strong in Proterozoic examples, but the morphogenetic influence of responsible mat biotas is less clearly known. Microfossils are infrequently preserved in morphologically complex stromatolites, and even when they do occur it is often unclear that the preserved fossils are those of populations responsible for stromatolite accretion. Thus, biological influence in stromatolite morphogenesis must be determined from the indirect evidence of lamina morphology (the generating curve of stromatolites), microstructure, and petrographic texture.

Microstructure and texture, in particular, are often thought to be under strong biological control (3, 6, 33). To the extent that this is true, stromatolites can be used as indicators of community heterogeneity and distribution among Proterozoic environments, greatly augmenting the data available from microfossils. However, this approach must be substantiated (or repudiated) by fabric analyses of recent stromatolites. Is there a one-to-one relationship between community composition and microstructure? Are specific microbial populations associated with diagnostic fabrics that are interpretable in the geological record (see, e.g., references 3 and 33)? This question is especially interesting as applied to eucaryotic components of microbial communities—can the presence of eucaryotic mat populations be inferred from textures preserved in mat carbonates?

Microbial trace fossils also occur as microborings produced by endolithic microorganisms in ooids, pisoids, lithified stromatolitic surfaces, and carbonate hardgrounds (see, e.g., reference 12). In the Backlundtoppen ooids of Svalbard, microborings of several types occur. These correspond to cellularly preserved endolithic microfossils in the same rocks, suggesting that the use of microborings as direct proxies for individual taxa is feasible in Proterozoic rocks.

Although this chapter is concerned principally with microorganisms, I cannot conclude a discussion of trace fossils without some mention of meiofauna. Body and trace fossils of macroscopic animals first appear in uppermost Proterozoic sedimentary rocks, but several authors (see, e.g., references 8, 36, and 41) have argued for a long metazoan prehistory during which the forerunners of the Ediacaran and later faunas existed as populations of very small individuals. Our best chance of testing this hypothesis lies in the recognition of distinctive microbial mat or other sedimentary fabrics by meiofauna (see, e.g., reference 2). As is the case for fabrics associated with specific mat-building communities, studies of possible Proterozoic meiofaunal fabrics must be related to observations made on present-day systems.

Biochemistry

It is well known that some metabolic processes impart geochemical signatures to sedimentary rocks. Because they are not easily destroyed by diagenesis or mild metamorphism, isotopic indicators are particularly valuable paleobiological guides to ancient metabolism (37). Stable carbon isotopes in co-occurring organic matter and carbonates attest to the presence of autotrophic carbon fixation and, under appropriate conditions, methanogensis and methylotrophy (16, 21); isotopic ratios of sulfur in sedimentary sulfates and sulfides record dissimilatory sulfate reduction; and nitrogen isotopes can provide clues to the operation of a biological nitrogen cycle.

Although this isotopic evidence of metabolism is important, it is really only at issue for Archean rocks. All of the processes mentioned above appear to have been present by the beginning of the Proterozoic Eon. Does this mean that there is nothing new to be learned paleobiologically by the isotopic analysis of Proterozoic rocks? Not at all. Des Marais et al. (chapter 17 of this volume) have shown that important information about carbon flow in specific modern and ancient environments is encoded in carbon isotopic ratios. Secular changes in carbon isotopic ratios also appear to provide a promising means of "measuring the pulse" of the global carbon cycle through time, which, in turn, can suggest the presence and timing of evolutionarily important changes in environmental parameters such as pO_2 and pCO_2 (29; chapter 17 of this volume). Secular patterns in sulfur isotopic ratios, oxygen isotopic ratios, strontium isotopic ratios, strontium abundance in carbonates, and paleosol development, and rare earth element patterns in sedimentary rocks also have a strong potential for revealing patterns of tectonic and environmental change in the Proterozoic Eon.

In carbonates from the Draken Conglomerate and Backlundtoppen formations, stable carbon isotopic ratios for both carbonates and co-occurring organic matter are enriched by 5 to 7‰ PDB, suggesting (in association with other data) that global rates of organic carbon burial were unusually high at the time this formation was deposited (29). Draken and Backlundtoppen carbonates contain $^{87}Sr/^{86}Sr$ ratios that are unusually low by Phanerozoic standards (L. A. Derry, L. Keto, S. Jacobsen, A. H. Knoll, and K. Swett, *Geochim. Cosmochim. Acta,* in press), an observation that is consistent with stratigraphic data indicating abundant rifting and incipient continental breakup during this period. Sulfur isotopic data are not yet available for the Svalbard formations, but the few data available from other late Proterozoic sequences suggest that at the time of Draken and Backlundtoppen sedimentation, the sulfate concentration in seawater was +16.5 to +20‰, rather near its Phanerozoic mean (I. B. Lambert and T. H. Donnelly, *in* H. K. Herbert, ed., *Stable Isotope and Fluid Processes in Mineralization*, in press). Interestingly, during the late Proterozoic era, the secular curves of carbon and sulfur isotopes appear not to have been closely coupled in the way that they have been over the past 570 million years (Lambert and Donnelly, in press). Collectively, these data reveal a picture of marked tectonic and biogeochemical change during the late Proterozoic era, with important paleobiological consequences (29; A. H. Knoll, *Geol. Soc. Am. Abstr. Progr.* **19**:730, 1987).

I have obviously been tempted to interpret some isotopic data from Upper Proterozoic sediments in global terms, but this interpretation can be sustained only if values for specific facies can be related to trends in the world ocean. To this end, comparative geochemical research on the isotopic and other geochemical signals preserved in sediments from specific facies is needed. How, for example, do isotopic ratios vary as a function of environment in the modern Bahamas, and how does this constrain the interpretation of data obtained from homologous facies in the Draken and Backlundtoppen formations? In their classic study of carbon and oxygen isotopes from modern Bahamian sediments, Lowenstam and Epstein (30) found that carbon isotopic ratios in ooids were enriched by as much as 5% PDB, whereas associated muds were in general isotopically lighter and more variable. On the other hand, carbonates from other depositional environments do not record such heavy values; this variation, as well as the fact that Proterozoic carbonates may have been deposited mainly along ocean margins, provides fertile ground for comparative geochemistry aimed at understanding environmental evolution. Can one identify specific "vital" effects that will allow paleobiological interpretation of more subtle paleoenvironmental variations in isotopic signals? Once again, careful ancient-modern comparisons promise to reveal much of interest to both geologists and microbiologists.

Molecular analyses of Precambrian organic matter have a chequered history (see e.g., references 17 and 32), but recent studies of little-altered Proterozoic sequences indicate that this approach has the potential to dramatically augment our knowledge of Proterozoic mat and other ecosystems. Well-preserved Middle and Upper Proterozoic sediments contain a variety of taxonomically and physiologically interpretable biomarker molecules (22; R. Summons, T. G. Powell, and C. J. Boreham, *Geochim. Cosmochim. Acta*, in press). For example, preliminary analysis of samples from the Draken Conglomerate Formation has revealed the presence of phytane, pristane, abundant *n*-alkanes, and a strong dominance of low-molecular-weight (C_{16} to C_{19}) compounds, methylalkanes, hopanoids, and steranes (R. Summons, unpublished data). This profile is similar to those obtained from other Upper Proterozoic kerogens (R. Summons, personal communication). It suggests a significant contribution to local primary production by both cyanobacteria and algae, possibly including red as well as green cyanobacteria and algae. Waxy hydrocarbons and biomarkers specific for higher plants and animals are notable for their absence. Such compounds are characteristic features of younger kerogens, but they do not occur in rocks over 700 million years old.

At present, the principal limitation of fossil biomarker analysis is not the restricted preservation of primary molecular signals in Proterozoic rocks; studies completed to date indicate that a paleobiologically significant molecular record is preserved in many kerogens whose thermal history has been mild. It is the incomplete understanding of the natural products chemistry of many living taxa and the postmortem fate of these molecules that most strongly restricts paleobiological interpretation.

CONCLUSIONS

The central theme of this paper has been the importance of environment in the interpretation of the Proterozoic biological record. The detailed comparison of sediments, stromatolites, and microfossils from late Proterozoic formations in Svalbard and East Greenland with modern counterparts in the Bahamas is not only increasing the precision with which we can interpret the former, but is also suggesting new questions to ask about the latter. For example, new comparative investigations of Bahamian with Svalbard and East Greenland microorganisms have already revealed the presence of at least two hitherto undescribed living cyanobacteria which we discovered as a result of finding their fossil counterparts in specific sedimentary facies. Ongoing microbiological, biogeochemical, and sedimentological studies of Bahamian sediments are certain to provide new insights into the nature of Draken and Backlundtoppen microbial communities. Other ancient-modern comparisons can be equally informative: for example, comparisons between modern sabkhas from the Persian Gulf with environmentally similar Proterozoic successions from the Belcher Islands, northwestern Greenland, northern Australia, and elsewhere (see, e.g., references 11, 18, and 34) and those between coastal hypersaline lakes such as occur in Baja California and the classical Bitter Springs microfossil beds (38). Through such analyses, we will obtain a far better understanding of how biotas have changed over time, as well as new insights into how environments themselves have evolved. The latter will be of particular importance in the interpretation of microbial phylogeny.

It is clear that the present arsenal of approaches to Proterozoic biology leaves much unseen or poorly resolved; however, the range of data that can be obtained is impressive, and new ways of looking at the record may well improve the available picture substantially. Limits to the interpretation of Proterozoic sedimentary successions exist, but they have not yet been reached. They will not be reached until we know far more about the present.

ACKNOWLEDGMENTS. I thank M. R. Walter, R. V. Burne, J. Bauld, I. Lambert, and R. Summons for helpful discussions. D. Des Marais provided helpful comments on an early draft of the manuscript.

This paper was written during the term of a Guggenheim Fellowship at the Baas Becking Geobiological Laboratory, Canberra, Australia; I am grateful to both institutions. This paper is based on research supported by National Science Foundation grant BSR-85-16328 and National Aeronautics and Space Administration grant NAGW-893.

LITERATURE CITED

1. **Aizenshtat, Z., G. Lipiner, and Y. Cohen.** 1984. Biogeochemistry of carbon and sulfur cycle in the microbial mats of the Solar Lake (Sinai), p. 281–312. *In* Y. Cohen, R. W. Castenholz, and H. O. Halvorson (ed.), *Microbial Mats: Stromatolites.* Alan R. Liss, Inc., New York.
2. **Awramik, S. M., D. S. McMennamin, C. Yin, Z. Zhao, Q. Ding, and S. Zhang.** 1985. Prokaryotic and eukaryotic microfossils from a Proterozoic/Phanerozoic transition in China. *Nature* (London) **315**:655–658.
3. **Bertrand-Sarfati, J.** 1976. An attempt to classify Late Precambrian stromatolite microstructures, p. 251–259. *In* M. R. Walter (ed.), *Stromatolites.* Elsevier/North-Holland Publishing Co., Amsterdam.
4. **Burne, R. V., and L. S. Moore.** 1987. Microbialites: organosedimentary deposits of benthic microbial communities. *Palaios* **2**:241–254.
5. **Drouet, F., and W. Daily.** 1956. Revision of the coccoid Myxophyceae. *Butler Univ. Bot. Stud.* **12**:1–218.
6. **Gebelein, C. D.** 1974. Biologic control of stromatolite microstructure: implications for Precambrian time stratigraphy. *Am. J. Sci.* **274**:575–598.
7. **Gerasimerko, L. M., and I. N. Krylov.** 1985. Postmortem alterations of cyanobacteria in the algal-bacterial films in the hot springs of Kamchatka. *Dokl. Akad. Nauk SSSR* **272**:201–203. (In Russian.)
8. **Glaessner, M. F.** 1984. *The Dawn of Animal Life: a Biohistorical Study.* Cambridge University Press, Cambridge.
9. **Golubic, S., and E. S. Barghoorn.** 1977. Interpretation of microbial fossils with special reference to the Precambrian, p. 1–14. *In* E. Flügel (ed.), *Fossil Algae.* Springer-Verlag KG, Berlin.
10. **Golubic, S., and J. W. Focke.** 1978. *Phormidium hendersonii* Howe: identity and significance of a modern stromatolite building organism. *J. Sediment. Petrol.* **48**:751–764.
11. **Golubic, S., and H. J. Hofmann.** 1976. Comparison of modern and mid-Precambrian Entophysalidaceae (Cyanophyta) in stromatolitic algal mats: cell division and degradation. *J. Paleontol.* **50**:1074–1082.
12. **Golubic, S., R. D. Perkins, and K. J. Lukas.** 1975. Boring microorganisms and microborings in carbonate substrates, p. 229–269. *In* R. W. Frey (ed.), *The Study of Trace Fossils.* Springer-Verlag KG, Berlin.
13. **Green, J., A. H. Knoll, S. Golubic, and K. Swett.** 1987. Paleobiology of distinctive benthic microfossils from the Upper Proterozoic Limestone-Dolomite "Series", East Greenland. *Am. J. Bot.* **74**:928–940.
14. **Hardie, L. A. (ed.).** 1977. *Sedimentation on the Modern Carbonate Tidal Flats of Northwest Andros Island, Bahamas.* Johns Hopkins University Press, Baltimore.
15. **Harris, P. M.** 1983. The Joulters ooid shoal, Great Bahama Bank, p. 132–141. *In* T. Peryt (ed.), *Coated Grains.* Springer-Verlag KG, Berlin.
16. **Hayes, J. M.** 1983. Geochemical evidence bearing on the origin of aerobiosis, a speculative hypothesis, p. 291–301. *In* J. W. Schopf (ed.), *Earth's Earliest Biosphere, Its Origin and Evolution.* Princeton University Press, Princeton, N.J.
17. **Hayes, J. M., I. R. Kaplan, and K. W. Wedeking.** 1983. Precambrian organic geochemistry, preservation of the record, p. 93–134. *In* J. W. Schopf (ed.), *Earth's Earliest Biosphere, Its Origin and Evolution.* Princeton University Press, Princeton, N.J.
18. **Hofmann, H. J.** 1976. Precambrian microflora, Belcher Islands: significance and systematics. *J. Paleontol.* **50**:1040–1073.
19. **Horodyski, R. J.** 1977. *Lyngbya* mats at Laguna Mormona, Baja California, Mexico: comparison

with Proterozoic stromatolites. *J. Sediment. Petrol.* 47:1305–1320.

20. **Horodyski, R. J., B. Bloesser, and S. vonder Haar.** 1977. Laminated algal mats from a coastal lagoon, Laguna Mormona, Baja California, Mexico. *J. Sediment. Petrol.* 47:680–696.
21. **Irwin, H., C. Curtis, and N. Coleman.** 1977. Isotopic evidence for source of diagenetic carbonates formed during burial of organic-rich sediments. *Nature* (London) **269**:209–213.
22. **Jackson, M. J., T. G. Powell, R. E. Summons, and I. P. Sweet.** 1986. Hydrocarbon shows and petroleum source rocks in sediments as old as 1.7 $\times 10^9$ years. *Nature* (London) **322**:727–729.
23. **Knoll, A. H.** 1981. Paleoecology of Late Precambrian microfossil assemblages, p. 17–54. *In* K. J. Niklas (ed.), *Paleobotany, Paleoecology, and Evolution.* Praeger, New York.
24. **Knoll, A. H.** 1982. Microfossils from the Late Precambrian Draken Conglomerate, Ny Friesland, Svalbard. *J. Paleontol.* **56**:755–790.
25. **Knoll, A. H.** 1985. The distribution and evolution of microbial life in the Late Proterozoic era. *Annu. Rev. Microbiol.* **39**:391–417.
26. **Knoll, A. H., and E. S. Barghoorn.** 1975. Precambrian eukaryotic organisms: a reassessment of the evidence. *Science* **190**:52–54.
27. **Knoll, A. H., and S. Golubic.** 1979. Anatomy and taphonomy of a Precambrian algal stromatolite. *Precambrian Res.* **10**:115–151.
28. **Knoll, A. H., S. Golubic, J. Green, and K. Swett.** 1986. Organically preserved microbial endoliths from the late Proterozoic of East Greenland. *Nature* (London) **321**:856–857.
29. **Knoll, A. H., J. M. Hayes, J. Kaufman, K. Swett, and I. Lambert.** 1986. Secular variation in carbon isotope ratios from Upper Proterozoic successions of Svalbard and East Greenland. *Nature* (London) **321**:832–838.
30. **Lowenstam, H. A., and S. Epstein.** 1957. On the origin of sedimentary aragonite needles of the Great Bahama Bank. *J. Geol.* **65**:364–375.
31. **Margulis, L., B. D. D. Grosovsky, J. F. Stolz, E. J. Gong-Collins, S. Lenk. D. Read, and A. López-Cortés.** 1983. Distinctive microbial structures and the pre-Phanerozoic fossil record. *Precambrian Res.* **20**:443–477.
32. **McKirdy, D. M., and J. H. Hahn.** 1982. Composition of kerogen and hydrocarbons in Precambrian rocks, p. 123–154. *In* H. D. Holland and M. Schidlowski (ed.), *Mineral deposits and the Evolution of the Biosphere.* Springer-Verlag KG, Berlin.
33. **Monty, C. L. V.** 1976. The origin and development of cryptalgal fabrics, p. 193–249. *In* M. R. Walter (ed.), *Stromatolites.* Elsevier/North-Holland Publishing Co., Amsterdam.
34. **Muir, M. D.** 1979. A sabkha model for deposition of part of the Proterozoic McArthur Group of the Northern Territory, and implications for mineralization. *Bur. Miner. Resour. J. Aust. Geol. Geophys.* 4:149–162.
35. **Rippka, R., J. Deruelles, J. B. Waterbury, M. Herdman, and R. Y. Stanier.** 1979. Generic assignments, strain histories and properties of pure cultures of cyanobacteria. *J. Gen. Microbiol.* **111**: 1–61.
36. **Runnegar, B.** 1982. The Cambrian explosion: animals or fossils? *J. Geol. Soc. Aust.* **29**:395–411.
37. **Schidlowski, M., J. M. Hayes, and I. R. Kaplan.** 1983. Isotopic inferences of ancient biochemistries: carbon, sulfur, hydrogen, and nitrogen, p. 149–186. *In* J. W. Schopf (ed.), *Earth's Earliest Biosphere, Its Origin and Evolution.* Princeton University Press, Princeton, N.J.
38. **Southgate, P. N.** 1986. Depositional environment and mechanism of preservation of microfossils, upper Proterozoic Bitter Springs Formation, Australia. *Geology* 14:683–686.
39. **Stal, L. J., and W. E. Krumbein.** 1985. Isolation and characterization of cyanobacteria from a marine microbial mat. *Bot. Mar.* **28**:351–365.
40. **Walter, M. R. (ed.)** 1976. *Stromatolites.* Elsevier/North-Holland Publishing Co., Amsterdam.
41. **Walter, M. R., and G. R. Heys.** 1985. Links between the rise of the metazoa and the decline of stromatolites. *Precambrian Res.* **29**:149–174.
42. **Waterbury, J. B., and R. Y. Stanier.** 1978. Patterns of growth and development in pleurocapsalean cyanobacteria. *Microbiol. Rev.* **42**:2–44.
43. **Wilson, C. B.** 1961. The Upper Middle Hecla Hoek rocks of Ny Friesland, Spitsbergen. *Geol. Mag.* **98**:89–116.
44. **Zhang, Z.** 1986. Solar cyclicity in the Precambrian rock record. *Palaeontology* **29**:101–111.

INDEX